D1920525

Edited by
Xiao-Feng Wu

Transition Metal-Catalyzed Heterocycle Synthesis via C-H Activation

Related Titles

Crabtree, R.H.

The Organometallic Chemistry of the Transition Metals, Sixth Edition
6 Edition

2014
Print ISBN: 978-1-118-13807-6
ePub ISBN: 978-1-118-78824-0
WOL obook PDF ISBN: 978-1-118-78830-1
Adobe PDF ISBN: 978-1-118-78844-8
Vital eBook XML ISBN: 978-1-118-89498-9
CourseSmart ISBN: 978-1-118-89501-6

Shi, Z. (ed.)

Homogeneous Catalysis for Unreactive Bond Activation

2015
Print ISBN: 978-1-118-45223-3
WOL obook PDF ISBN: 978-1-118-78898-1
ePub ISBN: 978-1-118-78899-8
Adobe PDF ISBN: 978-1-118-78903-2

Bates, R.R.

Organic Synthesis Using Transition Metals. 2e
2 Edition

2012
Print ISBN: 978-1-119-97894-7
ePub ISBN: 978-1-118-31366-4
MobiPocket ISBN: 978-1-118-31367-1
Adobe PDF ISBN: 978-1-119-94285-6
ISBN: 978-1-119-94286-3

Yudin, A.K. (ed.)

Catalyzed Carbon-Heteroatom Bond Formation

2011
Print ISBN: 978-3-527-32428-6
ISBN: 978-3-527-63338-8
Adobe PDF ISBN: 978-3-527-63339-5
ePub ISBN: 978-3-527-63340-1
MobiPocket ISBN: 978-3-527-64173-4

Eicher, T., Hauptmann, S., Speicher, A.

The Chemistry of Heterocycles
Structure, Reactions, Synthesis, and Applications
3 Edition

2012
Print ISBN: 978-3-527-32868-0

Metal-Catalyzed Cross-Coupling Reactions and More

2014
Print ISBN: 978-3-527-33154-3
ISBN: 978-3-527-65558-8
MobiPocket ISBN: 978-3-527-65559-5
ePub ISBN: 978-3-527-65560-1
Adobe PDF ISBN: 978-3-527-65561-8

Alvarez-Builla, J., Vaquero, J.J., Barluenga, J. (eds.)

Modern Heterocyclic Chemistry

2011
Print ISBN: 978-3-527-33201-4
ISBN: 978-3-527-63773-7

Pincer and Pincer-Type Complexes
Applications in Organic Synthesis and Catalysis

2014
Print ISBN: 978-3-527-33442-1
ISBN: 978-3-527-68130-3
MobiPocket ISBN: 978-3-527-68131-0
Adobe PDF ISBN: 978-3-527-68132-7
ePub ISBN: 978-3-527-68133-4

Burke, A.J., Silva Marques, C.

Catalytic Arylation Methods
From the Academic Lab to Industrial Processes

2015
Print ISBN: 978-3-527-33518-3
ISBN: 978-3-527-67270-7
Adobe PDF ISBN: 978-3-527-67274-5
MobiPocket ISBN: 978-3-527-67284-4
ePub ISBN: 978-3-527-67285-1

Ananikov, V.P. (ed.)

Understanding Organometallic Reaction Mechanisms and Catalysis
Computational and Experimental Tools

2014
Print ISBN: 978-3-527-33562-6
ISBN: 978-3-527-67821-1
ePub ISBN: 978-3-527-67822-8
MobiPocket ISBN: 978-3-527-67823-5
Adobe PDF ISBN: 978-3-527-67824-2

Murakami, M., Chatani, N. (eds.)

Cleavage of Carbon-Carbon Single Bonds by Transition Metals

2016
Print ISBN: 978-3-527-33632-6
ISBN: 978-3-527-68009-2
Adobe PDF ISBN: 978-3-527-68010-8
ePub ISBN: 978-3-527-68011-5
MobiPocket ISBN: 978-3-527-68012-2

Ackermann, L., Gunnoe, T., Habgood, L. (eds.)

Catalytic Hydroarylation of Carbon-Carbon Multiple Bonds

2017
Print ISBN: 978-3-527-34013-2
Adobe PDF ISBN: 978-3-527-69763-2
WOL obook PDF ISBN: 978-3-527-69764-9
ePub ISBN: 978-3-527-69765-6
MobiPocket ISBN: 978-3-527-69766-3

Edited by
Xiao-Feng Wu

Transition Metal-Catalyzed Heterocycle Synthesis via C-H Activation

WILEY-VCH
Verlag GmbH & Co. KGaA

Editor

Dr. Xiao-Feng Wu
Albert-Einstein-Str. 29a
18059 Rostock
Germany

All books published by **Wiley-VCH** are carefully produced. Nevertheless, authors, editors, and publisher do not warrant the information contained in these books, including this book, to be free of errors. Readers are advised to keep in mind that statements, data, illustrations, procedural details or other items may inadvertently be inaccurate.

Library of Congress Card No.: applied for

British Library Cataloguing-in-Publication Data
A catalogue record for this book is available from the British Library.

Bibliographic information published by the Deutsche Nationalbibliothek
The Deutsche Nationalbibliothek lists this publication in the Deutsche Nationalbibliografie; detailed bibliographic data are available on the Internet at <http://dnb.d-nb.de>.

© 2016 Wiley-VCH Verlag GmbH & Co. KGaA, Boschstr. 12, 69469 Weinheim, Germany

All rights reserved (including those of translation into other languages). No part of this book may be reproduced in any form – by photoprinting, microfilm, or any other means – nor transmitted or translated into a machine language without written permission from the publishers. Registered names, trademarks, etc. used in this book, even when not specifically marked as such, are not to be considered unprotected by law.

Print ISBN: 978-3-527-33888-7
ePDF ISBN: 978-3-527-69191-3
ePub ISBN: 978-3-527-69193-7
Mobi ISBN: 978-3-527-69194-4
oBook ISBN: 978-3-527-69192-0

Cover Design Schulz Grafik-Design, Fußgönheim, Germany
Typesetting SPi Global, Chennai, India
Printing and Binding Markono Print Media Pte Ltd, Singapore

Printed on acid-free paper

Dedicated to my wife and children
Qing-Yuan Wei, Nuo-Yu Wu, and Nuo-Lin Wu
Thanks for their understanding and support!

Contents

List of Contributors

Kazuma Amaike
Nagoya University
Graduate School of Science
Department of Chemistry
Chikusa
Nagoya 464-8602
Japan

Matthias Beller
Leibniz Institute for Catalysis
(LIKAT Rostock)
Albert-Einstein-Street 29a
18059 Rostock
Germany

Kevin J. T. Carr
Heriot-Watt University
School of Engineering and
Physical Sciences
Institute of Chemical Sciences
Room 2.16
William Perkin Building
Edinburgh EH14 4AS
UK

Feng Chen
Peking University
State Key Laboratory of Natural
and Biomimetic Drugs
Xue Yuan Road 38
Beijing 100191
China

Jianbin Chen
Leibniz-Institut für Katalyse an
der Universität Rostock e.V.
Leibniz Institute for Catalysis
(LIKAT Rostock)
Albert-Einstein-Street 29a
Rostock 18059
Germany

Chien-Hong Cheng
National Tsing Hua University
Department of Chemistry
101 Sec 2, Kuang-Fu Road
Hsinchu 30013
Taiwan

Wei Ding
Central China Normal University
College of Chemistry
152 Luoyu Road
Wuhan 430079
China

Wei Guo
Central China Normal University
College of Chemistry
152 Luoyu Road
Wuhan 430079
China

Lin He
Leibniz-Institut für Katalyse an der Universität Rostock e.V.
Leibniz Institute for Catalysis (LIKAT Rostock)
Albert-Einstein-Street 29a
Rostock 18059
Germany

Fang Hu
Zhejiang University
Department of Chemistry
38 Zheda Road
Hangzhou 310027
China

Hanmin Huang
State Key Laboratory for Oxo Synthesis and Selective Oxidation
Lanzhou Institute of Chemical Physics
Chinese Academy of Sciences
18 Tianshui Middle Road
Lanzhou 730000
China

Kenichiro Itami
Nagoya University
Graduate School of Science
Department of Chemistry
Chikusa
Nagoya 464-8602
Japan

Xiaolei Ji
State Key Laboratory for Oxo Synthesis and Selective Oxidation
Lanzhou Institute of Chemical Physics
Chinese Academy of Sciences
18 Tianshui Middle Road
Lanzhou 730000
China

Ning Jiao
Peking University
State Key Laboratory of Natural and Biomimetic Drugs
Xue Yuan Road 38
Beijing 100191
China

Bin Li
Nankai University
State Key Laboratory of Element-Organic Chemistry
College of Chemistry
94 Weijin Road
Tianjin 300071
China

and

Nankai University
Collaborative Innovation Center of Chemical Science and Engineering
94 Weijin Road
Tianjin 300071
China

Jin-Heng Li
Hunan University
State Key Laboratory of Chemo/Biosensing and Chemometrics
College of Chemistry and Chemical Engineering
Changsha 410082
China

Lixin Li
State Key Laboratory for Oxo Synthesis and Selective Oxidation
Lanzhou Institute of Chemical Physics
Chinese Academy of Sciences
18 Tianshui Middle Road
Lanzhou 730000
China

Bin Liu
Zhejiang University
Department of Chemistry
38 Zheda Road
Hangzhou 310027
China

Zhanxiang Liu
Zhejiang University
Department of Chemistry
ZJU-NHU United R&D Center
38 Zheda Road
Hangzhou 310027
China

Liang-Qiu Lu
Central China Normal University
College of Chemistry
152 Luoyu Road
Wuhan 430079
China

Stuart A. Macgregor
Heriot-Watt University
School of Engineering and
Physical Sciences
Institute of Chemical Sciences
Room 2.16
William Perkin Building
Edinburgh EH14 4AS
UK

Claire L. McMullin
Heriot-Watt University
School of Engineering and
Physical Sciences
Institute of Chemical Sciences
Room 2.16
William Perkin Building
Edinburgh EH14 4AS
UK

Krishnamoorthy Muralirajan
National Tsing Hua University
Department of Chemistry
101
Sec 2
Kuang-Fu Road
Hsinchu 30013
Taiwan

Yasushi Obora
Kansai University
Faculty of Chemistry
Materials
and Bioengineering
Department of Chemistry and
Materials Engineering
3-3-35, Yamate
Suita
Osaka 564-8680
Japan

Bing-Feng Shi
Zhejiang University
Department of Chemistry
38 Zheda Road
Hangzhou 310027
China

Ren-Jie Song
Hunan University
State Key Laboratory of
Chemo/Biosensing and
Chemometrics
College of Chemistry and
Chemical Engineering
Changsha 410082
China

Baiquan Wang
Nankai University
State Key Laboratory of
Element-Organic Chemistry
College of Chemistry
94 Weijin Road
Tianjin 300071
China

and

Nankai University
Collaborative Innovation Center
of Chemical Science and
Engineering
94 Weijin Road
Tianjin 300071
China

Xiao-Feng Wu
Leibniz-Institut für Katalyse an
der Universität Rostock e.V.
Leibniz Institute for Catalysis
(LIKAT Rostock)
Albert-Einstein-Street 29a
Rostock 18059
Germany

Wen-Jing Xiao
Central China Normal University
College of Chemistry
152 Luoyu Road
Wuhan 430079
China

Yuanjing Xiao
East China Normal University
Shanghai Key Laboratory of
Green Chemistry and Chemical
Processes
Department of Chemistry
School of Chemistry and
Molecular Engineering
3663 N. Zhongshan Road
Shanghai 200062
China

Junichiro Yamaguchi
Nagoya University
Graduate School of Science
Department of Chemistry
Chikusa
Nagoya 464-8602
Japan

Yudong Yang
Sichuan University
Key Laboratory of Green
Chemistry and Technology of the
Ministry of Education
College of Chemistry and State
Key Laboratory of Biotherapy
West China Medical School
29 Wangjiang Road
Chengdu 610064
China

Naohiko Yoshikai
Nanyang Technological
University
Division of Chemistry and
Biological Chemistry
School of Physical and
Mathematical Sciences
21 Nanyang Link
Singapore 637371
Singapore

Jingsong You
Sichuan University
Key Laboratory of Green
Chemistry and Technology of the
Ministry of Education
College of Chemistry and State
Key Laboratory of Biotherapy
West China Medical School
29 Wangjiang Road
Chengdu 610064
China

Ting-Ting Zeng
Central China Normal University
College of Chemistry
152 Luoyu Road
Wuhan 430079
China

Junliang Zhang
East China Normal University
Shanghai Key Laboratory of
Green Chemistry and Chemical
Processes
Department of Chemistry
School of Chemistry and
Molecular Engineering
3663 N. Zhongshan Road
Shanghai 200062
China

Yuhong Zhang
Zhejiang University
Department of Chemistry
ZJU-NHU United R&D Center
38 Zheda Road
Hangzhou 310027
China

Foreword 1

Heterocyclic compounds, a huge and structurally very diverse class of organic compounds, are of utmost importance for many fields such as pharmaceuticals, agrochemicals, and materials. Whereas an enormous armory of synthetic methods for their formation exists, still many limitations and unsolved problems exist. The recent use of transition-metal-catalyzed C–H activation methods for the formation of heterocycles represents an exciting alternative. Not only can the use of cheaper starting materials and the more efficient, greener formation of the desired heterocycles be a consequence, but original disconnections might also allow the formation of completely new products, reaching out to new product space – a promising source of innovation. The last few years have seen a rapid development of C–H activation chemistry and catalyst systems are getting more and more practical and attractive. Consequently, applications in industry and academia will soon become everyday reality.

This multiauthor book provides the reader with a great overview on the most important and exciting applications of C–H activation in the formation of heterocycles. It covers a broad range of topics, ranging from Pd- and Rh-catalyzed C–H activation, the workhorses of the field, to the use of earth abundant metals, an important trend in modern C–H activation research. In addition, heterogeneous C–H activation is also included, as is the detailed mechanistic analysis and the application in natural product synthesis. Xiao-Feng Wu and his multiauthor team of international leaders in the field did a great job and their book will serve the synthetic community well! It is not difficult to foresee that C–H activation, in the formation of heterocycles and beyond, will become increasingly important in the near future.

Münster
July 2015

Frank Glorius

Foreword 2

Heterocycles represent key structural motifs in numerous fundamental and applied areas, ranging *inter alia* from material sciences or coordination chemistry to crop protection and pharmaceutical industries. Traditionally, the assembly of heterocycles largely involved the interconversion of preexisting functional groups, the installation of which often requires lengthy synthetic operations. Since these approaches, thus, produce major amounts of undesired by-products and waste, the focus has shifted during the past decade toward the development of methods that capitalize upon the activation of otherwise inert C–H bonds. Hence, metal-catalyzed C–H functionalizations have revolutionized the efficiency of heterocycle syntheses, leading to an overall streamlining of the construction of increasingly complex target structures. Particularly, oxidative transformations were identified as versatile tools for the effective *de novo* synthesis of, among others, oxygen-, nitrogen-, and phosphorous-containing heterocycles.

The synthetic utility of the C–H activation approach is, however, not limited to the assembly of heterocycles. Indeed, the past decade has gained an enormous momentum in establishing tools for the step-economical functionalization of heterocycles. Furthermore, in a more general context, heterocycles are not only of relevance as substrates for C–H functionalizations but are also mandatory as ligands and/or as site selectivity ensuring directing groups in C–H activation chemistry.

Organic substrates of interest for direct functionalizations usually bear numerous C–H bonds of comparable dissociation energies. Therefore, a major obstacle in achieving synthetically useful C–H functionalizations is constituted by fully controlling the site selectivity in intermolecular transformations. While this represents a formidable challenge in direct arene functionalizations, heteroarenes often display C–H bonds with increased kinetic acidity, mostly proximal to an electronegative heteroatom. Hence, recent years have witnessed a major growth in protocols for metal-catalyzed C–H functionalizations that proceed in a (concerted metalation) deprotonation fashion. For instance, earth abundant base metals, such as copper, nickel, or cobalt catalysts, allowed for site-selective C–H functionalizations of various five-membered heteroarenes, provided that rather strong bases were utilized. Since the functional group tolerance of such reactions is often compromised, a powerful alternative was established by carboxylate-assisted

C–H metalations, largely with the aid of 4d transition metal catalysts based on palladium, rhodium, or ruthenium. Thereby, challenging heteroarene functionalizations proved ultimately viable under exceedingly mild reaction conditions.

From a historical perspective, it is noteworthy that several C–H functionalization protocols were in fact initially developed for heterocyclic substrates. For instance, one of the arguably most powerful tools in C–H functionalization chemistry, the palladium(0)-catalyzed direct arylation with organic electrophiles, was originally devised for the site-selective diversification of indoles. This pioneering report nicely illustrated the power of controlling the site selectivity in heteroarene C–H functionalizations. Thus far, most of the advances in heterocyclic C–H activation were accomplished through the cleavage of $C(sp^2)$–H bonds. However, major recent progress was constituted by the functionalization of thermodynamically less stable but frequently more difficult to activate $C(sp^3)$–H bonds.

With an ever-increasing toolbox of metal-catalyzed C–H activation procedures at hand, it does not come as a surprise that elegant applications of this strategy to more complex settings have appeared in the recent years. Arguably, the full power of the C–H activation approach has, thus, very recently been highlighted by step-economical syntheses of bioactive natural products as well as the bioorthogonal, late-stage diversification of heteroarenes in biooligomers, such as nucleosides or oligopeptides.

The book summarizes all aspects of metal-catalyzed C–H functionalizations of relevance to the synthesis and diversification of heterocycles, which should prove invaluable for scientists working in academia as well as practitioners in an industrial environment. Thus, this book is expected to be found in libraries and on the bookshelves of chemists who enjoy chemical synthesis, catalysis, and medicinal chemistry.

In consideration of the crucial importance of heterocyclic chemistry in many applied areas, along with the step-economical nature of C–H activations, further exciting developments are expected in this rapidly evolving research area. In addition to novel disconnection approaches for the *de novo* heterocycle syntheses, the need for more environmentally benign, economical catalysts and reagents will need to be addressed. Only by doing so the transition from academic research to industrial applications on scale can be realized in order to unleash the full potential of catalytic C–H activation chemistry.

Göttingen
July 2015

Lutz Ackermann

Preface

Heterocycles' synthesis is in the central of organic chemistry in the past, current, and will definitely be in the future as well. Numerous methodologies have been established for the preparation of all kinds of heterocyclic compounds. Generally, traditional protocols require the usage of preactivated compounds as their substrates which demand tedious steps. Come into the twenty-first century, taking into consideration of sustainable development, many procedures based on C–H activation have been developed. Almost all the transition metals have been tested as catalysts in heterocycles' synthesis via C–H activation. The most obvious advantage of these procedures is "straightforward"; no preactivation steps are needed. Under this background, the idea of this book comes into my mind. As mentioned in the title, the commonly applied transition metal catalysts in C–H activation reactions have been covered. With this book, we hope to further promote the development of heterocyclic chemistry.

30 June 2015 *Xiao-Feng Wu*

1
Computational Studies of Heteroatom-Assisted C–H Activation at Ru, Rh, Ir, and Pd as a Basis for Heterocycle Synthesis and Derivatization

Kevin J. T. Carr, Stuart A. Macgregor, and Claire L. McMullin

1.1
Introduction

Modern computational chemistry is a key tool by which insight into organometallic reaction mechanisms can be gained. The ability to characterize short-lived intermediates and transition states provides an ideal complement to experiment, where such information is often extremely difficult, if not impossible, to obtain. Recent years have seen great advances in understanding the mechanisms of C–H bond activation, and this area was the subject of several major reviews toward the end of the last decade [1, 2]. More recently, the focus has shifted to how C–H activation can be integrated into catalytic cycles for useful organic transformations. The C–H bond activation event is itself mechanistically diverse, with oxidative addition (OA), σ-bond metathesis (SBM), and electrophilic activation (EA) all potentially available at late transition metal centers, depending on the metal, its oxidation state, and the coordination environment. C–H activation assisted by a heteroatom base (typically a carboxylate or carbonate) falls into the last of these categories and forms the focus of this chapter. More recently, computational studies of catalytic cycles for C–H activation and functionalization have become more common. These reveal the complexity of what are usually multiple-step processes, and calculations are particularly well placed to test different mechanistic possibilities. Such studies are most effectively pursued through a close interaction between experiment and computation, and increasingly this is allowing for a more quantitative assessment of computed reaction mechanisms.

As well as progress in mechanistic understanding, the last 10 years have seen important developments in the computational methodologies available to model transition metal reactivity. While density functional theory (DFT) remains the core method of choice, the ability to model larger systems that more closely reflect experiment has highlighted the known shortcomings of DFT in describing dispersion interactions. These long-range, stabilizing interactions are individually weak, but their cumulative effect in large systems can be significant. Methods to incorporate this component include its separate calculation (e.g.,

Transition Metal-Catalyzed Heterocycle Synthesis via C—H Activation, First Edition. Edited by Xiao-Feng Wu.
© 2016 Wiley-VCH Verlag GmbH & Co. KGaA. Published 2016 by Wiley-VCH Verlag GmbH & Co. KGaA.

with Grimme's D3 parameter set [3]), use of functionals that include a treatment of dispersion (e.g., B97-D [4], ωB97X-D [5]), or use of a Minnesota functional (e.g., M06 or its variants [6]) where the functional parameterization (e.g., to reproduce molecular structures from crystallographic data) captures dispersion effects without explicitly identifying them. The validity of such approaches is seen, for example, in marked improvements in calculated L_nM-PR_3 dissociation energies [7]. These developments make the choice of functional for the study of C–H activation and functionalization especially critical when larger models are employed in calculations.[1)]

This chapter will focus on how computational chemistry has provided insights into heteroatom-assisted C–H activation and functionalization at Ru, Rh, Ir, and Pd metal centers. These will include reactions that both construct new heterocycles and those that introduce new substituents into an already intact heterocyclic skeleton. The text will first cover Pd and will then treat Ru, Rh, and Ir together, providing in each case a brief background on computational work on heteroatom-assisted C–H activation. More recent developments will then be considered covering the literature since our last review [1b], that is, from 2009 until March 2015. In many cases, work on nonheterocyclic substrates is included to illustrate important general points regarding the C–H activation mechanism. Unless otherwise stated, DFT calculations have been employed throughout and only the major functional employed in each study will be highlighted, along with any solvent used in brackets. The notation DFT2//DFT1 will be used to indicate cases where a second functional (DFT2) has been used to recompute the energy of a system optimized with an initial functional (DFT1). The original papers should be consulted for full computational details.

1.2 Palladium

1.2.1 Intramolecular Heteroatom-Assisted C–H Activation

1.2.1.1 Early Computational Studies

Computational work under this heading stems from 2005 when Davies, Macgregor, and coworkers reported BP86 calculations on the cyclometalation of dimethylbenzylamine at $Pd(OAc)_2$ [8], a system that had been the subject of detailed experimental studies in the 1980s by Ryabov and coworkers (Figure 1.1) [9]. Starting from $Pd(OAc)_2(Me_2NCH_2Ph)$, **1**, a two-step process was characterized involving initial displacement of one arm of the κ^2-OAc ligand by the incoming *ortho*-C–H bond of the benzyl substituent. This forms an agostic intermediate, **2**, that polarizes the C–H bond and sets up an H-bond to the

1) High-level wavefunction methods promise the accurate treatment of these long-range electron correlation effects; however, such methods are still too computationally demanding for routine use on large model systems.

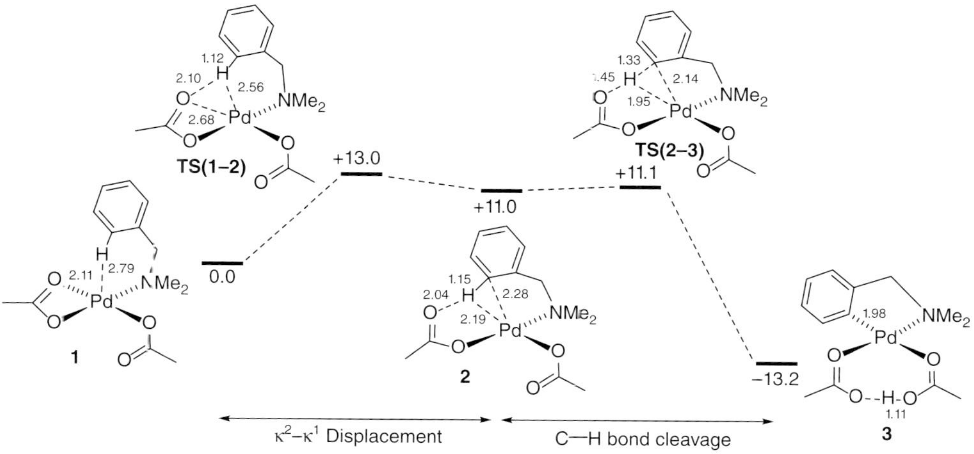

Figure 1.1 Computed reaction profile for C–H activation in $Pd(OAc)_2(Me_2NCH_2Ph)$. Energies are in kcal/mol and include a correction for zero-point energies; selected distances in Å [8].

pendant arm of the OAc ligand. C–H cleavage then proceeds with a minimal barrier to give the cyclometalated product, **3**, as an HOAc adduct. The overall barrier was computed to be 13 kcal mol^{-1} with the initial $\kappa^2-\kappa^1$ displacement of acetate via **TS(1 – 2)** corresponding to the highest point on the profile. Alternative mechanisms including oxidative addition or proton transfer onto the Pd-bound acetate oxygen were found to entail considerably higher barriers. The implication of an agostic intermediate, as opposed to a Wheland (or arenium)-type species, was significant as it indicates a fundamental difference to electrophilic aromatic substitution (S_EAr) processes. This type of reaction was subsequently characterized as proceeding via "ambiphilic metal–ligand assistance" (AMLA), a term that stresses the synergic role of the electron-deficient metal center and the nearby intramolecular base in promoting facile C–H bond activation [1b]. Both AMLA-6 and AMLA-4 processes have been defined, depending on whether a six-membered or four-membered transition state is involved. The role of an agostic intermediate indicates AMLA C–H activation is not (unlike S_EAr) restricted to aromatic C–H bonds, and indeed the equivalent $C(sp^3)$–H activation of $Me_2NCH_2CH_2CH_3$ was predicted to proceed with a barrier of only 20 kcal mol^{-1}.

Shortly afterward, Maseras, Echavarren, and coworkers published a B3LYP study on C–H activation in Pd(2-phenethylphenyl)(X)(PH_3) species (**4**, X = Br or HCO_3, Figure 1.2), as model intermediates in $Pd(OAc)_2$-catalyzed intramolecular C–C coupling [11]. The Pd–aryl bond is formed via oxidative addition of an aryl bromide precursor and acts as a directing group. With X = HCO_3, C–H activation proceeds by a one-step proton abstraction via **TS4A** in a similar way to that seen at $Pd(OAc)_2(Me_2NCH_2Ph)$, albeit without an agostic intermediate being located. With X = Br, proton abstraction was much less accessible; however, in this case, external deprotonation by a bicarbonate anion (**TS4B**) was found to

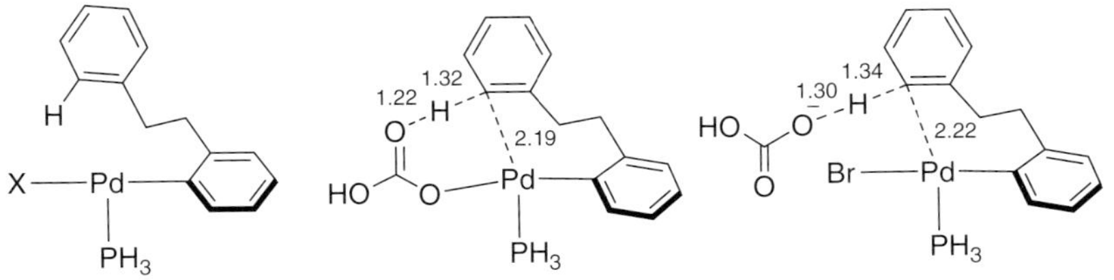

X = Br, HCO_3
4
TS4A
$\Delta G^{\ddagger} = 23.5$
TS4B
$\Delta G^{\ddagger} = 17.4$

Figure 1.2 C–H activation transition states derived from **4**, involving internal (**TS4A**) and external (**TS4B**) deprotonation. Computed free energies in kcal mol^{-1} and key distances in Å [10].

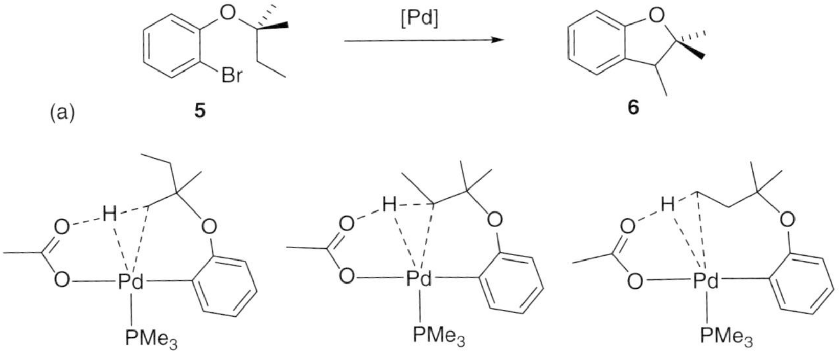

(a) **5** **6**

TS7A **TS7B** **TS7C**
(b) $\Delta G^{\ddagger} = 27.0$ $\Delta G^{\ddagger} = 32.5$ $\Delta G^{\ddagger} = 33.1$

Figure 1.3 (a) Pd-catalyzed formation of dihydrobenzofurans and (b) alternative transition states for C(sp^3)–H activation with computed free energies relative to Pd(Ar)(OAc)(PMe_3), **7**, in kcal mol^{-1} [12].

be competitive. A follow-on study showed that both intra- and intermolecular deprotonation mechanisms correctly capture the accelerating effect of electron-withdrawing substituents (F, Cl) on the aryl ring, as seen experimentally [10]. Such patterns are again inconsistent with a S_EAr mechanism.

The importance of agostic interactions was also seen in the ability to cleave C(sp^3)–H bonds as part of the Pd-catalyzed synthesis of dihydrobenzofurans (**5** → **6**, Figure 1.3a) reported by Fagnou and coworkers [12]. B3LYP calculations on a model Pd(Ar)(OAc)(PMe_3) species, **7**, characterized a series of concerted, inner-sphere deprotonation mechanisms for reaction at the three distinct C(sp^3)–H bonds of the 2-methylbutoxy substituent (Figure 1.3b). Reaction at one of the Me positions that results in a six-membered palladacycle is favored via **TS7A** over activation of either the CH_2 or CH_3 groups of the ethyl substituent via **TS7B** and **TS7C**, respectively, the latter forming a seven-membered palladacycle. These one-step C–H activation processes were consistent with the "concerted metalation deprotonation" (CMD) concept that Fagnou and coworkers were developing at this time. Lledós and Urriolabeitia found similar B3LYP-computed processes accounted for observed selectivities in the cyclometalation of stabilized iminophosphoranes, $Ar_3P{=}NC(O)Ph$ at $Pd(OAc)_2$, where *exo* activation via an agostic intermediate is kinetically favored over a one-step *endo* activation [13].

Overall, these early studies of carboxylate-assisted intramolecular C–H activation established the key features of these AMLA/CMD processes, where an electron-deficient metal center works in concert with a pendant carboxylate base to promote C–H activation. This is most evident when an agostic intermediate is involved and such species also rationalize how these systems can also perform $C(sp^3)$–H bond activation. Whether C–H activation is achieved as a one- or two-step process appears rather system dependent. Alternative mechanisms, for example, proton transfer onto a halide ligand, oxidative addition, or AMLA-4 processes involving proton transfer onto the inner (Pd-bound) oxygen of the carboxylate were all ruled out. Likewise, no evidence for S_EAr processes had been reported. Subsequent work was set against this background and considered the various other parameters that may affect the C–H bond activation process.

1.2.1.2 The Role of the Base, Solvent, and Additives on Pd-Mediated Intramolecular C–H Activation

The precise role of base additives in promoting C–H activation has been the subject of several computational studies. Clot and Baudoin used B3PW91(DMF) calculations to study the reactions of 2-halo-*tert*-butylbenzene (**8**) at $\{Pd(PR_3)\}$ species to form cyclobutarenes (**9**) (Figure 1.4a). These studies compared a model system (with R = Me) to those with the bulky phosphines used experimentally (R = ^{t}Bu [14, 15] or Cyp [16]). With R = Me and ^{t}Bu, the initial Ar–Br oxidative addition and subsequent C–H activation and C–C reductive elimination steps were all compared in the presence of three different bases (OAc, HCO_3^-, and CO_3^{2-}). Experimentally, use of CO_3^{2-} proved most efficient. Calculations on the PMe_3 model failed to capture this preference for carbonate and, moreover,

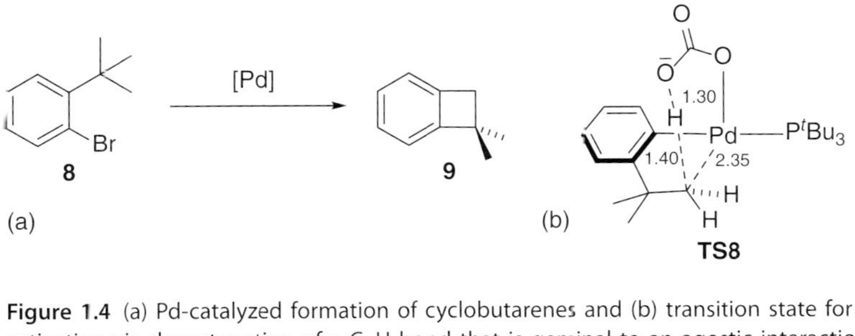

Figure 1.4 (a) Pd-catalyzed formation of cyclobutarenes and (b) transition state for $C(sp^3)$–H activation via deprotonation of a C–H bond that is geminal to an agostic interaction; key distances in Å [14].

did not identify C–H activation as the rate-limiting step as suggested by a significant experimental k_H/k_D kinetic isotope effect (KIE). In contrast, with the {Pd(P^{t}Bu$_3$)}/CO_3^{2-} model, a novel variation on the C–H activation mechanism was characterized whereby the bulk of the P^{t}Bu$_3$ ligand promotes an isomer with carbonate *trans* to the agostic C–H of the substrate (Figure 1.4). From this position, the deprotonation of a geminal C–H bond can be accessed via **TS8** and is promoted by the greater acidity of this position due to the neighboring agostic interaction [17]. With this mechanism, C–H activation is more accessible with CO_3^{2-} than with acetate or bicarbonate, and this C–H activation also becomes rate limiting.

The role of the base was also considered by Rousseaux and coworkers in a joint experimental and computational study on Pd(Ar)(Br)(PMe$_3$)$_2$ (**10**, Ar = 2-(dimethylcarbamoyl)phenyl), a model for the Pd(Ar)(OPiv)(PCy$_3$)$_2$ (OPiv = pivalate, ^{t}BuCO$_2^-$) used experimentally (see Figure 1.5) [18]. Efficient catalysis required both pivalate and carbonate additives, and B3LYP(toluene) calculations suggest that the carboxylate base is important not only for the C–H activation but also in promoting phosphine loss from the Pd center (**10** → **11**, $\Delta G = -0.6$ kcal mol^{-1}). C–H activation is computed to be strongly endergonic, and so the initial HOAc adduct formed, **12**, should rapidly revert back to **11**. This is avoided by the intervention of the carbonate additive which performs a rapid deprotonation, thus allowing the reaction to proceed to products. Kinetic studies show a zeroth-order dependence in [substrate], first order in [Pd], and saturation kinetics at [OPiv]:[Pd] > 3 : 1. The weak coordinating ability of the carbamoyl moiety is also important, as more basic amine substrates will displace one arm of the κ^2-OAc ligand in **11** to form a stable bidentate adduct from which C–H activation is much less accessible.

Additional factors, including the role of specific solvation and of some commonly used additives, have been included in the computational modeling of

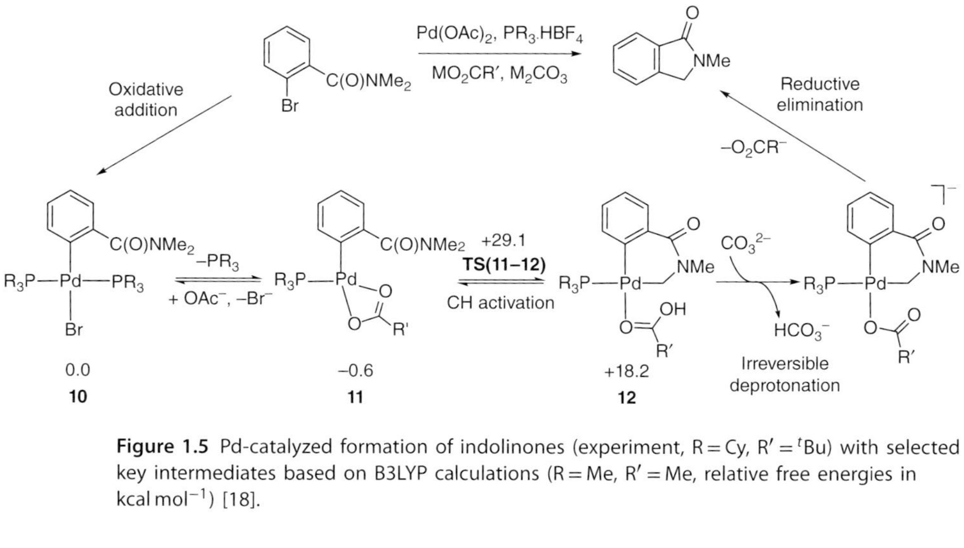

Figure 1.5 Pd-catalyzed formation of indolinones (experiment, R = Cy, R′ = ^{t}Bu) with selected key intermediates based on B3LYP calculations (R = Me, R′ = Me, relative free energies in kcal mol^{-1}) [18].

intramolecular C–H activation. Martínez and coworkers considered the role of the acidic medium in the cyclometalation of *N*-methylbenzylimine at $Pd(OAc)_2$ in acetic acid [19]. A dicationic intermediate, $[Pd(HOAc)_3(MeN{=}CHPh)]^{2+}$, was invoked to account for the rate acceleration observed in detailed kinetic studies performed in acidic media. B3LYP(MeOH) calculations (where MeOH was used to provide a dielectric similar to acetic acid) suggest C–H activation proceeds as a two-step process, the first involving displacement of an acetic acid ligand by the approaching *ortho*-C–H bond. The liberated acetic acid molecule then performs the deprotonation of the resultant agostic intermediate and leaves as an H_2OAc^+ cation (see **TS13**, Figure 1.6a). The overall barrier for this process is only 25 kJ mol^{-1}, with the deprotonation via **TS13** corresponding to the highest point on the profile.

The intervention of methanol solvent has also been considered in the $C(sp^2)$–H functionalization of *N*-methoxybenzamide at $Pd(OAc)_2$. M06(MeOH)//B3LYP calculations suggest that after *N*-deprotonation C–H activation proceeds with a computed barrier of 19.6 kcal mol^{-1}. However, a related transition state featuring an intervening MeOH molecule as a proton shuttle is competitive, at +22.4 kcal mol^{-1} (see **TS14**, Figure 1.6b) [20]. A similar role for external bicarbonate as a proton shuttle has been suggested by Wang, Wang, and coworkers for C–H activation of an anionic alanine amide species at Pd^{2+} (M06-L(DCE)//BP86; see **TS15**, Figure 1.6c) [21].

Supporting ligands can also play the role of the heteroatom base in C–H activation, as seen for an *N*-acyl monoprotected amino acid (MPAA) ligand studied by Houk, Yu, Wu, and coworkers (M06(SMD)//B3LYP calculations; see Figure 1.7a) [22]. This study also highlighted the flexibility of the directing group paradigm, with a remote cyano moiety providing access to an unusual *meta* selectivity in the reaction of substrate **16** with ethyl acrylate to give **17**. A combination of mass spectrometry and calculations provided evidence for a monomeric active species and, of several C–H activation mechanisms tested computationally, that proceeding via **TS16** in which the pendant *N*-acyl protecting group effects the deprotonation was most accessible ($\Delta G^{\ddagger} = 23.6$ kcal mol^{-1}). This is also consistent with the experimentally observed *meta* selectivity with barriers for activation

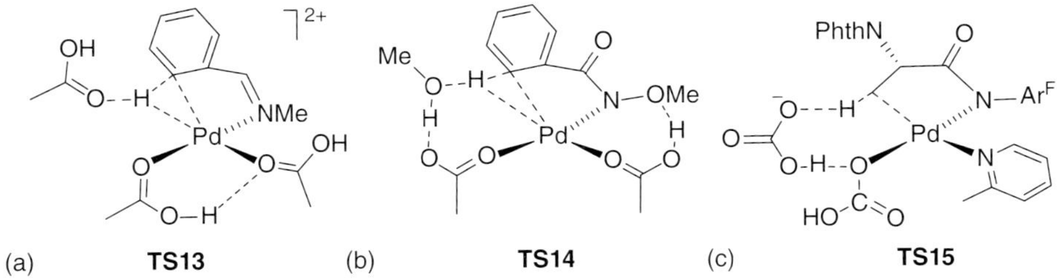

Figure 1.6 Cyclometalation transition state structures for (a) *N*-methylbenzylimine at $Pd(OAc)_2$ in acetic acid [19]; (b) *N*-methoxybenzamide at $Pd(OAc)_2$ in methanol [20]; and (c) an anionic alanine amide species at $Pd(2\text{-}Me\text{–}C_5H_4N)(HCO_3)$ (PhthN = phthalimido; $Ar^F = 4\text{-}C_6F_4\text{–}CF_3$) [21].

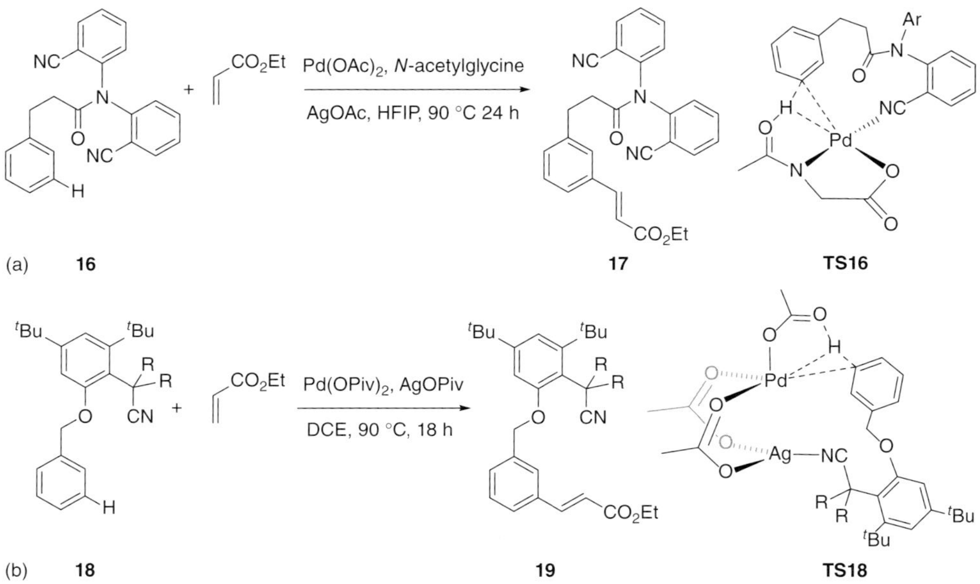

Figure 1.7 *Meta* selectivity via remote CN directing groups: (a) Pd-catalyzed alkenylation of **16–17** in the presence of an MPAA coligand via computed C–H activation transition state **TS16** featuring deprotonation by the *N*-acyl group (Ar = 2-C_6H_4–CN) [22] and (b) Pd/Ag-catalyzed alkenylation of **18–19** via computed heterobimetallic C–H activation transition state **TS18** [23].

at the *ortho* and *para* positions being slightly higher, by 0.9 and 0.5 kcal mol^{-1}, respectively. In a related study, the *meta*-selective C–H activation of **18** by a $Pd(OAc)_2$/AgOAc catalytic system was probed with M06(DCE)//B3LYP calculations (Figure 1.7b) [23]. A heterobimetallic $PdAg(OAc)_3$ system proved most compatible with experimental observations, with the remote CN group binding to Ag and directing the {Pd(OAc)} moiety to a *meta*-C–H bond (**TS18**). A computed barrier of 24.8 kcal mol^{-1} compares with barriers of 27.8 and 28.3 kcal mol^{-1} for *ortho*- and *para*-C–H activation, respectively. Other forms for the active species were considered but had higher computed barriers for *meta*-C–H activation ($Pd(OAc)_2$, 36.0 kcal mol^{-1}; $Pd_2(OAc)_4$, 29.3 kcal mol^{-1}; $Pd_3(OAc)_6$, 37.9 kcal mol^{-1}); moreover, for the monomeric and trimeric systems, an *ortho* selectivity was predicted.

Sunoj, Schaefer, and Anand have also considered the role of AgOAc and CsF additives in the $Pd(OAc)_2$-catalyzed amination of *N*-arylbenzamides with *N*-oxy- and *N*-chloromorpholines [24] (**20–21**, Figure 1.8) [25]. Using M06(DCE) calculations, a standard acetate-assisted C–H activation of the *N*-deprotonated substrate has a computed barrier of +28.1 kcal mol^{-1}. C–H activation at a range of monomeric as well as homo- and heterobimetallic species was then defined in which both additives were included. CsF is suggested to promote the initial *N*-deprotonation by forming HF; moreover, interaction of the Cs^+ cation with

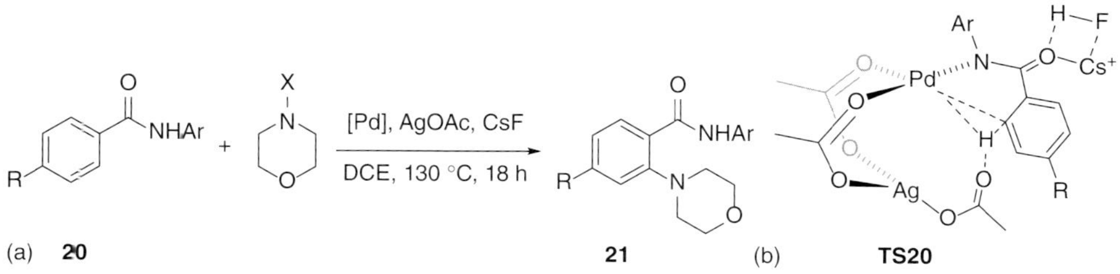

Figure 1.8 (a) Pd-catalyzed amination of N-arylbenzamides (experiment, R = ^{t}Bu, Ar = 4-$C_6F_4CF_3$, X = Cl, OAc,OBz) and (b) computed CsF-stabilized heterobimetallic transition state (R = Ar = H) [25].

both the acetate and HF can lead to significant stabilization in the subsequent C–H activation. The lowest-energy transition state was heterobimetallic **TS20**, which was computed to be 2.6 kcal mol^{-1} more stable than the isolated reactants. In all cases, the key elements of an AMLA/CMD C–H activation remain intact, that is, interaction with the Pd^{2+} metal center and efficient deprotonation by a nearby carboxylate ligand. Experimentally, the direct amination reaction requires heating to 130 °C for 18 h for efficient catalysis, and so the low barriers to C–H activation appear out of step with this. Presumably, the rate-determining step lies elsewhere in this catalytic cycle (e.g., the N–X activation or C–N coupling steps). In addition, the speciation of the "CsF" and "AgOAc" additives is particularly challenging. In this case, these are formulated as the solvent adducts CsF(DCE)$_6$ and AgOAc(DCE)$_2$, respectively. Such explicit solvation appears to be absent in the computed transition states, however.

A similar role for CsF in promoting *N*-deprotonation of amino acids at Pd(Ph)(I)(PCy$_3$) has been proposed by Musaev, Yu, and coworkers using M06-L(toluene) calculations [26]. A second equivalent of "a molecule" of CsF is then implicated in the subsequent C(sp^3)–H activation where F^- plays the role of an external base to effect the C–H activation. These steps are part of a computed pathway for the direct arylation of amino acids in which the largest barrier is around 10 kcal mol^{-1}. This reaction requires heating to 100 °C, suggesting a disconnect between the computed pathway and the semiquantitative energetics implied by experiment. As above, the speciation of an ionic solid such as CsF in an aromatic solvent is especially challenging.

1.2.1.3 Intramolecular C–H Activation of Heterocyclic Substrates

Martínez, Urriolabeitia, and coworkers have studied the regioselective C–H activation of (Z)-2-aryl-4-arylidene-5(4*H*)-oxazolones, **22**, at Pd(OAc)$_2$ in acidic media [27]. As seen in related studies on *N*-methylbenzylimine [19], reactivity is enhanced by increased acidity and favors the arylidene position (i.e., C–H^a) to give, unusually, a six-membered palladacycle, isolated as a dimer. B3LYP(MeCO$_2$H) calculations (X = Y = H) showed reduced barriers to C–H activation upon sequential protonation of the Pd active species and that the dicationic intermediate, $[Pd(MeCO_2H)_3(\kappa\text{-}N\text{-}\mathbf{22})]^{2+}$, provided the best agreement with experimental activation data. A two-step C–H activation process was characterized involving initial displacement of one acetic acid ligand and formation of an agostic intermediate followed by deprotonation by the external acetic acid (similar to **TS13**, Figure 1.6). The kinetic preference for C–H^a bond activation reflects a lower degree of reorganization involved in this process. Barriers were further reduced when computed with the analogous CF_3CO_2H system, consistent with enhanced reactivity seen experimentally. The nature of these two-step C–H activations does vary, however, with the higher transition state corresponding to C–H bond cleavage in $MeCO_2H$ but changing to the formation of the agostic intermediate in F_3CCO_2H.

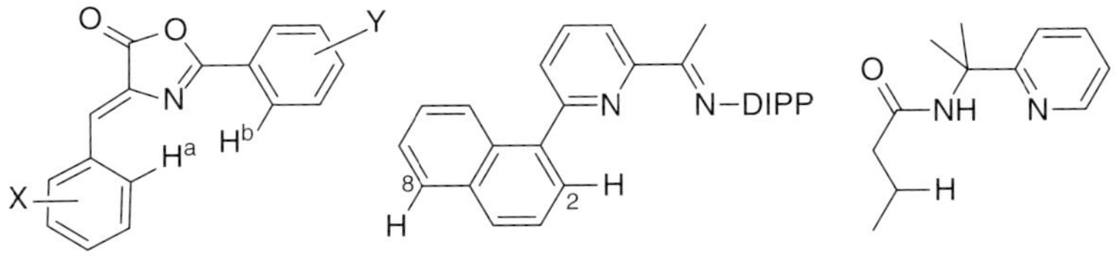

22 **23** **24**

Regioselective formation of six-membered palladacycles was also observed with naphthyl moieties fitted with a bidentate *N,N*-pyridylarylimine directing group (**23**, DIPP = 2,6-$C_6H_3{}^iPr_2$). BP86-D3(toluene) calculations on the full experimental system defined a two-step C–H activation process that is favored at the C8-position both kinetically ($\Delta\Delta G^{\ddagger} = 1.8\ \text{kcal mol}^{-1}$) and thermodynamically ($\Delta\Delta G = 3.5\ \text{kcal mol}^{-1}$) over a one-step process at the C2-position [28]. With monodentate 2-naphthylpyridine, both the kinetic and thermodynamic preference swaps to the 2-position, consistent with experimental observations and indicating that the additional imine anchor is indeed responsible for the change in selectivity. $Pd(OAc)_2$-mediated C(sp^3)–H bond activation of alkylamides fitted with a removable 2-pyridylisopropyl directing group (**24**) has been characterized with a barrier of 23.6 kcal mol^{-1} with M06-2X//B3LYP calculations [29].

Houk, Yu, and coworkers have investigated stereoselective C(sp^3)–H activation as part of the Pd(II)-catalyzed iodination and acetoxylation of oxazoline derivatives **25** (see Figure 1.9) [30]. M06//B3LYP calculations indicated the nature of the resting state to be crucial in controlling reactivity. With a bulky substrate ($R^1 = R^2 = {}^tBu$, **25a**), this is the $Pd_3(OAc)_6$ trimer and C–H activation has an overall barrier of 26.2 kcal mol^{-1}. In contrast, when R^1 = Et and $R^2 = {}^iPr$ (**25b**), a rather stable monomer, $Pd(OAc)_2(\kappa\text{-}N\text{-}\mathbf{25b})_2$, is formed from which C–H

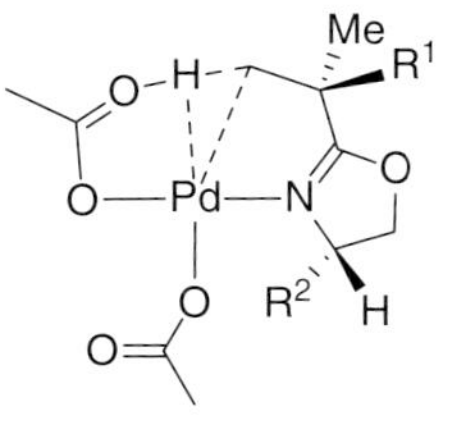

Figure 1.9 Computed C–H activation transition state for C(sp^3)–H bond activation of oxazolines ($R^1 = R^2 = {}^tBu$, **25a**; R^1 = Et, $R^2 = {}^iPr$, **25b**), highlighting the preferred *anti* arrangement for **25a** [30].

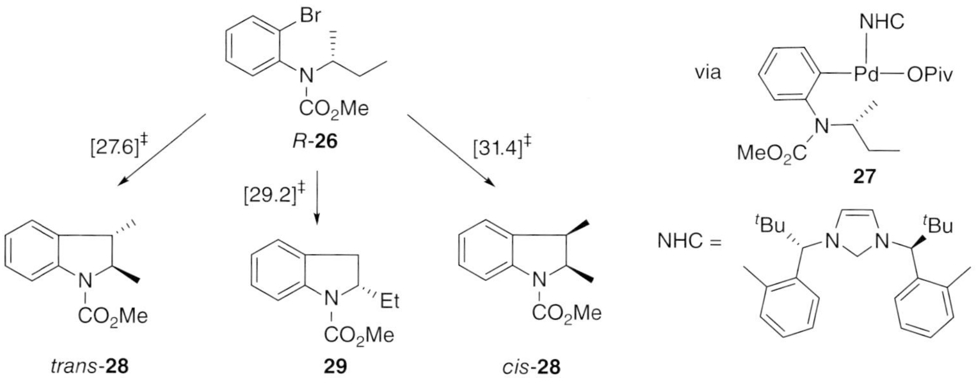

Figure 1.10 Enantioselective indoline formation from *R*-**26** via Pd–NHC intermediate **27**. Free energies (at 413 K) for the enantio-determining C–H activation transition states leading to products *cis-/trans*-**28** and **29** are indicated in kcal mol^{-1} [31a].

activation entails a barrier of 38.4 kcal mol^{-1}. The highly stereoselective reaction with **25a** is traced to a preferred *anti* arrangement of the bulky ^{t}Bu substituents in the transition state, this being 2.3 kcal mol^{-1} more stable than alternative *syn* forms. This $\Delta\Delta G^{\ddagger}$ equates to a 96% diastereomeric excess, de (cf. 82% experimentally), while with **25b** $\Delta\Delta G^{\ddagger} = 0.8$ kcal mol^{-1}, giving a calculated de of 17% (cf. 0% experimentally).

Further evidence of the ability of DFT calculations to capture the subtle energetics associated with enantioselective processes was seen in an M06-L(xylenes) study of the asymmetric synthesis of indolines reported by Kündig and coworkers (see Figure 1.10) [31]. For example, with indoline *R*-**26**, Ar–Br activation at the Pd(NHC) catalyst followed by Br/OPiv substitution accesses intermediate **27**. From here C–H activation favors the methylene position and specifically the C–H bond that gives rise to the *trans* isomer of the product (*trans*-**28**). C–H activation of the methyl group (leading to **29**) is then preferred to activation of the methylene C–H bond leading to *cis*-**28**. Computed free energy barriers with *S*-**26** and a benzyl analog also match the experiment product distributions. The origin of the enantioselectivity is traced to destabilizing interactions with the NHC ligand, and similar results were obtained with the B97-D functional.

1.2.2 Intermolecular C–H Activation

1.2.2.1 Early Computational Studies

The first computational study of heteroatom-assisted intermolecular C–H activation was reported in 2000 by Sakaki and coworkers, who studied the C–H activation of benzene and methane at $M(O_2CH)_2$ (M = Pd, Pt) at the MP4(SDQ) level [32]. For benzene, this involved a two-step process with $\kappa^2-\kappa^1$ displacement

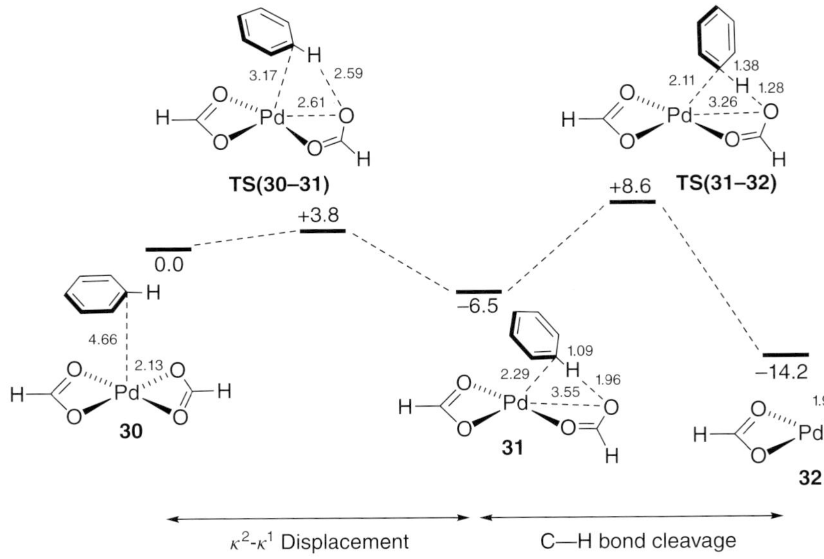

Figure 1.11 Computed reaction profile for C-H activation of benzene at $Pd(\kappa^2\text{-}CO_2H)_2$; MP4(SDQ) energies are in kcal mol^{-1} and selected distances in Å [32].

of one formate ligand to give an intermediate σ-complex (**31**, Figure 1.11), which features a short H···O contact of 1.96 Å (Figure 1.11). This facilitates the subsequent heterolytic cleavage of the C–H bond.

The observation that fluorinated arenes are particularly effective substrates in the $Pd(OAc)_2$-catalyzed direct arylation of aryl halides prompted Fagnou and coworkers to consider the mechanism of C–H activation [33]. Using B3LYP calculations, several mechanisms were considered at a $Pd(Ph)(X)(PH_3)$ model species (X = Br, HCO_3), and a concerted metalation with proton abstraction by the intramolecular bicarbonate base found to be most accessible (see Figure 1.12 where **TS33** was computed with the full P^tBu_2Me ligand). The absence of any Pd···F interactions indicated a truly nondirected C–H activation process. Moreover, no pathway relating to a S_EAr process could be located, and this, along with the greater reactivity associated with the more acidic C–H bonds of fluoroarenes, suggested a different mechanistic interpretation was required for such processes. The term "concerted metalation deprotonation," was later

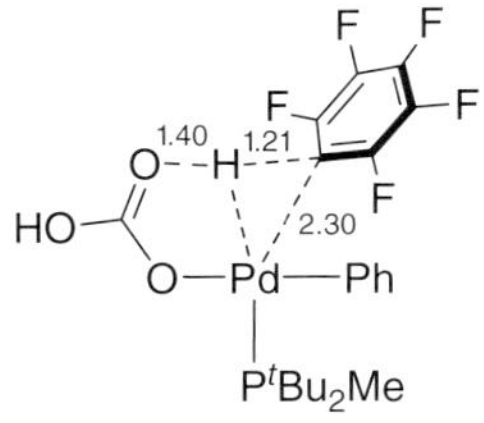

Figure 1.12 Computed transition state for C–H bond activation of C_6F_5H at $Pd(Ph)(HCO_3)(P^tBu_2Me)$, with selected distances in Å [33].

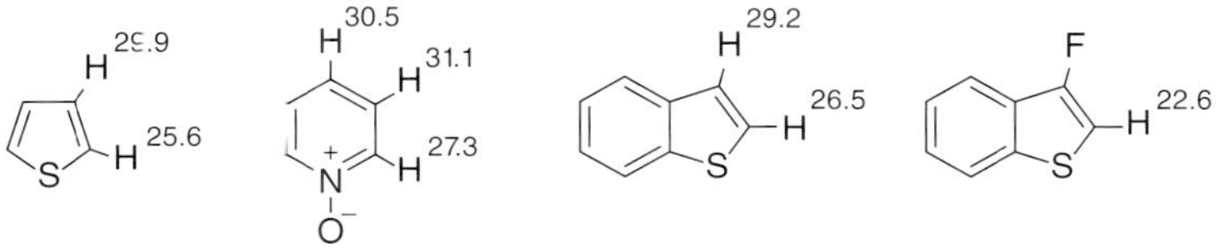

Figure 1.13 Computed free energy activation barriers (kcal mol^{-1}) for C–H activation of selected heterocycles at Pd(Ph)(OAc)(PMe_3) [35].

adopted by this group. A follow-on study showed a slightly reduced computed barrier with OPiv as the base (24.9 kcal mol^{-1}) compared to OAc (26.2 kcal mol^{-1}) [34].

The CMD concept was generalized in a 2008 B3LYP study by Gorelsky, Lapointe, and Fagnou that considered a wide range of (hetero)aromatic substrates. In particular, electron-rich arenes, hitherto thought to react via S_EAr processes, were shown instead to react via the CMD mechanism [35]. Moreover, the CMD pathway correctly reproduces the observed regioselectivities, including preferential reaction at the 2-position of thiophene and the *ortho* position of pyridine *N*-oxide (Figure 1.13). The accelerating effect of electron-withdrawing substituents is again apparent, for example, in the lower barrier for C2–H activation of 3-fluorobenzothiophene compared to benzothiophene. The overall trends in reactivity and the regioselectivities predicted in this paper were subsequently verified experimentally [36].

This and subsequent papers by Gorelsky [37] employed the activation strain model [38] to analyze trends in the CMD activation barriers (see Figure 1.14 for C–H activation of benzene). In this approach, the activation energy, ΔE_{act}, for C–H activation at Pd(Ph)(OAc)(PMe_3) is decomposed into two terms, ΔE_{dist}, the sum of the energies required to distort the individual reaction components to

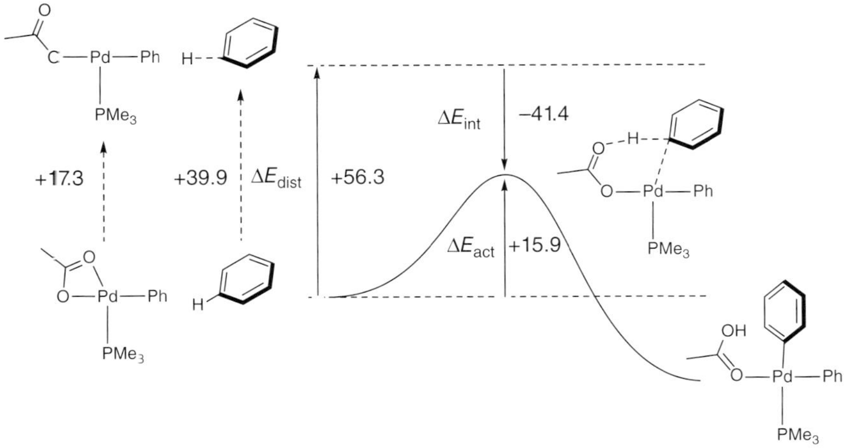

Figure 1.14 The activation strain model illustrated for C–H activation of benzene at Pd(Ph)(κ^2-OAc)(PMe_3) with component energies indicated in kcal mol^{-1} [37a].

their geometries seen in the transition state, and ΔE_{int}, the energy of interaction between these two fragments. This permitted a wide range of substrates to be categorized into one of three classes, depending on which was the dominant contribution to the overall barrier. For Class I substrates, ΔE_{dist} dominates, and in particular the distortion associated with the substrate C–H bond; this destabilizing terms tends to be lower with more acidic C–H bonds, such as in C_6F_5H and other electron-deficient substrates such as heterocyclic *N*-oxides. For Class II species, ΔE_{int} dominates, and this stabilizing term is enhanced by more nucleophilic substrates. Finally, for Class III substrates, variations in both ΔE_{dist} and ΔE_{int} must be considered. Computed reaction profiles indicate the possibility of precomplexation of the substrate at Pd, as either π-bound, heteroatom-bound, or C–H-bound adducts. In no case was a Wheland intermediate located and with the very electron-deficient C_6F_5H no precursor was seen. Gorelsky has summarized this body of work in a recent review, including an overview of C–H activation in heterocyclic *N*-oxides, azoles, thiophenes, and F- and Cl-substituted arenes [37b].

The direct arylation of polyfluorinated arenes, Ar^F–H, has also been considered by Eisenstein, Perutz, and coworkers [39]. Starting from a Pd(Ph)(κ^1-OAc)(DMA)(PMe_3) adduct (DMA = dimethylacetamide), B3PW91(DMA) calculations provided energetics for DMA/arene substitution, C–H activation, loss of AcOH, and the final C–C coupling and loss of the biaryl product. The computed trends were found to depend on the number of *ortho*-F substituents present. DMA/arene substitution is always endergonic but is most accessible with substrates with no *ortho*-F for which η^2-CH=CH adducts are located. With two *ortho*-F weak Pd · · · C–H adducts are seen, where this interaction is supported by H bonding to the pendant arm of the κ^1-OAc ligand. Two *ortho*-F substituents also promote the C–H activation step such that this group of substrates has the lowest overall barriers to CH activation. After loss of HOAc, C–C reductive coupling was then found to be disfavored by *ortho*-F substituents. The trends in C–H activation are shown to correlate well with both C–H bond acidities (as judged by computed heterolytic bond dissociation energies) and the homolytic bond strength of the new Pd–Ar bond in the Pd(Ar^F)(Ph)(κ^1-OAc)(PMe_3) intermediates. This latter factor also serves to increase barriers for C–C coupling. Overall, the C–H activation step is more sensitive to the F substituents as the effects of both the C–H acidity and the Pd–Ar^F bond strengths reinforce each other; in contrast, for the C–C coupling, the need to break a stronger Pd–Ar^F bond is counteracted by the formation of a stronger Ph–Ar^F bond. Ess and coworkers have also emphasized the role of the Pd–Ar bond strength on the regioselectivities of (hetero)arene activation at Pd(Ph)(κ^2-OAc)(PMe_3) (M06 calculations) [40]. Gorelsky has since questioned this in his recent review [37b] and reemphasized the role of the ΔE_{int} term, noting the lack of relationship between this and the computed Pd–Ar^F heterolytic dissociation energies. However, as the Pd–Ar^F bond strength is only one component of the ΔE_{int} term, these two interpretations are not necessarily contradictory.

1.2.2.2 Direct Functionalization via C–H Activation of Heterocyclic Substrates

Synthetic protocols are now in place for a range of direct functionalizations with heterocyclic substrates, and the field is still expanding rapidly. This activity is matched by parallel computational studies, where in many cases the full catalytic cycle is considered. In the following, the discussion will focus on the computational understanding of the C–H activation step, but the importance of this step within the context of the overall catalytic cycle will also be discussed, where appropriate.

Gorelsky and Lapointe modeled C–H activation at the 5-position of 2-substituted thiophenes at a Pd(Ph)(κ^2-OAc)(PMe_3) model system, with computed barriers (B3LYP) referenced against the parent unsubstituted thiophene [37a]. A correlation with the substrate's nucleophilic character as gauged by Hammett σ_p parameters was found, and a correlation between σ_p and both ΔE_{act} and ΔE_{dist} was also seen. Lower barriers were computed with both electron-withdrawing substituents (CN, CO_2Me) and certain electron-donating groups (*N*-pyrrolidine), and these implied higher reactivities were corroborated by experimental competition studies. That both electron-withdrawing and electron-donating groups can be more reactive reflects the counterbalancing effects of enhanced acidity (less favorable ΔE_{int}, more favorable ΔE_{dist}) versus enhanced nucleophilicity (more favorable ΔE_{int}, less favorable ΔE_{dist}). The relationship between σ_p and the components of the activation strain model is significant: negative gradient Hammett plots are often taken to be indicative of S_EAr character, but this study shows they can also be compatible with heteroatom-assisted C–H activation processes. Further B3LYP calculations confirmed that introduction of an electron-withdrawing chloro substituent into substituted thiophenes or *N*-methylindoles not only enhances reactivity but can also induce useful regioselectivities in the subsequent C–H functionalization steps [41]. This is synthetically attractive as the Cl substituent can both be readily and regioselectively introduced and then (after C–H functionalization) easily removed by conventional means.

The role of different *para*-substituted triarylphosphines, P(4-$C_6H_4R)_3$ (R = H, OMe, F, and CF_3), in controlling trends in the direct arylation of thiazole and 2-methylthiophene substrates has been investigated by Ozawa and coworkers using M06-2X(THF)//B3LYP calculations [42]. Experimentally, the rate of reaction at Pd(Ar)(κ^2-OAc){P(4-$C_6H_4R)_3$} intermediates (Ar = 2,6-$C_6H_3Me_2$) follows opposite trends for thiazole (R = CF_3 > F > H > OMe) and 2-methylthiophene (R = OMe > H > F > CF_3). Calculations (Ar = Ph, Figure 1.15) trace this to a change in the rate-determining steps between these two systems. With thiazole, C–H activation via **TS34** is rate limiting, and the barrier for this process reflects variation in the energies of the *N*-bound precursors that are stabilized by more electron-withdrawing phosphines. In contrast, with 2-methylthiophene, C–C reductive coupling is rate limiting (**TS35**), and this process is promoted by electron-donating phosphines.

The unusual C4-selective arylation of 2,3-disubstituted thiophenes reported by Itami and coworkers [43] has been the subject of two theoretical studies.

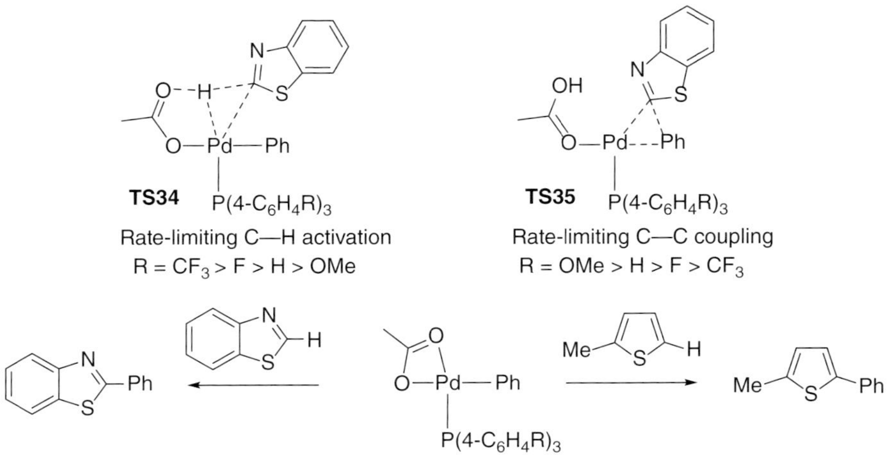

Figure 1.15 Different reactivity trends in the direct arylation of thiazole and 2-methylthiophene [42].

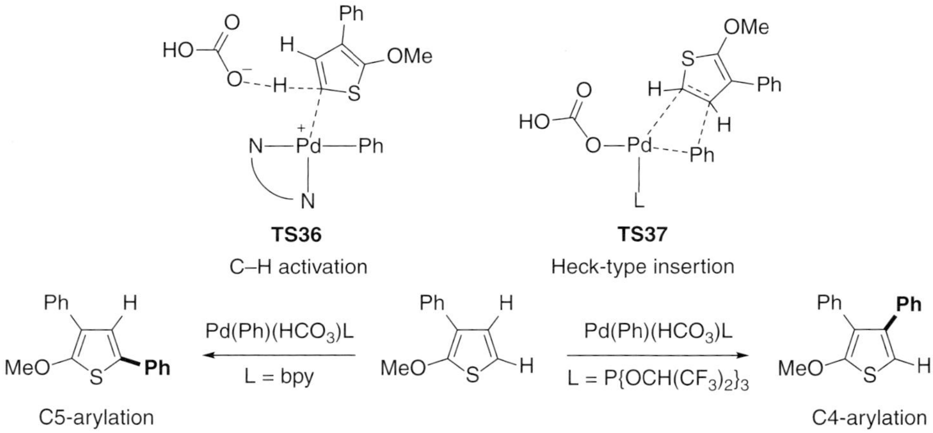

Figure 1.16 C4 versus C5 selectivity in the direct arylation of 2-phenyl-3-methoxythiophene [45].

Experimentally, the regioselectivity is controlled by the choice of coligand, with 2,2′-bipyridine (bpy) giving C5-arylation and $P\{OCH(CF_3)_2\}_3$ giving C4-arylation. Grimme, Itami, Studer, and coworkers used dispersion-corrected DFT (B2PLYP-D3//TPSS-D3) to model the reaction of $[Pd(bpy)Ph]^+$ with the parent thiophene and showed that insertion into the Pd–Ph bond had an intrinsic preference for phenyl transfer onto the C4-position [44]. Fu and coworkers considered the reactions of 2-phenyl-3-methoxythiophene at $Pd(Ph)(\kappa^1\text{-}CO_3H)(bpy)$ and $Pd(Ph)(\kappa^2\text{-}CO_3H)\{P(OR)_3\}$ (R = OMe, $CH(CF_3)_2$) using an M06(toluene)//B3LYP protocol [45]. In the bpy system, C–H activation at C5 proceeds with a computed barrier of 29.9 kcal mol^{-1} via **TS36** (see Figure 1.16). The bidentate nature of

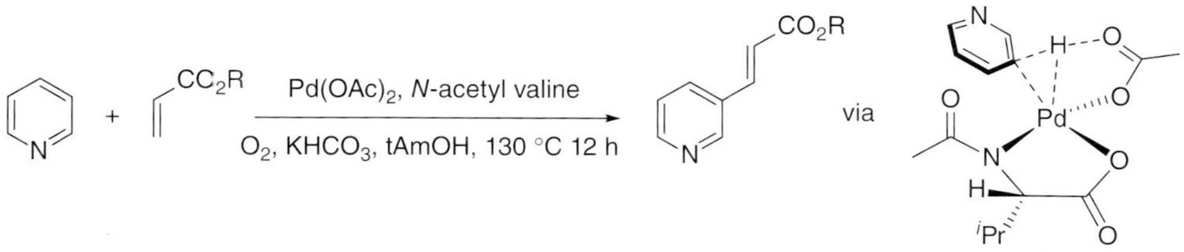

Figure 1.17 Pd-catalyzed *meta*-selective alkenylation of pyridine in the presence of an MPAA coligand (experiment R = ^{n}Bu; computed R = Et) [46].

the bpy ligand requires the initial displacement of the bicarbonate ligand by the incoming thiophene, with the bicarbonate then effecting an external deprotonation. The alternative Heck-type insertion of the thiophene favors Ph transfer onto the 4-position but has a slightly higher barrier of 31.6 kcal mol^{-1}. Initial calculations on a Pd(Ph)(κ^2-CO_3H){P(OMe)$_3$} model gave a similar outcome, and it was only by using the full P{OCH(CF_3)$_2$} ligand in the calculations that the swap in regioselectivity was reproduced, with barriers of 36.7 kcal mol^{-1} for C–H activation at C5 and 34.4 kcal mol^{-1} for insertion with Ph transfer to C4 via **TS37**. Experimentally, the reaction proceeds at 130 °C in 12 h. The authors postulate that specific H-bonding interactions in the transition state control the C4 regioselectivity and that this is only introduced once the full P{OCH(CF_3)$_2$} ligand is considered.

The *meta*-selective C–H activation of pyridine and other electron-deficient monosubstituted arenes has been modeled by Wu, Zeng, and coworkers using B3LYP(1-pentanol) calculations [46]. This step is part of a Pd(OAc)$_2$-catalyzed Heck alkenylation using MPAA coligands (see Figure 1.17). The calculations employed *N*-acetyl-valine which upon *N*-deprotonation and in the presence of pyridine forms Pd(κ^1-OAc)(NC_5H_5){κ-*N*,*O*-{AcONCH(^{i}Pr)CO_2}} as the precursor to C–H activation. Rearrangement from an *N*-bound to a π-bound pyridine adduct ($\Delta G = +17.1$ kcal mol^{-1}) precedes C–H activation, and for the *meta* position, this proceeds through a CMD transition state, **TS38**, with an overall barrier of 31.0 kcal mol^{-1}. This pathway proves slightly more accessible than either *ortho*-C–H activation ($\Delta G^{\ddagger} = 32.4$ kcal mol^{-1}) or *para*-C–H activation ($\Delta G^{\ddagger} = 31.9$ kcal mol^{-1}). Note that the potential noninnocence of the *N*-acetyl group in C–H activation has been subsequently highlighted (see Figure 1.7) [22]. Reprotonation then facilitates *N*-decoordination of the MPAA ligand, and this is thought to facilitate the functionalization steps (modeled with ethylacrylate). A related study by Zhang and coworkers considered monosubstituted arenes [47].

Gorelsky has modeled the C–H activation step in the direct arylation of both free and Cu(I)-bound azoles at a Pd(Ph)(κ^2-OAc)(PMe$_3$) intermediate [37c]. These two reactions provide complementary regioselectivities, with reaction at C5 favored for the free azoles but C2-arylation favored in the presence of Cu(I) salts. With *N*-methylimidazole and oxazole, the formation of *N*-bound

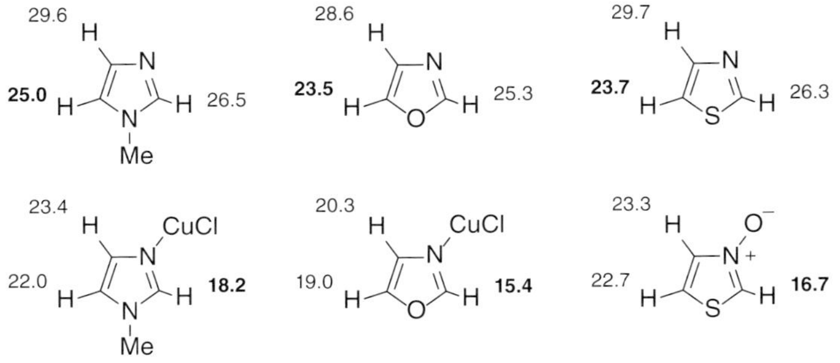

Figure 1.18 Computed free energy activation barriers (kcal mol^{-1}) for C–H activation at Pd(Ph)(κ^2-OAc)(PMe$_3$) for *N*-methylimidazole and oxazole (with and without bound CuCl), thiozole, and thiazole *N*-oxide [37c].

CuCl adducts leads to a significant lowering of the computed barriers to C2–H activation, consistent with the observed swap in regioselectivity (see Figure 1.18). The Lewis acidic CuCl additive switches the behavior of the azole from a Class II ligand (substrate nucleophilicity dominant) [48] to a Class I species (C–H bond acidity dominant). Formation of the thiazole *N*-oxide has a similar effect with the B3LYP-computed barrier for C2–H bond activation being 16.7, 9.6 kcal mol^{-1} lower than the free thiazole. Gorelsky and DeBoef also considered C–H activation of *N*-alkylindoles as a step in the direct C2-arylation of these species [49]. A CMD mechanism is again found to accommodate electronically diverse substituents in the C6-position with changes in computed barriers interpreted in terms of the Class II behavior of this substrate; regioselectivities were also correlated to changes in the distribution of the highest occupied molecular orbitals of these species.

Both Sakaki [50] and Sanford and Schoenebeck [51] have studied the Pd-mediated C–H oxidative coupling of benzo[h]quinoline with arenes. The latter study focused on changes in regioselectivity observed with 1,3-dimethoxybenzene (DMB), depending on whether acetate (*meta*/*meta*) or carbonate (*ortho*/*para*) is used. M06-L(DMSO/DMB)//ωB97X-D calculations on both monomeric and dimeric intermediates featuring a cyclometalated benzo[h]quinoline ligand suggest that C–H activation of 1,3-dimethoxybenzene proceeds through a CMD transition state with the same *ortho/para* selectivity irrespective of whether acetate or carbonate is used (the latter being modeled as "$NaCO_3^-$"). In contrast, the subsequent C–C coupling step (which is promoted by benzoquinone) proceeds with *meta/meta* selectivity, again for both bases. The observed swap in selectivity arises as a change in the base also changes the identity of the rate-determining step. With carbonate, this is the C–H activation, which leads to *ortho/para* selectivity. However, with acetate, the C–C reductive elimination transition state lies marginally above that for C–H activation, and so the *meta/meta* selectivity is favored.

The Pd(OAc)$_2$-catalyzed oxidative coupling of *N*-methylimidazole (as a model for xanthines) with thiophene has been modeled by Hu, You, and coworkers

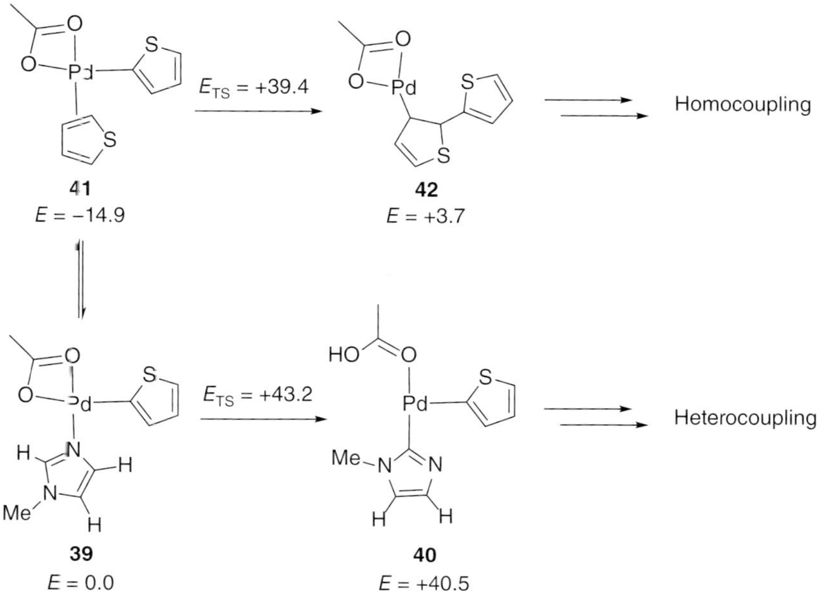

Figure 1.19 Computed key stationary points with energies in kcal mol^{-1} for the reactions of thiophene and *N*-methylimidazole at a Pd(OAc)$_2$ catalyst [52].

using B3LYP calculations [52]. Pd(OAc)$_2$ is considered to be the active species at which thiophene readily undergoes C–H activation at the 2-position with a barrier of only 9.6 kcal mol^{-1} (see Figure 1.19). This process is described as a S$_E$Ar reaction, although the computed transition state appears similar to many CMD-like processes in the literature. Binding of *N*-methylimidazole then forms an intermediate, **39**, from which C–H activation at the 2-position of the *N*-methylimidazole proceeds with a barrier of 43.2 kcal mol^{-1} to give **40**. Alternatively, the substitution of *N*-methylimidazole by a second thiophene gives **41** from which a Heck-type insertion of the C2=C3 bond of this second thiophene into the extant Pd-thiophenyl bond has been characterized. This process has a slightly lower barrier of 39.4 kcal mol^{-1} relative to **39**, but would imply homocoupling via **42**, contrary to experimental observations. Further study of this discrepancy would be of interest. Experimentally, the reaction proceeds in 1,4-dioxane at 120 °C and requires 20 h. Shaik and Kosuch have also commented on the importance of concentration effects in controlling homo- versus heterocoupling in oxidative coupling reactions [53].

The speciation of the Pd(OAc)$_2$ catalyst in such reactions is a key point. For example, Hartwig and Tan have shown that the Pd(OAc)$_2$-catalyzed direct arylation of 2-bromotoluenes and benzene proceeds more efficiently in DMA solvent in the absence of added phosphine ligands [54]. A preformed dianionic dimer, $[ArPdBr_2]_2^{2-}$, was also active. B3LYP calculations showed that C–H activation of benzene at Pd(Ph)(κ^2-OPiv)(DMA) has a computed free energy barrier of 31 kcal mol^{-1} compared to a barrier of 42 kcal mol^{-1} at Pd(Ph)(κ^2-OPiv)(P^{t}Bu$_3$).

It is postulated that steric hindrance around the Pd center disfavors reaction in the presence of bulky phosphines and that "ligandless" C–H activation may be accessible for these unactivated substrates.

The importance of heterocyclic *N*-oxides in synthesis has prompted computational and experimental mechanistic studies of their reactivity. Campeau and coworkers performed both kinetic studies and supporting B3LYP calculations on the direct arylation of pyridine *N*-oxide [55]. Under catalytic conditions, the coupling of *para*-nitropyridine *N*-oxide with 5-bromo-*m*-xylene was shown to be zeroth order in the aryl bromide, first order in the pyridine *N*-oxide, and half order in the Pd catalyst (whether supplied as either $Pd(OAc)_2/[^tBu_3PH][PF_6]$ or $Pd(P^tBu_3)_2$). Zero-order kinetics in $[P^tBu_3]$ were also observed, and side-by-side experiments comparing pyridine *N*-oxide and pyridine *N*-oxide-d_5 revealed a k_H/k_D KIE of 3.3. These outcomes suggest a mechanism involving facile loss of P^tBu_3 and Ar–Br activation, followed by rate-determining C–H activation of the pyridine *N*-oxide, with the formation of a dimeric resting state between these two events. B3LYP calculations confirmed that C–H activation at the 2-position of pyridine *N*-oxide at $Pd(Ph)(\kappa^2\text{-}CO_2R)(PMe_3)$ proceeds through an inner-sphere CMD process with computed barriers of 34.9 kcal mol^{-1} (R = OH) and 34.1 kcal mol^{-1} ($R = CH_3$). The effects of a range of different substituents around the pyridine ring were also successfully rationalized computationally. Further stoichiometric reactions of *para*-nitropyridine *N*-oxide were also performed to assess the role of base, starting with a preformed $Pd(Ph)Br(P^tBu_3)$ reactant. In this case, use of K_2CO_3 alone gave very low yields, but 30% conversion was seen with KOPiv, either with or without K_2CO_3. Further runs showed that achieving the observed catalytic activities required excess K_2CO_3 but only a substoichiometric amount of KOPiv. Thus, K_2CO_3 is proposed to act as a terminal proton acceptor to avoid the buildup of pivalic acid in solution. This interpretation was supported by stoichiometric runs using $Pd(Ph)(\kappa^2\text{-}OAc)(P^tBu_3)$, **43**, which gave the same yield whether run with or without K_2CO_3. It was noticeable, however, that the product yield in these runs (about 48%) was lower than expected from the performance of the catalytic process.

Picking up on this point, Hartwig and coworkers revisited this system and showed that **43** is in fact inactive toward pyridine *N*-oxide [56]. Instead, a cyclometalated dimer, $[Pd(OAc)(^tBu_2PCMe_2CH_2)]_2$ (**44**, formed via decomposition of **43**), was proposed to be the active species. Kinetic studies showed direct arylation at **44** to be first order in pyridine *N*-oxide and half order in the dimer. These results indicate opening of dimer **44** to give the cyclometalated monomer, $Pd(\kappa^2\text{-}OAc)(^tBu_2PCMe_2CH_2)$, **45**, which then effects the C–H activation. The inactivity of the monomeric bromo-analog, $Pd(Br)(^tBu_2PCMe_2CH_2)$, suggested that C–H activation proceeds via a CMD pathway involving acetate rather than the cyclometalated arm of the phosphine ligand. Further runs in the presence of both **43** and **44** showed a zeroth-order dependence on [**43**], confirming its non-involvement in the C–H activation. Finally, B3LYP calculations demonstrated a reduced C–H activation barrier of 27 kcal mol^{-1} at **45** compared to 33 kcal mol^{-1} at **43**. Based on these results, a cooperative mechanism shown in Figure 1.20

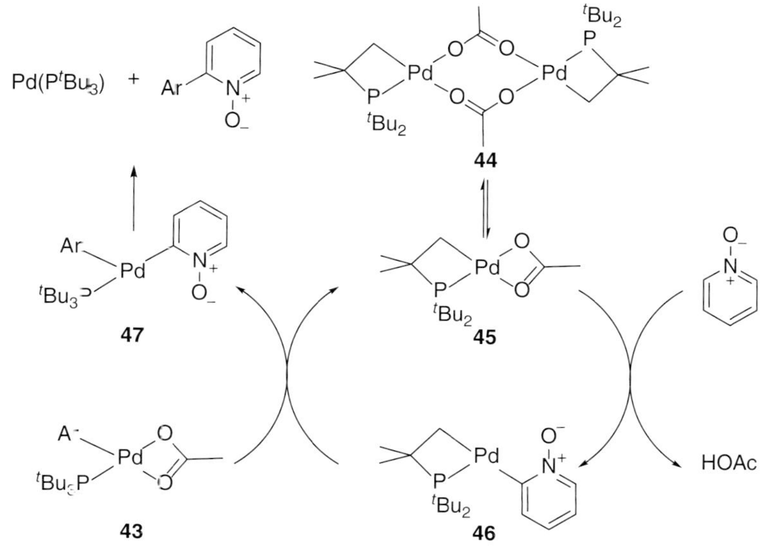

Figure 1.20 Proposed mechanism for the direct arylation of pyridine *N*-oxides in the Pd(OAc)$_2$/P^{t}Bu$_3$ system. Figure adapted from reference [56].

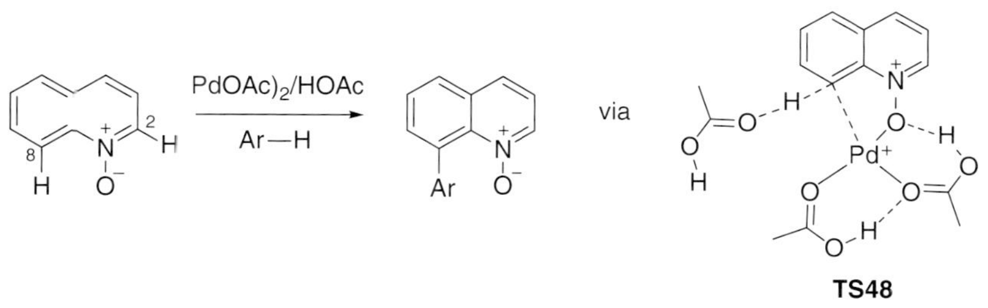

Figure 1.21 Direct, C8-selective arylation of QNO by Pd(OAc)$_2$ in acetic acid and the proposed computed transition state [57].

was proposed. In this, the pyridine *N*-oxide undergoes C–H activation at **45** to generate a heteroaryl ligand in **46**. This then undergoes exchange with the OAc ligand in **43** to regenerate **45** and form Pd(Ar)(Ar′)(P^{t}Bu$_3$), **47**. Reductive elimination from **47** releases the biaryl product and Pd(P^{t}Bu$_3$), with the latter reforming **44**/**45** under the catalytic conditions.

Larionov and coworkers have described the Pd(OAc)$_2$-catalyzed direct arylation of quinolone *N*-oxides (QNOs) in acetic acid, a process that proceeds selectively at the 8-position (Figure 1.21) [57]. B3LYP(HOAc) calculations suggest this process occurs at dicationic [Pd(HOAc)$_3$(κ-*O*–QNO)]$^{2+}$, with C–H activation involving an agostic intermediate formed via displacement of one HOAc molecule that then acts as an external base. This is a similar scenario to that described by Martíniz, Urriolabeitia, and coworkers [27]. A barrier of

24.7 kcal mol^{-1} is computed via a five-membered transition state, **TS48**; in comparison, activation at the 2-position involves a four-membered transition state and hence a much higher barrier of 40.4 kcal mol^{-1}. C–H activation is also endergonic, consistent with the observation of H/D exchange experimentally. Comparison with the equivalent reactions at a Pd(Ph)(κ^2-OAc)(PMe_3) model intermediate shows C–H activation at the C2-position would be favored ($\Delta G^{\ddagger} = 29$ kcal mol^{-1}).

1.3 Ruthenium, Rhodium, and Iridium

Much of the insight gained from computational and mechanistic studies into heteroatom-assisted C–H activation at Pd(II) metal centers has proven to be applicable to Ru(II), Rh(III), and Ir(III) systems. Therefore, only a brief overview of the most relevant computational studies of these processes will be presented. Computational studies of Ru-, Rh-, and Ir-catalyzed C–H functionalization reactions will then be described. Most of these target the synthesis or derivatization of heterocycle systems and also include a treatment of the initial C–H activation step.

1.3.1 Intramolecular Heteroatom-Assisted C–H Activation

In 2006, Davies, Macgregor, and coworkers investigated the cyclometalation of Me_2NCH_2Ph by $[IrCl_2Cp^*]_2$ in the presence of NaOAc [58]. Acetate plays a dual role in this process, by both inducing dimer opening and acting as a base in the subsequent C–H activation. Initial BP86 calculations on a model $[Ir(Me_2NCH_2Ph)(\kappa^2\text{-OAc})Cp]^+$ system characterized a concerted C–H activation incorporating dissociation of one Ir–O bond and deprotonation by the pendant acetate arm (i.e., an AMLA-6 process). This process was favored over either deprotonation by the Ir-bound oxygen (AMLA-4) or an oxidative addition to an Ir(V)-hydride. Later [58b], a slightly lower-energy two-step pathway featuring an agostic intermediate was characterized, with rate-limiting $\kappa^2-\kappa^1$ displacement of acetate at +13.4 kcal mol^{-1}. Lower barriers were computed with weaker bases, RCO_2^- (R = Me > Ph > HO > CCl_3 > CF_3), reflecting the fact that a weak base will also be a poor ligand and so facilitate the rate-limiting $\kappa^2-\kappa^1$ carboxylate displacement step. The cyclometalation of *N*-methyl-2-pyrrole imine was also considered [59].

Maseras, Dixneuf, and coworkers subsequently studied the cyclometalation of 2-phenylpyridine at the $[RuCl_2(p\text{-cymene})]_2$ dimer in the presence of *N*-heterocyclic carbene ligands [60]. B3LYP calculations were performed using a *cis*-$RuCl_2$(IMe)(2-Ph–C_5H_4N) model system (where IMe = 1,3-dimethylimidazol-2-ylidene) with bicarbonate base. Although this species showed a distinct agostic interaction *trans* to the IMe ligand, a concerted

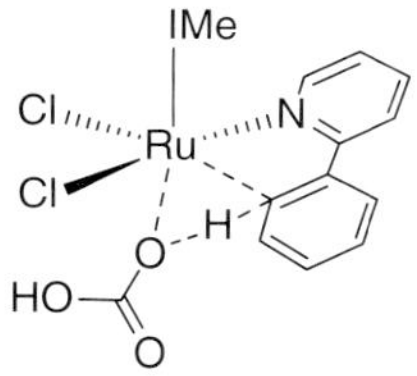

TS49

Figure 1.22 C–H activation transition state for cyclometalation of 2-phenylpyridine at {*cis*-Ru(Cl)$_2$(IMe)} in the presence of bicarbonate base [60].

oxidative addition to a Ru(IV)-hydride proved very unfavorable. Instead, coordination of bicarbonate is preferred from which a low-energy intramolecular deprotonation can be accessed with a barrier of only 13.9 kcal mol^{-1} via **TS49** (Figure 1.22). This involves the Ru-bound oxygen atom and a four-membered (i.e., AMLA-4) transition state.

More recently, Davies, Macgregor, and coworkers highlighted the importance of dispersion effects when seeking a more quantitative assessment of observed data [61]. Experimentally, *N*-alkylimines (**H-L$_{1-5}$**) and 2-phenylpyridine (**H-L$_6$**) react at [MCl$_2$Cp*]$_2$/NaOAc (M = Rh, Ir) to give well-defined cyclometalated species as their chloride adducts (Figure 1.23a). Moreover, H/D exchange experiments indicate irreversible C–H activation at Ir (kinetic control) but reversible C–H activation at Rh (thermodynamic control). Competition experiments also gave different trends in the reactivity of these species at each metal.

These features were studied with BP86-D3 calculations, in either MeOH (Rh) or CH_2Cl_2 (Ir) solvent. Computed profiles for the reactions of the 2-thiophenyl derivative **H-L$_3$** are shown in Figure 1.24. OAc-induced opening of the [MCl$_2$Cp*]$_2$ dimer followed by substitution of one anionic ligand by **H-L$_3$** forms **IV$_M$**. For Rh, a two-step C–H activation is characterized via an agostic intermediate, **V$_{Rh}$**, to give the cyclometalated HOAc adduct **VI$_{Rh}$**. Cl/HOAc substitution leads to the observed product **VII$_{Rh}$**. The energetics of this process ($\Delta G_{calc} = -4.0$ kcal mol^{-1}; $\Delta G^{\ddagger}_{calc} = 11.8$ kcal mol^{-1}) indicate reversible C–H activation, consistent with the observed H/D exchange. For Ir, the reaction is both more exergonic ($\Delta G_{calc} = -14.2$ kcal mol^{-1}) and has a much higher barrier ($\Delta G^{\ddagger}_{calc} = 18.4$ kcal mol^{-1}), consistent with an irreversible reaction under kinetic control. Moreover, C–H activation proceeds in a single step, corresponding to $\kappa^2-\kappa^1$ displacement of OAc. Experimentally, the Rh system is less efficient and takes much longer to achieve reasonable yields. The lower barrier for Rh indicates this is not a kinetic effect but rather reflects a slow approach to equilibrium that ultimately favors product formation.

Importantly, use of both an extended basis set (with diffuse functions) and the D3 correction proved essential to reproduce the favorable thermodynamics expected for all metal/substrate combinations. This reflects the stabilization derived from having the bulky substrate adjacent to the {MCp*} fragment and in turn stresses the importance of using the full experimental system in the calculations. With this approach, the more subtle trends in the relative reactivity of the six substrates were also reasonably captured. Thus, for Rh (Figure 1.23a, thermodynamic control), the trend in ΔG_{calc} largely follows the increase in

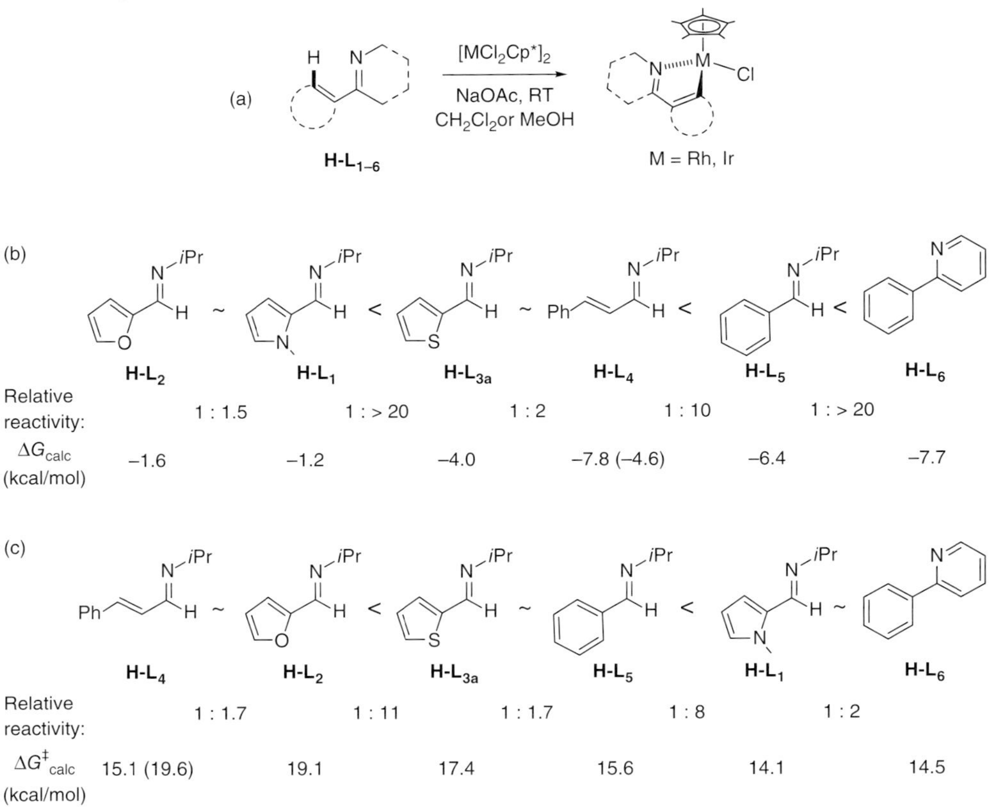

Figure 1.23 (a) Cyclometalation of *N*-alkylimines (**H-L$_{1-5}$**) and 2-phenylpyridine (**H-L$_6$**) at $[MCl_2Cp^*]_2$ (M = Rh, Ir); relative experimental and computed reactivities for substrates **H-L$_{1-6}$** at $[MCl_2Cp^*]_2$, (b) M = Rh, and (c) M = Ir. Computed data (kcal mol^{-1}) give the overall free energy changes, ΔG_{calc}, for M = Rh and calculated activation barriers, $\Delta G^{\ddagger}_{calc}$, for M = Ir [61].

substrate reactivity, whereas for Ir (Figure 1.23b, kinetic control), increased substrate reactivity is mirrored in a decrease in $\Delta G^{\ddagger}_{calc}$. One exception was **H-L$_4$**, the reactivity of which was overestimated for both metals. This is thought to relate to the different shape of this substrate (which features a vinyl CH=CHPh substituent rather than the aromatic substituents in **H-L$_{1-3}$** and **H-L$_5$**) and the propensity of the dispersion correction to emphasize intramolecular stabilization over interaction with the environment (e.g., solvent). A different degree of dispersion stabilization therefore occurs upon binding **H-L$_4$** to the metal fragment. To explore this issue, a model substrate **H-L$_{4'}$** featuring a smaller $CH{=}CH_2$ substituent was computed (values in parenthesis in Figure 1.23), and this gave both a reduced ΔG_{calc} at Rh and an increased $\Delta G^{\ddagger}_{calc}$ at Ir.

Around the same time, Zheng and coworkers [62] modeled the acetate-assisted cyclometalation of *N*-phenylbenzaldimines at $[MCl_2Cp^*]_2$ dimers (M = Rh, Ir). B3LYP(MeOH) calculations found cyclometalation at Ir to be favored

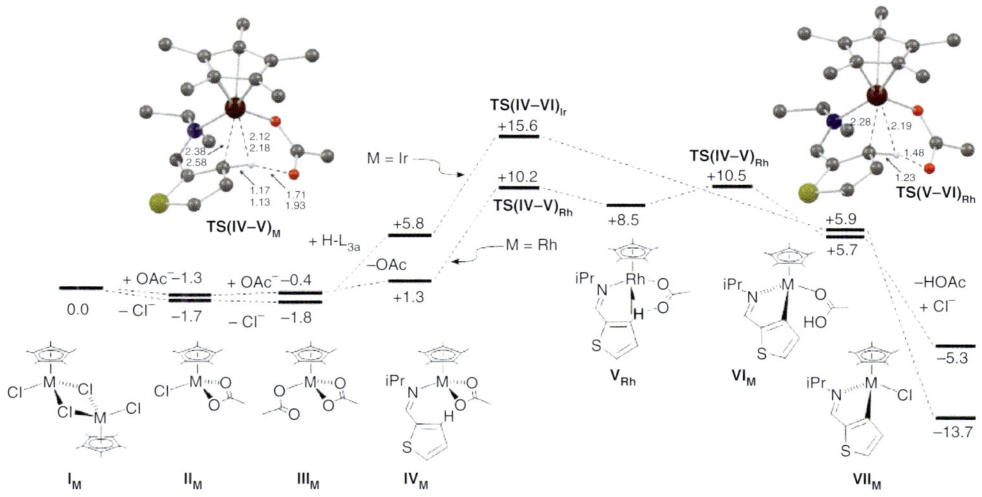

Figure 1.24 Computed free energy reaction profiles (kcal mol^{-1}) for C–H activation of **H-L$_3$** at $[MCl_2Cp^*]_2$ for M = Ir (in CH_2Cl_2) and M = Rh (in MeOH). Computed C–H activation transition states are shown with key distances in Å (Rh, plain text; Ir, italics) and nonparticipating H atoms omitted for clarity.

thermodynamically (Rh, $\Delta G_{calc} = -0.5$ kcal mol^{-1}; Ir, $\Delta G_{calc} = -8.4$ kcal mol^{-1}), but, contrary to the above study, the C–H activation barrier was higher for Rh (Rh, $\Delta G^{\ddagger}{}_{calc} = 26.4$ kcal mol^{-1}; Ir, $\Delta G^{\ddagger}{}_{calc} = 19.7$ kcal mol^{-1}). Experimentally, the reaction is slower with Rh [63]. The effect of phenyl substituents, R, positioned *meta* to the directing group, was also assessed. When R = CF_3 or Me, C–H activation is computed to be kinetically favored at the less hindered *ortho*-C–H bond, as seen experimentally. With R = OMe, the two pathways are computed to be in competition, while for R = F, activation of the adjacent *ortho*-C–H bond is favored (perhaps reflecting the "*ortho*-effect" of fluorine [64]). Experimentally, a mixture of products is seen when R = OMe and F.

1.3.2
Intermolecular C–H Activation

Extensive computational work has been published on the intermolecular C–H bond activation of benzene at Ru(II), Rh(III), and Ir(III) metal centers fitted with a heteroatomic base as a ligand. When this base is a κ^2-carboxylate ligand, the C–H mechanism has been shown to be similar to analogous reactions at Pd(II) carboxylates. Computational studies of benzene C–H activation have been reported at Ir(κ^2-CO_3)Cp(PMe_3) [37a], [Ir(κ^2-OAc)Cp(PMe_3)]$^+$[65], and Ir(acac)$_2$(κ^2-OAc) (acac = acetylacetonoate) [66]. García-Melchor, Gorelsky, and Woo also compared the C–H activations of benzene, 2-methoxythiophene, 2,3-dimethylfuran, and *N*-methylindoles at a [Ir(Ph)(κ^2-CO_3)(PMe_3)(py)] catalyst and found reactivity trends were accommodated within a CMD-type mechanism [67].

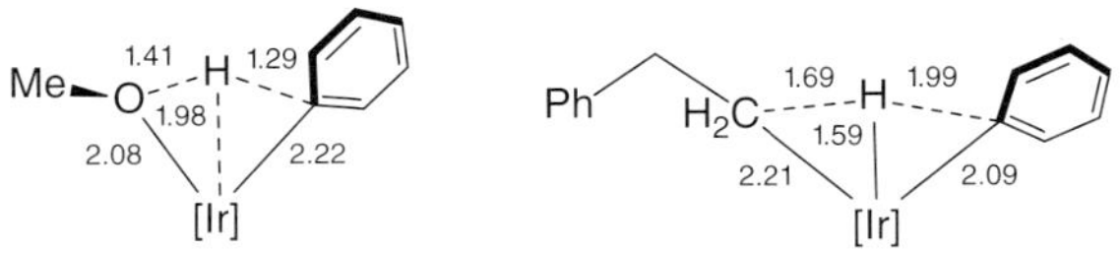

Figure 1.25 Comparison of OMe-assisted and σ-bond metathesis transition states at [Ir]R species ([Ir] = Ir(acac)$_2$; R = OMe, **TS50** [68]; R = PhCH$_2$CH$_2$, **TS51** [69]). Key selected distances are given in Å.

In addition to processes involving a bidentate base, C–H activation can also be facilitated by a monodentate ligand, X (X = OR, NR$_2$), acting as a heteroatomic base. This results in a net 1,2-addition of the C–H bond over the M–X single bond via a four-membered transition state. Two key computational studies characterized this process. Firstly, Oxgaard, Goddard, and coworkers studied the C–H activation of benzene at Ir(acac)$_2$(OMe) using B3LYP calculations [68]. Orbital analysis of the computed pathway via **TS50** (Figure 1.25) highlighted the role of the oxygen lone pair in promoting a proton transfer. This contrasts with a related σ-bond metathesis process (**TS51**) [69], where a pair of electrons in an M–C bond is used to accept the transferring proton. These authors categorized the heteroatom-assisted process as an "internal electrophilic substitution" (IES).

Shortly afterward, Gunnoe, Cundari, and coworkers reported B3LYP calculations on benzene C–H activation at a series of isoelectronic d^6 [(Tab)M(X)(PH$_3$)]^q model species (Tab = tris(azo)borate; X = OH, OPh, or NHPh and including M = Ru (q = 0) and Ir (q = 1)) [70]. Analysis of the four-membered transition state using the atoms-in-molecules approach revealed an absence of a bond critical point between the transferring hydrogen and the metal center. This is again distinct from σ-bond metathesis in related Ir(III) systems where a direct Ir–H interaction is seen. The availability of a lone pair of the heteroatomic base, X, is again the key feature, leading the authors to characterize this process as in intramolecular deprotonation. The interaction of the electron-deficient metal center also renders the C–H bond more acidic and so facilitates deprotonation.

Ess, Periana, Goddard, and coworkers subsequently compared four-membered and six-membered C–H activation transition states in a B3LYP study on the C–H activation of benzene at Ir(acac)$_2$(X) species (X = OH, OAc) [66]. An activation strain model was used to compare the performance of hydroxide and acetate, where the latter could access both four-membered and six-membered transition states, depending on whether the proton transfers onto the Ir-bound or pendant oxygen, respectively. C–H activation via a six-membered process is favored and is associated with a lower ΔE_{dist} term compared to the more constrained four-membered transition states. The acetate and hydroxide four-membered transition states are very similar.

More recently, further computational studies of the C–H activation of benzene at Ir [71] and Rh [72] have appeared, while Jones and coworkers compared aryl C(sp^2)–H bond activation and alkyl C(sp^3)–H bond activation

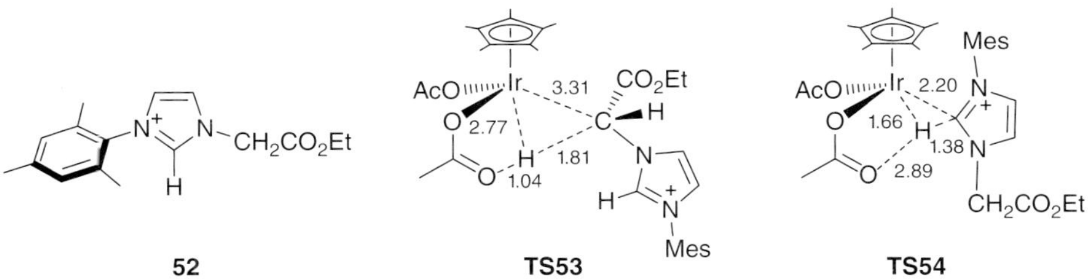

52 **TS53** **TS54**

$\Delta G^{\ddagger}_{calc}$ = +41.6 kcal/mol $\Delta G^{\ddagger}_{calc}$ = +44.7 kcal/mol

Figure 1.26 Key distances (Å) and relative free energies (kcal mol^{-1}) in transition states for C(sp^3)–H activation (**TS54**) and C(sp^2)–H activation (**TS55**) of an α-imidazolium ester, **53**, at $Ir(OAc)_2Cp^*$.

of isopropylbenzene at $[M(bdmpza)Cl_3]^-$ complexes (M = Rh, Ir; bdmpza = *bis*-(3,5-dimethylpyrazol-1-yl)acetate; B3LYP(TFA)) [73]. Competing C(sp^2)–H and C(sp^3)–H bond activation processes of α-pyridinium and α-imidazolium esters (Figure 1.26, **52**) at $[IrCl_2Cp^*]_2$/NaOAc were reported by Cross and coworkers. M06(CH_2Cl_2)//B3LYP calculations showed the observed C(sp^3)–H activation to be kinetically favored over the thermodynamically preferred C(sp^2)–H activation [74]. Both processes involve a nondirected C–H activation and initiate from H-bonded adducts formed at Ir(κ^1-OAc)$_2$Cp*. Although the computed barriers of >40 kcal mol^{-1} are rather high for room temperature processes, the different transition state geometries show some interesting features: C(sp^3)–H activation is dominated by interaction with the base (**TS53**: C···H = 1.81 Å, O···H = 1.04 Å, Ir···H = 2.77 Å), while the metal plays a more prominent role in the C(sp^2)–H activation (**TS54**: C···H = 1.38 Å, O···H = 2.89 Å, Ir···H = 1.66 Å). This is consistent with the AMLA concept in which the complementary involvement of base and metal combines to accommodate the activation of a wide range of different C–H bonds [59].

1.3.3
C–H Activation and Functionalization

A range of studies have now appeared modeling complete catalytic cycles based on an initial C–H bond activation and the subsequent functionalization steps. The functionalization processes involve reactions with an insertion partner (an alkyne, an alkene, or a carbene/nitrene source) and lead to the formation of new heterocyclic rings or direct replacement of the C–H bond with a new C–R group. In many cases, the focus of the computational studies is on these functionalization processes as well as the reoxidation steps (particularly with internal oxidants). The C–H activation in these catalytic processes is typically of the intramolecular AMLA-6 type, although exceptions do exist, as will be detailed later. Input from parallel experimental studies, including the determination of k_H/k_D KIEs, the observation of H/D exchange, competition experiments, and the isolation of key intermediates both complement and provide benchmark data for the computational modeling. In the following, figures will show the model systems used in the calculations and relevant experimental results and conditions will be given in the text.

1.3.3.1 Heterocycle Formation with Internal Oxidants

In 2011, Guimond and coworkers published the first computational study of Rh(III)-catalyzed heterocycle formation based on initial C–H activation and subsequent functionalization [75]. The coupling of benzamides with alkynes to give isoquinolones was considered, with an OR group (OR = OMe, OPiv) installed on the benzamide nitrogen to act as an internal oxidant. B3LYP(MeOH) calculations were performed for the model reaction of *N*-acetoxybenzamide with acetylene at a model $Rh(OAc)_2Cp$ catalyst (see Figure 1.27). After *N*-deprotonation and loss of HOAc, directed C–H activation proceeds with an overall barrier of

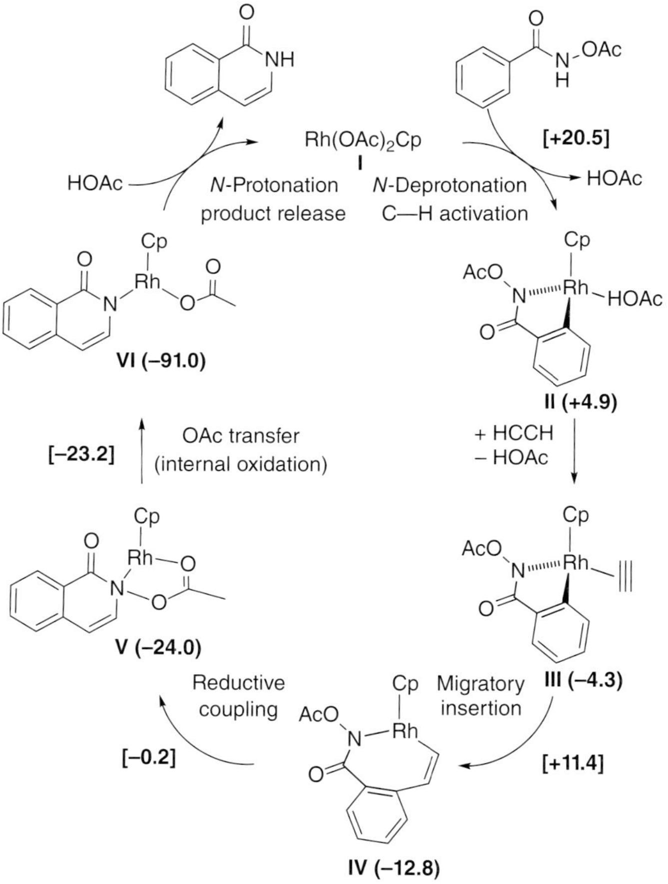

Figure 1.27 Computed catalytic cycle for the coupling of *N*-acetoxybenzamide with acetylene at $Rh(OAc)_2Cp$ [75]. Computed free energies of intermediates and transition states are given in $kcal\,mol^{-1}$, with the latter indicated in square brackets, and are quoted relative to the reactants at $0.0\,kcal\,mol^{-1}$.

20.5 kcal mol^{-1}. The alternative C–H activation of the neutral (i.e., *N*-protonated) substrate was considerably less accessible. Facile HOAc/HCCH substitution, insertion, and $N(sp^3)$–$C(sp^2)$ reductive coupling then led to intermediate **V** in which the isoquinolone is bound to Rh (Rh–N = 2.13 Å) and the *N*-acetate group bridges the Rh–N bond (N–O = 1.49 Å; Rh–O = 2.21 Å). N–O bond cleavage with transfer of acetate then occurs with a minimal barrier of 0.8 kcal mol^{-1} with concomitant formal Rh(I) to Rh(III) oxidation. Protonolysis by HOAc then releases the isoquinolone product and completes the catalytic cycle. The calculations suggest rate-limiting C–H activation, consistent with the observation of a large k_H/k_D KIE when an *N*-(pivaloyloxy)benzamide substrate was used experimentally. However, as no KIE is observed with *N*-methoxybenzamide, the nature of the rate-limiting step is clearly substrate dependent.

An important complement to this study was published by Xia and coworkers in 2012 who considered the mechanism of 3,4-dihydroquinolone formation via the coupling of alkenes with PhC(O)NH(OR), where the OR group dictates whether $N(sp^3)$–$C(sp^3)$ coupling (OR = OPiv) or β-H elimination occurs (OR = OMe; see Figure 1.28a for the case of ethene) [76]. M06(MeOH) calculations compared profiles for the reactions of both benzamides with ethene at a $Rh(OAc)_2Cp^*$ catalyst. Very similar N–H and C–H activation profiles are computed in each case, with overall barriers of about 18 kcal mol^{-1} and the endergonic formation of the HOAc adducts ($\Delta G_{calc} \approx +4$ kcal mol^{-1}) indicating reversible C–H activation. After HOAc/ethene substitution, insertion forms a conformationally flexible seven-membered rhodacycle ($\mathbf{I}_{\mathbf{OR}}$, Figure 1.28b) from which a low-energy isomer can be accessed when OR = OPiv ($\mathbf{I'}_{\mathbf{OR}}$, Figure 1.28c) which features an interaction between the pendant oxygen and the metal center.

From these seven-membered rhodacycles, the mechanisms diverge. For OR = OMe (Figure 1.28b), facile β-H transfer followed by formal N–H reductive coupling produces the Rh(I) intermediate $\mathbf{III}_{\mathbf{OMe}}$. Reaction with HOAc then facilitates the cleavage of the N–O bond ($\Delta G^{\ddagger}_{calc} = 13.6$ kcal mol^{-1}), forming MeOH and the Rh(III) species $\mathbf{IV}_{\mathbf{OMe}}$. A transition state for reoxidation via OMe transfer onto Rh in $\mathbf{III}_{\mathbf{OMe}}$ is 22.4 kcal mol^{-1} higher in energy. A second HOAc then protonates the amide nitrogen in $\mathbf{IV}_{\mathbf{OMe}}$, releasing the product and regenerating the catalyst. A transition state for alternative direct C–N coupling from $\mathbf{I}_{\mathbf{OMe}}$ was located, but at +43.8 kcal mol^{-1} it is too high to be competitive. In contrast, N–O bond cleavage in $\mathbf{I'}_{\mathbf{OPiv}}$ (Figure 1.28c) with transfer of the OPiv group onto Rh ($\Delta G^{\ddagger}_{calc} = 14.3$ kcal mol^{-1}) is now favored over β-H transfer and N–H reductive coupling ($\Delta G^{\ddagger}_{calc} = 23.2$ kcal mol^{-1}, $\mathbf{I}_{\mathbf{OR}}$ to $\mathbf{III}_{\mathbf{OPiv}}$, Figure 1.28b). The OPiv group is thought to play several roles in this mechanistic swap: the extra Rh–O interaction in the seven-membered rhodacycle $\mathbf{I'}_{\mathbf{OPiv}}$ disfavors the initial β-H transfer step, while the greater bulk of the OPiv unit also destabilizes the alkene intermediate, $\mathbf{II}_{\mathbf{OPiv}}$, making β-H transfer endergonic. The Rh–O interaction in $\mathbf{I'}_{\mathbf{OPiv}}$ also facilitates N–O bond cleavage and affects the internal oxidation to give a Rh(V) nitrene intermediate ($\mathbf{V}_{\mathbf{OPiv}}$, Rh–N = 1.871 Å). This species then undergoes facile C–N reductive coupling and protonolysis to complete the cycle.

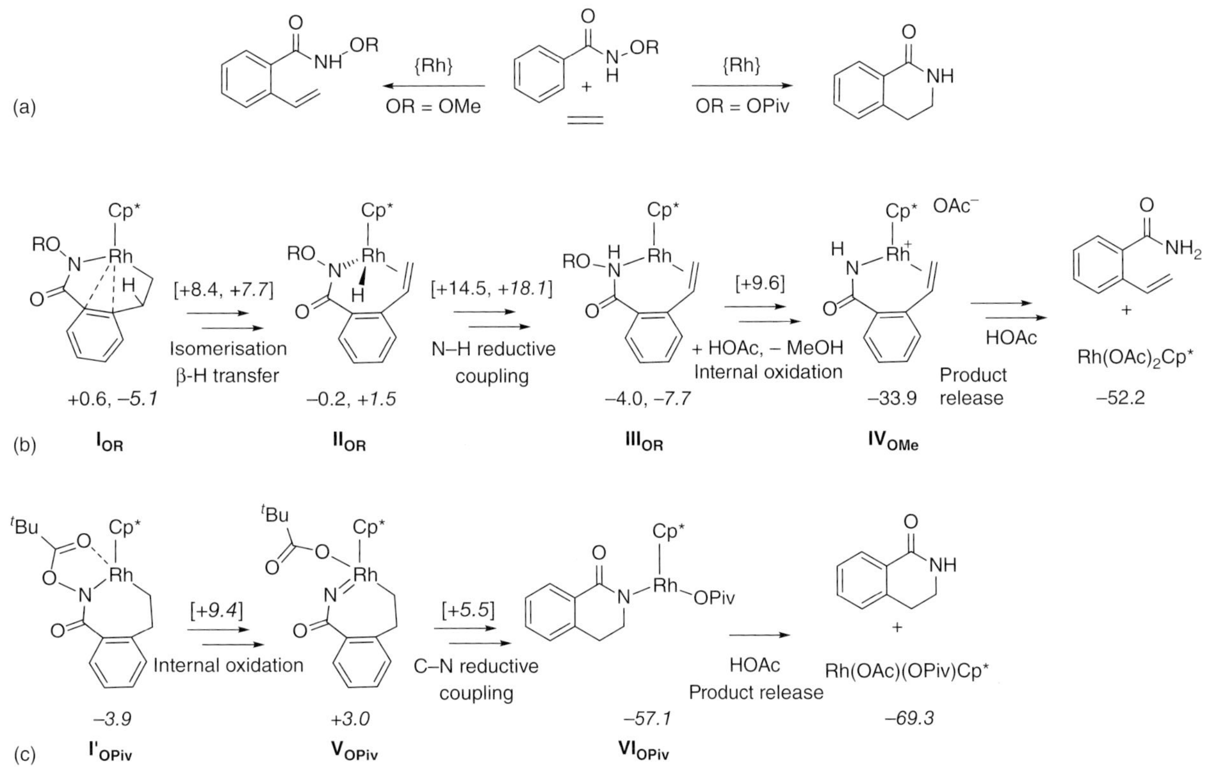

Figure 1.28 (a) Reactions of PhC(O)NH(OR) (OR = OMe, OPiv) with ethene at $Rh(OAc)_2Cp^*$ [76]. Key stationary points (kcal mol^{-1}; free energies quoted relative to the reactants at 0.0 kcal mol^{-1}) for (b) alkenylation for OR = OMe and OPiv (in italics); data for the onward reaction of $\mathbf{III}_{OR}$ for OMe only; and (c) $C(sp^3)$–N coupling for OR = OPiv. Double arrows indicate several steps are involved with the energy of the highest transition state between the two minima indicated in square brackets.

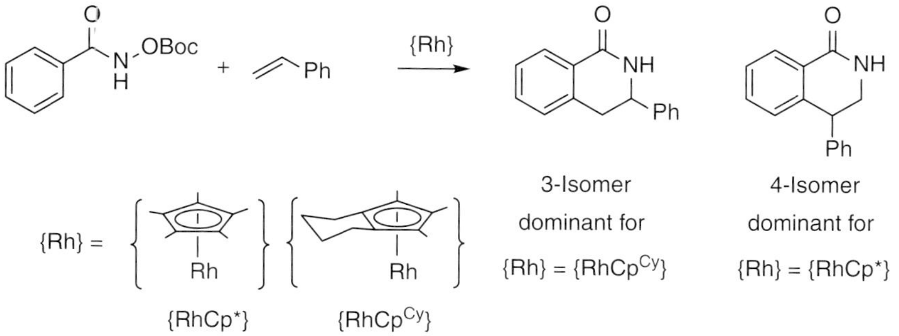

Figure 1.29 Regioselective dihydroisoquinolone formation with Rh-cyclopentadienyl catalysts [77].

The regioselectivity of dihydroisoquinoline formation between PhC(O)NHO-Boc (OBoc=OC(O)^{t}Bu) and styrene has been studied with ωB97X-D(MeOH) calculations (Figure 1.29) [77]. Experimentally, {RhCp*} catalysts favor the 4-isomer and {RhCp^{Cy}} catalysts favor the 3-isomer. Calculations based on the pathway defined by Xia above indicate very similar transition state energies for the insertion and internal oxidation steps. This made quantitative predictions about selectivity difficult, and so the computed data were used as a basis for kinetic simulations. These predicted a 76 : 24 ratio in favor of the 3-isomer with {RhCp^{Cy}}, in excellent agreement with experiment. For {RhCp*}, the computed ratio is 49 : 51, indicating that the 4-isomer does become relatively more favored, although this is underestimated compared to experiment where a 16 : 84 ratio is seen. The authors point out that a 0.5 kcal mol^{-1} shift in energy would reproduce this result, highlighting the challenge to theory when attempting to model such subtle selectivity effects.

Ma and coworkers have considered the related coupling of *N*-acetoxybenzamide with allene-enes to construct eight-membered lactam rings (Figure 1.30) [78]. Using a Rh(CO_3)Cp* catalyst, M06(MeOH) calculations suggest that the carbonate can effect both the *N*-deprotonation and (as HCO_3^-) the subsequent C–H activation. The C–H activation barrier of 20.8 kcal mol^{-1} is the highest along the computed profile, consistent with efficient catalysis at room temperature and a significant k_H/k_D KIE. C–H activation was computed to be endergonic (ΔG_{calc} = +5.4 kcal mol^{-1}), although experimentally no H/D exchange was seen in the absence of the coupling partner. Coordination of the allene-ene permits insertion of first the distal allenic double bond and then the terminal alkene to form seven- and then nine-membered rhodacyclic intermediates (**55** and **56**, respectively). The profile proceeds as in Figure 1.28c, with the key OAc transfer to give a Rh(V) nitrene again shown to be favored over β-H transfer/N–H reductive coupling. A related study by Huang and Chen on the Rh$(OAc)_2$Cp*-catalyzed coupling of PhC(O)NHOR (OR = OPiv, OMe) with both cyclohexylallene and 1,1-dimethylallene was recently reported [79]. M06(MeOH)//B3LYP calculations confirm the observed regioselectivities: with cyclohexylallene, the 4-isomer

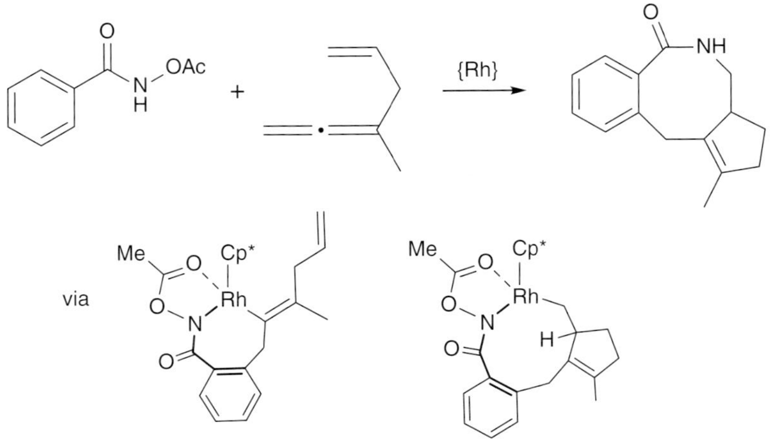

55 56

Figure 1.30 Rh-catalyzed lactam formation from *N*-acetoxybenzamide and 3-methylhexa-1,2,5-triene [78].

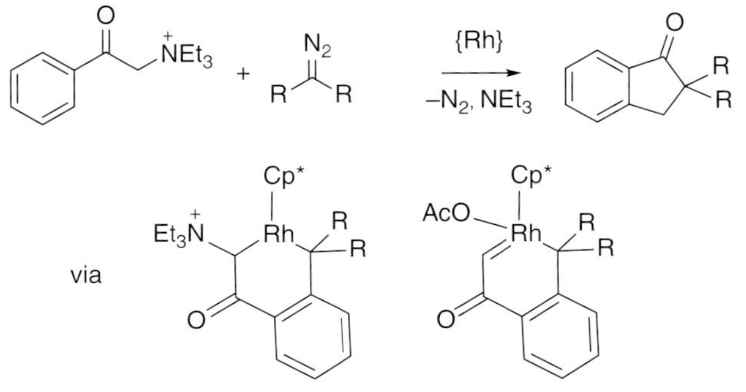

57 58

Figure 1.31 Rh-catalyzed benzocyclopentanone formation from benzophenone ammonium salts and α-diazoesters ($R = CO_2{}^iPr$) [80].

(cf. Figure 1.30) is favored, but with 1,1-dimethylallene, greater steric hindrance in the insertion transition state drives the formation of the 3-isomer.

Ammonium groups can also act as internal oxidants, with C–N bond cleavage resulting in loss of an amine and formation of carbocyclic products. Lan, Wan, Li, and coworkers used M06(MeCN) and B3LYP-D3(MeCN) calculations to model $Rh(OAc)_2Cp^*$-catalyzed benzocyclopentanone formation from benzophenone ammonium salts and α-diazoesters (Figure 1.31) [80]. Initial deprotonation and loss of HOAc forms an enolate intermediate from which C–H activation proceeds preferentially from the C-bound form ($\Delta G^{\ddagger}_{calc}$: M06 = 27.2 kcal mol^{-1}, B3LYP-D3 = 26.3 kcal mol^{-1}); barriers via the O-bound isomer were 2–3 kcal mol^{-1} higher. The subsequent steps involve diazoester coordination and N_2 loss followed by carbene insertion to give **57**. NEt_3 loss and coordination of OAc then gives **58** via a transition state at 25.2 kcal mol^{-1} (M06) or 20.7 kcal mol^{-1} (B3LYP-D3). **58** is a Rh(V) carbene complex that is isoelectronic with the previous Rh(V) nitrene species. Carbene insertion and protonolysis then complete the cycle. Overall, the initial C–H activation is rate limiting, consistent with an observed k_H/k_D KIE of about 4. C–H activation is, however, computed to be endergonic, which implies a reversibility not seen in attempted H/D exchange reactions. The authors postulate that this discrepancy may be due to the presence of CsOAc in the catalytic system that slows the reverse protonolysis.

N–N bond cleavage as a means of internal oxidation has been exploited in the coupling of 2-acetyl-1-arylhydrazines with alkynes to give indoles (Figure 1.32). This $Rh(OAc)_2Cp^*$-catalyzed reaction was studied by Chen and Lin who reported M06(DCE) energetics with an additional D3 correction [81]. After deprotonation, the terminal NAc position acts as a directing group for a two-step C–H activation via $\mathbf{I_{NAc}}$. Unusually, C–H bond cleavage from the agostic intermediate has a significant additional barrier of 10.5 kcal mol^{-1}, giving an overall barrier for C–H activation of 25.9 kcal mol^{-1}. HOAc/PhCCPh substitution gives $\mathbf{II_{NAc}}$, from which alkyne insertion (via a transition state at 29.2 kcal mol^{-1}) gives $\mathbf{III_{NAc}}$ at −5.1 kcal mol^{-1}. This isomerizes first via a facile 1,2-Rh shift to $\mathbf{IV_{NAc}}$ and then by an HOAc-assisted H^+ transfer to $\mathbf{V_{NAc}}$. C–N reductive coupling constructs the indole moiety ($\mathbf{VI_{NAc}}$) which is bound to a formally Rh(I) center and an NHC(O)Me group. The latter then transfers onto Rh to generate a Rh(III) species ($\mathbf{VII_{NAc}}$). Reaction with HOAc induces loss of H_2NOAc, protonolysis of the Rh–N bond, and release of the indole.

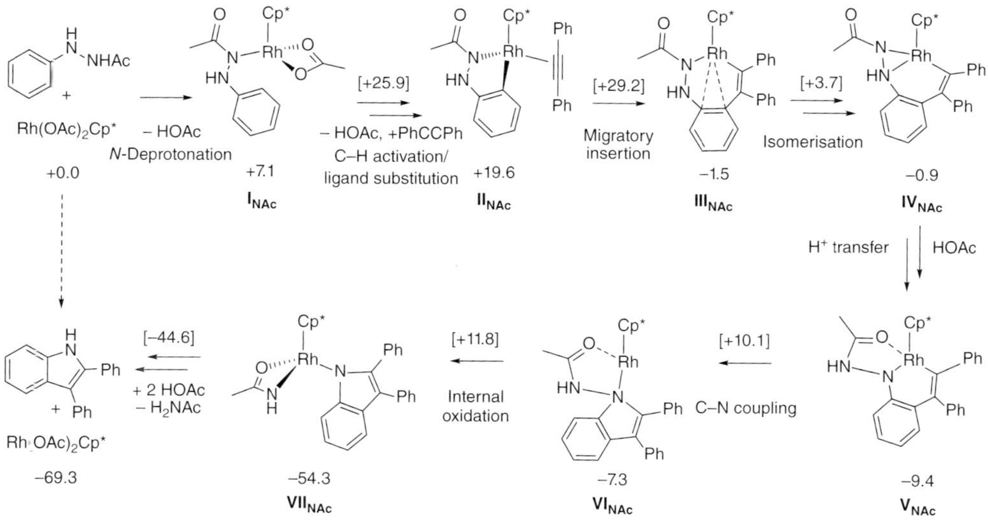

Figure 1.32 Key stationary points on free energy profiles (kcal mol^{-1}) for the $Rh(OAc)_2Cp^*$-catalyzed reaction of 2-acetyl-1-arylhydrazines with diphenylacetylene to give indoles [81]. Double arrows indicate several steps are involved, with the energy of the highest transition state between the two minima indicated in square brackets.

In contrast to previous pathways, no Rh(V) species are invoked in this process, but rather a Rh(III)–Rh(I)–Rh(III) pathway is proposed. Experimentally, the reaction proceeds at 70 °C [82] and a k_H/k_D KIE of 2.3 is observed. The rate-limiting alkyne insertion implied by the calculations therefore appears inconsistent with this; however, an endothermic (i.e., reversible) initial C–H activation step would also be expected to affect the overall rate [83]. The use of kinetic simulations to quantify these effects would be of interest. Calculations comparing the –NH(NHOAc) and –NH(OAc) groups suggest the former is a weaker oxidant that can promote C–N bond coupling with alkynes but not with alkenes.

1.3.3.2 Heterocycle Formation without Internal Oxidants

Davies, Macgregor, and coworkers have studied the $Rh(OAc)_2Cp^*$-catalyzed oxidative coupling of 5-methyl-3-phenylpyrazole and 4-octyne with BP86(DCE) and BP86-D3(DCE) calculations (Figure 1.33) [84]. Experimental H/D exchange studies indicate reversible C–H activation both in the absence (40% *ortho*-deuteration) and in the presence of the alkyne (7% incorporation). This indicates close competition between protonolysis of the initial cyclometalated intermediate, **II** (via **TS(I–II)**), and the onward reaction involving HOAc/alkyne substitution and migratory insertion (via **TS(III–IV)**). This competition was only correctly reproduced when the D3 dispersion correction was employed and primarily reflects changes in the HOAc/^{n}PrCC^{n}Pr substitution step (ΔG_{calc}:

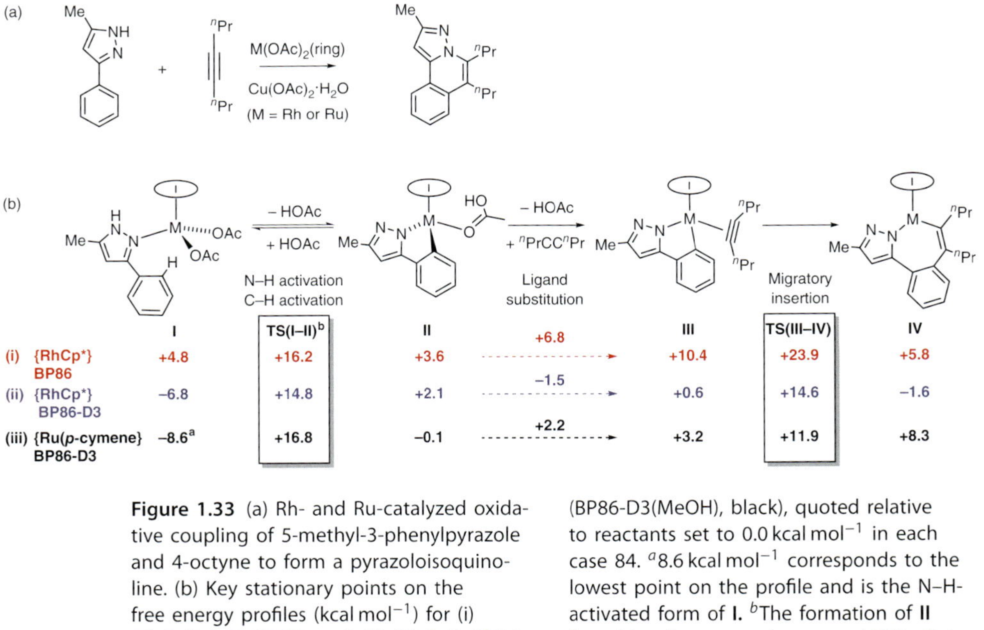

Figure 1.33 (a) Rh- and Ru-catalyzed oxidative coupling of 5-methyl-3-phenylpyrazole and 4-octyne to form a pyrazoloisoquinoline. (b) Key stationary points on the free energy profiles (kcal mol^{-1}) for (i) {RhCp*} (BP86(DCE), red), (ii) {RhCp*} (BP86-D3(DCE), blue), and (iii) {Ru(*p*-cymene)} (BP86-D3(MeOH), black), quoted relative to reactants set to 0.0 kcal mol^{-1} in each case 84. [a]8.6 kcal mol^{-1} corresponds to the lowest point on the profile and is the N–H-activated form of **I**. [b]The formation of **II** from **I** involves several steps and **TS(I–II)** is the highest point in this process.

BP86 = +6.8 kcal mol^{-1}; BP86-D3 = −1.5 kcal mol^{-1}). This is particularly sensitive to dispersion effects because of the different bulk of the two ligands involved. Alternative methods that include a treatment of dispersion effects (e.g., M06 or B97-D, or addition of the D3 correction to other pure and hybrid functionals) also performed well in this regard.

Comparison with the Ru(OAc)$_2$(*p*-cymene) catalyst proved instructive. For both metals, rate-limiting C–H activation was computed, with a higher barrier for Ru (25.4 kcal mol^{-1}) compared to Rh (21.8 kcal mol^{-1}). This is consistent with the higher temperature required for Ru experimentally (2-methylbutanol, 120 °C cf. DCE at 83 °C for Rh). For Rh, an observed k_H/k_D KIE of 2.7 ± 0.5 is consistent with the computed two-step C–H activation where the higher transition state, **TS(I−II)$_{Rh}$** (Figure 1.34), corresponds to C–H bond cleavage occurring after an agostic intermediate (a calculated KIE based on **TS(I−II)$_{Rh}$** was 5.5). In contrast, the observed k_H/k_D KIE at Ru was small (1.1 ± 0.2). In this case, however, calculations indicated a one-step C–H activation via **TS(I−II)$_{Ru}$**, which equates to the $\kappa^2-\kappa^1$ displacement of OAc. **TS(I−II)$_{Ru}$** therefore features minimal C–H bond elongation and the computed KIE is only 1.2, in good agreement with experiment. An important conclusion from this study is that the lack of a significant k_H/k_D KIE does not *necessarily* mean that C–H activation is not rate limiting; it may simply be that the C–H activation has a very early transition state geometry.

Lledós, Urriolabeitia, and coworkers have used M06(MeOH) calculations to model the oxidative coupling of benzylamines with 2-butyne at [Ru(Cl)$_2$(*p*-cymene)]$_2$ (Figure 1.35) [85]. The presence of protic N–H bonds in benzylamine promotes the formation of H-bonded ion pairs upon OAc dissociation, and these proved more reactive than non-ion-paired species where OAc was modeled as fully dissociated in solution. Thus, after OAc/amine substitution at Ru(OAc)$_2$(*p*-cymene) to give **I·OAc**, an outer-sphere C–H activation is computed via a transition state at 24.6 kcal mol^{-1}, 2.5 kcal mol^{-1} below a more conventional intramolecular process. Although this external deprotonation cannot strictly

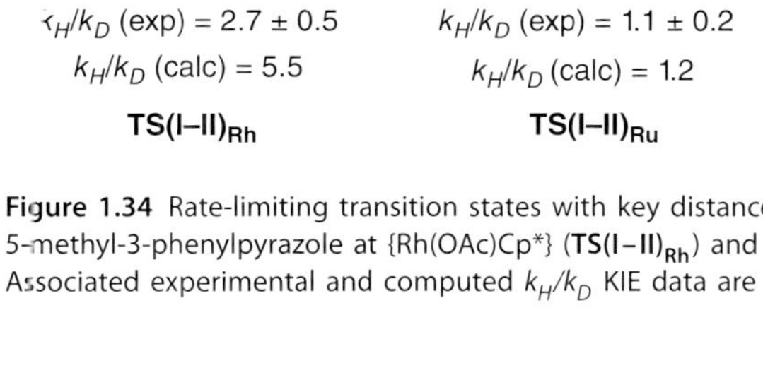

Figure 1.34 Rate-limiting transition states with key distances (Å) for the C–H activation of 5-methyl-3-phenylpyrazole at {Rh(OAc)Cp*} (**TS(I−II)$_{Rh}$**) and {Ru(OAc)(*p*-cymene)} (**TS(I−II)$_{Ru}$**). Associated experimental and computed k_H/k_D KIE data are also shown [84].

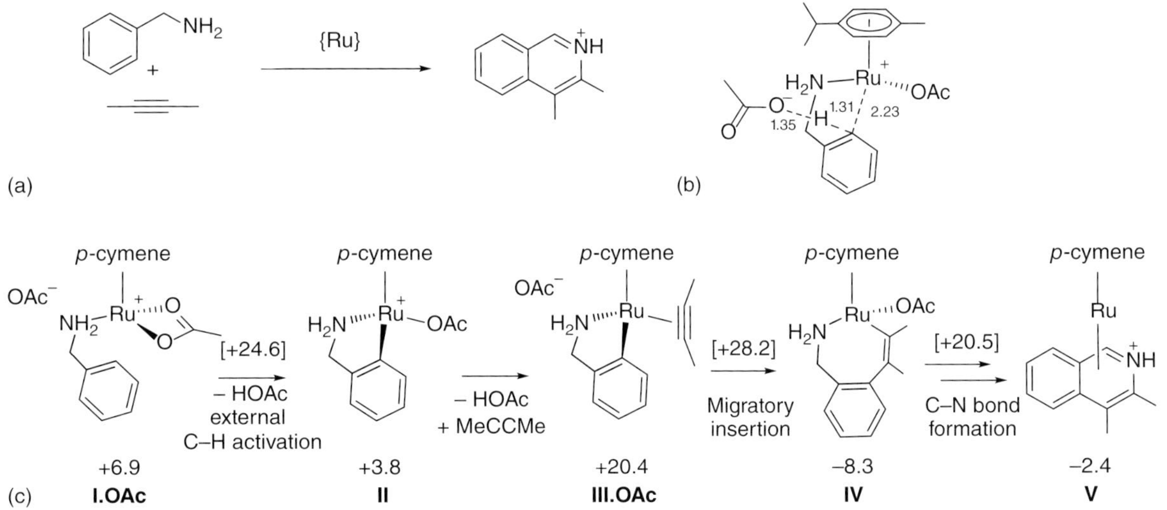

Figure 1.35 (a) Ru-catalyzed oxidative coupling of benzylamines with 2-butyne to form isoquinolones; (b) transition state for C–H activation via external deprotonation with key distances (Å); and (c) key stationary points on the free energy profile (kcal mol^{-1}; energies quoted relative to the reactants at 0.0 kcal mol^{-1}). Double arrows indicate several steps are involved, with the energy of the highest transition state between the two minima indicated in square brackets [85].

be termed *AMLA* (as this implies a coordinated ligand acting as a base), it does display some common features, namely, the interaction of the C–H bond with an electron-deficient metal center which activates it to deprotonation by a base. "CMD" would describe this concerted process appropriately. After loss of HOAc, OAc/MeCCMe substitution gives a further ion pair, **III·OAc**, that undergoes insertion to a seven-membered ruthenacycle, which is then trapped by OAc to form a saturated intermediate, **IV**. Of several routes considered for C–N bond formation, a process involving *N*-decoordination and nucleophilic attack of the free amine arm at the bound Ru-alkenyl moiety was found to be most accessible, with an overall barrier relative to **IV** of 28.8 kcal mol^{-1}; OAc is again H bonded to the amine during this process. The protonated 1,2-dihydroisoquinoline is thus formed bound to a Ru(0) metal center (**V**). Overall, no clear rate-determining process could be identified, with insertion and C–N bond formation both having similar barriers of about 29 kcal mol^{-1}, both slightly higher than C–H activation. Experimentally, temperatures of about 100 °C are required. Related profiles were computed for 1-naphthylmethyl, 2-methylallylamine, and 2-thiophenemethylamine.

The $Ir(OAc)_2Cp^*$-catalyzed oxidative coupling of benzoic acids with alkynes has been studied with BP86(MeOH) calculations by Ison and coworkers (Figure 1.36a) [86]. These confirmed the previously reported result [58a] that C–H activation proceeds via acetate assistance and not OA. The subsequent insertion of 4-octyne insertion is thought to be rate limiting and has a transition state at +29.6 kcal mol^{-1}. This is perhaps rather high for a reaction that completes after 1 h at 60 °C, and it would be interesting to see how this barrier would be affected by a treatment of dispersion effects. After insertion, C–O reductive coupling then gives the isocoumarin product bound as an η^4-ligand at an Ir(I) center. Oxidation via reaction with AgOAc to give an Ir(II)OAc intermediate (and a Ag atom) is proposed to be necessary to release the organic product, with reaction with a second equivalent of AgOAc then regenerating the $Ir(III)(OAc)_2$ catalyst.

The related synthesis of phosphaisocoumarins via the Rh-catalyzed coupling of phenylphosphonates and alkynes (Figure 1.36b) was modeled by Zhao and coworkers with B3LYP-D3(^{t}BuOH) calculations [87]. Using a model system ($Rh(OAc)_2Cp$, PhP(O)(OMe)(OH), and 2-butyne), the authors defined a pathway with an energy span [88] of 36.8 kcal mol^{-1}, which increases to 50.2 kcal mol^{-1} if

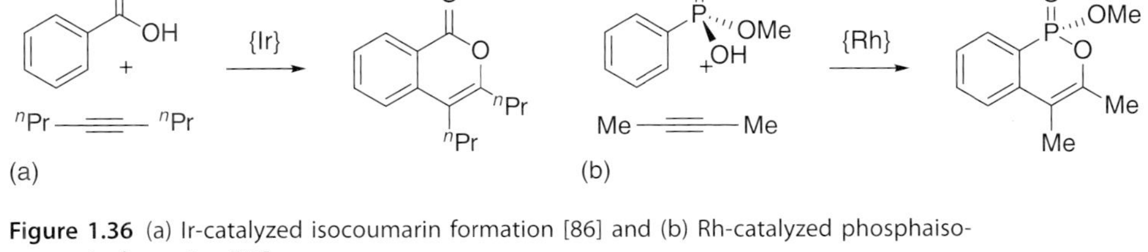

Figure 1.36 (a) Ir-catalyzed isocoumarin formation [86] and (b) Rh-catalyzed phosphaisocoumarin formation [87].

a low-energy carbonate intermediate, $Rh(CO_3)Cp\{PhP(O)(OMe)(OH)\}$, is taken into account. The computed rate-limiting transition state involves C–C coupling. It is difficult to reconcile these computed results with a process that operates efficiently at 90 °C and has an observed k_H/k_D KIE of 5.3 [89]. Use of a truncated model was addressed with tests comparing Cp and Cp* and gave similar results for the rate-limiting C–C coupling. However, these were performed with 2-butyne, an alkyne that is not used experimentally. The importance of combining the full steric bulk of all substrates and catalysts to construct a reasonable model has already been stressed [84].

1.3.4
Alkenylation and Amination

The first computational contribution to this area came from Huang, Fu, and coworkers with their 2013 study of Rh(III)-catalyzed oxidative Heck coupling of phenol carbamates with alkenes (see Figure 1.37) [90]. Using an M06(THF)//B3LYP protocol, they modeled the reaction of *m*-tolyldimethylcarbamate with ethyl acrylate at $[Rh(OAc)Cp^*]^+$, supported by preliminary work on a Cp model. An initial adduct formed via the carbamate oxygen (**I**, 0.0 kcal mol^{-1}) provides access to a one-step C–H activation at the less hindered *ortho* position with a barrier of 22.2 kcal mol^{-1} to give, after loss of HOAc, a six-membered metalacycle, **II**, at +7.9 kcal mol^{-1}. Ethyl acrylate can then undergo 2,1-insertion into the Rh–C bond with an overall barrier of 11.9 kcal mol^{-1} resulting in an eight-membered rhodacycle, **III**, at −5.5 kcal mol^{-1}. β-H transfer is then uphill ($\Delta G^{\ddagger} = 16.9$ kcal mol^{-1}; $\Delta G = +16.8$ kcal mol^{-1}), but it is assumed that the Rh(I)-H intermediate, **IV**, can then release the coupled product in the presence of $Cu(OAc)_2$ oxidant. C–H activation is therefore computed to be rate limiting, consistent with an intramolecular k_H/k_D KIE of 3.1. Experimentally, the modeled reaction requires 24 h at 110 °C [91]. An alternative mechanism based on initial

(a)

(b)

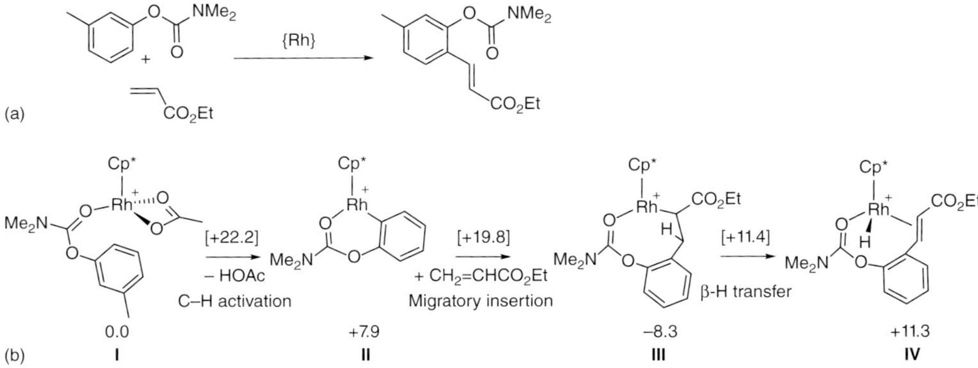

Figure 1.37 (a) Rh-catalyzed alkenylation of *m*-tolyldimethylcarbamate with ethyl acrylate and (b) key stationary points on the free energy profile (kcal mol^{-1}; energies quoted relative to the reactants at 0.0 kcal mol^{-1}) [90].

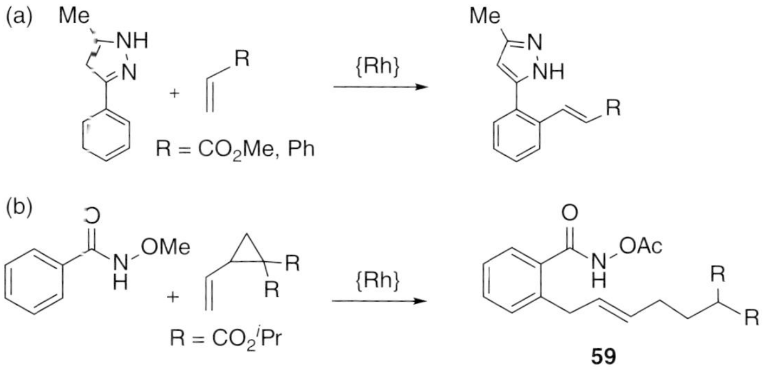

Figure 1.38 Rh-catalyzed alkenylation reactions: (a) 3-phenylpyrazole with H_2C=CHR (R = Ph, CO_2Me) [92] and (b) *N*-methoxybenzamide with dimethyl-2-vinylcyclopropane-1,1-dicarboxylate [93].

C–H activation of ethyl acrylate was also considered and found to have a higher barrier of 25.1 kcal mol^{-1}.

The related coupling of 5-methyl-3-phenylpyrazole with both methyl acrylate and styrene at $Rh(OAc)_2Cp^*$ was modeled by Davies, Macgregor, and coworkers using a BP86-D3(DCE) method (Figure 1.38a) [92]. Starting from alkene adducts of their previously computed cyclometalated intermediate [84], a clear preference for 2,1-insertion was computed for both alkenes, with barriers of about 20 kcal mol^{-1} for methyl acrylate and about 23 kcal mol^{-1} for styrene. Insertion produces conformationally flexible rhodacycles from which β-H transfer to both the *cis* and *trans* alkene products was characterized, with formation of the *trans* isomers being both kinetically and thermodynamically favored. Overall, for methyl acrylate, similar barriers for insertion and the preceding C–H activation were computed, whereas the higher insertion barrier for styrene makes this process rate limiting in this case. Using $M06(CF_3CH_2OH)$ calculations, Huang, Wang, and coworkers showed that the reaction of dimethyl-2-vinylcyclopropane-1,1-dicarboxylate with *N*-methoxybenzamide (Figure 1.38b) proceeds via rate-limiting 2,1-insertion with an overall barrier of 26.1 kcal mol^{-1}. This is preceded by a reversible C–H activation with a barrier of 21.4 kcal mol^{-1} [93]. Experimentally, a small k_H/k_D KIE of 1.7 was obtained from a parallel experiment. In this case, subsequent β-H elimination and ring opening of the pendant cyclopropane moiety gives access to a highly functionalized (*E*)-but-2-en-1-yl-dimethylmalonate moiety, **59**.

The direct C–H amination of *N-tert*-butylbenzamide with organic azides catalyzed by Gp 9 $M(OAc)_2Cp^*$ species has been modeled by Chang, Musaev, and coworkers with M06(DCE) calculations (Figure 1.39) [94]. Experimentally, $[MCl_2Cp^*]_2$ precatalysts are employed along with an $AgSbF_6$ additive, and the latter is assumed to effect the necessary C–H activation (presumably via an EA mechanism, i.e., without any specific base assistance) to arrive at a $[M(BA)Cp]^+$ intermediate (where BA is an *O*-bound, C–H-activated *N-tert*-butylbenzamide ligand). The reaction proceeds by coordination of PhN_3 followed by loss of N_2 to

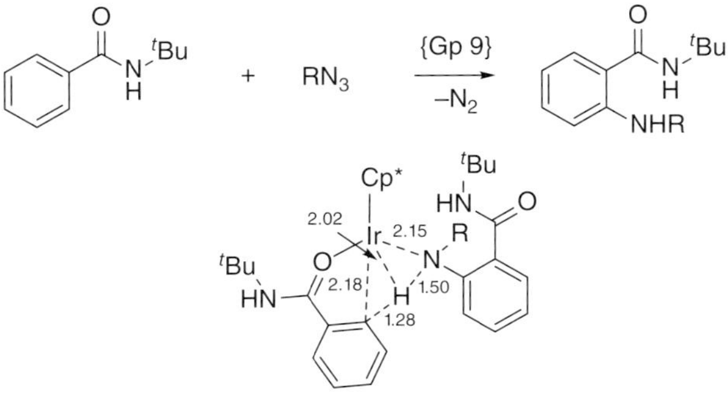

Figure 1.39 Gp 9-catalyzed amination of *N-tert*-butylbenzamide with organic azides, along with the structure of the AMLA-4 C–H activation transition state, **TS60**, with key distances in Å.

form a M(V) nitrene that can then insert into the M–C bond. Protodemetalation then proceeds via C–H activation of the benzamide substrate, in which the amido nitrogen acts as an internal base (an AMLA-4 step; see **TS60**, Figure 1.39). This releases the amination product and regenerates the $[M(BA)Cp]^+$ active species. When M = Rh, significant barriers are computed for N_2 loss (30.9 kcal mol^{-1}) and the final C–H activation step (40.6 kcal mol^{-1}). The equivalent barriers for Ir are 29.5 and 33.9 kcal mol^{-1}, respectively, with the reduced barriers to C–H activation arising from enhanced charge density on the nitrene nitrogen in this case. Experimentally, the Ir system is more promising with the reaction of ArN_3 (Ar = 3,5-$(CF_3)_2C_6H_3$) proceeding in 47% yield (90 min, 65 °C); in contrast, Rh gives an 8% yield under these conditions. A significant k_H/k_D KIE is seen observed for the Ir system and $\Delta H^{\ddagger}{}_{exp}$ is 20.3 kcal mol^{-1}, in reasonable agreement with the computed value of 17.9 kcal mol^{-1}. In contrast, the small value of $\Delta S^{\ddagger}{}_{exp}$ (1.4 cal mol^{-1} K^{-1}) is at odds with the strongly associative mechanism implied computationally. For the Co(III) analog, barriers of 31.7 kcal mol^{-1} (N_2 loss) and 51.9 kcal mol^{-1} (C–H activation) are computed, and indeed the Co system is an ineffective catalyst experimentally.

1.4 Conclusions

This chapter has surveyed the contribution of computational chemistry in understanding the mechanisms of heteroatom-assisted C–H bond activations at Pd as well as Ru, Rh, and Ir metal centers. The underlying basis of these processes is now well understood with the combination of an electron-deficient metal center and a chelating base providing a powerful means to cleave C–H bonds, especially when a directing group is employed to facilitate this process. Intermolecular C–H activation of (hetero)arenes is now well documented, particularly at Pd(II) where the

CMD concept has provided significant insight into both activities and selectivities. This process is now being exploited in Pd-catalyzed direct arylation chemistry and is proving an effective means to produce functionalized heterocycles. Challenges in the mechanistic understanding of such processes still remain, however, not least in defining the true nature of the Pd active species and the role of additives. Intramolecular C–H activation in the presence of a directing group is providing the basis for heterocycle synthesis via oxidative coupling reactions, particularly with alkynes and alkenes, and most commonly at Rh(III). The computation of full catalytic cycles based on C–H activation and functionalization is becoming more prevalent, and the discipline is well placed to play an active role in furthering efforts in this area. In doing so, the use of more realistic models that better reflect the systems used experimentally will be important, and this is made possible by the availability of increased computational power. However, such models bring with them the dual challenges of a correct treatment of weak interactions and an assessment of conformational flexibility. Close interaction with experiment is still vital in order to benchmark computational protocols against experimental data, and the most valuable studies in this chapter have been based on a close alignment of experiment and computation. Improved mechanistic insight therefore demands input from experiment in the form of quantitative experimental data, while computational chemists must be stringent in testing their methods against such data. With this approach, the role of computation in integrating C–H activation into ever-more ambitious catalytic schemes has a vibrant future.

Acknowledgments

We thank Prof Dai Davies (University of Leicester) and his group for many fruitful discussions and the EPSRC for financial support through awards EP/J002712/1 (KJTC) and EP/J021911/1 (CLM).

References

1. (a) Balcells, D., Clot, E., and Eisenstein, O. (2010) *Chem. Rev.*, **110**, 749–823; (b) Boutadla, Y., Davies, D.L., Macgregor, S.A., and Poblador-Bahamonde, A.I. (2009) *Dalton Trans.*, 5820–5831; (c) Perutz, R.N. and Sabo-Etienne, S. (2007) *Angew. Chem. Int. Ed.*, **46**, 2578–2592; (d) Vastine, B.A. and Hall, M.B. (2009) *Coord. Chem. Rev.*, **253**, 1202–1218; (e) Lapointe, D. and Fagnou, K. (2010) *Chem. Lett.*, **39**, 1119–1126.
2. For recent related reviews see: (a) Guan, W., Sayyed, F.B., Zeng, G., and Sakaki, S. (2014) *Inorg. Chem.*, **53**, 6444–6457; (b) Pahls, D.R., Allen, K.E., Goldberg, K.I., and Cundari, T.R. (2014) *Organometallics*, **33**, 6413–6419.
3. Grimme, S., Antony, J., Ehrlich, S., and Krieg, H. (2010) *J. Chem. Phys.*, **132**, 154104.
4. Grimme, S. (2006) *J. Comput. Chem.*, **27**, 1787–1799.
5. Chai, J.-D. and Head-Gordon, M. (2008) *Phys. Chem. Chem. Phys.*, **10**, 6615–6620.
6. Zhao, Y. and Truhlar, D. (2008) *Theor. Chem. Acc.*, **120**, 215–241.
7. Minenkov, Y., Occhipinti, G., and Jensen, V.R. (2009) *J. Phys. Chem. A*, **113**, 11833–11844.

8. Davies, D.L., Donald, S.M.A., and Macgregor, S.A. (2005) *J. Am. Chem. Soc.*, **127**, 13754–13755.
9. Ryabov, A.D., Sakodinskaya, I.K., and Yatsimirsky, A.K. (1985) *J. Chem. Soc., Dalton Trans.*, 2629–2638.
10. García-Cuadrado, D., de Mendoza, P., Braga, A.A.C., Maseras, F., and Echavarren, A.M. (2007) *J. Am. Chem. Soc.*, **129**, 6880–6886.
11. García-Cuadrado, D., Braga, A.A.C., Maseras, F., and Echavarren, A.M. (2006) *J. Am. Chem. Soc.*, **128**, 1066–1067.
12. Lafrance, M., Gorelsky, S.I., and Fagnou, K. (2007) *J. Am. Chem. Soc.*, **129**, 14570–14571.
13. Aguilar, D., Bielsa, R., Contel, M., Lledós, A., Navarro, R., Soler, T., and Urriolabeitia, E.P. (2008) *Organometallics*, **27**, 2929–2936.
14. Chaumontet, M., Piccardi, R., Audic, N., Hitce, J., Peglion, J.-L., Clot, E., and Baudoin, O. (2008) *J. Am. Chem. Soc.*, **130**, 15157–15166.
15. Kefalidis, C.E., Baudoin, O., and Clot, E. (2010) *Dalton Trans.*, **39**, 10528–10535.
16. Rousseaux, S., Davi, M., Sofack-Kreutzer, J., Pierre, C., Kefalidis, C.E., Clot, E., Fagnou, K., and Baudoin, O. (2010) *J. Am. Chem. Soc.*, **132**, 10706–10716.
17. Häller, L.J.L., Page, M.J., Macgregor, S.A., Mahon, M.F., and Whittlesey, M.K. (2009) *J. Am. Chem. Soc.*, **131**, 4604–4605.
18. Rousseaux, S., Gorelsky, S.I., Chung, B.K.W., and Fagnou, K. (2010) *J. Am. Chem. Soc.*, **132**, 10692–10705.
19. Aullón, G., Chat, R., Favier, I., Font-Bardia, M., Gómez, M., Granell, J., Martínez, M., and Solans, X. (2009) *Dalton Trans.*, 8292–8300.
20. Anand, M. and Sunoj, R.B. (2012) *Organometallics*, **31**, 6466–6481.
21. Dang, Y., Qu, S., Nelson, J.W., Pham, H.D., Wang, Z.-X., and Wang, X. (2015) *J. Am. Chem. Soc.*, **137**, 2006–2014.
22. Cheng, G.-J., Yang, Y.-F., Liu, P., Chen, P., Sun, T.-Y., Li, G., Zhang, X., Houk, K.N., Yu, J.-Q., and Wu, Y.-D. (2014) *J. Am. Chem. Soc.*, **136**, 894–897.
23. Yang, Y.-F., Cheng, G.-J., Liu, P., Leow, D., Sun, T.-Y., Chen, P., Zhang, X., Yu, J.-Q., Wu, Y.-D., and Houk, K.N. (2014) *J. Am. Chem. Soc.*, **136**, 344–355.
24. Yoo, E.J., Ma, S., Mei, T.S., Chan, K.S.L., and Yu, J.Q. (2011) *J. Am. Chem. Soc.*, **133**, 7652–7655.
25. Anand, M., Sunoj, R.B., and Schaefer, H.F. III, (2014) *J. Am. Chem. Soc.*, **136**, 5535–5538.
26. (a) Musaev, D.G., Figg, T.M., and Kaledin, A.L. (2014) *Chem. Soc. Rev.*, **43**, 5009–5031; (b) Figg, T.M., Wasa, M., Yu, J.-Q., and Musaev, D.G. (2013) *J. Am. Chem. Soc.*, **135**, 14206–14214.
27. Roiban, G.-D., Serrano, E., Soler, T., Aullón, G., Grosu, I., Cativiela, C., MartÍnez, M., and Urriolabeitia, E.P. (2011) *Inorg. Chem.*, **50**, 8132–8143.
28. Cross, W.B., Hope, E.G., Lin, Y.-H., Macgregor, S.A., Singh, K., Solan, G.A., and Yahya, N. (2013) *Chem. Commun.*, **49**, 1918–1920.
29. Chen, F.-J., Zhao, S., Hu, F., Chen, K., Zhang, Q., Zhang, S.-Q., and Shi, B.-F. (2013) *Chem. Sci.*, **4**, 4187–4192.
30. Giri, R., Lan, Y., Liu, P., Houk, K.N., and Yu, J.-Q. (2012) *J. Am. Chem. Soc.*, **134**, 14118–14126.
31. (a) Katayev, D., Larionov, E., Nakanishi, M., Besnard, C., and Kündig, E.P. (2014) *Chem. Eur. J.*, **20**, 15021–15030; (b) Larionov, E., Nakanishi, M., Katayev, D., Besnard, C., and Kündig, E.P. (2013) *Chem. Sci.*, **4**, 1995–2005.
32. Biswas, B., Sugimoto, M., and Sakaki, S. (2000) *Organometallics*, **19**, 3895–3908.
33. Lafrance, M., Rowley, C.N., Woo, T.K., and Fagnou, K. (2006) *J. Am. Chem. Soc.*, **128**, 8754–8756.
34. Lafrance, M. and Fagnou, K. (2006) *J. Am. Chem. Soc.*, **128**, 16496–16497.
35. Gorelsky, S.I., Lapointe, D., and Fagnou, K. (2008) *J. Am. Chem. Soc.*, **130**, 10848–10849.
36. Lapointe, D., Markiewicz, T., Whipp, C.J., Toderian, A., and Fagnou, K. (2011) *J. Org. Chem.*, **76**, 749–759.
37. (a) Gorelsky, S.I., Lapointe, D., and Fagnou, K. (2012) *J. Org. Chem.*, **77**, 658–668; (b) Gorelsky, S.I. (2013) *Coord. Chem. Rev.*, **257**, 153–164; (c) Gorelsky, S.I. (2012) *Organometallics*, **31**, 794–797.
38. van Zeist, W.-J. and Bickelhaupt, F.M. (2010) *Org. Biomol. Chem.*, **8**, 3118–3127.

39. Guihaumé, J., Clot, E., Eisenstein, O., and Perutz, R.N. (2010) *Dalton Trans.*, **39**, 10510–10519.
40. Petit, A., Flygare, J., Miller, A.T., Winkel, G., and Ess, D.H. (2012) *Org. Lett.*, **14**, 3680–3683.
41. Liégault, B., Petrov, I., Gorelsky, S.I., and Fagnou, K. (2010) *J. Org. Chem.*, **75**, 1047–1060.
42. Wakioka, M., Nakamura, Y., Hihara, Y., Ozawa, F., and Sakaki, S. (2014) *Organometallics*, **33**, 6247–6252.
43. (a) Yanagisawa, S., Ueda, K., Sekizawa, H., and Itami, K. (2009) *J. Am. Chem. Soc.*, **131**, 14622–14623; (b) Ueda, K., Yanagisawa, S., Yamaguchi, J., and Itami, K. (2010) *Angew. Chem. Int. Ed.*, **49**, 8946–8949.
44. Steinmetz, M., Ueda, K., Grimme, S., Yamaguchi, J., Kirchberg, S., Itami, K., and Studer, A. (2012) *Chem. Asian J.*, **7**, 1256–1260.
45. Tang, S.-Y., Guo, Q.-X., and Fu, Y. (2011) *Chem. Eur. J.*, **17**, 13866–13876.
46. Cong, X., Tang, H., Wu, C., and Zeng, X. (2013) *Organometallics*, **32**, 6565–6575.
47. Zhang, S., Shi, L., and Ding, Y. (2011) *J. Am. Chem. Soc.*, **133**, 20218–20229.
48. Campeau, L.-C., Bertrand-Laperle, M., Leclerc, J.-P., Villemure, E., Gorelsky, S., and Fagnou, K. (2008) *J. Am. Chem. Soc.*, **130**, 3276–3277.
49. Potavathri, S., Pereira, K.C., Gorelsky, S.I., Pike, A., LeBris, A.P., and DeBoef, B. (2010) *J. Am. Chem. Soc.*, **132**, 14676–14681.
50. Ishikawa, A., Nakao, Y., Sato, H., and Sakaki, S. (2010) *Dalton Trans.*, **39**, 3279–3289.
51. Sanhueza, I.A., Wagner, A.M., Sanford, M.S., and Schoenebeck, F. (2013) *Chem. Sci.*, **4**, 2767–2775.
52. Xi, P., Yang, F., Qin, S., Zhao, D., Lan, J., Gao, G., Hu, C., and You, J. (2010) *J. Am. Chem. Soc.*, **132**, 1822–1824.
53. Meir, R., Kozuch, S., Uhe, A., and Shaik, S. (2011) *Chem. Eur. J.*, **17**, 7623–7631.
54. Tan, Y. and Hartwig, J.F. (2011) *J. Am. Chem. Soc.*, **133**, 3308–3311.
55. Sun, H.-Y., Gorelsky, S.I., Stuart, D.R., Campeau, L.-C., and Fagnou, K. (2010) *J. Org. Chem.*, **75**, 8180–8189.
56. Tan, Y., Barrios-Landeros, F., and Hartwig, J.F. (2012) *J. Am. Chem. Soc.*, **134**, 3683–3686.
57. Stephens, D.E., Lakey-Beitia, J., Atesin, A.C., Ateşin, T.A., Chavez, G., Arman, H.D., and Larionov, O.V. (2015) *ACS Catal.*, **5**, 167–175.
58. (a) Davies, D.L., Donald, S.M.A., Al-Duaij, O., Macgregor, S.A., and Polleth, M. (2006) *J. Am. Chem. Soc.*, **128**, 4210–4211; (b) Boutadla, Y., Davies, D.L., Macgregor, S.A., and Poblador-Bahamonde, A.I. (2009) *Dalton Trans.*, 5887–5893.
59. Davies, D.L., Donald, S.M.A., Al-Duaij, O., Fawcett, J., Little, C., and Macgregor, S.A. (2006) *Organometallics*, **25**, 5976–5978.
60. Özdemir, I., Demir, S., Çetinkaya, B., Gourlaouen, C., Maseras, F., Bruneau, C., and Dixneuf, P.H. (2008) *J. Am. Chem. Soc.*, **130**, 1156–1157.
61. Carr, K.J.T., Davies, D.L., Macgregor, S.A., Singh, K., and Villa-Marcos, B. (2014) *Chem. Sci.*, **5**, 2340–2346.
62. Li, J., Hu, W., Peng, Y., Zhang, Y., Li, J., and Zheng, W. (2014) *Organometallics*, **33**, 2150–2159.
63. Li, L., Brennessel, W.W., and Jones, W.D. (2009) *Organometallics*, **28**, 3492–3500.
64. Clot, E., Eisenstein, O., Jasim, N., Macgregor, S.A., McGrady, J.E., and Perutz, R.N. (2011) *Acc. Chem. Res.*, **44**, 333–348.
65. Davies, D.L., Macgregor, S.A., and Poblador-Bahamonde, A.I. (2010) *Dalton Trans.*, **39**, 10520–10527.
66. Ess, D.H., Bischof, S.M., Oxgaard, J., Periana, R.A., and Goddard, W.A. III, (2008) *Organometallics*, **27**, 6440–6445.
67. García-Melchor, M., Gorelsky, S.I., and Woo, T.K. (2011) *Chem. Eur. J.*, **17**, 13847–13853.
68. Oxgaard, J., Tenn, W.J. III, Nielsen, R.J., Periana, R.A., and Goddard, W.A. III, (2007) *Organometallics*, **26**, 1565–1567.
69. Oxgaard, J., Muller, R.P., Goddard, W.A., and Periana, R.A. (2004) *J. Am. Chem. Soc.*, **126**, 352–363.
70. Cundari, T.R., Grimes, T.V., and Gunnoe, T.B. (2007) *J. Am. Chem. Soc.*, **129**, 13172–13182.
71. (a) Meier, S.K., Young, K.J.H., Ess, D.H., Tenn, W.J. III, Oxgaard, J., Goddard,

W.A. III, and Periana, R.A. (2009) *Organometallics*, **28**, 5293–5304; (b) Bischof, S.M., Ess, D.H., Meier, S.K., Oxgaard, J., Nielsen, R.J., Bhalla, G., Goddard, W.A. III, and Periana, R.A. (2010) *Organometallics*, **29**, 742–756.

72. Webster-Gardiner, M.S., Fu, R., Fortman, G.C., Nielsen, R.J., Gunnoe, T.B., and Goddard, W.A. III, (2015) *Catal. Sci. Technol.*, **5**, 96–100.
73. Rhinehart, J.L., Manbeck, K.A., Buzak, S.K., Lippa, G.M., Brennessel, W.W., Goldberg, K.I., and Jones, W.D. (2012) *Organometallics*, **31**, 1943–1952.
74. Cross, W.B., Razak, S., Singh, K., and Warner, A.J. (2014) *Chem. Eur. J.*, **20**, 13203–13209.
75. Guimond, N., Gorelsky, S.I., and Fagnou, K. (2011) *J. Am. Chem. Soc.*, **133**, 6449–6457.
76. Xu, L., Zhu, Q., Huang, G., Cheng, B., and Xia, Y. (2012) *J. Org. Chem.*, **77**, 3017–3024.
77. Wodrich, M.D., Ye, B., Gonthier, J.F., Corminboeuf, C., and Cramer, N. (2014) *Chem. Eur. J.*, **20**, 15409–15418.
78. Wu, S., Zeng, R., Fu, C., Yu, Y., Zhang, X., and Ma, S. (2015) *Chem. Sci.*, **6**, 2275–2285.
79. Xing, Z., Huang, F., Sun, C., Zhao, X., Liu, J., and Chen, D. (2015) *Inorg. Chem.*, **54**, 3958–3969.
80. Yu, S., Liu, S., Lan, Y., Wan, B., and Li, X. (2015) *J. Am. Chem. Soc.*, **137**, 1623–1631.
81. Chen, W.-J. and Lin, Z. (2015) *Organometallics*, **34**, 309–318.
82. Zhao, D., Shi, Z., and Glorius, F. (2013) *Angew. Chem. Int. Ed.*, **52**, 12426–12429.
83. Simmons, E.M. and Hartwig, J.F. (2012) *Angew. Chem. Int. Ed.*, **51**, 3066–3072.
84. Algarra, A.G., Cross, W.B., Davies, D.L., Khamker, Q., Macgregor, S.A., McMullin, C.L., and Singh, K. (2014) *J. Org. Chem.*, **79**, 1954–1970.
85. Ruiz, S., Villuendas, P., Ortuño, M.A., Lledós, A., and Urriolabeitia, E.P. (2015) *Chem. Eur. J.*, **21**, 8626–8636.
86. Frasco, D.A., Lilly, C.P., Boyle, P.D., and Ison, E.A. (2013) *ACS Catal.*, **3**, 2421–2429.
87. Liu, L., Wu, Y., Wang, T., Gao, X., Zhu, J., and Zhao, Y. (2014) *J. Org. Chem.*, **79**, 5074–5081.
88. Kozuch, S. and Shaik, S. (2011) *Acc. Chem. Res.*, **44**, 101–110.
89. (a) Park, Y., Seo, J., Park, S., Yoo, E.J., and Lee, P.H. (2013) *Chem. Eur. J.*, **19**, 16461–16468; (b) Seo, J., Park, Y., Jeon, I., Ryu, T., Park, S., and Lee, P.H. (2013) *Org. Lett.*, **15**, 3358–3361.
90. Zhang, Q., Yu, H.-Z., Li, Y.-T., Liu, L., Huang, Y., and Fu, Y. (2013) *Dalton Trans.*, **42**, 4175–4184.
91. Gong, T.-J., Xiao, B., Liu, Z.-J., Wan, J., Xu, J., Luo, D.-F., Fu, Y., and Liu, L. (2011) *Org. Lett.*, **13**, 3235–3237.
92. Algarra, A.G., Davies, D.L., Khamker, Q., Macgregor, S.A., McMullin, C.L., Singh, K., and Villa-Marcos, B. (2015) *Chem. Eur. J.*, **21**, 3087–3096.
93. Wu, J.-Q., Qiu, Z.-P., Zhang, S.-S., Liu, J.-G., Lao, Y.-X., Gu, L.-Q., Huang, Z.-S., Li, J., and Wang, H. (2015) *Chem. Commun.*, **51**, 77–80.
94. Figg, T.M., Park, S., Park, J., Chang, S., and Musaev, D.G. (2014) *Organometallics*, **33**, 4076–4085.

2
Pd-Catalyzed Synthesis of Nitrogen-Containing Heterocycles

Lixin Li, Xiaolei Ji, and Hanmin Huang

2.1
Introduction

Nitrogen-containing heterocycles (Figure 2.1) are prevalently used in pharmaceuticals, agrochemicals, functionalized materials, and bioactive natural products [1]. Driven by this prevalence, intense effort has been devoted to the synthesis of azaheterocycles. As such, a lot of traditional transition-metal-catalyzed methods have been extensively developed to synthesize such kinds of compounds by many leading chemists, such as Hegedus [2], Fischer [3], Larock [4], and Buchwald [5], who undoubtedly made a pioneering contribution in this area. However, until the last decade, many chemists started to pay attention to the synthesis of nitrogen heterocycles via transition-metal-catalyzed C–H activation [6].

2.2
General Consideration on Palladium Chemistry

Palladium can successfully affect an extraordinary number of C–C and C–N bond formation reactions, and thus many name reactions, such as Kumada, Stille, Suzuki, Sonogashira, and Negishi cross-coupling reaction, Heck reaction, and Wacker reaction, have been established under relatively mild reaction conditions on the basis of palladium catalysis. Furthermore, the catalytic amount of palladium loading can be used in many reactions and tolerate several functional groups, which avoids protection chemistry successfully and reduces the number of steps of synthetic routes. Therefore, palladium was found of such wide utility in many synthesis strategies. Palladium compounds exist in three oxidation states, such as Pd(0), Pd(II), and Pd(IV). Every oxidation state exhibits different chemistry properties. In common, Pd(0) as nucleophile is able to insert chemical bond of some organic compounds, such as aryl, allylic, or vinylic halides or triflates and aminals. Then Pd(0) can be easily oxidized to the Pd(II) state, subsequently, which can transform to Pd(0) to undergo reductive elimination. Based on the transformation ability of Pd(0) and Pd(II), this chemistry can be synthetically useful for the synthesis of heterocycles.

Transition Metal-Catalyzed Heterocycle Synthesis via C—H Activation, First Edition. Edited by Xiao-Feng Wu.
© 2016 Wiley-VCH Verlag GmbH & Co. KGaA. Published 2016 by Wiley-VCH Verlag GmbH & Co. KGaA.

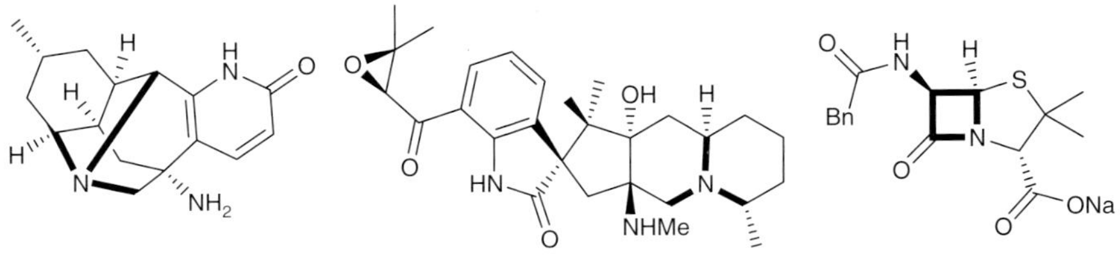

Figure 2.1 Nitrogen-containing heterocycles.

Palladium(II) complexes are extremely important in organopalladium chemistry, as we know; they are easily stored and handled due to the fact that they are soluble in most common organic solvents and stable in air. Palladium(II) complexes are typically electrophilic, so they can coordinate with many common organic substrates which are electron-rich species, such as olefins, arenes, alkynes, and some alkanes containing heteroatom. Some of the most useful Pd(II) chemistry is based on the rapidness and reversible formation of Pd(II) complexes with olefins and alkynes; then the complexes have two paths to react: (i) undergo subsequent attack by nucleophiles and (ii) activate C–H via palladacycle. Based on this, it is a powerful and useful tool for constructing the nitrogen-containing heterocycles through palladium-catalyzed C–H activation. This chemistry will be the focus of this chapter.

2.3 Heterocycle Synthesis via C(sp^3)–H Activation

The most efficient way to form many saturated N-heterocycles would be to construct a C–N bond by substituting an H atom on a C(sp^3)–H group with N. Because of the thermodynamic and kinetic stability of the C–H bond, the work of direct C–H activation posed a challenge. However, this perspective examines the different methods and strategies that have been developed for the direct C–H activation of both activated (allylic, benzylic) and inert (alkyl) C(sp^3)–H bonds (Scheme 2.1), with a particular focus on applications of these methods to the synthesis of nitrogen heterocycles [6d].

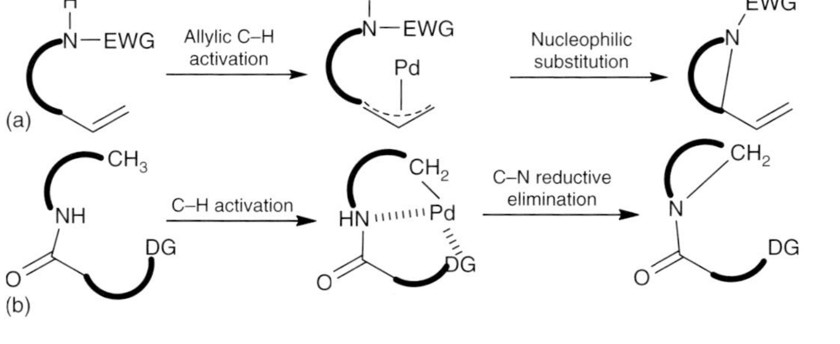

Scheme 2.1 C(sp^3)–H activation strategies: (a) the allylic C–H activation and (b) the unactivated C–H activation. DG = directing group.

2.3.1
Heterocycle Synthesis via Activated C(sp^3)–H Bonds

The pioneer work of a catalytic, intramolecular allylic C–H activation was reported by Larock and coworkers in 1996 [7]; the reaction underwent the cyclization of tosylamides **1** bearing alkene substituents to afford indoline products. It was suggested that π-allylic palladium intermediates were likely formed during the cyclization (Scheme 2.2).

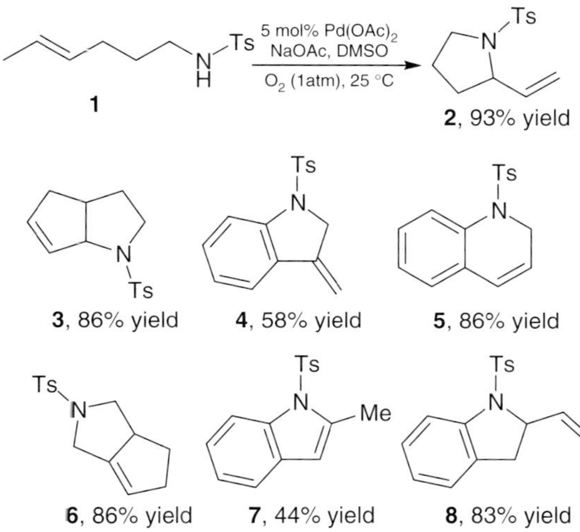

Scheme 2.2 Larock's allylic C–H activation.

The similar method was employed by Broggini and coworkers in 2004 [8]; a divergent synthesis of quinazolin-4-ones and 1,4-benzodiazepin-5-ones via Pd(II)-catalyzed intramolecular C–H activation of tosylated *N*-allyl-anthranilamides was documented (Scheme 2.3). Both formations of six- and

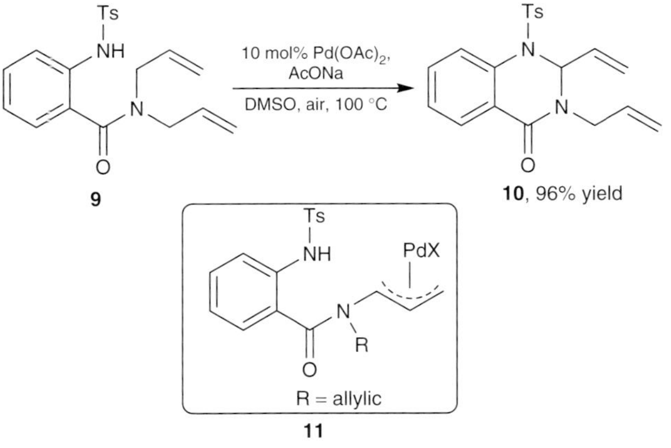

Scheme 2.3 Broggini's allylic C–H activation.

seven-membered ring were available in high yields depending on the different reaction conditions. The literature data reported a possible mechanism for amination reaction: Pd(II) activation of the carbon–carbon double bond followed by nucleophilic attack of the amine to the π-coordinated olefin **11**.

Thereafter, a serial of palladium-catalyzed intramolecular C–H activation of allylic C–H bonds (Scheme 2.4) was reported by White [9], who first developed a novel route for accessing chiral syn-1,2-amino alcohols enabled through the discovery of a Pd/sulfoxide-catalyzed diastereoselective allylic C–H activation reaction of chiral homoallylic *N*-tosyl carbamates. The author discovered that the addition of bis-sulfoxide ligands (**L-1**) to $Pd(OAc)_2$ promoted allylic C–H cleavage of α-olefins versus oxypalladation in the presence of weak oxygen nucleophiles (i.e., carboxylic acids). Evidences presented that the mechanism proceeds via Pd(II)/sulfoxide-promoted allylic C–H cleavage to furnish a π-allylic palladium intermediate followed by counterion-promoted functionalization with the tethered *N*-tosyl carbamate nucleophile.

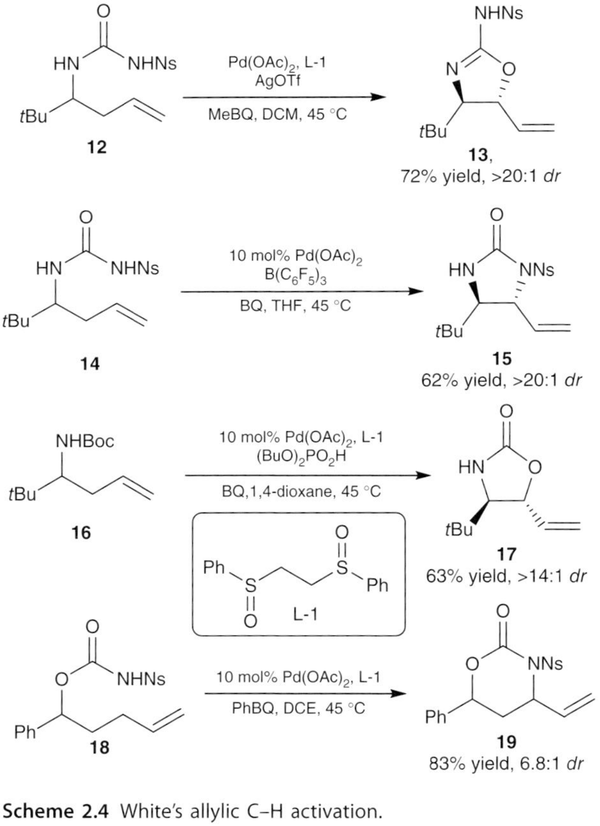

Scheme 2.4 White's allylic C–H activation.

Then, a macrolactonization reaction (Scheme 2.5) was reported by White in 2006 [10]. The strategy provided medicinal chemists with a useful tool to prepare macrolactonization of ω-alkenoic acids via Pd-catalyzed C–H activation

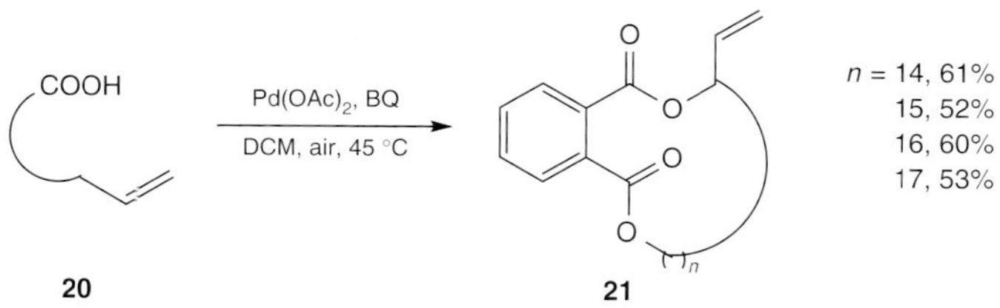

Scheme 2.5 A macrolactonization reaction via allylic C–H activation.

(Scheme 2.5). The cyclization was efficiently performed to furnish 14- to 17-membered macrocyclic benzolactons in moderate yield with outstanding regioselectivities. A possible mechanism proceeded via Pd-templated π-allyl carboxylate intermediate.

Poli reported a similar method to synthesize nitrogen-containing heterocycles via π-allyl-palladium intermediate [11]. It was worth noting that when the reaction is conducted in acetic acid rather than in methylene dichloride or terahydrofuran, a strong accelerating effect occurs in direct intramolecular allylic C(sp^3)–H cyclization (Scheme 2.6). Stoichiometric tests and computational Density Functional Theory analysis of the palladium reoxidation step provided an overview of the structural and energetic role of acetic acid in increasing the efficacy of the entire catalytic cycle (Scheme 2.6).

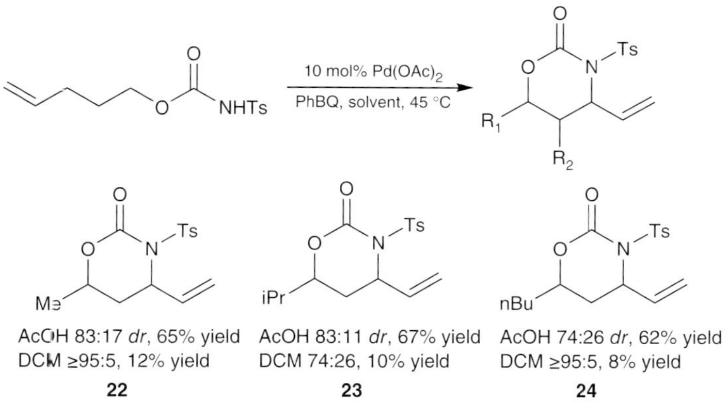

Scheme 2.6 Poli's allylic C–H activation.

Liu and coworkers reported a base-modulated regioselective Pd-catalyzed intramolecular aerobic oxidative amination of alkenes for five- or seven-membered ring products by varying the reaction conditions [12].

2.3.2 Heterocycle Synthesis via Unactivated C(sp^3)–H Bonds

In contrast with allylic C–H bonds, the alkane C–H bonds are unactivated; although a few significant accomplishments have been made in this area, the corresponding catalytic activation of unactivated C(sp^3)–H bonds is rare. Reaction mechanism described here is proposed to coordinate with the direct group and then undergo C–H activation through palladium cyclic and C–N reductive elimination from palladium complexes formed *in situ*. In common, C (sp^2)–N reductive elimination is the elementary step responsible for the release of the product, but the corresponding process is rare for alkylpalladium amido complexes. This is partly due to the tendency of alkylpalladium complexes would undergo facile β-hydride elimination instead of C(sp^3)–N reductive elimination. However, a few isolated examples of palladium-catalyzed unactivated C(sp^3)–H for heterocycle synthesis have been documented.

In 2009, Glorius reported on the first realization of this strategy in palladium-catalyzed synthesis of valuable indoline products from many differently substituted anilides [13]. Although other acyl groups, such as pivaloyl, benzoyl, or trifluoroacetyl, were tested, the *N*-acetyl group was found to be the most effective group in this transformation. The tolerance of functional group was found to be extraordinarily broad, with the exception of acidic or basic functional groups, such as carboxylic acids or basic amines (Scheme 2.7).

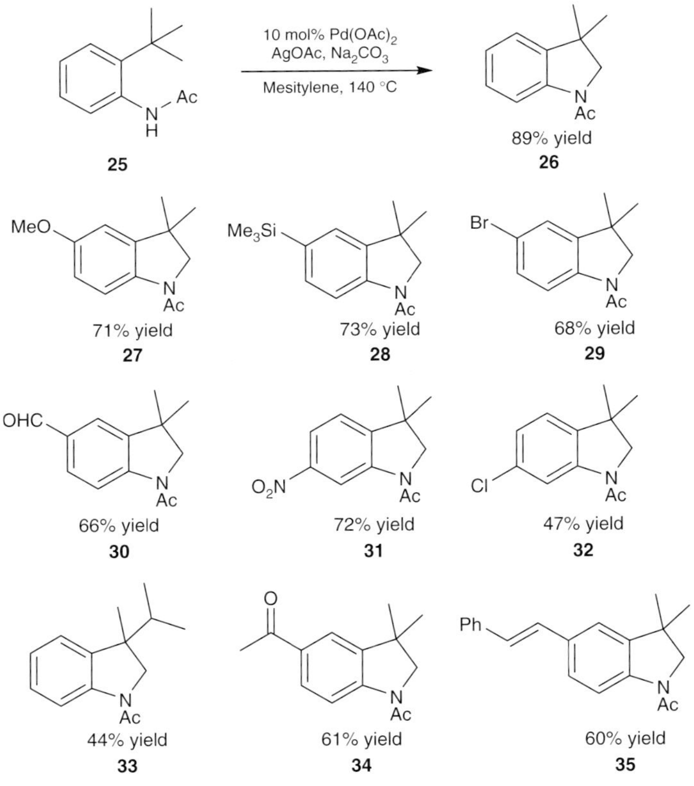

Scheme 2.7 Synthesis of indolines by C(sp^3)–H activation.

It is easy to find out the difficulty associated with C(sp^3)–H as compared to the C(sp^2)–H activation by a competition experiment with biaryl substrate (Scheme 2.8); However, a slight preference was found for the functionalization of C–H bonds at the activated position through the intramolecular competition between an activated benzylic position and two unactivated methyl groups (Scheme 2.9).

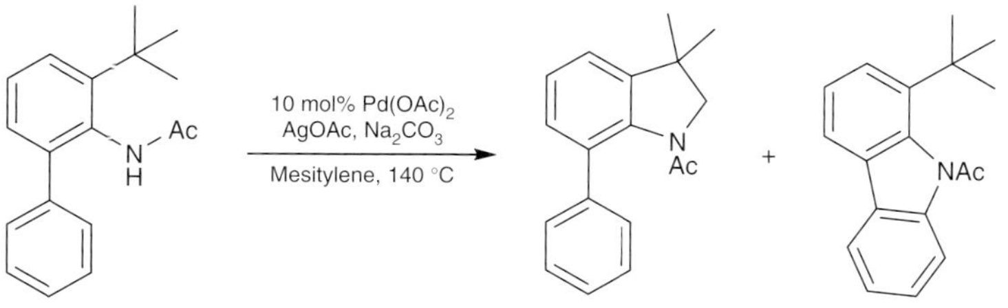

36 **37**, 0% **38**, 39% yield

Scheme 2.8 The relative reactivity of activated versus unactivated C–H bonds.

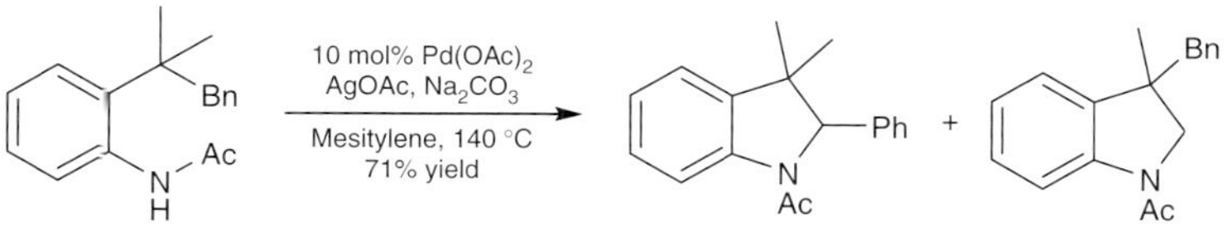

39 **40** 52:48 **41**

Scheme 2.9 Intramolecular competition experiment.

A possible reaction mechanism for the palladium-catalyzed cyclization of *N*-(2-*tert*-butylphenyl)acetamide is shown in Scheme 2.10. The acetanilide moiety acting as a directing group can coordinate to the palladium(II) catalyst, replacing the acetate ligand, and then the activation of the unactivated C–H bond of an alkyl group can take place to give intermediate. Subsequent reductive elimination of palladium(II) leads to the formation of indoline product.

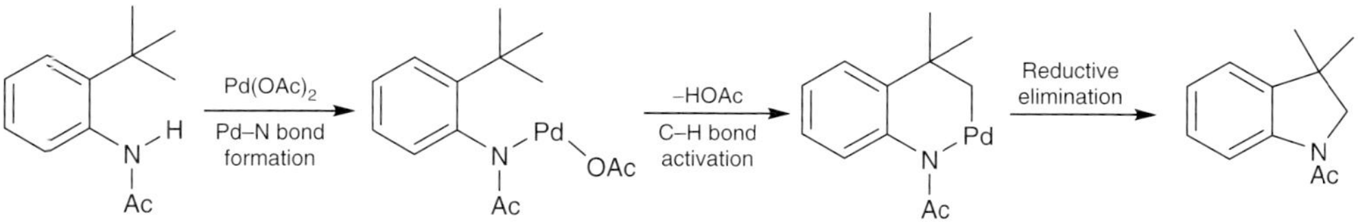

42 **43** **44** **45**

Scheme 2.10 Proposed mechanism for indoline formation.

Directing groups as an important tool have been applied in the palladium(II)-catalyzed C–H activation. Except for the acetate ligand, others also have been employed. An initial report by Daugulis in 2005 introduced the picolinamide group for the regioselective arylation of γ-C(sp^3)–H groups [14]. Since then, the picolinamide group has been successfully employed in a number of diverse transformations.

In 2012, Chen and coworkers developed the palladium-catalyzed intramolecular amination of C(sp^3)–H at γ and δ positions to synthesize a series of nitrogen-containing heterocycles [15], including azetidine (Scheme 2.11), pyrrolidines, and indolines. With the optimal conditions, using catalytic $Pd(OAc)_2$, an oxidant ($PhI(OAC)_2$, 2.5 equiv.) and an acid additive (AcOH) in toluene at 110 °C, Chen tested them on other picolinamide substrates bearing primary γ-C(sp^3)–H bonds; surprisingly, the seemingly unfavorable four-membered azetidine was obtained as the major product. It is possible that a Pd(IV) intermediate was formed via $PhI(OAC)_2$ oxidation of the palladacycle intermediate because the subsequent C–N and C–O reductive elimination pathways would lead to the formation of the cyclized and acetoxylated product. It was also noteworthy that no β-H elimination product was detected under the aforementioned reaction conditions.

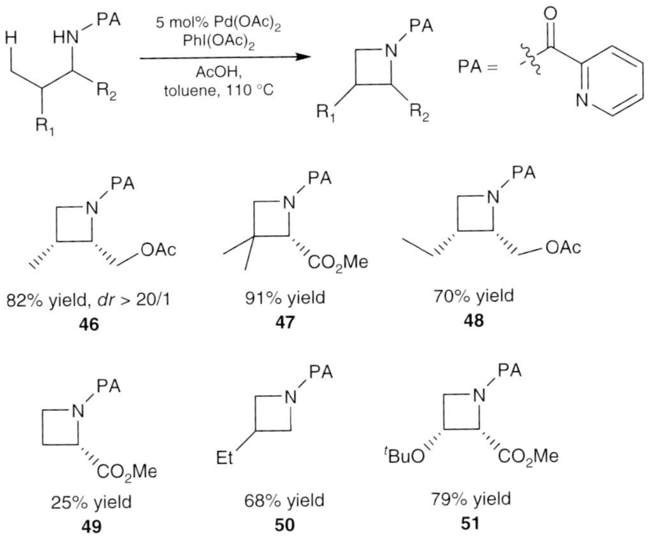

Scheme 2.11 Azetidine formation.

The same strategy was employed for the synthesis of pyrrolidines (Scheme 2.12) from substrates containing δ-methyl C–H bonds. The author believed that compared with the ring contraction from a five-membered palladacycle to a four-membered azetidine product, the formation of a five-membered pyrrolidine product from a six-membered palladacycle intermediate would be much more favorable. However, substrates lacking substituent at the γ-positions provide only low yields of the pyrrolidine products, which suggest the possible importance of the Thorpe–Ingold effect to the efficiency of the transformation.

In the same year, Daugulis reported a palladium-catalyzed method for heterocycles by employing a picolinamide as directing group [16], $PhI(OAc)_2$ as

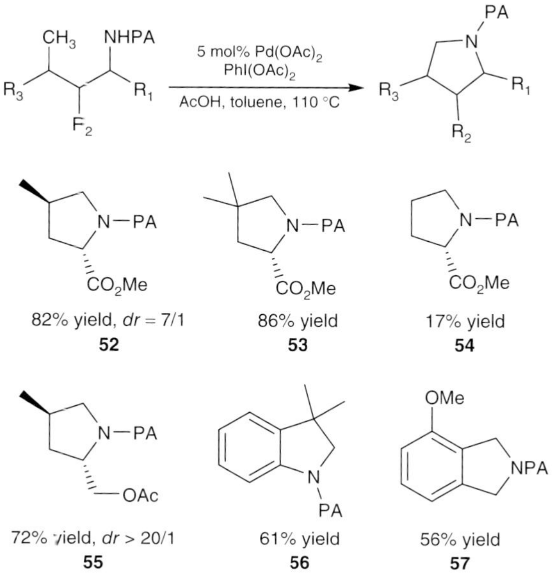

Scheme 2.12 Synthesis of pyrrolidine via C–H activation.

oxidant, and toluene as solvent at 80–120 °C. The cyclization, which forms a five-membered ring pyrrolidine product, was tolerant of some functional groups, including esters and aromatic chlorides and bromides (Scheme 2.13). The control

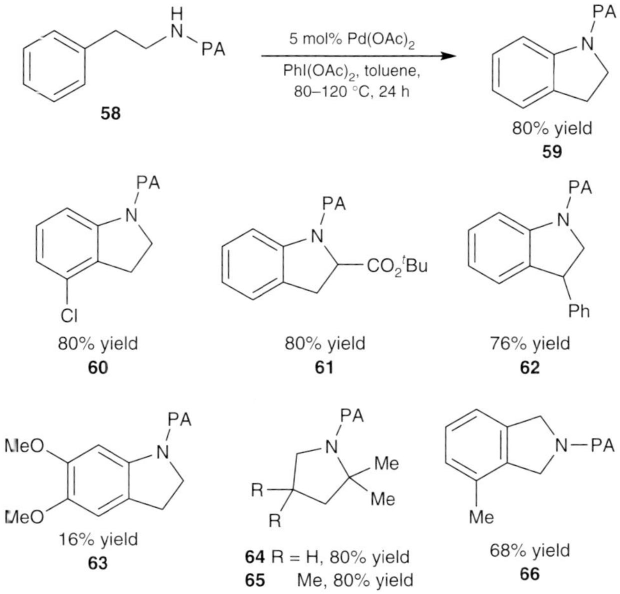

Scheme 2.13 Heterocycle synthesis via C–H activation.

experiments showed that a picolinamide directing group is essential for achieving cyclization (Scheme 2.14).

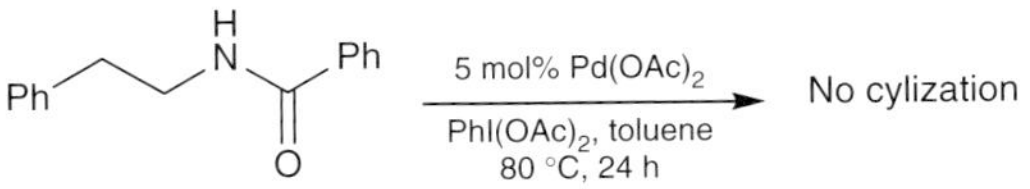

67

Scheme 2.14 Controlled experiment.

So far, several aforementioned intramolecular strategies have been established that all employ highly activated nitrogen source such as carbamates, sulfamates, amides, or azides to achieve C–H functionalization. Although all of them are powerful techniques for the introduction of nitrogen-based functionalities into aliphatic molecule, these methods are unable to provide a direct route toward tertiary amines, which are crucial structural motifs in numerous alkaloids. Therefore, the development of methods for direct N–H/C–H coupling of secondary amine and unactivated alkanes is highly desirable.

The road less traveled to amination has been achieved by Gaunt and coworkers [17]; using a palladium-catalyzed C–H activation reaction, fully aliphatic unactivated amines were converted into aziridines. Additionally, a secondary amine moiety was engaged as an efficient group for C–H functionalization of alkanes.

Most directed aliphatic C–H activation processes underwent a five-membered-ring cyclometalated intermediate. However, the article described a palladium-catalyzed C–H bond activation mode which proceeds through a remarkable four-membered-ring cyclopalladation pathway (Scheme 2.15).

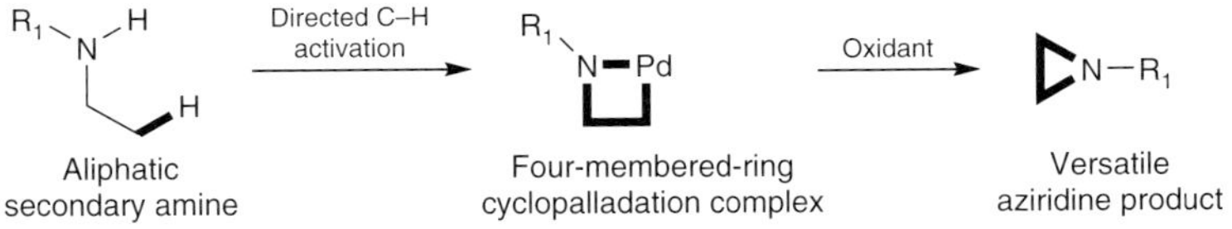

Scheme 2.15 Heterocycle synthesis via four-membered-ring cyclopalladation complex.

Gaunt and coworkers discovered that sterically demanding cyclic secondary amines underwent facile methyl group activation in the presence of palladium acetate, forming strained four-membered-ring palladacycle. Under an oxidative condition, the palladacycle was converted into aziridine through reductive elimination of palladium, leading to the formation of a C–N bond (Scheme 2.16). This strategy represents the first example of using unactivated secondary aliphatic amines to convert a C–H into a C–N bond. The transformation was demonstrated for a wide range of morpholinone-based substrates, showing both its generality and its high tolerance toward various functional groups.

Furthermore, the utility of four-membered-ring-palladacyclic intermediates was extended to the carbonylation of aliphatic C–H bonds (Scheme 2.17).

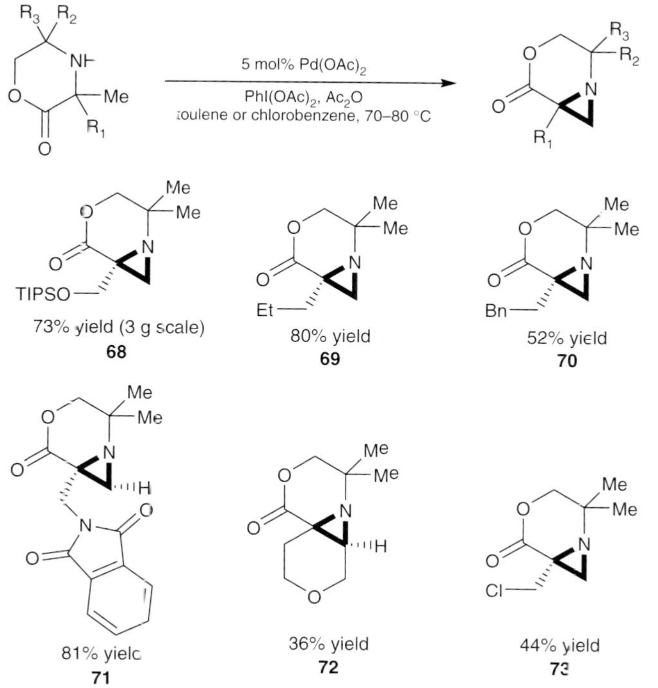

Scheme 2.16 Scope of aziridination reaction.

Importantly, this method demonstrates a broad substrate scope, as various cyclic and acyclic secondary amines were converted into β-lactams. Indeed, the strategy greatly provided synthetic and medicinal chemists with a powerful tool to solve a long-standing problem of producing C–N bonds in fully aliphatic systems through a C–H activation approach.

In 2010, the Pd(II)-catalyzed β-H carbonylation under CO (1 atm), with amide-directed C–H activation and insertion of CO into the Pd–C bond, was achieved by the Yu group [18]. The novel method was found to be effective for the synthesis of 1,4-dicarbonyl compounds that are widespread in biologically important natural products (Scheme 2.18).

2.4 Heterocycles via C(sp^2)–H Activation

Comparing to C(sp^3)–H, the activation of C(sp^2)–H bonds was not difficult. Buchwald developed a method in which the combination of an amide and an unactivated arene [19], under palladium catalysis (Pd(OAc)$_2$), can be used to efficiently produce a series of carbazoles with the aid of a directing functional group. The derivatives of carbazoles have important photophysical and biological

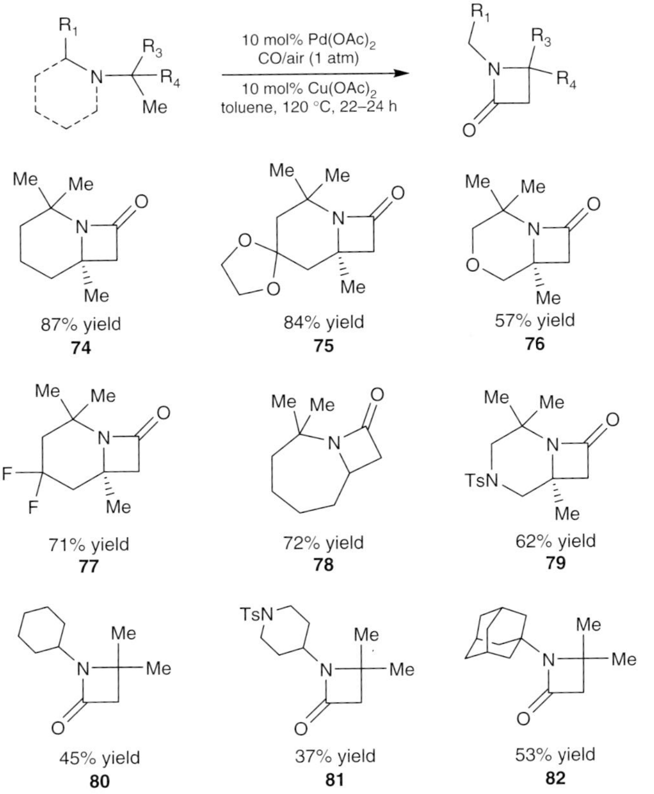

Scheme 2.17 Scope of the C–H carbonylation reaction.

properties. A plausible reaction pathway shows the formation of a six-membered palladacycle whose reductive elimination leads to product and Pd(0). The latter was reoxidized to Pd(II) by $Cu(OAc)_2$, and the reduced Cu species was in turn reoxidized to Cu(II) by oxygen (Scheme 2.19).

After two years, Inamoto developed an approach to the synthesis of 3-aryl/alkylindazoles [20], by means of a $Pd(OAc)_2$-catalyzed C–H amination reactions of hydrazine compounds. This method features the use of novel combinations of such $Pd(OAc)_2/Cu(OAc)_2/AgOCOCF_3$, which successfully effect catalytic C–H activation followed by amination to give the cyclized products. The cyclization proceeded in milder reaction conditions (50 °C) and hence tolerated various functional groups such as alkoxycarbonyl and cyano groups and halogen atoms (Scheme 2.20). The data showed that the cyclization may be controlled by both steric and electronic factors on the arene.

Subsequently, Gaunt reported a Pd(II)-catalyzed intramolecular C–H bond amination strategy and its application to direct carbazole formations [21].

The author believed that Pd(II)/Pd(0) process occurs via coordination of the metal center to the amide, carbopalladation, and reductive elimination to form

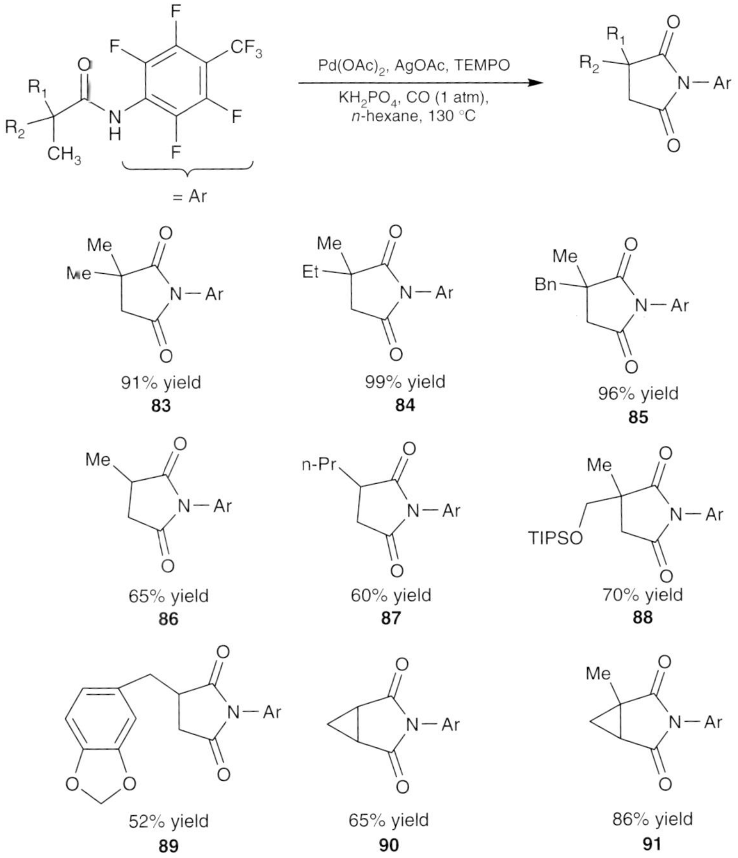

Scheme 2.18 Pd-catalyzed C(sp^3)–H carbonylation.

the C–N bond but requires rather forcing reaction conditions (Scheme 2.21). Reductive elimination from a high oxidation state Pd(IV) center would more readily facilitate C–N bond formation, so the cyclization could occur under ambient conditions.

With an optimized reaction condition, a catalytic $Pd(OAc)_2$, 1 equiv. $PhI(OAc)_2$ and AcOH, room temperature, the reaction can tolerate substrates broadly; it was noticed that electron-rich substrates reacted faster than electron-deficient molecules (Scheme 2.22). *N*-Glycosyl carbazoles, due to the prevalence of this motif in a range of natural products, were accessed successfully through this method (Scheme 2.23).

Ma and coworkers reported the palladium-catalyzed direct amination of phenylalanine derivatives for the synthesis of dipeptides containing indoline-2-carboxylate via 2-methoxyiminoacyl (MIA) as a directing group and an fluorine source as the oxidant [22]. This method features high efficiency and wide functional group tolerance (Scheme 2.24).

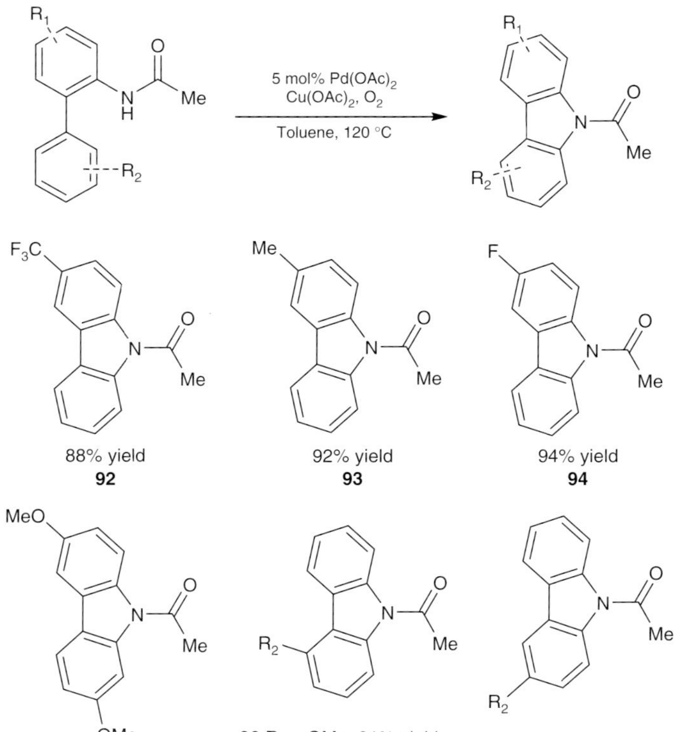

88% yield **92**

92% yield **93**

94% yield **94**

88% yield **95**

96 R_2 = OMe, 81% yield
97 CF_3, 81% yield
98 F, 78% yield

99 R_2 = OMe, 41% yield
100 Me, 82% yield

Scheme 2.19 Carbazole synthesis via C(sp^2)–H activation.

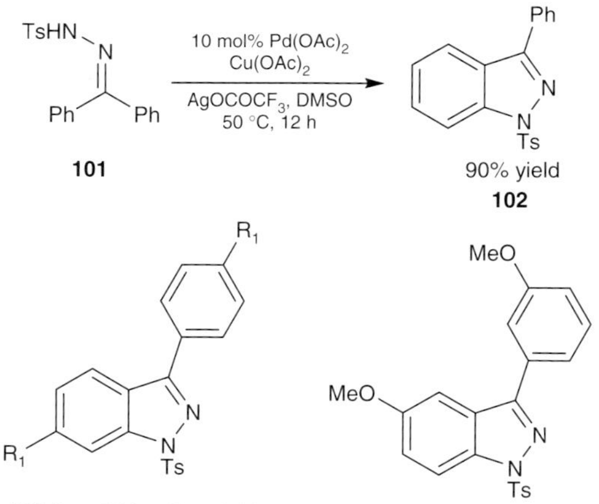

101

90% yield **102**

103 R_1 = OMe, 13% yield
104 = Me, 73% yield

96% yield **105**

Scheme 2.20 Indazole synthesis via C(sp^2)–H activation.

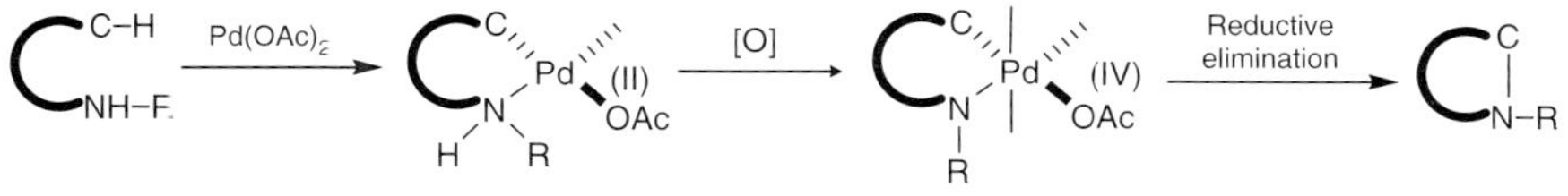

Scheme 2.21 Proposed mechanistic hypothesis for C(sp^2)–H activation.

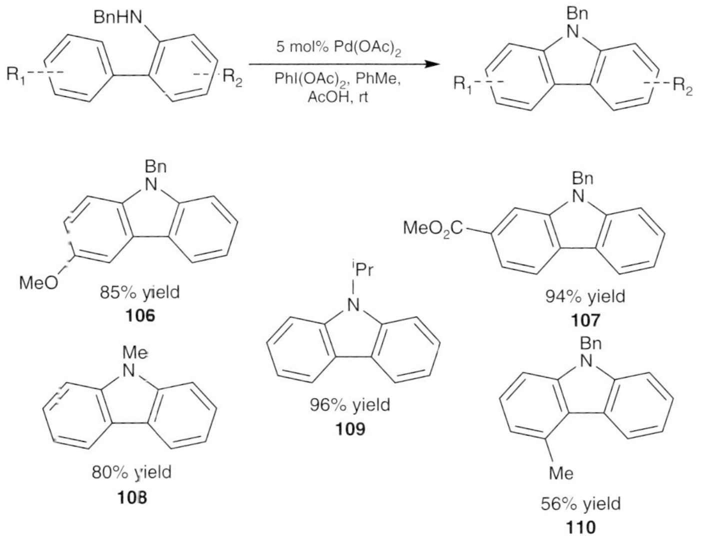

Scheme 2.22 Carbazole synthesis via C(sp^2)–H activation.

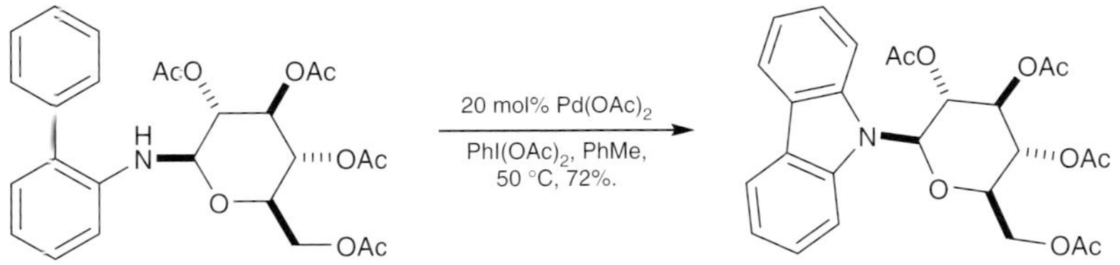

Scheme 2.23 Formation of *N*-glycosyl carbazoles.

Chang and coworkers developed Pd-catalyzed cyclization of *N*-(2-halobenzyl)-substituted pyrroles or phenyl variants to afford condensed hetero- or carbocycles, respectively, via tandem activation of benzyl halide and aromatic C–H bond [23]. The Pd-catalyzed cyclization protocol was successfully applied to a broad range of *N*-(2-chlorobenzyl)pyrroles to furnish pyrroloindoles. Interestingly, the reaction of substrates bearing electron-deficient groups on the pyrrole was also highly efficient to give products which are difficult to obtain using the conventional Friedel–Crafts approach (Scheme 2.25).

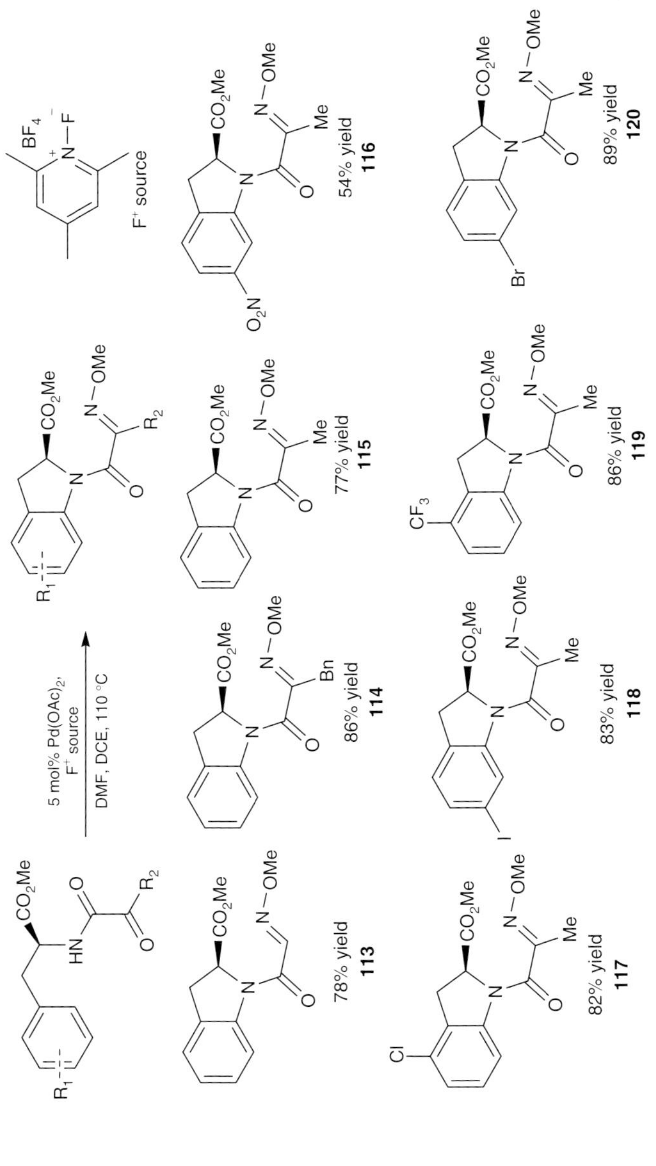

Scheme 2.24 Indoline synthesis via C(sp^2)–H activation.

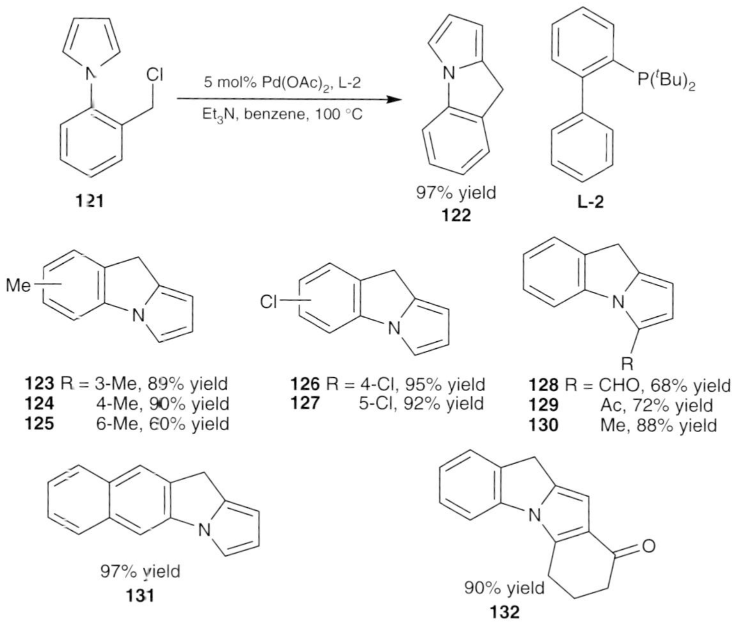

Scheme 2.25 Synthesis of condensed pyrroloindoles via intramolecular C(sp^2)–H activation.

A proposed mechanism was described as oxidative addition of benzyl halide to Pd(0) takes place leading to a Pd(II) species, the plausible palladacycle **134** for the final reductive elimination formed via either C–H activation or electrophilic aromatic substitution (Scheme 2.26).

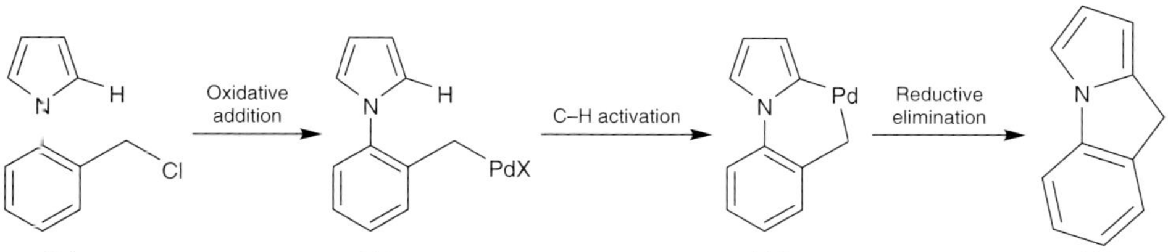

Scheme 2.26 Proposed mechanistic pathways.

2.5 Conclusions

In this chapter, we have presented some examples for the efficient synthesis of nitrogen-containing heterocycles. In the last two decades, many novel synthetic routes have been developed to prepare nitrogen-containing heterocycles via the chemical transformation of palladium-catalyzed C–H activation. Many of these methods are sufficiently mild and functional group compatible for applications in nitrogen-containing heterocycle synthesis. However, there are also lots of works that need to be done for the Pd-catalyzed activation of unactivated C–H bonds in higher selectivity and yield. Undoubtedly, continued progress toward this goal will inspire new application of C–H activation in the synthesis of nitrogen heterocycles.

References

1. Takano, S. and Ogasawara, K. (1989) *Alkaloids*, **36**, 225–252.
2. (a) Hegedus, L.S., Allen, G.F., and Waterman, E.L. (1976) *J. Am. Chem. Soc.*, **98**, 2674–2676; (b) Hegedus, L.S. (1988) *Angew. Chem. Int. Ed. Engl.*, **27**, 1113–1126.
3. (a) Fischer, E. and Jourdan, F. (1883) *Ber. Dtsch. Chem. Ges.*, **16**, 2241–2245; (b) Fischer, E. and Hess, O. (1884) *Ber. Dtsch. Chem. Ges.*, **17**, 559–568.
4. (a) Larock, R.C. and Yum, E.K. (1991) *J. Am. Chem. Soc.*, **113**, 6689–6690; (b) Larock, R.C., Yum, E.K., and Refvik, M.D. (1998) *J. Org. Chem.*, **63**, 7652–7662.
5. (a) Wagaw, S., Yang, B.H., and Buchwald, S.L. (1998) *J. Am. Chem. Soc.*, **120**, 6621–6622; (b) Rutherford, J.L., Rainka, M.P., and Buchwald, S.L. (2002) *J. Am. Chem. Soc.*, **124**, 15168–15169.
6. (a) Taber, D.F. and Tirunahari, P.K. (2011) *Tetrahedron*, **67**, 7165–7210; (b) Beccali, E.M., Broggini, G., Martinelli, M., and Scottocornola, S. (2007) *Chem. Rev.*, **107**, 5318–5365; (c) Zeni, G. and Larock, R.C. (2006) *Chem. Rev.*, **106**, 4644–4680; (d) Jeffrey, J.L. and Sarpong, R. (2013) *Chem. Sci.*, **4**, 4092–4106; (e) Ramirez, T.A., Zhao, B., and Shi, Y. (2012) *Chem. Soc. Rev.*, **41**, 931–942; (f) Li, H., Li, B.J., and Shi, Z.J. (2011) *Catal. Sci. Technol.*, **1**, 191–206; (g) Roizen, J.L., Harvey, M.E., and Bois, J.D. (2012) *Acc. Chem. Res.*, **45**, 911–922.
7. (a) Larock, R.C., Hightower, T.R., Hasvold, L.A., and Peterson, K.P. (1996) *J. Org. Chem.*, **61**, 3584–3585; (b) Sun, W.W., Cao, P., Mei, R.Q., Li, Y., Ma, Y.L., and Wu, B. (2014) *Org. Lett.*, **16**, 480–483; (c) Li, B., Wang, S.Q., Liu, B., and Shi, B.F. (2015) *Org. Lett.*, **17**, 1200–1203.
8. (a) Beccalli, E.M., Broggini, G., Paladino, G., Penoni, A., and Zoni, C. (2004) *J. Org. Chem.*, **69**, 5627–5630; (b) Abbiati, G., Beccalli, E.M., Broggini, G., and Zoni, C. (2003) *J. Org. Chem.*, **68**, 7625–7628.
9. (a) Rice, G.T. and White, M.C. (2009) *J. Am. Chem. Soc.*, **131**, 11707–11711; (b) Strambeanu, I.I. and White, M.C. (2013) *J. Am. Chem. Soc.*, **135**, 12032–12037; (c) Osberger, T.J. and White, M.C. (2014) *J. Am. Chem. Soc.*, **136**, 11176–11181; (d) Frauhoffer, K.J. and White, M.C. (2007) *J. Am. Chem. Soc.*, **129**, 7274–7276; (e) Nishikawa, Y., Kimura, S., Kato, Y., Yamazaki, N., and Hara, O. (2015) *Org. Lett.*, **17**, 888–891.
10. Fraunhoffer, K.J., Prabagaran, N., Sirois, L.E., and White, M.C. (2006) *J. Am. Chem. Soc.*, **128**, 9032–9033.
11. Nahra, F., Liron, F., Prestat, G., Mealli, C., Messaoudi, A., and Poli, G. (2009) *Chem. Eur. J.*, **15**, 11078–11082.
12. Wu, L., Qiu, S., and Liu, G. (2009) *Org. Lett.*, **11**, 2707–2710.
13. Neumann, J.J., Rakshit, S., Droge, T., and Glorius, F. (2009) *Angew. Chem. Int. Ed.*, **48**, 6892–6895.
14. Zaitsev, V.G., Shabashov, D., and Daugulis, O. (2005) *J. Am. Chem. Soc.*, **127**, 13154–13155.
15. He, C., Zhao, Y., Zhang, S., Lu, C., and Chen, G. (2012) *J. Am. Chem. Soc.*, **134**, 3–6.
16. Nadres, E.T. and Daugulis, O. (2012) *J. Am. Chem. Soc.*, **134**, 7–10.
17. McNally, A., Haffemayer, B., Collins, B.S.L., and Gaunt, M. (2014) *Nature*, **510**, 129–133.
18. Yoo, E.J., Wasa, M., and Yu, J.Q. (2010) *J. Am. Chem. Soc.*, **132**, 17378–17380.
19. (a) Tsang, W.C.P., Zheng, N., and Buchwald, S.L. (2005) *J. Am. Chem.*

Soc., **127**, 14560–14561; (b) Tsang, W.C.P., Munday, R.H., Brasche, G., Zheng, N., and Buchwald, S.L. (2008) *J. Org. Chem.*, **73**, 7603–7610; (c) Mei, T.S., Leow, D., Xiao, H., Laforteza, B.N., and Yu, J.Q. (2013) *Org. Lett.*, **15**, 3058–3061; (d) Youn, S.W., Bihn, J.H., and Kim, B.S. (2011) *Org. Lett.*, **13**, 3738–3741; (e) He, G., Lu, C., Zhao, Y., Nack, W.A., and Chen, G. (2012) *Org. Lett.*, **14**, 2944–2947; (f) Yang, W., Chen, J., Huang, X., Ding, J., Liu, M., and Wu, H. (2014) *Org. Lett.*, **16**, 5418–5421; (g) Mei, T.S., Wang, X., and Yu, J.Q. (2009) *J. Am. Chem. Soc.*, **131**, 10806–10807; (h) Chng, L.L., Yang, J., Wei, Y., and Ying, J.Y. (2014) *Chem. Commun.*, **50**, 9049–9052.

20. Inamoto, K., Saito, T., Katsuno, M., Sakamoto, T., and Hiroya, K. (2007) *Org. Lett.*, **9**, 2931–2934.
21. Jordan-Hore, J.A., Johansson, C.C.C., Gulias, M., Beck, E.M., and Gaunt, M.J. (2008) *J. Am. Chem. Soc.*, **130**, 16184–16186.
22. He, Y.P., Zhang, C., Fan, M., Wu, Z., and Ma, D. (2015) *Org. Lett.*, **17**, 496–499.
23. (a) Hwang, S.J., Cho, S.H., and Chang, S. (2008) *J. Am. Chem. Soc.*, **130**, 16158–16159; (b) Watanabe, T., Qishi, S., Fujii, N., and Ohno, H. (2008) *Org. Lett.*, **10**, 1759–1762; (c) Ren, H. and Knochel, P. (2006) *Angew. Chem. Int. Ed.*, **45**, 3462–3465.

3
Pd-Catalyzed Synthesis of Oxygen-Containing Heterocycles

Yudong Yang and Jingsong You

3.1
Introduction

Oxygen-containing heterocycles are prevalent structural motifs in many natural products, biologically active molecules, pharmaceuticals, agrochemicals, organic functional materials, and valuable intermediates for organic synthesis (Figure 3.1) [1, 2]. In the past few decades, enormous endeavor was devoted into the synthesis of oxacycles, and substantial methods have been developed in this course [3]. Although these reactions typically occur in high yields with a good functional group compatibility, they also often suffered from multistep procedures, harsh reaction conditions, and the requirement of prefunctionalized starting materials. Thus, there is still ample room to improve the synthetic routes to provide these structures.

From the viewpoint of green chemistry, high efficiency, and step economy, transition metal-catalyzed direct functionalization of C–H bonds which are ubiquitous in organic compounds represents an ideal strategy in organic synthesis. Distinctly different from the traditional transition metal-catalyzed coupling reactions with the cleavage of C–X bonds (X = OTf, I, Br, Cl, $B(OH)_2$, etc.), transition metal-catalyzed C–H functionalization avoids the prefunctionalization of the starting materials and hence shortens the synthetic route and reduces the undesired wastes discharged [4]. Thus it should be highly attractive and potentially useful in the construction of oxacycles. However, to use this tactic as a practical tool in organic synthesis, two fundamental challenges must be overcome: activation of the inert C–H bonds [5] and controlling the site selectivity in molecules [6]. In the past few decades, numerous transition metals have been studied to address the first challenge by reaction with the C–H bonds to form more reactive C–M bonds which could be further converted to other functional groups [7, 8]. For the second challenge, several strategies have also been developed, including the utility of intrinsic electronic nature of substrates [9] and directing group which could coordinate to the metal center and selectively deliver the catalyst to the proximal C–H bond [10]. Accompanied with these achievements, a broad set of oxygen-containing heterocycles have

Transition Metal-Catalyzed Heterocycle Synthesis via C—H Activation, First Edition. Edited by Xiao-Feng Wu.
© 2016 Wiley-VCH Verlag GmbH & Co. KGaA. Published 2016 by Wiley-VCH Verlag GmbH & Co. KGaA.

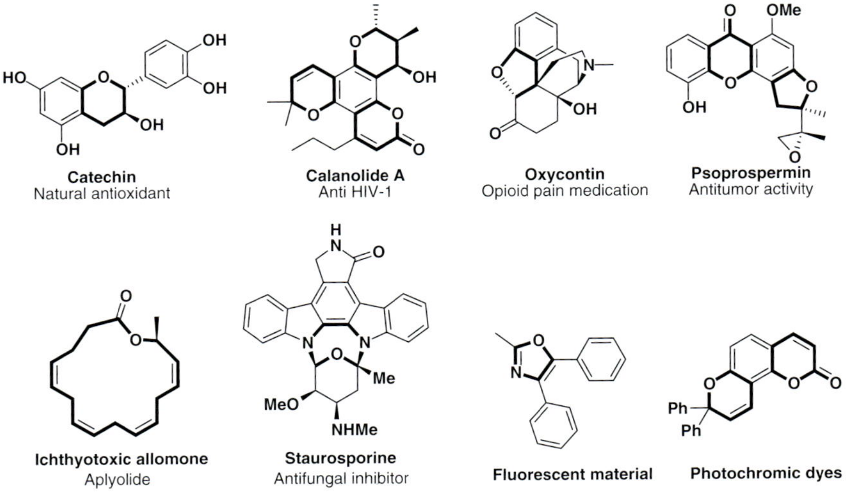

Figure 3.1 Selected examples of functional molecules containing oxacycles.

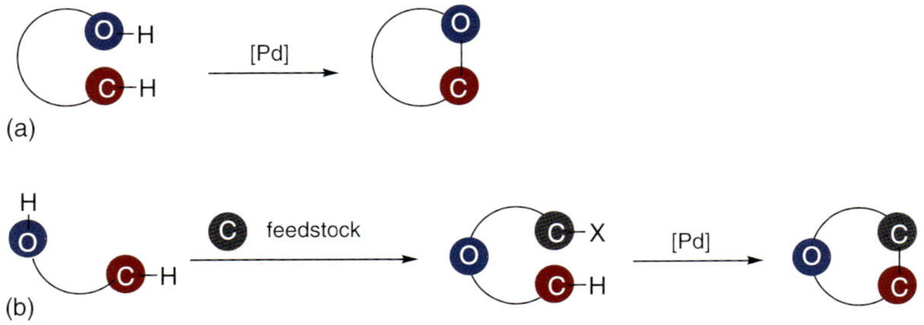

Scheme 3.1 General pathways for the synthesis of oxacycles. (a) C–H activation/C–O formation and (b) C–H activation/C–C formation.

been synthesized via C–H activation/C–C formation or C–H activation/C–O formation (Scheme 3.1).

In this chapter, we will focus on the synthesis of oxygen-containing heterocycles by palladium-catalyzed C–H activation. The section is roughly divided into C–C bond and C–O bond formations according to the type of bond formed on the site which undergoes C–H activation. And the details are described according to the (i) type of transformations involved, (ii) ring size of oxacycles, (iii) class of substrates, and (iv) directing groups.

3.2 Palladium-Catalyzed C–H Activation/C–C Formation to Construct Oxacycles

Palladium-catalyzed synthesis of oxygen-containing heterocycles involving a C–H activation/C–C formation process has been substantially investigated in the past few decades. Palladium-catalyzed arylation, olefination, alkylation, and carbonylation have all been developed as efficient approaches to afford the desired oxacycles. These reactions predominately occur on an aromatic C–H bond, and thus benzo-oxacycles are often obtained, including benzofurans, coumarins, isocoumarins, benzosultones, and benzopyrans.

3.2.1 Palladium-Catalyzed C–H Bond Arylation

Palladium-catalyzed C–H arylation is perhaps the most rigorously studied C–H bond functionalization reaction up to now [4, 8a, 11]. A broad set of benzofused oxacycles have been synthesized by this approach. Both aromatic halides and arenes with the innate hydrogen remaining could be used as the coupling partners to arylate the inert C–H bond.

With aromatic bromides as the substrates, in 2009, Gevorgyan reported a concise procedure for the synthesis of silicon-containing oxacycles via the palladium-catalyzed intramolecular C–H arylation. And the resulted silyl oxacycles could be subsequently desilylated into the corresponding *ortho*-arylated phenols easily (Scheme 3.2) [12].

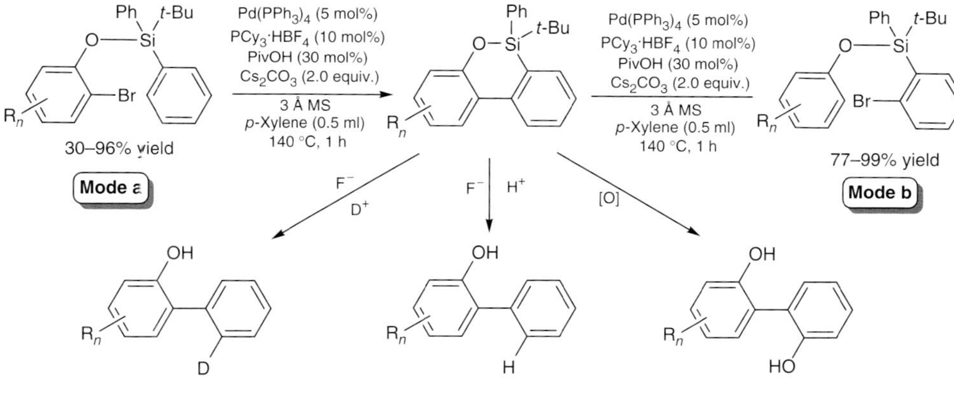

Scheme 3.2 Pd-catalyzed intramolecular arylation of phenol derivatives.

Sultones are the internal esters of hydroxy sulfonic acids and sulfur analogs of lactones. The biological activities of sultones are concerned with toxicological, skin sensitization, and antiviral properties [13]. In 2009, Majumdar developed a $Pd(PPh_3)_4$-catalyzed intramolecular C–H arylation reaction of benzenesulfonic acid 2-bromophenyl esters to afford polycyclic sultones, which are generally synthesized by elimination of the corresponding hydroxyl sulfonic acid derivatives, in up to 90% yield (Scheme 3.3, path a) [14]. TBAB was found to be critical in this reaction, and no reaction occurred in the absence of this additive. An electrophilic palladation mechanism was proposed for this transformation.

Subsequently, Doucet and coworkers reported a palladium-catalyzed intramolecular direct arylation of 2-bromobenzenesulfonic acid derivatives

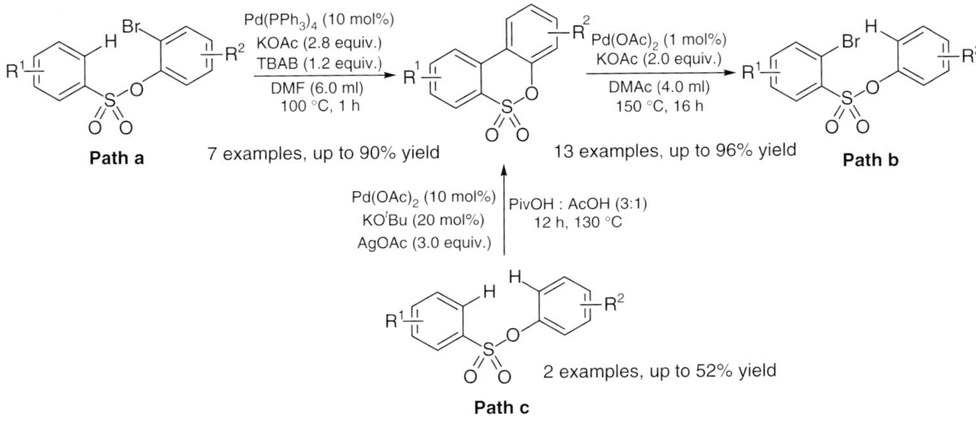

Scheme 3.3 Pd-catalyzed intramolecular C–H arylation reaction to construct sultones.

to synthesize sultones under phosphine-free conditions with up to 96% yield (Scheme 3.3, path b) [15]. The arylation occurred regioselectively on the C(sp^2)–H bond rather than C(sp^3)–H bond to afford six-membered oxacycles exclusively. Compared with Majumdar's work, higher temperature and prolonged time are required in this method. In addition, palladium-catalyzed synthesis of sultones involving a double C(sp^2)–H functionalization has also been achieved by Laha and coworkers (Scheme 3.3, path c) [16].

An example of the construction of dibenzofurans by palladium-catalyzed C–H arylation was reported by Itatani in 1974 (Scheme 3.4) [17]. Compared with previous methods [18], this reaction could obviate the tedious synthetic route and is suitable for the preparation of substituted dibenzofurans with high efficiency. It should be noted that substituents on the benzene nucleus are believed to reduce the interaction between molecules and favor intramolecular coupling rather than intermolecular dimerization.

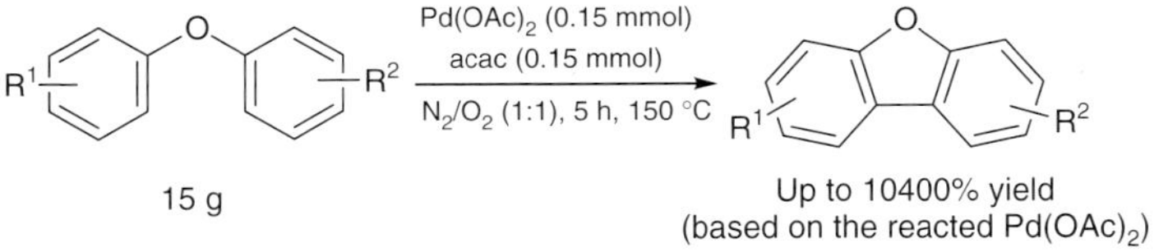

Scheme 3.4 Pd-catalyzed synthesis of dibenzofurans via intramolecular C–H arylation.

Medium-ring-containing biaryls are generally difficult to access by classical routes; thus development of new approach to this structure is highly attractive. In 2011, Greaney reported the intramolecular dehydrogenative coupling of indoles and other arenes to synthesize medium-ring heterocycles [19]. In the investigation of this double C–H activation reaction, three oxazapane derivatives were obtained in good yields by the catalysis of $Pd(OAc)_2$ (Scheme 3.5). The result of intramolecular kinetic isotope effect (KIE) study implied that this reaction might proceed via a concerted metalation–deprotonation (CMD) mechanism. In addition, the possibility of electrophilic palladation mechanism is excluded due to the observed selectivities for electron-poor sites in competition experiments.

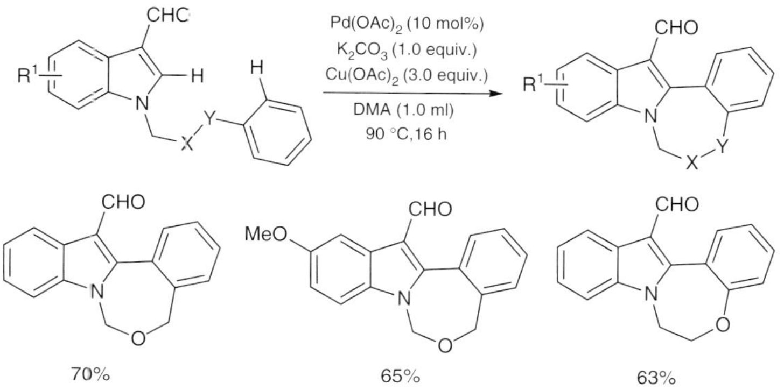

Scheme 3.5 Pd-catalyzed intramolecular C–H arylation of indoles.

3.2.2 Palladium-Catalyzed C–H Olefination

Palladium-catalyzed C–H olefination is also a very useful protocol to synthesize oxygen-containing heterocycles, since it allows to functionalize the substrates with an unsaturated olefinic moiety which would constitute a π-conjugated structure or be elaborated further. Two pathways are often encountered in this approach: (i) addition of the resulted arylpalladium complex to alkynes and subsequent protonation or transformation and (ii) addition of the arylpalladium complex to olefins and following palladium hydride elimination [7b, 20]. The current methods are mainly focused on the functionalization of aromatic C–H bonds to form benzo-oxacycles.

Phenol and its derivatives are the most common substrates for the construction of oxacycles in this approach. In 2000, Fujiwara and coworkers reported a palladium-catalyzed intramolecular addition reaction of aryl alkynoates to construct coumarins in a mixture of trifluoroacetic acid and CH_2Cl_2 at room temperature (Scheme 3.6) [21]. This method is simple and efficient to give the desired oxygen-containing heterocycles in high yields. However, this palladium-catalyzed C–H olefination reaction is only constrained to the electron-rich aromatic compounds owing to the inherent limitation of electrophilic metalation process.

As a further modification of Fujiwara's work, Kitamura and coworkers realized the construction of coumarins by palladium-catalyzed C–H olefination of phenols with acrylates and subsequent intramolecular esterification (Scheme 3.7)

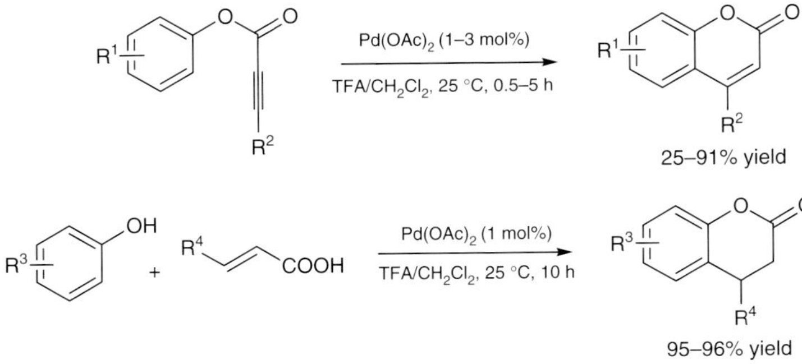

Scheme 3.6 Synthesis of coumarins and 3,4-dihydrocoumarins.

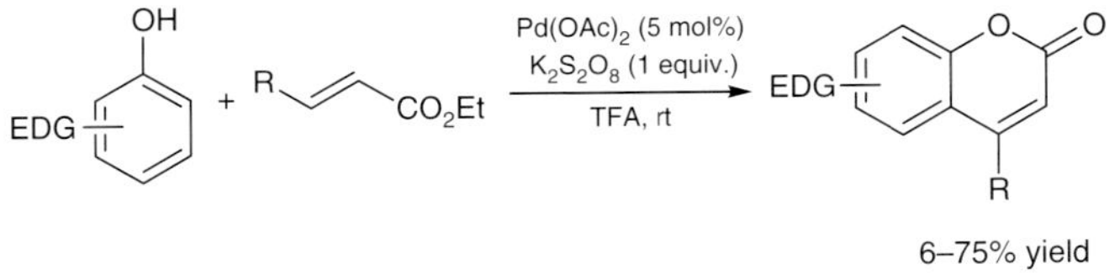

Scheme 3.7 Pd(II)-catalyzed reaction of phenols with acrylates.

[22]. $K_2S_2O_8$ was optimized to oxidize the generated Pd(0) species and accomplished the catalytic cycles. However, only limited substrates were reported, and the yields were low to moderate.

In 2013, Hong and coworkers reported a one-pot protocol to achieve similar structures by a sequential Pd(II)-catalyzed dehydrogenation-oxidative Heck-cyclization process [23]. Using 20 mol% $Pd(TFA)_2$, 1.0 equiv. of $Cu(OAc)_2$, the cyclohexanones could easily react with electron-deficient olefins to afford coumarins in PivOH under an O_2 atmosphere with modest to high yields (Scheme 3.8). It is noted that the product scope is somewhat extended to the coumarins with electron-withdrawing groups on the aromatic rings. In addition, the obtained coumarin products could be further alkenylated to provide an extended π-system.

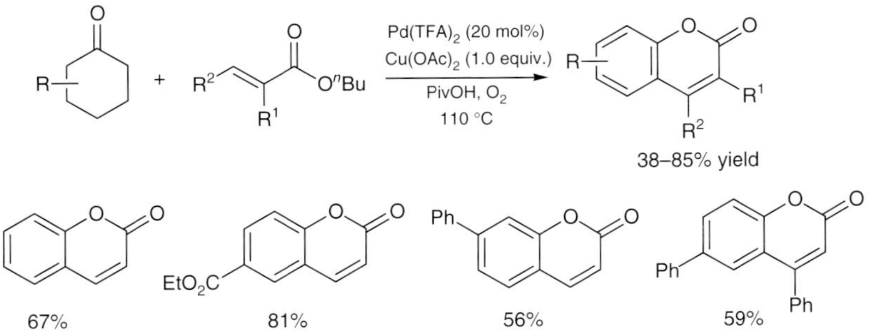

Scheme 3.8 The one-pot process of sequential dehydrogenation-oxidative Heck cyclization.

In contrast to the earlier reactions with an acid as solvent, in the presence of 10 mol% $Pd(CH_3CN)_4(BF_4)_2$, 2.0 equiv. of $Cu(OAc)_2$, and 0.8 equiv. of NaOPiv, phenols could react with electron-withdrawing olefins to afford coumarin derivatives in mesitylene in moderate to good yields (Scheme 3.9) [24]. Phenols with all electronic characters including electron neutral, rich, and poor are suitable for this reaction. In addition, estrone and *N*-phth-Me-D-tyrosine were functionalized by this method to synthesize more complex molecules.

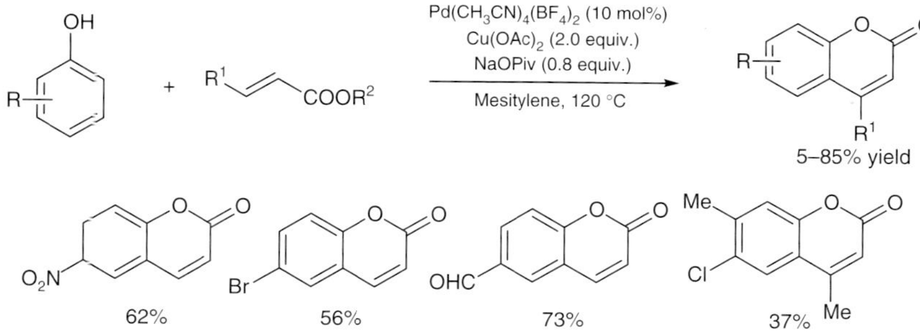

Scheme 3.9 Palladium-catalyzed directed C–H alkenylation of phenols.

Flavone- and coumarin-fused benzopyrans are privileged structural motifs in many biologically active molecules such as cyclomorusin and cycloheterophyllin [25]. Recently, Hong and coworkers reported a tandem alkenylation/C–O cyclization reaction to achieve these structural features by the selective activation of C3–H with palladium [26]. In the presence of 20 mol% $Pd(acac)_2$, 3 equiv. of $Cu(OAc)_2$, 2 equiv. of Cs_2CO_3, and 20 mol% Al_2O_3, a fairly broad range of flavone- and coumarin-fused benzopyran derivatives could be obtained in *t*-BuOH with moderate to high yields (Scheme 3.10). Al_2O_3 in this reaction was supposed to coordinate to the chromone substrate and thus enhance the reactivity.

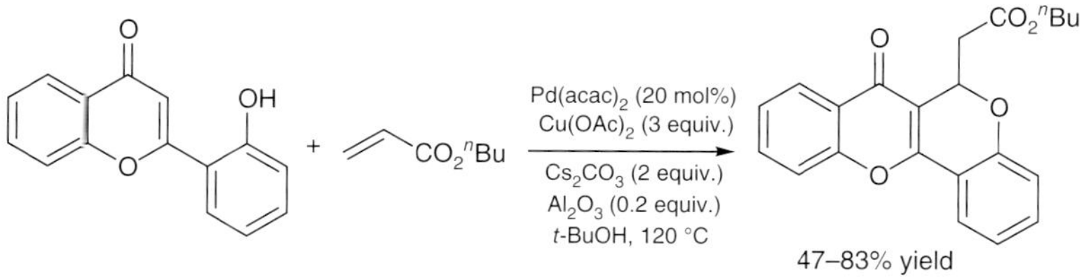

Scheme 3.10 Pd(II)-catalyzed C–H alkenylation/C–O cyclization of flavones.

It is known that alkynes could readily interact with transition metals to form η^2-metal complexes and then undergo further elaboration [27]; thus alkynes should be potentially useful in the *ortho*-aromatic C–H olefination to form oxygen-containing heterocycles. In 2012, Minami and Hiyama reported a palladium-catalyzed cycloaddition of alkynyl aryl ethers to give internal alkynes

(Scheme 3.11) [28]. With 5 mol% $Pd(OAc)_2$, 10 mol% PCy_3, 5 mol% Zn, and toluene as solvent, this reaction could proceed smoothly at 90°C to give the corresponding chromenes in moderate to excellent yields. It should be noted that a bulky silyl substituent on the terminated position of alkynoxy group is essential for this transformation owing to the less sterically hindered trimethylsilyl giving no product. More recently, the same group also accomplished the construction of chromenes with allenes under similar conditions [29].

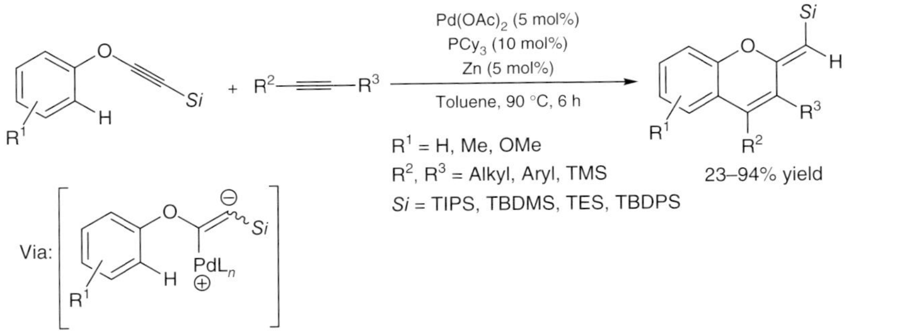

Scheme 3.11 Palladium-catalyzed cycloaddition of alkynyl aryl ethers with internal alkynes.

A concise route to oxygen-containing helical tetrasubstituted alkenes via a palladium-catalyzed domino reaction with Sonogashira coupling/carbopalladation/C–H functionalization sequence has been developed by Tietze and coworkers (Scheme 3.12) [30]. In this work, the *in situ*-generated vinyl palladium intermediate from alkyne plays a vital role in the C–H activation. The reaction is efficient, and a mixture of *Z*- and *E*-isomers of the helical alkenes was afforded under the typical condition (44–96% yields, $E/Z = 1.5{:}1$ to 1:1.6).

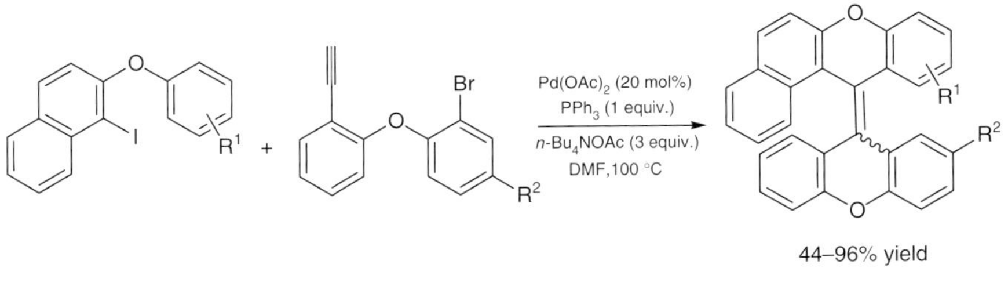

Scheme 3.12 Palladium-catalyzed domino reaction.

Besides six-membered heterocycles, the construction of five-membered analogs by C–H olefination of phenol derivatives has also been reported. In 2011, Wang and coworkers reported a palladium-catalyzed oxidative cyclization of 3-phenoxy acrylates to construct benzofurans via an intramolecular aromatic C–H olefination (Scheme 3.13) [31]. In the presence of 5 mol% $Pd(OAc)_2/PPh_3$ and 2.0 equiv. of CF_3CO_2Ag, benzofurans were obtained in good yields in

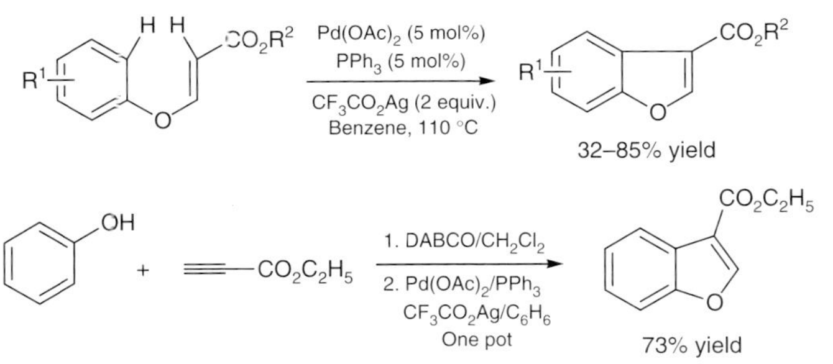

Scheme 3.13 Palladium-catalyzed synthesis of benzofurans.

benzene under air. The authors also successfully developed a one-pot protocol to synthesize benzofurans by the reaction of propiolates and phenols.

Then Wang and coworkers extended the substrate range to bromoalkynes by a phenoxide addition/C–H olefination protocol (Scheme 3.14) [32]. With this reaction, a broad set of (*Z*)-2-bromovinyl phenyl ethers could be obtained with excellent regio- and stereoselectivity and subsequently undergo intramolecular cyclization to afford 2-aryl- or 2-alkyl-substituted benzofurans in good yields. Both electron-withdrawing and electron-donating groups are tolerant in this reaction, yet the phenols with electron-withdrawing groups at *para* positions are more favorable. Hence, this method is considered to be an efficient alternative to Sonogashira coupling/5-endo-dig cyclization of 2-halophenols and alkynes to synthesize benzofurans.

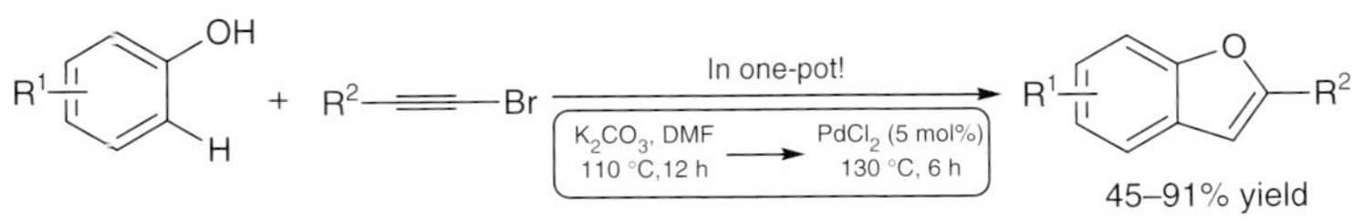

Scheme 3.14 Palladium-catalyzed one-pot synthesis of benzofurans.

Further extension of this strategy to the unactivated internal alkynes was realized by Sahoo and coworkers (Scheme 3.15) [33]. Using 5 mol% $Pd_2(dba)_3$, 10 mol% 1,10-phenanthroline or bathophenanthroline, and 2.0 equiv. of $Cu(OAc)_2$, phenols could readily react with internal alkynes to afford 2,3-substituted benzofurans in good to excellent yields. In addition, several complex phenol derivatives of pharmaceutical importance were also investigated in this work, including *N*-boc-L-tyrosine, estrone, umbelliferone, and vanillideneacetone.

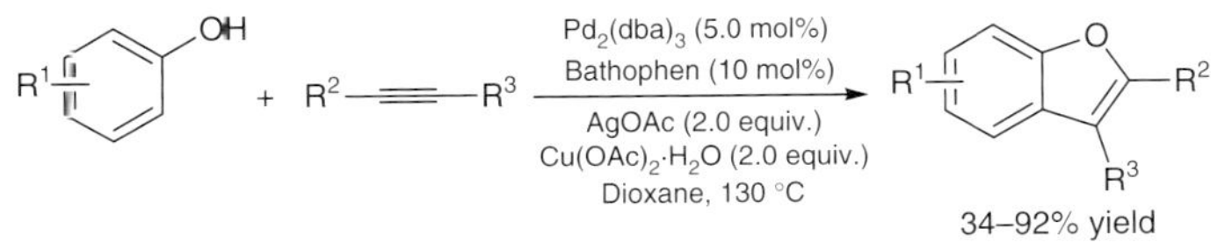

Scheme 3.15 Palladium-catalyzed oxidative annulation of phenols and alkynes.

The aforementioned methods to construct benzofurans are all achieved by activation of C(sp^2)–H bond. In 2012, based on the alkynoxy-directed synthesis of 2-methylidene-2*H*-chromene derivatives [28], Hiyama and coworkers developed a palladium-catalyzed hydrobenzylation of *ortho*-tolyl alkynyl ethers to construct 2-methylidene-2,3-dihydrobenzofurans via the activation of benzylic C(sp^3)–H bond [34].

Besides phenols, other substrates have also been used to construct oxygen-containing heterocycles involving a C–H alkenylation process. In 1998, Miura and coworkers described an oxidative cross-coupling of benzoic acids with acrylate esters to synthesize phthalide and isocoumarin derivatives in the catalysis of palladium and copper under air conditions [35]. The ratio of these two products is highly dependent on the structures of substrates. In 2013, the Lee group achieved the construction of similar structures with Ag_2O as the oxidant under the catalysis of $Pd(OAc)_2$ (Scheme 3.16) [36]. The *ortho* position was found to be critical for the selectivity of the reaction. When the *ortho*-substituted substrates are employed, phthalides would be obtained, whereas isocoumarins are afforded when the innate hydrogen remains. It is noteworthy that this protocol is efficient only when the electron-rich or electron-neutral benzoic acids are attempted; thus it is proposed to proceed through a S_EAr mechanism.

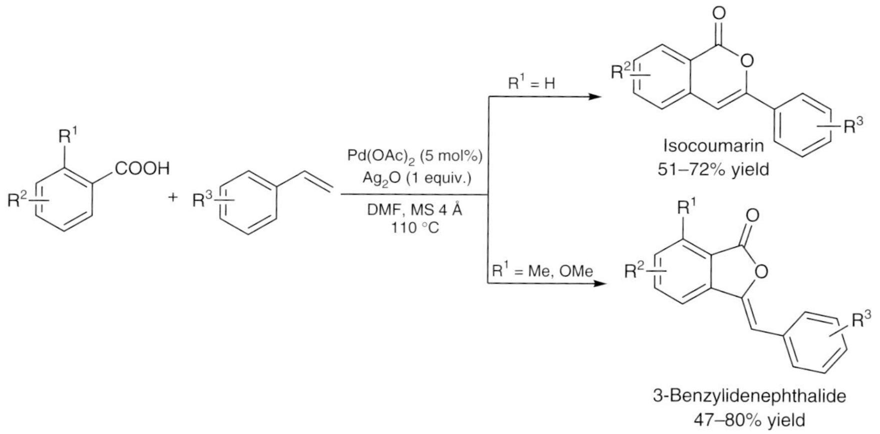

Scheme 3.16 Pd-catalyzed oxidative coupling of benzoic acids and vinyl arenes.

Hydroxyl is a weak directing group for coordination to the transition metals, and the alcohol is also known to be easily oxidized or decomposed by Pd(II) complex; thus it is envisioned difficult to implement the direction of C–H activation [37]. In 2010, Yu's group reported a Pd(II)-catalyzed cascade C–H olefination/oxidative cyclization reaction with the direction of a hydroxy group (Scheme 3.17) [38]. In the assistant of an amino acid, a diversity of primary, secondary, and tertiary alcohol could be converted to the corresponding pyran derivatives with exocyclic alkene moieties.

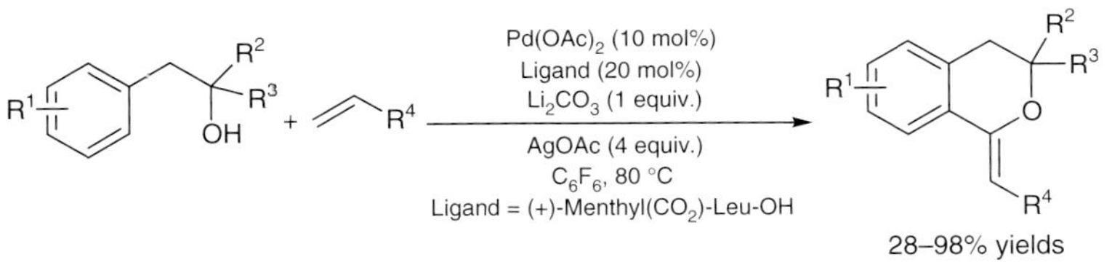

Scheme 3.17 Pd(II)-catalyzed hydroxyl-directed C–H olefination.

3.2.3
Palladium-Catalyzed C–H Alkylation

One of the often encountered reaction in this field is the Catellani reaction which is a useful method for the polyfunctionalization of aromatic molecules with assistance of norbornene [39]. In 2002, Lautens reported a palladium-catalyzed sequential alkylation/alkenylation reaction to synthesize benzoxepines from aryl iodides and bromoenoates [40]. Later, the same group developed a palladium-catalyzed three-component domino reaction to synthesize polysubstituted oxacycles under similar reaction conditions [41]. In 2007, Lautens and coworkers realized the construction of oxygen-containing tricycles based on the Catellani reaction again (Scheme 3.18) [42]. In the presence of 10 mol% $Pd(OAc)_2$, 22 mol% PPh_3, 5 equiv. Cs_2CO_3, and 3 equiv. of norbornene, a variety of symmetric and asymmetric oxygen-containing tricycles could be synthesized from a terminal olefin and an aryl iodide having two oxygen-tethered primary alkyl bromides. The reaction is very soon and ring size could vary from five to seven members. However, a high temperature of 190°C and microwave assistance are required in this methodology.

After a stereochemical and mechanistic investigation of palladium-catalyzed tandem intramolecular *ortho*-alkylation/Heck reaction of secondary iodides, this

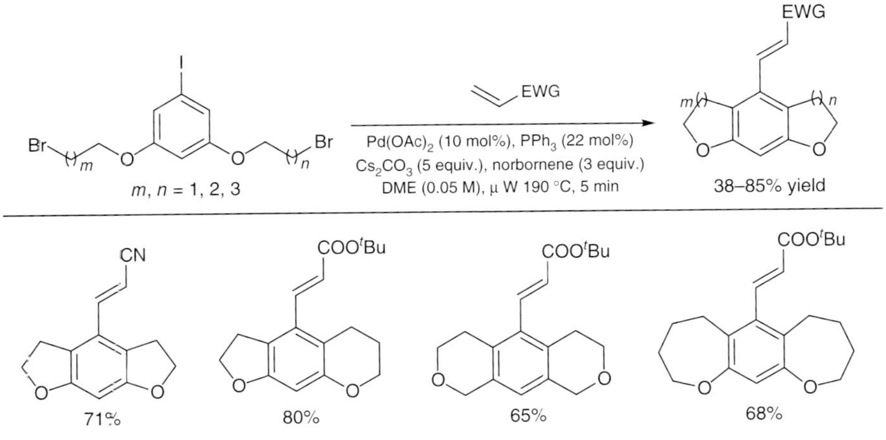

Scheme 3.18 Formation of oxygen-containing tricyclic heterocycles.

topic was further explored. In 2009, Lautens and coworkers further extended this chemistry to construct polycyclic oxacycles by the intramolecular or intermolecular *ortho*-alkylation of aromatic C–H bonds with secondary alkyl iodides and bromides (Scheme 3.19) [43].

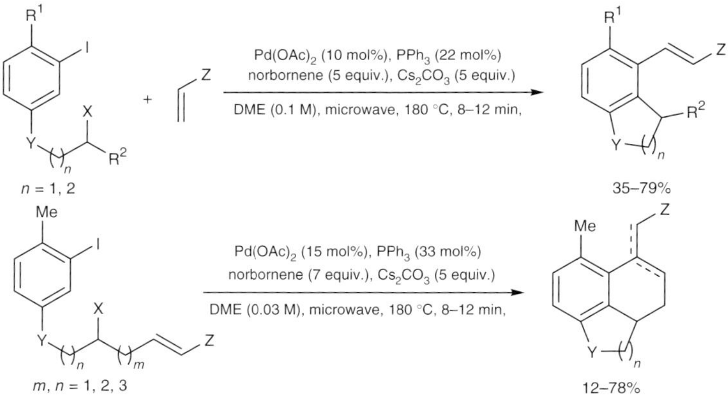

Scheme 3.19 Construction of polycyclic oxacycles based on Catellani reaction.

A carboxylic acid-directed palladium-catalyzed alkylation of benzoic acids with alkyl halides to provide benzolactones was reported by Yu and coworkers in 2009 (Scheme 3.20) [44]. The reaction is rather general to a wide scope of substrates and gives the desired products in moderate to excellent yields.

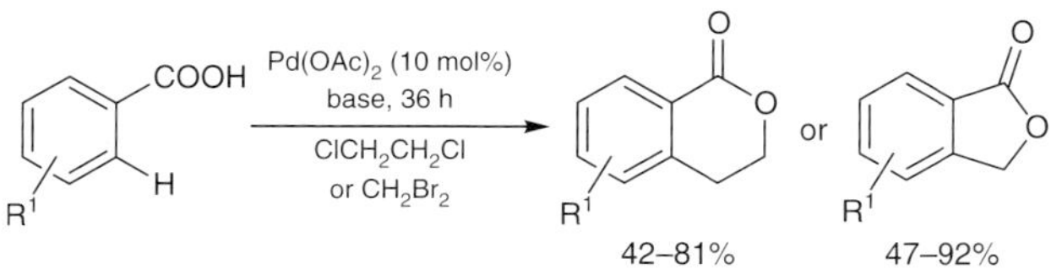

Scheme 3.20 Palladium-catalyzed synthesis of benzolactones.

3.2.4
Palladium-Catalyzed C–H Carbonylation and Carboxylation

In 2008, Yu and coworkers reported Pd(II)-catalyzed C–H carboxylation of benzoic and phenylacetic acid derivatives to form 1,2- and 1,3-dicarboxylic acids (Scheme 3.21) [45]. In this investigation, several cyclic anhydrides were found stable enough to be isolated, and the others were subsequently converted into the corresponding dicarboxylic acids under the catalytic conditions.

Different to Yu's work which was directed by the carboxy group [45], in 2012, Guan and coworkers described an alkyl amino-directed palladium-catalyzed C–H carbonylation of *N*-alkylanilines to synthesize isatoic anhydrides (Scheme 3.22) [46]. Using 5 mol% $Pd(OAc)_2$, 2.2 equiv. of $Cu(OAc)_2$, and 0.2 equiv. of KI, isatoic

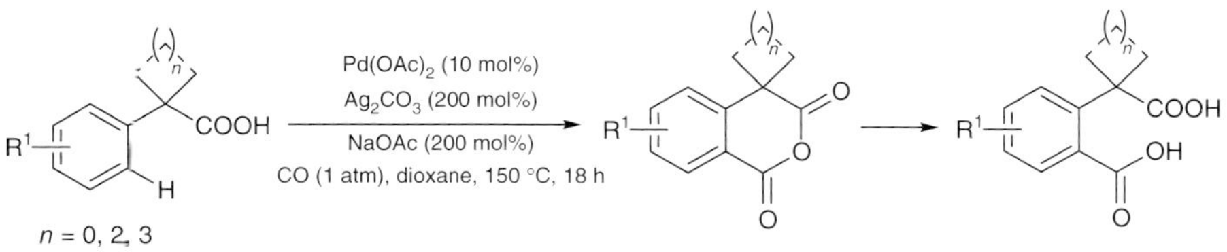

Scheme 3.21 Pd-catalyzed C–H carbonylation of benzoic and phenylacetic acid derivatives.

anhydrides could be readily prepared from *N*-alkylanilines in high yields under the CO atmosphere. This new carbonylation reaction displayed a broad substrate scope as well as good functional group tolerance and proved to be a quite general methodology to provide isatoic anhydrides.

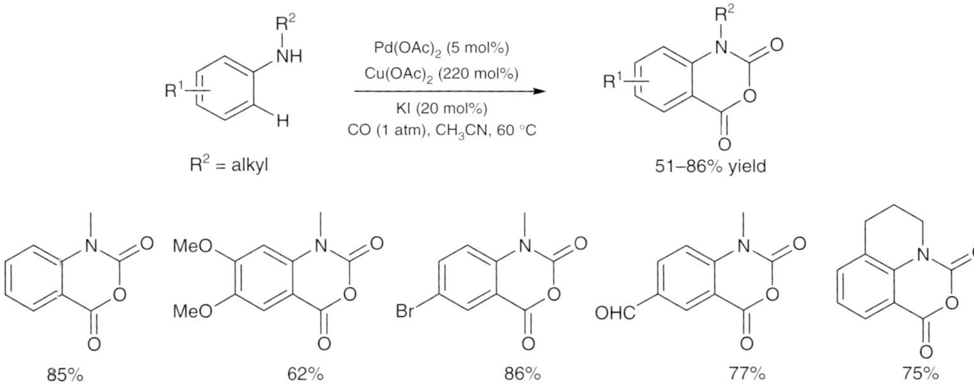

Scheme 3.22 Pd-catalyzed C–H carbonylation synthesis of isatoic anhydrides.

Besides the cyclic acid anhydrides, the oxaphosphorinanone oxides, which are bioisostere of acid anhydrides and inhibitors of β-lactamase [47], have also been synthesized by palladium-catalyzed C–H carbonylation of phosphonic and phosphinic acids (Scheme 3.23) [48]. In the presence of 5 mol% $Pd(OAc)_2$, 1.5 equiv. of $PhI(OAc)_2$, and 1.5 equiv. of AgOAc, benzylphosphonic and phosphinic acid derivatives could give the corresponding oxaphosphorinanone oxides in moderate to high yields by incorporation of a carbonyl group.

In addition, the palladium-catalyzed C–H carbonylation is also useful to construct pyranone derivatives. In 2011, Yu's group reported a Pd(II)-catalyzed *ortho*-C–H carbonylation reaction of phenethyl alcohol derivatives to give 1-isochromanones by connecting the aryl moiety and alcohol with incorporated carbonyl group (Scheme 3.24) [49]. A wide range of functionalities such as methoxy, fluoro, chloro, and bromo are tolerated under the typical condition and afford various 1-isochromanones in moderate to good yields. The *N*-protected amino acids are supposed to reduce the rate of Pd(II) reduction in the presence of CO in this work.

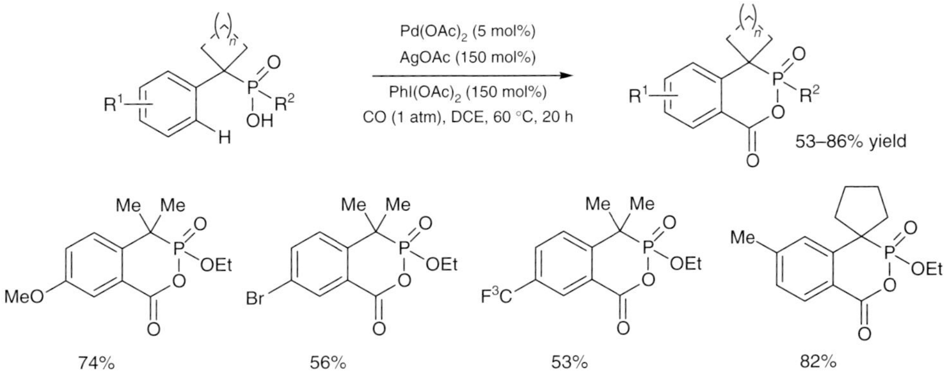

Scheme 3.23 Pd-catalyzed C–H carbonylation synthesis of oxaphosphorinanone oxides.

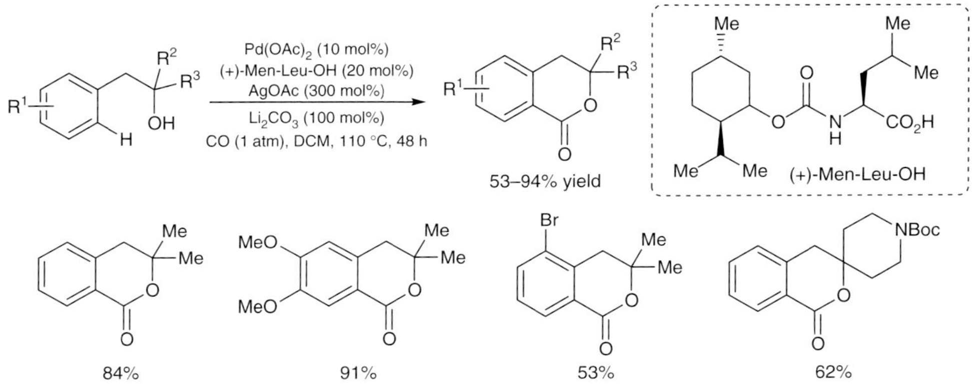

Scheme 3.24 Pd-catalyzed synthesis of oxaphosphorinanone oxides via C–H carbonylation.

In 2013, Shi and coworkers described the first example of phenolic hydroxy group-directed C–H carbonylation to prepare lactones (Scheme 3.25) [50]. Using 5 mol% $Pd(OAc)_2$, 10 mol% $Cu(OAc)_2$, 2 equiv. of Na_2CO_3, and 0.5 equiv. of PivOH, dibenzopyranones could be synthesized from 2-arylphenols under an atmospheric pressure of CO and air. The reaction is of wide substrate scope, whereas the steric hindrance and electronic character of phenolic aromatic ring are critical to the yield.

Later on, Cheng and Chuang reported an extra acid- and base-free condition to achieve similar structures in the catalysis of $Pd(OAc)_2$ at lower temperature (Scheme 3.26) [51]. The reaction is also tolerant to a variety of substrates to give the corresponding products in good yields. However, the phenol moieties equipped with a methyl or strongly electron-withdrawing groups are not compatible in this condition.

More recently, Jiang and coworkers used 3-arylquinolin-4(1*H*)-one as the substrates and realized the construction of 6*H*-isochromene[4,3-*c*]-quinolin-6-ones by palladium-catalyzed C–H carbonylation and cyclization (Scheme 3.27) [52].

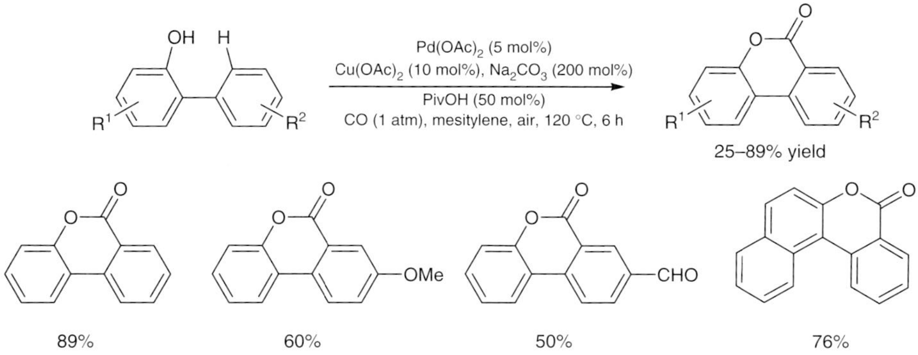

Scheme 3.25 Pd-catalyzed C–H carbonylation of phenol to prepare lactones.

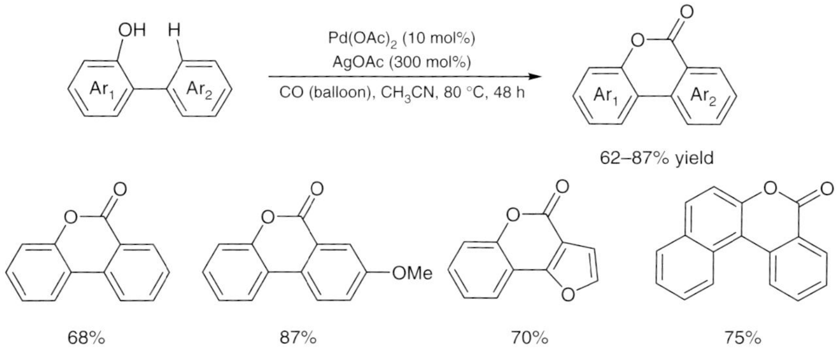

Scheme 3.26 Pd-catalyzed C–H carbonylation of phenol to prepare lactones.

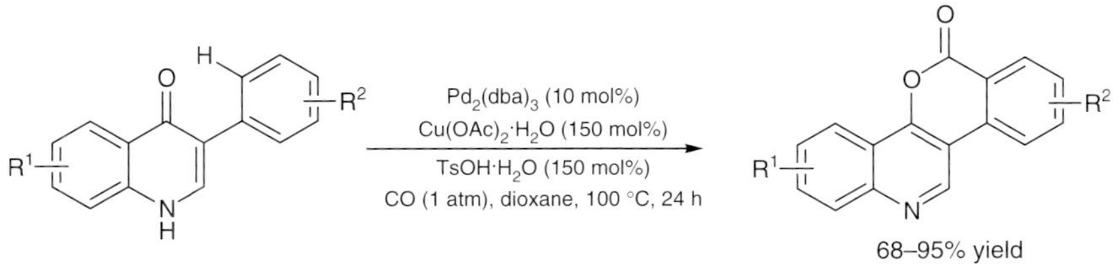

Scheme 3.27 Pd-catalyzed C–H carbonylation to prepare polycyclic heterocyclic compounds.

Xanthone scaffold is envisioned as a privileged structure in many biologically active molecules [53], and it is conventionally synthesized through the cyclodehydration of 2,2′-dihydroxybenzophenones and electrophilic cycloacylation of 2-aryloxybenzoic acids [54]. To refine these synthetic routes, Lei and coworkers described a Pd-catalyzed double C–H activation/carbonylation of diaryl ethers to afford xanthones (Scheme 3.28) [55]. The method is general, and a wide range of diaryl ethers bearing different substituents worked well under the acidic condition.

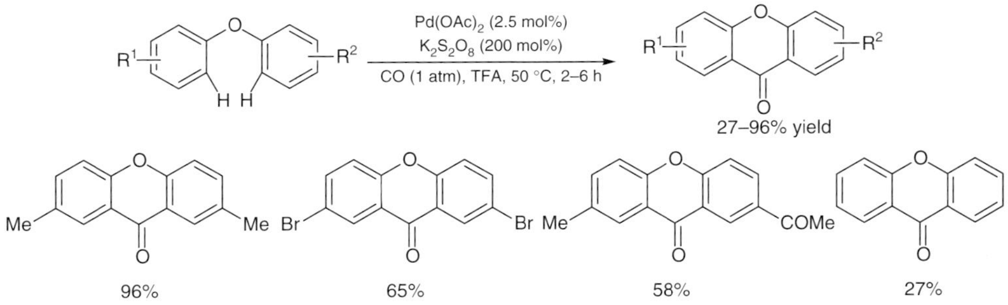

Scheme 3.28 Pd-catalyzed C–H carbonylation of diaryl ethers to prepare xanthones.

Besides the palladium-catalyzed carbonylation of aromatic C–H bond with CO, the carboxylation of alkenyl C–H bond to give oxygen-containing heterocycles has also been reported. In 2013, Iwasawa revealed a palladium-catalyzed carboxylation of alkenyl C–H bonds of 2-hydroxystyrenes to prepare coumarins (Scheme 3.29) [56]. Using 5 mol% $Pd(OAc)_2$, 3.0 equiv. of Cs_2CO_3, and diglyme as solvent, a wide range of 2-hydroxystyrenes could smoothly convert to corresponding coumarins in satisfactory yields under atmospheric pressure of CO_2.

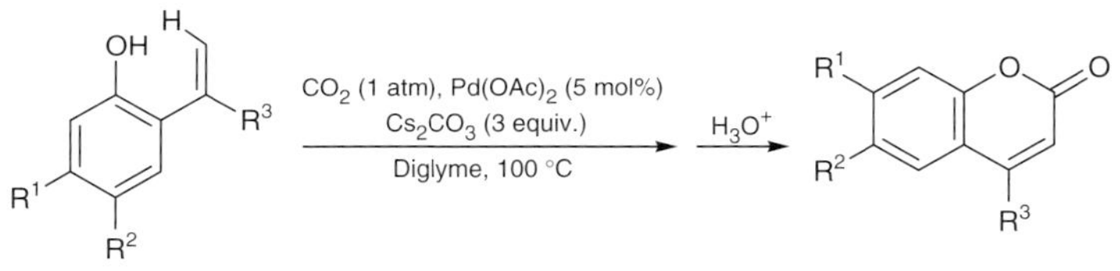

Scheme 3.29 Pd-catalyzed C–H carboxylation of 2-hydroxystyrenes to prepare coumarins.

3.3 Palladium-Catalyzed C–H Activation/C–O Formation to Construct Oxacycles

Palladium-catalyzed C–O formation via C–H activation offers us another opportunity to construct oxygen-containing heterocycles. Compared with the extensively investigated palladium-catalyzed C–H activation/C–C formation, this reaction is much less developed which would be partly attributed to the weak coordinating ability of oxygenic groups [10a] and the reluctant C–O reductive elimination [57, 58]. These methods are generally achieved with alcohols, phenols, and carboxylic acids and their phosphorous analogs through intramolecular cyclization. The oxygenic functional groups on the substrates are not only used to direct the C–H activation but also turned into a part of the desired oxacycles; thus it is highly efficient. In addition, the development of this approach also in turn promoted the progress of oxygen-directing transition metal-catalyzed C–H functionalization.

3.3.1 Palladium-Catalyzed C–O Bond Formation via C(sp²)–H Activation

To construct dihydrobenzofuran rings, in 2010, Yu and coworkers developed a palladium-catalyzed C–H activation/C–O cyclization reaction of phenylethanol derivatives (Scheme 3.30) [59]. Using 5 mol% $Pd(OAc)_2$, 1.5 equiv. of $PhI(OAc)_2$, and 1.5 equiv. of Li_2CO_3, the phenylethanols could be readily cyclized to form dihydrobenzofurans in moderate to high yields. Both the tertiary and secondary alcohols are suitable to this methodology, albeit lower yield for secondary alcohol presumably owing to the potential oxidation.

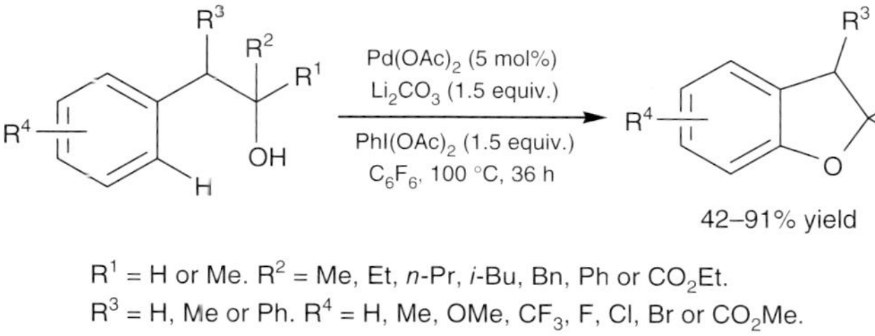

Scheme 3.30 Pd-catalyzed C–H activation/C–O cyclization of aliphatic alcohol.

Later, Yu and Davies further elaborated this fascinating reaction in the enantioselective synthesis of highly functionalized 2,3-dihydrobenzofurans (Scheme 3.31) [60]. Through a rhodium-catalyzed enantioselective intermolecular C–H insertion and following palladium-catalyzed C–H activation/C–O cyclization sequence, a variety of 2,3-dihydrobenzofurans were synthesized with a high regio-, diastereo-, and enantioselectivity.

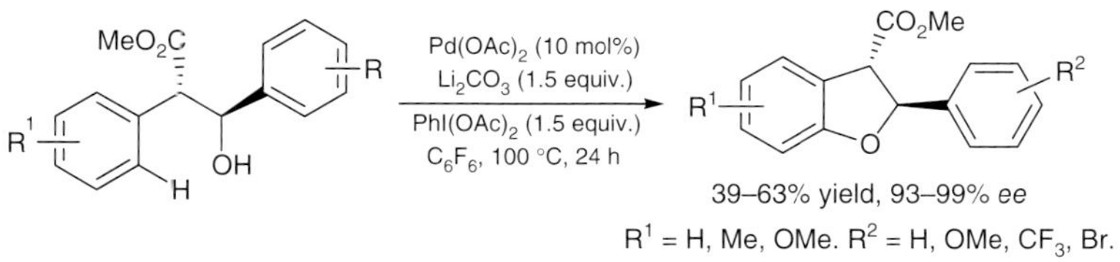

Scheme 3.31 Enantioselective synthesis of highly functionalized 2,3-dihydrobenzofurans.

Considering the similarity of the hydroxyl group in alcohols and silanols, Gevorgyan developed a silanol-directed palladium-catalyzed C–H activation/C–O cyclization of phenoxy silanols [61]. The obtained oxasilacycles were further desilylated with TBAF to give catechols. In a follow-up work, the same group disclosed a palladium-catalyzed C–H oxygenation with benzylsilanols to achieve oxasilacycles under similar conditions (Scheme 3.32) [62]. The obtained oxasilacycles proved to be valuable synthetic intermediates and could be converted to a variety of useful compounds.

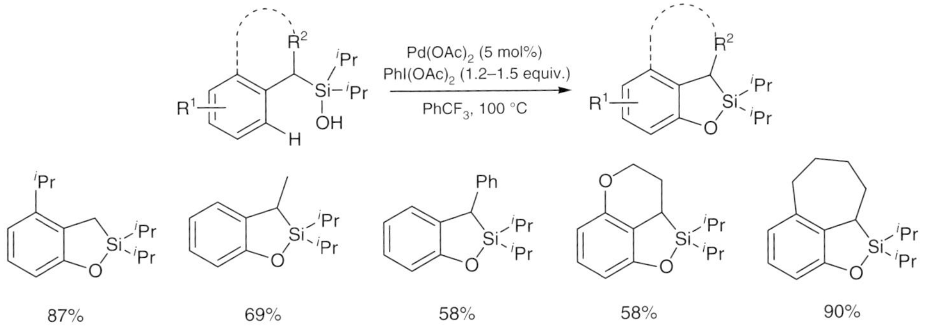

Scheme 3.32 Pd-catalyzed silanol-directed C–H oxygenation.

The construction of dibenzofurans by the C–O bond cyclization of 2-phenols has also been reported. Using 5 mol% $Pd(OAc)_2$, 10 mol% IPr, 10 mol% 4,5-diazafluoren-9-one, 0.5 equiv. of MesCOONa, and 2 equiv. of K_2CO_3, Liu realized the synthesis of dibenzofurans by phenol-directed intramolecular C–O cyclization based on a Pd(0)/Pd(II) catalytic cycle (Scheme 3.33, path a) [63]. Later on, the Yoshikai group reported a similar reaction to achieve dibenzofurans with *tert*-butyl peroxybenzoate as an oxidant (Scheme 3.33, path b) [64]. Different from Liu's work, this reaction is accomplished through a Pd(II)/Pd(IV) catalytic cycle, and the rate-limiting step is the C–H bond cleavage rather than C–O reductive elimination.

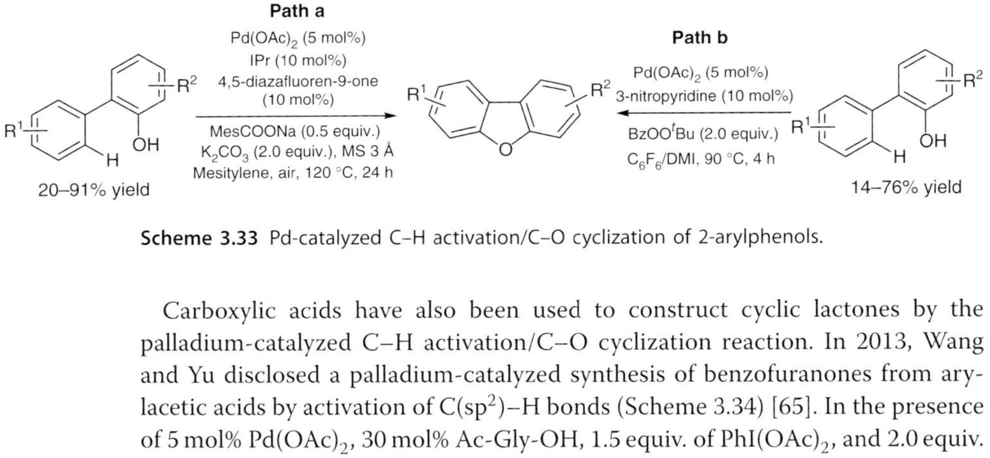

Scheme 3.33 Pd-catalyzed C–H activation/C–O cyclization of 2-arylphenols.

Carboxylic acids have also been used to construct cyclic lactones by the palladium-catalyzed C–H activation/C–O cyclization reaction. In 2013, Wang and Yu disclosed a palladium-catalyzed synthesis of benzofuranones from arylacetic acids by activation of C(sp^2)–H bonds (Scheme 3.34) [65]. In the presence of 5 mol% $Pd(OAc)_2$, 30 mol% Ac-Gly-OH, 1.5 equiv. of $PhI(OAc)_2$, and 2.0 equiv. of KOAc, the corresponding lactones could be afforded in modest to excellent yields in *t*-BuOH. Furthermore, by optimization of the ligand to Boc-Ile-OH, a variety of diphenylacetic acids could undergo lactonization with excellent enantioselectivities. It should be noted that either *ortho*-substituted group at the phenyl ring or hydrogen at the α-position would block the reaction.

Almost the same time, Shi and coworkers independently reported a direct lactonization of 2-arylacetic acids through Pd(II)-catalyzed C–H activation/C–O

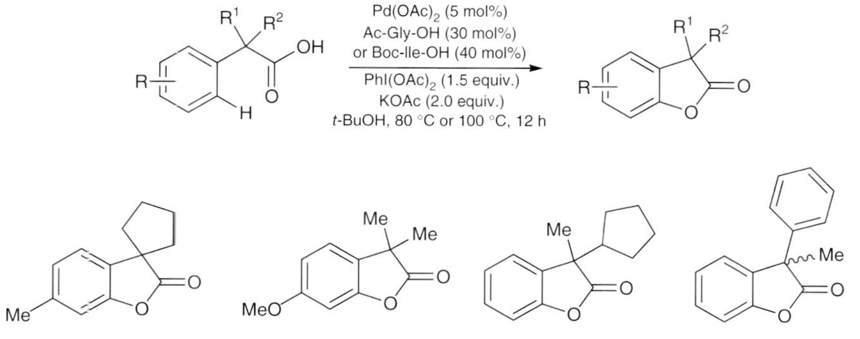

94% yield　94% yield　89% yield　78% yield, 95% ee

Scheme 3.34 Pd(II)-catalyzed enantioselective C–H activation/C–O bond formation.

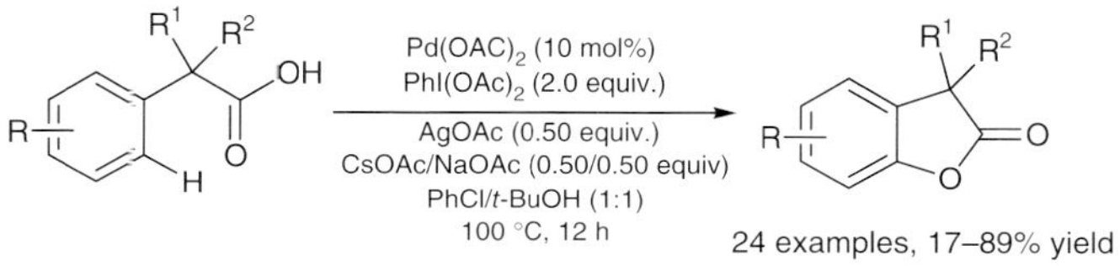

Scheme 3.35 Pd(II)-catalyzed C–H activation/C–O cyclization to benzofuranones.

cyclization (Scheme 3.35) [66]. Compared with Yu's work [65], although phenylacetic acids with hydrogen at the α-position are also not suitable to this reaction, the *ortho*-substituted substrates work well under the standard condition.

Based on the ligand-promoted C–H activation strategy [67], Wang developed a practical palladium-catalyzed biaryl lactone synthesis (Scheme 3.36) [68]. A diverse set of biaryl carboxylic acids could be cyclized to form the desired

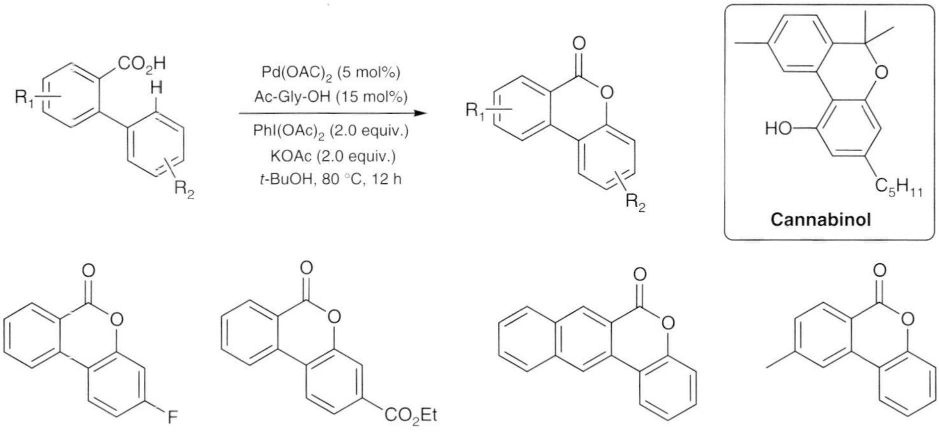

81%　74%　80%　94%

Scheme 3.36 Pd(II)-catalyzed C–H activation/C–O cyclization of biphenyl carboxylic acid.

oxygen-containing heterocycles in moderate to excellent yields. In addition, the utility of this methodology was further demonstrated in a total synthesis of the natural product cannabinol from commercially available starting materials.

An efficient synthetic method to prepare phosphorus benzolactone analogs based on the palladium-catalyzed C–H activation/C–O cyclization reaction has been achieved by Lee and coworkers (Scheme 3.37) [69]. It should be noted that both C(sp^2)–H bond and benzylic C(sp^3)–H bond could be activated under the standard conditions.

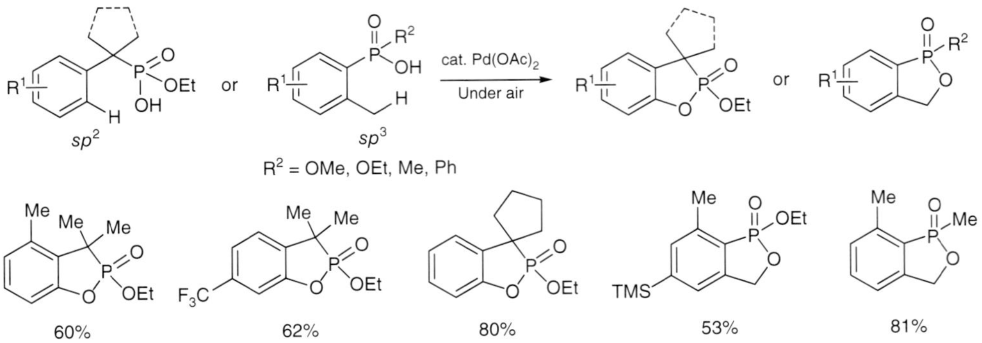

Scheme 3.37 Pd-catalyzed C(sp^2 and sp^3)–H activation/C–O bond formation.

In addition, through palladium-catalyzed C–H activation/C–O cyclization of *in situ*-generated enamines, Jiang and coworkers developed an efficient method to construct oxazole derivatives from readily available amides and ketones in one pot (Scheme 3.38) [70].

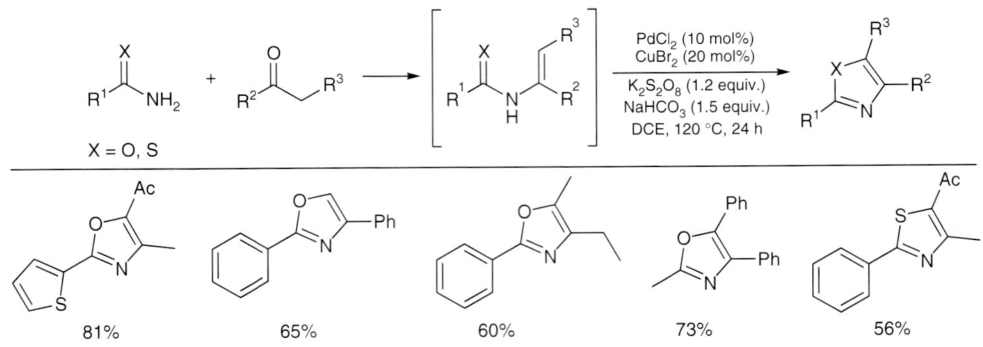

Scheme 3.38 Synthesis of oxazole and thiazole derivatives.

3.3.2 Palladium-Catalyzed C–O Bond Formation via Allylic C–H Activation

In 1977, Hegedus reported a concise method to construct isocoumarins from 2-alkenylbenzoic acids [71]. Although these reactions mainly depend on the use of a stoichiometric amount of palladium salt, several 3-methylisocoumarins could also be obtained by a catalytic manner in the presence of copper acetate and oxygen. To develop a general, catalytic route to unsaturated lactones, in 1993, Larock and coworkers reported a $Pd(OAc)_2$-catalyzed cyclization of alkenoic acids to five- and six-membered lactones (Scheme 3.39) [72]. The methodology is very mild, and a broad scope of mono-, di-, and trisubstituted alkenes could be converted to the desired products in high yields.

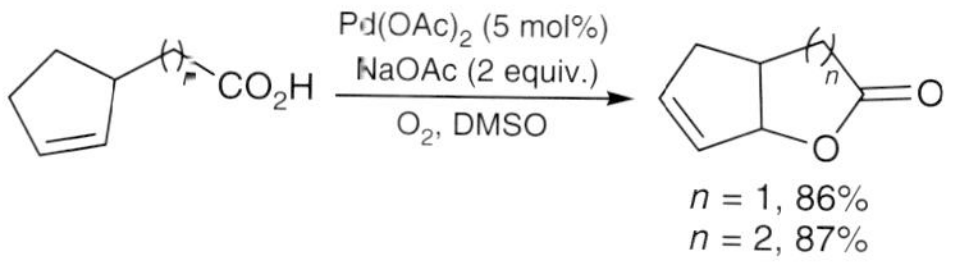

Scheme 3.39 Palladium-catalyzed cyclization of alkenoic acids to synthesize lactones.

In 2005, the White group further extended this chemistry to the construction of macrocyclic lactones [73]. In the presence of 10–20 mol% of $Pd(OAc)_2$/phenyl bis-sulfoxide and 2 equiv. of BQ, a wide scope of 14- to 17-membered macrocyclic benzolactones could be afforded in moderate yields with outstanding regioselectivities under air. With the combination of $Pd(OAc)_2$/bis-sulfoxide and phosphoric acid, White and coworkers achieved the synthesis of oxazolidinones from *N*-Boc amines in good yields and excellent diastereoselectivities via allylic C–H cleavage (Scheme 3.40) [74].

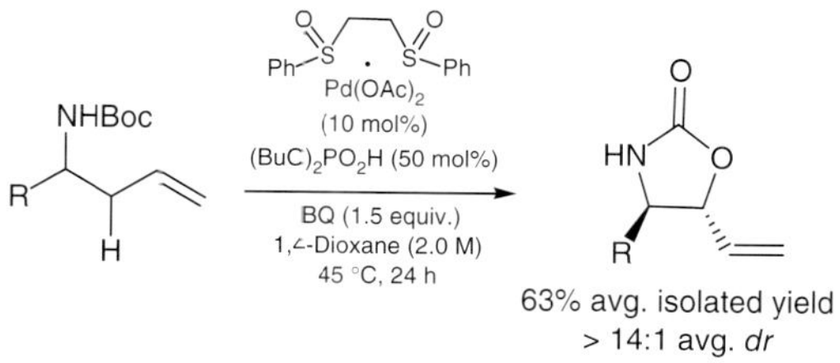

Scheme 3.40 Synthesis of oxazolidinones via allylic C–H oxidation.

Almost the same time, White and coworkers also disclosed a Pd(II)/bis-sulfoxide-catalyzed allylic C–H activation/C–O cyclization reaction to afford chromans, isochromans, and pyrans with chromium complex as the cocatalyst. In this work, phenols rather than carboxylic acids were used as the nucleophiles. However, the reaction is limited to the terminal olefins (Scheme 3.41) [75].

An enantioselective oxidative cyclization of 4-alkenoic acids to give γ-lactone derivatives via palladium-catalyzed allylic C–H activation was also disclosed by Sasai in 2011 (Scheme 3.42) [76]. In the presence of 10 mol% $Pd(OAc)_2$ and 15 mol% spiro bis(isoxazoline) ligand, a variety of five-membered ring lactones could be constructed in moderate to excellent yields. However, the enantioselectivity is highly depended on the structures of substrates.

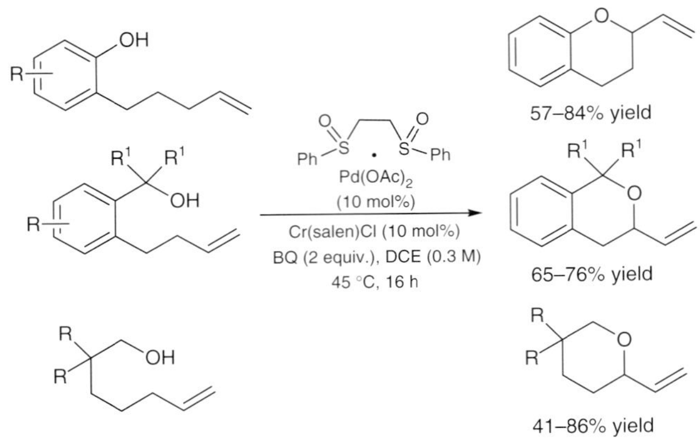

Scheme 3.41 Synthesis of chromans, isochromans, and pyrans via allylic C–H oxidation.

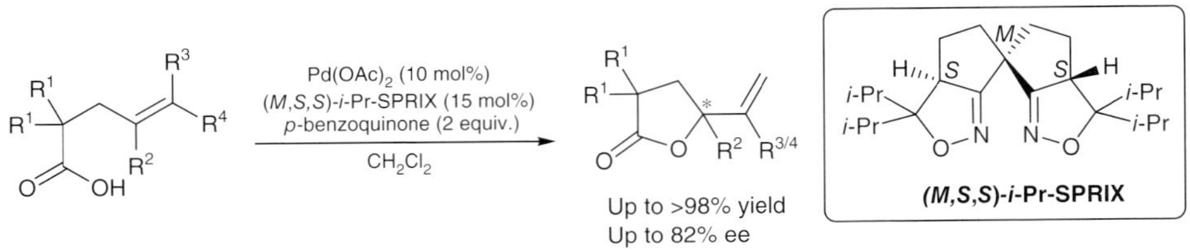

Scheme 3.42 Oxidative cyclization of 4-alkenoic acid derivatives.

3.4 Conclusions

Based on the rapid development of C–H activation in the past few decades, palladium-catalyzed synthesis of oxygen-containing heterocycles via C–H activation has been obtained a significant progress; in turn, these achievements also promoted the advance of transition metal-catalyzed C–H functionalization. A broad scope of oxacycles has been synthesized via this approach. Some of these structures are even difficult to access through the conventional methods and hence highlight the palladium-catalyzed C–H activation a powerful and efficient method to achieve novel transformations.

Despite the great success and potential of palladium-catalyzed C–H activation in the synthesis of oxygen-containing heterocycles, a number of issues still exist. To date, most of the reactions are focused on the activation of aromatic C–H bonds, and only a few works on the construction of oxacycles via activation of C(sp^3)–H bonds have been reported. In addition, these processes also require high loading of transition metal catalysts, which leaves the current methods a long way to go for the practical application. Furthermore, from the viewpoint of atom economy and green chemistry, the use of excess metal salts or hypervalent iodine reagents as the oxidant would produce undesired wastes and impair the appeal of this strategy. Future studies would be focused on the development of the reactions involving the activation of $C(sp^3)$–H bonds, reducing the amount of catalysts and exploiting cleaner oxidants such as molecular oxygen.

References

1. (a) Majumdar, K.C. and Chattopadhyay, S.K. (eds) (2011) *Heterocycles in Natural Product Synthesis*, Wiley-VCH Verlag GmbH, Weinheim; (b) Peng, X.-S., Fan, R., and Wu, J. (2014) in *Progress in Heterocyclic Chemistry*, vol. 26 (eds G.W. Gribble and J.A. Joule), Elsevier, Boston, MA, pp. 193–235; (c) Santos, C.M.M. and Silva, A.M.S. (2014) in *Progress in Heterocyclic Chemistry*, vol. 26 (eds G.W. Gribble and J.A. Joule), Elsevier, Boston, MA, pp. 463–520.
2. (a) Gulevich, A.V., Dudnik, A.S., Chernyak, N., and Gevorgyan, V. (2013) *Chem. Rev.*, **113**, 3084–3213; (b) Properzi, R. and Marcantoni, E. (2014) *Chem. Soc. Rev.*, **43**, 779–791.
3. (a) Pirenty, A., Moreau, X., and Campagne, J.-M. (2006) *Chem. Rev.*, **106**, 911–939 (b) Shu, X.-Z., Shu, D., Schienebeck, C.M., and Tang, W. (2012) *Chem. Soc. Rev.*, **41**, 7698–7711; (c) Krause, N. and Winter, C. (2011) *Chem. Rev.* **111**, 1994–2009; (d) Sambiagio, C., Marsden, S.P., Blacker, A.J., and McGowan, P.C. (2014) *Chem. Soc. Rev.*, **43**, 3525–3550; (e) Cavitt, M.A., Phun, L.H., and France, S. (2014) *Chem. Soc. Rev.*, **43**, 804–818.
4. Yu, J.-Q. and Shi, Z. (eds) (2010) *C-H Activation Topics in Current Chemistry*, vol. 292, Springer-Verlag, Berlin.
5. Shen, C., Zhang, P., Sun, Q., Bai, S., Hor, T.S.A., and Liu, X. (2015) *Chem. Soc. Rev.*, **44**, 291–314.
6. Dick, A.R. and Sanford, M.S. (2006) *Tetrahedron*, **62**, 2439–2463.
7. (a) Dyker, G. (1999) *Angew. Chem. Int. Ed.*, **38**, 1698–1712; (b) Yeung, C.S. and Dong, V.M. (2011) *Chem. Rev.*, **111**, 1215–1292; (c) Ackermann, L. (2011) *Chem. Rev.* **111**, 1315–1345.
8. (a) Chen, X., Engle, K.M., Wang, D.-H., and Yu, J.-Q. (2009) *Angew. Chem. Int. Ed.*, **48**, 5094–5115; (b) Franzoni, I. and Mazet, C. (2014) *Org. Biomol. Chem.*, **12**, 233–241.
9. (a) Newhouse, T. and Barana, P.S. (2011) *Angew. Chem. Int. Ed.*, **50**, 3362–3374; (b) Kuhl, N., Hopkinson, M.N., Wencel-Delord, J., and Glorius, F. (2012) *Angew. Chem. Int. Ed.*, **51**, 10236–10254; (c) Rossi, R., Bellina, F., Lessi, M., and Manzini, C. (2014) *Adv. Synth. Catal.*, **356**, 17–117.
10. (a) Lyons, T.W. and Sanford, M.S. (2010) *Chem. Rev.*, **110**, 1147–1169; (b) Engle, K.M., Mei, T.-S., Wasa, M., and Yu, J.-Q. (2012) *Acc. Chem. Res.*, **45**, 788–802; (c) Neufeldt, S.R. and Sanford, M.S. (2012) *Acc. Chem. Res.*, **45**, 936–946; (d) Zhang, F. and Spring, D.R. (2014) *Chem. Soc. Rev.*, **43**, 6906–6919.
11. Ackermann, L., Vicente, R., and Kapdi, A.R. (2009) *Angew. Chem. Int. Ed.*, **48**, 9792–9826.
12. Huang, C. and Gevorgyan, V. (2009) *J. Am. Chem. Soc.*, **131**, 10844–10845.
13. Mondal, S. (2012) *Chem. Rev.*, **112**, 5339–5355.
14. Majumdar, K.C., Mondal, S., and Ghosh, D. (2009) *Tetrahedron Lett.*, **50**, 4781–4784.
15. Bheeter, C.B., Bera, J.K., and Doucet, H. (2012) *Adv. Synth. Catal.*, **354**, 3533–3538.
16. Laha, J.K., Jethava, K.P., and Dayal, N. (2014) *J. Org. Chem.*, **79**, 8010–8019.
17. Shiotuni, A. and Itatani, H. (1974) *Angew. Chem. Int. Ed. Engl.*, **13**, 471.
18. Parham, W.E. and Strassburg, R.W. (1961) *J. Org. Chem.*, **26**, 4749–4753.
19. Pintori, D.G. and Greaney, M.F. (2011) *J. Am. Chem. Soc.*, **133**, 1209–1211.
20. (a) Jia, C., Kitamura, T., and Fujiwara, Y. (2001) *Acc. Chem. Res.*, **34**, 633–639; (b) Beccalli, E.M., Broggini, G., Martinelli, M., and Sottocornola, S. (2007) *Chem. Rev.*, **107**, 5318–5365; (c) Karimi, B., Behzadnia, H., Elhamifar, D., Akhavan, P.F., Esfahani, F.K., and Zamani, A. (2010) *Synthesis*, 1399–1427; (d) Le Bras, J. and Muzart, J. (2011) *Chem. Rev.*, **111**,

1170–1214; (e) Zhou, L. and Lu, W. (2014) *Chem. Eur. J.*, **20**, 634–642.

21. Jia, C., Piao, D., Oyamada, J., Lu, W., Kitamura, T., and Fujiwara, Y. (2000) *Science*, **287**, 1992–1995.
22. Aoki, S., Oyamada, J., and Kitamura, T. (2005) *Bull. Chem. Soc. Jpn.*, **78**, 468–472.
23. Kim, D., Mina, M., and Hong, S. (2013) *Chem. Commun.*, **49**, 4021–4023.
24. Zhang, X.-S., Li, Z.-W., and Shi, Z.-J. (2014) *Org. Chem. Front.*, **1**, 44–49.
25. (a) Achmad, S.A., Hakim, E.H., Juliawaty, L.D., Makmur, L., Suyatno, A.N., and Ghisalberti, E.L. (1996) *J. Nat. Prod.*, **59**, 878–879; (b) Zheng, Z.-P., Cheng, K.-W., To, J.T.-K., Li, H., and Wang, M. (2008) *Mol. Nutr. Food Res.*, **52**, 1530–1538; (c) Wei, B.-L., Weng, J.-R., Chiu, P.-H., Hung, C.-F., Wang, J.-P., and Lin, C.-N. (2005) *J. Agric. Food. Chem.*, **53**, 3867–3871; (d) Likhitwitayawuid, K., Supudompol, B., Sritularak, B., Lipipun, V., Rapp, K., and Schinazi, R.F. (2005) *Pharm. Biol.*, **43**, 651–657.
26. Kim, Y., Moon, Y., Kang, D., and Hong, S. (2014) *Org. Biomol. Chem.*, **12**, 3413–3422.
27. Mamane, V., Hannen, P., and Fürstner, A. (2004) *Chem. Eur. J.*, **10**, 4556–4575.
28. Minami, Y., Shiraishi, Y., Yamada, K., and Hiyama, T. (2012) *J. Am. Chem. Soc.*, **134**, 6124–6127.
29. Minami, Y., Kanda, M., and Hiyama, T. (2014) *Chem. Lett.*, **43**, 181–183.
30. Tietze, L.F., Hungerland, T., Eichhorst, C., Düfert, A., Maaß, C., and Stalke, D. (2013) *Angew. Chem. Int. Ed.*, **52**, 3668–3671.
31. Li, C., Zhang, Y., Li, P., and Wang, L. (2011) *J. Org. Chem.*, **76**, 4692–4696.
32. Wang, S., Li, P., Yu, L., and Wang, L. (2011) *Org. Lett.*, **13**, 5968–5971.
33. Kuram, M.R., Bhanuchandra, M., and Sahoo, A.K. (2013) *Angew. Chem. Int. Ed.*, **52**, 4607–4612.
34. Minami, Y., Yamada, K., and Hiyama, T. (2013) *Angew. Chem. Int. Ed.*, **52**, 10611–10615.
35. Miura, M., Tsuda, T., Satoh, T., Pivsa-Art, S., and Nomura, M. (1998) *J. Org. Chem.*, **63**, 5211–5215.
36. Nandi, D., Ghosh, D., Chen, S.-J., Kuo, B.-C., Wang, N.M., and Lee, H.M. (2013) *J. Org. Chem.*, **78**, 3445–3451.
37. (a) Hennings, D.D., Iwasa, S., and Rawal, V.H. (1997) *J. Org. Chem.*, **62**, 2–3; (b) Satoh, T., Kawamura, Y., Miura, M., and Nomura, M. (1997) *Angew. Chem. Int. Ed. Engl.*, **36**, 1740–1742; (c) Ferreira, E.M. and Stoltz, B.M. (2001) *J. Am. Chem. Soc.*, **123**, 7725–7726; (d) Jensen, D.R., Pugsley, J.S., and Sigman, M.S. (2001) *J. Am. Chem. Soc.*, **123**, 7475–7476.
38. Lu, Y., Wang, D.-H., Engle, K.M., and Yu, J.-Q. (2010) *J. Am. Chem. Soc.*, **132**, 5916–5921.
39. (a) Catellani, M., Frignani, F., and Rangoni, A. (1997) *Angew. Chem. Int. Ed. Engl.*, **36**, 119–122; (b) Catellani, M. and Cugini, F. (1999) *Tetrahedron*, **55**, 6595–6602; (c) Catellani, M., Mealli, C., Motti, E., Paoli, P., Perez-Carreno, E., and Pregosin, P.S. (2002) *J. Am. Chem. Soc.*, **124**, 4336–4346; (d) Catellani, M. (2002) in *Handbook of Organopalladium Chemistry for Organic Synthesis* (eds E.I. Negishi and A. de Meijere), John Wiley & Sons, Inc., Hoboken, NJ, pp. 1479–1489; (e) Tsuji, J. (2004) *Palladium Reagents and Catalysis–New Perspectives for the 21st Century*, John Wiley & Sons, Inc., Hoboken, NJ, pp. 409–416.
40. Lautens, M., Paquin, J.-F., and Piguel, S. (2002) *J. Org. Chem.*, **67**, 3972–3974.
41. Pache, S. and Lautens, M. (2003) *Org. Lett.*, **5**, 4827–4830.
42. Alberico, D., Rudolph, A., and Lautens, M. (2007) *J. Org. Chem.*, **72**, 775–781.
43. Rudolph, A., Rackelmann, N., Turcotte-Savard, M.-O., and Lautens, M. (2009) *J. Org. Chem.*, **74**, 289–297.
44. Zhang, Y.-H., Shi, B.-F., and Yu, J.-Q. (2009) *Angew. Chem. Int. Ed.*, **48**, 6097–6100.
45. Giri, R. and Yu, J.-Q. (2008) *J. Am. Chem. Soc.*, **130**, 14082–14083.
46. Guan, Z.-H., Chen, M., and Ren, Z.-H. (2012) *J. Am. Chem. Soc.*, **134**, 17490–17493.
47. Silvaggi, N.R., Kaur, K., Adediran, S.A., Pratt, R.F., and Kelly, J.A. (2004) *Biochemistry*, **43**, 7046–7053.
48. Shin, S., Jeong, Y., Jeon, W.H., and Lee, P.H. (2014) *Org. Lett.*, **16**, 2930–2933.

49. Lu, Y., Leow, D., Wang, X., Engle, K.M., and Yu, J.-Q. (2011) *Chem. Sci.*, **2**, 967–971.
50. Luo, S., Luo, F.-X., Zhang, X.-S., and Shi, Z.-J. (2013) *Angew. Chem. Int. Ed.*, **52**, 10598–10601.
51. Lee, T.-H., Jayakumar, J., Cheng, C.-H., and Chuang, S.-C. (2013) *Chem. Commun.*, **49**, 11797–11799.
52. Ji, F., Li, X., Wu, W., and Jiang, H. (2014) *J. Org. Chem.*, **79**, 11246–11253.
53. Peres, V., Nagem, T.J., and de Oliveira, F.F. (2000) *Phytochemistry*, **55**, 683–710.
54. Pinto, M.M.M. and Castanheiro, R.A.P. (2009) *Curr. Org. Chem.*, **13**, 1215–1240.
55. Zhang, H., Shi, R., Gan, P., Liu, C., Ding, A., Wang, Q., and Lei, A. (2012) *Angew. Chem. Int. Ed.*, **51**, 5204–5207.
56. Sasano, K., Takaya, J., and Iwasawa, N. (2013) *J. Am. Chem. Soc.*, **135**, 10954–10957.
57. Hartwig, J.F. (2007) *Inorg. Chem.*, **46**, 1936–1947.
58. (a) Widenhoefer, R.A., Zhong, H.A., and Buchwald, S.L. (1997) *J. Am. Chem. Soc.*, **119**, 6787–6795; (b) Widenhoefer, R.A. and Buchwald, S.L. (1998) *J. Am. Chem. Soc.*, **120**, 6504–6511; (c) Aranyos, A., Old, D.W., Kiyomori, A., Wolfe, J.P., Sadighi, J.P., and Buchwald, S.L. (1999) *J. Am. Chem. Soc.*, **121**, 4369–4378.
59. Wang, X., Lu, Y., Dai, H.-X., and Yu, J.-Q. (2010) *J. Am. Chem. Soc.*, **132**, 12203–12205.
60. Wang, H., Li, G., Engle, K.M., Yu, J.-Q., and Davies, H.M.L. (2013) *J. Am. Chem. Soc.*, **135**, 6774–6777.
61. Huang, C., Ghavtadze, N., Chattopadhyay, B., and Gevorgyan, V. (2011) *J. Am. Chem. Soc.*, **133**, 17630–17633.
62. Huang, C., Ghavtadze, N., Godoi, B., and Gevorgyan, V. (2012) *Chem. Eur. J.*, **18**, 9789–9792.
63. Xiao, B., Gong, T.-J., Liu, Z.-J., Liu, J.-H., Luo, D.-F., Xu, J., and Liu, L. (2011) *J. Am. Chem. Soc.*, **133**, 9250–9253.
64. Wei, Y. and Yoshikai, N. (2011) *Org. Lett.*, **13**, 5504–5507.
65. Cheng, X.-F., Li, Y., Su, Y.-M., Yin, F., Wang, J.-Y., Sheng, J., Vora, H.U., Wang, X.-S., and Yu, J.-Q. (2013) *J. Am. Chem. Soc.*, **135**, 1236–1239.
66. Yang, M., Jiang, X., Shi, W.-J., Zhu, Q.-L., and Shi, Z.-J. (2013) *Org. Lett.*, **15**, 690–693.
67. (a) Wang, D.-H., Engle, K.M., Shi, B.-F., and Yu, J.-Q. (2010) *Science*, **327**, 315–319; (b) Engle, K.M., Wang, D.-H., and Yu, J.-Q. (2010) *J. Am. Chem. Soc.*, **132**, 14137–14151; (c) Ohmatsu, K., Imagawa, N., and Ooi, T. (2014) *Nat. Chem.*, **6**, 47–51.
68. Li, Y., Ding, Y.-J., Wang, J.-Y., Su, Y.-M., and Wang, X.-S. (2013) *Org. Lett.*, **15**, 2574–2577.
69. Eom, D., Jeong, Y., Kim, Y.R., Lee, E., Choi, W., and Lee, P.H. (2013) *Org. Lett.*, **15**, 5210–5213.
70. Zheng, M., Huang, L., Huang, H., Li, X., Wu, W., and Jiang, H. (2014) *Org. Lett.*, **16**, 5906–5909.
71. Korte, D.E., Hegedus, L.S., and Wirth, R.K. (1977) *J. Org. Chem.*, **42**, 1329–1336.
72. Larock, R.C. and Hightower, T.R. (1993) *J. Org. Chem.*, **58**, 5298–5300.
73. Fraunhoffer, K.J., Prabagaran, N., Sirois, L.E., and White, M.C. (2006) *J. Am. Chem. Soc.*, **128**, 9032–9033.
74. Osberger, T.J. and White, M.C. (2014) *J. Am. Chem. Soc.*, **136**, 11176–11181.
75. Ammann, S.E., Rice, G.T., and White, M.C. (2014) *J. Am. Chem. Soc.*, **136**, 10834–10837.
76. Takenaka, K., Akita, M., Tanigaki, Y., Takizawa, S., and Sasai, H. (2011) *Org. Lett.*, **13**, 3506–3509.

4
Pd-Catalyzed Synthesis of Other Heteroatom-Containing Heterocycles

Zhanxiang Liu and Yuhong Zhang

4.1
Introduction

Recently, a number of different protocols have enormously widened the application of Pd-catalyzed reactions, promoting a shift from a general interest in cross-coupling processes toward new and more challenging direct C–H functionalization studies on unactivated substrates. In fact, new procedures that pave the way to minimal waste production, thus meaning cheaper and improved atom economic reactions, are welcome. Pd-catalyzed C–H bond activation reactions for the synthesis of heterocycles have gained considerable attention over the past decades because these processes do not require substrates bearing functionalized groups [1, 2]. Besides various nitrogen- and oxygen-containing heterocycles, Pd-catalyzed C–H bond activation reactions have applications in the synthesis of sulfur, phosphorus, and silicon heterocycles. These special heterocycles are widely used materials in various fields, such as catalysis, pharmaceuticals, agrichemicals, and materials chemistry [3–5]. In this chapter, we discuss the recent developments in Pd-catalyzed reactions which form sulfur, phosphorus, and silicon-containing heterocyclic cores based on C–H bond activation strategies.

4.2
Sulfur-Containing Heterocycles

Sulfur-containing heterocycles has been attracted much attention of the chemist due to their potential biological activity, pharmaceutical significance, and wide application in functional materials [3]. Driven by the importance of these structures, great efforts have been devoted toward synthesizing these crucial structures. In recent years, transition metal-catalyzed C–H bond activation to directly synthesize the sulfur-based heterocycles has been numerously studied.

Transition Metal-Catalyzed Heterocycle Synthesis via C—H Activation, First Edition. Edited by Xiao-Feng Wu.
© 2016 Wiley-VCH Verlag GmbH & Co. KGaA. Published 2016 by Wiley-VCH Verlag GmbH & Co. KGaA.

4.2.1
Benzo[*b*]thiophenes

In 2008, a breakthrough was achieved by Doi and coworkers, who employed thioenols to synthesize of multisubstituted benzo[*b*]thiophenes by using palladium catalyst such as $PdCl_2$ or $PdCl_2$(cod) (Scheme 4.1) [6]. Surprisingly, this transformation proceeded more efficiently in the absence of reoxidants, and they found DMSO is crucial for high conversion in this process. A tentative catalytic cycle could be proposed (Scheme 4.2). Disulfide **B**, formed from the corresponding thioenol **A**, undergoes oxidative addition to palladium, leading to complex **C**. The complex **C** achieves the corresponding product through two possible pathways: the first one is electrophilic attack of the aryl ring at the palladium center to form the six-membered palladacycle **D**, followed by reductive elimination to provide benzo[*b*]thiophene, and the other way is electrophilic attack at the sulfur atom to directly obtain the products.

The palladium-catalyzed double C–H activation of simple benzyl phenyl sulfoxides to synthesize dibenzothiophenes (DBTs) was described by Antonchick and coworkers (Scheme 4.3) [7]. They proposed that the sulfoxide functionalized as a directing group to undergo the first *ortho*-C–H activation in the benzyl phenyl

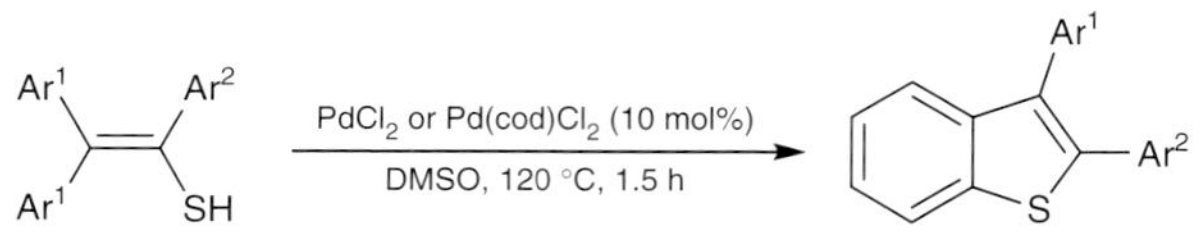

Scheme 4.1 Pd-catalyzed direct synthesis of benzo[*b*]thiophenes from thioenols.

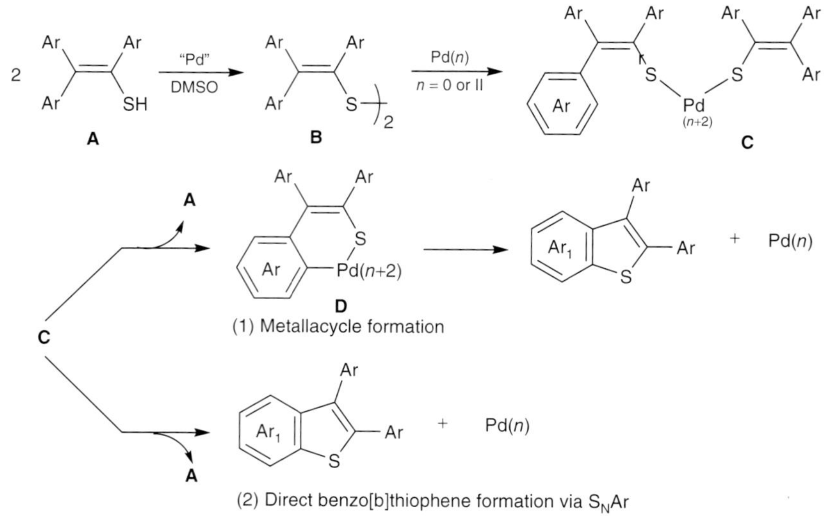

Scheme 4.2 The proposed mechanism.

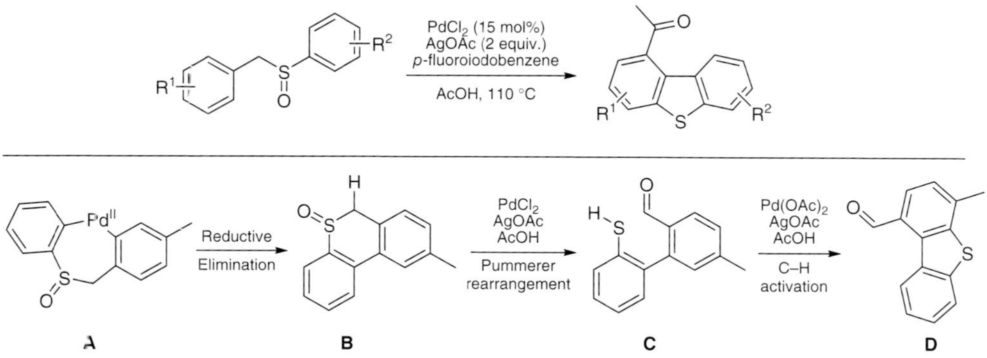

Scheme 4.3 Palladium-catalyzed synthesis of dibenzothiophenes from aryl sulfoxides.

ring. The second C–H activation gave the intermediate **A**, which underwent the reductive elimination to give the sulfur-bridged six-membered polycycle **B**. However, cycle **B** was unstable under the acidic reaction condition and took place the Pummerer rearrangement to produce the thiophenyl derivative **C**, which underwent the palladium-catalyzed cyclization to afford the DBTs. It is unusual that aryl iodide played important role in this transformation and the reaction was sluggish without the use of aryl iodide.

The Pd-catalyzed intramolecular direct arylation of the easily accessible bromo-diaryl sulfoxides to access DBT derivatives was reported by Colobert and coworkers (Scheme 4.4) [8]. This reaction provided a practical approach to construct highly valuable dibenzothiophene-based scaffolds, which have numerous applications such as bioactive compounds including agrochemicals, organic electronics, and materials chemistry. In addition, various diaryl sulfoxides bearing electron-withdrawing and electron-donating groups in the *ortho*, *meta*, or *para* position were well tolerated.

Scheme 4.4 Palladium-catalyzed arylation of 2-bromo-diaryl sulfoxides.

The Pd-catalyzed intramolecular oxidative C–H/C–H coupling of diaryl sulfides to form DBTs was developed by Zhou and coworkers (Scheme 4.5) [9]. This approach afforded various DBTs in moderate to good yields with tolerance of a broad range of substrates. Remarkably, the benzo[1,2-*b*:4,5-*b*′]bis[*b*]benzothiophene was successfully synthesized by this protocol, which could be used as an organic semiconductor for field-effect transistors. Furthermore, they found that diaryl sulfides bearing electron-rich substituents

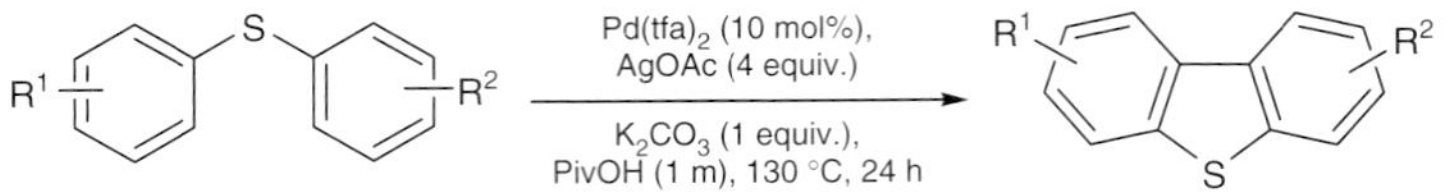

Scheme 4.5 Synthesis of dibenzothiophenes by Pd-catalyzed dual C–H activation.

worked better than those with electron-deficient substituents, which indicated that electrophilic aromatic substitution (SEAr) would be responsible for C–H activation.

The sulfur-bridged six-membered polycycle **B** was prepared by Zhang and coworkers by using the less acidic reaction conditions. They found that when the amount of acetic acid was decreased to 4 equiv. the Pummerer rearrangement could be thoroughly suppressed and the five-, six-, and seven-membered sulfur-bridged polycycles could be isolated by using the different sulfoxide substrates (Scheme 4.6a) [10]. The solvent had a significant effect on the reaction efficiency, and 1, 1, 2, 2-tetrachloroethane) gave the best results. A sulfide-bridged six-membered pyrene–thienoacene molecule was synthesized readily using this method, and the preliminary study revealed its excellent fluorescence performance to give a photoluminescence quantum yield as high as 0.48 (Scheme 4.6b).

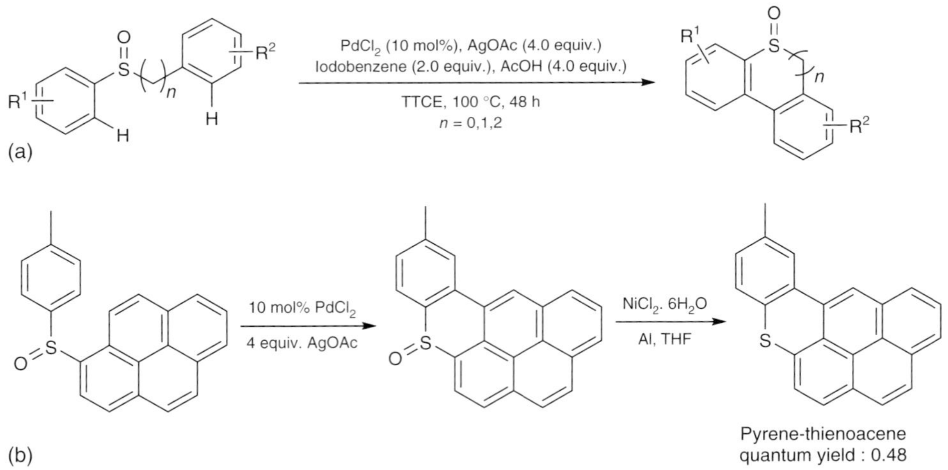

Scheme 4.6 Synthesis of sulfur-bridged polycycles via Pd-catalyzed dehydrogenative cyclization.

The Pd-catalyzed C–S/C–H bond activation of simple and diverse bromothiophenes with alkynes via ring rearrangement to prepare sulfur-based heterocycles and fulvenes was described by Duan and coworkers (Scheme 4.7) [11]. This novel transformation provided a new approach to afford not only the DBTs but also the thiopyrans.

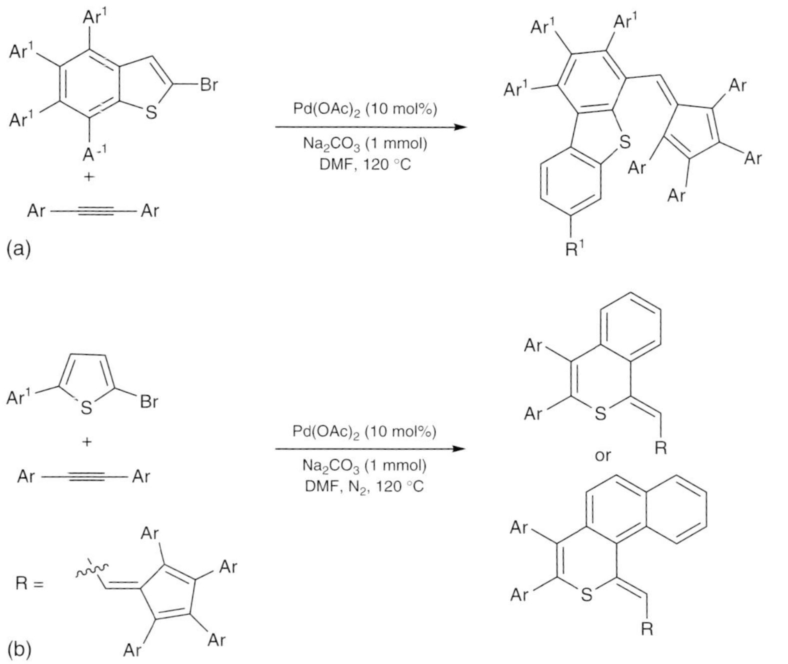

Scheme 4.7 Pd-catalyzed regioselective C–S bond cleavage of thiophenes.

The proposed mechanistic pathway of this transformation was shown in Scheme 4.8. Initially, oxidative addition of Pd(0) to bromobenzothiophene gives a thienylpalladium(II) intermediate **A** followed by *cis*-carbopalladation with alkyne to form a vinylic palladium intermediate **B** which then undergoes C–S bond activation and cycloaddition with alkyne to give the complex **C**. The complex **C** affords a six-membered palladacycle **D** through cyclopalladation with the neighboring Ph group. Finally, the subsequent reductive elimination obtains the DBTs with regeneration of Pd(0) species. In the thiopyran case (Scheme 4.8, bottom), the Pd–sulfur species **E** reacts with alkyne through carbothiolation to form a seven-membered palladacycle **F**, which produces thiopyran.

4.2.2 Benzothiazoles

Doi and coworkers reported the palladium-catalyzed C–H functionalization/C–S bond formation of thiobenzanilides to the synthesis of 2-substituted benzothiazoles (Scheme 4.9) [12]. The corresponding benzothiazole framework represents a privileged structure that exhibits the broad range of biological activities [13]. They found that the addition of Bu_4NBr greatly improved the efficiency of this reaction and the reaction could smoothly proceed under the palladium catalysts such as $PdCl_2$ or $PdCl_2$(cod) as well as $PdBr_2$.

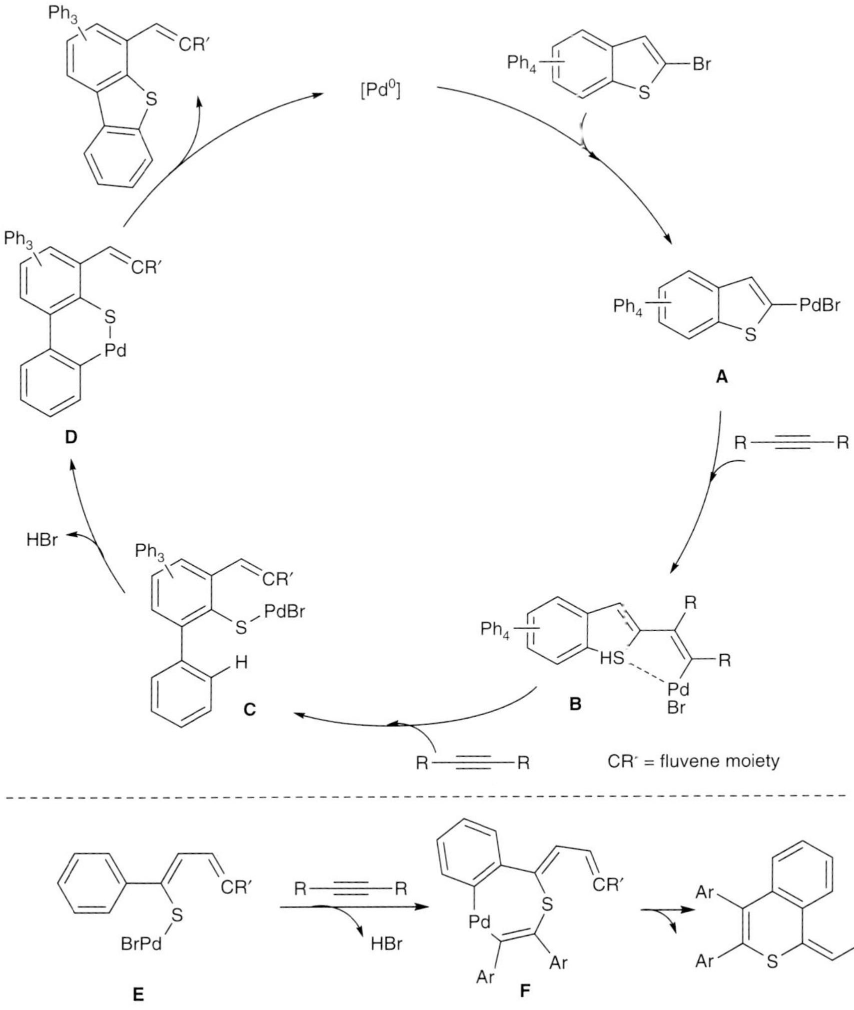

Scheme 4.8 The proposed mechanism.

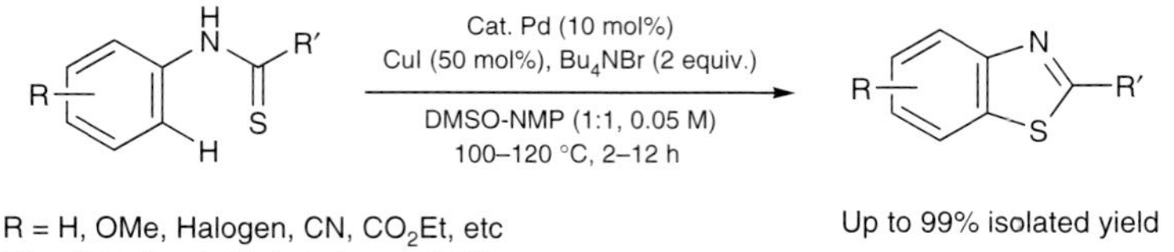

R = H, OMe, Halogen, CN, CO_2Et, etc
R′ = phenyl, substituted phenyl, NHBn
"Pd" = $PdCl_2$, $PdCl_2$(cod), $PdBr_2$

Up to 99% isolated yield

Scheme 4.9 Palladium-catalyzed synthesis of 2-substituted benzothiazoles.

Furthermore, this transformation could be successfully extended to synthesize the 2-aminobenzothiazole from *N*-substituted phenylthiourea. Subsequently, Batey and coworkers successfully employed an unusual O_2/MnO_2 reoxidation system for Pd-catalyzed cyclization of thioureas, leading to 2-aminobenzothiazoles (Scheme 4.10) [14]. They proposed that the mechanism of this reaction undergone electrophilic palladation pathway. A similar work was reported by Doi [15], Kondo [16], and coworkers, respectively.

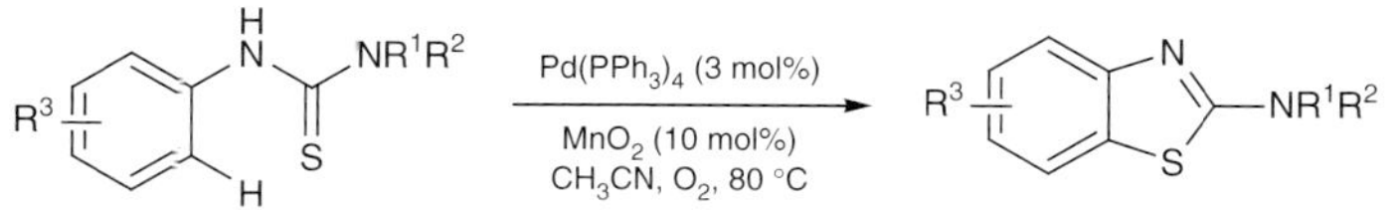

Scheme 4.10 Palladium-catalyzed synthesis of 2-aminobenzothiazoles.

The Pd- and Cu-catalyzed intramolecular C–S linkage of 2-fluoro, 2-chloro, 2-bromo, and 2-iodo aryl thioureas to regioselectively form 2-aminobenzothiazoles was reported by Patel and coworkers (Scheme 4.11) [17]. While Pd prefers predominantly C–H activation strategy, Cu favors a dehalogenative pathway during the formation of 2-aminobenzothiazoles. In the absence of halogens, Pd mostly undergoes the C–H activation pathway, but Cu is unproductive. However, Cu favors a dehalogenative path with less reactive halogens such as –F and –Cl. On the contrary, Pd achieved exclusively C–H activated product in the case of –Cl substrates and a mixture of C–H activated and dehalogenated products with fluoro. In addition, only the dehalogenated products were obtained by both Pd and Cu catalyst of 2-bromo and 2-iodo aryl thioureas.

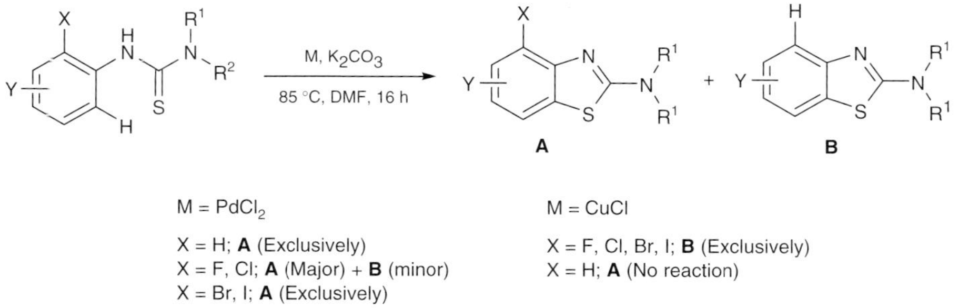

Scheme 4.11 Pd- and Cu-catalyzed regioselective synthesis of 2-aminobenzothiazoles.

The initial work for the synthesis of a wide range of sugar-based benzothiazoles from readily accessible glycosyl thioureas through palladium-catalyzed C–S coupling reaction was developed by Liu and coworkers (Scheme 4.12) [18]. The sugar-based benzothiazole products are a new class of molecules promising for many biological applications. They also found that these fluorescent glycoconjugates can readily enter mammalian tumor cells and exhibit potential in vitro

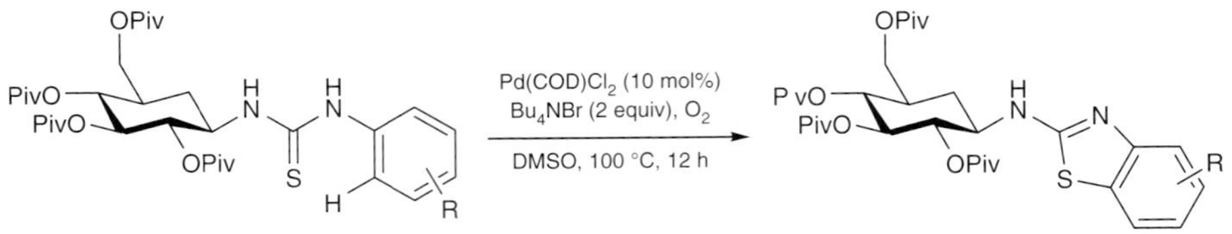

Scheme 4.12 Synthesis of sugar-based benzothiazoles through C–S coupling.

antitumor activity. The key step of this transformation involved the coordination to the Pd center with the assistance of pivaloyl carbonyl group, and the presence of intramolecular hydrogen bonding plays important roles in the efficiency and chemoselectivity of reaction.

Wu and coworkers explored a novel one-pot approach to 2-trifluoromethylbenzothiazoles based on the reaction of the sodium hydrosulfide hydrate using $PdCl_2$ as the sole catalyst in DMSO through the trifluoromethylthiobenzamides intermediate (Scheme 4.13) [19]. A wide range of 2-trifluoromethylbenzothiazoles bearing electron-donating or electron-withdrawing groups were gave in moderate to high yields, including the carbonyl group, the cyano group, the trifluoromethyl group, and halogens.

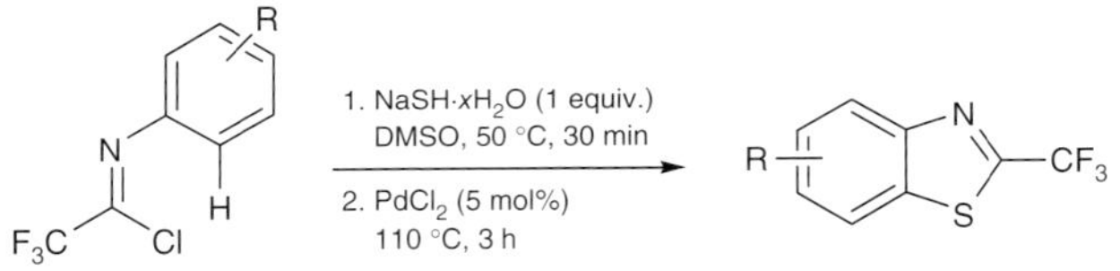

Scheme 4.13 Palladium-catalyzed synthesis of 2-trifluoromethy benzothiazoles.

4.2.3 Sultones

Sultones are the internal esters of hydroxy sulfonic acids and are the sulfur analogs of lactones. Sultones are demanded scaffolds in medicinal chemistry research. Biological studies on sultones are mainly concerned with their toxicological, skin sensitization, and antiviral activities [20]. Sultones are synthetically useful heterocycles which can react with a variety of compounds to introduce the alkylsulfonic acid function and therefore used as sulfoalkylating agents [21]. There have been several new developments for the synthesis of sultones which have also been applied in the total synthesis of natural products. In recent years, the palladium-catalyzed direct arylation of several aromatics via a C–H bond activation using aryl halides has led to successes. An intramolecular version of this reaction has allowed the synthesis of several biaryls via the formation of five- to seven-membered rings. Thus, the sultones should be synthesized by C–H activation via two pathways (Scheme 4.14).

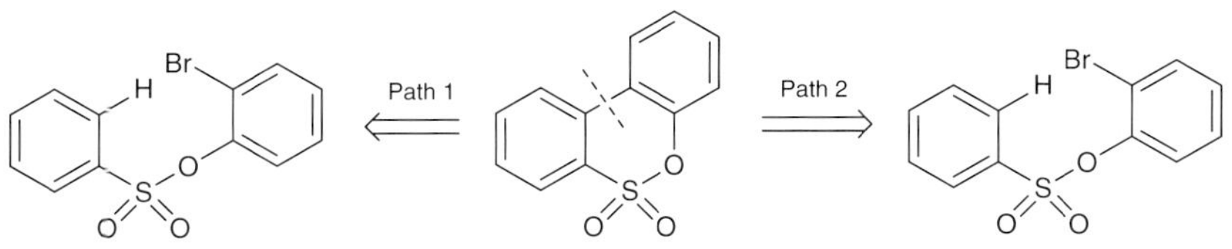

Scheme 4.14 Synthesis of sultones via Pd-catalyzed intramolecular direct arylation.

In 2009, Majumdar and coworkers developed a new efficient route for the synthesis of polycyclic sultones via ligand-free Pd-catalyzed intramolecular coupling reaction (Scheme 4.15) [22]. Using the optimized condition [$Pd(PPh_3)_4$/TBAB/KOAc/DMF/100 °C], benzenesulfonic acid 2-bromophenyl esters were successfully cyclized to afford the sultone in 90% yield.

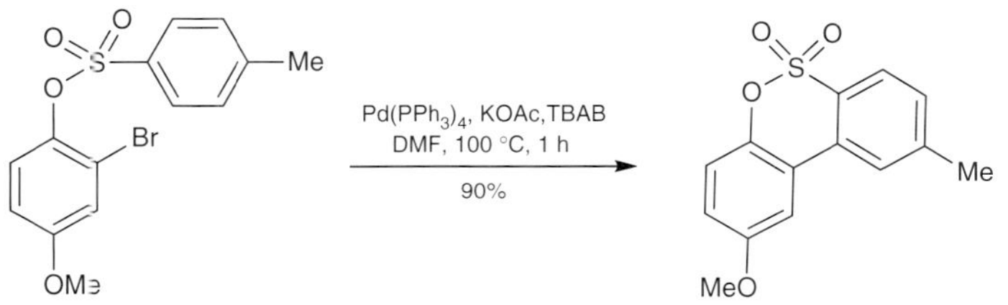

Scheme 4.15 Pd-catalyzed intramolecular coupling reaction of benzenesulfonic acid 2-bromophenyl esters.

Doucet and coworkers reported the intramolecular coupling reaction of 2-bromobenzenesulfonic acid phenyl esters in the presence of a phosphine-free palladium catalyst (Scheme 4.16) [23]. One mole percentage of phosphine-free Pd $(OAc)_2$ as the catalyst associated to KOAc promoted the intramolecular direct arylation via an sp^2 C–H bond functionalization of several 2-bromobenzenesulfonic acid derivatives. The nature of the substituents on the phenol moiety was found to have a huge influence on the reaction. Electron-donating substituents favored the reaction, whereas electron-withdrawing substituents are unfavorable. As a very wide variety of phenols are commercially available, such a cyclization would provide a very simple access to functionalized sultones.

Majumdar *et al.* reported tricyclic sultone derivatives starting from unactivated vinylic systems by the implementation of the palladium-catalyzed

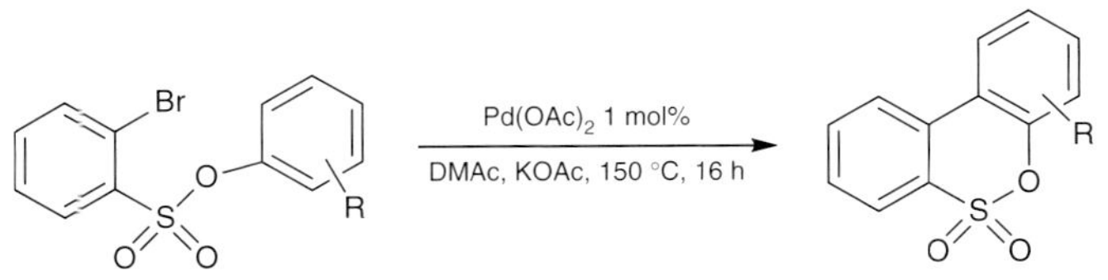

Scheme 4.16 Pd-catalyzed intramolecular coupling reaction of 2-bromobenzenesulfonic acid phenyl esters.

intramolecular Heck reaction [24]. The intramolecular Heck reaction of the 2′-vinylphenyl-2-bromobenzene sulfonates was carried out using $Pd(OAc)_2$ as catalyst, KOAc as base, and tetra-*n*-butylammonium bromide (TBAB) as promoter in dry DMF at 80 °C under nitrogen atmosphere, giving the sultones in 79–91% yield (Scheme 4.17). Using the same methodology, they also extended their work to the synthesis of polycyclic sultones.

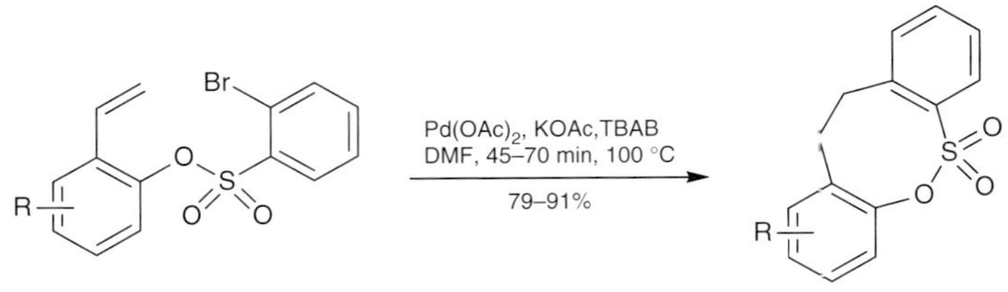

Scheme 4.17 Pd-catalyzed synthesis of tricyclic sultones.

4.2.4
Sultams

Sultams (cyclic sulfonamides), sulfonamides in which the S–N bond is part of a ring, have emerged as privileged structures in drug discovery due to their diverse biological properties [25, 26].

An efficient synthetic method for the synthesis of polycyclic sultams through palladium-catalyzed intramolecular ligand-free coupling reaction was described by Majumdar and coworkers (Scheme 4.18) [27]. The proposed mechanistic pathway of this transformation was shown in Scheme 4.19. The initial oxidation of *N*,*N*-bis(2-bromobenzyl)benzenesulfonamide to Pd(0) to form arylpalladium intermediate **A** followed by adding to the double bond of the electron-rich benzyl bromide to give the intermediate **B**, which give rise to the intermediate **C** by hydrogen elimination. This intermediate **C** may again add to the double bond of the electron-poor aromatic ring of the sulfonamide and may finally give the product sultam.

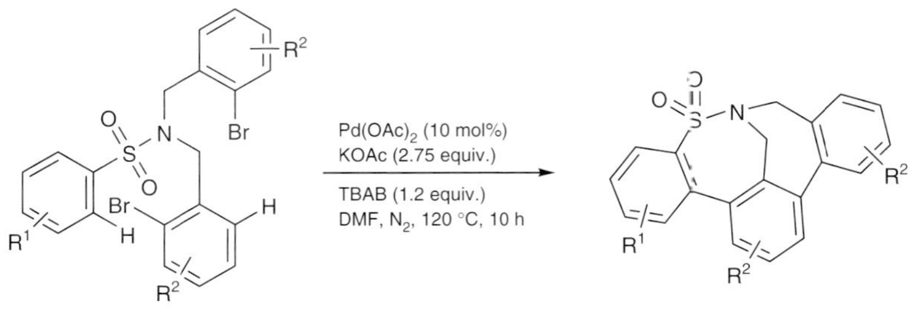

Scheme 4.18 Pd-catalyzed intramolecular coupling reaction to synthesize polycyclic sultams.

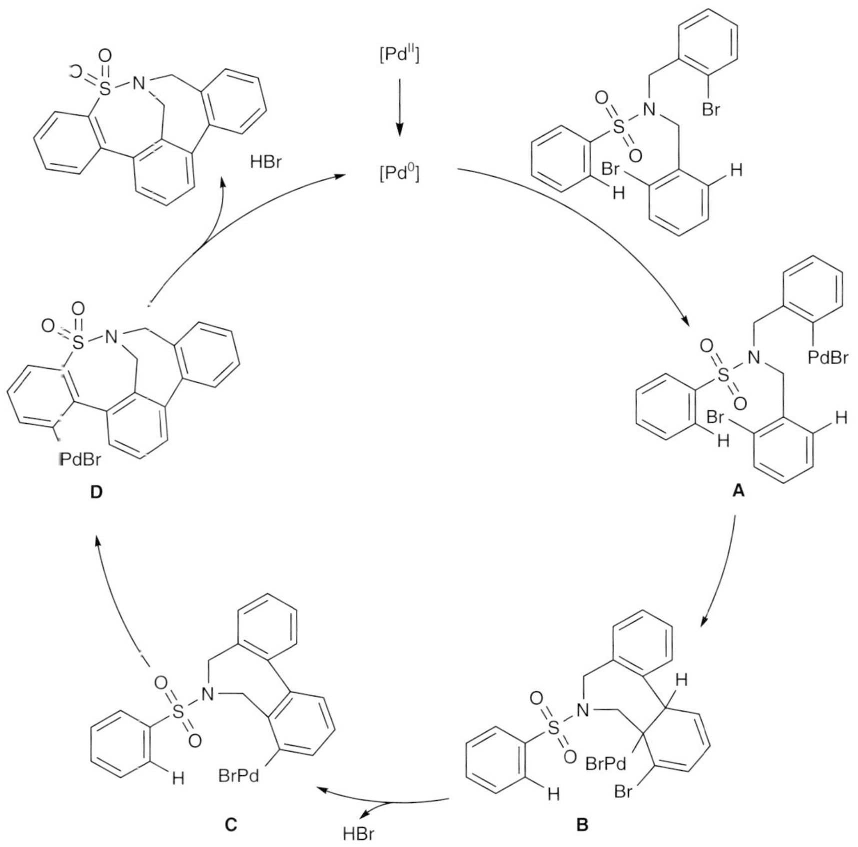

Scheme 4.19 The proposed mechanism.

Later on, the similar work for synthesis of skeletally diverse benzofused sultams based on a central α-halo benzene sulfonamide building block was developed by Hanson and coworkers (Scheme 4.20) [28]. This transformation employed commercially available α-haloaryl sulfonyl chlorides to form five-, six-, and seven-membered sultams in good to excellent yields.

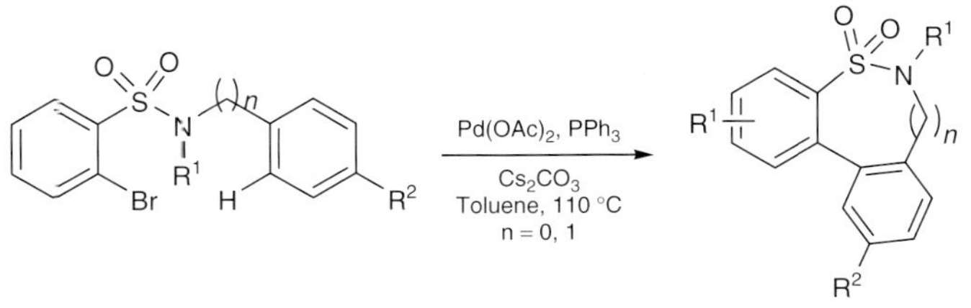

Scheme 4.20 Synthesis of skeletally diverse benzofused sultams based on a central α-halo benzene sulfonamide.

4.3 Phosphorus-Containing Heterocycles

Phosphorus-containing heterocycles are a less numerous and less accessible class of compounds in comparison with accessible nitrogen or sulfur heterocycles [29]. However, interest in this class of compound, especially toward the ones with an endocyclic P–C bond, has increased due to their recent use in wide variety of areas as model objects in fundamental research [30], as ligands for new catalysts [31], for modifying properties of materials [32] and as important building blocks for drug discovery [33].

4.3.1 P–C Heterocycles (Dibenzophosphole Oxides)

Phospholes are known to exhibit characteristic optical and electrochemical properties derived from the phosphorus-bridged 1,3-dienic π system [34]. Particular interest has recently been paid to their π-conjugated derivatives, such as nonfused phospholes, dibenzo[*b*,*d*]phospholes, benzo[*b*]phospholes, benzo[*c*]phospholes, and related compounds [35]. There are several methods to synthesize phospholes [36]. The classical method for the synthesis of phospholes is the reaction between the nucleophilic substitution of a P–X bond with a stoichiometric amount of an organometallic species such as organolithium or organomagnesium reagents (Scheme 4.21) [37].

Recently, C–H activation catalyzed by transition metals has been recognized as a challenging objective in organic synthesis. In 2011, Takai and coworkers established the first example of a Pd(0)-catalyzed C–H phosphorylation reaction in an

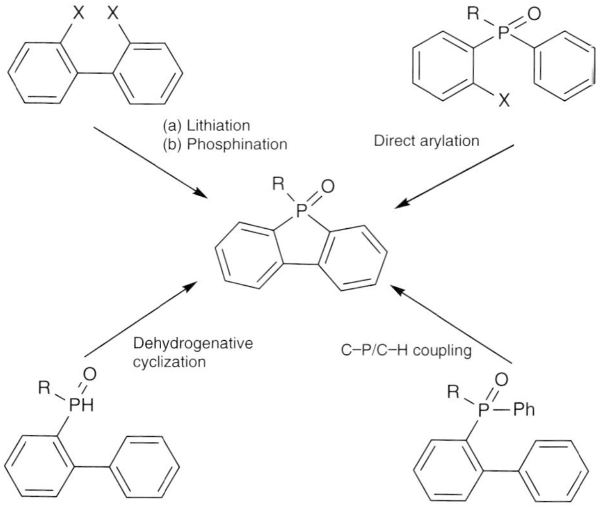

Scheme 4.21 The several methods for the synthesis of phospholes.

intramolecular fashion to synthesize dibenzophosphole oxides (Scheme 4.22) [38]. This reaction proceeds by P–H and C–H bond cleavage via dehydrogenation. By using a tethered phosphite as a directing group as well as the coupling partner, it successfully avoided the problem due to the strong coordinating property of the phosphorus coupling partners. This method could be applied to the synthesis of ladder-type dibenzophosphole oxide.

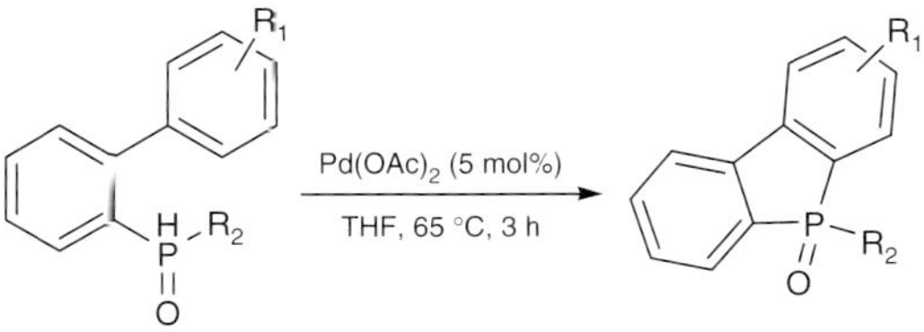

Scheme 4.22 Pd-catalyzed synthesis of dibenzophosphole oxides.

When employing a phosphine oxide with an electron-withdrawing group, a higher temperature and longer reaction time were necessary. The cyclization reaction was not inhibited by a substituent at the 2-position. It demonstrated that the factor of steric hindrance was not important. Additionally, the reaction proceeded when phosphine oxides with aliphatic or aromatic substituents on the phosphorus atom were used as substrates. The deuterium-labeling experiment was carried out. The kinetic isotope effect (KIE = 2.3) shows that the rate-determining step is C–H bond activation of the aromatic ring. The mechanism may involve Pd(0)/Pd(II) pathway (Scheme 4.23).

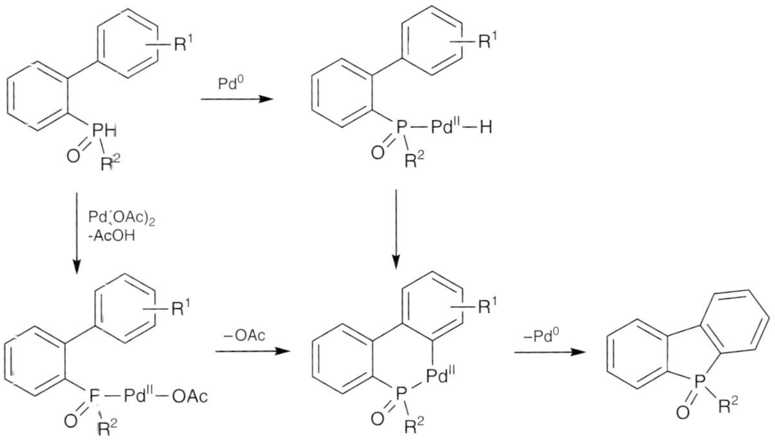

Scheme 4.23 Proposed mechanism for the formation of dibenzophosphole oxides.

In 2012, Takai and coworkers use a bis(biphenyl)phosphine oxide to synthesize a new phosphine oxide with a chiral phosphine center by dehydrogenative intramolecular cyclization via P–H and C–H bond activation under palladium catalysis (Scheme 4.24) [39]. The crystal X-ray structure analysis confirmed the absolute configuration of one enantiomer of the phosphine oxide. The corresponding phosphine, which also has a chiral center, was produced by reducing the phosphine oxide. It could be optically resolved by chiral HPLC column chromatography.

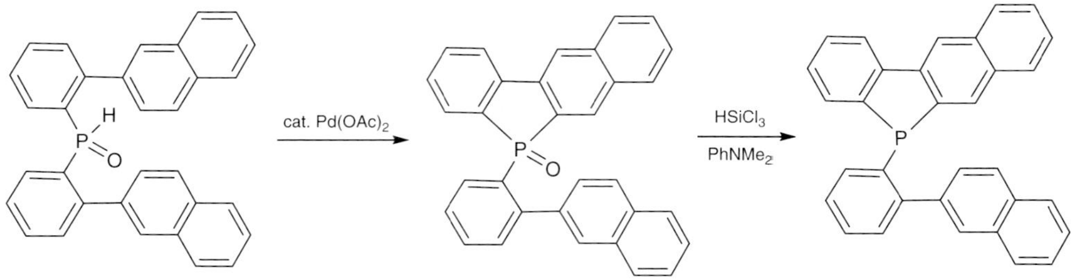

Scheme 4.24 Pd-catalyzed synthesis of a phosphine oxide with a chiral phosphorus center via C–H phosphination.

In 2014, Cui and coworkers developed a new method for the synthesis of dibenzophosphole oxides by palladium-catalyzed direct arylation reactions from the readily available *ortho*-halodiarylphosphine oxides (Scheme 4.25) [40]. The substrates with electron-rich and electron-poor functional groups can be tolerated to give the desired products in good to excellent yields. In addition, it was observed that slow addition of arylboronic acids is another critical factor to achieve satisfactory results. The utilities of this method were demonstrated by the facile access to extended *p*-conjugated systems of dibenzophosphole oxides combined with the second direct arylation or Suzuki–Miyaura coupling reactions in a tandem process. This strategy features simplicity, atom economics, high yields, and good functional group compatibility, providing an efficient and versatile entrance to a variety of dibenzophosphole oxides applicable in materials.

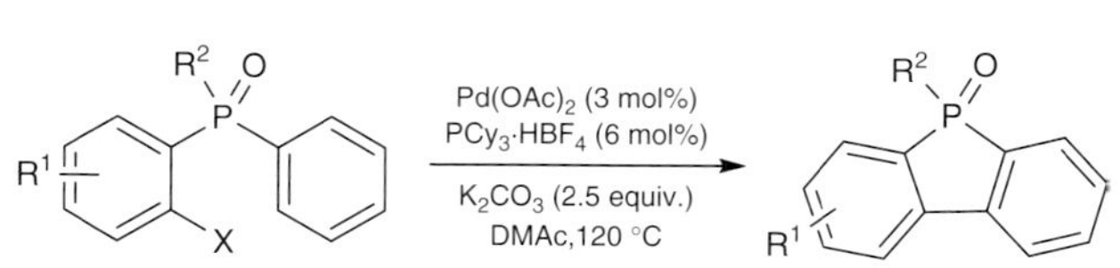

Scheme 4.25 Pd-catalyzed intramolecular direct arylation reactions of ortho-halodiarylphosphine oxides.

A palladium-catalyzed method for the synthesis of phospholes from triarylphosphines has been developed by Chatani and coworkers (Scheme 4.26) [41]. The reaction was found to occur in the presence of a catalytic quantity of $Pd(OAc)_2$ at 160 °C. The condition was very simple. The method was successfully

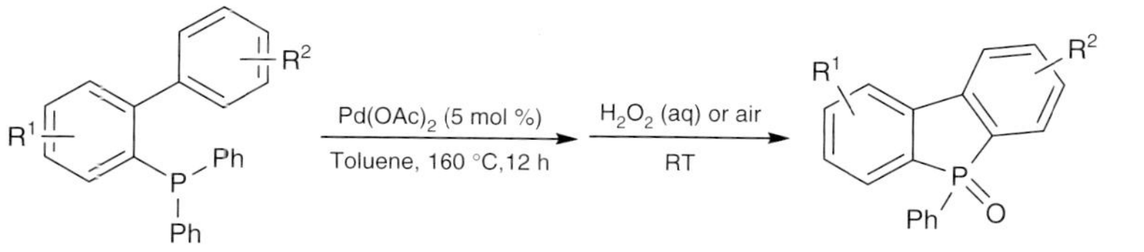

Scheme 4.26 Pd-catalyzed direct synthesis of phosphole derivatives from triarylphosphines.

applied to synthesize a diverse array of phospholes. The ether, amine, ketone, ester, nitrile, and halogen groups were tolerated.

A possible mechanism is depicted in Scheme 4.27. The reaction is initiated by the reaction of Pd(II) with **1a** to form the cyclopalladated complex **B**. Subsequent reductive elimination from **B** leads to the formation of phosphonium **C** along with Pd(0). The phosphonium **C** immediately undergoes to Pd(0) to provide phosphole **3a** and {PhPd(OAc)} (**D**) through cleavage of a C–P bond. Finally, **D** is protonated by AcOH, which is released in the initial cyclometalation step, to regenerate $Pd(OAc)_2$ (**A**).

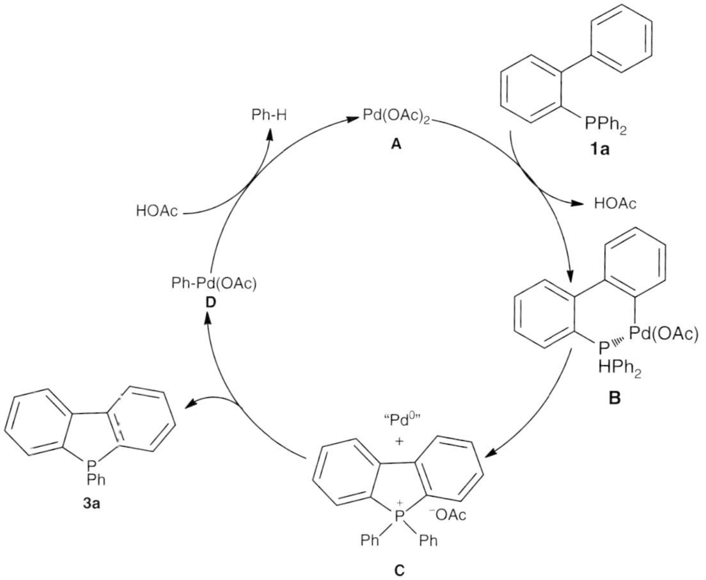

Scheme 4.27 A possible mechanism.

The advantages of this new method include (i) operational simplicity, (ii) direct use of simple starting materials, (iii) excellent functional group compatibility, and (iv) high modularity of the aromatic component to be incorporated. The application of this method to the synthesis of elaborated phosphole derivatives and other heterocyclic compounds is being actively investigated.

4.3.2
O–P=O Heterocycles

Pd-catalyzed activation/C–O bond formation for the synthesis of benzoxaphosphole oxides from phosphonic and phosphinic acids without prefunctionalization under aerobic conditions has been achieved by Lee and coworkers in 2013 (Scheme 4.28) [42]. For the reaction of C(sp^2) –H activation, the use of $Pd(OAc)_2$, $PhI(OAc)_2$, NaOAc, and DCE and a wide variety of benzoxaphosphole oxides were formed in good to excellent yields. It should be noted that acetoxylation of the C–H bond, which is usually proceeded with this oxidant, did not occur under the optimized reaction conditions. The reaction did not proceed without NaOAc because the corresponding salt that brings about C–H activation in a similar manner to that of benzoic acid substrates could not be produced *in situ*. Both electron-donating and electron-withdrawing groups are tolerated on the substituted aryl ring. Different from C(sp^2) –H activation, $(4\text{-MeO–}C_6H_4)_3P$ turned out to be the most effective ligand, and Ag_2CO_3 was determined to be the best oxidant for increasing the yields. The KIE studies indicated that the C–H bond cleavage at the *ortho* position of phosphonate is most likely involved with the rate-limiting step.

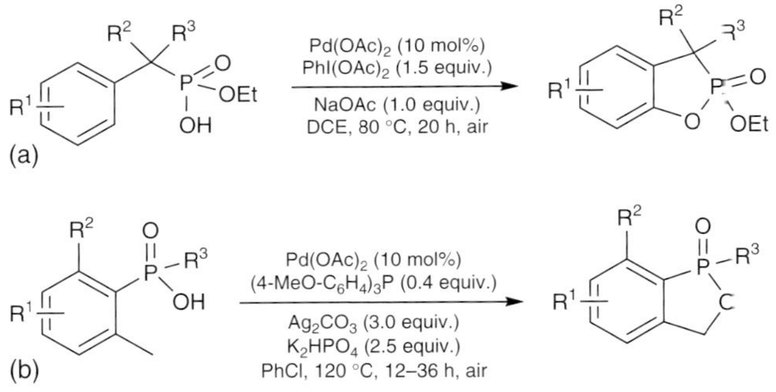

Scheme 4.28 Pd-catalyzed C–H activation/C–O bond formation.

In 2014, Lee and coworkers also reported an efficient phosphaannulation by Pd-catalyzed carbonylation of C–H bonds of phosphonic and phosphinic acids for the synthesis of oxaphosphorinanone oxides (Scheme 4.29) [43]. In the reaction, AgOAc and $PhI(OAc)_2$ gave the best result. In the transformation, ethyl hydrogen benzylphosphonate and methyl-substituted phosphonate were totally ineffective. It indicated that introduction of two substituents at the α-position was essential for successful carbonylation. Both electron-donating and electron-withdrawing groups on the aryl ring were tolerated.

A plausible mechanism for this catalytic carbonylation was described in Scheme 4.30. The KIE ($k_H/k_D = 3.76$) indicate that the C–H cleavage at the ortho position of benzylphosphonic acids is most likely involved with the rate-limiting step. The proposed catalytic cycle started by coordination of the OH group in the

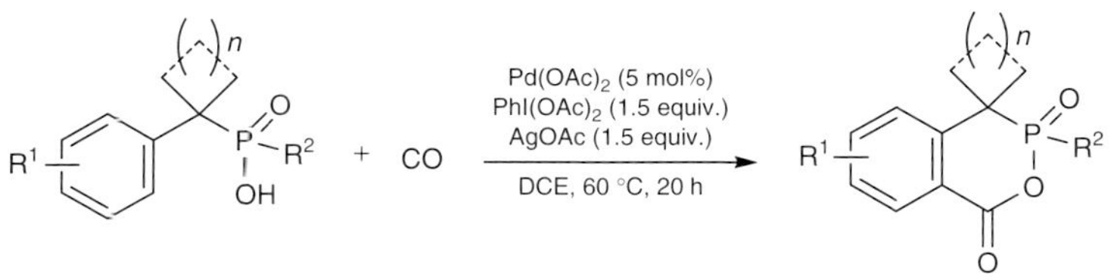

Scheme 4.29 Palladium-catalyzed carbonylation of C–H bonds of phosphonic and phosphinic acids.

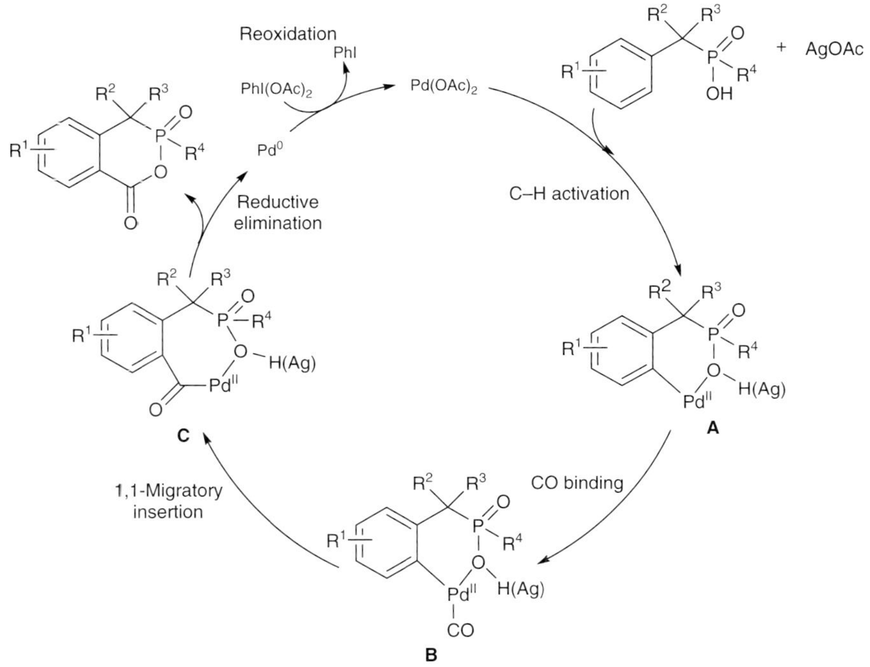

Scheme 4.30 Proposed mechanism of carbonylation.

phosphonic and phosphinic acids to Pd(II) catalyst to give Pd(II) phosphonate and phosphinate, respectively. Subsequent ortho-metalation to provide palladacycle intermediate **A**, CO binding, 1,1-migratory insertion, and reductive elimination affords the corresponding carbonylated products, oxaphosphorinanone oxides. However, the possibility of a mechanism of carbonylation via a Pd(II)/Pd(IV) process cannot be ruled out.

4.3.3 P–N Heterocycles

A novel protocol for the Pd-catalyzed *ortho*-arylation of aryl phosphinamide with boronic acid is reported by Han and coworkers (Scheme 4.31) [44]. By using

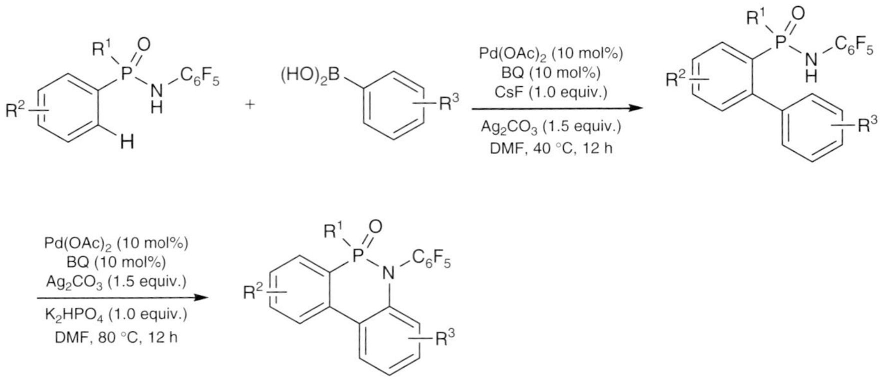

Scheme 4.31 Pd-catalyzed C–H intramolecular amination oriented by a phosphinamide group.

phosphinamide as a new directing group, the reaction proceeds efficiently under mild conditions at 80 °C. Mechanistic studies reveal that the reaction proceeds via a Pd^{II} to Pd^0 cycle. The phosphinamide group is also shown to be an effective orienting group for direct C–H amination.

4.4 Silicon-Containing Heterocycles

Silicon-containing heterocycles such as siloles or silines have received much attention in materials sciences because of their unique electronic properties, arising mainly from their low-lying LUMO [45]. Particular interest has been focused on the application of π-conjugated siloles to light-emitting materials [46]. Consequently, much effort has been devoted to the development of new methods for their efficient and selective synthesis.

4.4.1 Benzosiloles

Silicon-bridged biaryls (SBArs) have attracted growing attention in the fields of functional organic materials such as light-emitting diodes, field-effect transistors, and solar cells. The conventional synthesis of SBArs always involves dilithiation of the corresponding 2,2′-dihalobiaryls with dichlorosilanes. However, this protocol is limited to symmetrical substrates in most cases.

In 2008, Shimizu and coworkers developed a novel approach of palladium-catalyzed intramolecular coupling of readily available 2-(arylsilyl)aryl triflates (Scheme 4.32) [47]. The key to this reaction was the installation of bulky substituents on silicon and the use of Et_2NH as a base. Symmetrical and asymmetrical 9-silafluorenes, substituted with electron-donating and/or electron-withdrawing groups, such as NMe_2, OMe, CF_3, and CN, were synthesized from the

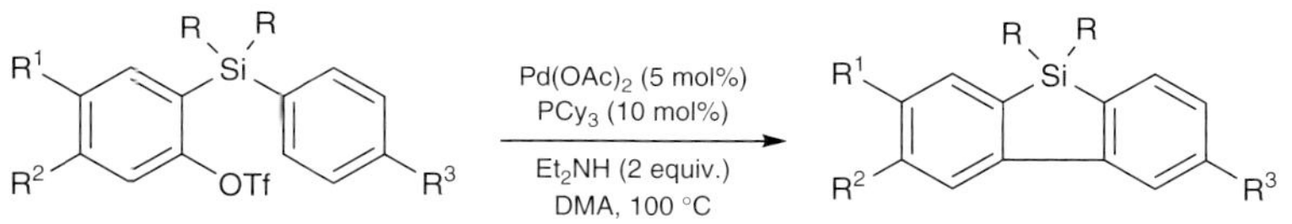

Scheme 4.32 Synthesis of functionalized 9-silafluorenes via palladium-catalyzed intramolecular direct arylation.

corresponding triflates in high to excellent yields. For a substrate incorporating CF_3 as substituent R^3, the reaction required 120 °C to proceed smoothly. In addition to 9-silafluorenes, this new protocol can be extended to the preparation of silicon-bridged benzene–heteroarenes. Benzothiophene, benzofuran, and indole moieties also underwent direct arylation, affording tetracyclic products in fair to high yields.

Among the SBArs reported here, silicon-bridged 2-phenylindole was found to exhibit noteworthy solid-state blue fluorescence. A solution of this heterocycle in cyclohexane was, upon irradiation with UV light ($\lambda_{ex} = 320$ nm), strongly fluorescent with an emission maximum triflate, which was subjected to the original conditions ($Pd(OAc)_2/2PCy_3$ 5 mol%, Et_2NH (2 equiv.), dimethylacetamide (DMA), 100 °C). Unexpectedly, Si-bridged 2-phenylindoles were isolated as a major product (62% yield) along with the expected product (8%) (Scheme 4.33) [48]. Then they screened the conditions and found that the use of 1,2-bis(diphenylphosphino)ethane (dppe) and Et_2NH in large excess in the presence of $Pd(OAc)_2$ as a catalyst was effective in suppressing the formation of the by-product so that Si-bridged 2-phenylindoles were isolated in 89% yield.

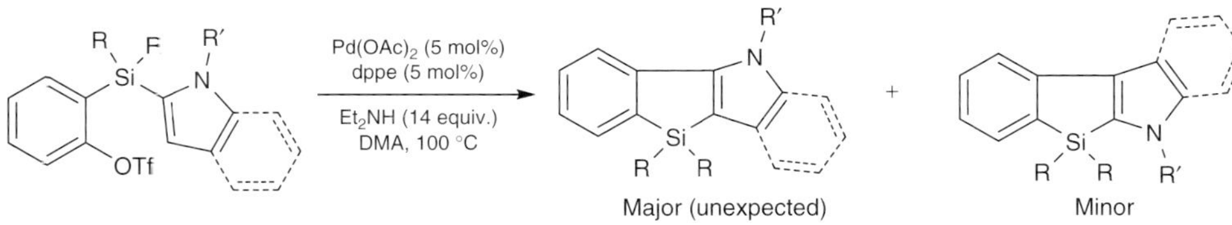

Scheme 4.33 Palladium-catalyzed intramolecular coupling of 2-[(2-pyrrolyl)silyl]aryl triflates through 1, 2-silicon migration.

A plausible mechanism was also proposed in Scheme 4.34. Arylpalladium **A** generated by the oxidative addition (OA) of 2-[(2-pyrrolyl)silyl]aryl triflates to a Pd(0) complex would undergo intramolecular electrophilic substitution (ES) at the 3-position of the pyrrole/indole to give **B** (*route a*), followed by migration of the Pd atom to the 2-position, giving rise to cationic intermediate **C**. The silicon–cation stabilizing effect may assist the migration to overcome the steric hindrance caused by the spirocyclic structure in **C**. Alternatively, direct palladation at the 2-position in **A** leading to **C** may be operative (*route b*). Subsequent 1,2-Si migration followed by deprotonation by a base and then reductive elimination (RE) furnishes the catalytic cycle to produce the product.

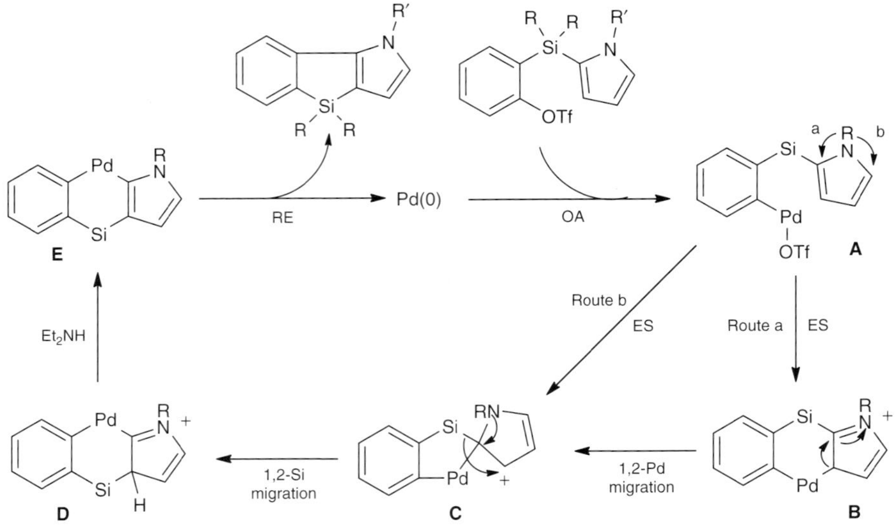

Scheme 4.34 A plausible mechanism.

However, the corresponding 2-silylfuran and -(benzo)thiophenes did not undergo silicon migration which suggested that the electron-donating ability of nitrogen should be one of the key factors for the realization of the novel intramolecular coupling. It was worth pointing out that chlorine tolerated the conditions to give chlorinated products in high yields, which made it possible to get further extension of the π-conjugated system by use of the chlorine functionality.

Considering this wide utility of the dibenzosilole structural motif in materials science, the preparation of enantioenriched chiral dibenzosiloles would be of high significance in view of their potential future applications. In 2012, Shintani and Hayashi groups developed the first example of Pd-catalyzed asymmetric synthesis of Si stereocenters through enantioselective C–H bond functionalization of prochiral 2-(arylsilyl)aryl triflates (Scheme 4.35) [49]. High chemo- and enantioselectivities were achieved by employing Josiphos-type ligand (R, Sp)-L_2 under mild conditions.

In the competition experiment, electron-deficient 4-trifluoromethyl aryl triflate reacted 21 times faster than electron-rich 4-methyl substrates. Intermolecular KIE $k_H/k_D = 1.5$ suggested that oxidative addition of aryl triflate to Pd(0) is most likely the turnover-limiting step of the catalytic cycle. A plausible mechanism was also proposed in Scheme 4.36.

4.4.2
Oxasiline and Azasiline

Gevorgyan and coworkers reported Pd-catalyzed intramolecular arylation using the common TBDPS and Br–TBDPS protecting group as an efficient aryl group

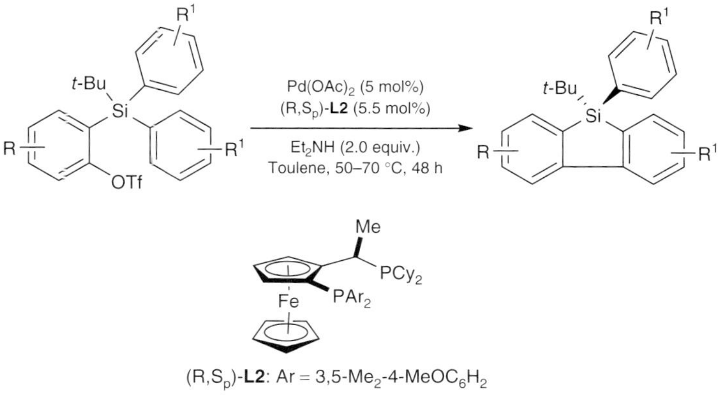

Scheme 4.35 Pd-catalyzed asymmetric synthesis of Si-stereogenic dibenzosiloles.

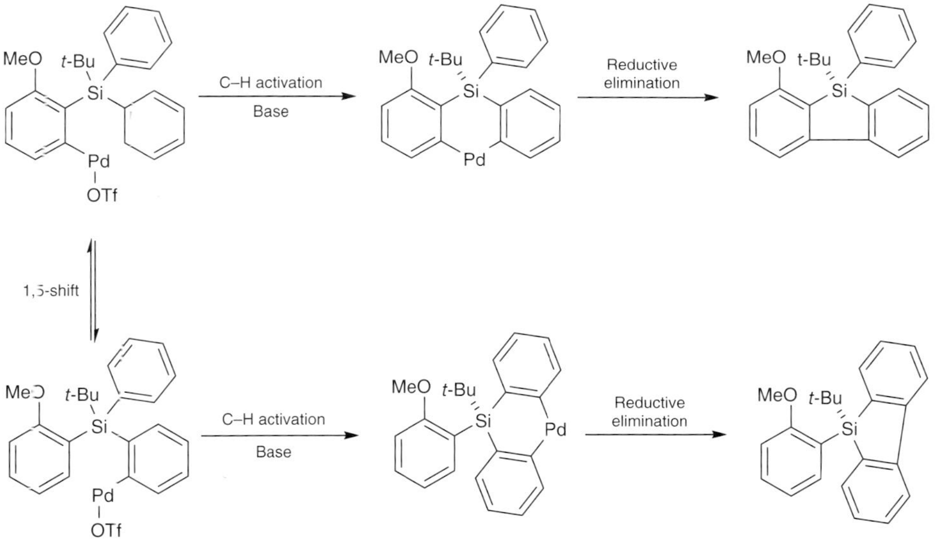

Scheme 4.36 Proposed reaction pathways to dibenzosilole and its isomer.

donor for phenols (Scheme 4.37) [50]. It was found that the latter protecting group was more efficient to give the products in higher yields than the former. Diverse substrates possessing MeO, Cl, F, NO_2, and CHO groups could perfectly tolerate under these reaction conditions. Moreover, N–Br–TBDPS-protected anilines also can undergo the Pd-catalyzed intramolecular arylation to give azasilacycle in good yield.

It was beneficial to employ this temporary silicon tether motif not only because of the ease of its deprotection but also because it provided easy access to deuterated biaryls and biphenols (Scheme 4.38).

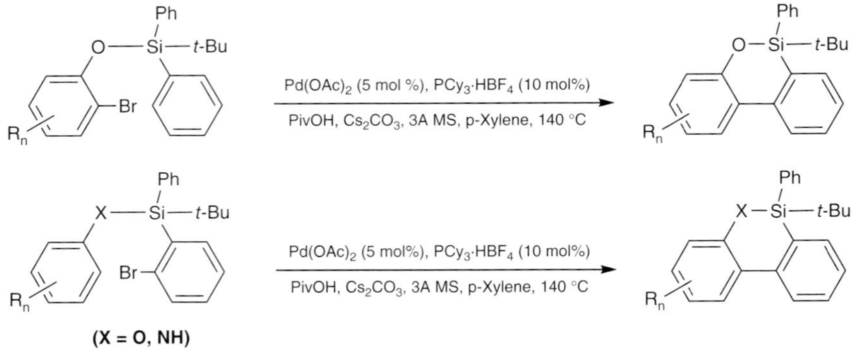

Scheme 4.37 Arylation of TBDPS-protected *o*-bromophenols.

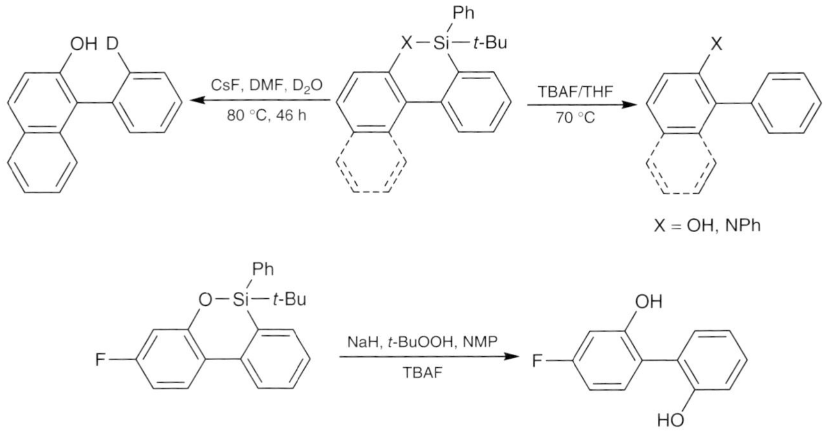

Scheme 4.38 The transformation of oxasilacycles and azasilacycle to arylated phenols and aniline.

4.5 Summary and Conclusions

Due to the importance of sulfur, phosphorus, and silicon heterocycles in functional materials chemistry, the development of efficient and straightforward methods for their synthesis is of paramount importance. In this chapter, we have sought to highlight recent advances in the construction of these heterocycles via palladium-catalyzed C–H activation followed by C–C or C–heteroatom bond formation. Since C–H bonds are ubiquitous in nature and organic molecules, direct bond construction formation via C–H functionalization will minimize the preactivation of substrates, alleviate the waste production, and shorten the synthetic steps. In a strategic level, it could provide novel and more efficient synthetic planning of valuable heterocyclic building blocks. However, these still require the expensive palladium catalysts in high loadings, together with numerous additives and reagents. The development of more reliable and efficient catalysis will make C–H functionalization ever more practical and useful in the sulfur, phosphorus, and silicon heterocycle synthesis. The discovery of new reactions, study of the underlying mechanism, and application in organic synthesis will continue to impact this field. It will provide the chemists with powerful tools to solve the challenges in synthetic organic chemistry.

References

1. For a recent review on synthesis of heterocycles via the Pd-catalyzed C–H activation, see Beccalli, E.M., Broggini, G., Fasana, A., and Rigamonti, M. (2011) *J. Organomet. Chem.*, **696**, 277.
2. For selected reviews on the synthesis of heterocycles through C–H bond activation strategy, see (a) Chen, X., Engle, K.M., Wang, D.-H., and Yu, J.-Q. (2009) *Angew. Chem. Int. Ed.*, **48**, 5094; (b) Gutekunst, W.R. and Baran, P.S. (2011) *Chem. Soc. Rev.*, **40**, 1976; (c) Stokes, B.J. and Driver, T.G. (2011) *Eur. J. Org. Chem.*, **2011**, 4071; (d) Mei, T.-S., Kou, L., Ma, S., Engle, K.M., and Yu, J.-Q. (2012) *Synthesis*, **44**, 1778; (e) Yamaguchi, J., Yamaguchi, A.D., and Itami, K. (2012) *Angew. Chem. Int. Ed.*, **51**, 8960; (f) Wencel-Delord, J. and Glorius, F. (2012) *Nat. Chem.*, **5**, 369; (g) Yoshikai, N. and Wei, Y. (2013) *Asian J. Org. Chem.*, **2**, 466.
3. Rance, D.J. (1989) in *Sulfur-Containing Drugs and Related Organic Compounds. Chemistry, Biochemistry and Toxicology*, vol. **1** (ed L.A. Damani), Ellis Horwood, Chichester, pp. 217–259.
4. (a) Börner, A. (ed) (2008) *Phosphorus Ligands in Asymmetric Catalysis: Synthesis and Applications*, vol. **1–3**, Wiley-VCH Verlag GmbH, Weinheim; (b) Bansal, R.K. (ed) (2010) *Phosphorus Heterocycles II*, Topics in Heterocyclic Chemistry, vol. **21**, Springer-Verlag, Berlin; (c) Peruzzini, M. and Gonsalvi, L. (eds) (2011) *Phosphorus Compounds: Advanced Tools in Catalysis and Material Sciences*, Springer-Verlag, Berlin.
5. (a) Weber, W.P. (1983) *Silicon Reagents for Organic Synthesis*, Springer-Verlag, Heidelberg; (b) Muzafarov, A.M. (2011) *Silicon Polymers*, Springer-Verlag, Heidelberg.
6. Inamoto, K., Arai, Y., Hiroya, K., and Doi, T. (2008) *Chem. Commun.*, 5529–5531.
7. Samanta, R. and Antonchick, A.P. (2011) *Angew. Chem. Int. Ed.*, **50**, 5217–5220.
8. Wesch, T., Berthelot-Bréhier, A., Leroux, F.R., and Colobert, F. (2013) *Org. Lett.*, **15**, 2490–2493.
9. Che, R., Wu, Z., Li, Z., Xiang, H., and Zhou, X. (2014) *Chem. Eur. J.*, **20**, 7258–7261.
10. Wang, B., Liu, Y., Lin, C., Xu, Y., Liu, Z., and Zhang, Y. (2014) *Org. Lett.*, **16**, 4574–4577.
11. Huang, H., Li, J., Lescop, C., and Duan, Z. (2011) *Org. Lett.*, **13**, 5252–5255.
12. Inamoto, K., Hasegawa, C., Hiroya, K., and Doi, T. (2008) *Org. Lett.*, **10**, 5147–5150.
13. Weekes, A.A. and Westwell, A.D. (2009) *Curr. Med. Chem.*, **16**, 2430–2440.
14. Joyce, L.L. and Batey, R.A. (2009) *Org. Lett.*, **11**, 2792–2795.
15. Inamoto, K., Hasegawa, C., Kawasaki, J., Hiroya, K., and Doi, T. (2010) *Adv. Synth. Catal.*, **352**, 2643–2655.
16. Inamoto, K., Nozawa, K., and Kondo, Y. (2012) *Synlett*, **23**, 1678–1682.
17. Sahoo, S.K., Banerjee, A., Chakraborty, S., and Patel, B.K. (2012) *ACS Catal.*, **2**, 544–551.
18. Shen, C., Xia, H., Yan, H., Chen, X., Ranjit, S., Xie, X., Tan, D., Lee, R., Yang, Y., Xing, B., Huang, K.-W., Zhang, P., and Liu, X. (2012) *Chem. Sci.*, **3**, 2388–2393.
19. Zhu, J., Chen, Z., Xie, H., Li, S., and Wu, Y. (2010) *Org. Lett.*, **12**, 2434–2436.
20. Buglass, A.J. and Tillet, J.G. (1991) in *The Chemistry of Sulfonic Acids, Esters and their Derivatives*, Chapter 19 (eds S. Patai and Z. Rappoport), John Wiley & Sons, Inc., New York, p. 789.

21. Mondal, S. (2012) *Chem. Rev.*, **112**, 5339–5355.
22. Majumdar, K.C., Mondal, S., and Ghosh, D. (2009) *Tetrahedron Lett.*, **50**, 4781–4784.
23. Bheeter, C.B., Bera, J.K., and Doucet, H. (2012) *Adv. Synth. Catal.*, **354**, 3533–3538.
24. (a) Majumdar, K.C., Chattopadhyay, B., and Sinha, B. (2008) *Synthesis*, **23**, 3857; (b) Majumdar, K.C., Chattopadhyay, B., and Sinha, B. (2009) *Lett. Org. Chem.*, **6**, 453.
25. Levy, L. (1992) *Drugs Future*, **17**, 451–454.
26. (a) Silvestri, R., Marfe, G., Artico, M., La Regina, G., Lavecchia, A., Novellino, E., Morgante, M., Di Stefano, C., Catalano, G., Filomeni, G., Abruzzese, E., Ciriolo, M.R., Russo, M.A., Amadori, S., Cirilli, R., La Torre, F., and Salimei, P.S. (2006) *J. Med. Chem.*, **49**, 5840; (b) Lebegue, N., Gallet, S., Flouquet, N., Carato, P., Pfeiffer, B., Renard, P., Leonce, S., Pierre, A., Chavatte, P., and Berthelot, P. (2005) *J. Med. Chem.*, **48**, 7363.
27. Majumdar, K.C., Mondal, S., and De, N. (2008) *Synlett*, **18**, 2851–2855.
28. Rayabarapu, D.K., Zhou, A., Jeon, K.O., Samarakoon, T., Rolfe, A., Siddiqui, H., and Hanson, P.R. (2009) *Tetrahedron*, **65**, 3180–3188.
29. (a) Quin, L. (1981) *The Heterocyclic Chemistry of Phosphorus System Based on the Phosphorus Carbon Bond*, John Wiley & Sons, Inc., New York; (b) Mader, M. and Bartlett, P. (1997) *Chem. Rev.*, **97**, 1281–1301; (c) Nyulszi, L. (2001) *Chem. Rev.*, **101**, 1229–1246.
30. (a) Toshimitsu, A., Saeki, T., and Tamao, K. (2001) *J. Am. Chem. Soc.*, **123**, 9210–9211; (b) Heinike, J., Gupta, N., Surana, A., Peulecke, N., Witt, B., Steinhauser, K., Bansal, R.K., and Jones, P.G. (2001) *Tetrahedron*, **57**, 9963–9972.
31. Birkholz, M.-N., Freixa, Z., and van Leeuwen, P.W.N.M. (2009) *Chem. Soc. Rev.*, **38**, 1099.
32. Krebs, F.C., Larsen, P.S., Larsen, J., Jacobsen, C.S., Boutton, C., and Thorup, N. (1997) *J. Am. Chem. Soc.*, **119**, 1208–1216.
33. (a) Martinek, T., Forro, E., Gunther, G., Sillanpaa, R., and Fulop, F. (2000) *J. Org. Chem.*, **65**, 316–321; (b) Bujard, M., Gouverneur, V., and Mioskowski, C. (1999) *J. Org. Chem.*, **65**, 2119–2123.
34. (a) Baumgartner, T. and Reau, R. (2006) *Chem. Rev.*, **106**, 4681; (b) Matano, Y. and Imahori, H. (2009) *Org. Biomol. Chem.*, **7**, 1258; Representative recent works: (c) Fukazawa, A. and Yamaguchi, S. (2009) *Chem. Asian J.*, **4**, 1386; (d) Ren, Y. and Baumgartner, T. (2011) *J. Am. Chem. Soc.*, **133**, 1328; (e) Ren, Y., Kan, W.H., Henderson, M.A., Bomben, P.G., Berlinguette, C.P., Thangadurai, V., and Baumgartner, T. (2011) *J. Am. Chem. Soc.*, **133**, 17014; (f) Matano, Y., Saito, A., Fukushima, T., Tokudome, Y., Suzuki, F., Sakamaki, D., Kaji, H., Ito, A., Tanaka, K., and Imahori, H. (2011) *Angew. Chem. Int. Ed.*, **50**, 8016; (g) Bruch, A., Fukazawa, A., Yamaguchi, E., Yamaguchi, S., and Studer, A. (2011) *Angew. Chem. Int. Ed.*, **50**, 12094.
35. For selected examples, see(a) Bock, T., Mohwald, H., and Mulhaupt, R. (2007) *Macromol. Chem. Phys.*, **208**, 1324; (b) Rusch-Salazar, L.A. and Sheares, V.V. (2003) *J. Polym. Sci., Part A: Polym. Chem.*, **41**, 2277; (c) Evans, O.R., Manke, D.R., and Lin, W. (2002) *Chem. Mater.*, **14**, 3866; (d) Jin, S. and Gonsalves, K.E. (1998) *Macromolecules*, **31**, 1010.
36. Dithienophosphole oxides: (a) Dienes, Y., Eggenstein, M., Karpati, T., Sutherland, T.C., Nyulaszi, L., and Baumgartner, T. (2008) *Chem. Eur. J.*, **14**, 9878–9889; Bis-phosphoryl-bridged stilbene: (b) Fukazawa, A., Hara, M., Okamoto, T., Son, E.-C., Xu, C., Tamao, K., and Yamaguchi, S. (2008) *Org. Lett.*, **10**, 913–916.
37. (a) Chen, R.-F., Fan, Q.-L., Zheng, C., and Huang, W. (2006) *Org. Lett.*, **8**, 203; (b) Geramita, K., McBee, J., and Tilley, T.D. (2009) *J. Org. Chem.*, **74**, 820; (c) Fukazawa, A., Kiguchi, M., Tange, S., Ichihashi, Y., Zhao, Q., Takahashi, T., Konishi, T., Murakoshi, K., Tsuji, Y., Staykov, A., Yoshizawa, K., and Yamaguchi, S. (2011) *Chem. Lett.*, **40**, 174.
38. Kuninobu, Y., Yoshida, T., and Takai, K. (2011) *J. Org. Chem.*, **76**, 7370–7376.
39. Kuninobu, Y., Origuchi, K., and Takai, K. (2012) *Heterocycles*, **85**, 3029.

40. Cui, Y., Fu, L., Cao, J., Deng, Y., and Jiang, J. (2014) *Adv. Synth. Catal.*, **356**, 1217–1222.
41. Baba, K., Tobisu, M., and Chatani, N. (2013) *Angew. Chem. Int. Ed.*, **52**, 11892–11895.
42. Eom, D., Jeong, Y., Kim, Y.R., Lee, E., Choi, W., and Lee, P.H. (2013) *Org. Lett.*, **15**, 5210.
43. Shin, S., Jeong, Y., Jeon, W.H., and Lee, P.H. (2014) *Org. Lett.*, **16**, 2930–2933.
44. Guan, J., Wu, G.-J., and Han, F.-S. (2014) *Chem. Eur. J.*, **20**, 3301–3305.
45. Yamaguchi, S. and Tamao, K. (1996) *Bull. Chem. Soc. Jpn.*, **69**, 2327–2334.
46. (a) Yamaguchi, S. and Tamao, K. (1998) *J. Chem. Soc., Dalton Trans.*, 3693; (b) Yamaguchi, S., Endo, T., Uchida, M., Izumizawa, T., Furukawa, K., and Tamao, K. (2000) *Chem. Eur. J.*, **6**, 1683–1692; (c) Yamaguchi, S. and Tamao, K. (2005) *Chem. Lett.*, **34**, 2; (d) Yamaguchi, S. and Xu, C. (2005) *J. Synth. Org. Chem. Jpn.*, **63**, 1115; (e) Yamaguchi, S., Xu, C., and Okamoto, T. (2006) *Pure Appl. Chem.*, **78**, 721.
47. Shimizu, M., Mochida, K., and Hiyama, T. (2008) *Angew. Chem. Int. Ed.*, **47**, 9760–9764.
48. Mochida, K., Shimizu, M., and Hiyama, T. (2009) *J. Am. Chem. Soc.*, **131**, 8350.
49. Shintani, R., Otomo, H., Ota, K., and Hayashi, T. (2012) *J. Am. Chem. Soc.*, **134**, 7305.
50. Huang, C. and Gevorgyan, V. (2009) *J. Am. Chem. Soc.*, **131**, 10844.

5
Rh-Catalyzed Synthesis of Nitrogen-Containing Heterocycles

Krishnamoorthy Muralirajan and Chien-Hong Cheng

5.1
Introduction

The synthesis of nitrogen-containing heterocycles has attracted great attention largely due to their prominence as building blocks for many therapeutically useful materials and the wide range of potential biological activities. In particular, five- and six-membered heterocycles, such as indoles, lactams, carbazoles, pyrroles, imidazoles, pyridines, isoquinolines, isoquinolones, and pyridones, are important classes of *N*-heterocycles that exist in many natural products, inhibitors, organic dyes, and pharmaceutical agents [1]. In addition, tailoring the properties of these compounds to their particular function requires synthetic methods capable of selectively introducing appropriate substituents onto the desired heterocyclic scaffolds. For this purpose, synthetic chemists continue to play a key role in the creation and functionalization of *N*-heterocyclic cores.

In this context, some important coupling reactions have been investigated for the formation of *N*-heterocycles. These reactions provide easy access to various *N*-heterocycles in high yields with different substitution patterns, but their substrate scope is limited by the need for prefunctionalization [2]. Transition metal-catalyzed C–H bond functionalization is an efficient alternative approach for the synthesis of *N*-heterocycles that avoids preactivation of the substrates and reduces the number of synthesis steps. As such several current reviews on the synthesis of *N*-heterocyclic compounds by transition metal-catalyzed C–H activation have been compiled [3].

Recently, a large number of novel transformations that involve rhodium (Rh)-catalyzed C–N bond-forming reactions have been developed for the construction of different *N*-heterocyclic rings. Such transformations frequently occur with high stereoselectivity and a broad array of substrates under mild conditions. Selection of a suitable catalyst for a preferred substrate can seem intimidating. However, a rational solution to this problem can be delivered considering the possible mechanistic cycles for the individual transformations.

Transition Metal-Catalyzed Heterocycle Synthesis via C—H Activation, First Edition. Edited by Xiao-Feng Wu.
© 2016 Wiley-VCH Verlag GmbH & Co. KGaA. Published 2016 by Wiley-VCH Verlag GmbH & Co. KGaA.

Rh can exist in different oxidation states; organic reactions via Rh-catalyzed C–H bond activation are generally controlled by the use of Rh(I) and Rh(III) species. Rh(I)-catalyzed C–H activation has received less attention in the past, whereas the Rh(III)-catalyzed directing group-assisted C–H activation of various aromatic substrates with alkynes and alkenes has been extensively investigated. In the latter case, it has been established by various research groups that the Rh(III) catalyst promotes selective C–H bond activation to form an organometallic complex with a C–Rh bond that can subsequently insert into various π components and result in an intermediate that undergoes reductive elimination to provide the corresponding *N*-heterocyclic compounds (Scheme 5.1) [3e, 4].

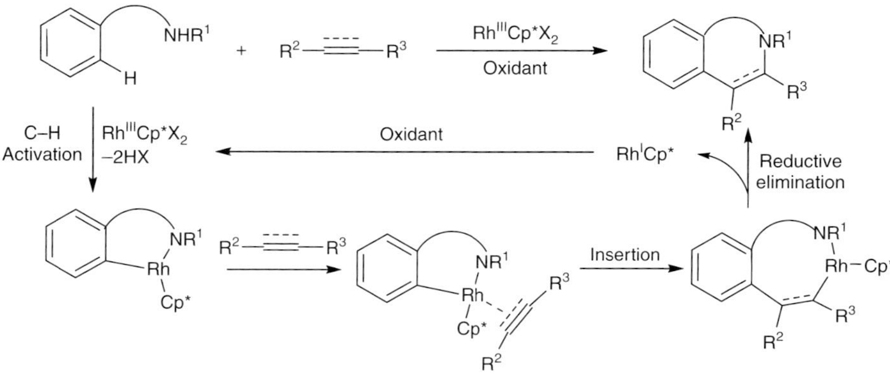

Scheme 5.1 A possible mechanism for the synthesis of *N*-heterocycles by C–H bond activation.

5.2 Synthesis of Five-Membered Nitrogen Heterocycles

5.2.1 Synthesis of Indoles

The indole scaffold is a common structural moiety found in several natural products and functional materials. For the synthesis of indole products, Rh-catalyzed C–H activation has been considered as a simple alternative strategy to complement traditional synthetic protocols. The first examples of Rh(III)-catalyzed oxidative coupling of acetanilides and internal alkynes were described by Fagnou in 2008 for the construction of indole derivatives **1** (Eq. (5.1)) [5a]. Heterocyclic substituents, such as thiophene and indole, on the alkyne were tolerated. In addition, unsymmetrical alkynes afforded high regioselectivity, but terminal alkynes failed to participate. The application of this method toward the synthesis of a paullone derivative, which shows significant biological activity, was also studied. Later, by using enynes as the functionalized alkyne substrates, 2-alkenylindoles were constructed (Eq. (5.2)) [5b,c].

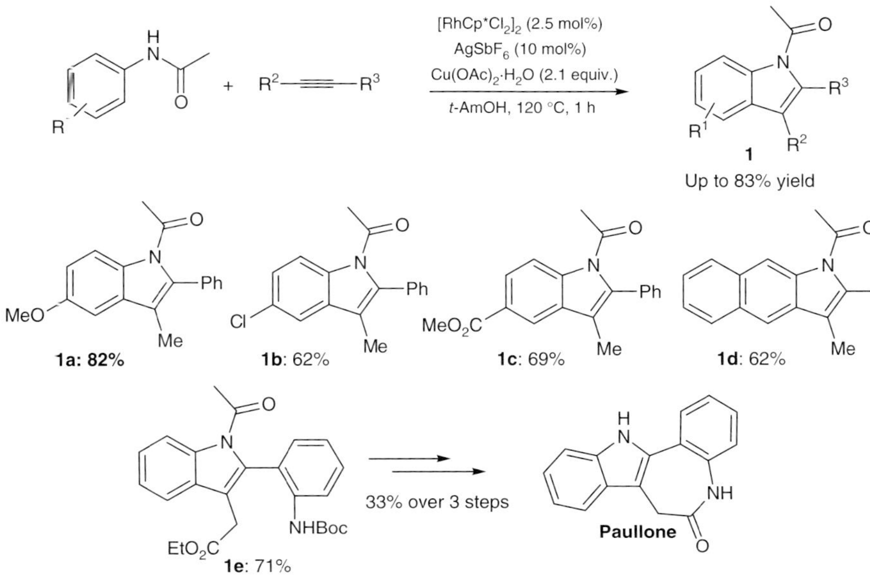

(5.1)

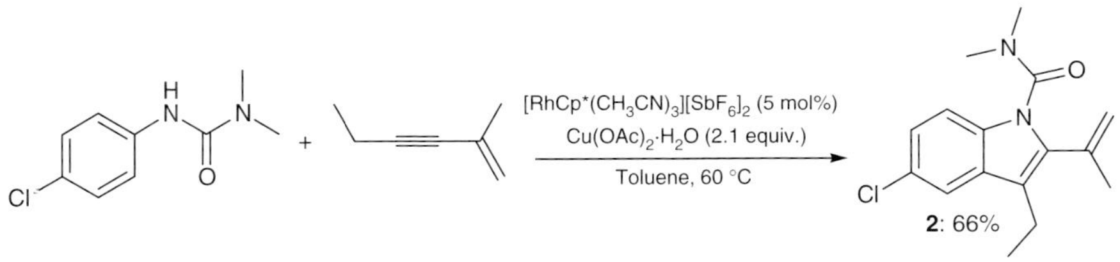

(5.2)

A possible reaction mechanism for this Rh-catalyzed oxidative coupling is shown in Scheme 5.2. First, C–H bond cleavage of the acetanilide leads to a six-membered rhodacycle intermediate **A2**. Next, coordination and insertion of the alkyne into the Rh–C bond afford intermediate **A3**, which undergoes reductive elimination to afford product **1** and Rh(I) species **A4**. The latter is reoxidized by Cu(II) to **A1** to regenerate the catalyst [5a–c].

In 2014, Rh(III)-catalyzed intramolecular annulation of alkyne-tethered acetanilides for the synthesis of fused tricyclic indoles **3** via C–H activation was developed by the groups of Liu, Jia, and Li [6a–c]. The reactions have a good substrate scope, utilize molecular oxygen as the cooxidant, and proceed with complete regioselectivity. It is interesting to note that only the sterically hindered *ortho*-C–H bond of the arene moiety participated in C–C bond formation (Eq. (5.3)). Similarly, the cyclization of anilides with allyl carbonates also proceeded to form 2-substituted indoles **4** (Eq. (5.4)) [6d].

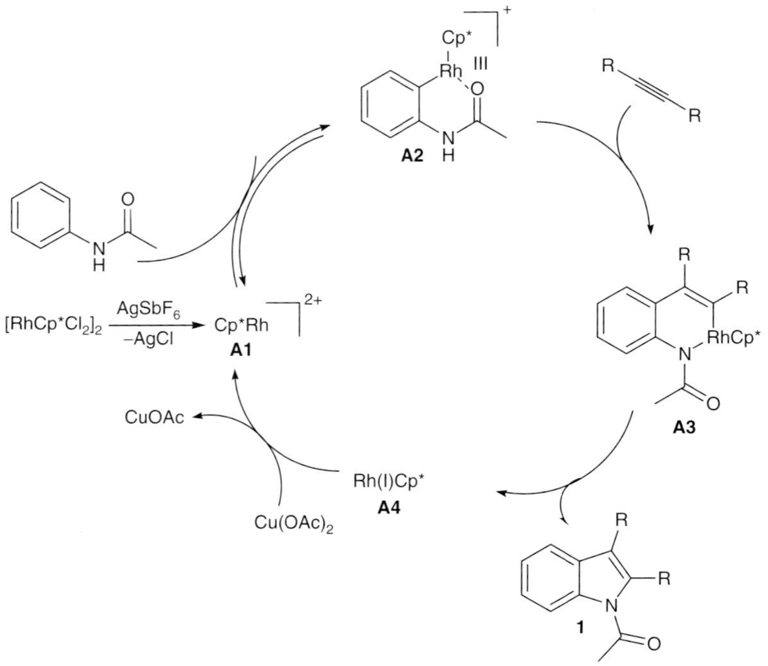

Scheme 5.2 Proposed catalytic mechanism for indole formation.

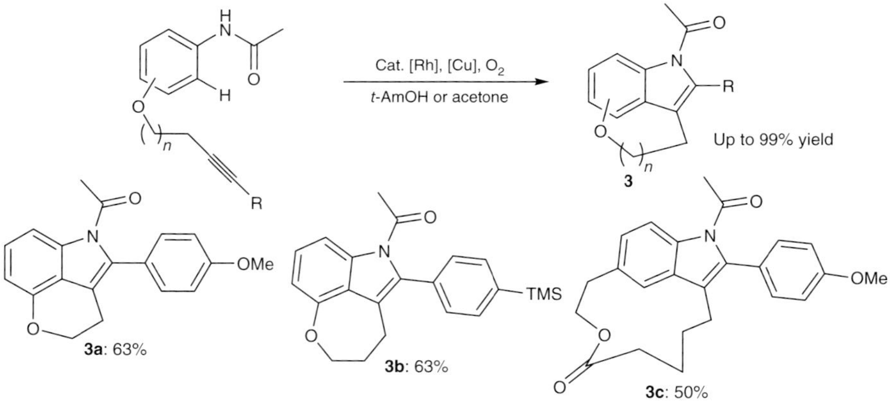

(5.3)

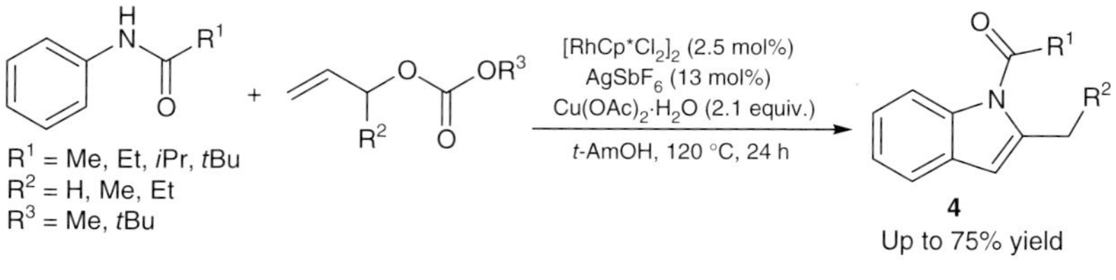

(5.4)

Various convenient methods for the synthesis of protected indoles have been reported by several groups [7a–e]. Interesting C–H oxidative coupling reactions of various arenes with alkynes are shown in Eqs. (5.5)–(5.9). These methods allow quick access to a number of functional molecules in good yields via an N–N bond oxidative cleavage or use of an external oxidant. Excellent regioselectivity was achieved by using asymmetrically substituted alkynes.

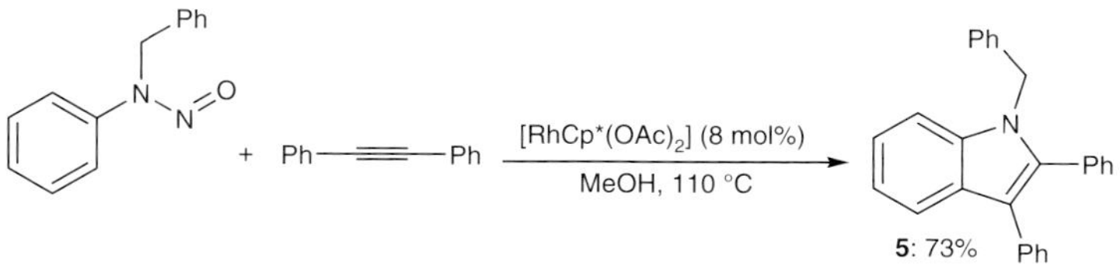

(5.5)

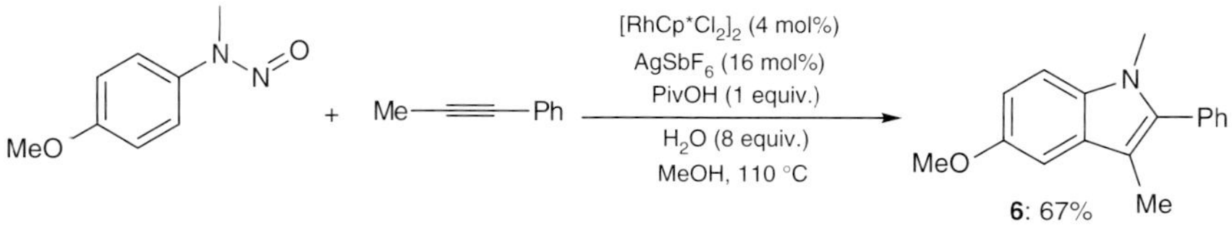

(5.6)

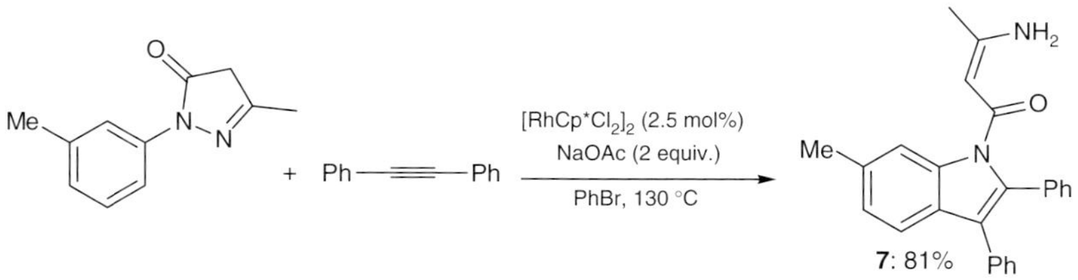

(5.7)

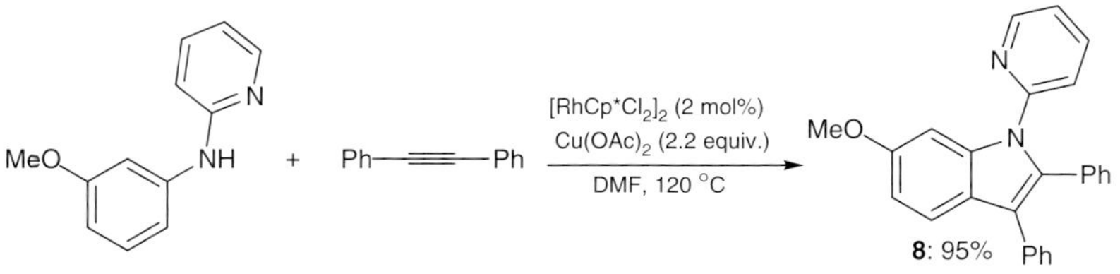

(5.8)

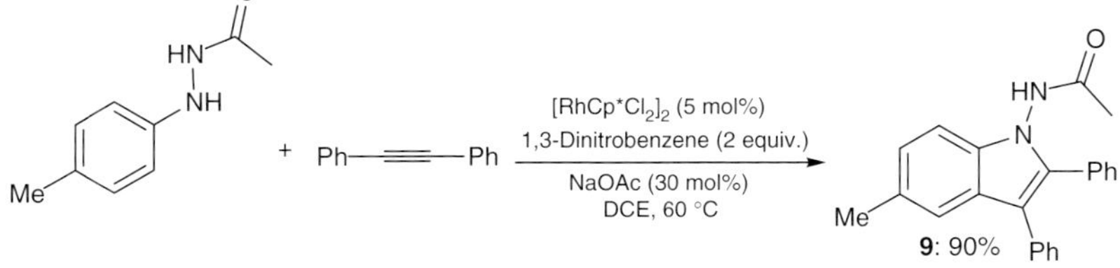

(5.9)

5.2.2
Synthesis of Isoindolines

Recently, Kim and coworkers developed an efficient strategy for the construction of isoindolines **10** via Rh(III)-catalyzed oxidative alkenylation and subsequent intramolecular cyclization of *N*-benzyltriflamides with olefins. Under these reaction conditions, various substituted electron-deficient alkenes were successfully employed in good product yields [8]. The authors proposed a mechanism to account for the catalytic reaction (Eq. (5.10)).

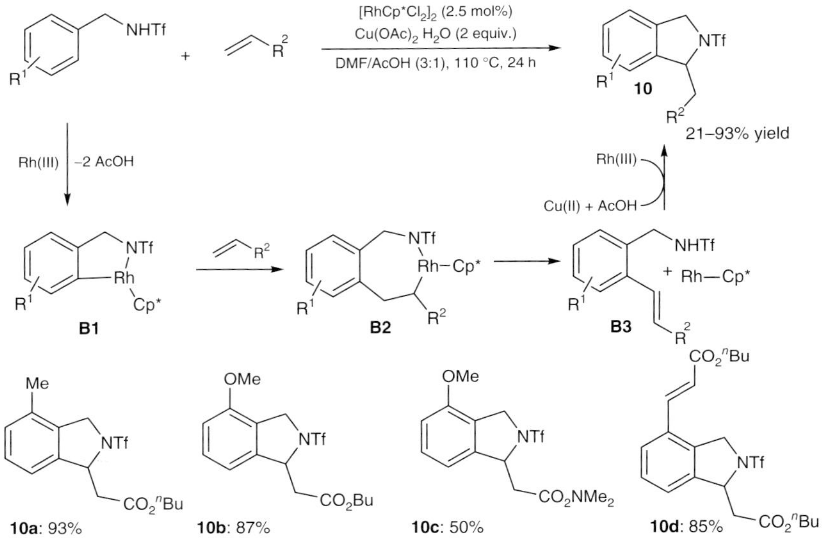

(5.10)

Miura *et al.* also demonstrated a Rh-catalyzed room-temperature dehydrogenative coupling of *α*,*α*-disubstituted benzylamines with alkenes. The reaction proceeded via an *ortho*-alkenylation directed by the free amino group and subsequent cyclization to produce isoindole-1-yl derivatives **11** (Eq. (5.11)). It is rare for catalytic processes to use an amino group as a directing group [9]. The reaction was unsuccessful when styrene was used as the coupling partner.

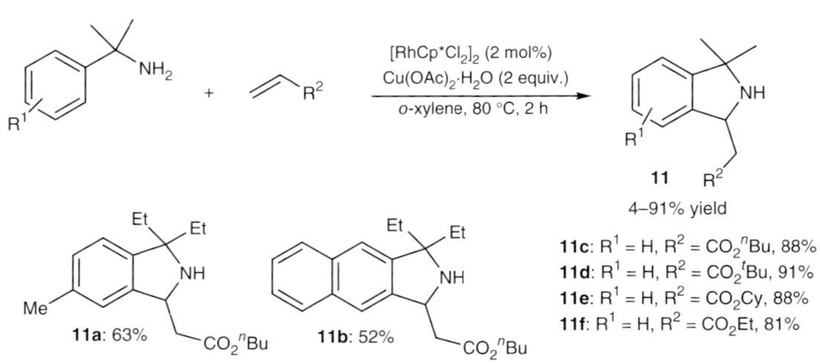

(5.11)

5.2.3
Synthesis of Unprotected Indoles

The classical "Fischer indole synthesis" has persisted as a commonly used procedure due to its working simplicity. However, this method suffers some drawbacks, such as the need for strong acids, poor selectivity with unsymmetrical ketones, and poor functional group tolerance. Huang *et al.* developed an alternative synthetic strategy for the synthesis of unprotected indoles: a tandem *ortho*-C–H activation and annulation by using triazenyl arenes and alkynes. This reaction involves a ring contraction by a 1,2-Rh shift and an N=N insertion (Eq. (5.12)) [10a]. Another interesting method involves the synthesis of indole derivatives by Rh-catalyzed aerobic C–H coupling of simple arylamines with alkynes. In this reaction the acetanilide generated *in situ* from acetic anhydride and aniline acts as a directing group (Eq. (5.13)) [10b].

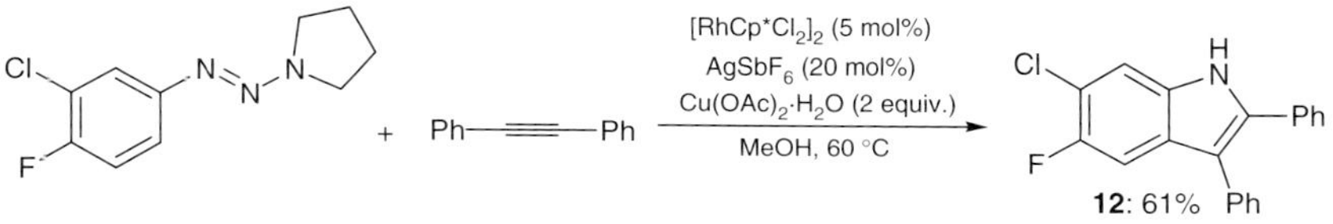

(5.12)

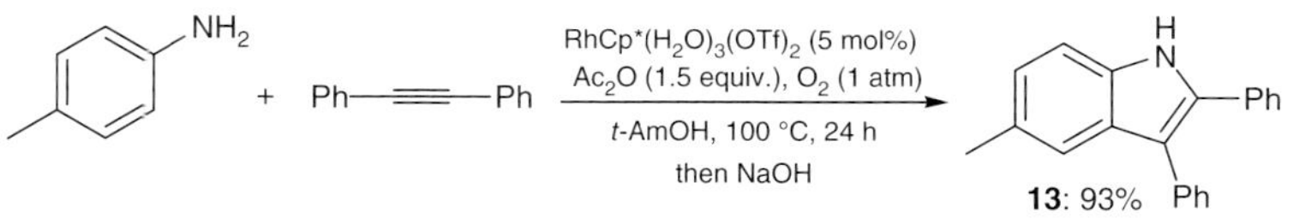

(5.13)

A highly regioselective protocol was reported that delivered numerous 3,4-disubstituted indole derivatives with excellent yields by hydrazine-directed Rh-catalyzed C–H activation, in which the protected hydrazine acts as a directing group and as an internal oxidant (Eq. (5.14)) [11a]. The reaction was promoted by an acetic acid additive. Similarly, a Rh(III)-catalyzed reaction of an arylhydrazine salt, a diethyl ketone, and an alkyne to synthesize a 3,4-disubstituted indole was also reported by our group. The reaction proceeds by in situ condensation of an arylhydrazine with diethyl ketone to form an arylhydrazone intermediate (Eq. (5.15)) [11b].

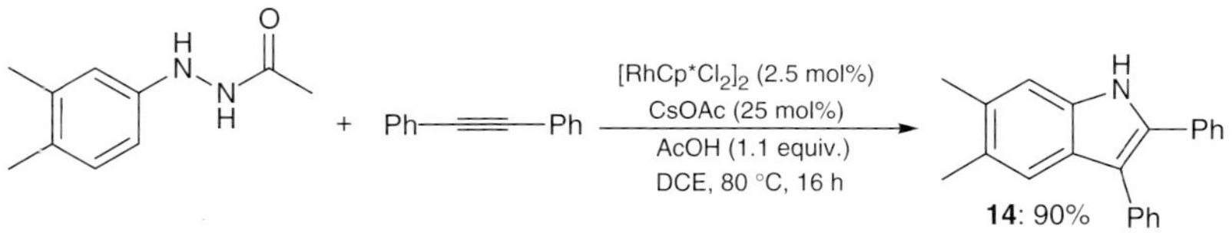

(5.14)

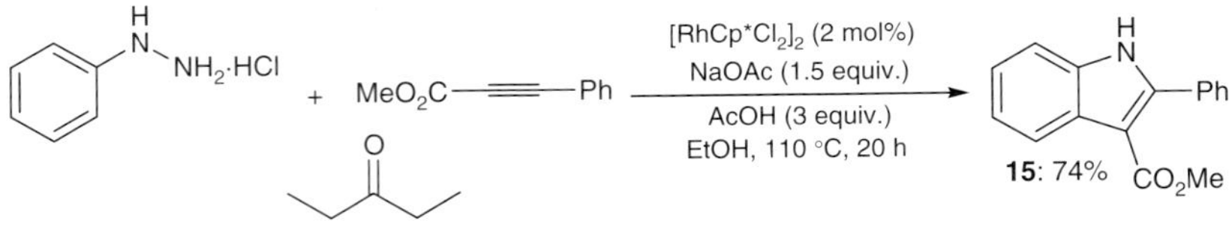

(5.15)

5.2.4 Synthesis of Indolines

Recently, the synthesis of indoline derivatives from acetanilides and allylic alcohols was achieved via Rh-catalyzed *ortho*-C–H bond activation, alkene insertion, and subsequent reductive elimination (Eq. (5.16)) [12a]. Similarly, aryldiazene carboxylates reacted with alkenes to afford diverse 1-aminoindolines without any external oxidant (Eq. (5.17)) [12b].

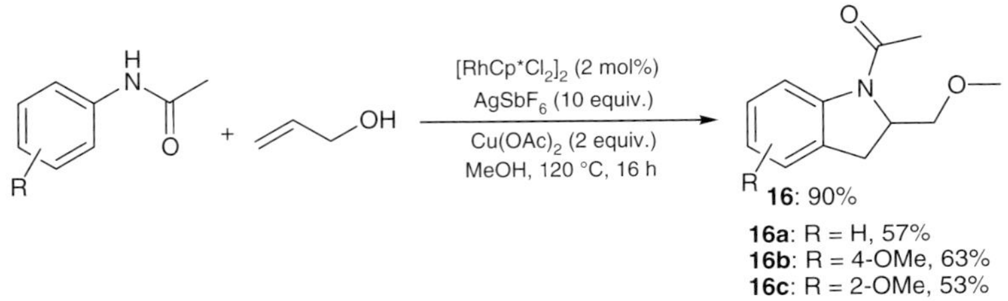

(5.16)

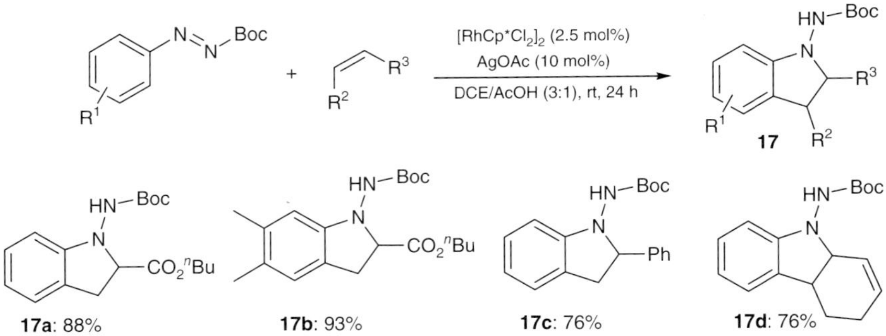

(5.17)

5.2.5 Synthesis of Indazoles

Lavis and Ellman reported an efficient method for the preparation of *N*-aryl-*2H*-indazoles **18** by Rh(III)-catalyzed C–H bond addition of azobenzenes to aldehydes without external oxidant (Eq. (5.18)) [13]. In this reaction, the regioselective activation of azobenzenes can be controlled either by electronic or by steric effects, and the azo moiety serves as a nucleophile to trap the initial aldehyde addition product. A wide range of aldehydes and azobenzenes participate in the reaction to afford differently substituted indazoles. An interesting feature of this reaction is that the *N*-aryl-2*H*-indazole derivatives can readily be converted to 2*H*-indazoles **19**.

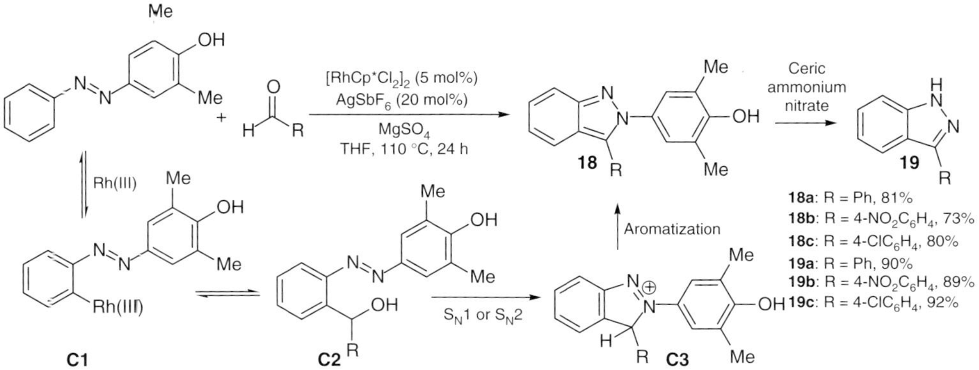

(5.18)

Glorius described a tandem Rh-catalyzed C–N bond and Cu-catalyzed N–N bond formation between arylimidates and organic azides to provide 1*H*-indazoles [14a]. The reaction is scalable and O_2 is used as the terminal oxidant (Eq. (5.19)). The 1*H*-indazoles were obtained in moderate to high yields with good functional group tolerance. The formation of the N–N bond in this process does not appear to be Rh catalyzed. Similarly, Kim and coworkers demonstrated a Rh(III)-catalyzed oxidative olefination of 1,2-disubstituted arylhydrazines with electron-deficient terminal alkenes and an intramolecular aza-Michael reaction to synthesize 2,3-dihydro-1*H*-indazoles in good yields (Eq. (5.20)) [14b].

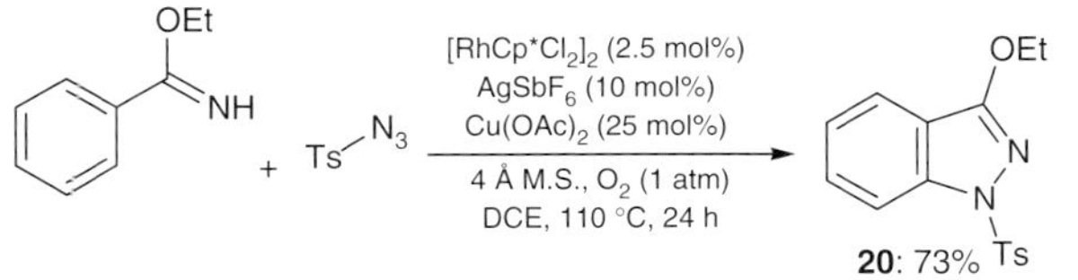

(5.19)

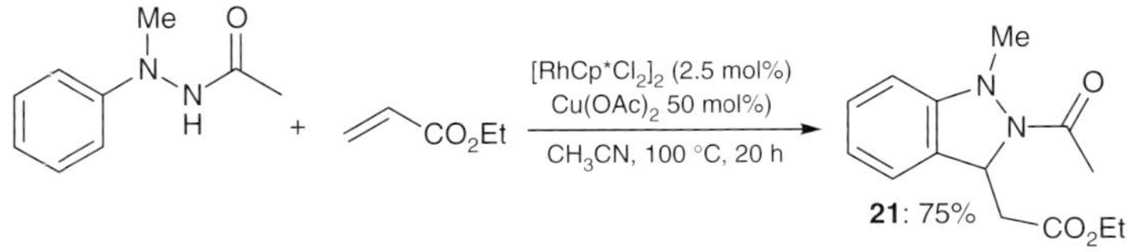

(5.20)

5.2.6 Synthesis of Isoxazoles

Xu and Zhang developed an efficient aryl C–H amination reaction by using *N*-methoxybenzamides and *N*-Boc-hydroxylamine for the synthesis of benzoisoxazolones **22** (Eq. (5.21)) [15]. In this reaction, the unstable nitroso intermediate was generated by in situ Cu(I)-catalyzed aerobic oxidation of *N*-Boc-hydroxylamine.

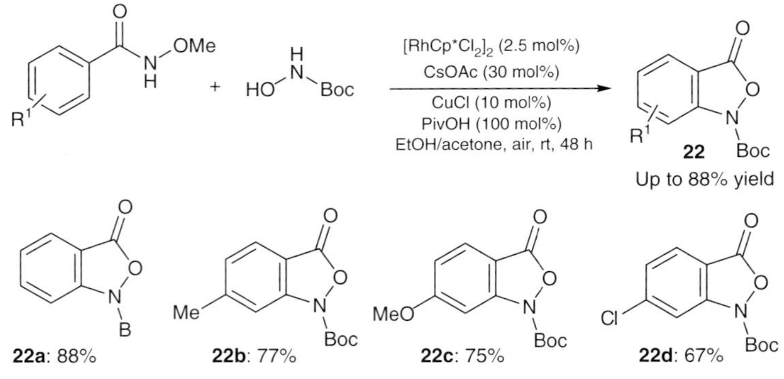

(5.21)

5.2.7
Synthesis of Pyrroles

Highly substituted pyrroles are important *N*-heterocyclic compounds found in a number of natural and synthetic bioactive compounds. In 2010, Glorius and coworkers reported a new method for the synthesis of pyrroles from enamines and alkynes by Rh-catalyzed sp^3 C–H bond activation (Eq. (5.22)) [16]. Several *N*-substituents on the enamides were examined, and the acetyl group was found to be critical for the reaction. Internal alkynes with aromatic substituents were successfully coupled. It is noteworthy that the reaction appears to proceed by an sp^3 C–H activation at the γ-position, which leads to a six-membered rhodacycle **D2** or **D3**. The resulting intermediate undergoes alkyne insertion and subsequent reductive elimination to afford **23**.

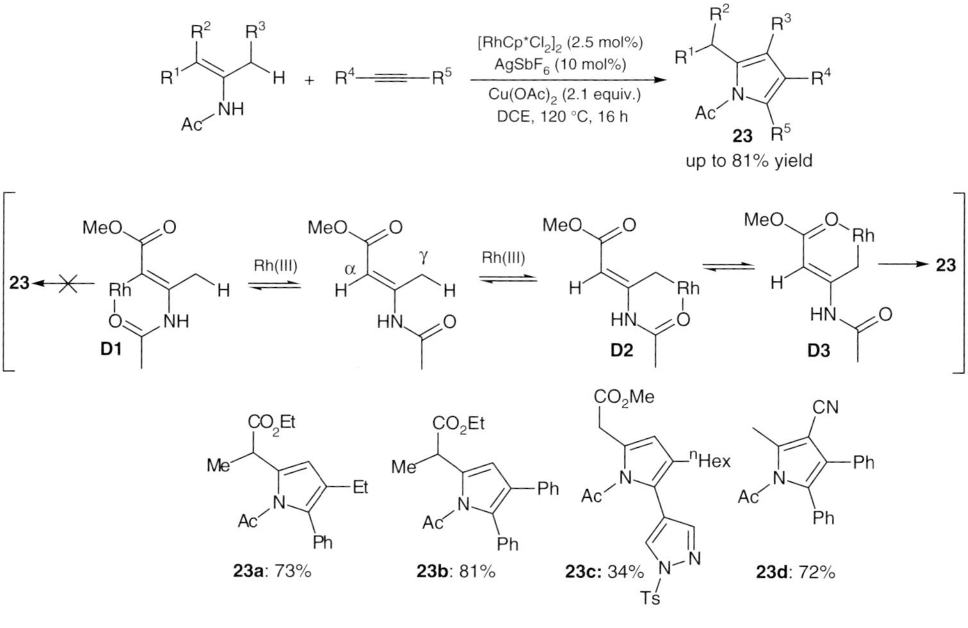

(5.22)

Interestingly, a similar method can be used for the synthesis of 3-borylated pyrroles **24** from enamides and internal alkyne *N*-methyliminodiacetic acid (MIDA) boronates (Eq. (5.23)) [17]. Product **24a** undergoes Suzuki–Miyaura coupling with bromobenzene to give **24c** in good yield, with a concomitant removal of the acetyl group.

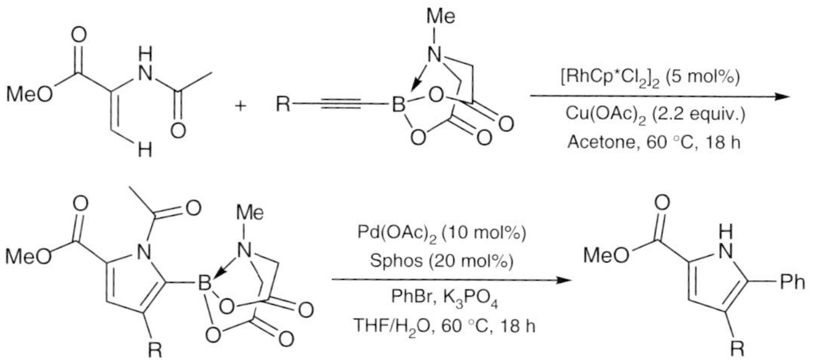

24a: R = Me, 82% (5:1)
24b: R = 4-Me-Ph, 85%, (5:1)
24c: R = Me, 91%

(5.23)

Similarly, Huestis and Fagnou reported a cationic Rh-catalyzed preparation of unsymmetrically substituted pyrroles from the coupling of enamides with enynes in good yield (Eq. (5.24)) [5c]. In this reaction, alkenes were selectively introduced at the C2 position of the corresponding pyrroles by using a Rh-catalyzed vinylic C–H activation approach.

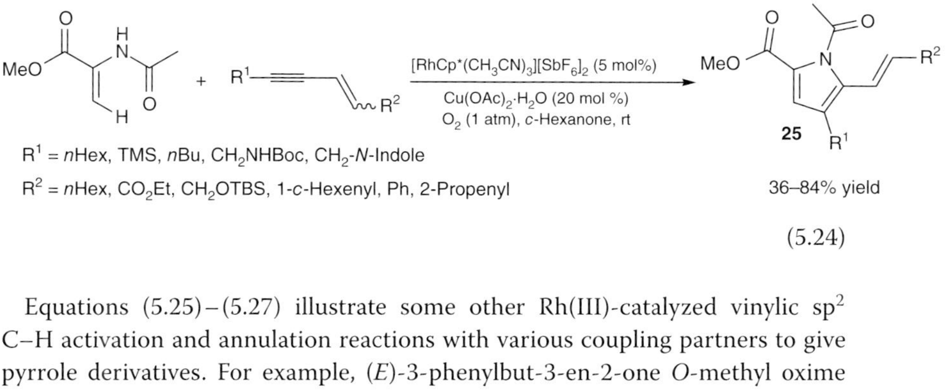

25

R^1 = *n*Hex, TMS, *n*Bu, CH$_2$NHBoc, CH$_2$-*N*-Indole
R^2 = *n*Hex, CO$_2$Et, CH$_2$OTBS, 1-*c*-Hexenyl, Ph, 2-Propenyl

36–84% yield

(5.24)

Equations (5.25)–(5.27) illustrate some other Rh(III)-catalyzed vinylic sp^2 C–H activation and annulation reactions with various coupling partners to give pyrrole derivatives. For example, (*E*)-3-phenylbut-3-en-2-one *O*-methyl oxime reacted with the *N*-tosyl imine of ethyl glyoxylate in DCE (1,2-dichloroethane) to afford pyrrole **26** in 78% yield [18a]. Similar reactions that involve vinylic sp^2 C–H and allylic C–H activation have also been reported [18b,c].

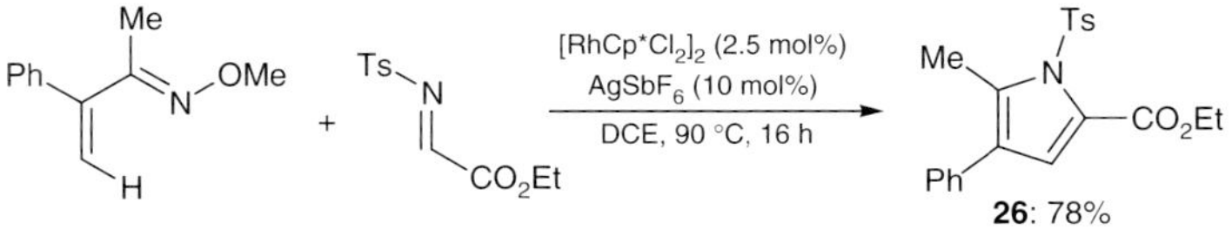

26: 78%

(5.25)

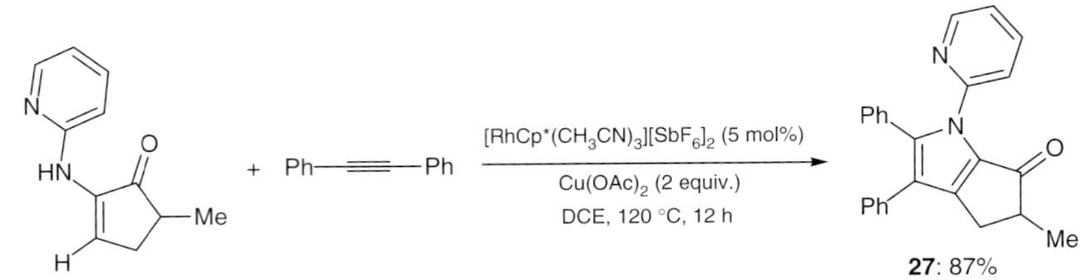

(5.26)

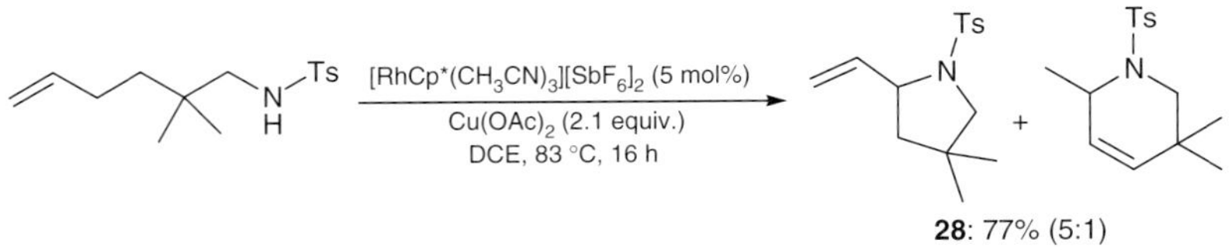

(5.27)

5.2.8 Synthesis of Isoindolin-1-ones

Glorius developed a Rh(III)-catalyzed oxidative *ortho*-olefination and *exo*-cyclization reaction for the synthesis of γ-butyrolactam derivatives [19a]. The resulting data showed that only products with a *Z*-configured exocyclic C=C bond are formed. A representative transformation, in which benzamide and *n*-butyl acrylate generate **29** in 62% yield, is shown in the following (Eq. (5.28)). Similarly, Zhou and Li developed a Rh(III)-catalyzed C–H bond activation and annulation of *O*-methyl oximes with isocyanates for the synthesis of 3-methyleneisoindolin-1-ones **30** (Eq. (5.29)) [19b].

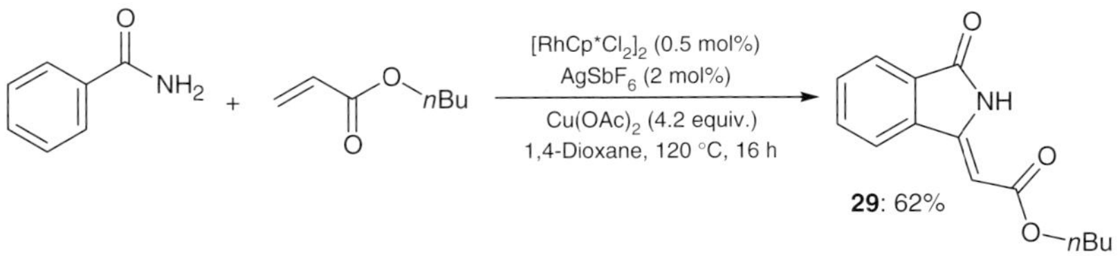

(5.28)

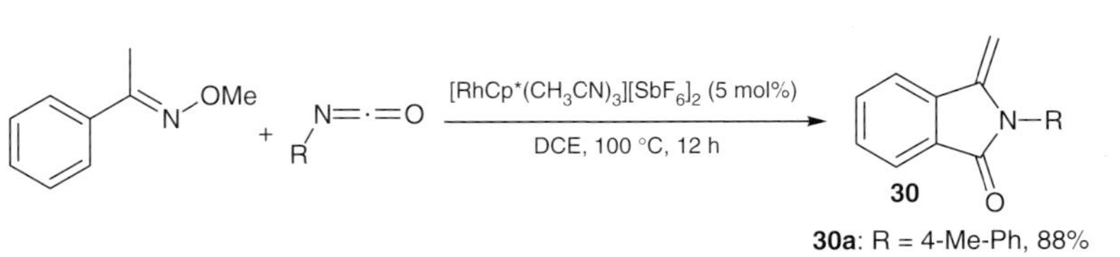

30a: R = 4-Me-Ph, 88%
30b: R = 4-OMe-Ph, 98%
30c: R = 4-Br-Ph, 80%
30d: R = 4-F-Ph, 75%

(5.29)

5.2.9
Synthesis of 3-Hydroxyisoindolin-1-ones

Kim and coworkers reported a Rh-catalyzed oxidative *ortho*-acylation of secondary benzamides by aryl aldehydes subsequent intramolecular cyclization to give 3-hydroxyisoindolin-1-ones **31** in good yields [20]. The reaction used Ag_2CO_3 as an oxidant (Eq. (5.30)).

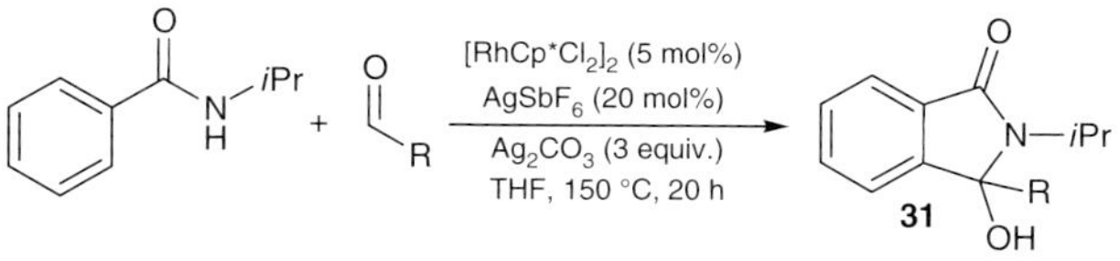

31a: R = 4-CF_3-Ph, 83%
31b: R = 4-CO_2Me-Ph, 78%
31c: R = 4-NO_2-Ph, 60%
31d: R = 4-CN-Ph, 40% (5.30)

5.2.10
Synthesis of 3-(Imino)isoindolinones

Falck and coworkers reported a Rh-catalyzed annulation of *N*-benzoylsulfonamides with isocyanides via C–H activation for the synthesis of 3-(imino)isoindolinones **32** in good yields [21]. The transformation effectively suppresses the competitive reaction and is compatible with a broad range of *N*-benzoylsulfonamides, as well as substituted isocyanides. A possible pathway is shown later (Eq. (5.31)). The catalytic cycle starts from the generation of five-membered rhodacycle **E1** by Rh(III)-catalyzed C–H activation and then 1,1-insertion of the isocyanide into the Rh–C bond to form intermediate **E3**. Reductive elimination affords annulated product **32** and Rh(I) which is oxidized by Cu(II) to regenerate the Rh(III) catalyst.

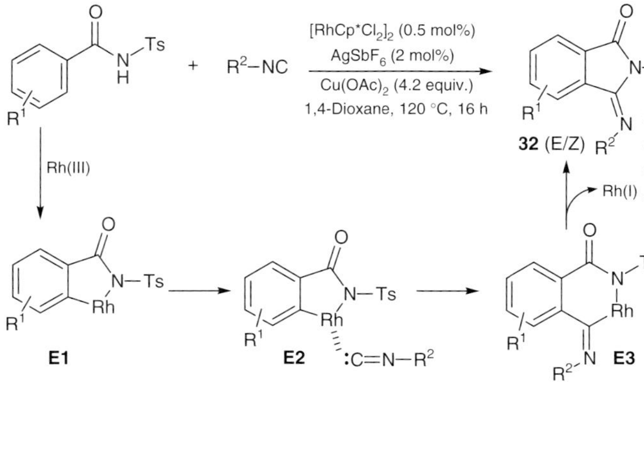

32a: R^1 = H, R^2 = Ph, 70%, 1:3
32b: R^1 = OMe, R^2 = Ph, 70%, 1:3
32c: R^1 = H, R^2 = *n*Pentyl, 54%, only *Z*
32d: R^1 = H, R^2 = *c*-Hexyl, 50%, only *Z*

(5.31)

Many different approaches have been described recently for the synthesis of five-membered cyclic lactams. A series of papers described the use of Rh(III)-catalyzed intermolecular oxidative insertion and C–N bond formation of amides and alkenes to provide *N*-heterocyclic derivatives are depicted in Eqs. (5.32)–(5.37). These approaches provide efficient routes for the construction of benzolactams, pyridine-fused lactams, 3,3-disubstituted isoindolinones, and oligocyclic lactams in good yields [22a–f].

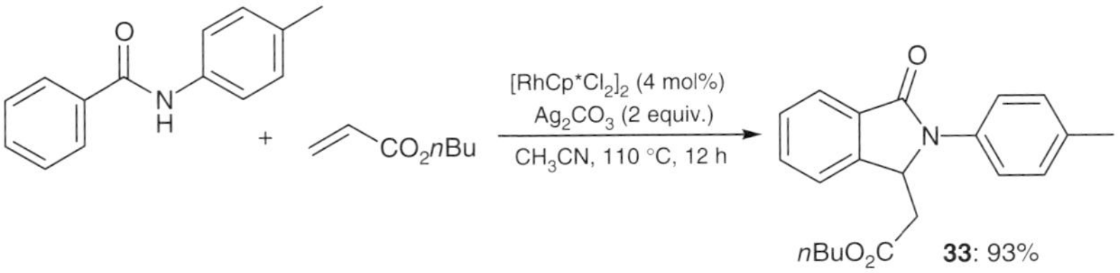

33: 93%

(5.32)

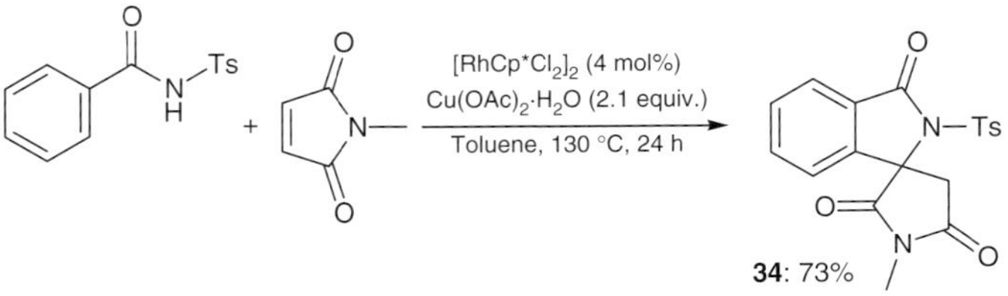

34: 73%

(5.33)

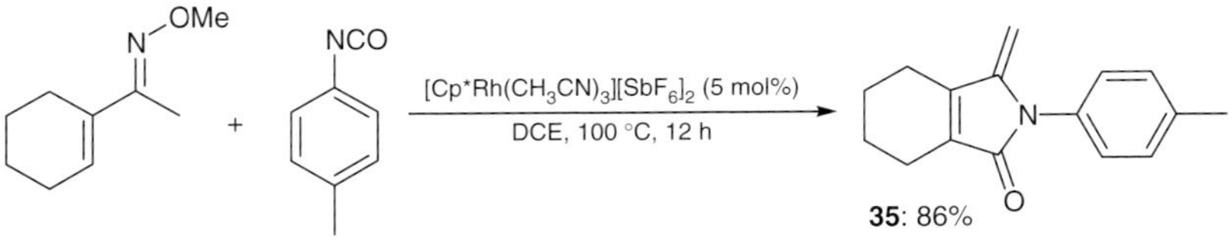

35: 86%

(5.34)

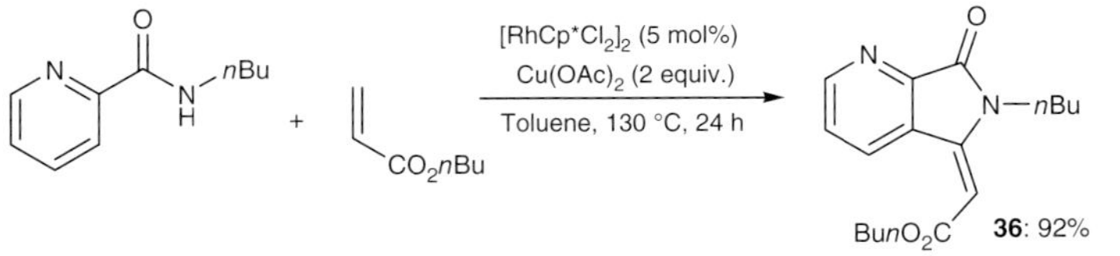

36: 92%

(5.35)

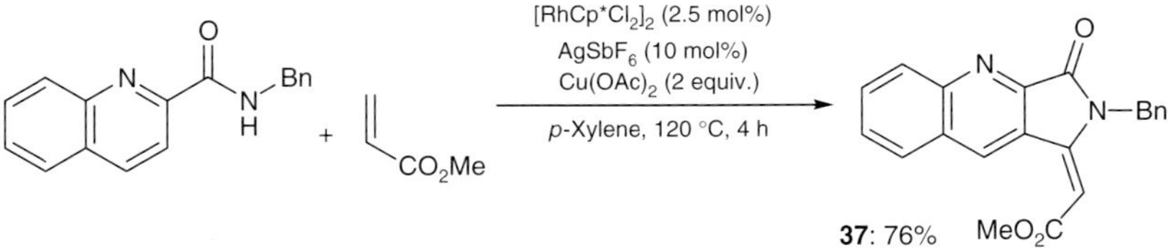

37: 76%

(5.36)

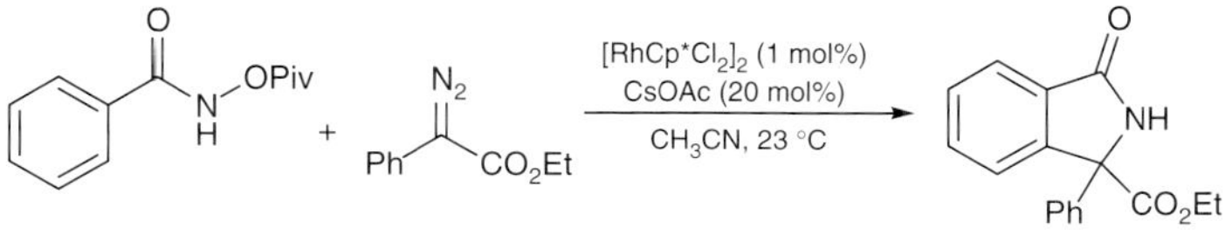

38: 96% (5.37)

5.2.11 Synthesis of Dihydrocarbazoles

The first example of Rh-catalyzed C–H coupling of 2-phenylpyridines with 7-azabenzonorbornadienes to provide *cis*-fused dihydrocarbazoles **39** was reported by Li and coworkers (Eq. (5.38)) [23]. The C–N bond forms via a double C–H activation, in which a seven-membered rhodacyclic complex is a key intermediate as a result of stereoselective insertion of the olefin into the Rh–C(aryl) bond. This strategy has also been employed for the synthesis of dihydrocarbazoles by using various directing groups.

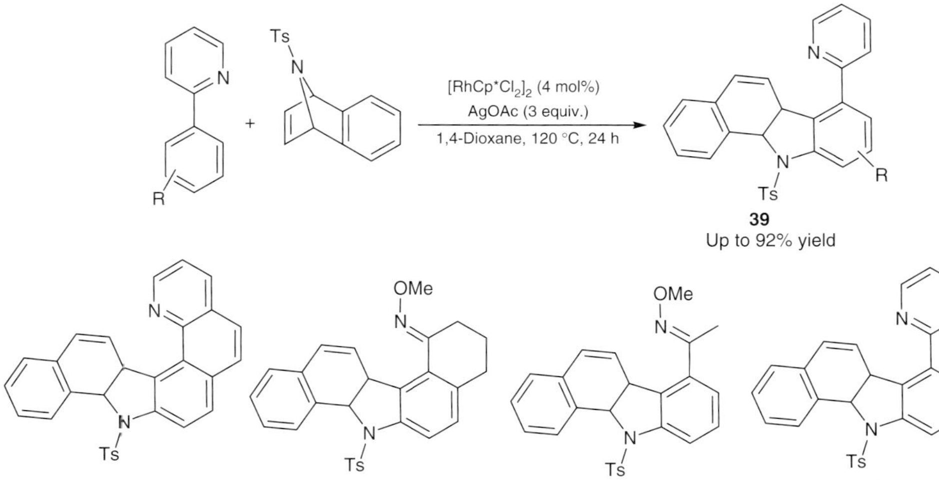

39
Up to 92% yield

39a: 88% **39b**: 52% **39c**: 60% **39d**: 38%

(5.38)

5.2.12 Synthesis of Sultams

Recently, Qian and Dong disclosed a highly efficient Rh-catalyzed three-component reaction of imines, alkynes, and aldehydes via C–H activation to afford highly fused polycyclic sultam skeletons [24]. This strategy allows the formation of four new bonds in a simple-to-perform, single-operation cascade of C–H activation/C=N insertion, C–H activation/C=O insertion, and cyclization (Eq. (5.39)).

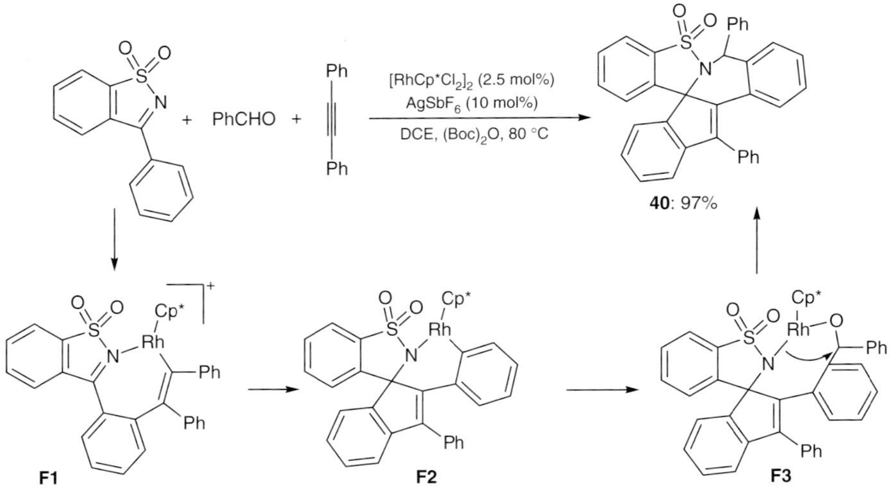

(5.39)

Another approach to the construction of benzofused sultams was also reported by Wang and Li groups [25]. The reaction provides a wide scope of *ortho*-alkylated benzofused sultam analogs in good yields (Eq. (5.40)). The *N*-acetyl group is important for rhodacycle formation and subsequent Michael addition reaction, which was finally removed under mild conditions to provide free NH sultams.

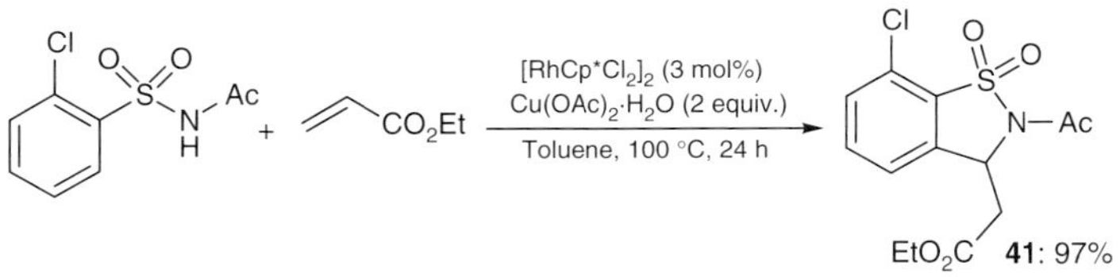

(5.40)

5.2.13
Synthesis of Phthalimides

The first example of Rh(III)-catalyzed oxidative carbonylation of aromatic amides by C–H/N–H activation to synthesize phthalimides, such as **42**, was reported by Rovis in 2011 [26a]. The presence of KH_2PO_4 is crucial to obtain the products in high yields (Eq. (5.41)). This reaction tolerates a variety of functional groups under standard conditions to afford phthalimides in excellent yields. Similarly, a Rh(III)-catalyzed direct C–H amidation of benzoic acids with isocyanates and subsequent intramolecular cyclization to give *N*-substituted phthalimides, such as **43**, is also known. This cascade cyclization reaction provided various phthalimides in 26–91% yield [26b]. In this case, NaOAc plays a vital role for the annulation and stimulation of the *ortho*-C–H bond activation of the benzoic acids (Eq. (5.42)).

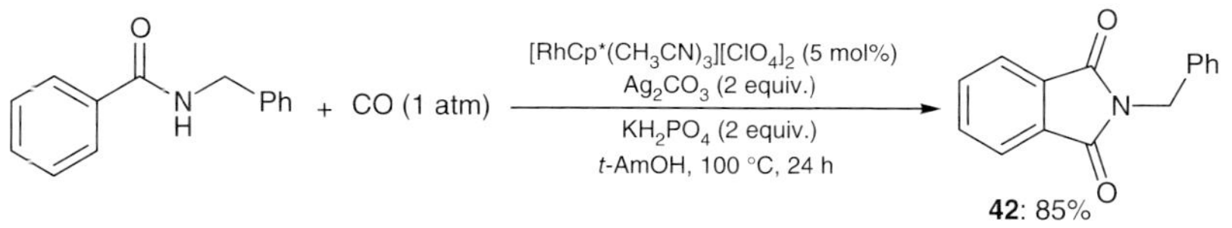

(5.41)

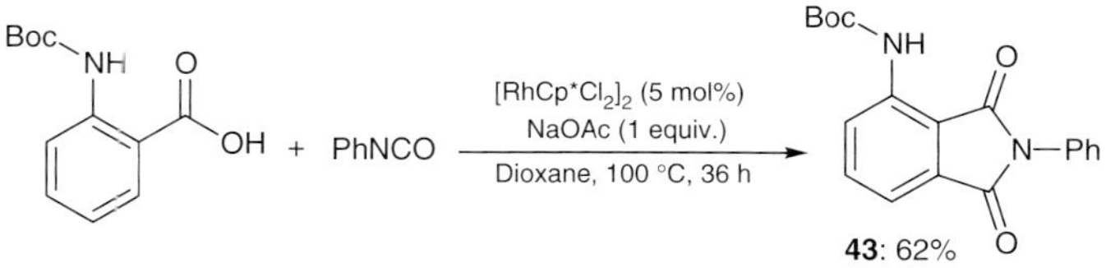

(5.42)

5.3
Synthesis of Six-Membered Nitrogen Heterocycles

5.3.1
Synthesis of Isoquinolines by Rh(I) Catalysis

Jun *et al.* demonstrated a Rh(I)-catalyzed cyclization of an *N*-benzyl aromatic ketimine with diphenylacetylene to provide isoquinoline **44** [27]. The chelation-assisted C–H activation strategy was employed for the first time for isoquinoline synthesis. However, the reaction required a high temperature (150 °C) and led to two different isoquinoline derivatives **44** and **44′**. Based on the experimental results, the authors proposed a plausible reaction mechanism that involved *ortho*-alkenylation, 6π-electrocyclization, intermolecular nucleophilic substitution, and dehydrogenative aromatization (Eq. (5.43)).

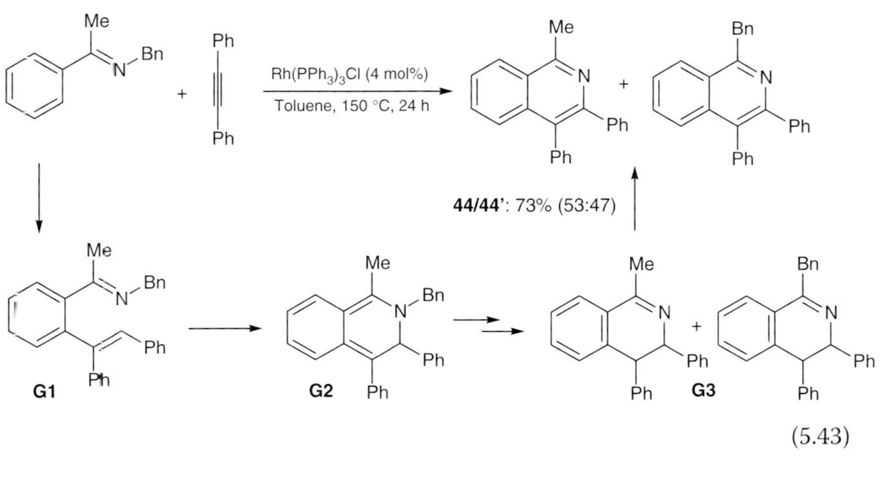

(5.43)

5.3.2
Synthesis of Isoquinolines by Rh(III) Catalysis

In 2009, Fagnou *et al.* discovered a Rh(III)-catalyzed synthesis of substituted isoquinolines **45** by oxidative annulation between *N*-*t*Bu aromatic aldimines and internal alkynes [28a]. The *t*-Bu leaving group was eliminated as isobutene in the reaction process and avoided the generation of a mixture of isoquinolines. However, the substrate scope was limited to aldimines, and a stoichiometric amount of $Cu(OAc)_2 \cdot H_2O$ was used as an external oxidant. Mechanistic studies omitted the *ortho*-alkenylation/6π-electrocyclization/oxidation pathway, and intermediate **H2** was crucial for the C–N reductive elimination to proceed (Eq. (5.44)). At the same time, Satoh and Miura groups also exploited a Rh(III)-catalyzed oxidative cyclization of N–H benzophenone imine and internal alkynes to give isoquinolines [28b]. Both aromatic and aliphatic alkynes were agreeable for this protocol, but a stoichiometric amount of the Cu(II) salt was required.

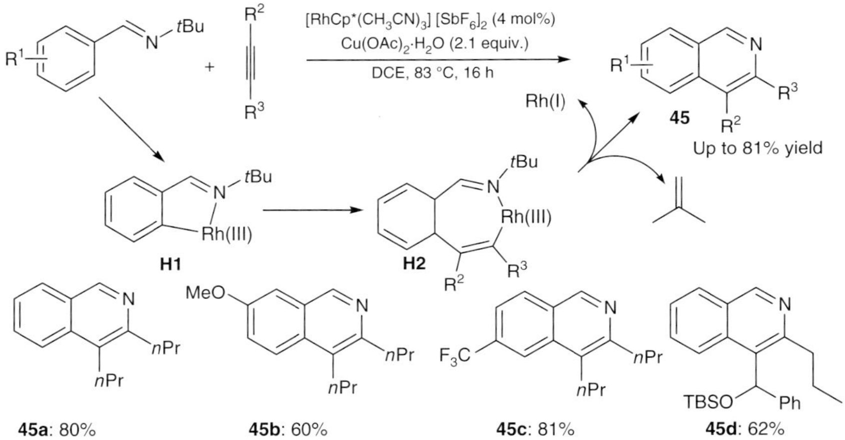

(5.44)

Several Rh-catalyzed annulation reactions were reported by other groups that allow the generation of a broad array of isoquinoline derivatives as illustrated by Eqs. (5.45)–(5.53). These reaction conditions are quite mild, and a wide range of arenes were coupled with alkynes to yield multisubstituted isoquinolines in good yields [29a–i].

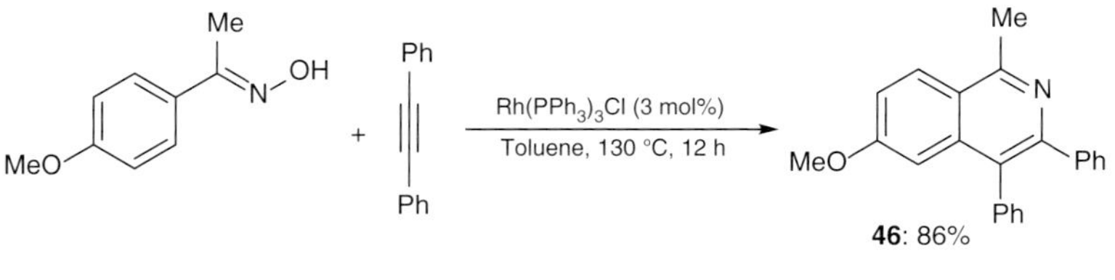

(5.45)

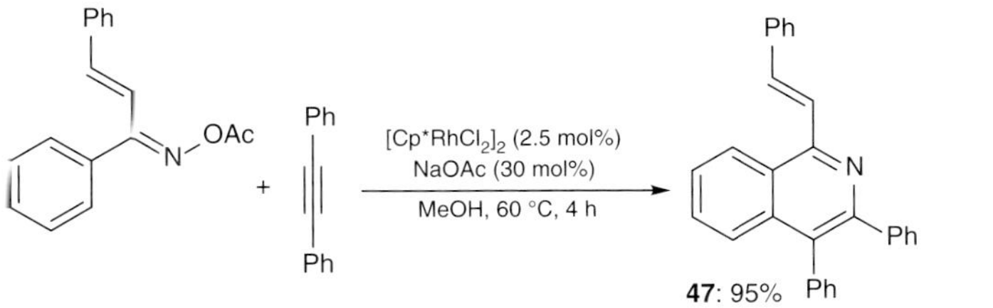

(5.46)

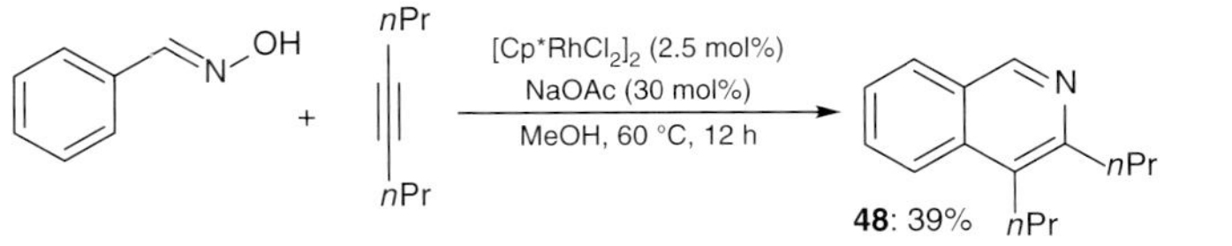

(5.47)

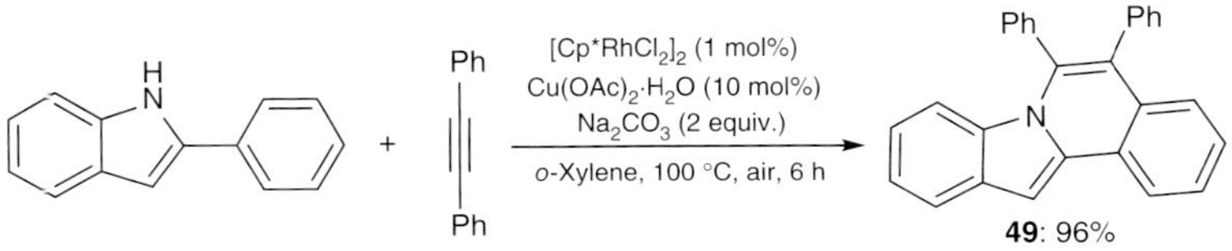

(5.48)

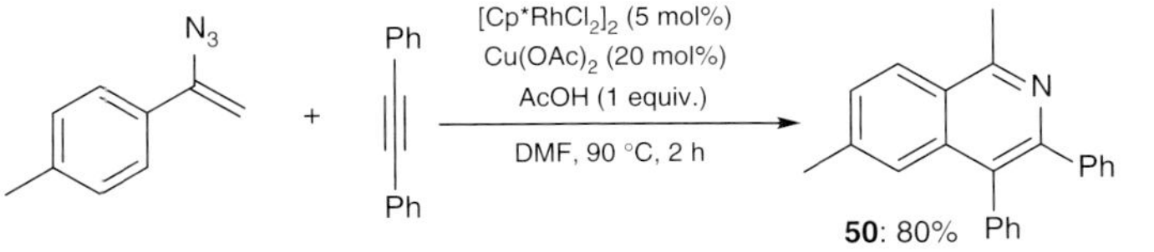

(5.49)

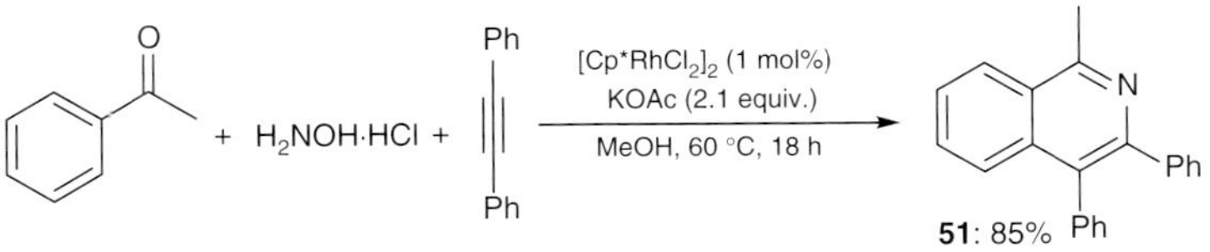

(5.50)

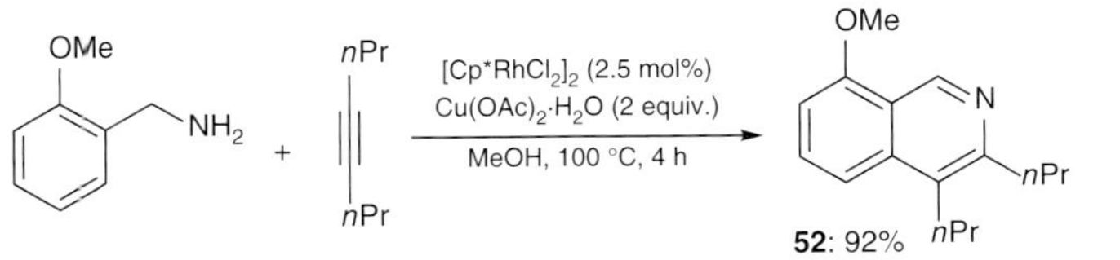

(5.51)

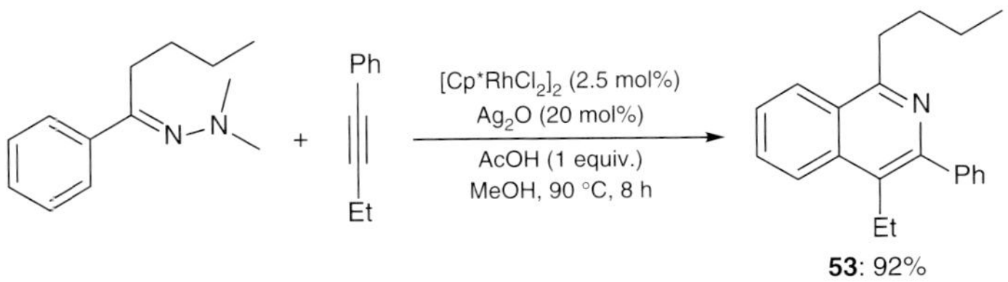

(5.52)

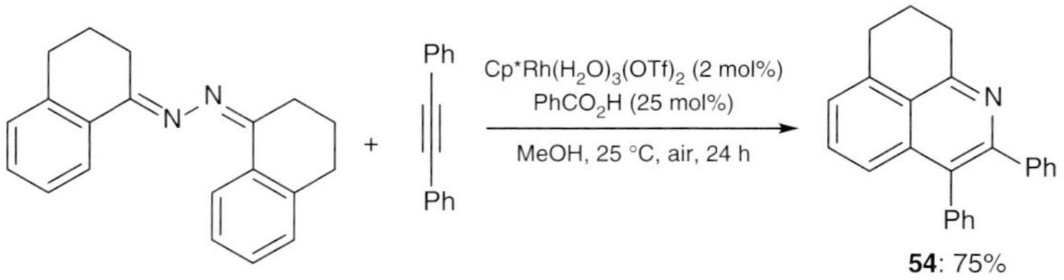

(5.53)

5.3.3
Synthesis of 1-Aminoisoquinolines

In 2011, Li *et al.* described a Rh(III)-catalyzed coupling of *N*-aryl and *N*-alkyl benzamidines with internal alkynes that afforded less accessible 1-aminoisoquinolines **55** [30]. In the presence of *ortho*-substituted benzamidines, the reaction occurred by dual oxidative cyclization and furnished indole–isoquinoline **56**. This unusual result occurred in the presence of an *ortho* steric effect after the formation of the isoquinoline moiety. The steric repulsion between the *ortho* substituent and *N*-phenyl ring forced these two groups apart (Eq. (5.54)).

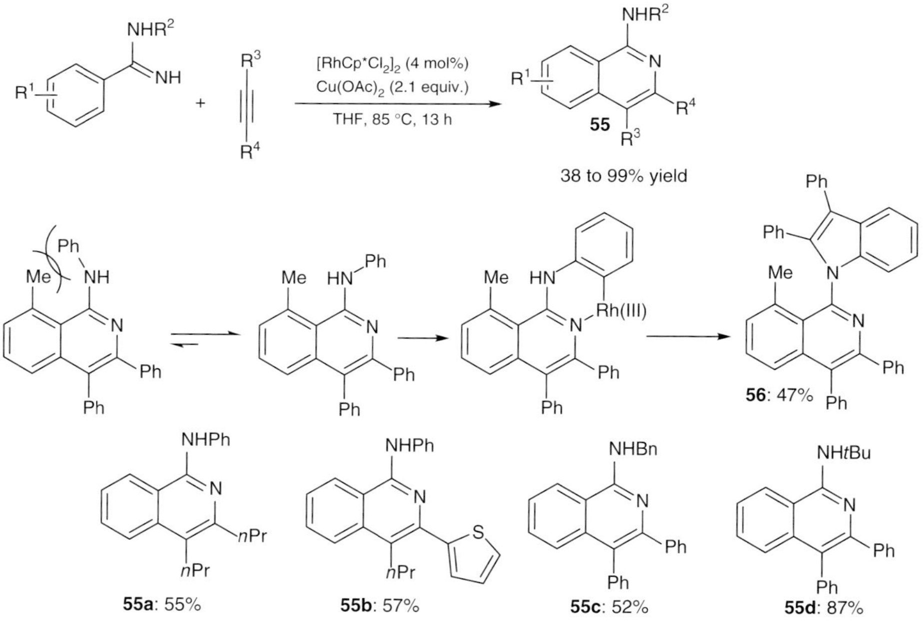

(5.54)

Similarly, Glorius and coworkers disclosed a Rh(III)-catalyzed C–H activation/cyclization/isomerization strategy to prepare isoquinolines with a complete regioselectivity from *O*-pivaloylketoximes and 1,3-dienes (Eq. (5.55)) [31]. Several electron-deficient 1,3-dienes could be efficiently transformed into

3-alkyl-substituted isoquinolines **57** under redox-neutral conditions, whereas the reaction of electron-neutral dienes proceeded in the presence of stoichiometric amounts of AgOAc to furnish the corresponding 3-alkenylated isoquinolines.

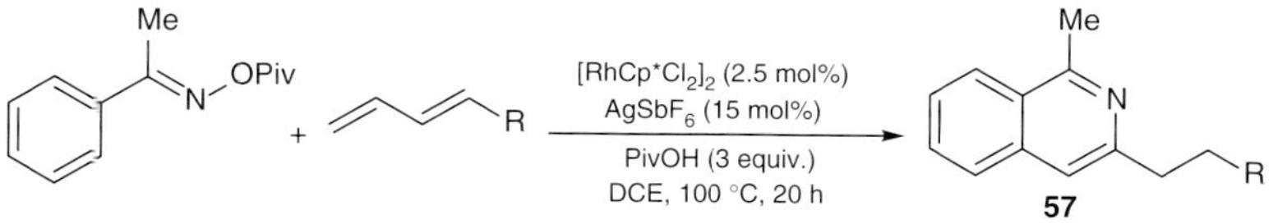

57

57a: R = COMe, 62%
57b: R = CO_2Me, 84%
57c: R = CN, 72%
57d: R = SO_2Ph, 66% (5.55)

5.3.4 Synthesis of Isoquinolones and Related Derivatives

In 2010, Fagnou and coworkers reported a cascade method for the synthesis of isoquinolones **58** by Rh-catalyzed annulation of *N*-methoxybenzamides with internal alkynes [32]. Both symmetrical and unsymmetrical alkynes were examined as the coupling partners (Eq. (5.56)). This external oxidant-free process used a N–O bond to unlock the C–N bond formation and catalyst release.

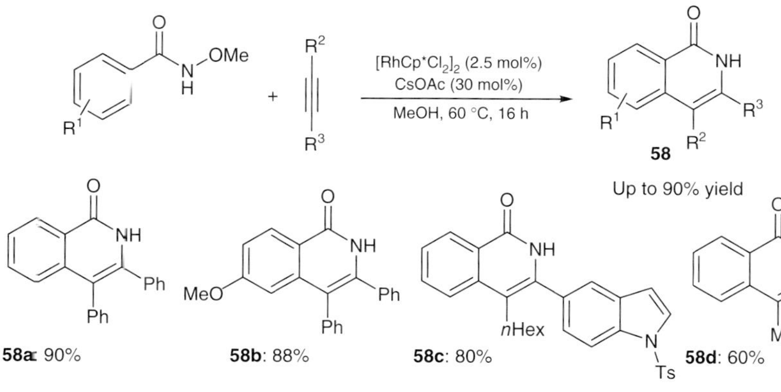

(5.56)

At the same time, similar methodology to succeed the catalytic C–H activation of benzamides under oxidative coupling conditions resulting in isoquinolones was also reported (Eqs. (5.57) and (5.58)). Both *N*-alkyl and *N*-aryl secondary benzamides could be employed, whereas primary benzamides reacted differently, and two alkyne units were oxidatively incorporated to give tricyclic products such as **60** [33].

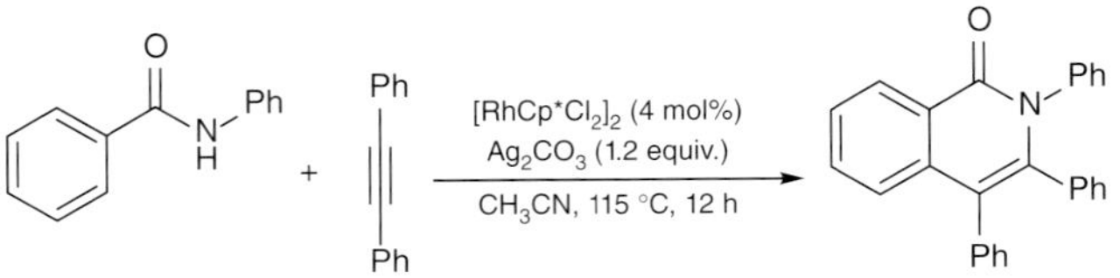

59: 95% (5.57)

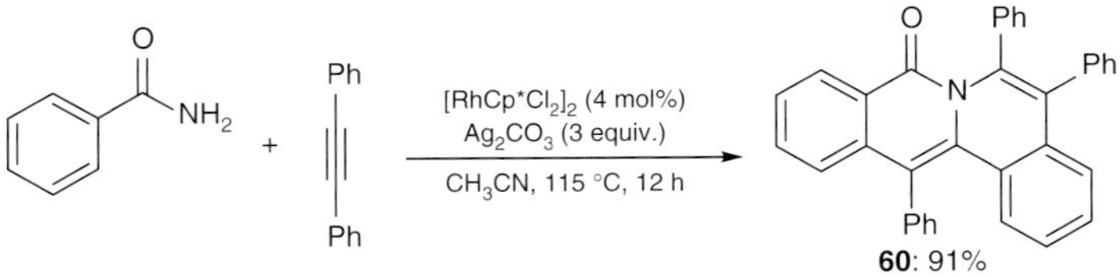

60: 91% (5.58)

The reaction was further extended to the annulation of *N*-pivaloxybenzamide with terminal alkynes to give 3-substituted isoquinolones **61** with good regioselectivity under mild conditions (Eq. (5.59)) [34]. If alkenes were used, the reaction gave 3,4-dihydroisoquinolones **62** in good yields (Eq. (5.60)); however, both diaryl- and dialkyl-substituted alkenes failed to react under the same reaction conditions.

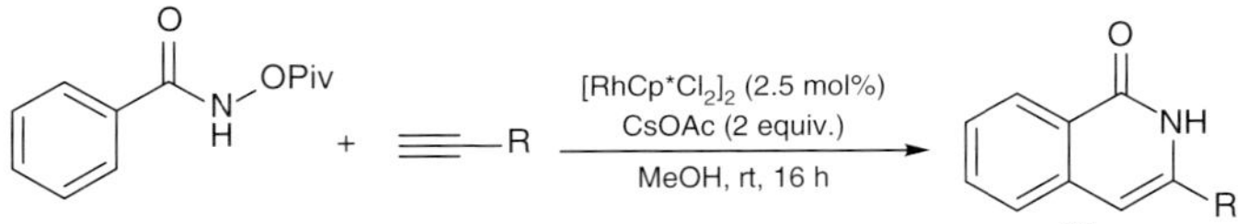

61
Up to 95% yield
61a: R = *n*-Hex, 92%
61b: R = TMS, 75%
61c: R = CO_2Me, 49%
61d: R = $(CH_2)_2OH$, 85% (5.59)

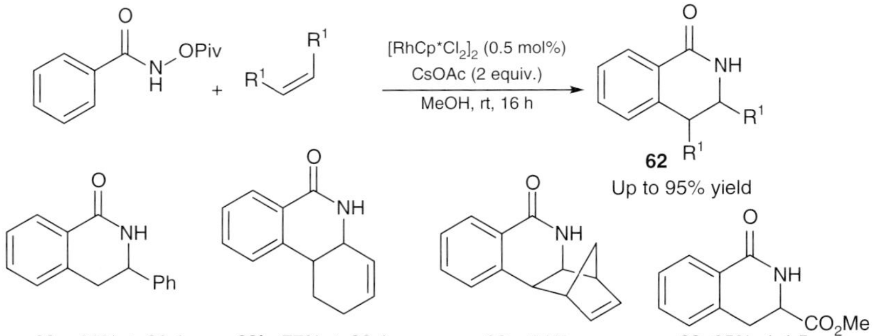

62
Up to 95% yield

62a: 90%, ≥ 20:1 **62b**: 77%, ≥ 20:1 **62c**: 91% **62**: 95%, 1:4.5

(5.60)

At this stage, many authors focused on different routes for the synthesis of hydroisoquinolone derivatives by utilizing Rh-catalyzed C–H activation and reductive elimination of amides. Only the control of reaction conditions and judicious selection of the alkene derivatives gave selective insertion with good yields. Selected results are shown in Eqs. (5.61)–(5.69). Notably, the presence of chiral Cp* ligands gave products with high regioselectivity and enantiomeric excess [35a–i].

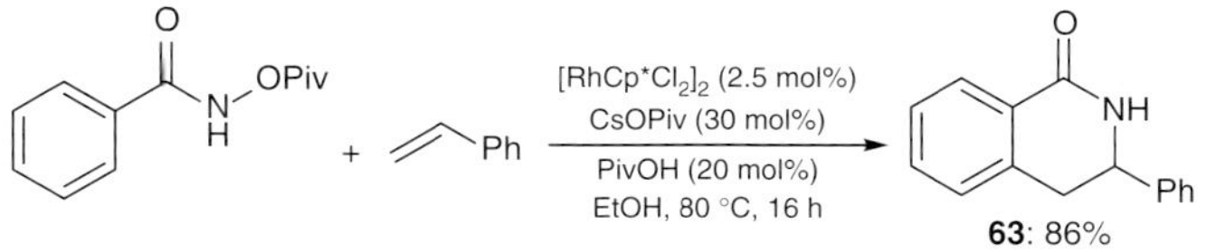

(5.61)

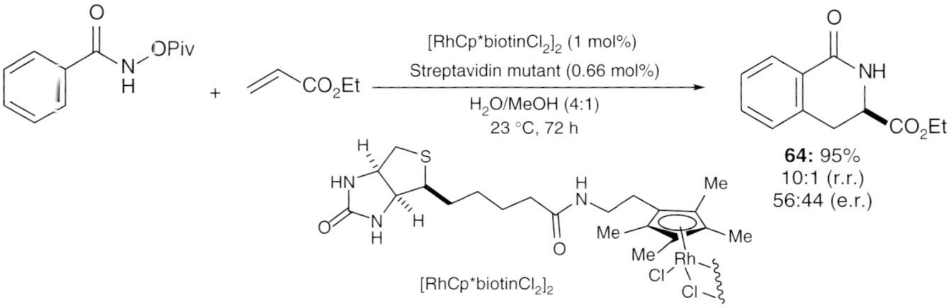

(5.62)

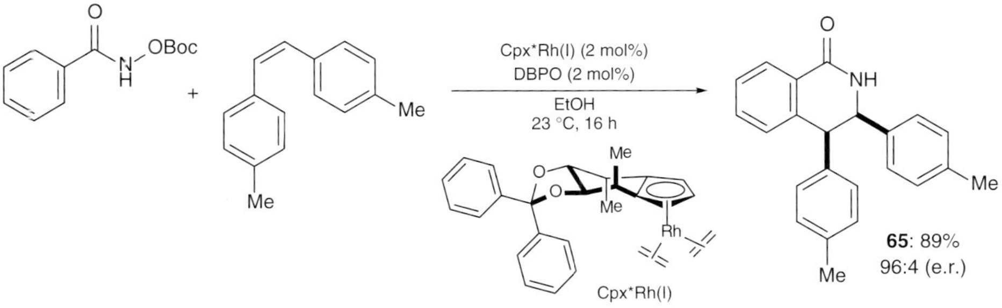

(5.63)

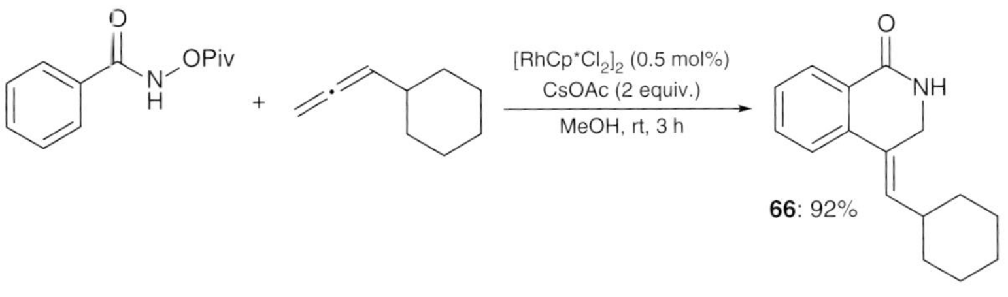

(5.64)

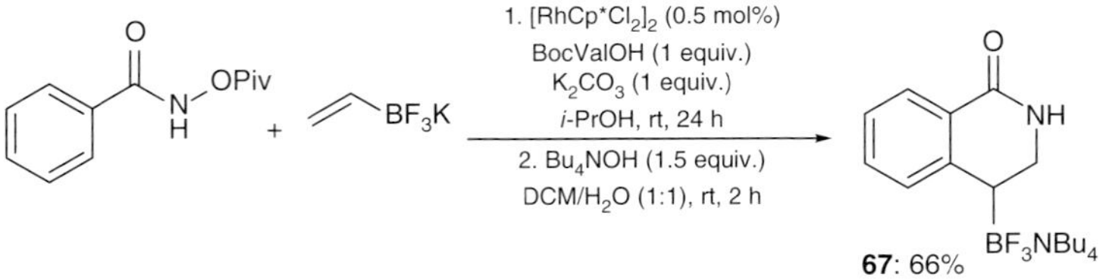

67: 66% (5.65)

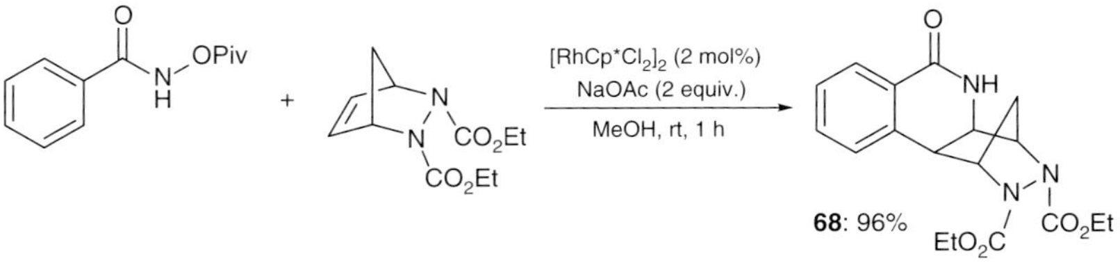

68: 96%

(5.66)

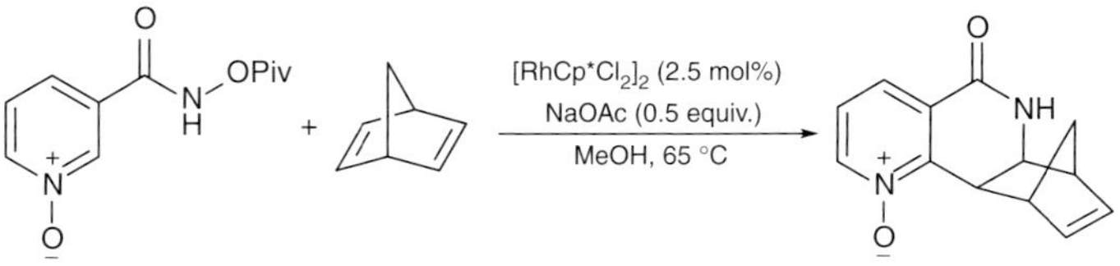

69: 63%, ≥20:1 (5.67)

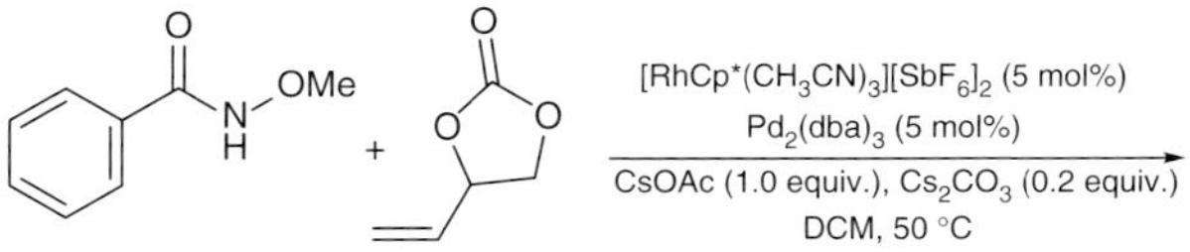

70: 72%

(5.68)

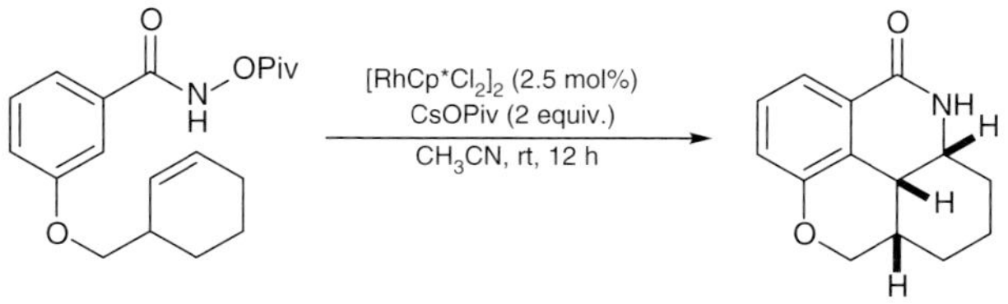

71: 50% (5.69)

An interesting Rh(III)-catalyzed intramolecular annulation of alkyne-tethered benzamides to give isoquinolones was recently developed by Park and coworkers [36a]. The reaction might involve intramolecular insertion of the alkyne moiety with rhodacycle **I1**, which leads to intermediate **I2** or **I2′**. Then, reductive elimination, followed by oxidative addition of the N–O bond, gives intermediate

I3 (Eq. (5.70)). Upon further protonation by acetic acid, product **72** is released, and the catalyst is regenerated. Similarly, Mascarenas and Gulias reported a Rh(III)-catalyzed intramolecular annulation of alkyne-tethered benzamides to give tricyclic isoquinolone derivatives in good yields (Eq. (5.71)) [36b].

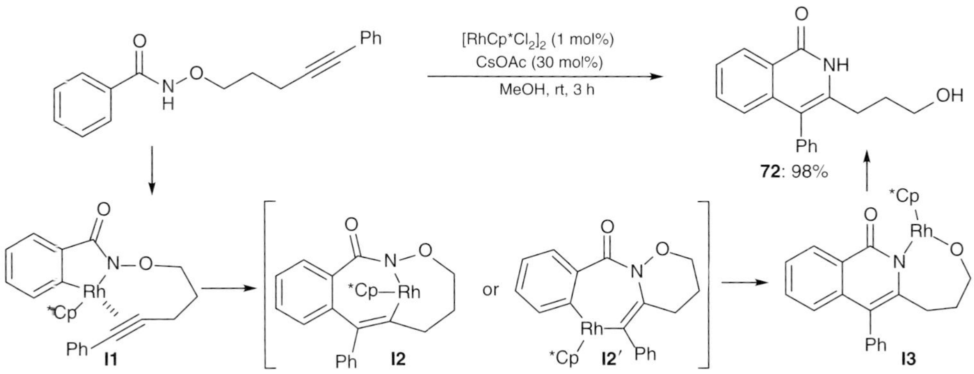

(5.70)

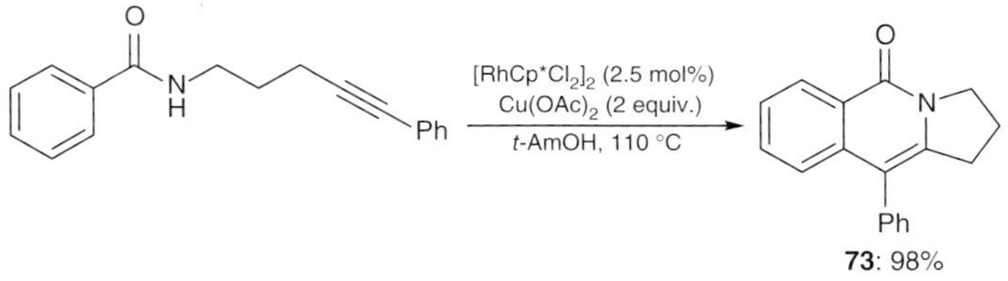

(5.71)

Recently, the preparation of 3,4-unsubstituted isoquinolones **74** by Rh-catalyzed C–H activation/annulation was established by Marsden *et al.* by using amides and vinyl acetate [37]. The vinyl acetate behaved as an acetylene equivalent, and a variety of 3,4-unsubstituted isoquinolin-1-(2*H*)ones were obtained in good yields (Eq. (5.72)).

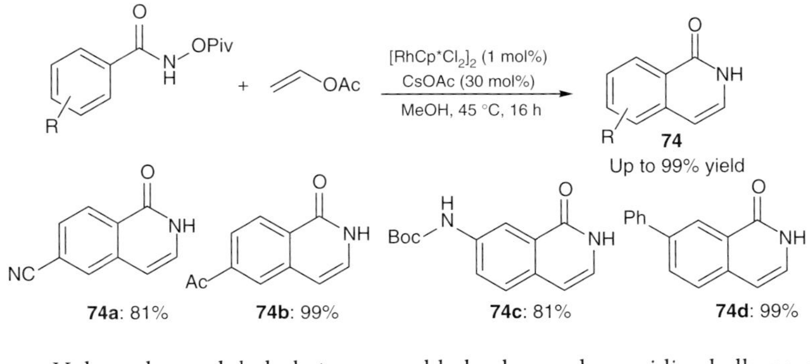

(5.72)

α-Halo and pseudohalo ketones could also be used as oxidized alkyne equivalents in a Rh(III)-catalyzed redox-neutral annulation to efficiently generate diverse isoquinolone structures (Eq. (5.73)) [38]. By this reaction, several 3-aryl- and 3-alkyl-substituted isoquinolones were produced in moderate to excellent yields; however, other *N*-substituted amides (H, Me, Ph, OPiv) and secondary mesylates were totally ineffective.

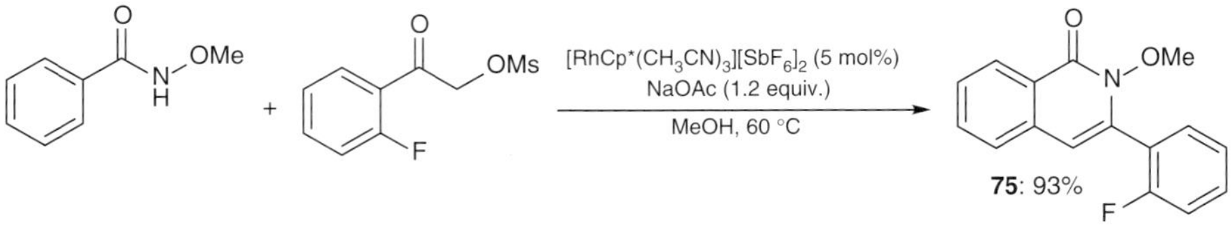

(5.73)

Tian and Lin disclosed a Rh(III)-catalyzed arylative cyclization reaction of 1,6-enynes with *N*-(pivaloyloxy)benzamides for the synthesis of a new type of tetracyclic isoquinolones **76** in quantitative yields [39]. Product formation fully depends on the O-substituted *N*-hydroxybenzamide substrate. For example, treatment of *N*-(pivaloyloxy)benzamide with 1,6-enynes in the presence of acetone afforded polycyclic products **76** by a Michael addition process (Eq. (5.74)), whereas the treatment of *N*-(methoxy)benzamide did not afford the Michael addition product.

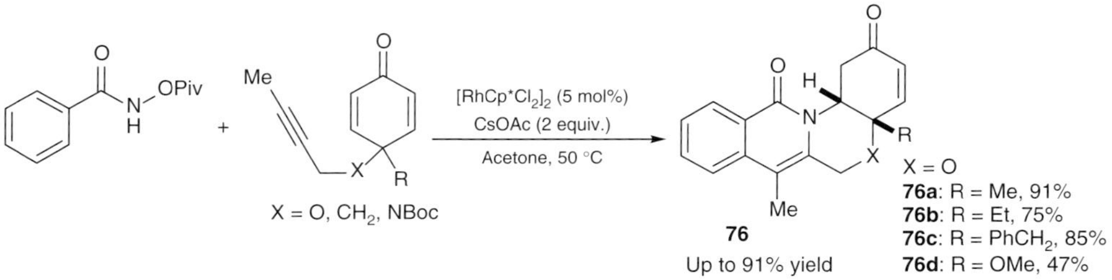

(5.74)

5.3.5 Synthesis of Phenanthridinones

In 2012, our group reported a Rh(III)-catalyzed regioselective synthesis of phenanthridinones **77** from *N*-methoxybenzamides and arylboronic acids via one-pot C–C/C–N bond formation under mild conditions [40a]. For example, treatment of *N*-methoxybenzamide and phenylboronic acid with $[RhCp^*Cl_2]_2$ and Ag_2O in methanol at 60 °C for 3 h afforded phenanthridinone in 94% yield. This is the first Rh-catalyzed dual oxidative C–H coupling reaction with two aromatic compounds to give a formal [4+2] cyclization product. In this reaction, the transformation is believed to proceed by a Rh^{III}/Rh^{I} catalytic cycle that is initiated by Ag_2O by transmetalation of the arylboronic acid (Eq. (5.75)). Similar methodology can be extended to aryltrialkoxysilanes for both coupling and annulation with *N*-methoxybenzamides (Eq. (5.76)) [40b].

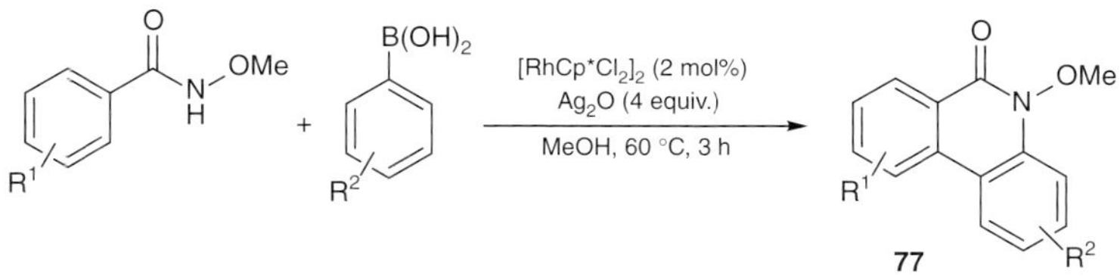

77
72 to 94% yield

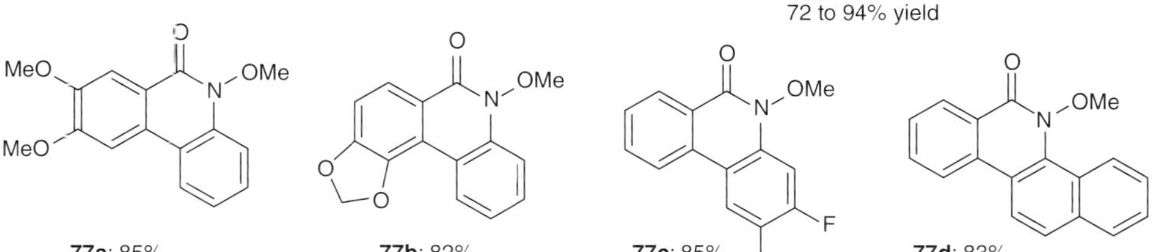

77a: 85% **77b**: 82% **77c**: 85% **77d**: 83%

(5.75)

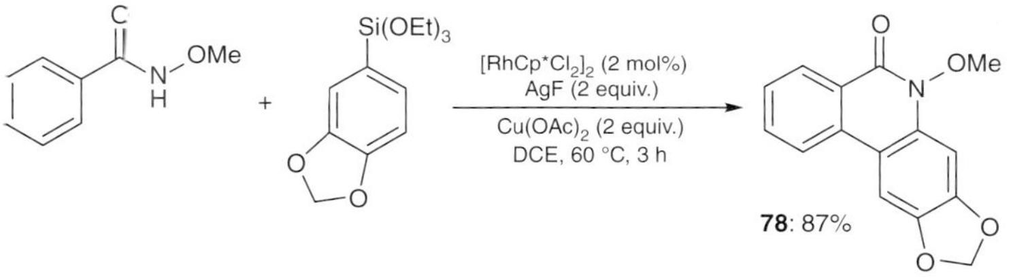

78: 87%

(5.76)

5.3.6
Synthesis of Quinolines

Hua *et al.* developed a one-pot synthesis of multisubstituted 2-aminoquinolines by the cyclization of 1-aryl tetrazoles with internal alkynes by Rh(III)-catalyzed double C–H bond activation and Cu(II)-mediated denitrogenation (Eq. (5.77)). In this reaction, copper plays a vital role in N–N bond cleavage [41].

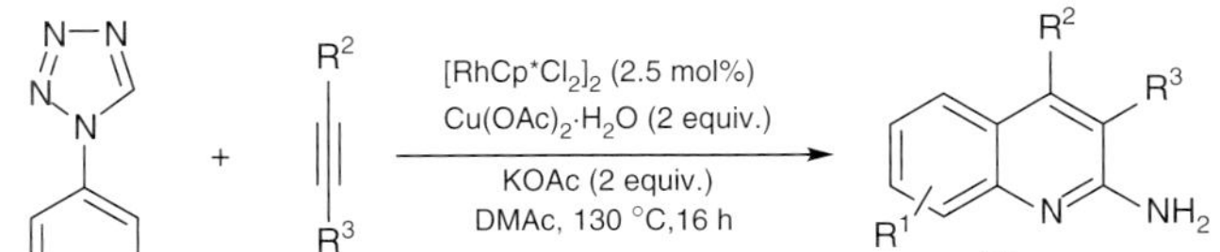

79
Up to 77% yield

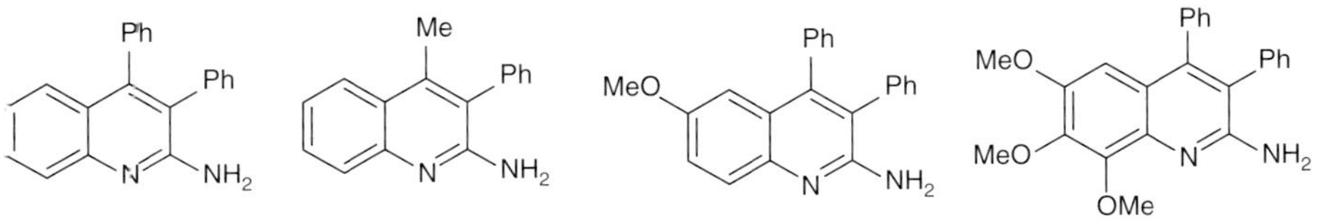

79a: 68% **79b**: 62% **79c**: 72% **79d**: 77%

(5.77)

5.3.7
Synthesis of Naphthyridines

An efficient one-pot synthesis of highly substituted naphthyridine-based polyheteroaromatic compounds by Rh(III)-catalyzed multiple C–H activation and multiple C–C/C–N bond formation was also reported by our group (Eq. (5.78)) [42]. This interesting new chemistry delivers highly substituted π-conjugated compounds, which can potentially be used in organic semiconductors and luminescent materials. Four possible rhodacycles (**J1–J4**) are likely involved in this reaction. The physical properties of these compounds, such as absorption and emission properties, were also studied. Product **80b** showed a very strong green photoluminescence with a narrow bandwidth of $\lambda = 75$ nm, and a strong deep-blue emission was observed for compound **81a**.

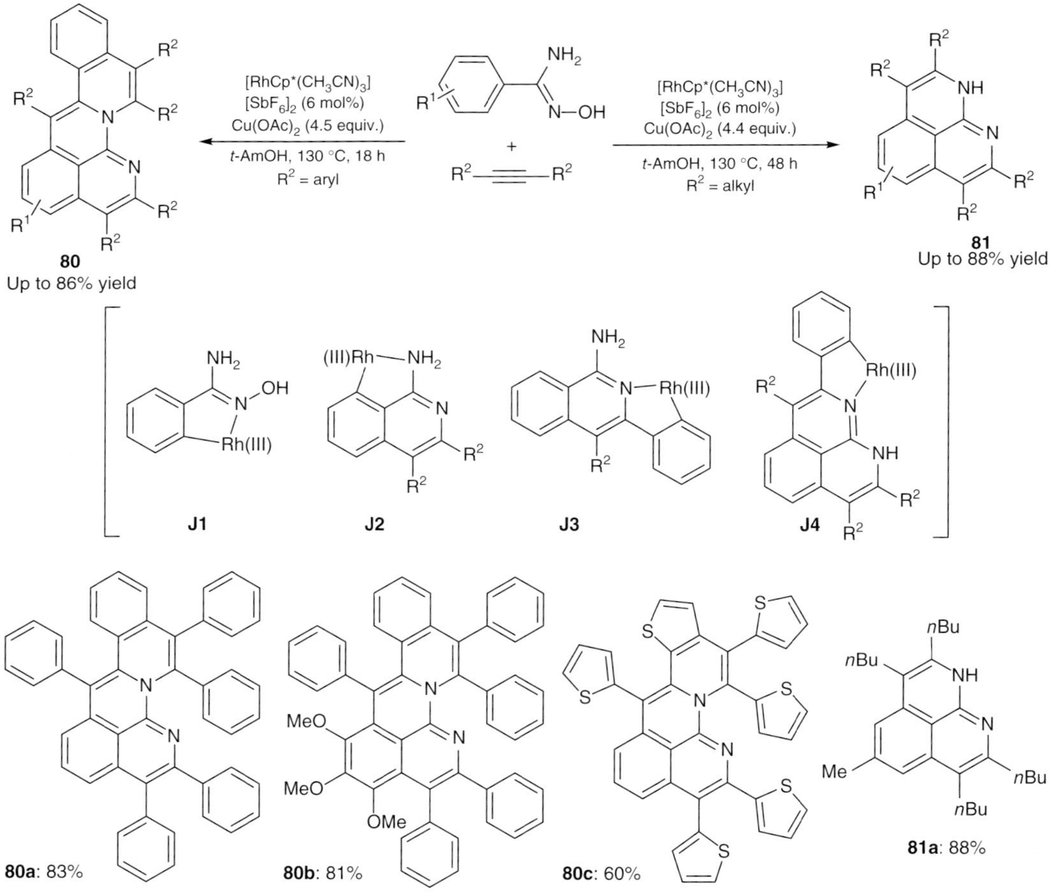

(5.78)

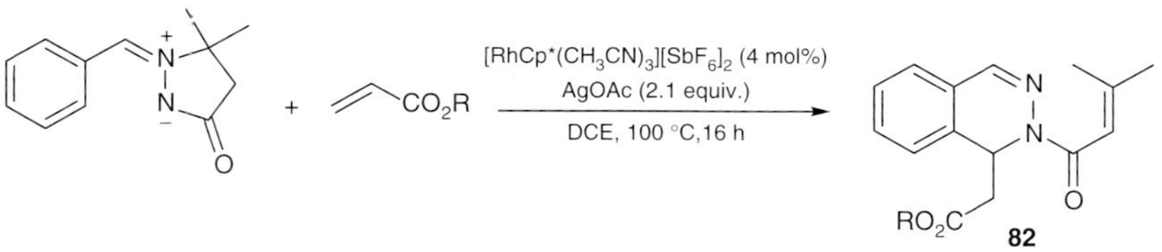

82
Up to 88% yield
82a: R = Bn, 88%
82b: R = Me, 77%
82c: R = Et, 71%

(5.79)

5.3.8
Synthesis of Phthalazines

A new Rh-catalyzed oxidative coupling of azomethine ylides with olefins to produce 1,2-dihydrophthalazines by selective C–H and C–N cleavage was reported by Li in 2012 [43a]. In this reaction, AgOAc proved to be an effective oxidant, and monoolefination/cyclization products **82** were isolated in higher yields (Eq. (5.79)). The Rh catalyst plays a vital role in this reaction not only involving in C–H activation but also promoting the cleavage of pyrazolidinone ring and Michael addition sequence. Related reaction of 1,2-dihydrophthalazine synthesis has also been reported (Eq. (5.80)) [43b].

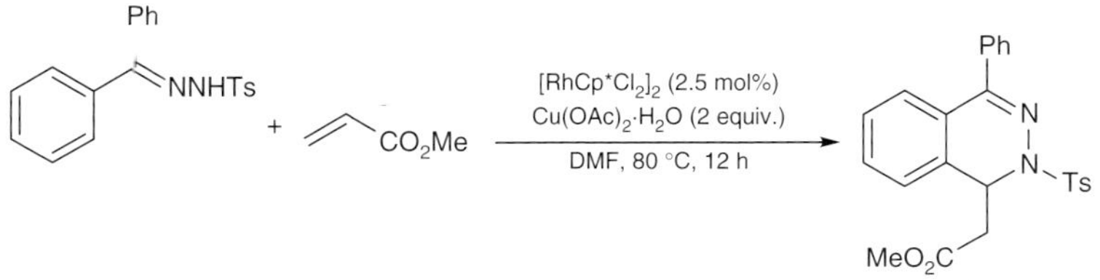

83: 93% (5.80)

5.3.9
Synthesis of Acridines and Phenazines

Ellman and coworkers reported a new [3+3] annulation reaction for the preparation of acridines and phenazines by Rh(III)-catalyzed C–H amination of aromatic imines and azobenzenes with aromatic azides. The reaction mechanism proceeds by C–H activation followed by intramolecular electrophilic aromatic substitution and aromatization to offer the respective products. Equations (5.81)–(5.83) illustrate that a broad range of acridines and phenazines can be generated with diverse functionality [44].

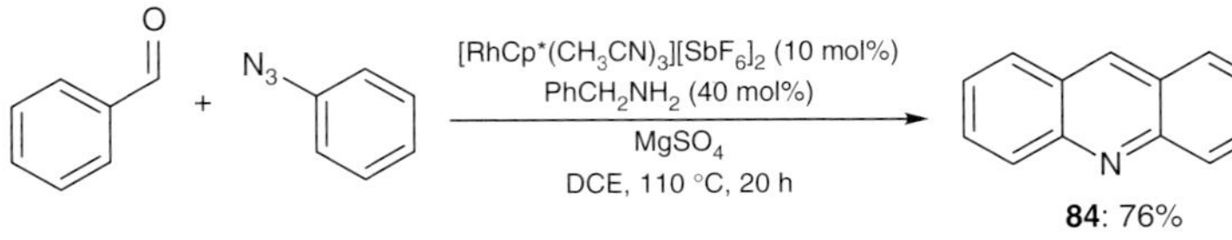

(5.81)

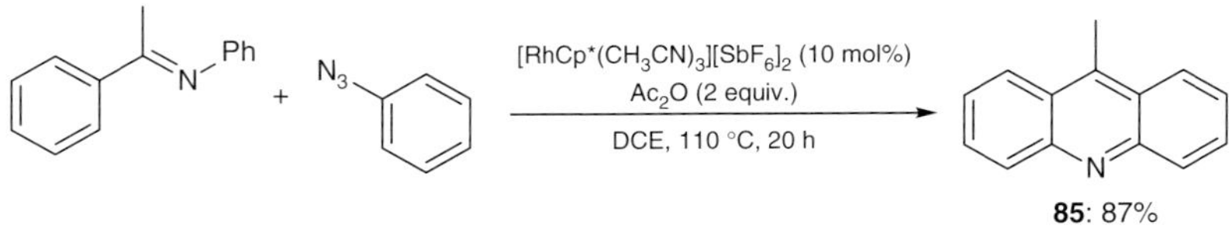

(5.82)

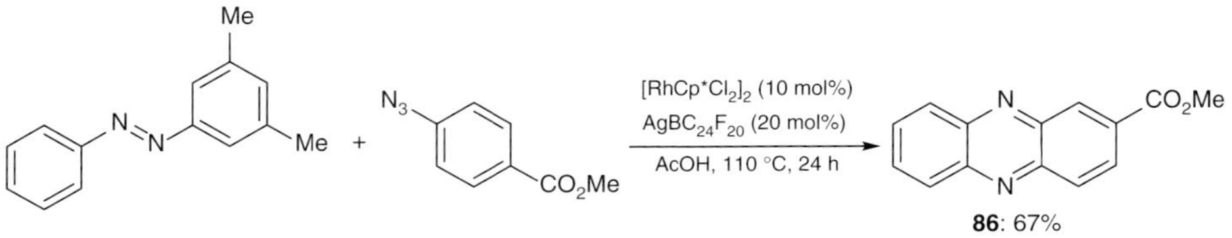

(5.83)

5.3.10 Synthesis of Cinnolines

You *et al.* developed a simple and convenient method for the synthesis of cinnolines **87** by Rh-catalyzed oxidative C–H activation/cyclization of azo compounds with alkynes [45]. In this reaction, *t*-Bu acts as a leaving group and also promotes the C–N reductive elimination; however, the reaction works only for aliphatic internal alkynes with moderate yield (Eq. (5.84)).

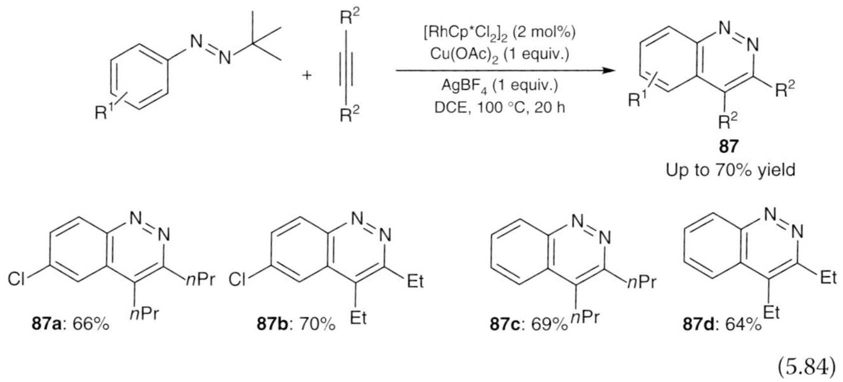

(5.84)

5.3.11
Synthesis of Isoquinolinones and Cinnolinones

The insertion of a Rh–C bond into alkynes or alkenes to give seven-membered metal intermediates has been extensively investigated, but reports of analogous Rh–C insertion with carbenoids to afford six-membered rhodacycles are rare in the literature. The reaction is started by electrophilic C–H activation, followed by coupling of the arylrhodium(III) complex with the diazomalonate. The reaction features a new route to α-aryl carbonyl compounds for specific N-heterocycles, as illustrated in Eqs. (5.85)–(5.87) [46a,b].

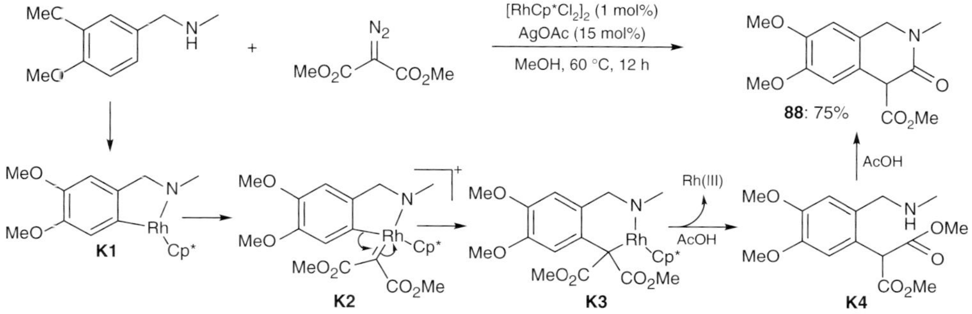

(5.85)

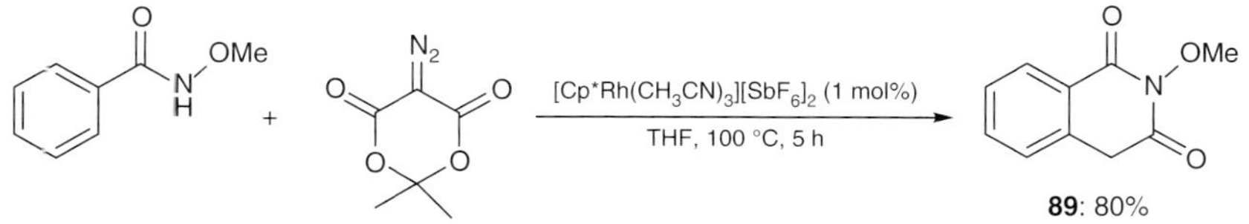

(5.86)

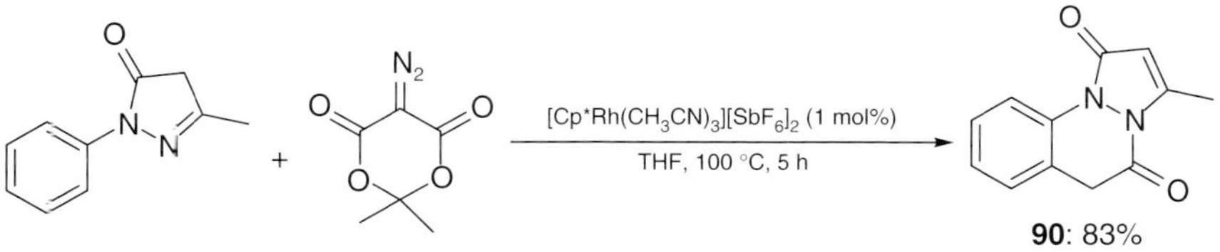

(5.87)

5.3.12
Synthesis of Dihydropyridines

An efficient Rh-catalyzed C–H alkenylation and electrocyclization sequence that provides easy access to highly substituted dihydropyridines **91** from α,β-unsaturated imines and alkynes was reported by Ellman (Eq. (5.88)) [47]. The alkyne scope was good: nonsymmetric alkynes gave the dihydropyridine products as single regioisomers in most cases. The 1,2-dihydropyridines can be oxidized efficiently to pyridines.

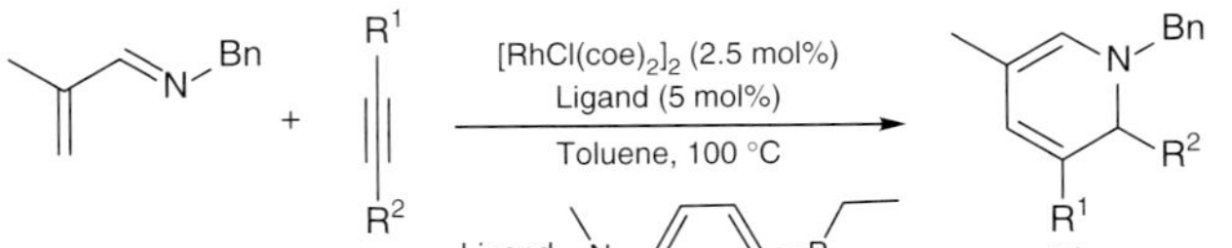

91

Up to 98% yield

91a: R^1, R^2 = Et, 98%
91b: R^1 = Me, R^2 = *i*Pr, 94%
91c: R^1 = Bn, R^2 = TMS, 81%
91d: R^1 = *n*Pr, R^2 = TMS, 50% (5.88)

5.3.13
Synthesis of Pyridines

In 2008, our group disclosed a novel method that involved the one-pot synthesis of multisubstituted pyridines **92** by Rh-catalyzed oxime-assisted alkenyl C–H bond functionalization of α,β-unsaturated oximes with alkynes [48]. The scope includes a variety of α,β-unsaturated oximes and symmetrical alkynes. This report is one of a few examples of Rh(I)-catalyzed alkenyl C–H bond functionalization. The mechanism is thought to occur by oxime-directed oxidative insertion of the Rh catalyst into the alkenyl C–H bond to form the hydrometalacycle **L1**. Hydrorhodation onto the alkyne then occurs to give **L2**, followed by reductive elimination to provide **L3**. Intermediate **L3** can undergo a 6π-electrocyclization and then a dehydration reaction to form pyridine (Eq. (5.89)). Overall, this Rh(I)-catalyzed reaction is a redox neutral.

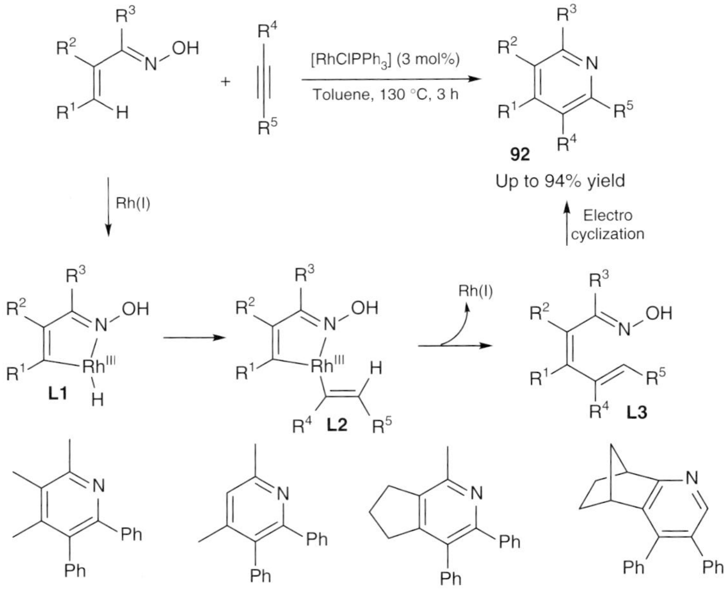

92a: 92% **92b**: 80% **92c**: 94% **92d**: 68% (5.89)

There has been a considerable focus on the development of reliable conditions for Rh-catalyzed C–N bond formation in the absence of metal oxidants. The first example of Rh(III)-catalyzed oxidative annulation reactions of α,β-unsaturated oximes and alkynes was described by Rovis in 2011 for the construction of pyridine derivatives (Eq. (5.90)) [49a]. Equations (5.91)–(5.96) illustrate subsequent modifications of these reactions for the synthesis of various substituted pyridines in good yields. Catalytic amounts of Rh, alkyne/alkene substrates, and a stoichiometric amount of a cooxidant, such as AgOAc or $Cu(OAc)_2 \cdot H_2O$, were required to facilitate catalyst turnover [49b–g].

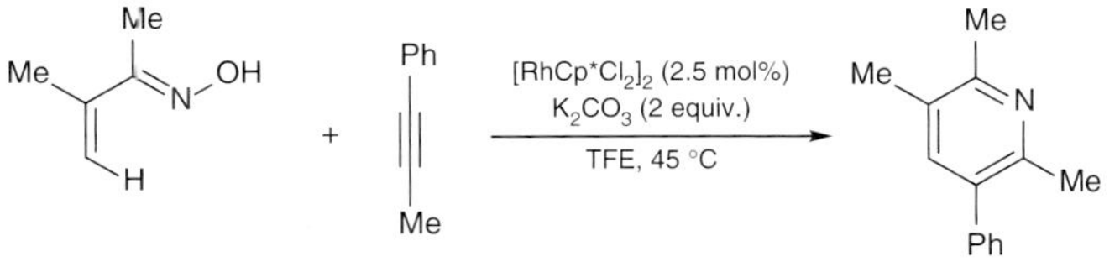

93: 87%, (2:1) (5.90)

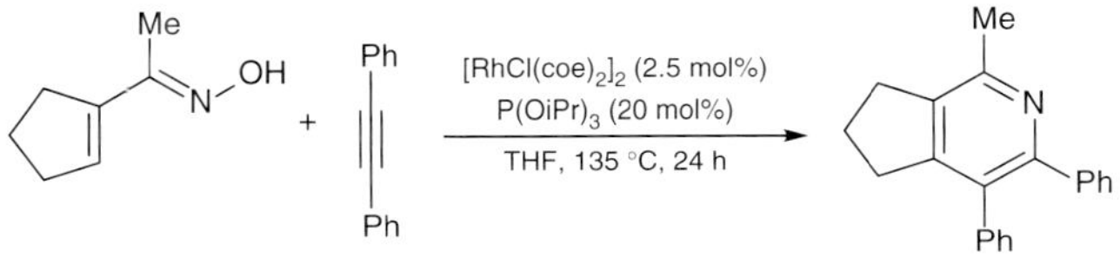

94: 79% (5.91)

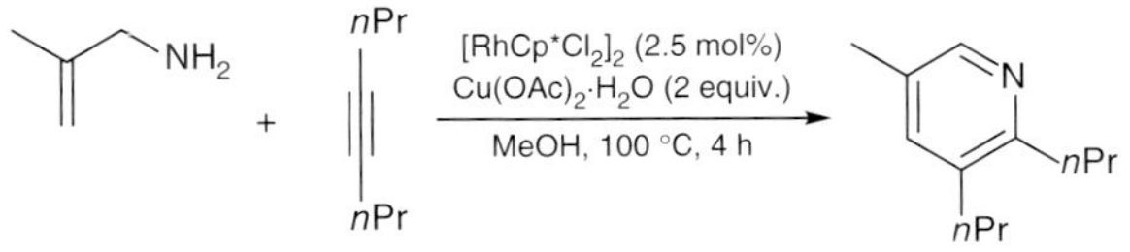

95: 92% (5.92)

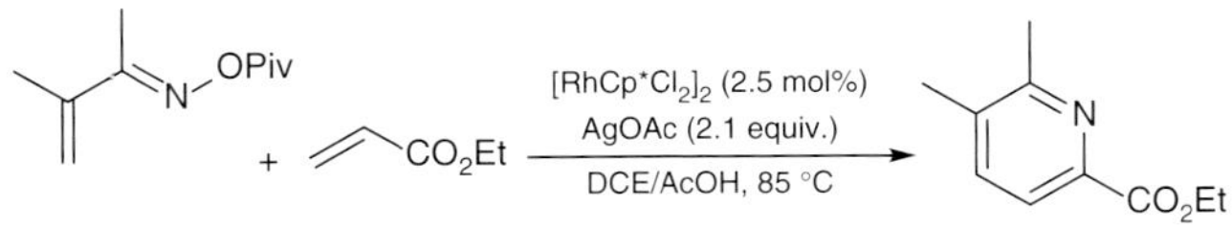

96: 92% (5.93)

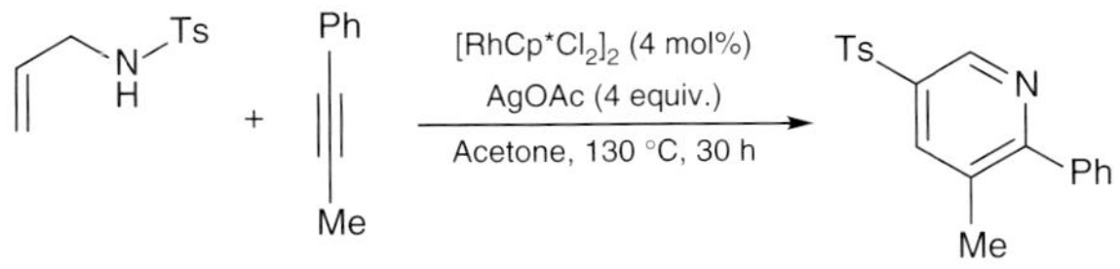

97: 81% (5.94)

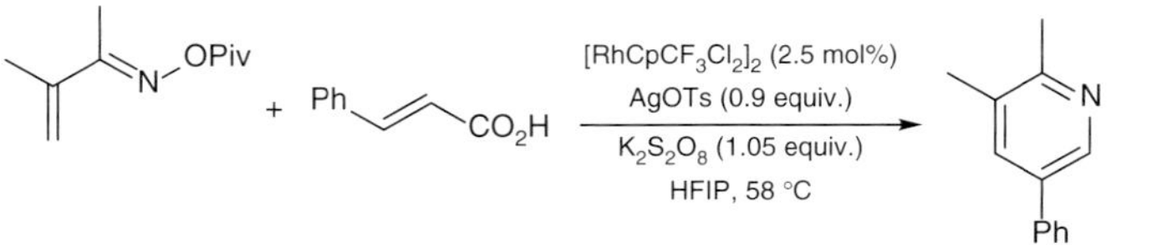

98: 75%, ≥20:1 (5.95)

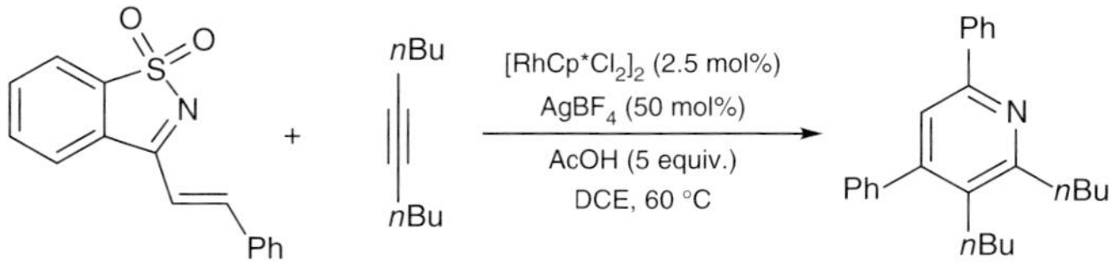

99: 68% (5.96)

5.3.14
Synthesis of Pyridones

Song, Li, and coworkers reported a Rh-catalyzed oxidative coupling of acrylamides with alkynes to generate 2-pyridones **100** (Eq. (5.97)) [50a]. Both diaryl- or aryl- and alkyl-substituted alkynes were suitable coupling partners for this reaction. The nature of the substituents plays a vital role in this reaction. When the steric hindrance around the olefin or *N*-substituent was increased, the reaction afforded *N*-acyl-indoles as the annulation products. Similarly, Hyster and Rovis reported an improved method for Rh-catalyzed highly regioselective synthesis of pyridones from the annulation of acrylamides and unsymmetrical alkynes [50b].

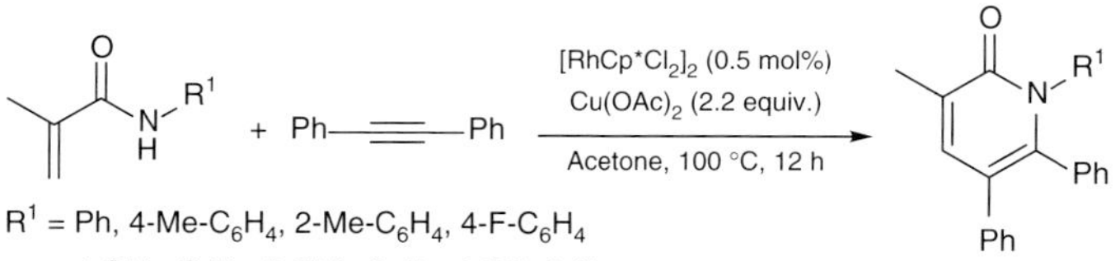

R^1 = Ph, 4-Me-C_6H_4, 2-Me-C_6H_4, 4-F-C_6H_4
4-OMe-C_6H_4, 3-OMe-C_6H_4, 4-CF_3-C_6H_4

100
48 to 94% yield (5.97)

5.3.15
Synthesis of Pyrimidinones

Cui and coworkers reported a Rh(III)-catalyzed C–H activation/cyclization of indoles and pyrroles at the C2 position with various alkyne and alkene substrates to give pyrimidin-2(1*H*)-ones and 3,4-dihydropyrimidin-2(1*H*)-ones **101** (Eq. (5.98)) [51]. The reaction tolerates a wide range of substituents on the aromatic rings. Moreover, both symmetrical and unsymmetrical alkynes/alkenes with various functional groups, such as hydroxyl, nitrile, and ester, also react efficiently to provide the expected *N*-heterocycles in good to excellent yields.

It is noteworthy that this protocol was applied to a concise synthesis of 5-HT_3-receptor antagonist **103** on a gram scale (Eq. (5.98)).

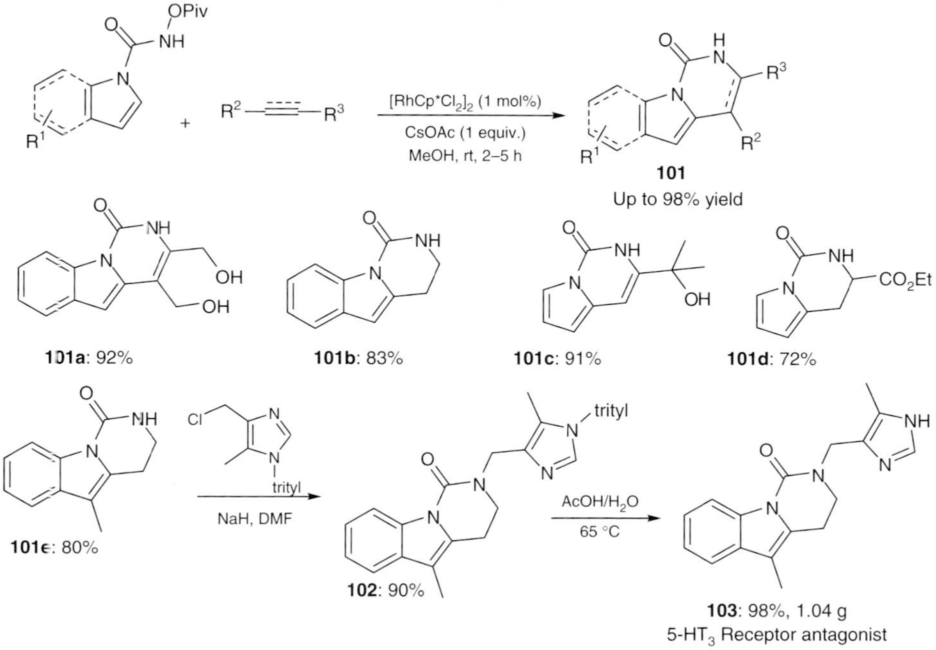

(5.98)

5.4 Synthesis of Quaternary Ammonium Salts

5.4.1 Synthesis of Isoquinolinium Salts

In 2008, Jones *et al.* disclosed a stoichiometric Rh(III)-mediated reaction of *N*-benzylidenemethylamine with dimethyl acetylenedicarboxylate to give the corresponding isoquinolinium salt [52a]. Inspired by this work, in 2012 we developed a Rh(III)-catalyzed one-pot synthesis of isoquinolinium salts **104** from benzaldehydes, primary amines, and internal alkynes by C–H activation and annulation [52b]. This was the first report for the synthesis of isoquinolinium salts by catalytic C–H activation. It is noteworthy that the current procedure was successfully applied to the total synthesis of isoquinolinone alkaloid oxychelerythrine **105** (Eq. (5.99)). In 2013, our group and Huang independently found that various isoquinolinium salts could be synthesized from aryl ketimines, 2-phenyl pyridines, and alkynes under similar reaction conditions, as shown in Eqs. (5.100)–(5.102) [52c–e].

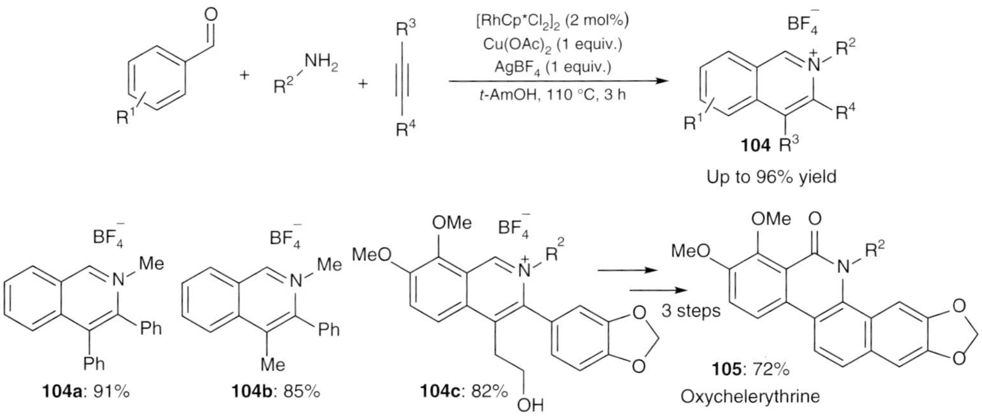

(5.99)

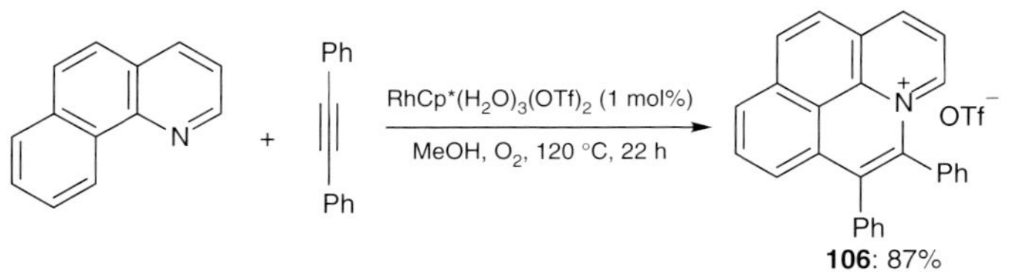

(5.100)

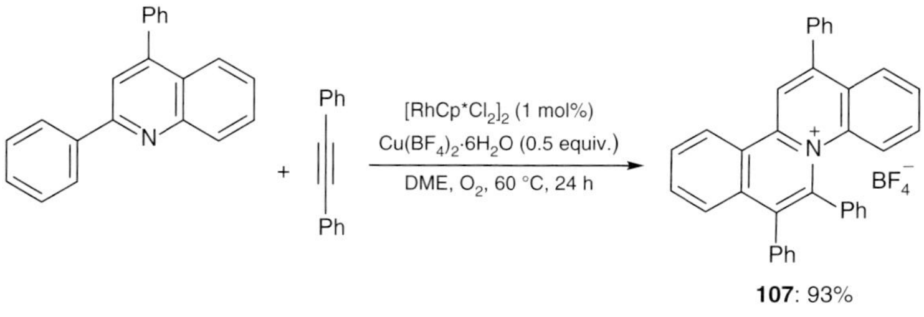

(5.101)

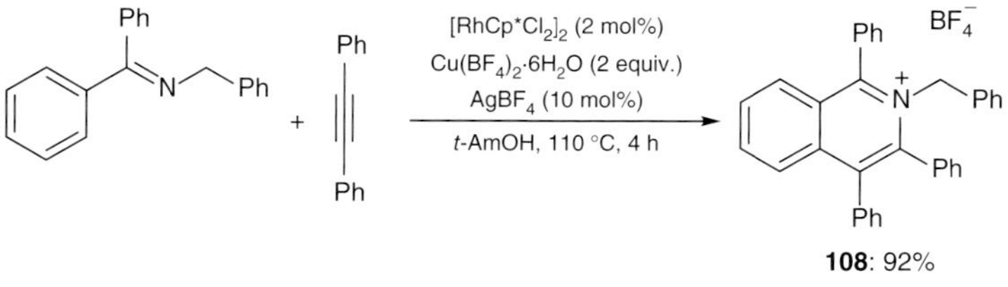

(5.102)

5.4.2
Synthesis of Quinolizinium and Pyridinium Salts

We also developed an interesting method for the synthesis of highly substituted quinolizinium salts in high yields with excellent functional group tolerance (Eq. (5.103)) [53a]. The reaction conditions can also be extended to synthesize pyridinium salts by a three-component reaction of ketones (Eq. (5.104))/aldehydes (Eq. (5.105)), amines, and alkynes [53b]. In this reaction, the *in situ*-generated imine acts as a directing group for vinylic C–H bond activation, and the presence of a counteranion was helpful to isolate quaternary ammonium salts. Unsymmetrical alkynes also proceed via [4+2] cyclization to form the corresponding pyridinium salts in good regioselectivity.

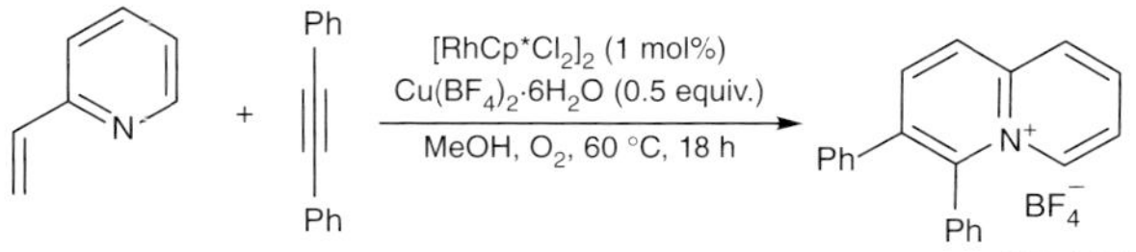

109: 91% (5.103)

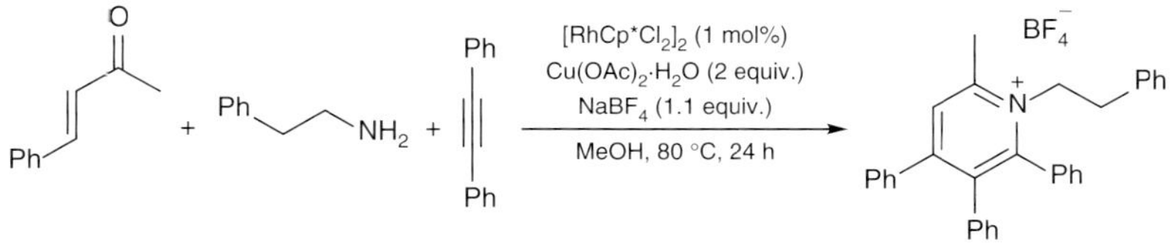

110: 90%

(5.104)

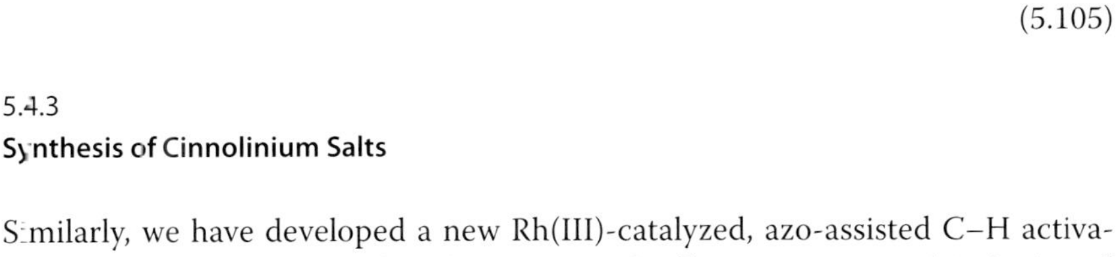

111: 85%

(5.105)

5.4.3
Synthesis of Cinnolinium Salts

Similarly, we have developed a new Rh(III)-catalyzed, azo-assisted C–H activation/annulation reaction of azobenzenes with alkynes to give multisubstituted cinnolinium salts **112** in good to excellent yields [54]. The reaction is compatible with various functional groups and proceeds under mild reaction conditions using air as the cooxidant. These cinnolinium salts have been further modified into three different *N*-heterocycles: indoles, indoloindoles, and cinnolines (Eq. (5.106)).

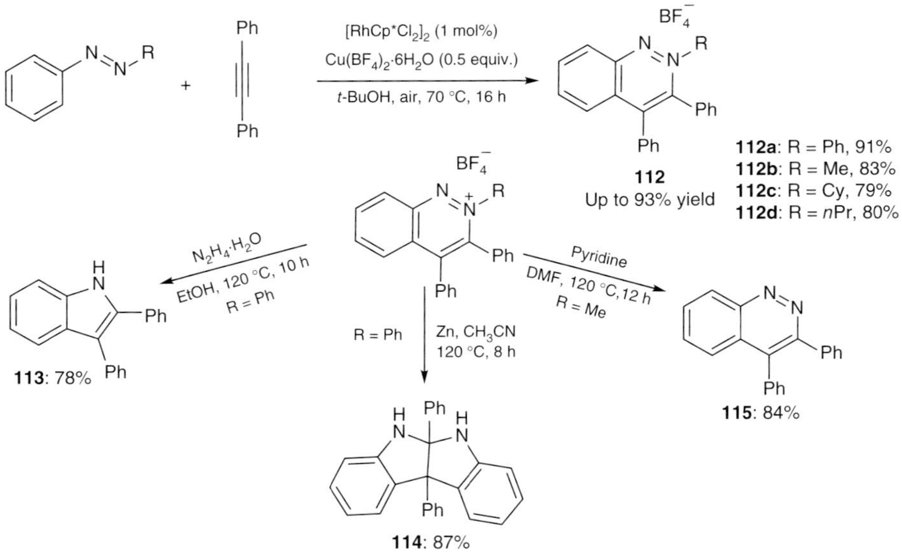

(5.106)

5.4.4
Synthesis of Isoquinoline *N*-Oxides and Pyridine *N*-Oxides

Recently, Glorius discovered a Rh(III)-catalyzed coupling of oximes with electron-deficient diazo compounds to give *N*-oxides of isoquinoline and pyridine [55]. The reaction proceeds with the liberation of H_2O and N_2 as by-products. In this reaction, the insertion of Rh–carbene into *ortho*-C–H bond followed by intramolecular dehydrative annulation plausibly accounts for the generation of *N*-oxides (Eqs. (5.107) and (5.108)). The diazo compounds functioned as equivalents of electron-deficient alkynes, enriching the product diversity of isoquinoline and pyridine synthesis.

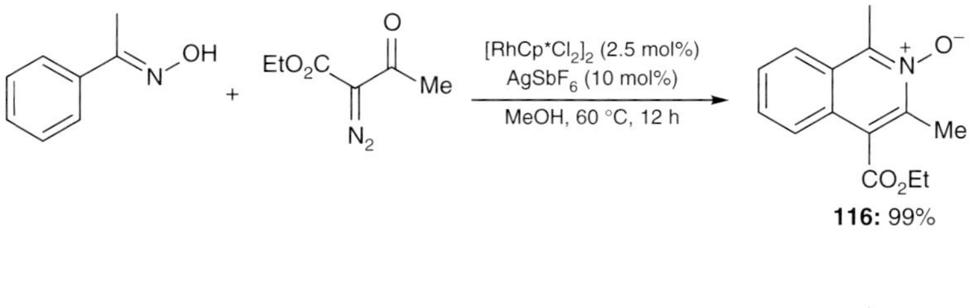

116: 99% (5.107)

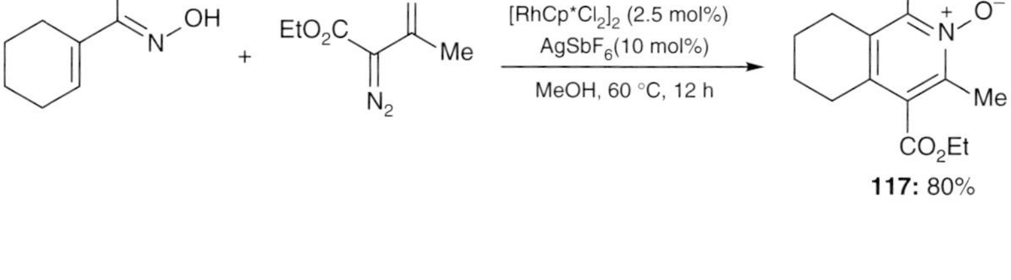

117: 80% (5.108)

5.5 Synthesis of Seven-Membered Nitrogen Heterocycles

5.5.1 Synthesis of Azepinones

A new Rh(III)-catalyzed intermolecular annulation of *N*-methoxybenzamides with *α*,*β*-unsaturated aldehydes or ketones to afford azepinones **118** was reported by Glorius *et al.* (Eq. (5.109)) [56a]. The reaction involves a tandem C–H activation, cyclization, and condensation sequence. The substrates with other *N*-substituents, such as H, phenyl, and OPiv, did not afford the corresponding products. Vinylcarbenoids were also used as the three-carbon component in the Rh(III)-catalyzed C–H activation and [4+3] cycloaddition with benzamides to access azepinones (Eq. (5.110)) [56b].

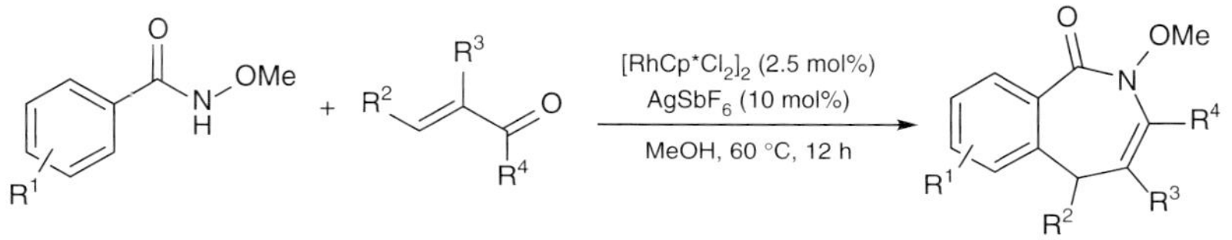

118
Up to 83% yield

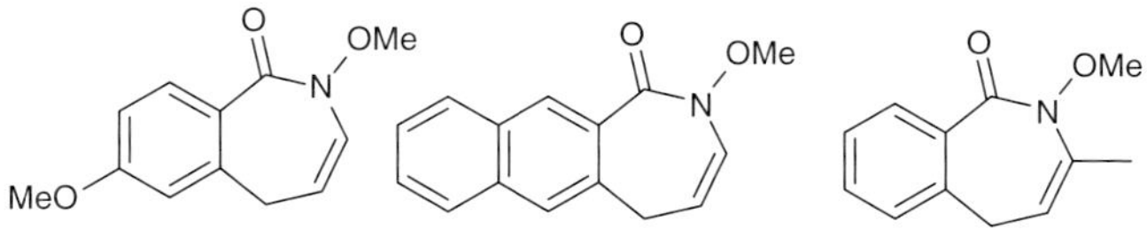

118a: 78% **118b**: 88% **118c**: 63%

(5.109)

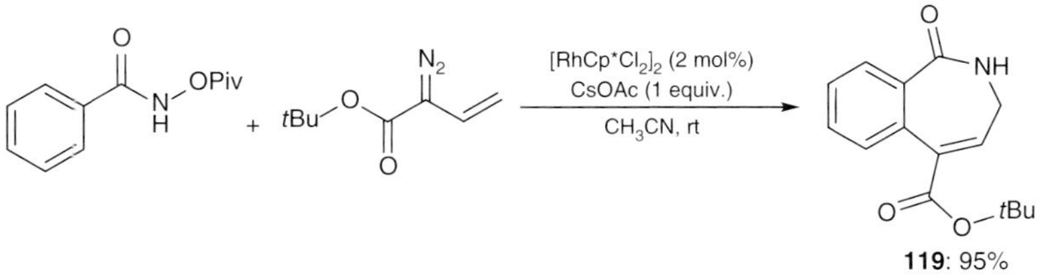

119: 95%

(5.110)

5.5.2 Synthesis of 1,2-Oxazepines

Chen and Zhao reported an interesting coupling reaction of *N*-phenoxyacetamides with *α*,*β*-unsaturated aldehydes to give 1,2-oxazepines **120** by C–H activation and [4+3] annulation (Eq. (5.111)) [57]. This simple [4+3] annulation method features mild reaction conditions and high product yields for various substituted

N-phenoxyacetamides and α,β-unsaturated aldehydes. However, use of cinnamaldehyde or α-methyl unsaturated aldehyde substrates resulted in no reaction. Finally, product **120d** underwent a reduction reaction under a H_2 atmosphere to afford the unusual benzofused chroman derivative **121** in 80% yield.

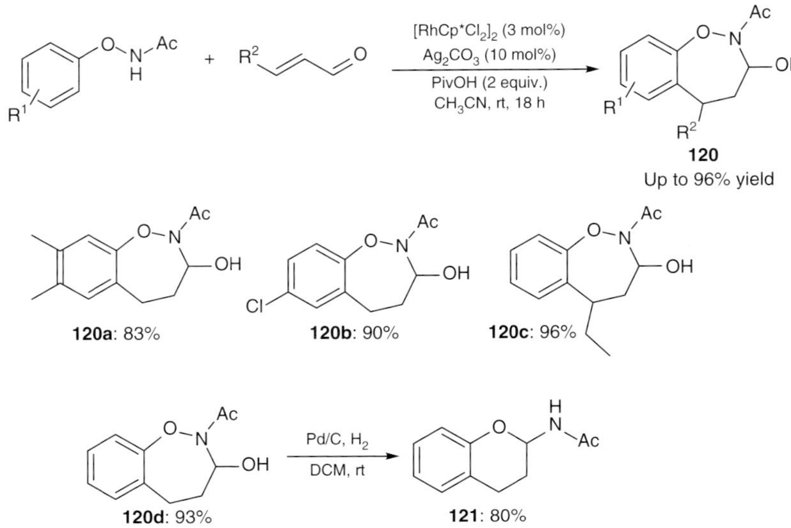

(5.111)

5.6
Summary and Conclusions

In recent years, research in the field of Rh catalysis has resulted in the development of numerous transformations that provide various nitrogen heterocycles. Most of the transformations result in the formation of a C–N bond with good to excellent yields, in addition to good stereocontrol. Despite the many developments that have been made in this arena, new findings are being reported with greater frequency. This research area will potentially remain a fruitful discipline for many years to come. In particular, the development of new Rh catalysts and Rh-catalyzed reactions will likely take a lead role, with more attention on the scope and application of the transformations. In addition, studies on the catalytic reaction mechanisms to understand the nature of the reactions will continue for some period.

References

1. For reviews on biologically active nitrogen heterocycles, see: (a) Kang, E.J. and Lee, E. (2005) *Chem. Rev.*, **105**, 4348–4378; (b) Saleem, M., Kim, H.J., Ali, M.S., and Lee, Y.S. (2005) *Nat. Prod. Rep.*, **22**, 696–716; (c) Bermejo,

A., Figadere, B., Zafra-Polo, M.-C., Barrachina, I., Estornell, E., and Cortes, D. (2005) *Nat. Prod. Rep.*, **22**, 269–309; (d) Bellina, F. and Rossi, R. (2006) *Tetrahedron*, **62**, 7213–7256; (e) Coldham, I. and Hufton, R. (2005) *Chem. Rev.*, **105**, 2765–2810; (f) Daugan, A., Grondin, P., Ruault, C., Gouville, A.D., Coste, H., Linget, J., Kirilovsky, J., Hyafil, F., and Labaudiniere, R. (2003) *J. Med. Chem.*, **46**, 4533–4542.

2. (a) Suzuki, A. (1998) in *Metal-Catalyzed Cross-Coupling Reactions* (eds F. Diederich and P.J. Stang), Wiley-VCH Verlag GmbH, New York, pp. 49–97; (b) Johansson Seechurn, C.C., Kitching, M.O., Colacot, T.J., and Snieckus, V. (2012) *Angew. Chem. Int. Ed.*, **51**, 5062–5086; (c) Yamaguchi, J., Yamaguchi, A.D., and Itami, K. (2012) *Angew. Chem. Int. Ed.*, **51**, 2–52; (d) Schmidt, A.W., Reddy, K.R., and Knolker, H.-J. (2012) *Chem. Rev.*, **112**, 3193–3328; (e) Chattopadhyay, B. and Gevorgyan, V. (2012) *Angew. Chem. Int. Ed.*, **51**, 862–872; (f) Qiu, G., Ding, Q., and Wu, J. (2013) *Chem. Soc. Rev.*, **42**, 5257–5269; (g) Suzuki, A. (1999) *J. Organomet. Chem.*, **576**, 147–168; (h) Han, F.-S. (2013) *Chem. Soc. Rev.*, **42**, 5270–5298.

3. (a) Alberico, D., Scott, M.E., and Lautens, M. (2007) *Chem. Rev.*, **107**, 174–238; (b) Lewis, J.C., Bergman, R.E., and Ellman, J.A. (2008) *Acc. Chem. Res.*, **41**, 1013–1025; (c) Campos, K.R. (2007) *Chem. Soc. Rev.*, **36**, 1069–1084; (d) Campeau, L.-C., Stuart, D.R., and Fagnou, K. (2007) *Aldrichim. Acta*, **40**, 35–41; (e) Song, G., Wang, F., and Li, X. (2012) *Chem. Soc. Rev.*, **41**, 3651–3678; (f) Cho, S.H., Kim, J.Y., Kwak, J., and Chang, S. (2011) *Chem. Soc. Rev.*, **40**, 5068–5083; (g) Sun, C.-L., Li, B.-J., and Shi, Z.-J. (2010) *Chem. Commun.*, **46**, 677–685; (h) Chen, X., Engle, K.M., Wang, D.-H., and Yu, J.-Q. (2009) *Angew. Chem. Int. Ed.*, **48**, 5094–5115; (i) Thansandote, P. and Lautens, M. (2009) *Chem. Eur. J.*, **15**, 5874–5883; (j) Kakiuchi, F. (2007) *Top. Organomet. Chem.*, **24**, 1–33; (k) Ackermann, L. (2007) *Top. Organomet. Chem.*, **24**, 35–60; (l) Satoh, T. and Miura, M. (2007) *Top. Organomet. Chem.*, **24**, 61–84; (m) Kalyani, D. and Sanford, M.S. (2007) *Top. Organomet. Chem.*, **24**, 85–116.

4. (a) Satoh, T. and Miura, M. (2010) *Chem. Eur. J.*, **16**, 11212–11222; (b) Patureau, F.W., Wencel-Delord, J., and Glorius, F. (2012) *Aldrichim. Acta*, **45**, 31–55; (c) Wu, Y., Wang, J., Mao, F., and Kwong, F.Y. (2014) *Chem. Asian J.*, **9**, 26–47; (d) Delord, J.W., Dröge, T., Liu, F., and Glorius, F. (2011) *Chem. Soc. Rev.*, **40**, 4740–4761; (e) Colby, D.A., Bergman, R.G., and Ellman, J.A. (2010) *Chem. Rev.*, **110**, 624–655; (f) Zhu, C., Wang, R., and Falck, J.R. (2012) *Chem. Asian J.*, **7**, 1502–1514; (g) Kuhl, N., Schroder, N., and Glorius, F. (2014) *Adv. Synth. Catal.*, **356**, 1443–1460; (h) He, R., Huang, Z.-T., Zheng, Q.-Y., and Wang, C. (2014) *Tetrahedron Lett.*, **55**, 5705–5713.

5. (a) Stuart, D.R., Bertrand-Laperle, M., Burgess, K.M.N., and Fagnou, K. (2008) *J. Am. Chem. Soc.*, **130**, 16474–16475; (b) Stuart, D.R., Alsabeh, P., Kuhn, M., and Fagnou, K. (2010) *J. Am. Chem. Soc.*, **132**, 18326–18339; (c) Huestis, M.P., Chan, L., Stuart, D.R., and Fagnou, K. (2011) *Angew. Chem. Int. Ed.*, **50**, 1338–1341.

6. (a) Zhang, X., Li, Y., Shi, H., Zhang, L., Zhang, S., Xu, X., and Liu, Q. (2014) *Chem. Commun.*, **50**, 7306–7309; (b) Taoa, P. and Jia, Y. (2014) *Chem. Commun.*, **50**, 7367–7370; (c) Zhou, B., Yang, Y., Tang, H., Du, J., Feng, H., and Li, Y. (2014) *Org. Lett.*, **16**, 3900–3903; (d) Cajaraville, A., López, S., Varela, J.A., and Saá, C. (2013) *Org. Lett.*, **15**, 4576–4579.

7. (a) Wang, C. and Huang, Y. (2013) *Org. Lett.*, **15**, 5294–5297; (b) Liu, B., Song, C., Sun, C., Zhou, S., and Zhu, J. (2013) *J. Am. Chem. Soc.*, **135**, 16625–16631; (c) Fan, Z., Song, S., Li, W., Geng, K., Xu, Y., Miao, Z.-H., and Zhang, A. (2015) *Org. Lett.*, **17**, 310–313; (d) Chen, J., Song, G., Pan, C.-L., and Li, X. (2010) *Org. Lett.*, **12**, 5426–5429; (e) Li, D.Y., Chen, H.J., and Liu, P.N. (2014) *Org. Lett.*, **16**, 6176–6179.

8. Mishra, N.K., Park, J., Sharma, S., Han, S., Kim, M., Shin, Y., Jang, J., Kwak, J.H.,

Jung, Y.H., and Kim, I.S. (2014) *Chem. Commun.*, **50**, 2350–2352.
9. Suzuki, C., Morimoto, K., Hirano, K., Satoh, T., and Miura, M. (2014) *Adv. Synth. Catal.*, **356**, 1521–1526.
10. (a) Wang, C., Sun, H., Fang, Y., and Huang, Y. (2013) *Angew. Chem. Int. Ed.*, **52**, 5795–5798; (b) Zhang, G., Yu, H., Qin, G., and Huang, H. (2014) *Chem. Commun.*, **50**, 4331–4334.
11. (a) Zhao, D., Shi, Z., and Glorius, F. (2013) *Angew. Chem. Int. Ed.*, **52**, 12426–12429; (b) Muralirajan, K. and Cheng, C.-H. (2014) *Adv. Synth. Catal.*, **356**, 1571–1576.
12. (a) Huang, L., Wang, Q., Qi, J., Wu, X., Huang, K., and Jiang, H. (2013) *Chem. Sci.*, **4**, 2665–2669; (b) Zhao, D., Vasquez-Cespedes, S., and Glorius, F. (2015) *Angew. Chem. Int. Ed.*, **54**, 1657–1661.
13. Lian, Y., Bergman, R.G., Lavis, L.D., and Ellman, J.A. (2013) *J. Am. Chem. Soc.*, **135**, 7122–7125.
14. (a) Yu, D.-G., Suri, M., and Glorius, F. (2013) *J. Am. Chem. Soc.*, **135**, 8802–8805; (b) Han, S., Shin, Y., Sharma, S., Mishra, N.K., Park, J., Kim, M., Kim, M., Jang, J., and Kim, I.S. (2014) *Org. Lett.*, **16**, 2494–2497.
15. Yang, W., Sun, J., Xu, X., Zhang, Q., and Liu, Q. (2014) *Chem. Commun.*, **50**, 4420–4422.
16. Rakshit, S., Patureau, F.W., and Glorius, F. (2010) *J. Am. Chem. Soc.*, **132**, 9585–9587.
17. Wang, H., Grohmann, C., Nimphius, C., and Glorius, F. (2012) *J. Am. Chem. Soc.*, **134**, 19592–19595.
18. (a) Lian, Y., Huber, T., Hesp, K.D., Bergman, R.G., and Ellman, J.A. (2013) *Angew. Chem. Int. Ed.*, **52**, 629–633; (b) Du, J., Zhou, B., Yang, Y., and Li, Y. (2013) *Chem. Asian J.*, **8**, 1386–1390; (c) Cochet, T., Bellosta, V., Roche, D., Ortholand, J.-Y., Greiner, A., and Cossy, J. (2012) *Chem. Commun.*, **48**, 10745–10747.
19. (a) Patureau, F.W., Besset, T., and Glorius, F. (2011) *Angew. Chem. Int. Ed.*, **50**, 1064–1067; (b) Zhou, B., Hou, W., Yang, Y., and Li, Y. (2013) *Chem. Eur. J.*, **19**, 4701–4706.
20. Sharma, S., Park, E., Park, J., and Kim, I.S. (2012) *Org. Lett.*, **14**, 906–909.
21. Zhu, C., Xie, W., and Falck, J.R. (2011) *Chem. Eur. J.*, **17**, 12591–12595.
22. (a) Wang, F., Song, G., and Li, X. (2010) *Org. Lett.*, **12**, 5430–5433; (b) Zhu, C. and Falck, J.R. (2012) *Chem. Commun.*, **48**, 1674–1676; (c) Hou, W., Zhou, B., Yang, Y., Feng, H., and Li, Y. (2013) *Org. Lett.*, **15**, 1814–1817; (d) Cai, S., Chen, C., Shao, P., and Xi, C. (2014) *Org. Lett.*, **16**, 3142–3145; (e) Martınez, A.M., Rodriguez, N., Arrayas, R.G., and Carretero, J.C. (2014) *Chem. Commun.*, **50**, 6105–6107; (f) Hyster, T.K., Ruhl, K.E., and Rovis, T. (2013) *J. Am. Chem. Soc.*, **135**, 5364–5367.
23. Qi, Z. and Li, X. (2013) *Angew. Chem. Int. Ed.*, **52**, 8995–9000.
24. Huang, J.-R., Song, Q., Zhu, Y.-Q., Qin, L., Qian, Z.-Y., and Dong, L. (2014) *Chem. Eur. J.*, **20**, 16882–16886.
25. Xie, W., Yang, J., Wang, B., and Li, B. (2014) *J. Org. Chem.*, **79**, 8278–8287.
26. (a) Du, Y., Hyster, T.K., and Rovis, T. (2011) *Chem. Commun.*, **47**, 12074–12076; (b) Shi, X.-Y., Renzetti, A., Kundu, S., and Li, C.-J. (2014) *Adv. Synth. Catal.*, **356**, 723–728.
27. Lim, S.-G., Lee, J.H., Moon, C.W., Hong, C.J.-B., and Jun, C.-H. (2003) *Org. Lett.*, **5**, 2759–2761.
28. (a) Guimond, N. and Fagnou, K. (2009) *J. Am. Chem. Soc.*, **131**, 12050–12051; (b) Fukutani, T., Umeda, N., Hirano, K., Satoh, T., and Miura, M. (2009) *Chem. Commun.*, 5141–5143.
29. (a) Parthasarathy, K. and Cheng, C.-H. (2009) *J. Org. Chem.*, **74**, 9359–9364; (b) Too, P.C., Wang, Y.-F., and Chiba, S. (2010) *Org. Lett.*, **12**, 5688–5691; (c) Zhang, X., Che, D., Zhao, M., Zhao, J., Jia, A., and Li, X. (2011) *Adv. Synth. Catal.*, **353**, 719–723; (d) Morimoto, K., Hirano, K., Satoh, T., and Miura, M. (2010) *Org. Lett.*, **12**, 2068–2071; (e) Wang, Y.-F., Toh, K.K., Lee, J.-Y., and Chiba, S. (2011) *Angew. Chem. Int. Ed.*, **50**, 5927–5931; (f) Zheng, L., Ju, J., Bin, Y., and Hua, R. (2012) *J. Org. Chem.*, **77**, 5794–5800; (g) Kim, D.-S., Park, J.-W., and Jun, C.-H. (2013) *Adv. Synth. Catal.*, **355**, 2667–2679; (h) Chuang, S.-C., Gandeepan, P., and Cheng, C.-H. (2013)

Org. Lett., **15**, 5750–5753; (i) Han, W., Zhang, G., Li, G., and Huang, H. (2014) *Org. Lett.*, **16**, 3532–3535.

30. Wei, X., Zhao, M., Du, Z., and Li, X. (2011) *Org. Lett.*, **13**, 4636–4639.
31. Zhao, D., Lied, F., and Glorius, F. (2014) *Chem. Sci.*, **5**, 2869–2873.
32. Guimond, N., Gouliaras, C., and Fagnou, K. (2010) *J. Am. Chem. Soc.*, **132**, 6908–6909.
33. Song, G., Chen, D., Pan, C.-L., Crabtree, R.H., and Li, X. (2010) *J. Org. Chem.*, **75**, 7487–7490.
34. Guimond, N., Gorelsky, S.I., and Fagnou, K. (2011) *J. Am. Chem. Soc.*, **133**, 6449–6457.
35. (a) Rakshit, S., Grohmann, C., Besset, T., and Glorius, F. (2011) *J. Am. Chem. Soc.*, **133**, 2350–2353; (b) Hyster, T.K., Knörr, L., Ward, T.R., and Rovis, T. (2012) *Science*, **338**, 500–503; (c) Ye, B. and Cramer, N. (2012) *Science*, **338**, 504–506; (d) Wang, H. and Glorius, F. (2012) *Angew. Chem. Int. Ed.*, **51**, 7318–7322; (e) Presset, M., Oehlrich, D., Rombouts, F., and Molander, G.A. (2013) *Org. Lett.*, **15**, 1528–1531; (f) Huckins, J.R., Bercot, E.A., Thiel, O.R., Hwang, T.-L., and Bio, M.M. (2013) *J. Am. Chem. Soc.*, **135**, 14492–14495; (g) Prakash, P., Jijy, E., Aparna, P.S., Viji, S., and Radhakrishnan, K.V. (2014) *Tetrahedron Lett.*, **55**, 916–920; (h) Zhang, S.-S., Wu, J.-Q., Liu, X., and Wang, H. (2015) *ACS Catal.*, **5**, 210–214; (i) Shi, Z., Boultadakis-Arapinis, M., Koester, D.C., and Glorius, F. (2014) *Chem. Commun.*, **50**, 2650–2652.
36. (a) Xu, X., Liu, Y., and Park, C.-M. (2012) *Angew. Chem. Int. Ed.*, **51**, 9372–9376; (b) Quinones, N., Seoane, A., Garcıa-Fandino, R., Mascarenas, J.L., and Gulias, M. (2013) *Chem. Sci.*, **4**, 2874–2879.
37. Webb, N.J., Marsden, S.P., and Raw, S.A. (2014) *Org. Lett.*, **16**, 4718–4721.
38. Yu, D.-G., de Azambuja, F., and Glorius, F. (2014) *Angew. Chem. Int. Ed.*, **53**, 2754–2758.
39. Fukui, Y., Liu, P., Liu, Q., He, Z.-T., Wu, N.-Y., Tian, P., and Lin, G.-Q. (2014) *J. Am. Chem. Soc.*, **136**, 15607–15614.
40. (a) Karthikeyan, J., Haridharan, R., and Cheng, C.-H. (2012) *Angew. Chem. Int. Ed.*, **51**, 12343–12347; (b) Senthilkumar, N., Parthasarathy, K., Gandeepan, P., and Cheng, C.-H. (2013) *Chem. Asian J.*, **8**, 2175–2181.
41. Zhang, L., Zheng, L., Guo, B., and Hua, R. (2014) *J. Org. Chem.*, **79**, 11541–11548.
42. Jayakumar, J., Parthasarathy, K., Chen, Y.-H., Lee, T.-H., Chuang, S.-C., and Cheng, C.-H. (2014) *Angew. Chem. Int. Ed.*, **53**, 9889–9892.
43. (a) Zhen, W., Wang, F., Zhao, M., Du, Z., and Li, X. (2012) *Angew. Chem. Int. Ed.*, **51**, 11819–11823; (b) Li, G., Ding, Z., and Xu, B. (2012) *Org. Lett.*, **14**, 5338–5341.
44. Lian, Y., Hummel, J.R., Bergman, R.G., and Ellman, J.A. (2013) *J. Am. Chem. Soc.*, **135**, 12548–12551.
45. Zhao, D., Wu, Q., Huang, X., Song, F., Lv, T., and You, J. (2013) *Chem. Eur. J.*, **19**, 6239–6244.
46. (a) Chan, W.-W., Lo, S.-F., Zhou, Z., and Yu, W.-Y. (2012) *J. Am. Chem. Soc.*, **134**, 13565–13568; (b) Shi, J., Zhou, J., Yan, Y., Jia, J., Liu, X., Song, H., Xu, H.E., and Yi, W. (2015) *Chem. Commun.*, **51**, 668–671.
47. Colby, D.A., Bergman, R.G., and Ellman, J.A. (2008) *J. Am. Chem. Soc.*, **130**, 3645–3651.
48. Parthasarathy, K., Jeganmohan, M., and Cheng, C.-H. (2008) *Org. Lett.*, **10**, 325–328.
49. (a) Hyster, T.K. and Rovis, T. (2011) *Chem. Commun.*, **47**, 11846–11848; (b) Neely, J.M. and Rovis, T. (2013) *J. Am. Chem. Soc.*, **135**, 66–69; (c) Martin, R.M., Bergman, R.G., and Ellman, J.A. (2012) *J. Org. Chem.*, **77**, 2501–2507; (d) Kim, D.-S., Park, J.-W., and Jun, C.-H. (2012) *Chem. Commun.*, **48**, 11334–11336; (e) Wang, D., Wang, F., Song, G., and Li, X. (2012) *Angew. Chem. Int. Ed.*, **51**, 12348–12352; (f) Neely, J.M. and Rovis, T. (2014) *J. Am. Chem. Soc.*, **136**, 2735–2738; (g) Zhang, Q.-R., Huang, J.-R., Zhang, W., and Dong, L. (2014) *Org. Lett.*, **16**, 1684–1687.
50. (a) Su, Y., Zhao, M., Han, K., Song, G., and Li, X. (2010) *Org. Lett.*, **12**, 5462–5465; (b) Hyster, T.K. and Rovis, T. (2011) *Chem. Sci.*, **2**, 1606–1610.

51. Zhang, Y., Zheng, J., and Cui, S. (2014) *J. Org. Chem.*, **79**, 6490–6500.
52. (a) Li, L., Brennessel, W.W., and Jones, W.D. (2008) *J. Am. Chem. Soc.*, **130**, 12414–12419; (b) Jayakumar, J., Parthasarathy, K., and Cheng, C.-H. (2012) *Angew. Chem. Int. Ed.*, **51**, 197–200; (c) Senthilkumar, N., Gandeepan, P., Jayakumar, J., and Cheng, C.-H. (2014) *Chem. Commun.*, **50**, 3106–3108; (d) Zhang, G., Yang, L., Wang, Y., Xie, Y., and Huang, H. (2013) *J. Am. Chem. Soc.*, **135**, 8850–8853; (e) Luo, C.-Z., Gandeepan, P., Jayakumar, J., Parthasarathy, K., Chang, Y.-W., and Cheng, C.-H. (2013) *Chem. Eur. J.*, **19**, 14181–14186.
53. (a) Luo, C.-Z., Gandeepan, P., and Cheng, C.-H. (2013) *Chem. Commun.*, **49**, 8528–8530; (b) Luo, C.-Z., Jayakumar, J., Gandeepan, P., Wu, Y.-C., and Cheng, C.-H. (2015) *Org. Lett.*, **17**, 924–927.
54. Muralirajan, K. and Cheng, C.-H. (2013) *Chem. Eur. J.*, **19**, 6198–6202.
55. Shi, Z., Koester, D.C., Boultadakis-Arapinis, M., and Glorius, F. (2013) *J. Am. Chem. Soc.*, **135**, 12204–12207.
56. (a) Shi, Z., Grohmann, C., and Glorius, F. (2013) *Angew. Chem. Int. Ed.*, **52**, 5393–5397; (b) Cui, S., Zhang, Y., Wang, D., and Wu, Q. (2013) *Chem. Sci.*, **4**, 3912–3916.
57. Duan, P., Lan, X., Chen, Y., Qian, S.-S., Li, J.J., Lu, L., Lu, Y., Chen, B., Hong, M., and Zhao, J. (2014) *Chem. Commun.*, **50**, 12135–12138.

6
Rh-Catalyzed Synthesis of Oxygen-Containing Heterocycles

Bin Liu, Fang Hu, and Bing-Feng Shi

6.1
Introduction

Oxygen-containing heterocycles are indispensable structural motifs not only because of their biological relevance but also as valuable synthetic organic building blocks. Due to its unique role in pharmaceuticals, agrochemicals, and other biologically active molecular scaffolds (Scheme 6.1), numerous strategies have been developed for the construction of oxygen-containing heterocycles [1]. From the viewpoint of atom and step economy [2], transition metal-catalyzed C–H activation reactions would be an ideal and promising strategy for the synthesis of these heterocycles [3, 4].

Rh-catalyzed C–H bond functionalization reactions employed both Rh(I) and Rhodium(III) catalysts, which are mechanistically different from each other [3k]. Rhodium(III) catalysts, in particular $[RhCp^*Cl_2]_2$ (Cp^* = pentamethylcyclopentadienyl), have been widely used in mediating the cleavage of $C(sp^2)$–H bonds and the subsequent coupling with unsaturated molecules, such as alkenes, alkynes, and carbonyl compounds [5, 6]. The present chapter aims to describe the recent advances in the synthesis of oxygen-containing heterocycles via Rh-catalyzed C–H functionalization. The examples in each section are organized according to the size of the ring, such as five-, six-, and seven-membered heterocycles. Each section is further divided into subsections according to whether the heterocyclics are prepared via intramolecular cyclization or intermolecular annulation. Some mechanistic studies will also be described.

6.2
Synthesis of Five-Membered Oxygen-Containing Heterocycles

6.2.1
Intermolecular Annulation

6.2.1.1 Phthalides

Phthalides are an important class of oxygen-containing heterocycles found in natural products and exhibit a broad spectrum of biological activities [7].

Transition Metal-Catalyzed Heterocycle Synthesis via C—H Activation, First Edition. Edited by Xiao-Feng Wu.
© 2016 Wiley-VCH Verlag GmbH & Co. KGaA. Published 2016 by Wiley-VCH Verlag GmbH & Co. KGaA.

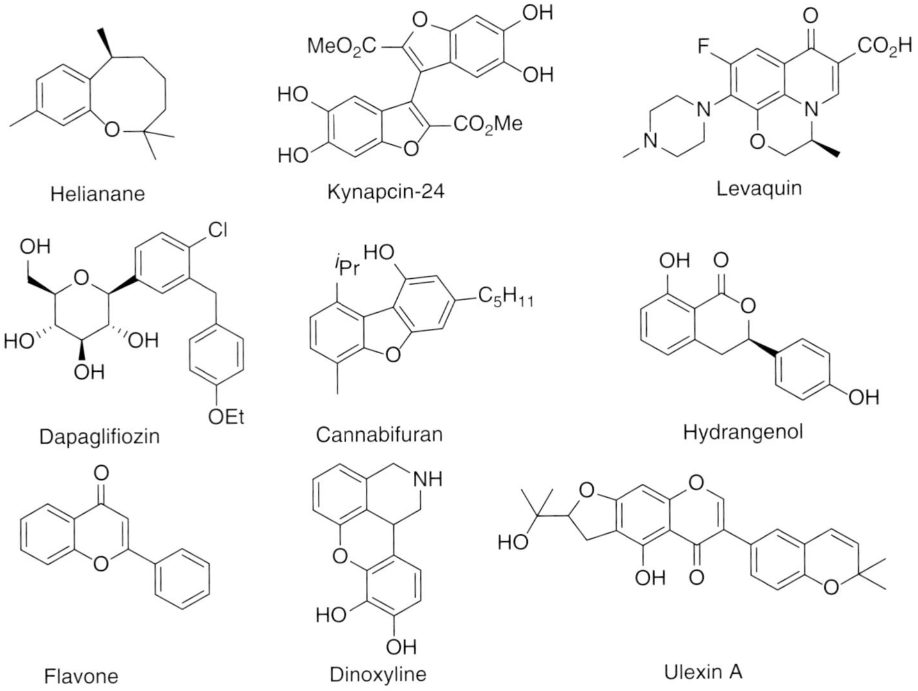

Scheme 6.1 Selected examples of natural products, pharmaceuticals, and biologically active compounds with oxygen-containing heterocycles.

Moreover, they are versatile building blocks for the synthesis of functionalized naphthalenes, anthracenes, and naphthacene natural products [8]. Recently, Rh-catalyzed C–H activation followed by nucleophilic addition to aldehydes emerged as a powerful alternative to access phthalides. In 2012, Li and coworkers developed a novel Rh(III)-catalyzed synthesis of three substituted phthalides from benzoic acids and aldehydes through carboxylate-directed *ortho*-C–H functionalization and subsequent intramolecular cyclization (Scheme 6.2a) [9]. In 2013, Gooßen and coworkers described the straightforward synthesis of 3-alkylidenephthalides from benzoic acids and aliphatic acids or anhydrides in the presence of $[Rh(cod)Cl]_2$ and CsF (Scheme 6.2b) [10].

Ellman and coworkers demonstrated that Rh(III)-catalyzed coupling of benzimidates with aldehydes furnished 3-arylphthalides in one pot with high yields (Scheme 6.3) [11]. The imidate group serves as a directing group for the *ortho*-C–H bond activation as well as cyclization of the resulting alcohol. The reaction was also applicable to aliphatic aldehydes by carefully tuning the substituents on the *N*-atom of imine.

Rh(III)-amine dual catalysis was developed by Seayad and coworkers in 2013 (Scheme 6.4) [12]. Both aryl- and alkyl-substituted phthalides were isolated in medium to high yields in the presence of $[RhCp^*Cl_2]_2$, $AgBF_4$, Ag_2CO_3, and 4-trifluoromethylaniline.

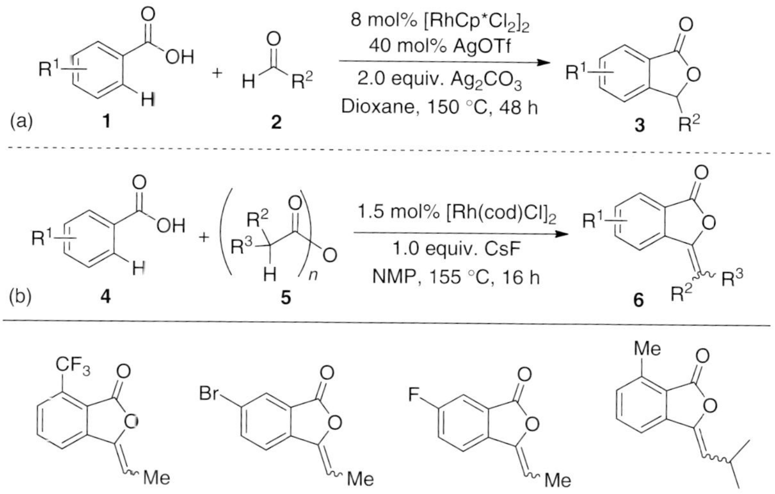

6a, 86%, $Z/E = 15:1$ **6b**, 64%, $Z/E > 20:1$ **6c**, 20%, $Z/E > 20:1$ **6d**, 98%, $Z/E > 20:1$

Scheme 6.2 (a) Synthesis of 3-substituted phthalides from aldehydes and aromatic acids and (b) synthesis of 3-alkylidenephthalides from benzoic acids.

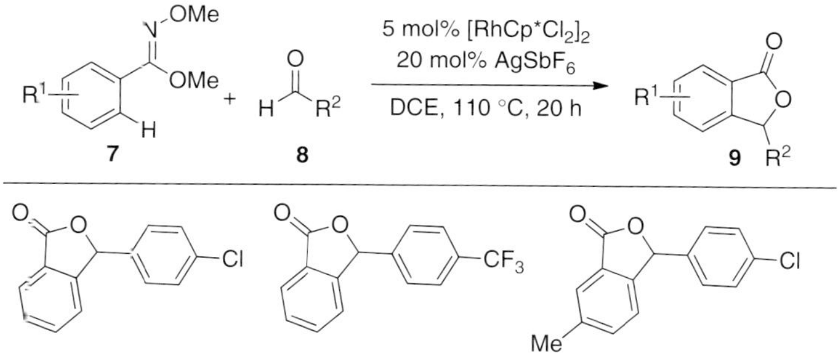

9a, 76% **9b**, 84% **9c**, 73%

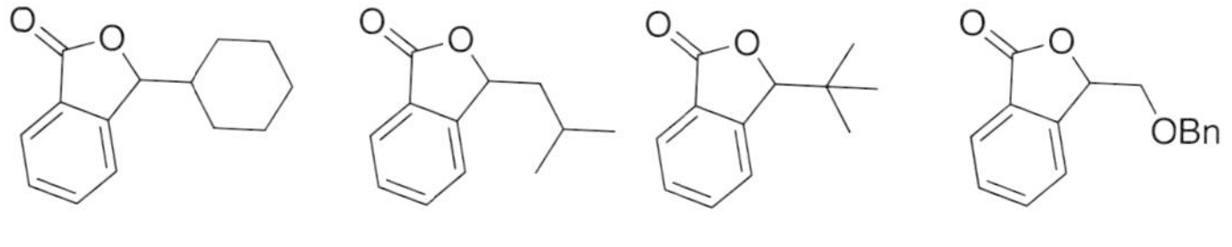

9d, 54%[a] **9e**, 53% **9f**, 52% **9g**, 61%

[a] 10 mol% $[RhCp^*Cl_2]_2$, 40 mol% $AgSbF_6$, 0.5 equiv K_2CO_3.

Scheme 6.3 Synthesis of phthalides from benzimidates and aldehydes.

6.2.1.2 Furans

Furans are very important five-membered aromatic heterocycle and broadly found in natural products and pharmaceuticals. They are also versatile precursors and synthetic intermediates in the preparation of cyclic and acyclic

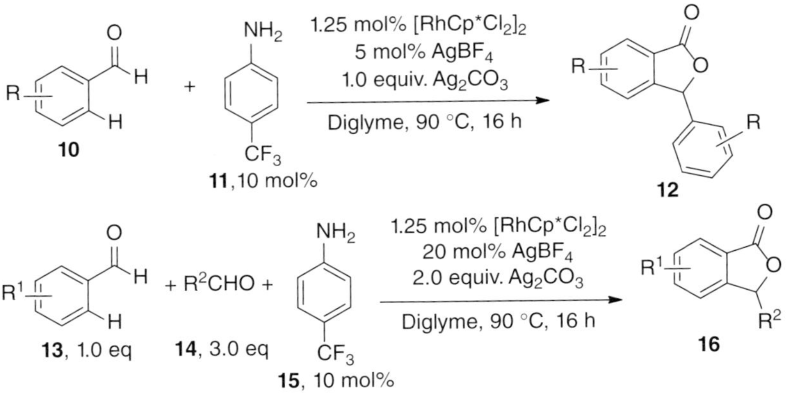

Scheme 6.4 Synthesis of phthalides by oxidative coupling of aldehydes.

molecules [13]. In 2013, the Lu group reported the synthesis of benzofurans via Rh(III)-catalyzed C–H annulation of *N*-phenoxyacetamides with internal alkynes (Scheme 6.5) [14]. O-NHAc (oxyacetamide) was used as the oxidizing directing groups. Deuterium was incorporated exclusively at the *ortho* position to the directing group after stirring **17a** in the presence

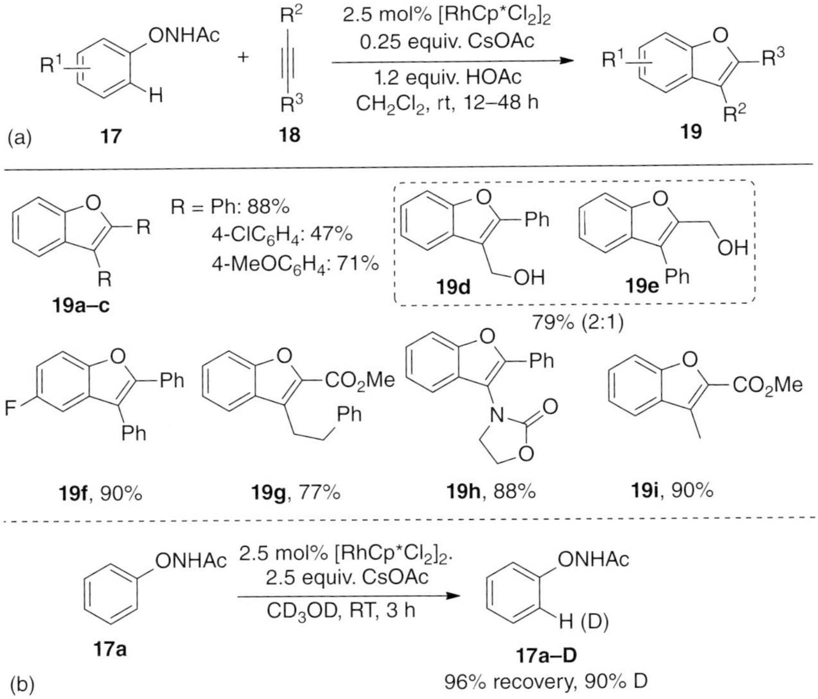

Scheme 6.5 Synthesis of benzofurans from *N*-phenoxyacetamides and alkynes.

of a catalytic amount of $[RhCp^{*}Cl_2]_2$ and 2.5 equiv. CsOAc in deuterated methanol at room temperature for 3 h (Scheme 6.5b). It is worth noting that no deuterium incorporation was found in the absence of CsOAc. These experiments indicated that the acetate anion is crucial to the cyclometalation step.

In 2013, the Ellman group reported the Rh(III)-catalyzed alkenyl C–H activation followed by the nucleophilic addition to aldehydes. Highly functionalized furans were synthesized efficiently in good yields (Scheme 6.6) [15].

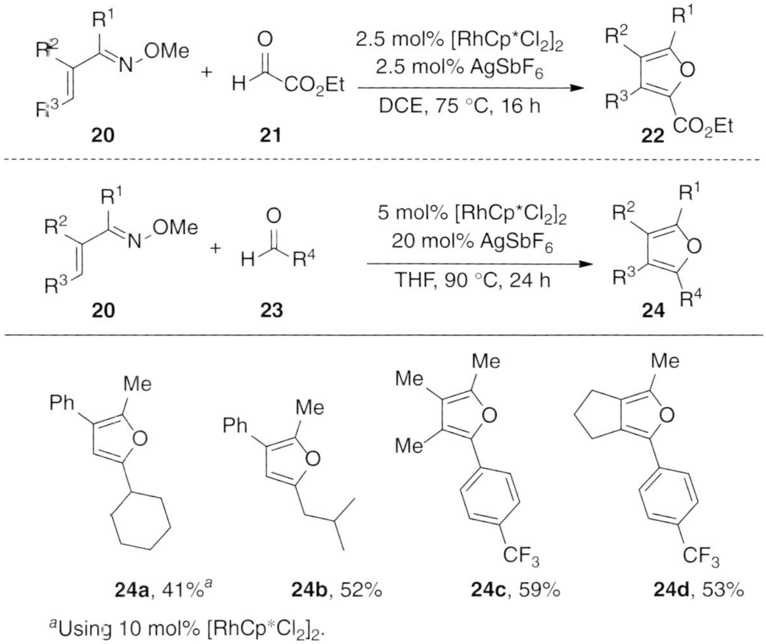

Scheme 6.6 Synthesis of furans via Rh(III)-catalyzed alkenyl C–H functionalization.

6.2.1.3 Other Five-Membered Oxygen-Containing Heterocycles

A Rh(III)-catalyzed dehydrogenative Heck reaction of salicylaldehydes bearing various functional groups with olefins was reported by Glorius and coworkers (Scheme 6.7) [16]. Among the electron-deficient alkenes investigated, ethyl acrylates exhibited optimal activity to give the corresponding products in good yields. Interestingly, when styrene was investigated under the present reaction conditions, only electron-deficient salicylaldehydes could afford the cyclization products.

The corresponding chalone was not observed when benzaldehyde **28** was utilized as the substrate (Scheme 6.8a). Moreover, 2-methoxybenzaldehyde underwent dehydrogenative Heck reaction only at *ortho* position of the aldehyde group (Scheme 6.8b). Therefore, they proposed that the presence of a hydroxy group in

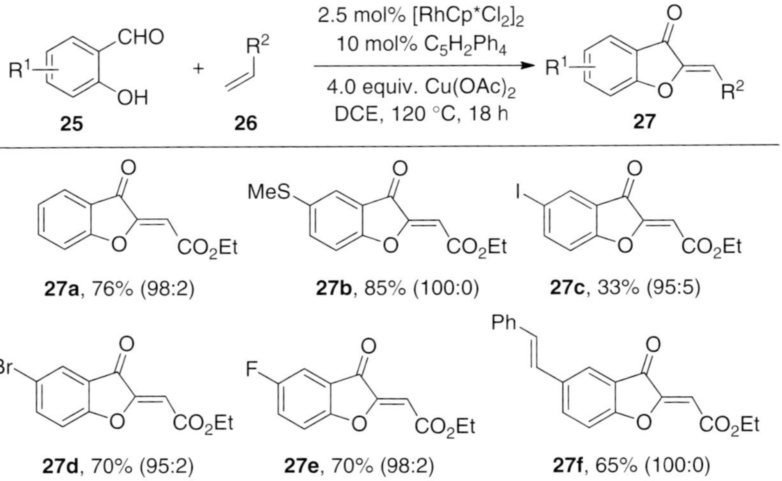

Scheme 6.7 Dehydrogenative Heck reaction of salicylaldehydes with electron-deficient olefins.

the substrate was critical for this transformation. The carbonyl group also plays a key role in this reaction because no desired products were detected when imines were subjected to the standard reaction conditions (Scheme 6.8c).

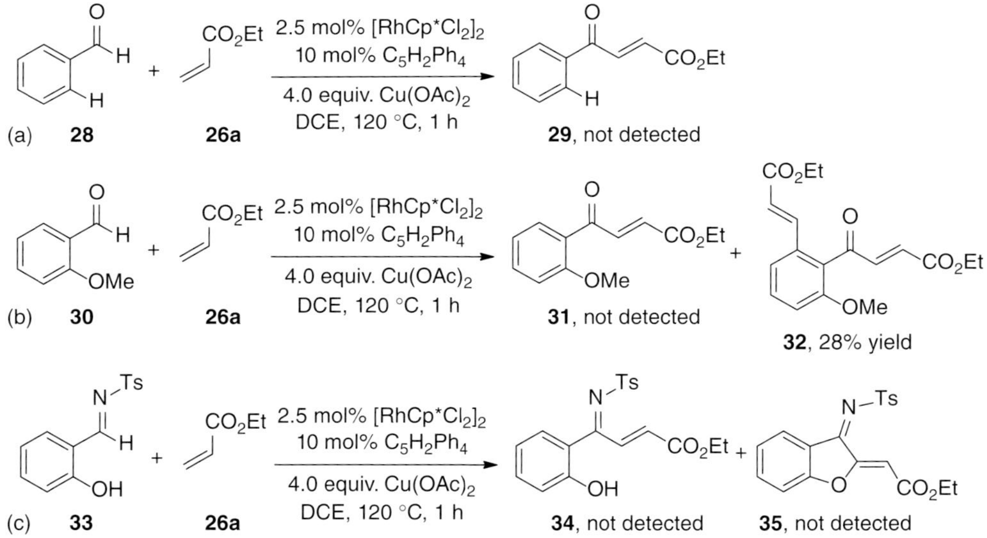

Scheme 6.8 Mechanism studies.

In the same year, a rhodium(III)-catalyzed cyclization by employing alkenes and arylphosphonic acid monoesters for the synthesis of benzoxaphosphole 1-oxides was reported by Lee and coworkers (Scheme 6.9) [17]. Using the combination of

$[RhCp^*Cl_2]_2$ as the catalyst and CH_3CN as the solvent, the cyclization showed broad scope to both internal alkynes and arylphosphonic acid monoesters. The observed isotopic effects ($k_H/k_D = 2.6$) indicated that the C–H bond cleavage step might be the rate-limiting step of this transformation.

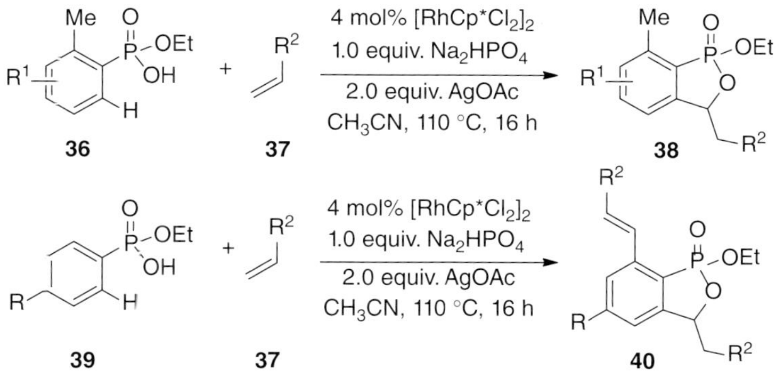

Scheme 6.9 Synthesis of benzoxaphosphole 1-oxides from arylphosphonic acid monoethyl esters and alkenes.

In 2014, O-NHAc (oxyacetamide) was used as a multitasking functional group to afford a fused dihydrofuran–dihydrooxazole scaffold bearing two adjacent quaternary stereogenic centers by Huang and coworkers (Scheme 6.10) [18]. Interestingly, single regioisomeric products were generated by using unsymmetrical internal alkynes as coupling partners.

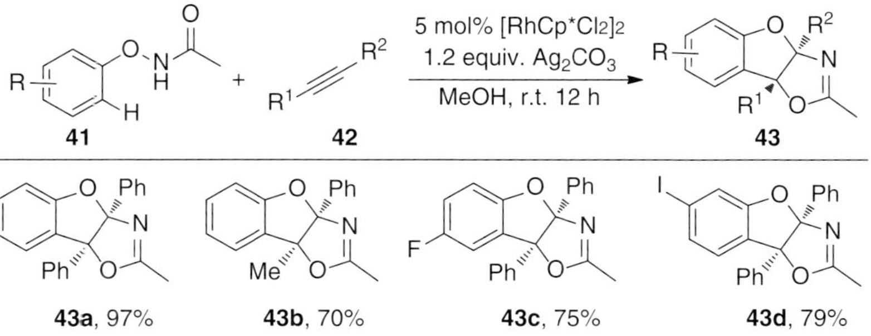

Scheme 6.10 Synthesis of dihydrobenzofuro[2,3-d]oxazoles from aryloxyacetamide and alkynes.

6.2.2 Intramolecular Cyclization

6.2.2.1 Dihydrobenzofurans

In 2001, Ellman and coworkers demonstrated the annulation of aryl ketimines with substituted olefins tethered to the *meta* position to provide the dihydrobenzofuran via $[(PPh_3)_3RhCl]$ catalyzed, aldimine-directed C–H activation [19]. They also developed enantioselective cyclization of aromatic imines with tethered 1,1-disubstituted alkenes through the combination of chiral phosphines and $[RhCl(coe)]_2$ [20]. A rhodium(III)-catalyzed distinct reaction pattern of tethered olefin-containing benzamides was developed by Rovis and coworkers (Scheme 6.11) [21]. This reaction might proceed through a seven-membered metallacycle **45**, which could then undergo hydroarylation, β-H elimination, or amidoarylation, depending on the substituents on amides.

The Glorius group independently reported a Rh(III)-catalyzed intramolecular C–H cyclization of tethered olefin-containing benzamides (Scheme 6.12) [22]. Noteworthy, a redox-neutral CONHPiv directing group was used.

More recently, Cramer and coworkers reported an asymmetric version of this chemistry to access functionalized chiral dihydrobenzofurans that possess a quaternary stereocenter using chiral rhodium catalyst (Scheme 6.13) [23]. They found that O-tethered substrate **54** was deuterated more quickly than *meta*-methyl-substituted derivative **55**. Furthermore, the reaction will take more time for both substrates when the chiral complex with a 1,2-disubstituted cyclopentadienyl ligand was replaced with achiral complex with a more hindered Cp* ligand. These differences between substrate **54** and **55** emphasize the significance of the alkoxy substituent as a secondary directing group for rhodium-catalyzed reactions.

6.2.2.2 Dibenzofuran

In 2013, Ryu and coworkers reported a rhodium-catalyzed arylation of 2-aryloxybenzoic acids through decarbonylative intramolecular C–H functionalization (Scheme 6.14) [24]. The existence of Ac_2O and KI could significantly enhance the reactivity. Based on the control experiments, a mixed anhydride will be produced through the condensation reaction of carboxylic acid and Ac_2O, which oxidize the rhodium(I) catalyst to rhodium(III) species.

6.3 Synthesis of Six-Membered Oxygen-Containing Heterocycles

6.3.1 Intermolecular Annulation

6.3.1.1 Chromenes

2*H*-Chromenes and their benzofused derivatives are ubiquitous in natural products and pharmaceuticals [25]. Moreover, the 2*H*-chromene skeleton is an important substructure moiety in a wide range of bioactive heterocyclic compounds, which have been used as traditional medicine for the treatment of hypertension and diabetes [26]. Consequently, tremendous efforts have been devoted to the synthesis of these compounds.

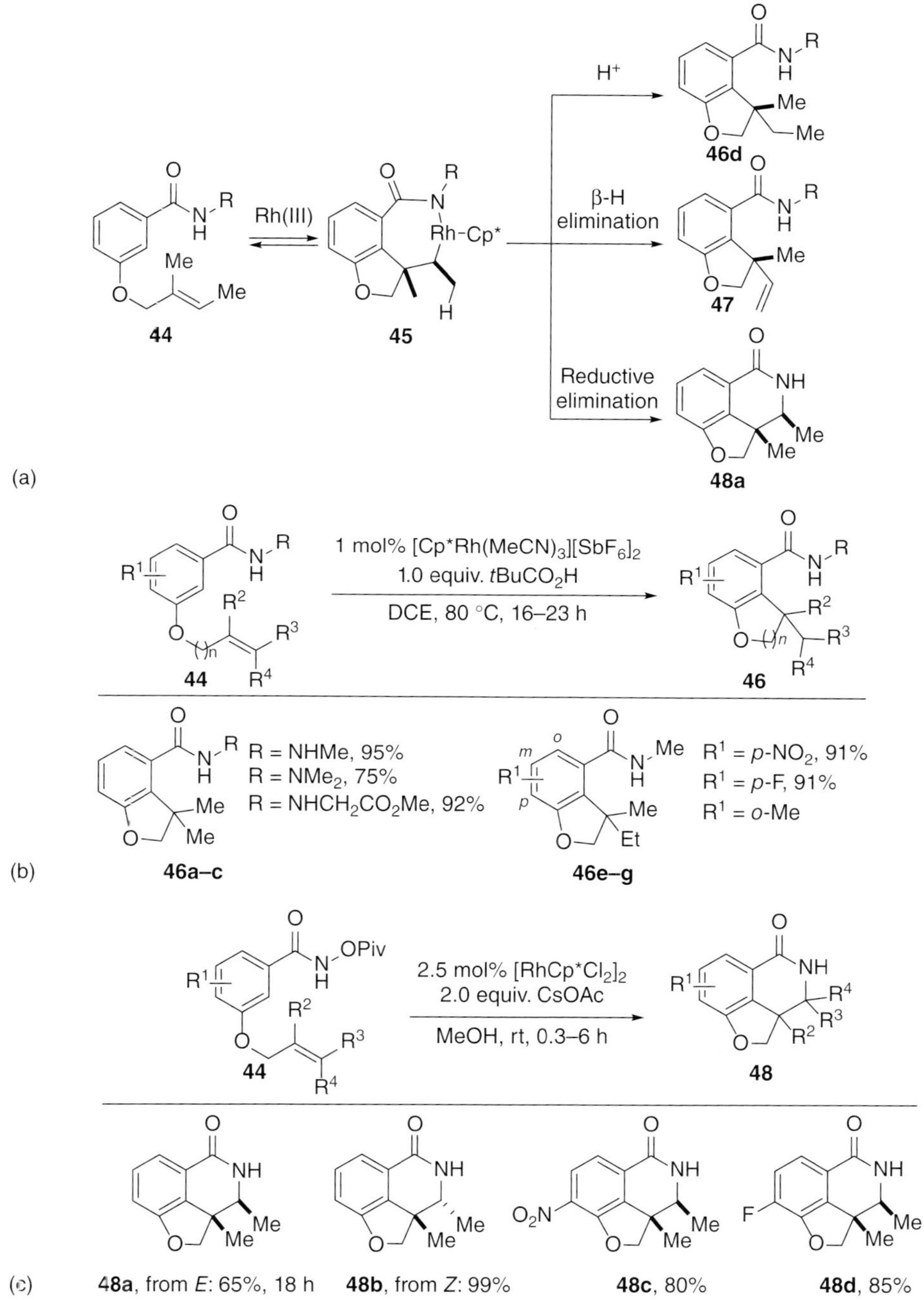

Scheme 6.11 Synthesis of dihydrobenzofuran via Rh(III)-catalyzed C–H functionalization of aromatic imines with tethered 1,1-disubstituted alkenes by Rovis. (a) Mechanistic hypothesis; (b) Rh(III)-catalyzed intramolecular hydroarylation; and (c) Rh(III)-catalyzed intramolecular amidoarylation.

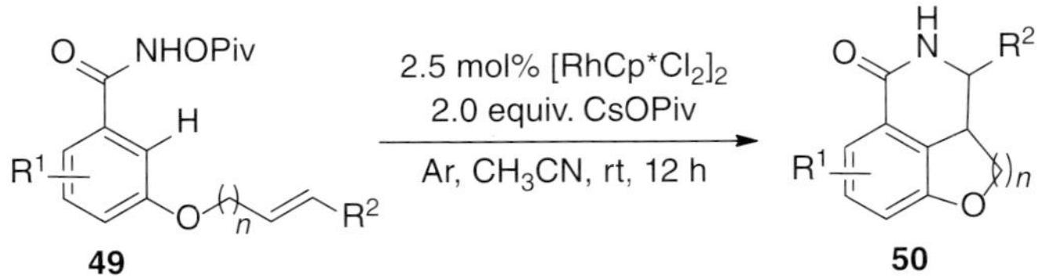

Scheme 6.12 Rh(III)-catalyzed intramolecular amidoarylations by Glorius.

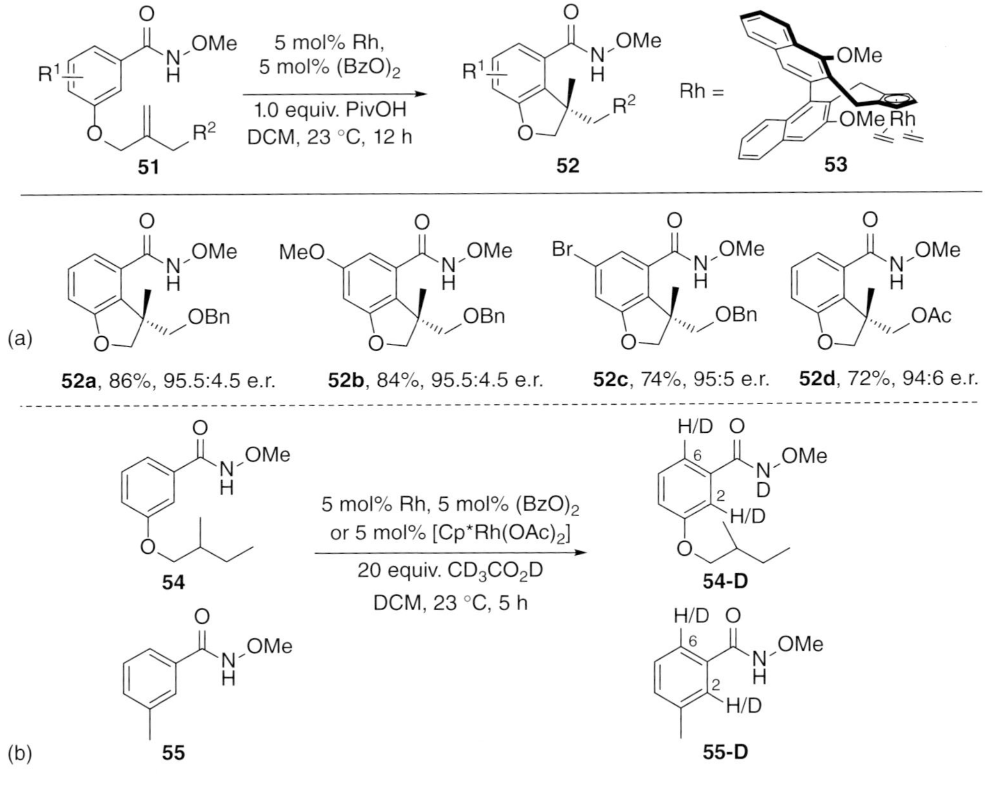

Scheme 6.13 Synthesis of chiral dihydrobenzofurans via Rh(III)-catalyzed enantioselective hydroarylation.

The synthesis of naphtho[1,8-*bc*]pyran derivatives and related compounds via the oxidative annulation of 1-naphthols with internal alkynes was first reported by the Miura group in 2010 (Scheme 6.15) [27]. A series of naphtho[1,8-*bc*]pyran derivatives was obtained in good yields with 1 mol% $[RhCp^*Cl_2]_2$ in the presence of *o*-xylene or DMF.

Shortly after, the same group reported a Rh(III)-catalyzed oxidative annulation of benzyl alcohols with internal alkynes for the synthesis of isochromenes (Scheme 6.16a) [28]. Interestingly, α,α-dimethylallyl alcohol also underwent the oxidative annulation reaction to give the corresponding products in good yields

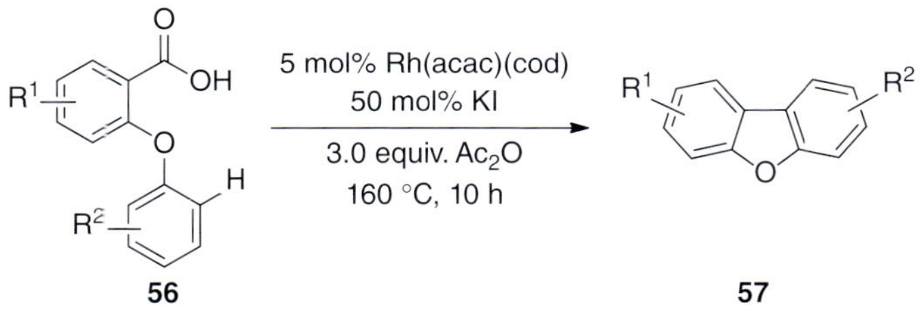

Scheme 6.14 Synthesis of dibenzofuran via decarbonylative C–H arylation of 2-aryloxybenzoic acids.

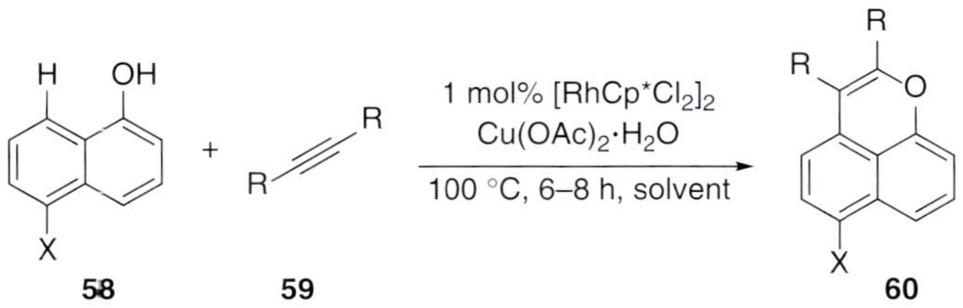

Scheme 6.15 Synthesis of naphtho[1,8-*bc*]pyran derivatives from 1-naphthols and alkynes.

via the alkenyl C(sp^2)–H activation. More recently, the oxidative annulation of tertiary benzyl alcohols with internal alkynes using a new type of rhodium(III) catalyst under ambient conditions was reported by Tanaka and coworkers (Scheme 6.16b) [29].

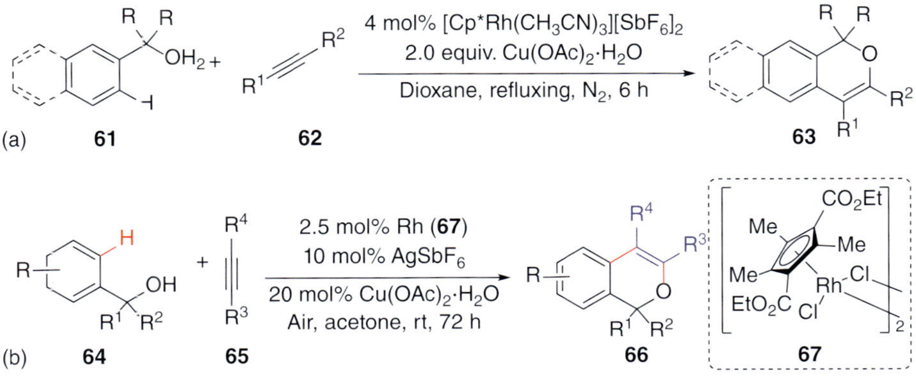

Scheme 6.16 Synthesis of isochromenes by oxidative annulation of benzyl alcohols with alkynes by Miura and Tanaka.

In 2012, Wang and coworkers reported the efficient synthesis of substituted naphtho[1,8-*bc*]pyrans by sequential cleavage of C(sp^2)–H/C(sp^3)–H and C(sp^2)–H/O–H bonds via hydroxyl-directed C–H bond activation (Scheme 6.17) [30]. In the presence of $[RhCp^*Cl_2]_2$ and $Cu(OAc)_2{\cdot}H_2O$, the reaction of benzoylacetonitriles with diphenylacetylene proceeded well to afford the desired annulation products in good yields. In addition to diphenylacetylene, the methyl-substituted diphenylacetylene reacted with benzoylacetonitrile smoothly, affording the corresponding naphtho[1,8-*bc*]pyrans **70i** in 84% yield. Interestingly, unsymmetrical alkynes, such as 1-phenyl-1-propyne, produced a single regioisomer **70j** in 61% yield.

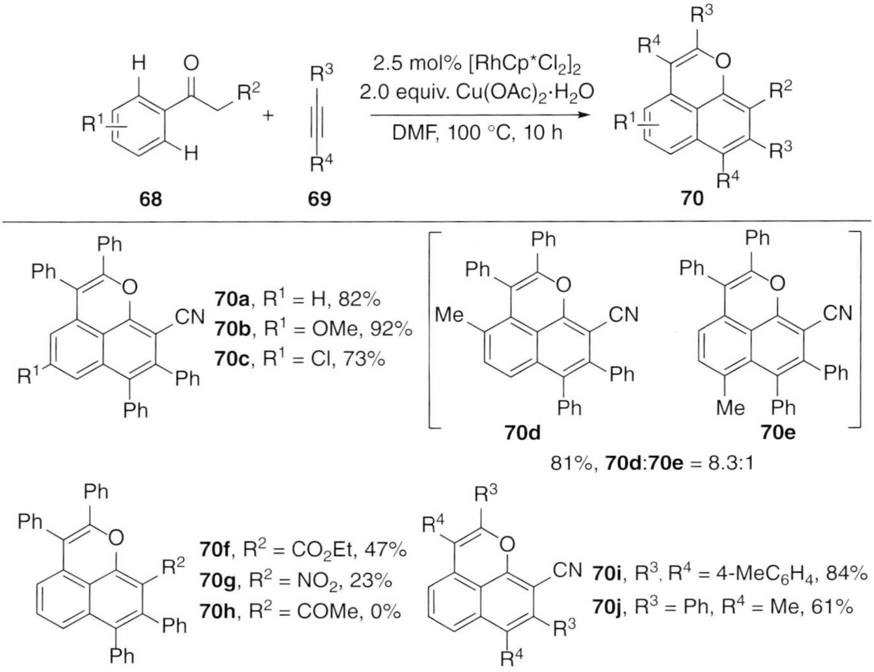

Scheme 6.17 Synthesis of naphtho[1,8-*bc*]pyrans via Rh(III)-catalyzed oxidative coupling of substituted benzoylacetonitriles with alkynes.

Lam and coworkers developed a novel strategy to enable 1,3-enynes to function as one-carbon oxidative annulation partners via 1,4-rhodium(III) migration (Scheme 6.18a) [31]. The reaction proceeded efficiently in 1,4-dioxane using $[RhCp^*Cl_2]_2$ as catalyst. A plausible mechanism of this annulation reaction was proposed in Scheme 6.18b. Initial coordination of the Rh(III) to enol **71a** was followed by cyclorhodation to provide the rhodacycle **A**. Rhodacycle **B** was afforded by the coordination and migratory insertion of the 1,3-enyne. Reductive elimination from **B** would give spiroindene F. However, the reversible protonolysis of **B** is possible in the presence of AcOH, which will provide the alkenylrhodium species **C**. Subsequently, a 1,4-rhodium migration of alkenylrhodium species **C** gave a new allylrhodium species **D**, which could interconvert with the π-allylrhodium species **E**. The benzopyran **73a** was obtained via the nucleophilic attack of the π-allylrhodium(III) moiety of **E** by the enol oxygen.

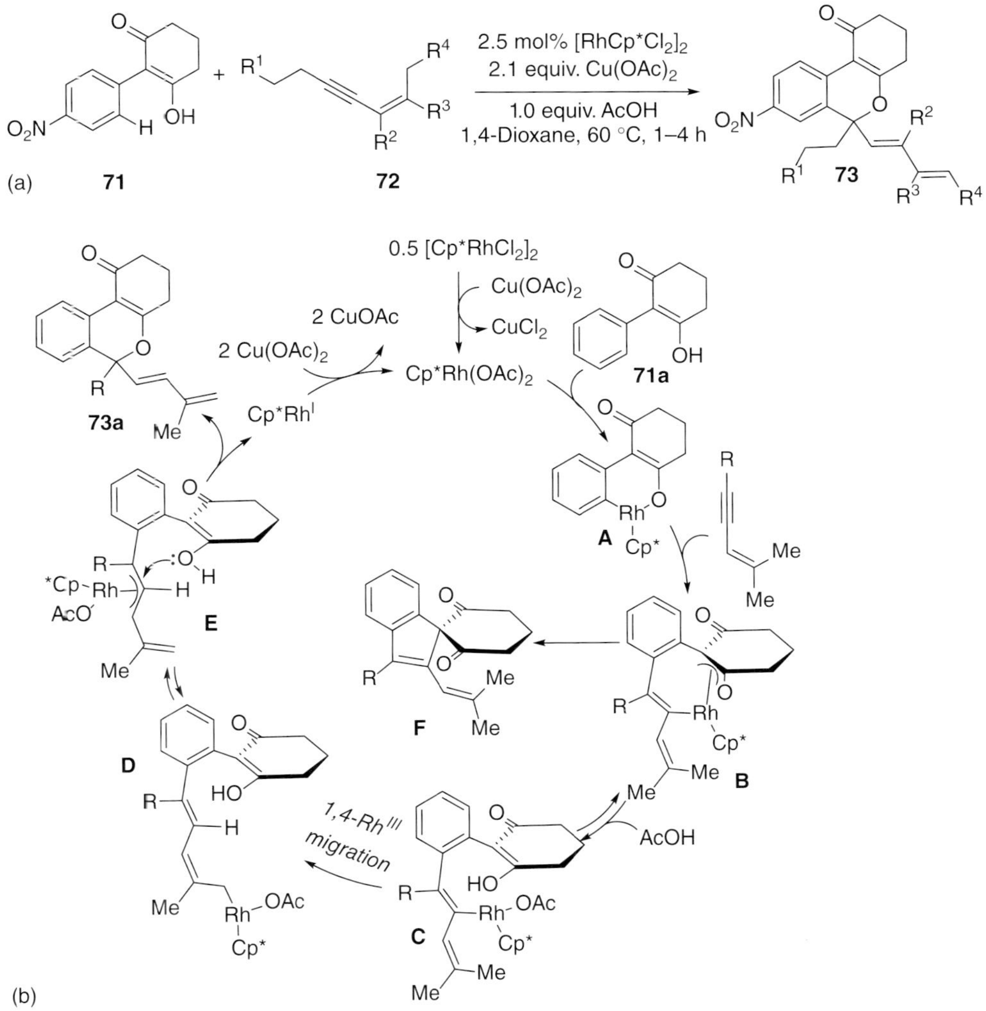

Scheme 6.18 Synthesis of benzopyrans from 2-aryl-3-hydroxy-2-cyclohexenone.

Zhang and coworkers reported the synthesis of tetracyclic naphthoxazoles via rhodium(III)-catalyzed annulation of naphthoquinone with internal alkynes (Scheme 6.19) [32]. A variety of substrates bearing synthetically useful functional groups were compatible with this protocol.

In 2014, Wang and coworkers discolesed a rhodium(III)-catalyzed annulation of *N*-phenoxyacetamide with cyclopropenes (Scheme 6.20), which represents the first example of using cyclopropene as a three-carbon unit in rhodium(III)-catalyzed C(sp^2)–H activation [33]. The reaction proceeded well at room temperature and showed broad substrate scope with respect to both cyclopropenes and *N*-phenoxyacetamides.

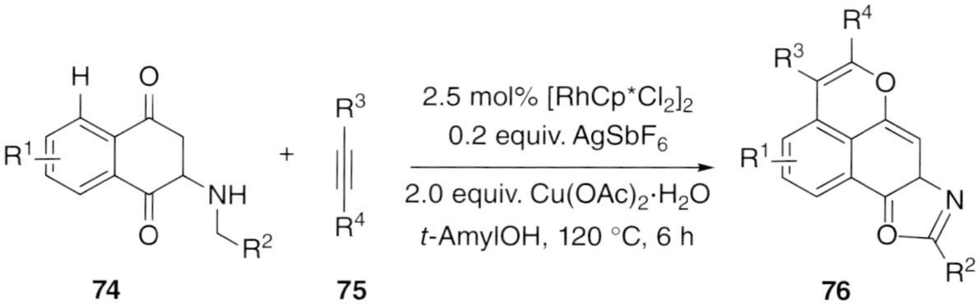

Scheme 6.19 Synthesis of tetracyclic naphthoxazoles from naphthoquinone.

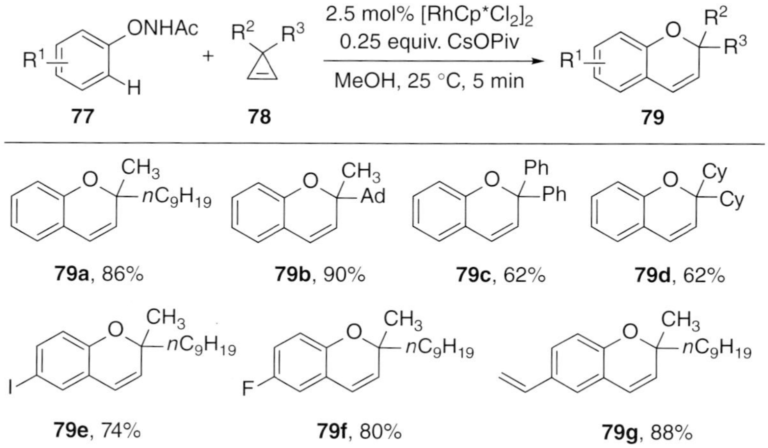

Scheme 6.20 Synthesis of 2*H*-chromene via rhodium(III)-catalyzed annulation of cyclopropenes with *N*-phenoxyacetamides.

A plausible reaction mechanism was proposed in Scheme 6.21 after a series of control experiments, and kinetic isotope effect experiments were conducted. First, coordination of [Cp*Rh(OPiv)$_2$] to phenoxyacetamide was followed by C–H bond cleavage to give the rhodacyclic intermediate **B**. Two pathways are possible for the formation of the eight-membered rhodacyclic intermediate **G**: β-carbon elimination of intermediate **D** or migratory insertion of intermediate **E**.

In 2015, the Gulías group reported a Rh(III)-catalyzed [5+1] annulations between 2-alkenylphenols and allenes, providing a practical entry to 2,2-disubstituted 2*H*-chromenes (Scheme 6.22) [34].

6.3.1.2 **Chromones**

In 2008, Miura, Satoh, and coworkers reported an interesting intermolecular oxidative coupling reaction between salicylaldehydes and internal alkynes to generate 2,3-disubstituted chromones in the presence of $[RhCl(cod)]_2/C_5H_2Ph_4$ as the catalyst and $Cu(OAc)_2{\cdot}H_2O$ as the oxidant (Scheme 6.23a) [35]. The Glorius group [16] described a Rh(III)-catalyzed dehydrogenative Heck reaction in the presence of 4 equiv. of $Cu(OAc)_2$ to give NO_2-substituted chromone **87** in

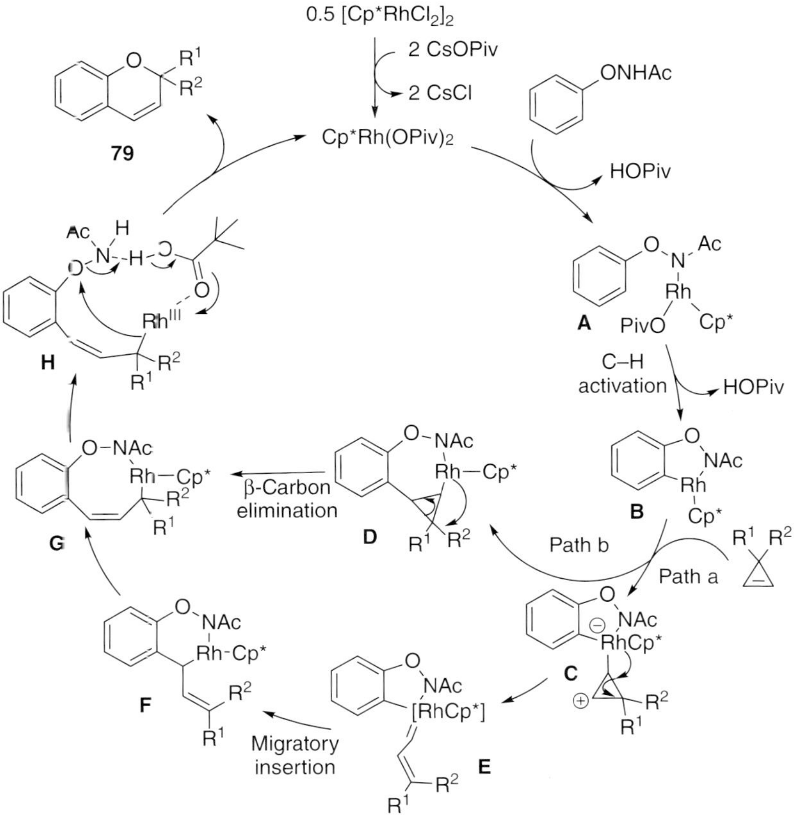

Scheme 6.21 Proposed mechanism.

44% yield (Scheme 6.23b). While chromane **88** was obtained in 34% yield in the presence of 2.0 equiv. of Ag_2CO_3.

6.3.1.3 Coumarin

Coumarin, which was first isolated from tonka bean (*Dipteryx odorata* Wild) in 1820 [36], have been demonstrated to possess anti-inflammatory, antiplatelet aggregation, antidote, antitumor, and AChE inhibitory activities [37]. As a consequence, significant efforts have been devoted to the synthesis of coumarins.

In 2007, the Miura group reported a Rh(III)-catalyzed oxidative coupling of benzoic acids with internal alkynes to the synthesis of isocoumarins via aromatic C–H activation (Scheme 6.24a) [38]. Importantly, the reaction of benzoic acids with alkynes takes place efficiently even with a reduced amount (5 mol%) of $Cu(OAc)_2{\cdot}H_2O$ under air (Scheme 6.24b) [5b]. The same group also developed the rhodium-catalyzed coupling of acrylic acids with alkynes to provide corresponding α-pyrone via vinylic C–H bond cleavage (Scheme 6.24c) [5c]. In 2015, Wen and coworkers described a Rh(III)-catalyzed synthesis of

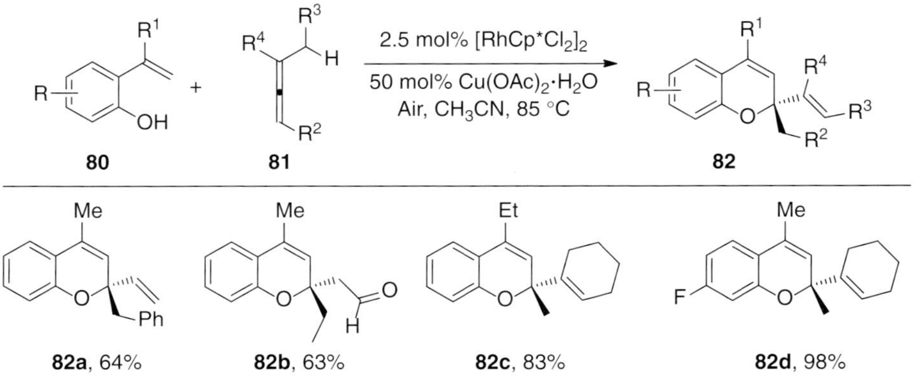

Scheme 6.22 Synthesis of 2,2-disubstituted 2*H*-chromenes from 2-alkenylphenols and allenes.

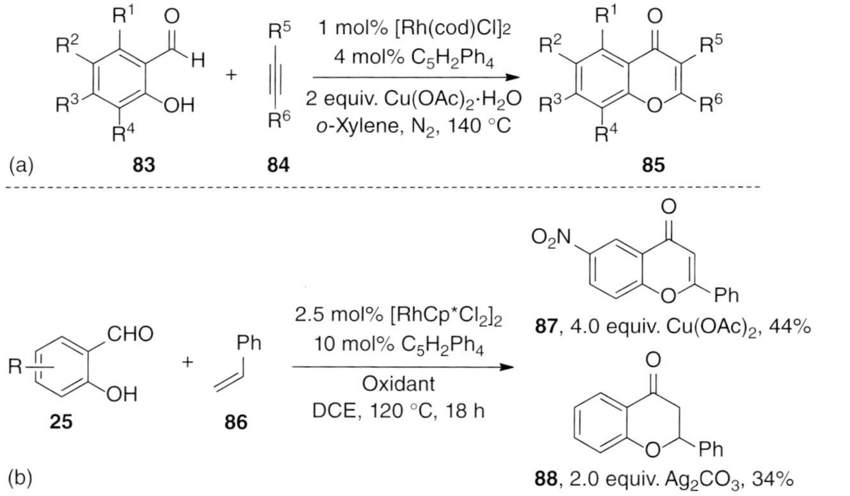

Scheme 6.23 (a) Synthesis 2,3-disubstituted chromones by oxidative coupling between salicylaldehydes and internal alkynes by Miura and Satoh and (b) synthesis of chromone and chromane from salicylaldehyde and styrene by Glorius.

3-substituted isocoumarins from benzoic acids with geminal-substituted vinyl acetates (Scheme 6.24d) [39].

In 2013, the synthesis of phosphaisocoumarins is achieved by Miura and coworkers via Rh(III)-catalyzed oxidative annulation of phenylphosphonic acids with alkynes (Scheme 6.25a) [40]. Shortly after, Lee and coworkers reported a similar result using phenylphosphinic or alkenylphosphonic acids to afford phosphaisocoumarins in good yields (Scheme 6.25b) [41].

In 2014, Liu and coworkers reported a Rh(III)-catalyzed annulation of O-benzoylhydroxylamines with internal alkynes to the preparation of isocoumarins

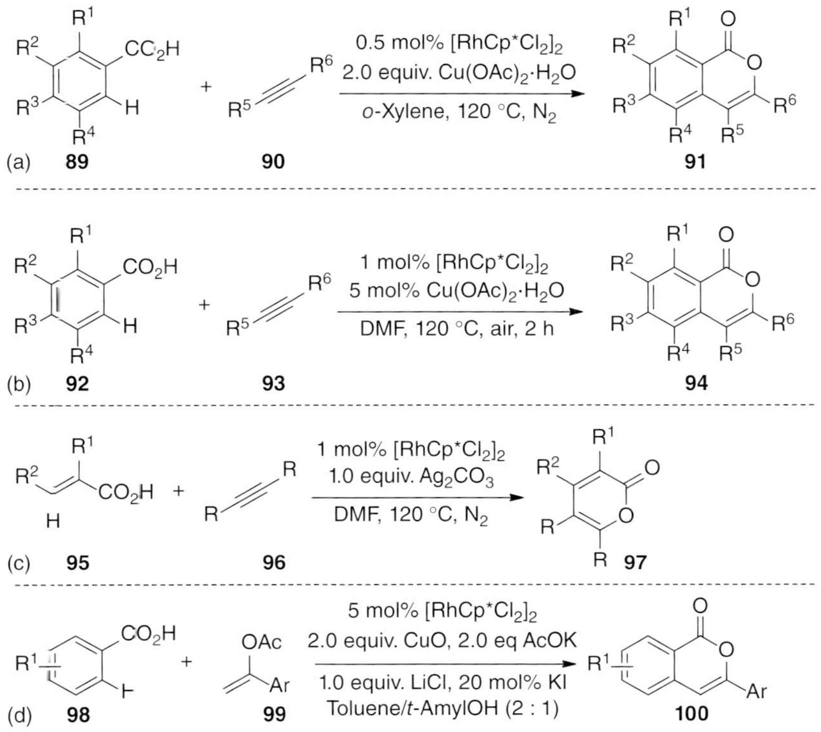

Scheme 6.24 (a) Synthesis of 3,4-diphenylisocoumarin via oxidative coupling of benzoic acids with alkynes; (b) synthesis of 3,4-diphenylisocoumarin under air; (c) synthesis of functionalized α-pyrone by oxidative coupling of substituted acrylic acids with alkynes; and (d) synthesis of 3-substituted isocoumarins from benzoic acids and geminal-substituted vinyl acetates.

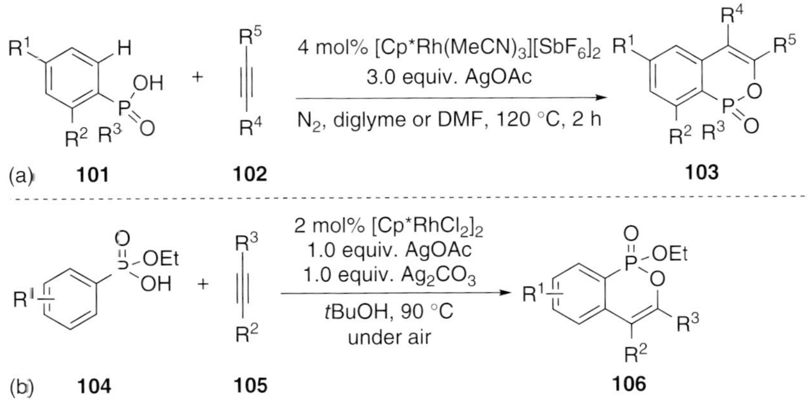

Scheme 6.25 (a) Synthesis of phosphaisocoumarins from phenylphosphinic acids and (b) synthesis of phosphaisocoumarins from arylphosphonic acid monoesters.

and α-pyrones (Scheme 6.26) [42]. The solvent effect plays a key role in this transformation, and TFE was found to be the best.

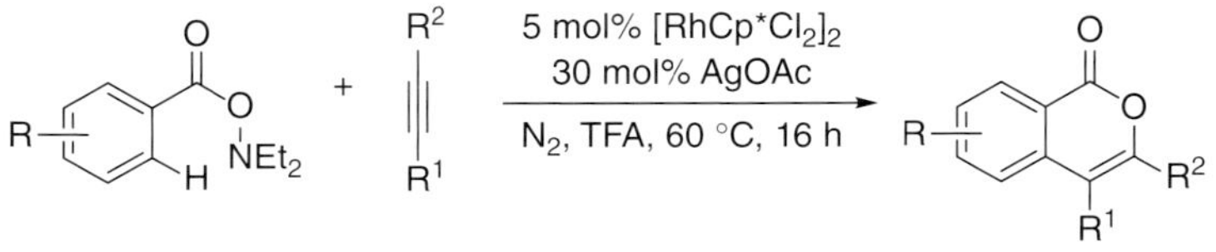

107 **108** **109**

Scheme 6.26 Synthesis of isocoumarins from *N,N*-diethyl-*O*-benzoylhydroxylamine and alkynes.

6.3.1.4 Other Six-Membered Oxygen-Containing Heterocycles

In 2013, Li and coworkers reported a Rh(III)-catalyzed annulation of 1-benzoylpyrrolidine with propargyl alcohols for the synthesis of (4-benzylidene) isochroman-1-ones (Scheme 6.27) [43]. Highly enantioenriched lactones could be isolated by employing the optically pure propargyl alcohols as the starting materials.

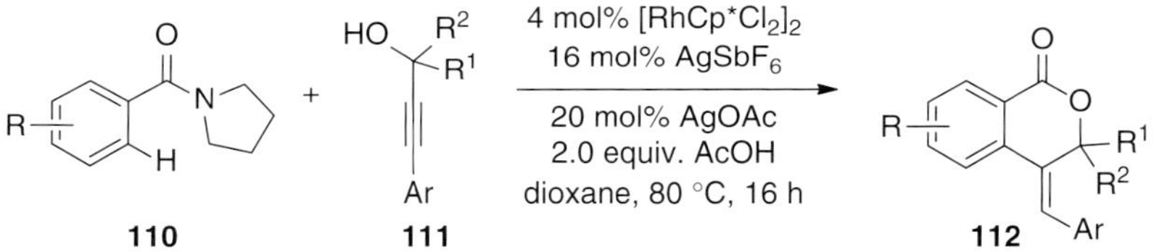

110 **111** **112**

Scheme 6.27 Synthesis of (4-benzylidene)isochroman-1-ones from benzamides and propargyl alcohols.

In 2014, the Glorius group developed a Rh(III)-catalyzed annulation of benzamides with 1,3-diynes to the synthesis of bisheterocycles (Scheme 6.28) [44]. Selective monoannulation was realized under room temperature and a lower catalyst loading. Oxygen-containing heterocycles were prepared by subsequent multiple annulation with acrylic acid and o-vinylphenol by alkenyl C–H activation.

6.3.2 Intramolecular Cyclization

In 2009, Urabe and coworkers communicated a $[Rh_2(tfa)_4]$-catalyzed isomerization and intramolecular cyclization of alkynyl ethers to the synthesis of dihydropyrans (Scheme 6.29) [45].

3,4-Dihydro-(*1H*)-benzopyranes were synthesized in excellent yields by Rovis through a Rh(III)-catalyzed intramolecular hydroarylation or amidoarylation (Scheme 6.30) [21].

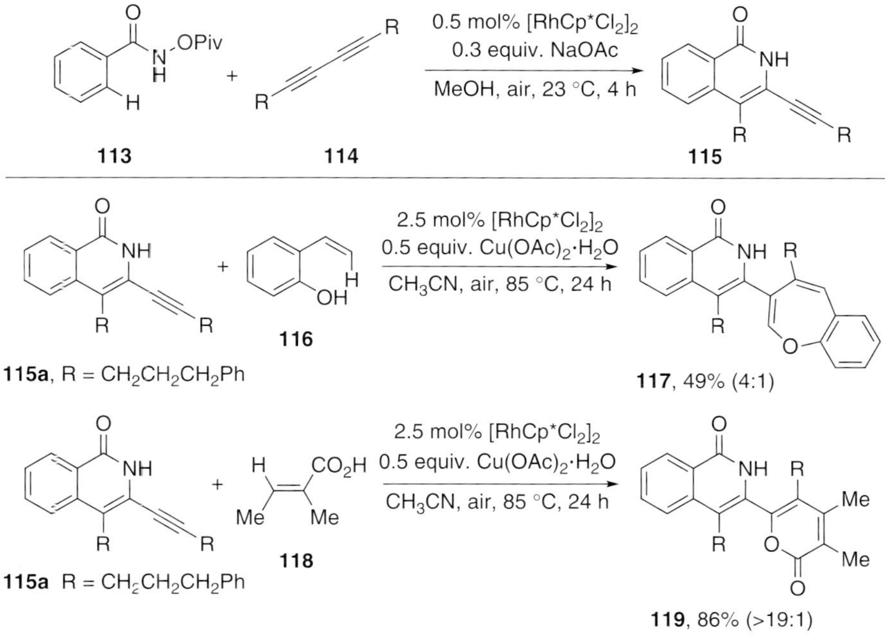

Scheme 6.28 Synthesis of diverse bisheterocycles.

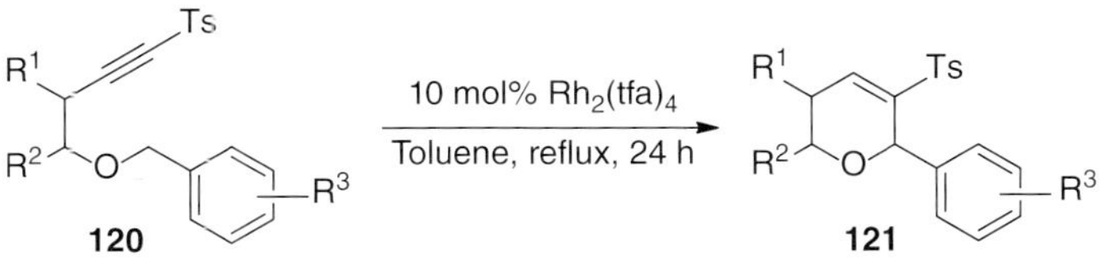

Scheme 6.29 Synthesis of dihydropyrans from acetylenic sulfones.

6.4 Synthesis of Seven-, Eight-, and Nine-Membered Oxygen-Containing Heterocycles

6.4.1 Intermolecular Annulation

In 2014, Mascareñas, Gulías, and coworkers reported the Rh(III)-catalyzed oxidative [5+2] annulation of 2-vinylphenols with alkynes (Scheme 6.31) [46]. The reaction took place with high regioselectivity when unsymmetrical aryl–alkyl alkynes were used as coupling partners with the product of phenyl tethered to the oxygen group generated predominantly.

An efficient synthesis of 1,2-oxazepines from N-phenoxyacetamides and α, β-unsaturated aldehydes under room temperature through Rh(III)-catalyzed intermolecular [4+3] annulation was reported by Zhao and coworkers (Scheme 6.32) [47]. It is worth noting that 1,2-oxazepine is readily transformed to chroman

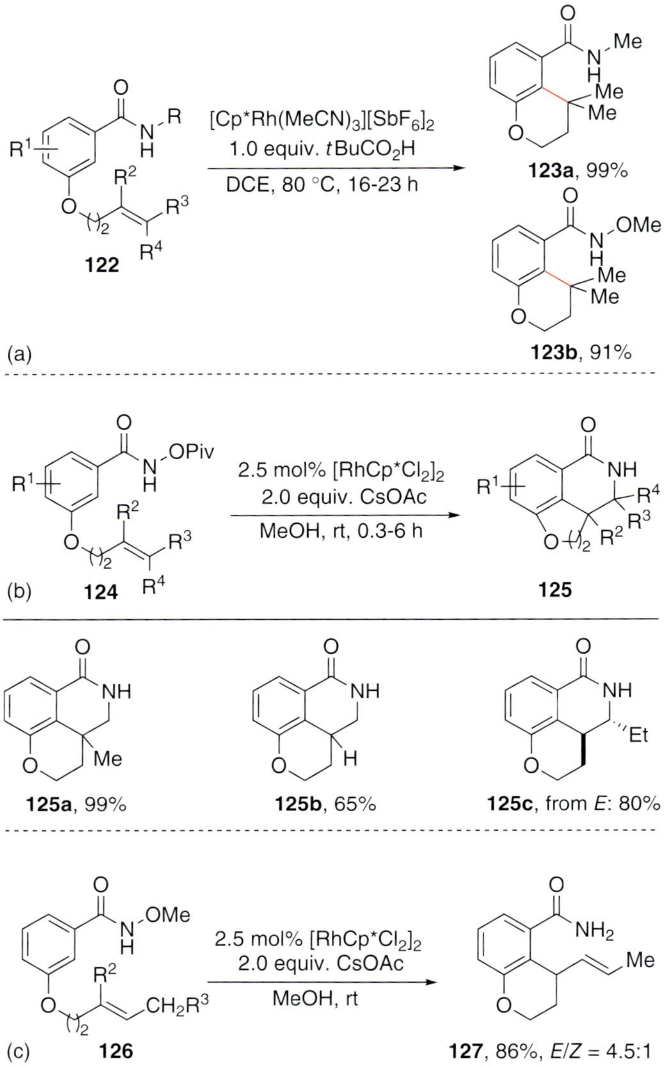

Scheme 6.30 Synthesis of dihydrobenzopyrans via intramolecular hydroarylation or amidoarylation. (a) Intermolecular hydroarylation; (b) intermolecular amidoarylation; and (c) intermolecular Heck-type reaction.

derivative, which are important building blocks in a number of biologically important molecules.

6.4.2
Intramolecular Cyclization

A Rh(III)-catalyzed intramolecular redox-neutral or oxidative cyclization strategy for the rapid synthesis of 3,4-fused indole skeletons from tethered alkynes was

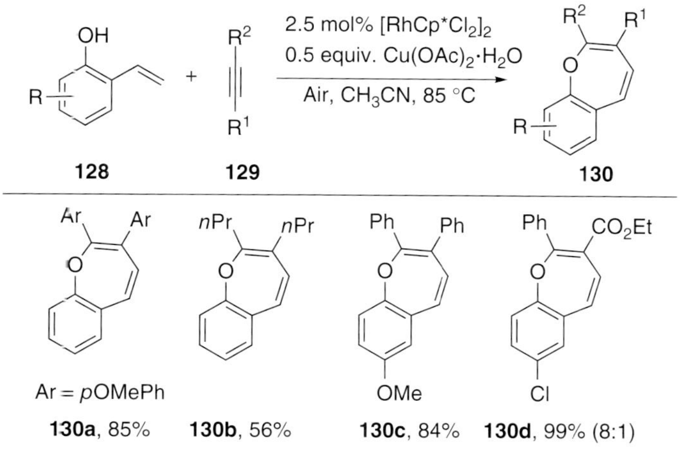

Scheme 6.31 Synthesis benzoxepines via Rh(III)-catalyzed annulation of *o*-vinylphenols with alkynes.

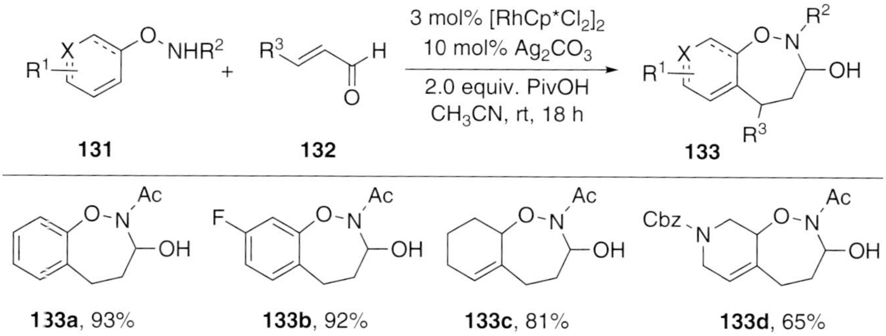

Scheme 6.32 Synthesis of 1,2-oxazepines from *N*-phenoxyacetamides and α,β-unsaturated aldehydes.

reported by Li and coworkers (Scheme 6.33) [48]. Interestingly, the scope of this intramolecular cyclization could be extended to generate 3,4-fused indoles with eight- to nine-membered rings.

6.5 Summary and Conclusions

Over the last decade, the rhodium-catalyzed C–H functionalization has emerged as a powerful approach for the synthesis of heterocycles, due to the obviation of the prefunctionalization of the starting materials. This review focused on rhodium-catalyzed C–H functionalization involving the formation of oxygen-containing heterocycles. Although the established methods have demonstrated

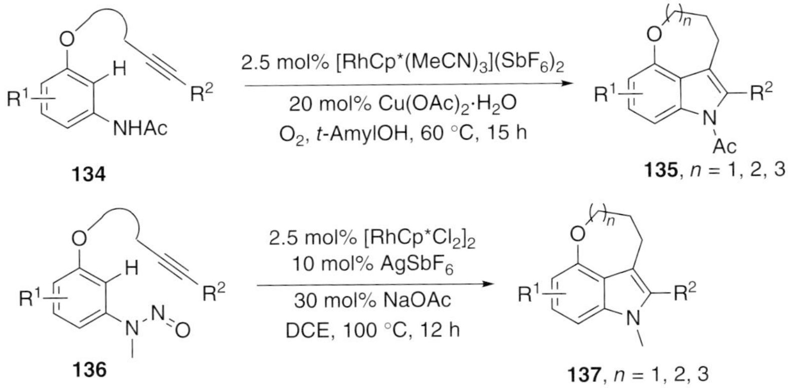

Scheme 6.33 Synthesis of 3,4-fused indole skeletons via intramolecular cyclization of tethered alkynes.

high reactivity, good selectivity, and broad substrate scope, significant challenges still remain: (i) these reactions are limited to the activation of C(sp^2)–H bonds and few examples of Rh-catalyzed C(sp^3)–H bonds activation have been reported and (ii) further studies are necessary to understand the mechanism of rhodium(III)-catalyzed C–H activation processes and exploit their potential. We anticipate that the development of new strategies and catalytic systems will further contribute to the synthesis of oxygen-containing heterocycles in the future.

References

1. For selected reviews, see: (a) Diederich, F. and Stang, P.J. (eds) (1998) *Metal-Catalyzed Cross-Coupling Reactions*, Wiley-VCH Verlag GmbH, New York; (b) Nakamura, I. and Yamamoto, Y. (2004) *Chem. Rev.*, **104**, 2127–2198; (c) Cacchi, S. and Fabrizi, G. (2005) *Chem. Rev.*, **105**, 2873–2920; (d) Nicolaou, K.C., Bulger, P.G., and Sarlah, D. (2005) *Angew. Chem. Int. Ed.*, **44**, 4442–4489; (e) Zeni, G. and Larock, R.C. (2006) *Chem. Rev.*, **106**, 4644–4680; (f) Chopade, P.R. and Louie, J. (2006) *Adv. Synth. Catal.*, **348**, 2307–2327; (g) Patil, N.T. and Yamamoto, Y. (2008) *Chem. Rev.*, **108**, 3395–3442; (h) Brandi, A., Cicchi, S., and Cordero, F.M. (2008) *Chem. Rev.*, **108**, 3988–4035; (i) Schmidt, A., Beutler, A., and Snovydovych, B. (2008) *Eur. J. Org. Chem.*, **2008**, 4073–4095; (j) Arndtsen, B.A. (2009) *Chem. Eur. J.*, **15**, 302–313; (k) Stokes, B.J. and Driver, T.G. (2011) *Eur. J. Org. Chem.*, **2011**, 4071–4088; (l) Cacchi, S., Fabrizi, G., and Goggiamani, A. (2011) *Org. Biomol. Chem.*, **9**, 641–652.
2. For perspectives on atom, step, and redox economy, see: (a) Trost, B.M. (1991) *Science*, **254**, 1471–1477; (b) Wender, P.A., Verma, V.A., Paxton, T.J., and Pillow, T.H. (2008) *Acc. Chem. Res.*, **41**, 40–49; (c) Burns, N.Z., Baran, P.S., and Hoffmann, R.W. (2009) *Angew. Chem. Int. Ed.*, **48**, 2854–2867.
3. For selected reviews on transition metal-catalyzed C–H activation, see: (a) Crabtree, R.H. (1985) *Chem. Rev.*, **85**, 245–269; (b) Shilov, A.E. and Shul'pin, G.B. (1997) *Chem. Rev.*, **97**, 2879–2932; (c) Jia, C., Kitamura, T., and Fujiwara, Y. (2001) *Acc. Chem. Res.*, **34**, 633–639; (d) Ritleng, V., Sirlin, C., and Pfeffer, M. (2002) *Chem.*

Rev., **102**, 1731–1770; (e) Labinger, J.A. and Bercaw, J.E. (2002) *Nature*, **417**, 507–514; (f) Kakiuchi, F. and Chatani, N. (2003) *Adv. Synth. Catal.*, **345**, 1077–1101; (g) Yu, J.-Q., Giri, R., and Chen, X. (2006) *Org. Biomol. Chem.*, **4**, 4041–4047; (h) Dick, A.R. and Sanford, M.S. (2006) *Tetrahedron*, **62**, 2439–2463; (i) Bergman, R.G. (2007) *Nature*, **446**, 391–393; (j) Hartwig, J.F. (2008) *Nature*, **455**, 314–322; (k) Colby, D.A., Bergman, R.G., and Ellman, J.A. (2010) *Chem. Rev.*, **110**, 624–655; (l) Lyons, T.W. and Sanford, M.S. (2010) *Chem. Rev.*, **110**, 1147–1169; (m) Yeung, C.S. and Dong, V.M. (2011) *Chem. Rev.*, **111**, 1215–1292; (n) Liu, C., Zhang, H., Shi, W., and Lei, A. (2011) *Chem. Rev.*, **111**, 1780–1824; (o) Engle, K.M., Mei, T.-S., Wasa, M., and Yu, J.-Q. (2012) *Acc. Chem. Res.*, **45**, 788–802; (p) Li, B.-J. and Shi, Z.-J. (2012) *Chem. Soc. Rev.*, **41**, 5588–5598; (q) Noisier, A.F.M. and Brimble, M.A. (2014) *Chem. Rev.*, **114**, 8775–8806; (r) Zhang, M., Zhang, Y., Jie, X., Zhao, H., Li, G., and Su, W. (2014) *Org. Chem. Front.*, **1**, 843–895; (s) Liu, C., Liu, D., and Lei, A. (2014) *Acc. Chem. Res.*, **47**, 3459–3470.

4. For selected reviews on heterocycle synthesis via transition metal-catalyzed C–H functionalization, see: (a) Thansandote, P. and Lautens, M. (2009) *Chem. Eur. J.*, **15**, 5874–5883; (b) Zhang, M. (2009) *Adv. Synth. Catal.*, **351**, 2243–2270; (c) Beccalli, E.M., Broggini, G., Fasana, A., and Rigamonti, M. (2011) *J. Organomet. Chem.*, **696**, 277–295.
5. For pioneering works, see: (a) Davies, D.L., Al-Duaij, O., Fawcett, J., Giardiello, M., Hilton, S.T., and Russell, D.R. (2003) *Dalton Trans.*, **21**, 4132–4138; (b) Ueura, K., Satoh, T., and Miura, M. (2007) *Org. Lett.*, **9**, 1407–1409; (c) Mochida, S., Hirano, K., Satoh, T., and Miura, M. (2009) *J. Org. Chem.*, **74**, 6295–6298; (d) Li, L., Brennessel, W.W., and Jones, W.D. (2008) *J. Am. Chem. Soc.*, **130**, 12414–12419; (e) Stuart, D.R., Bertrand-Laperle, M., Burgess, K.M.N., and Fagnou, K. (2008) *J. Am. Chem. Soc.*, **130**, 16474–16475; For a review, see: (f) Satoh, T. and Miura, M. (2010) *Chem. Eur. J.*, **16**, 11212–11222.
6. For recent reviews, see: (a) Song, G., Wang, F., and Li, X. (2012) *Chem. Soc. Rev.*, **41**, 3651–3678; (b) Colby, D.A., Tsai, A.S., Bergman, R.G., and Ellman, J.A. (2012) *Acc. Chem. Res.*, **45**, 814–825; (c) Patureau, F.W., Wencel-Delord, J., and Glorius, F. (2012) *Aldrichim. Acta*, **45**, 31–41; (d) Kuhl, N., Schroder, N., and Glorius, F. (2014) *Adv. Synth. Catal.*, **356**, 1443–1460.
7. (a) Lin, G., Chan, S.S.-K., Chung, H.-S., and Li, S.-L. (2005) in *Studies in Natural Products Chemistry*, vol. **32** (ed. A. Rahman), Elsevier, Amsterdam, pp. 611–669; (b) Beck, J.J. and Chou, S.C. (2007) *J. Nat. Prod.*, **70**, 891–900; (c) Xiang, M.-J. and Li, Z.-H. (2007) *Curr. Org. Chem.*, **11**, 833–844; (d) Mola, A.D., Palombi, L., and Massa, A. (2012) *Curr. Org. Chem.*, **16**, 2302–2320.
8. (a) Snieckus, V. (1990) *Chem. Rev.*, **90**, 879–933; (b) Mitchell, A.S. and Russell, R.A. (1995) *Tetrahedron*, **51**, 5207–5236; (c) Mal, D. and Pahari, P. (2007) *Chem. Rev.*, **107**, 1892–1918; (d) Hernández, E., Vélez, J.M., and Vlaar, C.P. (2007) *Tetrahedron Lett.*, **48**, 8972–8975; (e) Rathwell, K. and Brimble, M.A. (2007) *Synthesis*, 643–662; (f) Donner, C.D. (2013) *Tetrahedron*, **69**, 3747–3773.
9. Shi, X. and Li, C.-J. (2012) *Adv. Synth. Catal.*, **354**, 2933–2938.
10. Danoun, G., Mamone, P., and Gooßen, L.J. (2013) *Chem. Eur. J.*, **19**, 17287–17290.
11. Lian, Y., Bergman, R.G., and Ellman, J.A. (2012) *Chem. Sci.*, **3**, 3088–3092.
12. Tan, P.W., Juwaini, N.A.B., and Seayad, J. (2013) *Org. Lett.*, **15**, 5166–5169.
13. (a) Katritzky, A.R. (2008) *Comprehensive Heterocyclic Chemistry III*, 1st edn, Elsevier, Amsterdam; (b) Lipshutz, B.H. (1986) *Chem. Rev.*, **86**, 795–819.
14. Liu, G., Shen, Y., Zhou, Z., and Lu, X. (2013) *Angew. Chem. Int. Ed.*, **52**, 6033–6037.
15. Lian, Y., Huber, T., Hesp, K.D., Bergman, R.G., and Ellman, J.A. (2013) *Angew. Chem. Int. Ed.*, **52**, 629–633.
16. Shi, Z., Schröer, N., and Glorius, F. (2012) *Angew. Chem. Int. Ed.*, **51**, 8092–8096.

17. Ryu, T., Kim, J., Park, Y., Kim, S., and Lee, P.H. (2013) *Org. Lett.*, **15**, 3986–3989.
18. Chen, Y., Wang, D., Duan, P., Ben, R., Dai, L., Shao, X., Hong, M., Zhao, J., and Huang, Y. (2014) *Nat. Commun.*, **5**, 4610–4618.
19. Thalji, R.K., Ahrendt, K.A., Bergman, R.G., and Ellman, J.A. (2001) *J. Am. Chem. Soc.*, **123**, 9692–9693.
20. (a) Thalji, R.K., Ellman, J.A., and Bergman, R.G. (2004) *J. Am. Chem. Soc.*, **126**, 7192–7193; (b) Harada, H., Thalji, R.K., Bergman, R.G., and Ellman, J.A. (2008) *J. Org. Chem.*, **73**, 6772–6779.
21. Davis, T.A., Hyster, T.K., and Rovis, T. (2013) *Angew. Chem. Int. Ed.*, **52**, 14181–14185.
22. Shi, Z., Boultadakis-Arapinis, M., Koester, D.C., and Glorius, F. (2014) *Chem. Commun.*, **50**, 2650–2652.
23. Ye, B., Donets, P.A., and Cramer, N. (2014) *Angew. Chem. Int. Ed.*, **53**, 507–511.
24. Maetani, S., Fukuyama, T., and Ryu, I. (2013) *Org. Lett.*, **15**, 2754–2757.
25. Ellis, G.P. and Lockhart, I.M. (2009) in *The Chemistry of Heterocyclic Compounds: Chromenes, Chromanones, and Chromones*, vol. **31** (ed. G.P. Ellis), Wiley-VCH Verlag GmbH, New York, pp. 1–1196.
26. (a) Shibuya, H., Bohgaki, T., Matsubara, T., Watarai, M., Ohashi, K., and Kitagawa, I. (1999) *Chem. Pharm. Bull.*, **47**, 695–698; (b) Ohashi, K., Bohgaki, T., Matsubara, T., and Shibuya, H. (2000) *Chem. Pharm. Bull.*, **48**, 433–435.
27. Mochida, S., Shimizu, M., Hirano, K., Satoh, T., and Miura, M. (2010) *Chem. Asian J.*, **5**, 847–851.
28. Morimoto, K., Hirano, K., Satoh, T., and Miura, M. (2011) *J. Org. Chem.*, **76**, 9548–9551.
29. Burns, D.J. and Lam, H.W. (2014) *Angew. Chem. Int. Ed.*, **53**, 9931–9935.
30. Wang, M., Zhang, C., Sun, L.-P., Ding, C., and Zhang, A. (2014) *J. Org. Chem.*, **79**, 4553–4560.
31. Fukui, M., Hoshino, Y., Satoh, T., Miura, M., and Tanaka, K. (2014) *Adv. Synth. Catal.*, **356**, 1638–1644.
32. Tan, X., Liu, B., Li, X., Li, B., Xu, S., Song, H., and Wang, B. (2012) *J. Am. Chem. Soc.*, **134**, 16163–16166.
33. Zhang, H., Wang, K., Wang, B., Yi, H., Hu, F., Li, C., Zhang, Y., and Wang, J. (2014) *Angew. Chem. Int. Ed.*, **53**, 13234–13238.
34. Casanova, N., Seoane, A., Mascareñas, J.L., and Gulías, M. (2015) *Angew. Chem. Int. Ed.*, **54**, 2374–2377.
35. Shimizu, M., Tsurugi, H., Satoh, T., and Miura, M. (2008) *Chem. Asian J.*, **3**, 881–886.
36. Bruneton, J. (1999) *Pharmacognosy, Phytochemistry, Medicinal Plants*, 2nd edn, Intercept Ltd, Hampshire, pp. 263–277.
37. (a) Wu, L.H., Li, P., Wang, X., Zhuang, Z., Farzaneh, F., and Xu, R. (2010) *Pharm. Biol.*, **48**, 1344–1353; (b) Chia, Y.C., Chang, F.R., Wang, J.C., Wu, C.C., Chiang, M.Y., Lan, Y.H., Chen, K.S., and Wu, Y.C. (2008) *Molecules*, **13**, 122–128; (c) Chen, K.S., Wu, C.C., Chang, F.R., Chia, Y.C., Chiang, M.Y., Wang, W.Y., and Wu, Y.C. (2003) *Planta Med.*, **69**, 654–657; (d) Min, H.M., Aye, M., Taniguchi, T., Miura, N., Monde, K., Ohzawa, K., Nikai, T., Niwa, M., and Takaya, Y. (2007) *Tetrahedron Lett.*, **48**, 6155–6158; (e) Ito, C., Itoigawa, M., Onoda, S., Hosokawa, A., Ruangrungsi, N., Okuda, T., Tokuda, H., Nishino, H., and Furukawa, H. (2005) *Phytochemistry*, **66**, 567–572.
38. Ueura, K., Satoh, T., and Miura, M. (2007) *J. Org. Chem.*, **72**, 5362–5367.
39. Zhang, M., Zhang, J.-H., Han, T., Ruan, W., and Wen, T.-B. (2015) *J. Org. Chem.*, **80**, 620–627.
40. Unoh, Y., Hashimoto, Y., Takeda, D., Hirano, K., Satoh, T., and Miura, M. (2013) *Org. Lett.*, **15**, 3258–3261.
41. Seo, J., Park, Y., Jeon, I., Ryu, T., Park, S., and Lee, P.H. (2013) *Org. Lett.*, **15**, 3358–3361.
42. Li, X.G., Liu, K., Zou, G., and Liu, N.P. (2014) *Adv. Synth. Catal.*, **356**, 1496–1500.
43. Wang, F., Qi, Z., Sun, J., Zhang, X., and Li, X. (2013) *Org. Lett.*, **15**, 6290–6293.
44. Yu, D.-G., de Azambuja, F., Gensch, T., Daniliuc, C.G., and Glorius, F. (2014) *Angew. Chem. Int. Ed.*, **53**, 9650–9654.

45. Shikanai, D., Murase, H., Hata, T., and Urabe, H. (2009) *J. Am. Chem. Soc.*, **131**, 3166–3167.
46. Seoane, A., Casanova, N., Quiñones, N., Mascareñas, J.L., and Gulías, M. (2014) *J. Am. Chem. Soc.*, **136**, 834–837.
47. Duan, P., Lan, X., Chen, Y., Qian, S.-S., Li, J.J., Lu, L., Lu, Y., Chen, B., Hong, M., and Zhao, J. (2014) *Chem. Commun.*, **50**, 12135–12138.
48. Zhou, B., Yang, Y.-X., Tang, H.-Y., Du, J.-J., Feng, H.-J., and Li, Y.-C. (2014) *Org. Lett.*, **16**, 3900–3903.

7
Ruthenium-Catalyzed Synthesis of Heterocycles via C–H Bond Activation

Bin Li and Baiquan Wang

7.1
Introduction

Heterocycles are the largest and most diverse family of organic compounds, which represent important structural motifs found in natural products and pharmaceuticals [1]. The wide range of applications of heterocycles in biological and physiological materials requires more efficient methods for their synthesis.

Transition metal-catalyzed activation and functionalization of normally unreactive C–H bonds [2] have become an increasingly efficient and reliable approach for new carbon–carbon and carbon–heteroatom bond [3] formations. Thus this new strategy in organic synthesis allows the use of less expensive and more readily available starting materials to produce complex products in an atom- and step-economic fashion. Since the pioneer discoveries by Murai and coworkers on ruthenium(0)-catalyzed addition of aromatic C–H bonds to alkenes [4], various metal complexes have been employed as catalysts for direct functionalization of the ubiquitous C–H bond [2, 3]. Among these reports, novel methodologies utilizing transition metal-catalyzed C–H bond functionalization provide significant benefits toward the synthesis of a variety of heterocycles over the past decades.

Direct C–H bond functionalization via ruthenium complexes catalysis has been well explored [5]. However, the rapid progress for heterocycle synthesis via ruthenium-catalyzed C–H activation has been only achieved since 2011. By utilizing the less expensive and stable ruthenium(II) complex $[RuCl_2(p\text{-cymene})]_2$, the oxidative cyclization reactions with a series of unsaturated small molecules like alkenes and alkynes proved to be a fast access to heterocycles. Herein, this chapter is intended to provide an overview of reported examples for ruthenium-catalyzed C–H bond activation and cyclization for heterocycle synthesis till the end of 2014, including mechanism presentations

Transition Metal-Catalyzed Heterocycle Synthesis via C—H Activation, First Edition. Edited by Xiao-Feng Wu.
© 2016 Wiley-VCH Verlag GmbH & Co. KGaA. Published 2016 by Wiley-VCH Verlag GmbH & Co. KGaA.

for some typical transformations. Note that the reactions involving ruthenium-catalyzed carbene C–H insertion are considered to be out of the scope of this chapter.

7.2
Ruthenium-Catalyzed Heterocycle Synthesis via Intramolecular C–C Bond Formation Based on C–H Bond Activation

Jones and coworkers described the formation of 7-methyl indoles from ruthenium-catalyzed C(sp^3)–H bond activation of 2,6-xylylisocyanide in 1986 (Eq. (7.1)) [6]. The catalytic reaction proceeded by using ruthenium(II) complex $Ru(dmpe)_2H_2$ or $Ru(dmpe)_2$(2-naphthyl)H as catalyst (20 mol%) in C_6D_6 at high temperature (140 °C). Further studies indicated that the substrate scope is restricted to 2,6-disubstituted (2,6-dimethyl, 2-ethyl-6-methyl, 2,6-diethyl) isocyanides, since less substituted isocyanides only produced stoichiometric indole N–H oxidative addition adducts with [$Ru(dmpe)_2$].

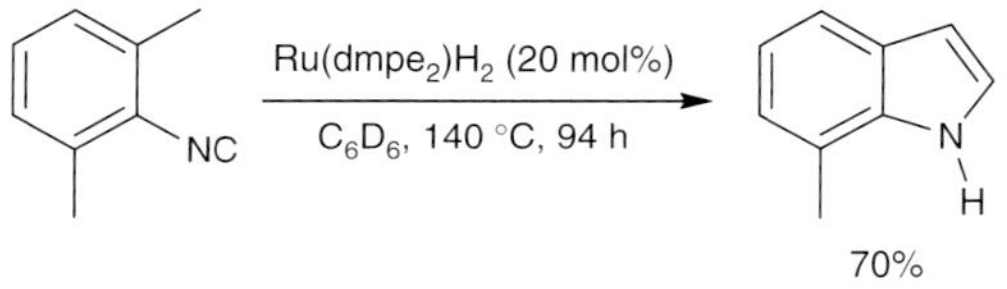

(7.1)

A proposed reaction mechanism is shown in Scheme 7.1. After coordination of isocyanide to [$Ru(dmpe)_2$] and dissociation of one end of a dmpe ligand, oxidative addition of a benzylic C–H bond gives rise to a six-membered metallacycle **1**. Then insertion of the isocyanide into the Ru–CH_2 bond and closure of the dmpe chelation generates intermediate **2**. Followed by tautomerism of the methylene hydrogen to the nitrogen and isomerization, the resulting intermediate *cis*-**3** undergoes reductive elimination to give the final product. The isolation and characterization of intermediate *trans*-**3** and its isomerization to *cis*-**3** were realized under less severe reaction conditions.

In 2013, Carreira and coworkers presented an attractive strategy for the synthesis of chiral pyrrolidones via ruthenium-catalyzed intramolecular hydrocarbamoylation of allylic formamides under CO atmosphere (Eq. (7.2)) [7]. A formal ruthenium-catalyzed intramolecular insertion into the formamide C–H bond and concomitant C–C bond formation by olefin hydrocarbamoylation were involved in the reaction, which made the reaction complete atom economy. The cyclization performed with a broad substrate scope. More interestingly, even homoallylic and bis-homoallylic formamide substrates were used, the reactions afforded five-membered nitrogen-containing heterocycles only.

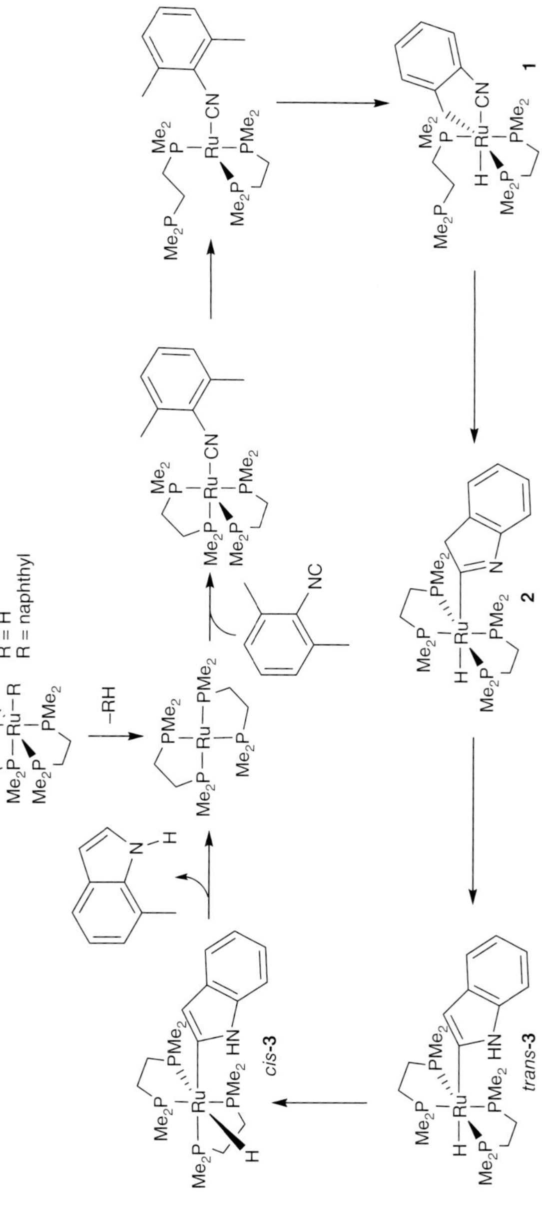

Scheme 7.1 Proposed mechanism for ruthenium-catalyzed indole synthesis from 2,6-xylylisocyanides.

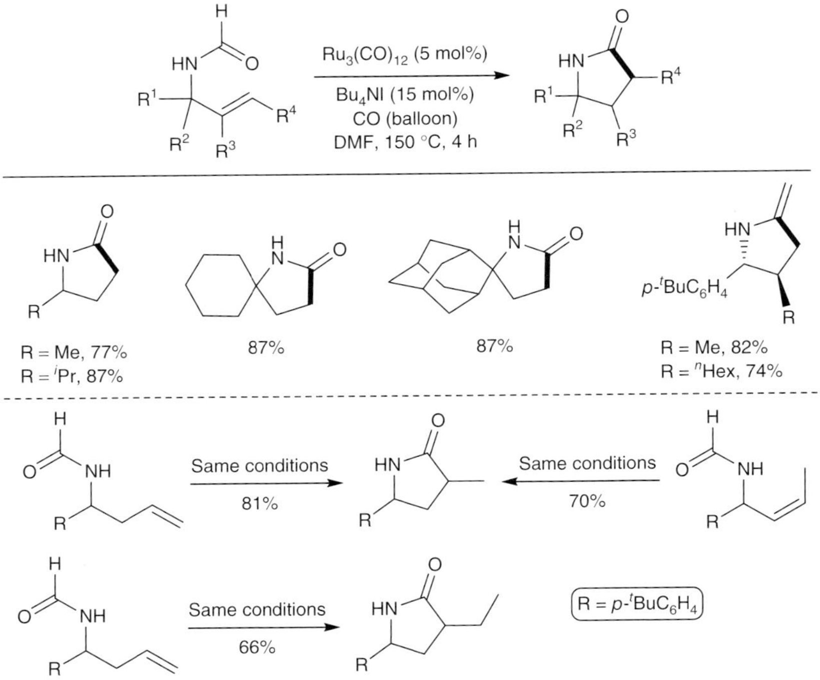

(7.2)

A possible mechanism involving initial insertion of the active ruthenium catalyst into the N–H bond of allylic formamide was proposed (Scheme 7.2). It was followed by reversible olefin insertion to form a ruthenacycle, which undergo β-hydride abstraction of the proximal formamide C–H bond and subsequent attack of the nucleophilic alkyl moiety onto the electrophilic carbonyl carbon to give intermediate **4**. After release of the product with reductive elimination, the active catalyst species was regenerated.

A ruthenium-catalyzed intramolecular olefin hydrocarbamoylation for the regiodivergent synthesis of indolin-2-ones and 3,4-dihydroquinolin-2-ones was disclosed by Chang and coworkers (Eq. (7.3)) [8]. The reactions underwent smoothly without requiring external CO atmosphere. In the presence of combined catalysts of $Ru_3(CO)_{12}/Bu_4NI$, a 5-exo-type cyclization proceeds favorably to form indolin-2-ones as a major product in good to excellent yields in DMSO/toluene cosolvent (catalytic system A). When the reaction was conducted in the absence of halide additives in *N*,*N*-dimethylacetamide (DMA)/PhCl (catalytic system B), 3,4-dihydroquinolin-2-ones were obtained in major in moderate to high yields via a 6-endo cyclization process. An excellent level of regioselectivity was observed with a variety of substrates to deliver 5-exo- or 6-endo-cyclized lactams.

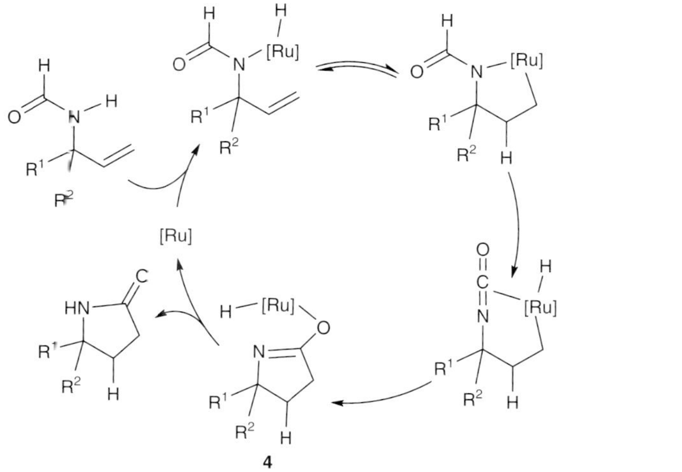

Scheme 7.2 Proposed mechanism for ruthenium-catalyzed pyrrolidone synthesis from allylic formamides.

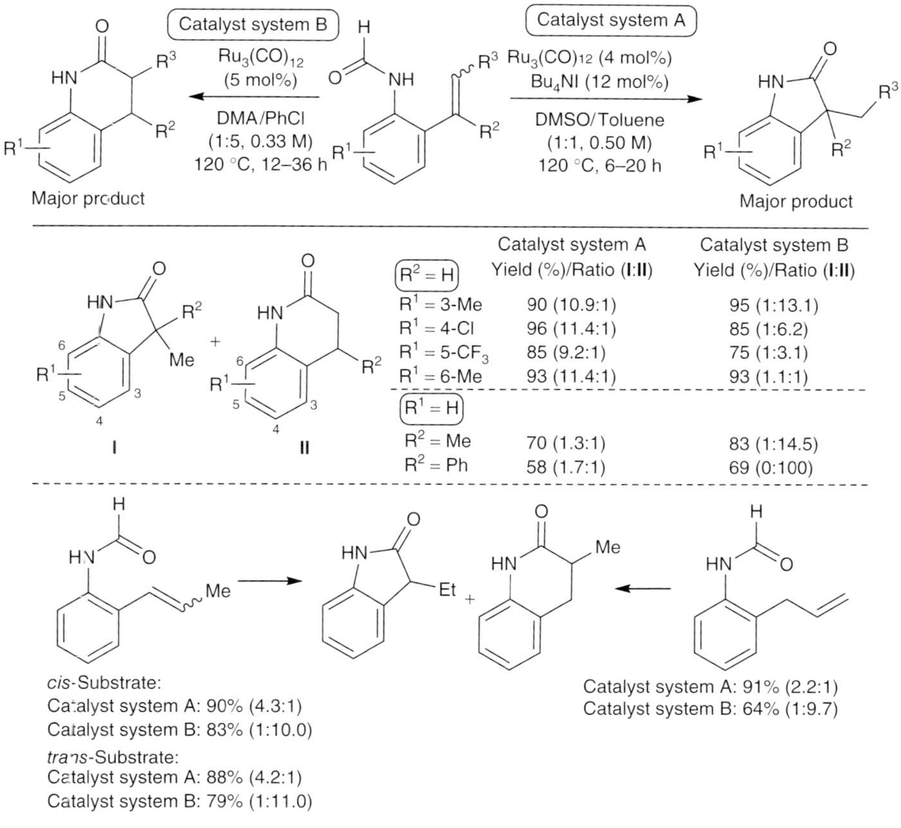

	Catalyst system A Yield (%)/Ratio (I:II)	Catalyst system B Yield (%)/Ratio (I:II)
R^2 = H		
R^1 = 3-Me	90 (10.9:1)	95 (1:13.1)
R^1 = 4-Cl	96 (11.4:1)	85 (1:6.2)
R^1 = 5-CF_3	85 (9.2:1)	75 (1:3.1)
R^1 = 6-Me	93 (11.4:1)	93 (1.1:1)
R^1 = H		
R^2 = Me	70 (1.3:1)	83 (1:14.5)
R^2 = Ph	58 (1.7:1)	69 (0:100)

(7.3)

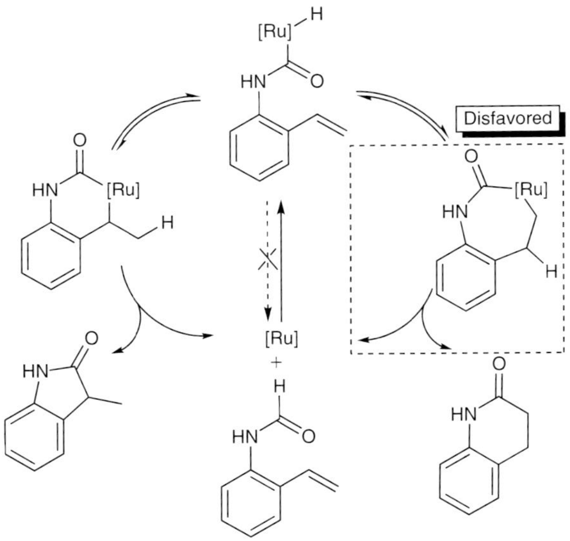

Scheme 7.3 Proposed mechanism for ruthenium-catalyzed intramolecular olefin hydrocarbamoylation through direct activation of the formyl C–H bond.

Two mechanistic pathways, which differed in the way of ruthenium-mediated initial cleavage of formyl C–H or amido N–H bond, were proposed for the catalytic cycle. As shown in Scheme 7.3, an irreversibly cleavage of formyl C–H bond by the active ruthenium complex was followed by reversible insertion of the olefin into the Ru–H bond, which afforded either six-membered or seven-membered ruthenacycle. After reductive elimination, indolin-2-ones or 3,4-dihydroquinolin-2-one was formed. According to isotopic studies, pathway leading to six-membered lactams is postulated to be less favored. Another cyclization process initiated by Ru-catalyzed oxidative addition of formyl N–H bond (Scheme 7.4) was similar to Carreira's proposal for their hydrocarbamoylation reaction of allylic formamides under similar ruthenium catalysis conditions [7]. The 6-endo cyclization process is proposed to be favored under the catalytic system B.

7.3 Ruthenium-Catalyzed Heterocycle Synthesis via Intramolecular C–N Bond Formation Based on C–H Bond Activation

Multisubstituted pyrazoles commonly are difficult to prepare with conventional methods. The synthesis of a variety of high diversified tri- and tetrasubstituted

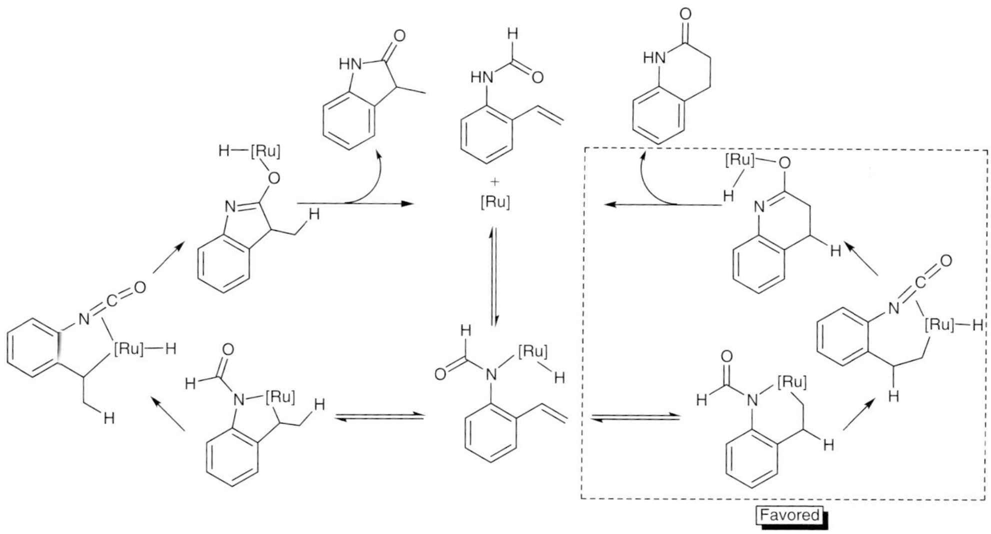

Scheme 7.4 Proposed mechanism for ruthenium-catalyzed intramolecular olefin hydrocarbamoylation through initial activation of the N–H bond.

pyrazoles was achieved by ruthenium(II)-catalyzed intramolecular oxidative C–N coupling from easily accessible starting materials (Eq. (7.4)) [9]. The C–H/N–H coupling transformation proceeded with excellent reactivity and high tolerance of functional groups. Notably, 1 atm of O_2 is employed as the oxidant for this catalytic process.

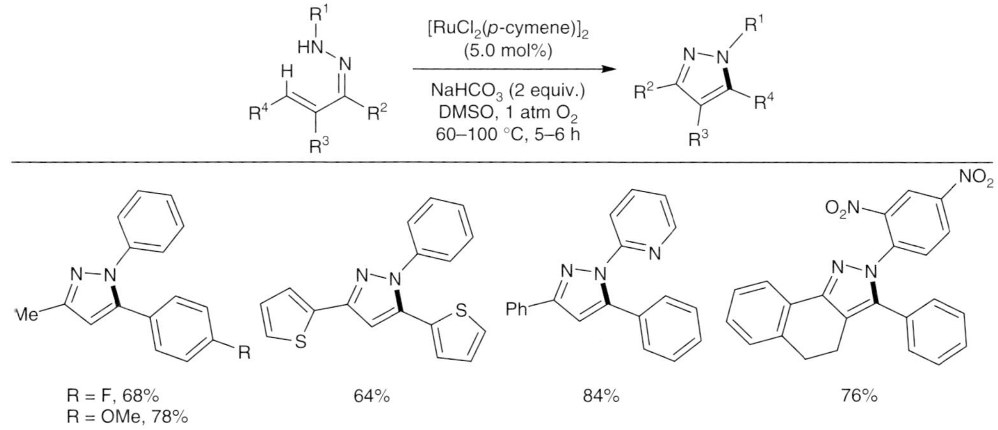

(7.4)

7.4
Ruthenium-Catalyzed Heterocycle Synthesis via Intermolecular C–C/C–O Bond Formation Based on C–H Bond Activation

7.4.1
Cyclization with Alkynes

Ruthenium(II) complex enabled aerobic oxidative annulation of substituted acids with alkynes was firstly described by the Jeganmohan group (Eq. (7.5)) [10]. The reactions were conducted in the presence of catalytic amount of silver and copper salt in DCE at 100 °C for 12 h to provide isocoumarin and α-pyrone derivatives in good to excellent yields. With the addition of $AgSbF_6$ and $Cu(OAc)_2 \cdot H_2O$, the catalytically active species was supposed to be the in situ generated cationic ruthenium(II) complex $[Ru(OAc)(p\text{-cymene})]^+[SbF_6]^-$. The addition of $AgSbF_6$ is important for this catalytic system in terms of controlling the regioselectivity of the alkyne insertion and suppressing the decarboxylative cyclization side reaction. The reaction is postulated to initiate with an *ortho*-C–H bond cycloruthenation to form five-membered ruthenacycle intermediate, followed by the alkyne regioselective insertion. Subsequent reductive elimination gives the desired isocoumarin product and regenerates the active ruthenium species for the next catalytic cycle with the aid of copper acetate.

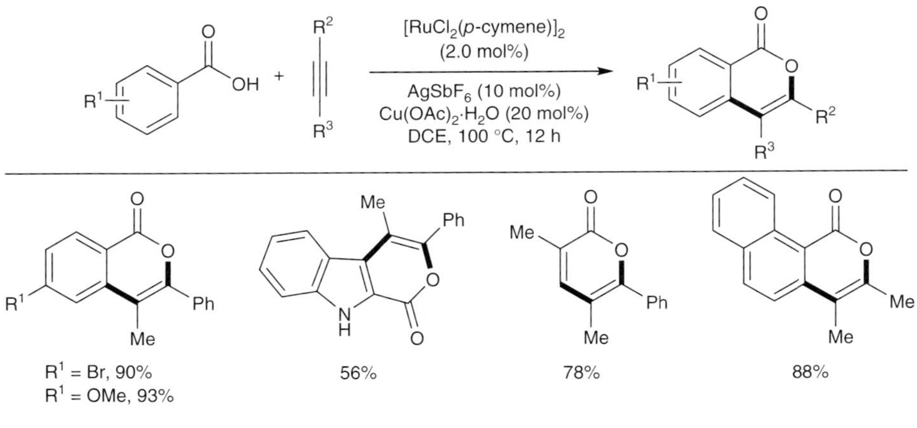

(7.5)

Ackermann and coworkers explored the same cyclization reactions, which were conducted with catalytic amount of KPF_6 and $Cu(OAc)_2 \cdot H_2O$ in *t*-AmOH at 120 °C for 16 h (Eq. (7.6)) [11]. They also proposed a similar catalytic cycle involving acetate-assisted rate-determining C–H bond metalation step according to mechanistic study.

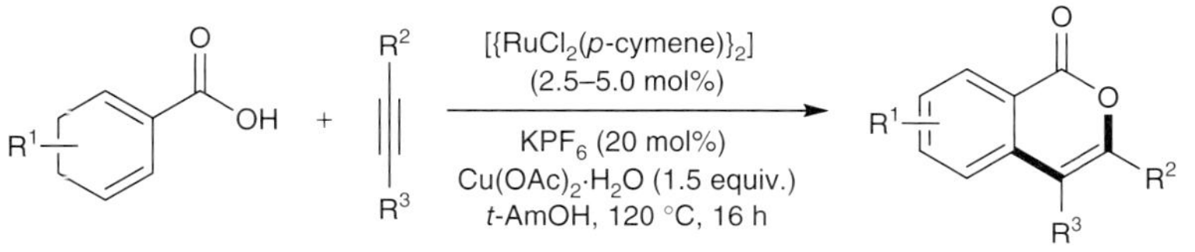

(7.6)

Hydroxyl group-directed oxidative annulations with alkynes for the production of fluorescent pyrans were reported (Eq. (7.7)) [12]. Not only naphthols but also 4-hydroxycoumarin and 4-hydroxy-substituted quinolin-2-one underwent this ruthenium(II)-catalyzed C–H/O–H bond functionalization process in a highly chemo- and regioselective manner. Competition reactions showed that electron-deficient alkynes are more reactive. Deuterium experiments also revealed a reversible C–H bond ruthenation step via carboxylate assistance.

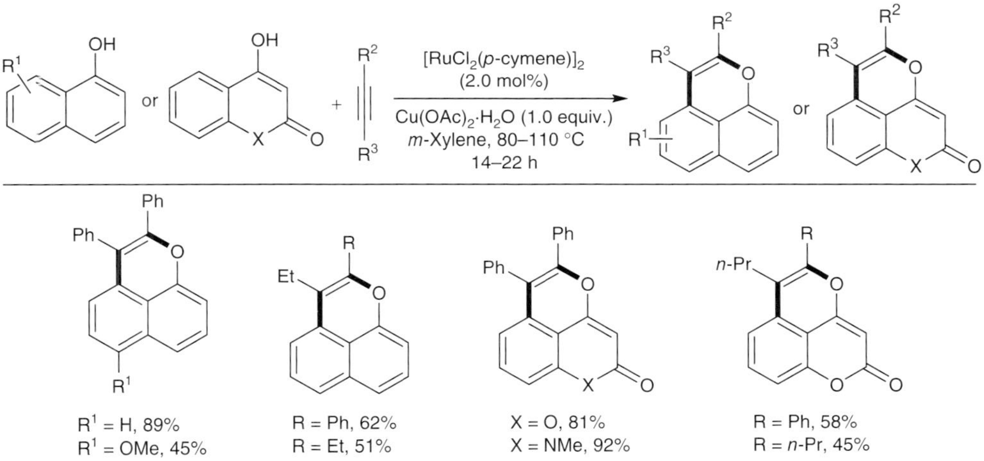

(7.7)

Unsymmetrical 2-aryl cyclic 1,3-dicarbonyl compounds which contain two distinct, nonadjacent sites for C–H bond functionalization were applied as substrates for the oxidative annulation reactions with alkynes. Interestingly, catalyst-controlled divergent C–H functionalization was realized in this transformation [13]. With the aid of ruthenium catalyst, functionalization of a hydrogen atom four bonds away from the oxygen of the directing group mainly occurred, leading to formation of benzopyran derivatives (Eq. (7.8)). In contrast, a palladium-based catalyst exclusively results in functionalization of a hydrogen atom five bonds away from the oxygen of the enol/enolate directing group, producing diverse substituted spiroindenes.

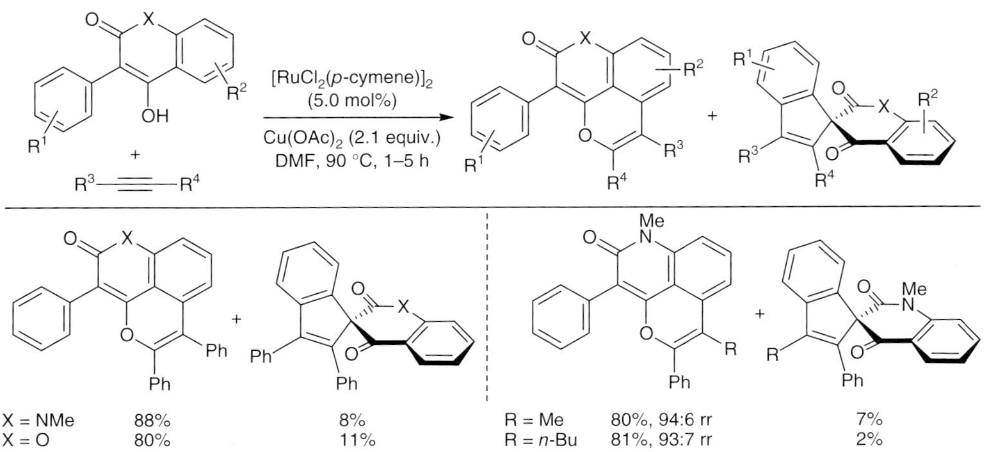

(7.8)

Oxidative cyclizations of phosphonic acid monoesters or phosphonic acids with alkynes via ruthenium(II)-catalyzed cleavage of C–H and O–H bond were achieved by Lee and coworkers (Eq. (7.9)) [14]. In the presence of [RuCl$_2$(*p*-cymene)]$_2$ (10–15 mol%)/KPF$_6$ (20 mol%), 1 equiv. of Ag$_2$CO$_3$ and AgOAc in *t*-BuOH, this methodology produced a wide range of phosphaisocoumarin derivatives in good to excellent yields under aerobic conditions. Vinyl phosphonic monoester was also suitable for this catalytic process by insertion of the alkyne into vinyl C–H and O–H bonds. Notably, the electronic effects of alkynes almost did not affect the present oxidative annulations. Furthermore, a rate-determining C–H bond metalation step was revealed to initiate the catalytic cycle.

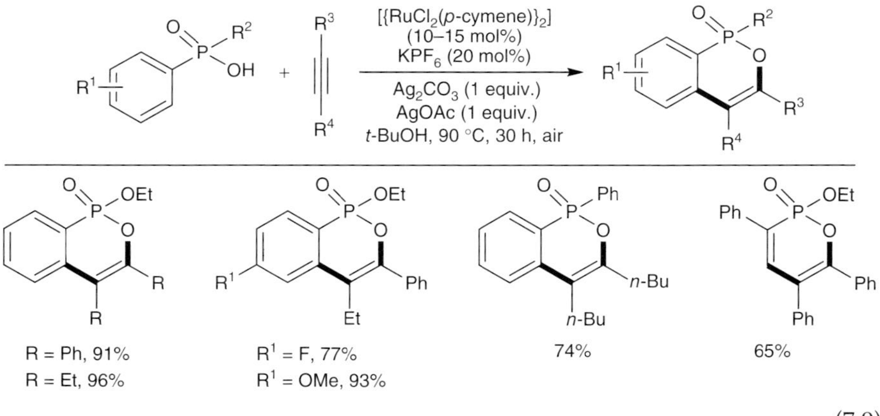

(7.9)

In 2014, Liu and Lu disclosed a Ru-catalyzed C–H functionalization with alkynes for the synthesis of benzofuran derivatives using –ONHPiv as an oxidizing directing group (Eq. (7.10)) [15]. The reaction showed good functional group tolerance and high regioselectivity. Low yields were obtained with the

N-phenoxypivalamide substrates bearing electron-withdrawing groups. Competition experiments revealed that the electron-deficient diarylalkynes were more reactive.

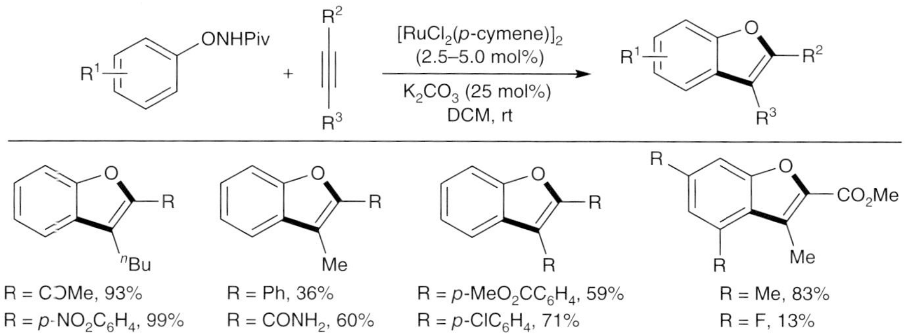

(7.10)

As shown in Scheme 7.5, the catalytic cycle initiates with an irreversible C–H cleavage by Ru(II) to yield a five-membered ruthenacycle intermediate. Subsequent alkyne insertion and protolysis led to intermediate **5**. Then the

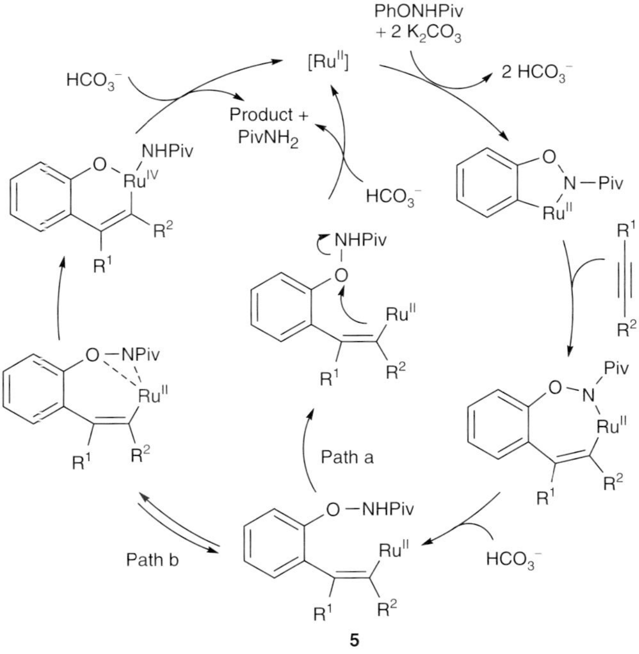

Scheme 7.5 Proposed mechanism for ruthenium-catalyzed benzofuran synthesis from *N*-phenoxypivalamide.

following two pathways might occur: (i) the C–O bond formation and the simultaneous O–N bond cleavage via intramolecular substitution and (ii) intramolecular oxidative addition and reductive elimination involving Ru(IV) intermediate.

Recently, Ackermann and coworkers reported aliphatic hydroxyl-directed oxidative annulation reactions of benzyl alcohols with alkynes to form isochromene derivatives (Eq. (7.11)) [16]. This C–H/O–H functionalization process performed smoothly by using $[RuCl_2(p\text{-cymene})]_2$ (5 mol%)/$AgPF_6$ (20 mol%) as catalyst and $Cu(OAc)_2 \cdot H_2O$ (20 mol%) as oxidant under an atmosphere of air. Various tertiary benzylic alcohols, α,α-dimethylallyl alcohol, and diverse internal alkynes are appropriate substrates for this transformation. The reaction occurred with moderate to high regioselectivity for unsymmetrical alkylarylacetylene, and an irreversible C–H metalation step is involved in the catalytic cycle.

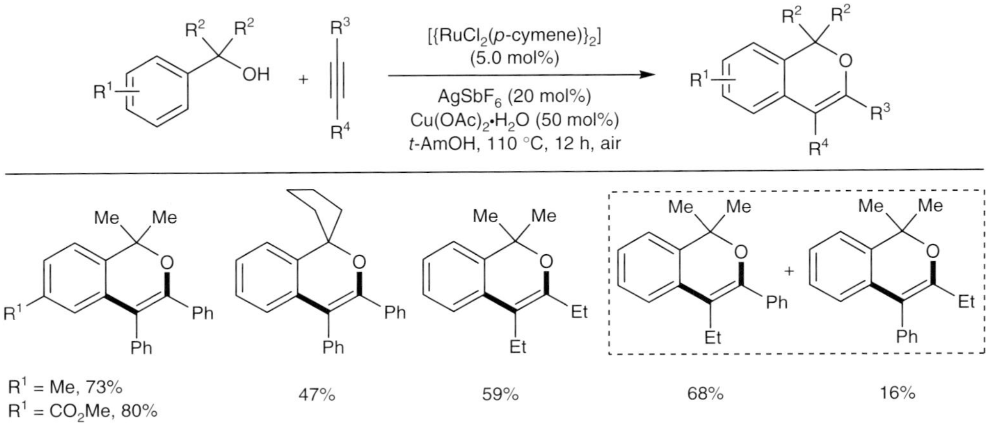

(7.11)

7.4.2
Cyclization with Alkenes

After the seminal work reported by Satoh and Miura in early 2011 on ruthenium-catalyzed oxidative vinylation of heteroarene carboxylic acids with alkenes [17], Ackermann demonstrated a ruthenium(II)-catalyzed cross-dehydrogenative C–H bond alkenylations of benzoic acid derivatives with acrylonitrile or alkyl acrylates. Following the oxidative C–H bond alkenylation reaction, subsequent intramolecular oxa-Michael reaction occurred leading to phthalides in good yields (Eq. (7.12)) [18]. The reactions took place with water as an environmentally benign medium under mild conditions.

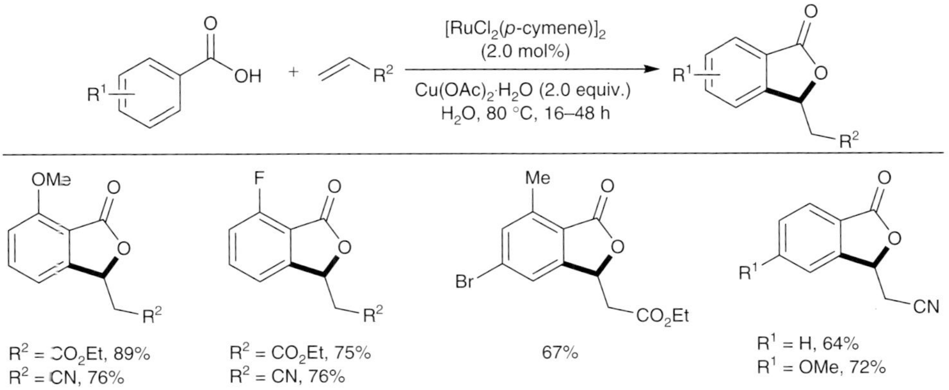

(7.12)

A ruthenium-catalyzed phthalide synthesis from the reaction of mandelic acids with acrylates was established by Wang and coworkers (Eq. (7.13)) [19]. A sequence of dialkenylation, decarboxylation, and subsequent intramolecular cyclization led to the final product. Of note, the aromatic rings of the mandelic acids with strong electron-withdrawing group or conjugated aryl group led to good yields of the corresponding products.

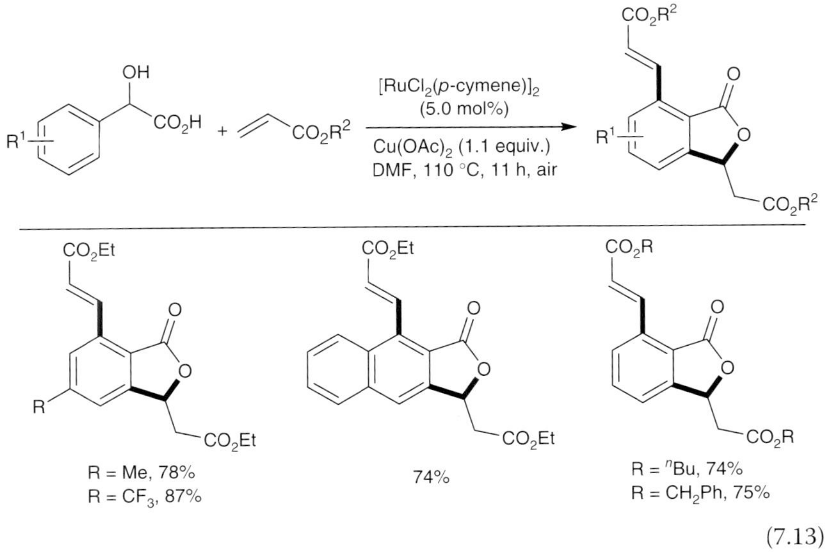

(7.13)

Lam and coworkers explored the direct C– bond alkenylation of 2-aryl-3-hydroxy-2-cyclohexenones (Eq. (7.14)) [20] after their studies on oxidative annulation of alkynes with these substrates [21]. The resulting benzopyrans were formed in the presence of $[RuCl_2(p\text{-cymene})]_2$ (2.5 mol%), $Cu(OAc)_2$ (2.1 equiv.), and K_2CO_3 (2.0 equiv.) in up to 78% yield via oxa-Michael-type addition of the enol oxygen atom to the alkenylated intermediate. The similar reaction could also proceed under the action of well-established palladium catalysis with more alkene substrates such as phenyl vinyl sulfone and methyl vinyl ketone.

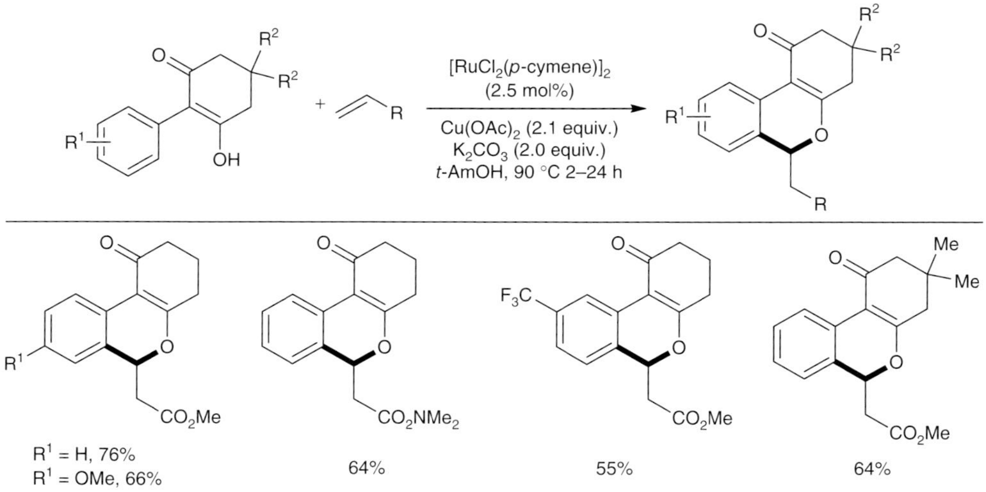

(7.14)

Catalyst-controlled divergent C–H functionalization was also achieved in the reaction of unsymmetrical 2-aryl cyclic 1,3-dicarbonyl compounds with electron-deficient terminal alkenes. In this context, ruthenium-based catalyst produced benzofurans selectively (Eq. (7.15)) [13]. In contrast, benzopyrans were formed in the palladium-catalyzed reactions.

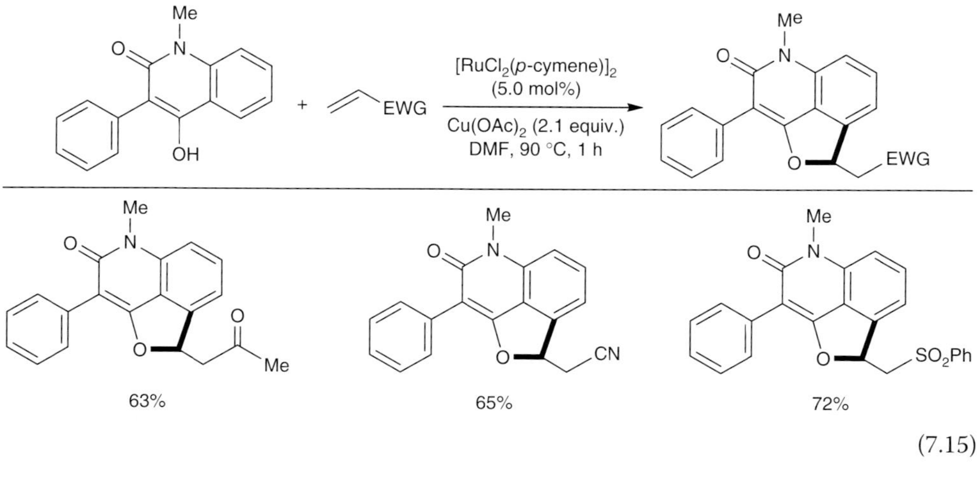

(7.15)

7.4.3
Cyclization with Carbon Monoxide

Murai and coworkers showed a ruthenium-catalyzed cyclocarbonylation of yne-aldehydes in 1998 (Eq. (7.16)) [22]. In the presence of a catalytic amount of $Ru_3(CO)_{12}$, the reaction of yne-aldehyde with CO (10 atm) in toluene at 160 °C gave a series of α,β-unsaturated bicyclic γ-butenolides in high yields. It is noteworthy that polyfunctional compounds could be formed in a single step by employing this method.

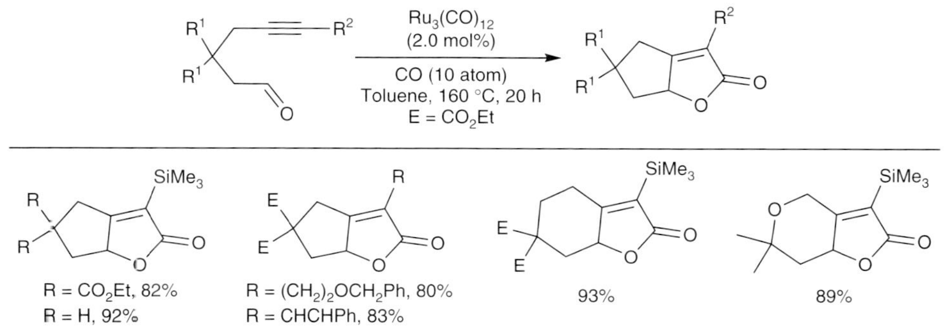

(7.16)

Two possible mechanisms were proposed. One is transition metal-catalyzed hetero-Pauson–Khand process. Another pathway was shown in Scheme 7.6, which was initiated by the oxidative addition of an aldehyde C–H bond to ruthenium.

A ruthenium-based complex enabled catalytic carbonylative C–H cyclization of 2-arylphenols was achieved by using balloon pressure of CO and O_2 (Eq. (7.17)) [23]. Under relatively mild reaction conditions, this methodology produced a variety of 6*H*-dibenzo[*b,d*]pyran-6-one derivatives in high yields with broad substrate scope. Competition experiment suggested that electron-rich substrates are more reactive. In addition, experiment with isotopically labeled substrate revealed that the C–H metalation step is reversible.

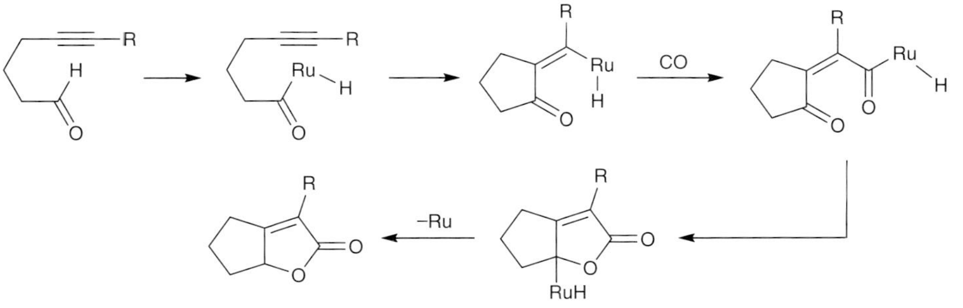

Scheme 7.6 Proposed mechanism for ruthenium-catalyzed cyclocarbonylation of yne-aldehydes.

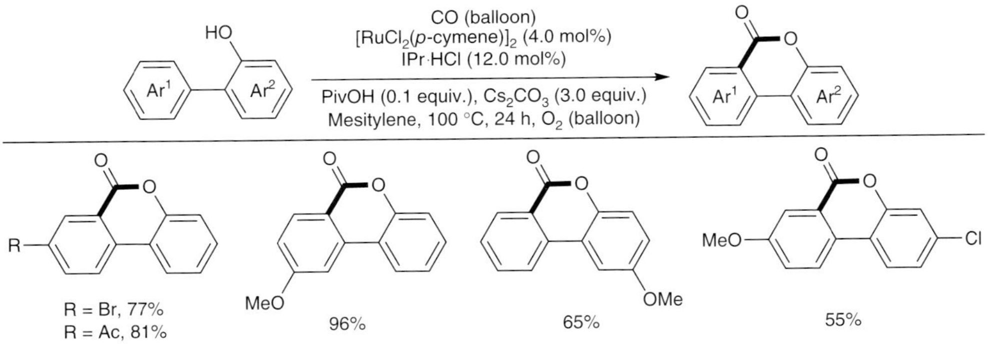

(7.17)

7.4.4
Cyclization with 1,2-Diol

Treatment of phenol with 1,2-diols and excess of cyclopentene (3 equiv.) in the presence of a well-defined cationic ruthenium hydride complex $[(C_6H_6)(PCy_3)(CO)RuH]^+BF_4^-$ (1 mol%) in toluene at 100 °C for 8–12 h led to the formation of benzofuran derivatives (Eq. (7.18)) [24]. The catalytic C–H coupling method exhibited a broad substrate scope, tolerated carbonyl and amine functional groups, obviated the use of any expensive and often toxic metal oxidants, and liberated water as the only by-product. Furthermore, excellent regioselective addition of the linear 1,2-diols were observed, which yielded the α-substituted benzofuran products exclusively. Such dehydrative C–H alkenylation and annulation reactions could be applied for a number of functionalized phenol and alcohol substrates of biological importance.

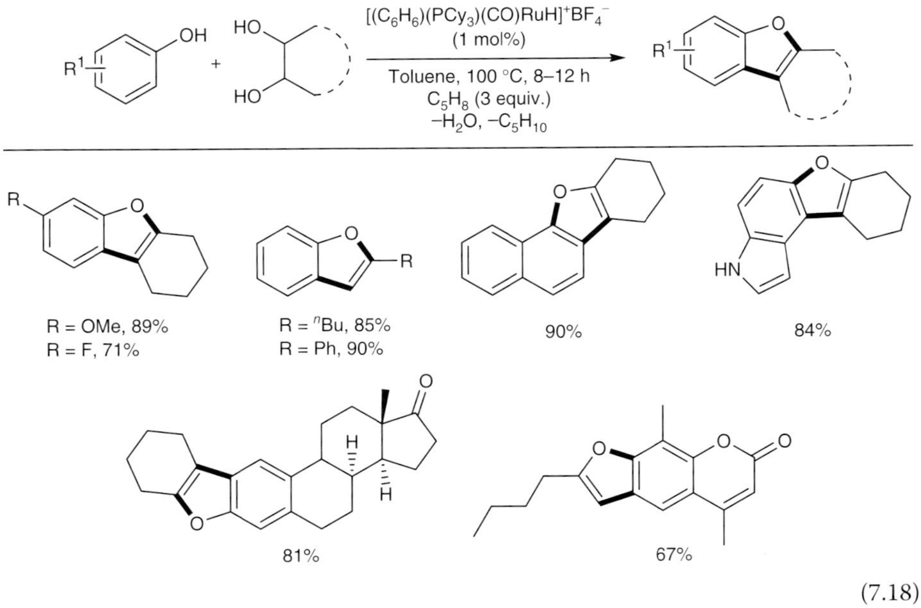

(7.18)

7.5
Ruthenium-Catalyzed Heterocycle Synthesis via Intermolecular C–C/C–N Bond Formation Based on C–H Bond Activation

7.5.1
Cyclization with Alkynes

The Nicholas group described that the annulation reactions of nitroaromatics with alkynes catalyzed by $[Cp^*Ru(CO)_2]_2$ under carbon monoxide regioselectively produced indole derivatives in 2002 (Eq. (7.19)) [25].

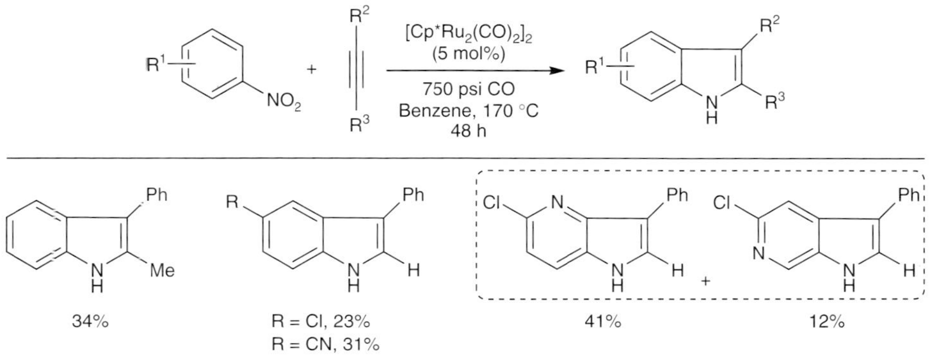

(7.19)

Almost the same time, they showed that indoles could also be formed from $[Cp^*Ru(CO)_2]_2$-catalyzed reductive annulation of nitrosoaromatics (ArNO) with alkynes (Eq. (7.20)) [26]. Although only moderate yields were observed for both methods, the transformations were conducted under neutral reaction conditions and exhibited high regioselectivity.

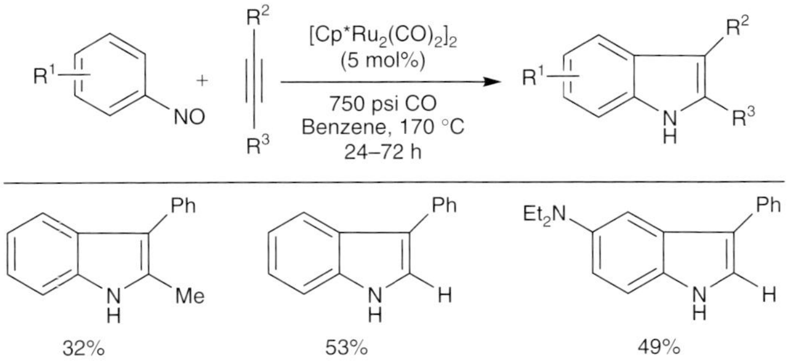

(7.20)

In 2005, Yi and coworkers reported a catalytic C–H bond activation and cyclization protocol by using the catalyst $Ru_3(CO)_{12}/NH_4PF_6$. This catalytic system offers an efficient method for the synthesis of tricyclic quinoline derivatives from regioselective coupling of benzocyclic amines with 2 equiv. of alkynes (Eq. (7.21)) [27]. Later, the catalytic coupling reactions of acyclic arylamines with terminal alkynes were also accomplished to deliver substituted quinoline and quinoxaline derivatives in up to 96% yield with $Ru_3(CO)_{12}/HBF \cdot OEt_2$ catalyst [28]. In the latter transformation, the arylamines with *meta*-substituent were found to promote the *ortho*-C–H bond activation of aniline, thus depressing the formation of hydroamination side product and increasing the efficiency of cyclization.

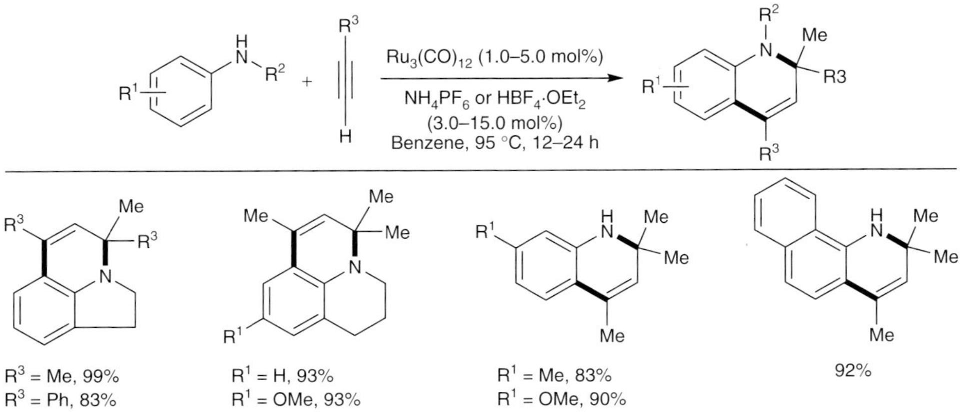

(7.21)

The reaction is postulated to proceed via an intramolecular migratory insertion of amine, followed by *ortho*-C–H bond activation and cyclization (Scheme 7.7). Notably, a catalytically active cationic ruthenium-acetylide complex has been isolated and characterized in this case.

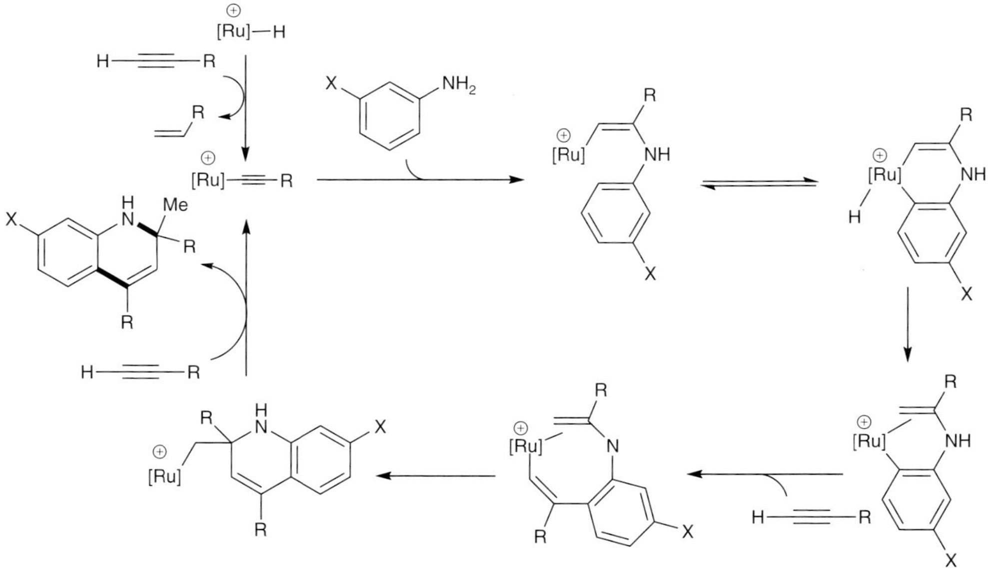

Scheme 7.7 Proposed mechanism for ruthenium-catalyzed cyclization of amines with alkynes.

Then until 2011, the Ackermann group reported a ruthenium(II)-catalyzed oxidative annulations of alkynes with benzamides (Eq. (7.22)) [29]. In this method, $Cu(OAc)_2 \cdot H_2O$ was used as the terminal oxidant to regenerate the ruthenium(II) catalyst in the catalytic cycle. This cyclization reaction proceeded via cleavage of both *ortho*-C–H and amide N–H bonds of benzamides to form isoquinolones in high chemo- and regioselectivity. The insertion of unsymmetrical alkylarylalkynes occurred with a connection of the electron-deficient alkyne carbon with nitrogen atom selectively. Moreover, electron-deficient alkynes exhibited higher reactivity in this transformation.

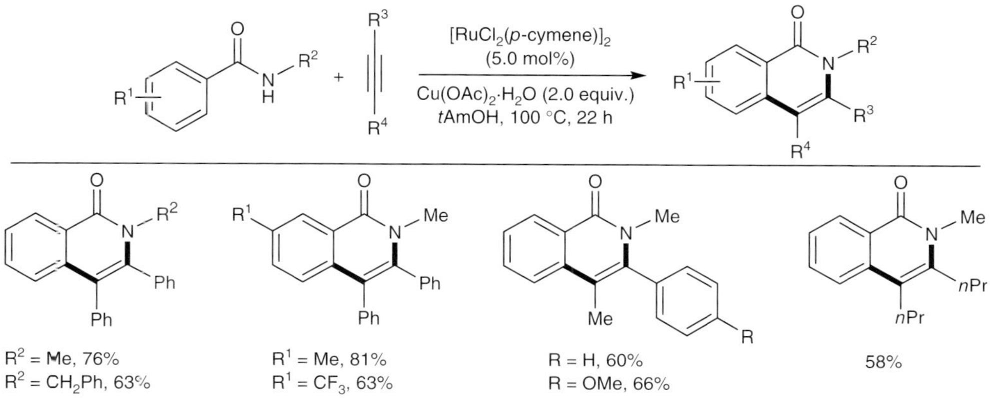

(7.22)

A rate-limiting C–H bond metalation step via acetate assistance was proposed to initiate the reaction. Then after alkyne insertion and subsequent reductive elimination, the final product was formed (Scheme 7.8).

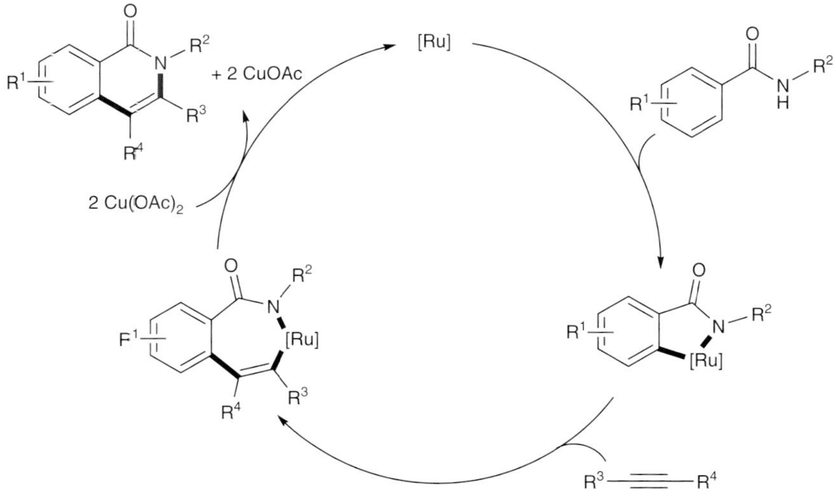

Scheme 7.8 Proposed mechanism for ruthenium-catalyzed cyclization of benzamides with alkynes.

The preparation of 2-pyridones through C–H/N–H bond functionalization and annulation of acrylamides with alkynes was achieved by employing the same ruthenium(II) catalytic system but using 1 equiv. of copper oxidant with good chemo- and regioselectivity (Eq. (7.23)) [30]. Notably, the ruthenium catalysts displayed improved substrate scope as compared to the similar reactions reported with rhodium catalyst.

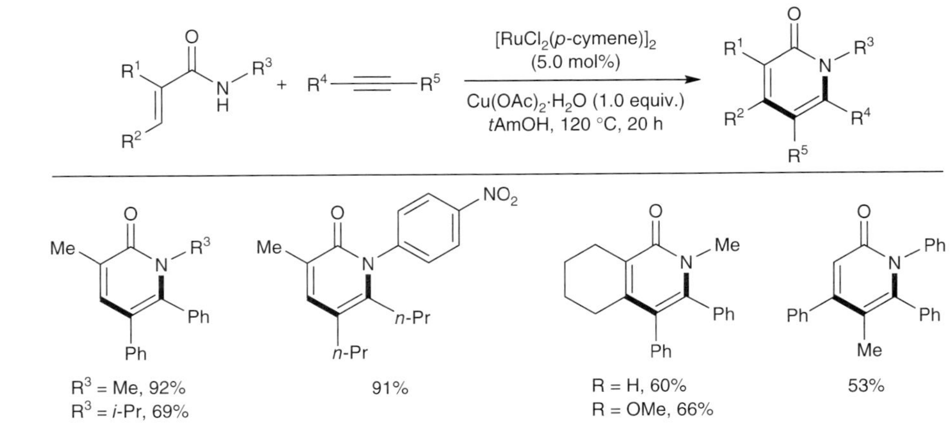

(7.23)

By introduction of an oxidizing directing group, Li and Wang demonstrated a regioselective ruthenium(II)-catalyzed isoquinolone synthesis through oxidative cyclization of *N*-methoxybenzamides with internal alkynes (Eq. (7.24)) [31]. This redox-neutral strategy circumvents the use of wasteful metal oxidants, and the reaction could take place at much milder conditions with broad substrate scope.

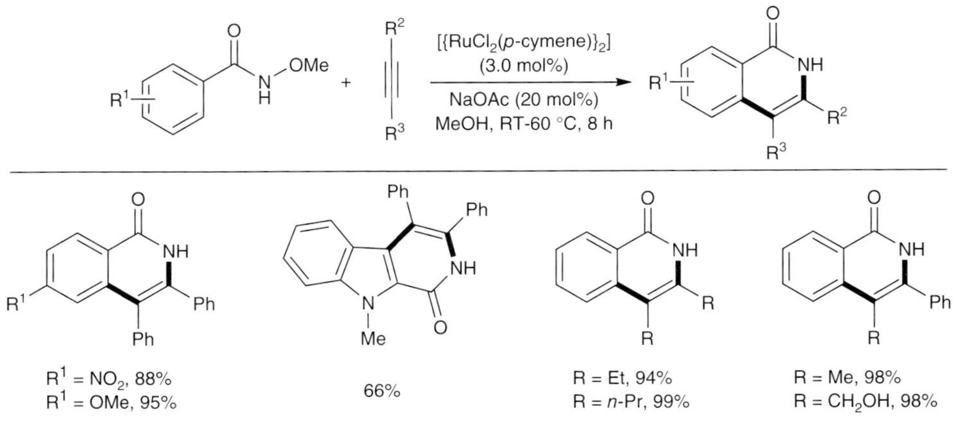

(7.24)

The same reaction can also be performed with water as solvent medium by the catalytic system of $[RuCl_2(p\text{-cymene})]_2$ associated with KO_2CMes (30 mol%) at 60 °C [32]. Free hydroxamic acid appeared also suitable for this annulation under slightly more drastic conditions at 60–100 °C and gave isoquinolone products in slightly low yield (Eq. (7.25)) [32a]. Later, the potassium salt of the electron-deficient carboxylic acid 3-$(F_3C)C_6H_4CO_2H$ proved to be a more efficient cocatalytic additive than KO_2CMes, setting the stage for effective dehydrative alkyne annulation by C–H/N–OH functionalization of hydroxamic acids with alkynes [33].

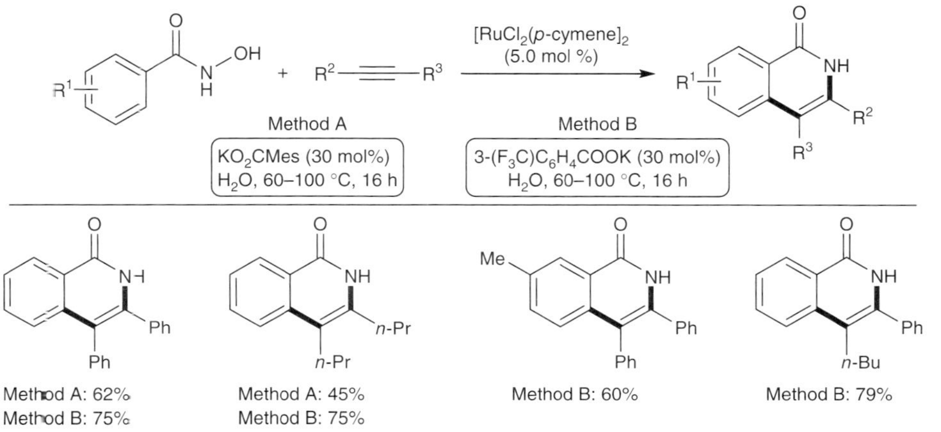

(7.25)

Ruthenium-catalyzed oxidative coupling of 2-aryl-substituted indoles and pyrroles with a variety of alkynes to generate structural analogs of bioactive marine alkaloids was reported (Eq. (7.26)) [34]. Ambient air was used as the terminal oxidant together with the addition of cocatalytic amounts of $Cu(OAc)_2 \cdot H_2O$ in this transformation. The annulation reactions favored the electron-deficient alkyne and the more acidic C–H bond. Mechanistic experiments showed that the catalytic process took place via a concerted, acetate-assisted deprotonative metalation step.

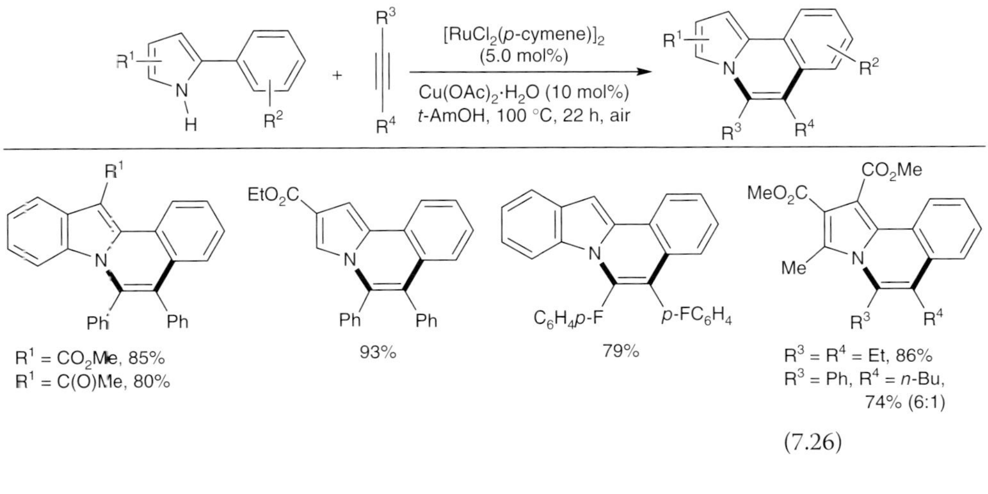

(7.26)

The same ruthenium catalytic system also enabled oxidative coupling of 2-arylbenzimidazoles with alkynes (Eq. (7.27)) [35a]. The benzimidazoisoquinoline derivatives were generated in toluene at reflux in good yields with high regioselectivity. Further studies showed that PEG-400 could be used as an alternate solvent medium to produce the desired products at room temperature in similar yields. More interestingly, the catalyst could be recycled for a few times with minimal loss of activity in the latter catalytic system. The related Ru-catalyzed oxidative annulation reactions of alkynes with 2-phenylimidazole and 2-arylphenanthroimidazoles were reported by the Wang group [35b] and the Hua [35c] group, respectively.

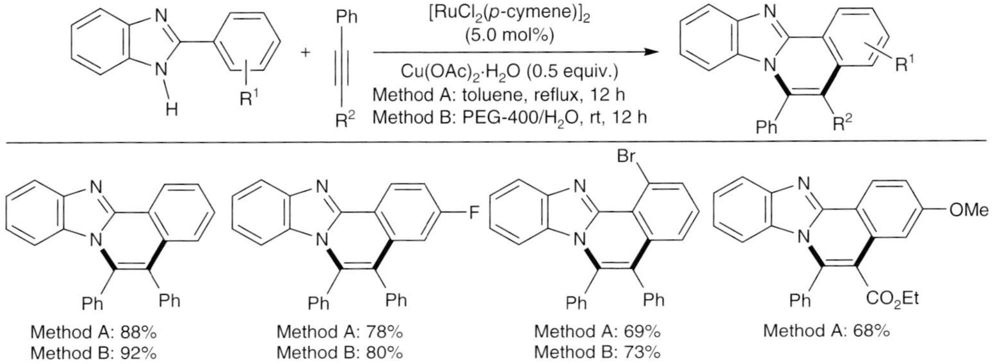

(7.27)

Under conditions of C–H/N–H bond functionalization, aryl-, heteroaryl-, and alkenyl-substituted 1H-pyrazoles underwent oxidative annulation with aryl and alkyl alkynes in high chemo- and regioselectivity in the presence of Ru(II)/$AgSbF_6$ catalyst (Eq. (7.28)) [36]. Aryl alkynes particularly bearing electron-donating substituents are more reactive in the present reaction system. A cationic ruthenium(II)-catalyzed reversible C–H bond metalation step was observed in the H/D exchange experiments.

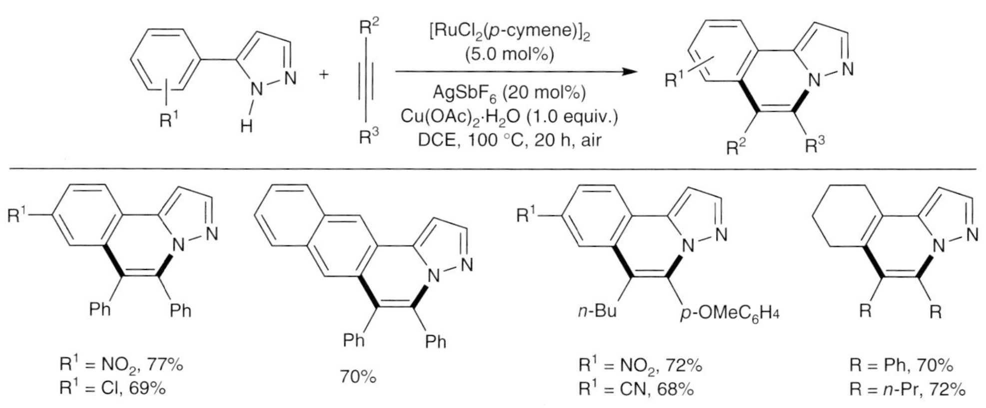

(7.28)

N-2-Pyrimidyl- (or pyridyl)-substituted anilines reacted with internal alkynes to generate indole derivatives under the conditions of cationic ruthenium(II) complexes enabled C–H bond functionalizations (Eq. (7.29)) [37]. This $[RuCl_2(p\text{-cymene})]_2/KPF_6$-catalyzed C–H/N–H bond cleavage process could perform most efficiently in water with ample scope, and the directing group can be easily removed by treatment with NaOEt in DMSO from indole. High-level regioselectivity was observed for alkylarylacetylenes with the alkyne carbon linked to the aryl group being selectively connected with the nitrogen atom. Notably, a rare six-membered ruthenacycle was proposed as a key intermediate for the cyclometalation step.

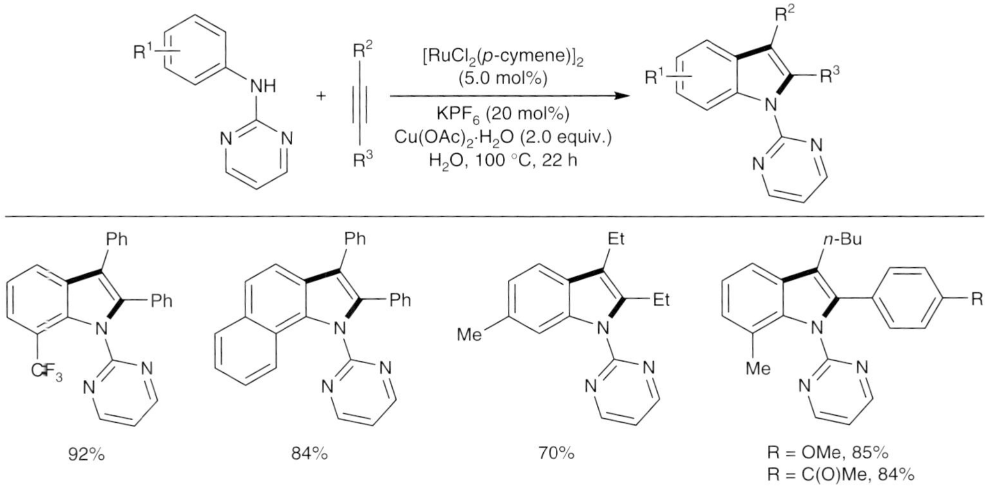

(7.29)

A detailed mechanistic study of the oxidative annulation coupling of isoquinolones with alkynes was investigated by Li and Wang (Scheme 7.9) [38]. The first step is an acetate-assisted cyclometalation to afford a five-membered ruthenacycle **6** via N–H/C–H bond cleavage of the isoquinolone. After the insertion of an alkyne into the Ru–C bond, a seven-membered ruthenacycle intermediate **7** is formed. Finally, oxidative coupling of the C–N bond occurs to generate a sandwich ruthenium(0) species **8**, which undergoes oxidation to release the desired dibenzo[*a*,*g*]quinolizin-8-one derivatives and regenerate the catalytically active Ru(II) complex with the aid of copper oxidant. It should be noted that all of the relevant intermediates were fully characterized and determined by single-crystal X-ray diffraction analysis.

Based on the results of the catalytic cycle study described previously, the catalytic reactions were subsequently investigated. The reaction occurred efficiently with $[RuCl_2(p\text{-cymene})]_2$ (7.5 mol%) as catalyst and $Cu(OAc)_2 \cdot H_2O$ (2.2 equiv.) as oxidant, together with 2 equiv. of Na_2CO_3 in high chemo- and regioselectivity (Eq. (7.30)). The resulting dibenzo[*a*,*g*]quinolizin-8-one derivatives was obtained with a broad substrate scope. 3,4-Diphenylisoquinolones

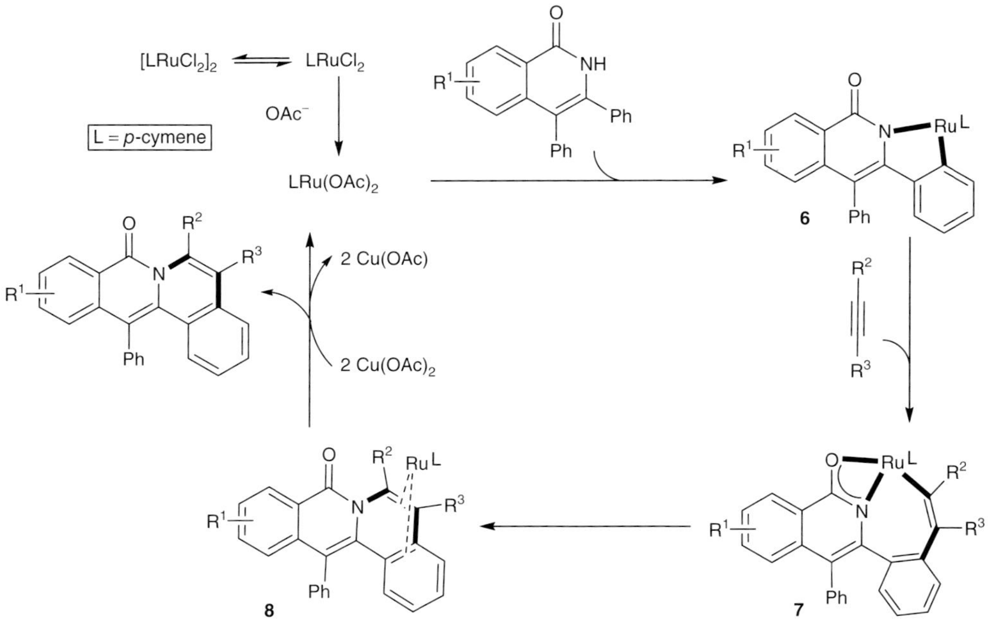

Scheme 7.9 Proposed mechanism for ruthenium-catalyzed oxidative annulations of isoquinolones with alkynes.

bearing electron-donating groups were less reactive than those bearing electron-withdrawing groups.

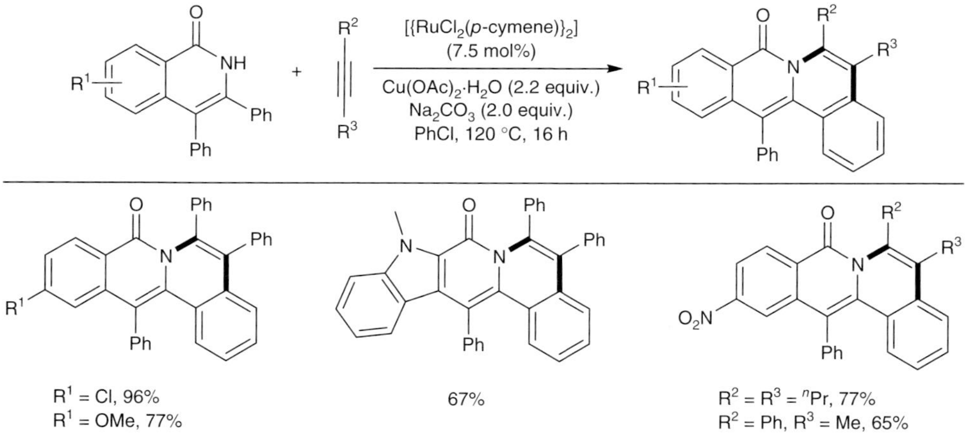

(7.30)

Under similar reaction conditions, a regioselective oxidative annulation of quinazolones with alkynes was reported by Peng and coworkers (Eq. (7.31)) [39]. A series of fused polycyclic heteroarenes were formed in moderate to high yields with a broad substrate scope. The substrates bearing electron-donating substituents on the 2-position of phenyl ring generally afforded better outcomes than electron-withdrawing ones. As to the mother aromatic ring of the substrate, the electron-deficient quinazolinones produced higher yields. Both the alkyl and aryl alkynes were suitable, and moderate regioselectivities were obtained for unsymmetric alkynes.

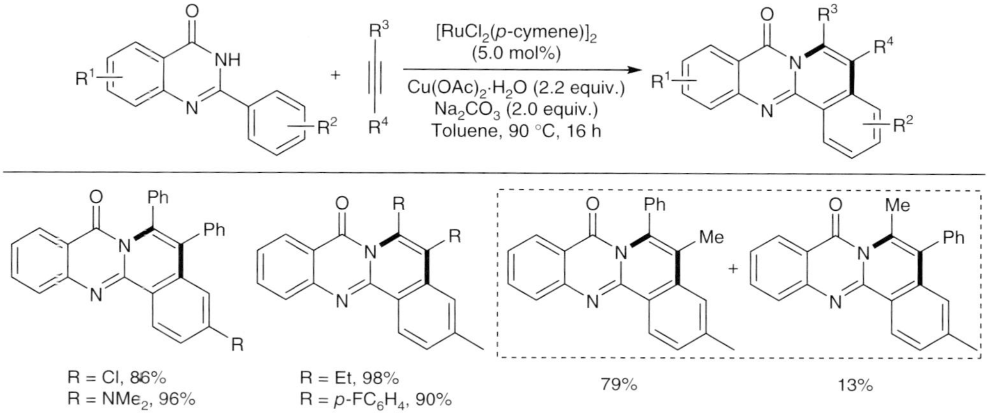

(7.31)

An efficient access to substituted isoquinolines via the oxidative cyclization of aromatic and heteroaromatic ketoximes with alkynes has been developed by Jeganmohan and coworkers (Eq. (7.32)) [40]. In the presence of 2.5 mol% of $[RuCl_2(p\text{-cymene})]_2$ and 25 mol% of NaOAc, the reaction occurred with high regioselectivity. A variety of unsymmetrical alkynes even including terminal alkynes are compatible for this transformation. Under the reaction conditions, the five-membered ruthenacycle key intermediate **9** was isolated from the stoichiometric reaction.

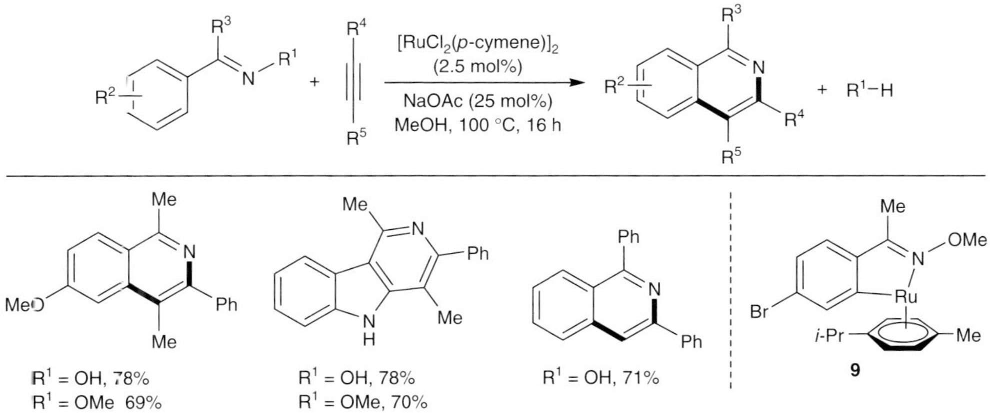

(7.32)

Later, the same reaction was also examined by Ackermann and coworkers with using the cocatalyst KPF_6 instead of NaOAc but in the presence of a higher catalyst loading (5 mol%) and longer reaction time (24 h) [41].

The ruthenium-catalyzed cyclization of substituted *N*-methoxy benzimidoyl halides with alkynes was also studied by the Jeganmohan group (Eq. (7.33)) [42]. In the presence of $[RuCl_2(p\text{-cymene})]_2$ (3.0 mol%) and CsOAc (25 mol%) in CF_3CH_2OH (TFE) at 100 °C for 16 h, 4-methyl and 4-*tert*-butyl *N*-methoxy benzimidoyl halides produced 1-haloisoquinolines, while *N*-methoxy benzimidoyl chloride and 4-halide substituted benzohydroximoyl chlorides provided 1-alkoxyisoquinolines. The electron-donating nature of the alkyl group on the aryl ring was supposed to suppress the nucleophilic addition of TFE at the imidoyl moiety for 4-methyl- and 4-*tert*-butyl-substituted substrates. Controlled experiments revealed that TFE was added to the imidoyl moiety after alkyne insertion. High regioselectivity was found for unsymmetric alkylarylalkynes with Ph group attached carbon of alkyne connecting to the nitrogen atom.

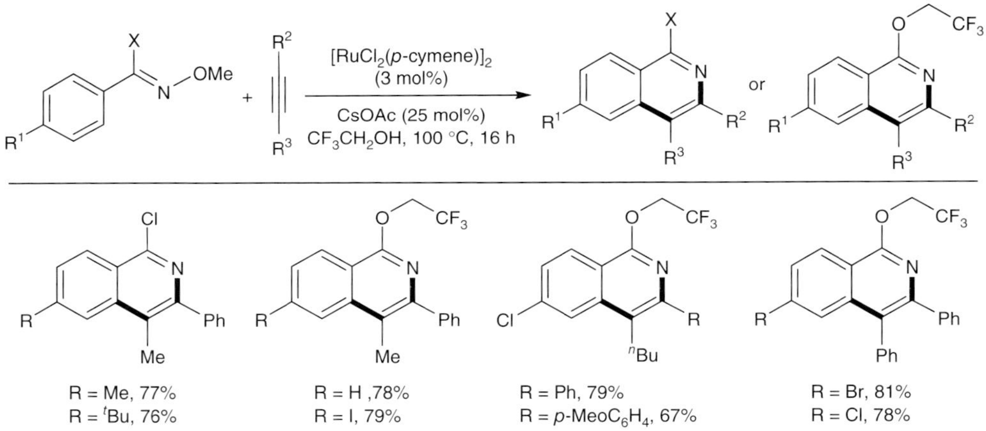

(7.33)

The Cheng group reported an efficient method for the synthesis of substituted isoquinolinium salts from one-pot three-component reaction of aryl aldehydes, amines, and alkynes using $[RuCl_2(p\text{-cymene})]_2/AgBF_4$ catalyst associated with the oxidant $Cu(BF_4)_2 \cdot 6H_2O$ (2 equiv.) in EtOH (Eq. (7.34)) [43]. This ruthenium-catalyzed C–H bond activation and annulation process exhibits good efficiency and high regioselectivity. The five-membered cationic ruthenacycle intermediate **10** was isolated and determined by single-crystal X-ray diffraction analysis as a 16-electron system which only contains a cyclometalated benzaldehyde imine and *p*-cymene ligand.

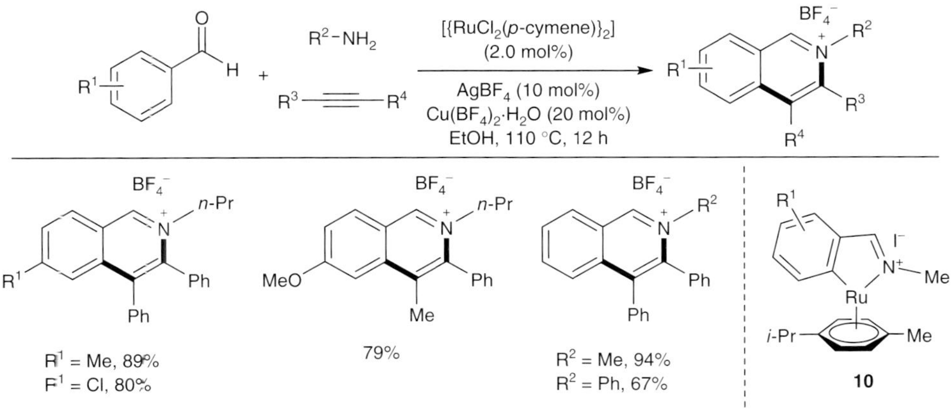

(7.34)

The aforementioned catalytic system was then applied for the oxidative cyclization of internal alkynes with 2-alkenylpyridines to synthesize quinolizinium salts (Eq. (7.35)) [44]. Interestingly, 2-heteroaryl pyridines bearing $C(sp^2)$–H bond at the 4-position with respect to the nitrogen atom were also suitable substrates for the annulation reactions. In this method, unsymmetrical alkynes afforded two regioisomeric products in excellent yield but moderate regioselectivity. The same reaction can also be catalyzed by $[Cp^*RhCl_2]_2$ catalyst but with 0.5 equiv. of $Cu(BF_4)_2 \cdot 6H_2O$ under an O_2 atmosphere. In some cases the less expensive ruthenium catalyst displayed complementary reactivity to the corresponding rhodium one.

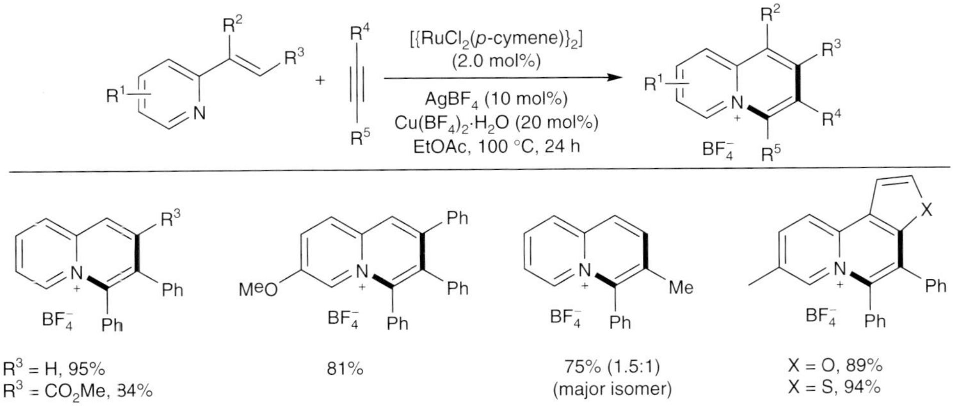

(7.35)

Oxidative annulation of enamides with alkynes via ruthenium(II)-catalyzed cleavage of $C(sp^2)$–H/N–H bonds was described by Li and Wang (Eq. (7.36)) [45]. The *N*-acetyl-substituted pyrroles were produced in high yield in the presence of 5 mol% of $[RuCl_2(p\text{-cymene})]_2$ catalyst and 0.5 equiv. of $Cu(OAc)_2 \cdot H_2O$ in DCE at 100 °C for 12 h. Interestingly, this process can afford *N*-unsubstituted pyrroles directly with the addition of $AgSbF_6$ and MeOH to the aforementioned catalytic system.

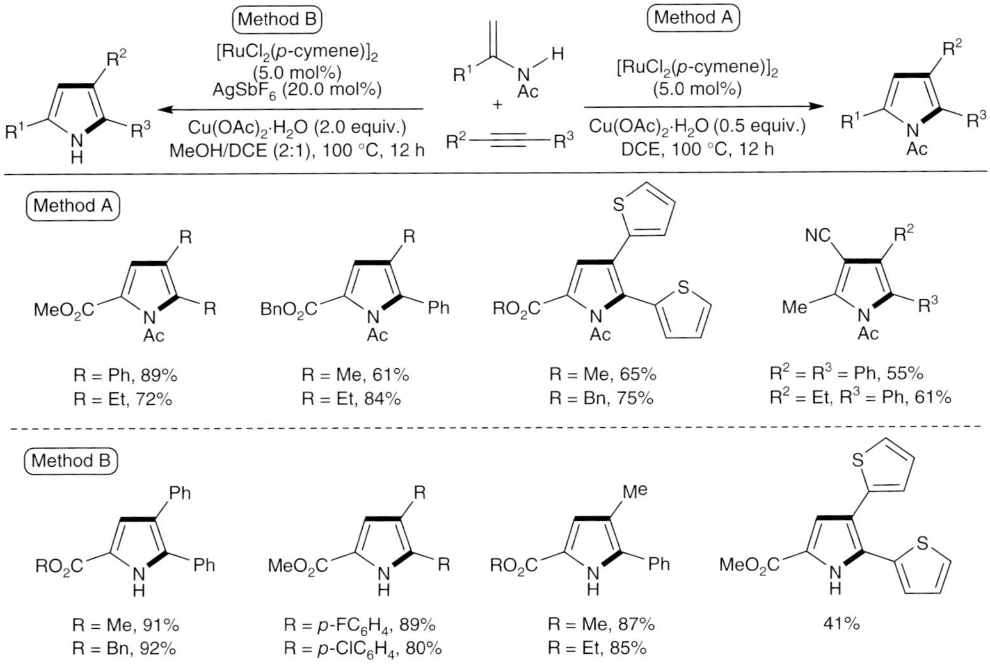

(7.36)

At the same time, Ackermann and coworkers also explored this ruthenium(II)-catalyzed alkene C–H bond functionalization to pyrrole synthesis in *t*-AmOH with 30 mol% of $Cu(OAc)_2 \cdot H_2O$ under ambient air conditions (Eq. (7.37)) [46].

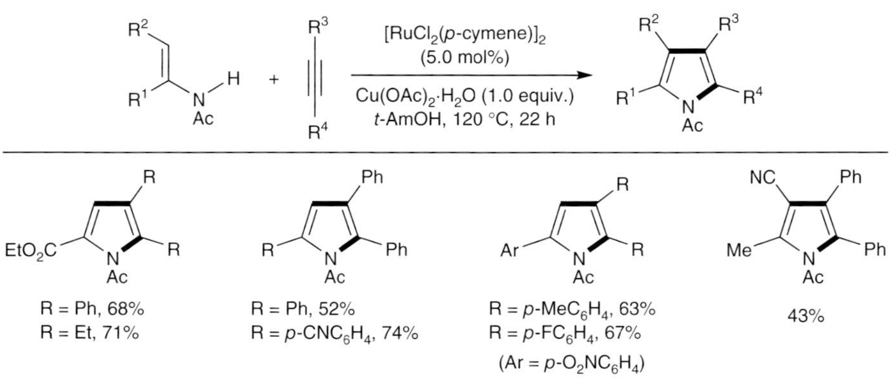

(7.37)

By employing a combination of catalytic amount of $[RuCl_2(p\text{-cymene})]_2$, $Cu(OAc)_2 \cdot H_2O$, and KPF_6 in acetic acid, isoquinolones were produced in good to excellent yields from the annulation of aromatic and heteroaromatic nitriles with alkynes (Eq. (7.38)) [47]. This method displayed high regioselectivity for both nitrile substrates and unsymmetrical alkylarylacetylenes; in the latter cases the aryl groups of the alkynes were installed adjacent to nitrogen atom selectively. It is postulated that a benzamide intermediate is involved in the catalytic cycle.

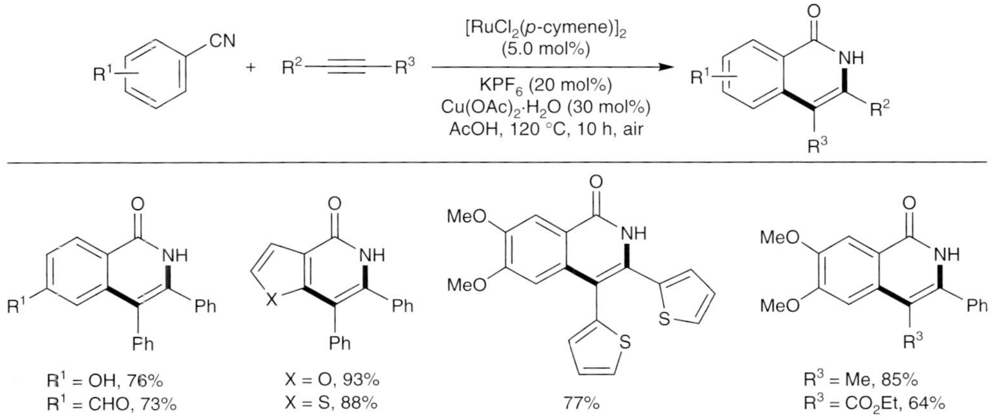

(7.38)

Aryl and heteroaryl amidines underwent oxidative C–H bond functionalization with alkynes with the aid of ruthenium catalysis [48]. In the presence of $[RuCl_2(p\text{-cymene})]_2/KPF_6$ as catalyst, the cyclization reactions led to the formation of a variety of 1-iminoisoindolines with high site selectivity, regioselectivity, and chemoselectivity (Eq. (7.39)).

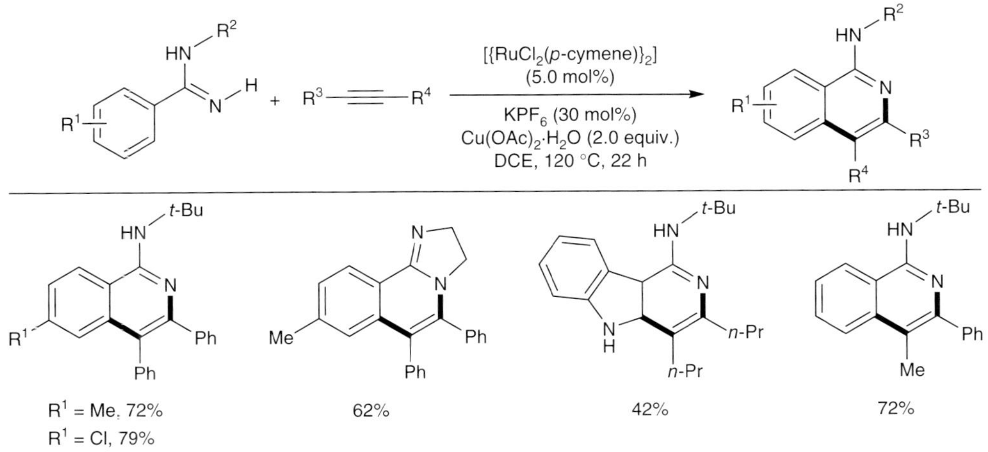

(7.39)

Unprotected primary amines serve as efficient directing groups in the cationic ruthenium(II)-catalyzed oxidative coupling of various benzylamines or heteroarylmethylamines with internal alkynes (Eq. (7.40)) [49]. A variety of isoquinolines, benzoisoquinolines, and fused heteroaryl[2,3-c] pyridines were generated from this catalytic C–H functionalization process. In some cases, products with insertion of two alkynes were also isolated.

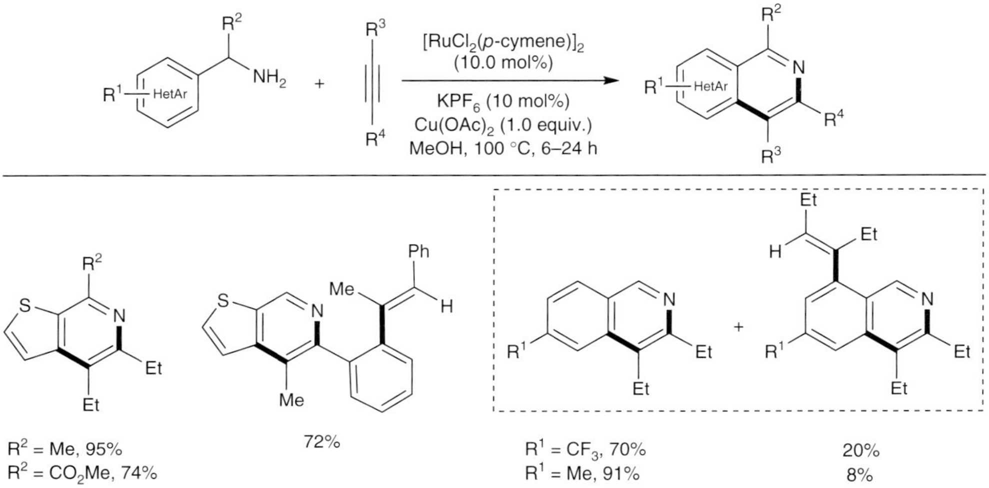

(7.40)

By using 8-aminoquinolinyl moiety as a bidentate directing group, *N*-quinolin-8-yl-benzamides annulated with alkynes under ruthenium(II)-catalyzed C–H activation to generate isoquinolones in open air (Eq. (7.41)) [50]. A similar reaction using 2-pyridinylmethylamine has been explored by using Ni(cod)$_2$ as catalyst [51]. Heteroaryl amides are also suitable for this catalytic process, and high regioselectivity is observed in the case of unsymmetrical alkylarylacetylenes. From the stoichiometric reaction of [RuCl$_2$(*p*-cymene)]$_2$, *N*-quinolin-8-ylbenzamide, and NaOAc, the *N,N*-coordinated ruthenium *N*-quinolin-8-ylbenzamide chloride species **11** was isolated. It can afford the isoquinolone derivative in high yield with alkyne, thus revealing the important role of the quinoline moiety in this reaction.

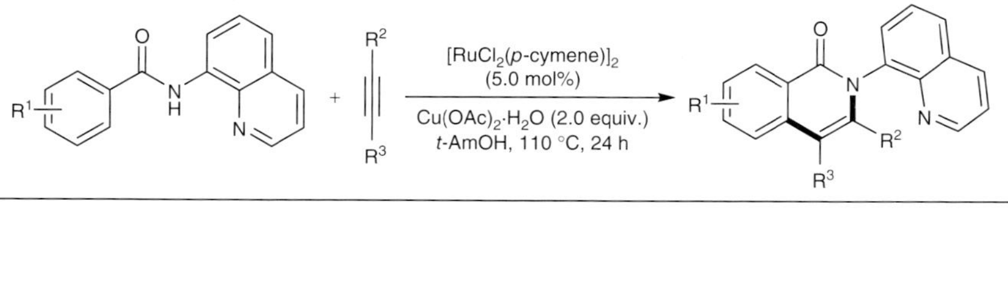

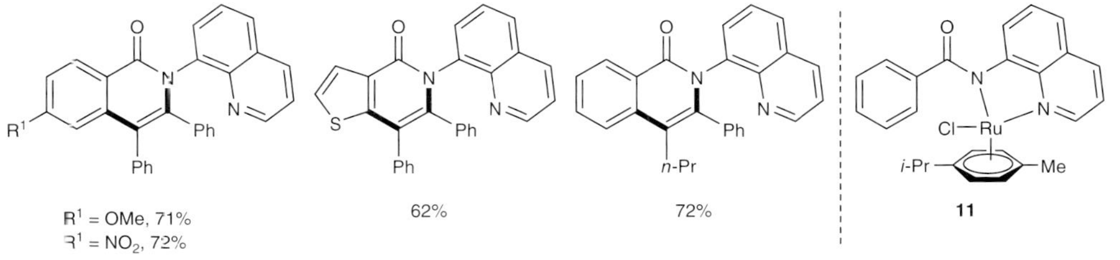

(7.41)

Isoquinolinones could also be formed from the annulation reactions of *o*-substituted *N*-benzoylated sulfoximine and diphenylacetylene via Ru(II)-catalyzed redox-neutral C–H bond functionalization (Eq. (7.42)) [52]. As a result, the formation of methyl phenyl sulfoxide was observed, and the methyl phenyl sulfoximine directing group can be regenerated.

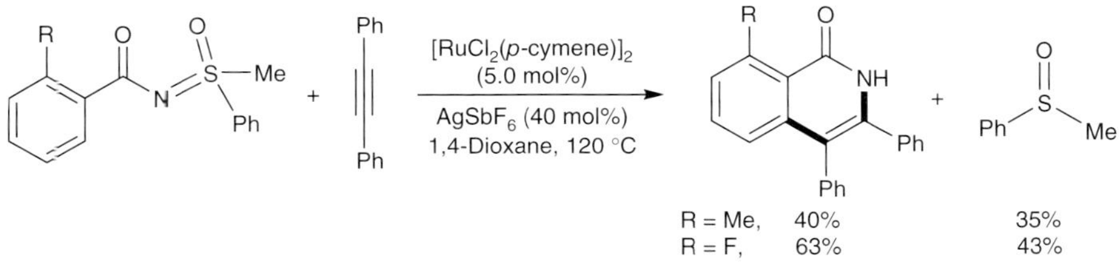

(7.42)

Under ruthenium-catalyzed *ortho*-C–H activation and intramolecular C–N bond formation, the condensation of iminophosphoranes (in situ generated from acyl azides and triphenylphosphine) with internal alkynes afforded a variety of isoquinolinone derivatives (Eq. (7.43)) [53]. The regioselective insertion of unsymmetrical alkynes led to an (aryl)C–N bond formation. Thiophene and indole-based acyl azides were also compatible for this transformation. A domino reaction sequence via coordination of ruthenium with *N*-atom of iminophosphoranes, *ortho*-cyclometalation, alkyne insertion, protonation, and reductive elimination was proposed for the catalytic cycle. Based on ^{1}H and ^{31}P NMR experiments, the involvement of benzamide during the reaction process was ruled out.

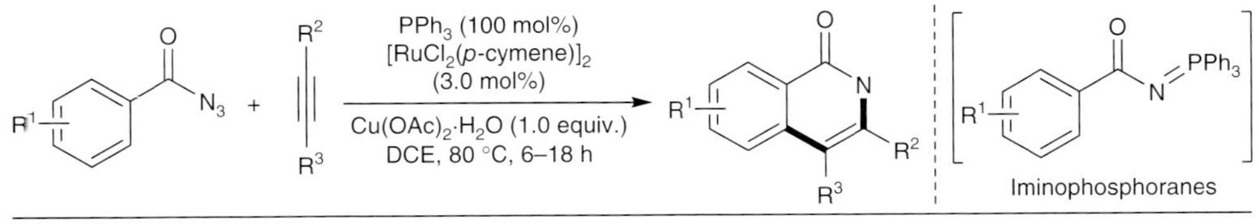

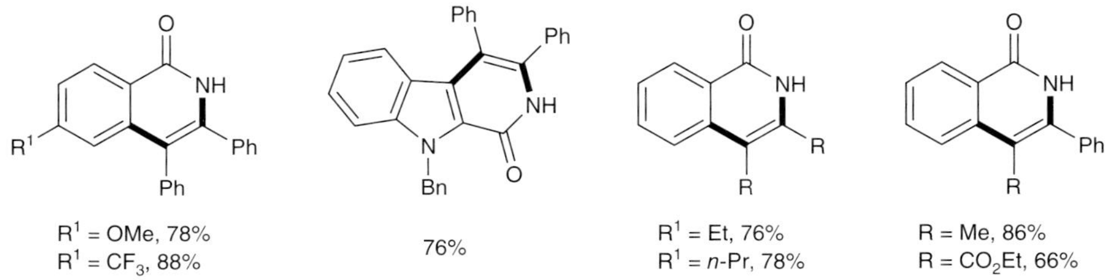

(7.43)

A ruthenium-catalyzed redox-neutral C–H activation reaction via cleavage of N–N bond is reported by Huang and coworkers (Eq. (7.44)) [54]. In this context, 3-(1*H*-indol-1-yl)propanamide derivatives were obtained in high yields by using pyrazolidin-3-one as an oxidative directing group. A broad substrate scope for both phenylpyrazolidin-3-ones and alkynes were well compatible with this catalytic transformation, especially the latter including commonly incompatible terminal alkynes. Meanwhile, excellent regioselectivity was demonstrated with single regioisomers of the indole products (2-aryl-3-alkylindoles for alkylarylacetylenes and 2-substituted indoles for terminal alkynes) obtained for unsymmetrical alkynes. Of note, the internal cleavage of the directing group resulted in installation of a pharmacologically significant *N*-propanamide functionality.

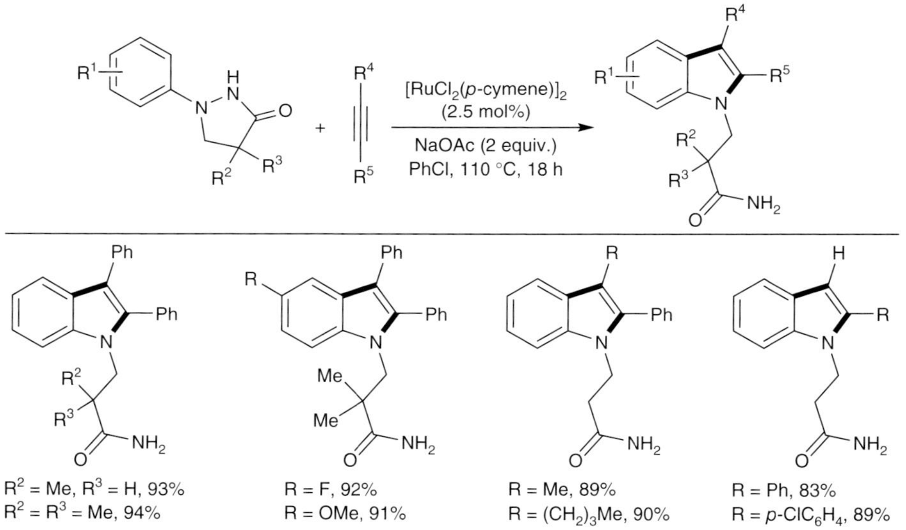

(7.44)

A Ru(II)–Ru(IV)–Ru(II) catalytic cycle was proposed (Scheme 7.10). After carboxylate-assisted cyclometalation and alkyne insertion, a Ru(IV) species **12** was formed by the cleavage of the cyclic N–N bond. Subsequent reductive elimination of **12** gives rise to the resulting indole product and then regenerates the catalytic active Ru(II) acetate species.

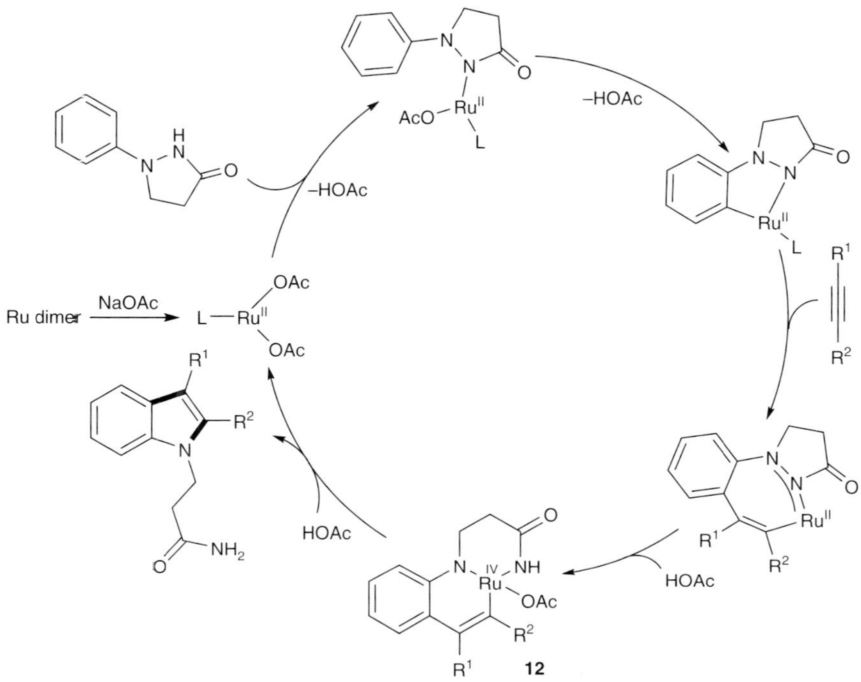

Scheme 7.10 Proposed mechanism for ruthenium-catalyzed 3-(1*H*-indol-1-yl)propanamide synthesis from phenylpyrazolidin-3-ones.

The cyclization of anilides with substituted propiolates provides 4-substituted-2-quinolinones in high yields under $[RuCl_2(p\text{-cymene})]_2/AgSbF_6$-catalyzed C–H activation in the presence of carboxylic acid (Eq. (7.45)) [55]. The reactions proceeded with high regioselectivity and good functional group tolerance. The electron-donating group-substituted anilides led to higher yields than electron-withdrawing group-substituted substrates. In the whole reaction process, organic acid plays multiple roles including acting as a proton source, accelerating *trans–cis* isomerization, and deacylation of anilide to aniline.

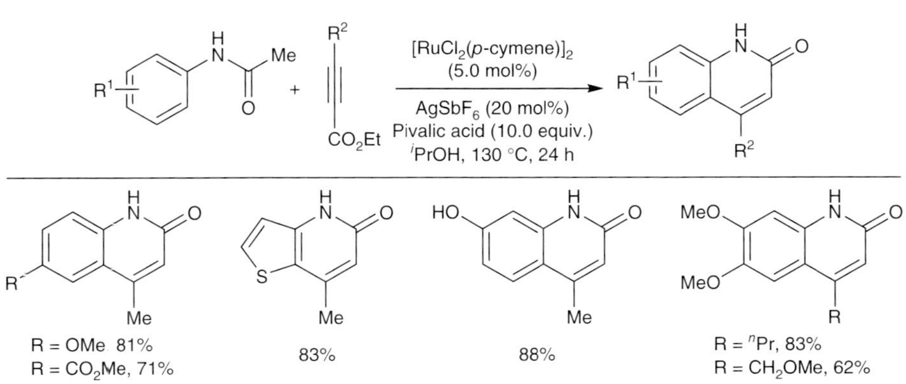

(7.45)

7.5.2
Cyclization with Alkenes

During the investigation of the ruthenium(II)-catalyzed *ortho*-alkenylation of arylpyrazoles with acrylates, Satoh and Miura found that benzanilide reacted with *n*-butyl acrylate in *o*-xylene to deliver lactam via oxidative alkenylation and subsequent intramolecular aza-Michael addition (Eq. (7.46)) [56].

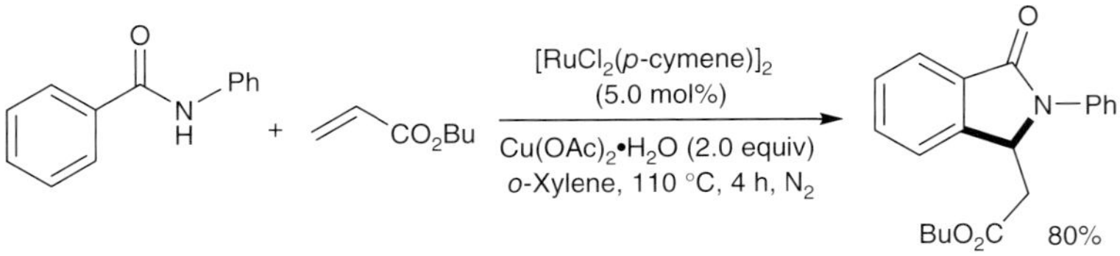

(7.46)

Similar cyclization product was also observed in the reaction of *N*-pentafluorophenyl benzamide with ethyl acrylate on activation with ruthenium(II) catalyst by Ackermann (Eq. (7.47)) [57].

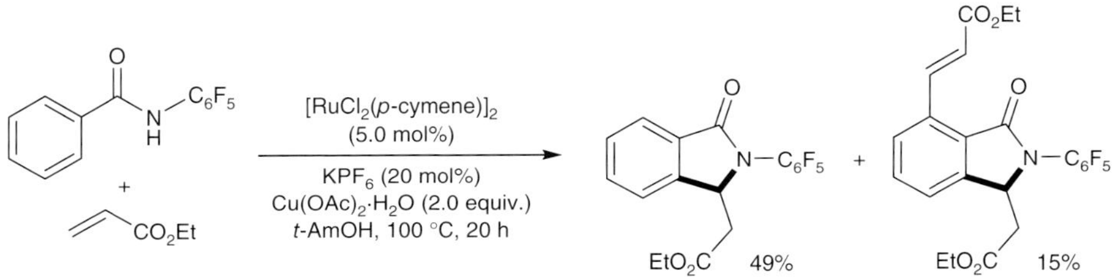

(7.47)

Under ruthenium catalysis, oxidative coupling of α,α-disubstituted benzylamines with acrylates can be performed efficiently at room temperature to produce (isoindol-1-yl)acetic acid derivatives (Eq. (7.48)) [58]. The reaction takes place accompanied by free amino group-directed *ortho*-alkenylation and successive intramolecular cyclization.

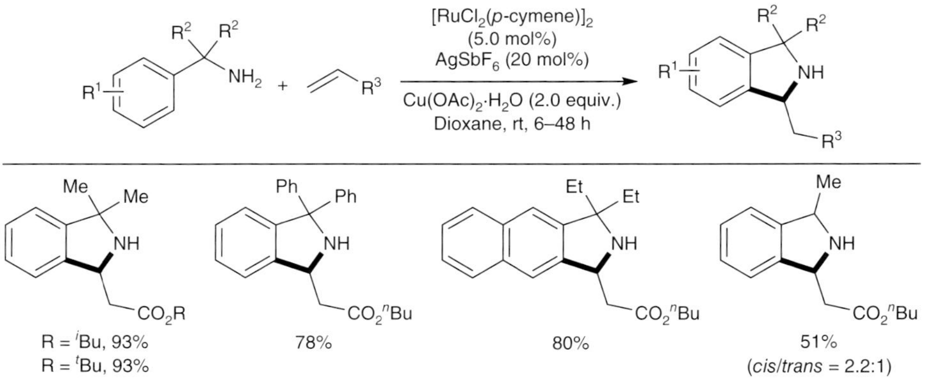

(7.48)

In 2014, Ackermann and coworkers reported the oxidative C–H alkenylation of sulfonamides. After ruthenium(II) catalyst enabled oxidative alkenylation of sulfonamides with acrylates at 120 °C for 18 h, the resulting *ortho*-alkenylated products underwent a chemoselective intramolecular aza-Michael reaction to yield the sultams by heating the reaction mixture at 150 °C for 5 h (Eq. (7.49)) [59].

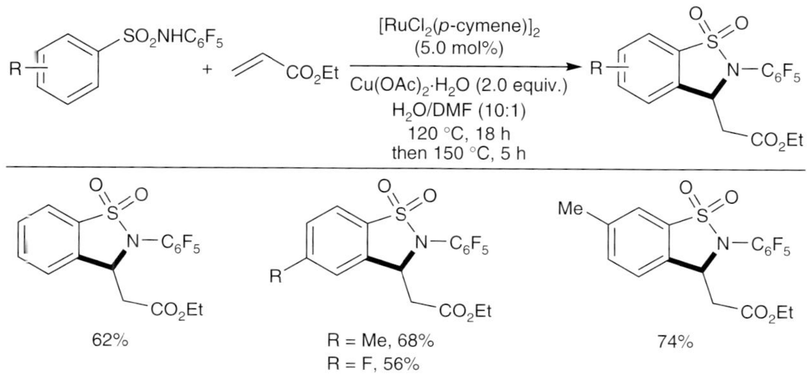

(7.49)

In the beginning of 2012, Li and Wang developed the first ruthenium-catalyzed C–H bond olefination of *N*-methoxybenzamides under redox-neutral conditions with the N–O bond acting as internal oxidant [60]. Using CONH(OMe) group as an oxidizing and directing group, the catalytic reaction is exclusively *ortho*- and mono-olefination selective. The alkenylation of *N*-methoxybenzamides with acrylate esters in CH_3OH worked smoothly to deliver the alkenylated products, but less active styrene and norbornadiene did not afford the corresponding olefinated benzamides under the same reaction conditions. Interestingly, the reactions with styrene and norbornadiene in solvent CF_3CH_2OH afforded 3,4-dihydroisoquinolinone derivatives in 47–87% yields (Eq. (7.50)).

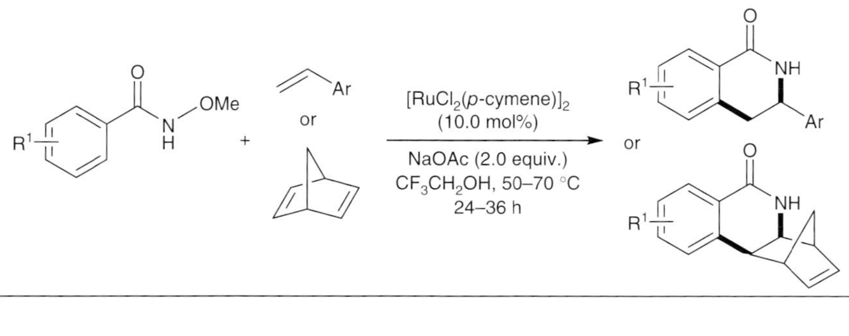

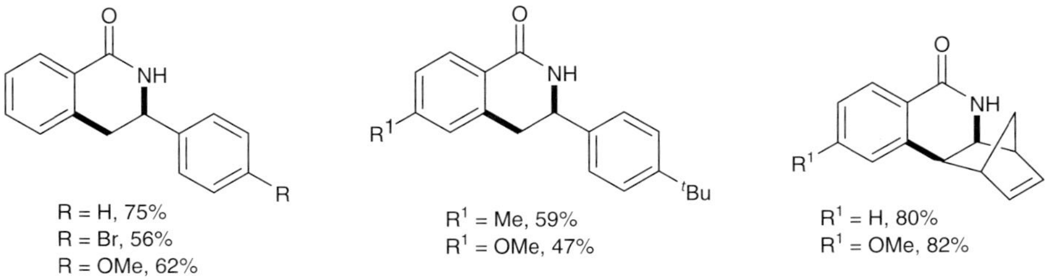

(7.50)

Mechanism studies suggested that the C–H bond metalation step is probably irreversible and involved in the rate-determining step. It was proposed that the reaction proceeds by an initial intermolecular carboruthenation of alkene via rate-determining C–H bond ruthenation and subsequently (i) β-hydride elimination of the potential intermediate of seven-membered ruthenacycle **13** (Pathway (a)) or (ii) C–N bond reductive elimination of the intermediate **9** for product formation (Pathway (b)) (Scheme 7.11).

Very recently, the cyclization of aromatic nitriles with alkenes was also explored. A variety of 3-methyleneisoindolin-1-ones were formed in high Z-stereoselective manner via in situ generated cationic Ru(II) catalyst (Eq. (7.51)) [61]. The high Z-stereoselectivity was attributed to the intramolecular hydrogen bonding.

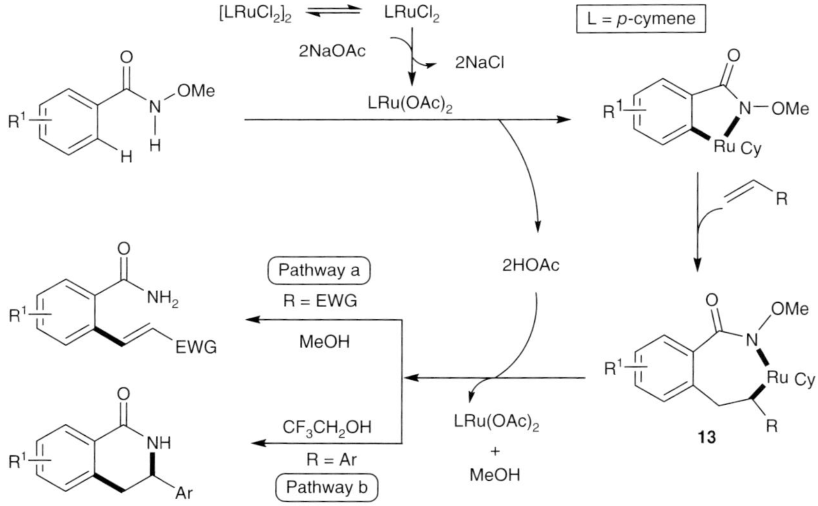

Scheme 7.11 Proposed mechanism for ruthenium-catalyzed oxidative alkenylation and cyclization of *N*-methoxybenzamides.

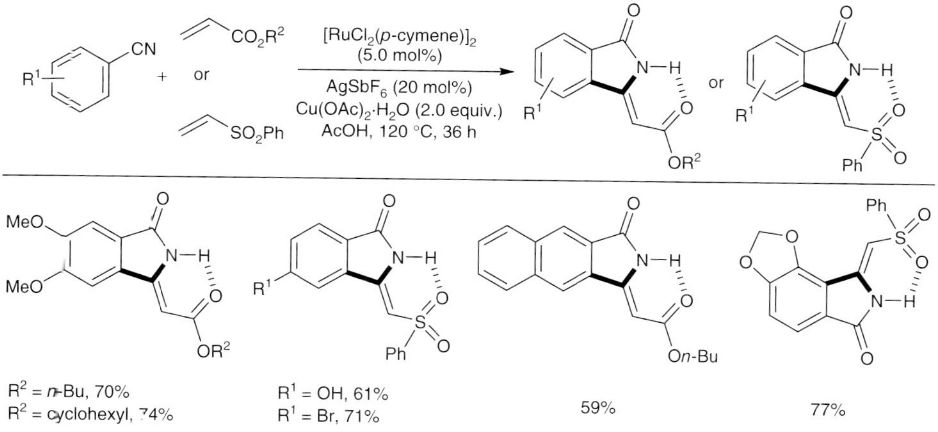

(7.51)

The whole cyclization process consists of three different catalytic reactions: (i) copper-catalyzed formation of a benzamide intermediate (catalytic cycle A), (ii) $[RuCl_2(p\text{-cymene})]_2/AgSbF_6$-catalyzed oxidative alkenylation of benzamide (catalytic cycle B), and (iii) Ru-catalyzed intramolecular aza-Michael addition and subsequent β-hydride elimination (catalytic cycle C) (Scheme 7.12).

By using ruthenium complexes derived from $[RuCl_2(p\text{-cymene})]_2/AgOAc$, oxidative alkenylations of amidines with substituted acrylates provided diversely substituted 1-iminoisoindolines (Eq. (7.52)) [48]. The resulting product was proposed to form in a process of oxidative alkenylation, intramolecular aza-Michael addition, and dehydrogenation.

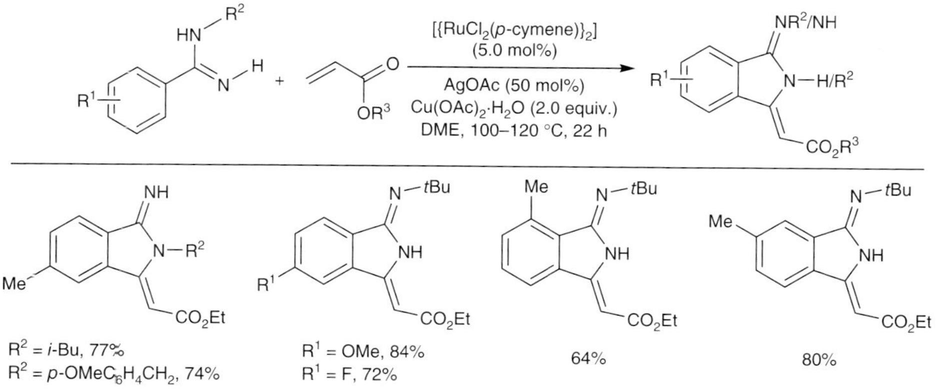

(7.52)

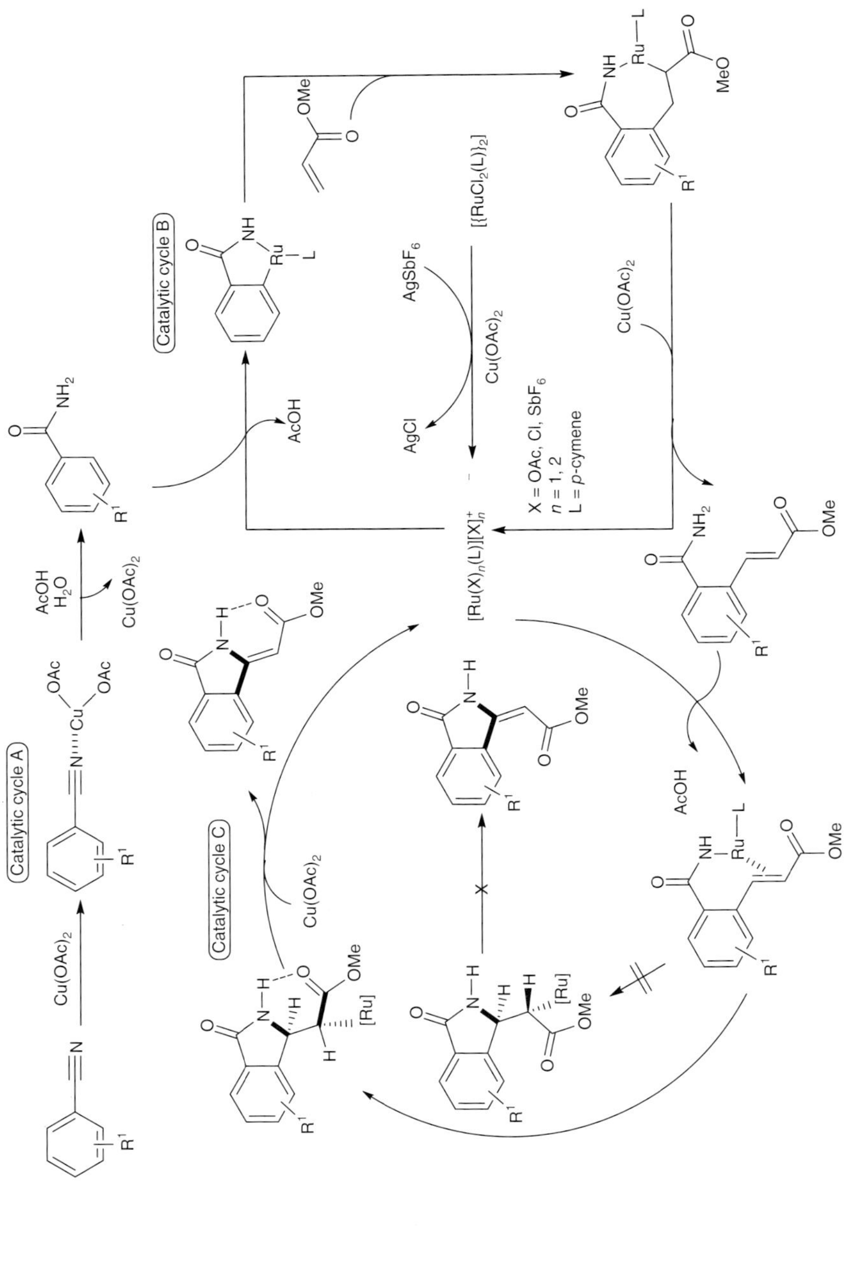

Scheme 7.12 Proposed mechanism for ruthenium-catalyzed oxidative alkenylation and cyclization of aromatic nitriles.

7.5.3 Cyclization with Carbon Monoxide

The Imhof group demonstrated a three-component coupling reaction of α,β-unsaturated imines with CO and alkenes catalyzed by $Ru_3(CO)_{12}$, which produced α,α-disubstituted β,γ-unsaturated γ-butyrolactams [62]. Similar reactions were also independently explored by Chatani and coworkers (Eq. (7.53)) [63]. Ethylene, vinylsilane, and norbornene could be used as the alkene partners, but the terminal alkenes were not compatible with the reaction system. The imine substrates bearing electron-withdrawing group showed low reactivity.

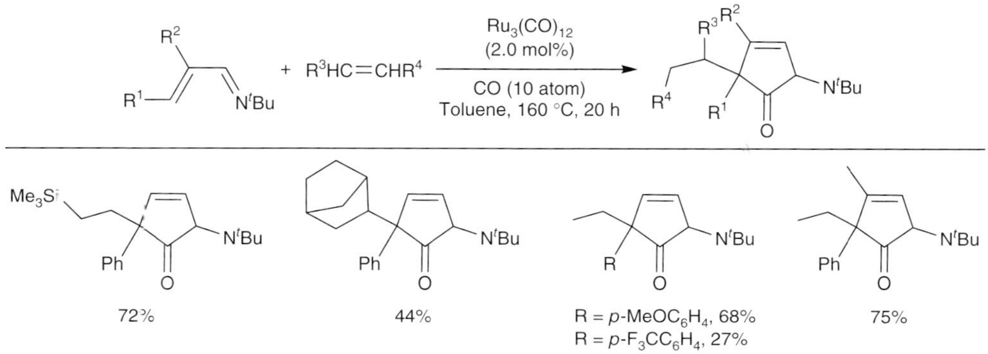

(7.53)

The reaction is proposed to proceed via a two-step sequence: (i) catalytic carbonylation of the olefinic C–H bond of α,β-unsaturated imines after insertion of ethylene and (ii) intramolecular nucleophilic attack of the imine nitrogen on the ketonic carbon followed by a 1,2-ethyl migration (Scheme 7.13). The initial formation of ethyl ketone was confirmed by the reaction of a cyclic unsaturated imine, which produced a β-aminocyclopentene derivative via an aldol-type condensation of the initially formed ketone (Eq. (7.54)).

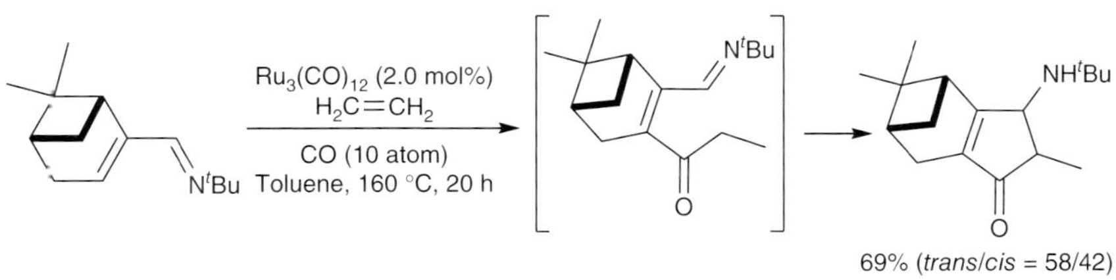

(7.54)

In 2009, Chatani and coworkers developed a ruthenium-catalyzed regioselective carbonylation of aromatic amides via C(sp^2)–H bond activation (Eq. (7.55)) [64]. It should be mentioned that the 2-pyridinylmethylamine moiety of the aromatic amides is required for this transformation to proceed. Under 10 atm of CO

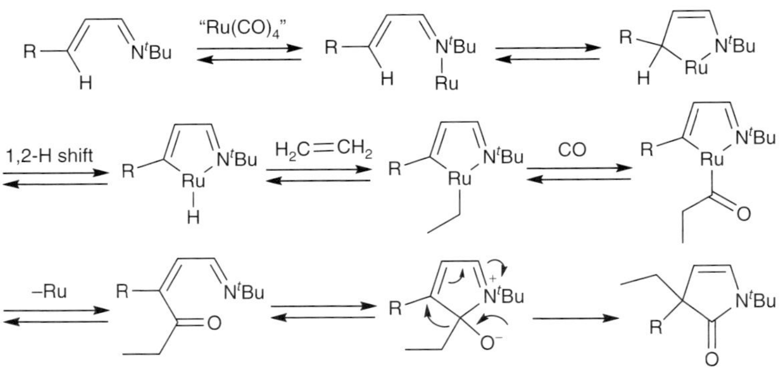

Scheme 7.13 Proposed mechanism for ruthenium-catalyzed three-component coupling reaction of α,β-unsaturated imines with CO and alkenes.

and 7 atm of ethylene together with small amount of water at 160 °C in toluene for 24 h, diverse phthalimides were formed in high yield. Interestingly, a dinuclear ruthenium complex **14**, in which 2-pyridinylmethylamine moiety coordinates to the ruthenium center in an *N,N*-fashion and the carbonyl oxygen coordinates to the other ruthenium center, was isolated from the stoichiometric reaction of amide with $Ru_3(CO)_{12}$. The dinuclear Ru(I) complex showed high catalytic activity under the standard reaction conditions but no reactivity without H_2O, thus indicating that complex **14** does not participate in the main catalytic cycle but rather exists in a resting state.

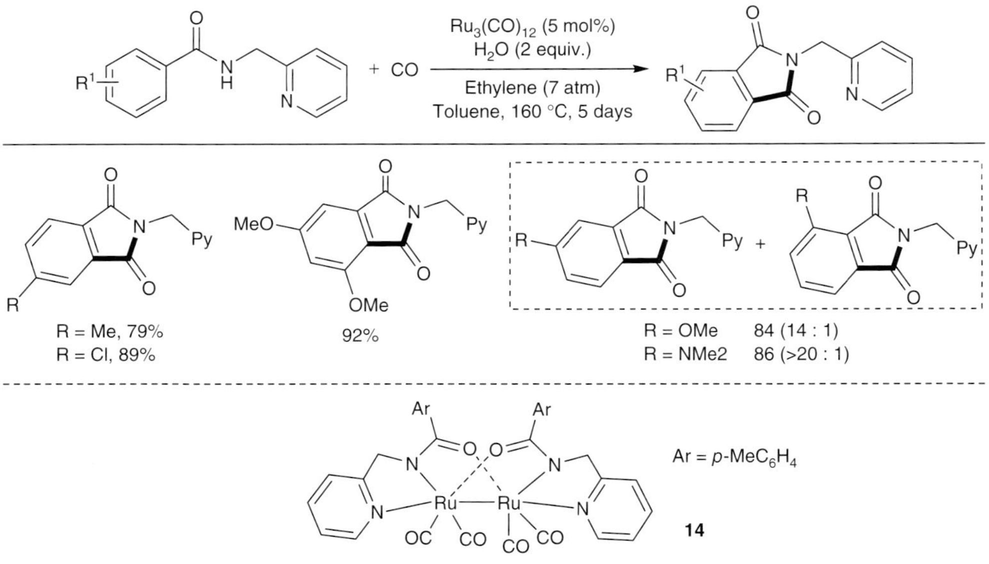

14

(7.55)

Two years later, a catalytic carbonylation of aliphatic amides by ruthenium-catalyzed C(sp^3)–H bond activation was reported by the same group (Eq. (7.56)) [65]. Under same reaction conditions but longer time (5 days), various substituted succinimides were obtained. The 2-pyridinylmethylamine bidentate system is also crucial for the success of the reaction. The carbonylation took place in a preference as the following sequence: C(sp^2)–H bond > cyclopropyl C(sp^3)–H bond > methyl C(sp^3)–H > methylene C(sp^3)–H bond.

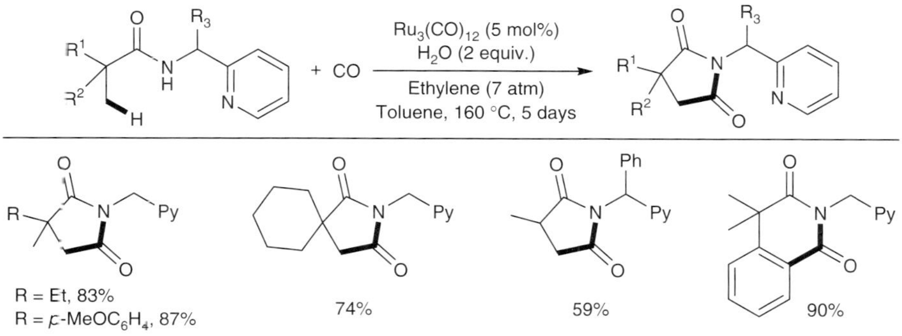

(7.56)

A dinuclear Ru(I) complex **15**, which is analog to **14**, was also isolated from the stoichiometric reaction of aliphatic amide with $Ru_3(CO)_{12}$. As shown in Scheme 7.14, the complex **15** is assumed to be reduced into a mononuclear species **16** with the aid of H_2O. Then followed by insertion of ethylene and

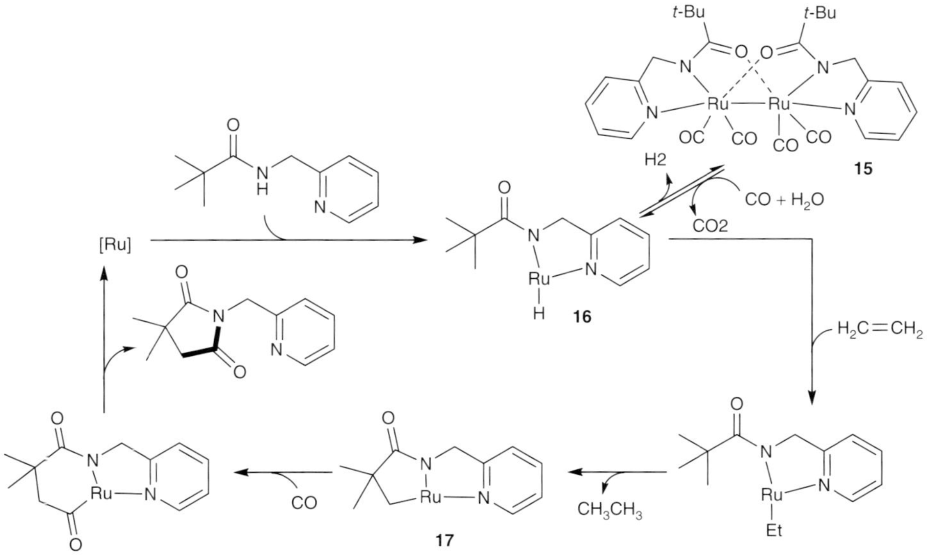

Scheme 7.14 Proposed mechanism for ruthenium-catalyzed carbonylation and cyclization of aliphatic amides.

release of ethane via irreversible C–H bond activation of one methyl group, the cyclometalated intermediate **17** is formed. Finally, CO insertion into Ru–C bond and subsequent reductive elimination lead to the final product with regeneration of the ruthenium catalyst.

7.5.4 Cyclization with Isocyanate

A novel route to synthesize phthalimide derivatives through ruthenium-catalyzed C–H bond functionalization of aromatic amides was developed by Ackermann and coworkers (Eq. (7.57)) [66]. This method is applicable to generate a potent COX-2 enzyme inhibitor in step-economical way. The reaction features by the insertion of a cycloruthenated species into a C–Het multiple bond of isocyanate and cleavage of pyrrolidinyl group. Electron-rich amides were found to favor the reaction, and an initial reversible C–H bond metalation step was also observed.

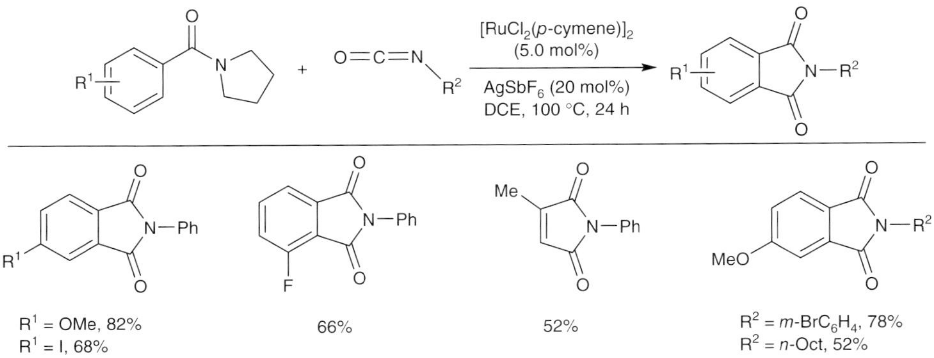

(7.57)

7.6 Summary and Conclusions

Ruthenium-catalyzed heterocycle synthesis through C–H bond activation and cyclization has received an increasing interest in the past few years. In this chapter the oxidative cyclization reactions with a series of unsaturated small molecules under ruthenium catalysis have been summarized. Among these reactions, the formations of heterocycles were realized by several synthetic pathways after initial C–H bond activation, which involved formation of intramolecular C–C bond, intramolecular C–N bond, intermolecular C–C/C–O bonds, and intermolecular C–C/C–N bonds. On the other hand, many challenges still need to overcome, such as heterocycle synthesis via ruthenium-catalyzed C(sp^3)–H activation. On the basis of the solid and exciting progress, no doubt that more reactive ruthenium catalysts and new innovative transformations will be discovered in this field in the near future.

References

1. (a) Joule, J.A. and Mills, K. (eds) (2000) *Heterocyclic Chemistry*, 4th edn, Blackwell, Oxford;(b) Eicher, T. and Hauptmann, S. (eds) (2003) *The Chemistry of Heterocycles*, Wiley-VCH Verlag GmbH, Weinheim;(c) Katrizky, A.R. and Pozharskii, A.F. (eds) (2000) *Handbook of Heterocyclic Chemistry*, 2nd edn, Pergamon Press, Amsterdam.
2. For selective reviews, see (a) Wencel-Delord, J. and Glorius, F. (2013) *Nat. Chem.*, **5**, 369–375; (b) Neufeldt, S.R. and Sanford, M.S. (2012) *Acc. Chem. Res.*, **45**, 936–946; (c) Song, G., Wang, F., and Li, X. (2012) *Chem. Soc. Rev.*, **41**, 3651–3678; (d) Kuhl, N., Hopkinson, M.N., Wencel-Delord, J., and Glorius, F. (2012) *Angew. Chem. Int. Ed.*, **51**, 10236–10254; (e) Yamaguchi, J., Yamaguchi, A.D., and Itami, K. (2012) *Angew. Chem. Int. Ed.*, **51**, 8960–9009; (f) Engle, K.M., Mei, T.-S., Wasa, M., and Yu, J.-Q. (2012) *Acc. Chem. Res.*, **45**, 788–802; (g) Colby, D.A., Tsai, A.S., Bergman, R.G., and Ellman, J.A. (2012) *Acc. Chem. Res.*, **45**, 814–825; (h) Ackermann, L. (2011) *Chem. Rev.*, **111**, 1315–1345; (i) Wencel-Delord, J., Dröge, T., Liu, F., and Glorius, F. (2011) *Chem. Soc. Rev.*, **40**, 4740–4761; (j) Wendlandt, A.E., Suess, A.M., and Stahl, S.S. (2011) *Angew. Chem. Int. Ed.*, **50**, 11062–11087; (k) Sun, C.-L., Li, B.-J., and Shi, Z.-J. (2011) *Chem. Rev.*, **111**, 1293–1314; (l) Herrmann, P. and Bach, T. (2011) *Chem. Soc. Rev.*, **40**, 2022–2038; (m) Yeung, C.S. and Dong, V.M. (2011) *Chem. Rev.*, **111**, 1215–1292; (n) Colby, D.A., Bergman, R.G., and Ellman, J.A. (2010) *Chem. Rev.*, **110**, 624–655; (o) Lyons, T.W. and Sanford, M.S. (2010) *Chem. Rev.*, **110**, 1147–1169; (p) Willis, M.C. (2010) *Chem. Rev.*, **110**, 725–748; (q) Bmalcells, D., Clot, E., and Eisenstein, O. (2010) *Chem. Rev.*, **110**, 749–823; (r) Yu, J.-Q. and Shi, Z.-J. (eds) (2010) *C-H Activation*, Topics in Current Chemistry, vol. 292, Springer, Berlin.
3. Selective reviews for carbon–carbon bond formation, see (a) Bugaut, X. and Glorius, F. (2011) *Angew. Chem. Int. Ed.*, **50**, 7479–7481; (b) Cho, S.H., Kim, J.Y., Kwak, J., and Chang, S. (2011) *Chem. Soc. Rev.*, **40**, 5068–5083; (c) Ashenhurst, J.A. (2010) *Chem. Soc. Rev.*, **39**, 540–548; (d) McGlacken, G.P. and Bateman, L.M. (2009) *Chem. Soc. Rev.*, **38**, 2447–2464; Selective reviews for carbon–heteroatom bond formation, see (e) Hartwig, J.F. (2011) *Chem. Soc. Rev.*, **40**, 1992–2002; (f) Enthaler, S. and Company, A. (2011) *Chem. Soc. Rev.*, **40**, 4912–4924; (g) Stokes, B.J. and Driver, T.G. (2011) *Eur. J. Org. Chem.*, **2011**, 4071–4088; (h) Mkhalid, I.A.I., Barnard, J.H., Marder, T.B., Murphy, J.M., and Hartwig, J.F. (2010) *Chem. Rev.*, **110**, 890–931.
4. Murai, S., Kakiuchi, F., Sekine, S., Tanaka, Y., Kamatani, A., Sonoda, M., and Chatani, N. (1993) *Nature*, **366**, 529–531.
5. For selective reviews, see (a) Ackermann, L. (2014) *Acc. Chem. Res.*, **47**, 281–295; (b) Kozhushkov, S.I. and Ackermann, L. (2013) *Chem. Sci.*, **4**, 886–896; (c) Arockiam, P.B., Bruneau, C., and Dixneuf, P.H. (2012) *Chem. Rev.*, **112**, 5879–5918; (d) Li, B. and Dixneuf, P.H. (2014) in *Ruthenium in Catalysis*, vol. 48 (eds C. Bruneau and P.H. Dixneuf), Springer, Heidelberg, pp. 119–194; (e) Bruneau, C. (2014) in *Ruthenium in Catalysis*, vol. 48 (eds C. Bruneau and P.H. Dixneuf), Springer, Heidelberg, pp. 195–236; (f) Sarkar, S.D., Liu, W., Kozhushkov, S.I., and Ackermann, L. (2014) *Adv. Synth. Catal.*, **356**, 1461–1479.
6. (a) Jones, W.D. and Kosar, W.P. (1986) *J. Am. Chem. Soc.*, **108**, 5640–5641; (b) Hsu, G.C., Kosar, W.P., and Jones, W.D. (1994) *Organometallics*, **13**, 385–396.

7. Armanino, N. and Carreira, E.M. (2013) *J. Am. Chem. Soc.*, **135**, 6814–6817.
8. Li, B., Park, Y., and Chang, S. (2014) *J. Am. Chem. Soc.*, **136**, 1125–1131.
9. Hu, J., Chen, S., Sun, Y., Yang, J., and Rao, Y. (2012) *Org. Lett.*, **14**, 5030–5033.
10. Chinnagolla, R.K. and Jeganmohan, M. (2012) *Chem. Commun.*, **48**, 2030–2032.
11. (a) Ackermann, L., Pospech, J., Graczyk, K., and Rauch, K. (2012) *Org. Lett.*, **14**, 930–933; (b) Deponti, M., Kozhushkov, S., Yufit, D.S., and Ackermann, L. (2013) *Org. Biomol. Chem.*, **11**, 142–148.
12. Thirunavukkarasu, V.S., Donati, M., and Ackermann, L. (2012) *Org. Lett.*, **14**, 3416–3419.
13. Dooley, J.D., Chidipudi, R.S., and Lam, H.W. (2013) *J. Am. Chem. Soc.*, **135**, 10829–10836.
14. Park, Y., Jeon, I., Shin, S., Min, J., and Lee, P.H. (2013) *J. Org. Chem.*, **78**, 10209–10220.
15. Zhou, Z., Liu, G., Shen, Y., and Lu, X. (2014) *Org. Chem. Front.*, **1**, 1161–1165.
16. Nakanowatari, S. and Ackermann, L. (2014) *Chem. Eur. J.*, **20**, 5409–5413.
17. Ueyama, T., Mochida, S., Fukutani, T., Hirano, K., Satoh, T., and Miura, M. (2011) *Org. Lett.*, **13**, 706–708.
18. Ackermann, L. and Pospech, J. (2011) *Org. Lett.*, **13**, 4153–4155.
19. Chen, L., Li, H., Yu, F., and Wang, L. (2014) *Chem. Commun.*, **50**, 14866–14869.
20. Chidipudi, S.R., Wieczysty, M.D., Khan, I., and Lam, H.W. (2013) *Org. Lett.*, **15**, 570–573.
21. Chidipudi, S.R., Khan, I., and Lam, H.W. (2012) *Angew. Chem. Int. Ed.*, **51**, 12115–12119.
22. Chatani, N., Morimoto, T., Fukumoto, Y., and Murai, S. (1998) *J. Am. Chem. Soc.*, **120**, 5335–5336.
23. Inamoto, K., Kadokawa, J., and Kondo, Y. (2013) *Org. Lett.*, **15**, 3962–3965.
24. Lee, D.-H., Kwon, K.-H., and Yi, C.S. (2012) *J. Am. Chem. Soc.*, **134**, 7325–7328.
25. Penoni, A. and Nicholas, K.M. (2002) *Chem. Commun.*, 484–485.
26. Penoni, A., Volkmann, J., and Nicholas, K.M. (2002) *Org. Lett.*, **4**, 699–701.
27. Yi, C.S., Yun, S.Y., and Guzei, I.A. (2005) *J. Am. Chem. Soc.*, **127**, 5782–5783.
28. Yi, C.S. and Yun, S.Y. (2005) *J. Am. Chem. Soc.*, **127**, 17000–17006.
29. Ackermann, L., Lygin, A.V., and Hofmann, N. (2011) *Angew. Chem. Int. Ed.*, **50**, 6379–6382.
30. Ackermann, L., Lygin, A.V., and Hofmann, N. (2011) *Org. Lett.*, **13**, 3278–3281.
31. (a) Li, B., Feng, H., Xu, S., and Wang, B. (2011) *Chem. Eur. J.*, **17**, 12573–12577; one substrate example of intramolecular annulation using oxidizing-directing-group strategy, see (b) Xu, X., Liu, Y., and Park, C.-M. (2012) *Angew. Chem. Int. Ed.*, **51**, 9372–9376.
32. (a) Ackermann, L. and Fenner, S. (2011) *Org. Lett.*, **13**, 6548–6551; (b) Kornhaaß, C., Kuper, C., and Ackermann, L. (2014) *Adv. Synth. Catal.*, **356**, 1619–1624.
33. Yang, F. and Ackermann, L. (2014) *J. Org. Chem.*, **79**, 12070–12082.
34. Ackermann, L., Wang, L., and Lygin, A.V. (2012) *Chem. Sci.*, **3**, 177–180.
35. (a) Kavitha, N., Sukumar, G., Kumar, V.P., Mainkar, P.S., and Chandrasekhar, S. (2013) *Tetrahedron Lett.*, **54**, 4198–4201; (b) Wang, R. and Falck, J.R. (2014) *J. Organomet. Chem.*, **759**, 33–36; (c) Zheng, L. and Hua, R. (2014) *J. Org. Chem.*, **79**, 3930–3936.
36. (a) Ma, W., Graczyk, K., and Ackermann, L. (2012) *Org. Lett.*, **14**, 6318–6321; (b) Algarra, A.G., Cross, W.B., Davies, D.L., Khamker, Q., Macgregor, S.A., McMullin, C.L., and Singh, K. (2014) *J. Org. Chem.*, **79**, 1954–1970.
37. Ackermann, L. and Lygin, A.V. (2012) *Org. Lett.*, **14**, 764–767.
38. Li, B., Feng, H., Wang, N., Ma, J., Song, H., Xu, S., and Wang, B. (2012) *Chem. Eur. J.*, **18**, 12873–12879.
39. Lu, H., Yang, Q., Zhou, Y., Guo, Y., Deng, Z., Ding, Q., and Peng, Y. (2014) *Org. Biomol. Chem.*, **12**, 758–764.
40. Chinnagolla, R.K., Pimparkar, S., and Jeganmohan, M. (2012) *Org. Lett.*, **14**, 3032–3035.
41. Kornhaa*β*, C., Li, J., and Ackermann, L. (2012) *J. Org. Chem.*, **77**, 9190–9198.
42. Chinnagolla, R.K., Pimparkar, S., and Jeganmohan, M. (2013) *Chem. Commun.*, **49**, 3703–3705.

43. (a) Parthasarathy, K., Senthilkumar, N., Jayakumar, J., and Cheng, C.-H. (2012) *Org. Lett.*, **14**, 3478–3481; Ru-catalyzed oxidative cyclization of benzophenone imine with diphenylacetylene to yield isoquinolinium salt, see (b) Senthilkumar, N., Gandeepan, P., Jayakumar, J., and Cheng, C.-H. (2014) *Chem. Commun.*, **50**, 3106–3108.
44. Luo, C.Z., Gandeepan, P., and Cheng, C.-H. (2013) *Chem. Commun.*, **49**, 8528–8530.
45. Li, B., Wang, N., Liang, Y., Xu, S., and Wang, B. (2013) *Org. Lett.*, **15**, 136–139.
46. Wang, L. and Ackermann, L. (2013) *Org. Lett.*, **15**, 176–179.
47. Reddy, M.C., Manikandan, R., and Jeganmohan, M. (2013) *Chem. Commun.*, **49**, 6060–6062.
48. Li, J., John, M., and Ackermann, L. (2014) *Chem. Eur. J.*, **20**, 5403–5408.
49. Villuendas, P. and Urriolabeitia, E.P. (2013) *J. Org. Chem.*, **78**, 5254–5263.
50. Allu, S. and Swamy, K.C.K. (2014) *J. Org. Chem.*, **79**, 3963–3972.
51. Shiota, H., Ano, Y., Aihara, Y., Fukumoto, Y., and Chatani, N. (2011) *J. Am. Chem. Soc.*, **133**, 14952–14955.
52. Yadav, M.R., Rit, R.K., Shankar, M., and Sahoo, A.K. (2014) *J. Org. Chem.*, **79**, 6123–6134.
53. Arigela, R.K., Kumar, R., Joshi, T., Maharo, R., and Kundu, B. (2014) *RSC Adv.*, **4**, 57749–57753.
54. Zhang, Z., Jiang, H., and Huang, Y. (2014) *Org. Lett.*, **16**, 5976–5979.
55. Manikandan, R. and Jeganmohan, M. (2014) *Org. Lett.*, **16**, 3568–3571.
56. Hashimoto, Y., Ueyama, T., Fukutani, T., Hirano, K., Satoh, T., and Miura, M. (2011) *Chem. Lett.*, **40**, 1165–1166.
57. Ackermann, L., Wang, L., Wolfram, R., and Lygin, A.V. (2012) *Org. Lett.*, **14**, 728–731.
58. Suzuki, C., Morimoto, K., Hirano, K., Satoh, T., and Miura, M. (2014) *Adv. Synth. Catal.*, **356**, 1521–1526.
59. Ma, W., Mei, R., Tenti, G., and Ackermann, L. (2014) *Chem. Eur. J.*, **20**, 15248–15251.
60. Li, B., Ma, J., Wang, N., Feng, H., Xu, S., and Wang, B. (2012) *Org. Lett.*, **14**, 736–739.
61. Reddy, M.C. and Jeganmohan, M. (2014) *Org. Lett.*, **16**, 4866–4869.
62. (a) Berger, D. and Imhof, W. (1999) *Chem. Commun.*, 1457–1458; (b) Berger, D. and Imhof, W. (2000) *Tetrahedron*, **56**, 2015–2023; (c) Göbel, A. and Imhof, W. (2001) *Chem. Commun.*, 593–594; (d) Imhof, W., Berger, D., Kötteritzsch, M., Rost, M., and Schönecker, B. (2001) *Adv. Synth. Catal.*, **343**, 795–801.
63. Chatani, N., Kamitani, A., and Murai, S. (2002) *J. Org. Chem.*, **67**, 7014–7018.
64. Inoue, S., Shiota, H., Fukumoto, Y., and Chatani, N. (2009) *J. Am. Chem. Soc.*, **131**, 6898–6899.
65. (a) Hasegawa, N., Charra, V., Inoue, S., Fukumoto, Y., and Chatani, N. (2011) *J. Am. Chem. Soc.*, **133**, 8070–8073; (b) Shibata, K., Hasegawa, N., Fukumoto, Y., and Chatani, N. (2012) *ChemCatChem*, **4**, 1733–1736; (c) Hasegawa, N., Shibata, K., Charra, V., Inoue, S., Fukumoto, Y., and Chatani, N. (2013) *Tetrahedron*, **69**, 4466–4472.
66. Sarkar, S.D. and Ackermann, L. (2014) *Chem. Eur. J.*, **20**, 13932–13936.

8
Cu-Catalyzed Heterocycle Synthesis

Feng Chen and Ning Jiao

8.1
Introduction

Heterocycles are important ubiquitous structures in biologically active molecules, natural products, and drugs. And they are also useful building blocks in organic synthesis. Many methods have been well developed to construct various N-, O-, S-containing heterocycles. The most common and powerful methods are intramolecular and intermolecular bond formations from halides precursor catalyzed by transition metals. Direct functionalization of C–H bonds into heterocycles catalyzed by transition metals are efficient and environmentally benign. In this chapter, we would focus on the application of non-noble metal copper catalysts in the formation of four-, five-, and six-membered-ring heterocycles via direct C–H functionalization using halogen-free-containing substrates.

8.2
Four-Membered-Ring Formation

Lactams (cyclic amides) and related compounds are important partial structures of natural products. The Beckmann rearrangement [1, 2], Schmidt reaction [3–5], cyclization of amino acids [6–8], and iodolactamization [9] are well-known synthetic methods for forming lactams. Direct C–H amidation as an alternative method to the synthesis of lactams is still rare. Very recently, Kuninobu, Kanai [10], and Ge [11] groups reported the intramolecular C(sp^3)–H and C(sp^2)–H amidation by oxidative cyclization with copper catalyst simultaneously. Kuninobu, Kanai, and coworkers realized the C(sp^3)–H amidation at a terminal methyl group or the internal benzylic position of an alkyl chain using a bidentate directing group 5-methoxyquinolyl with $Cu(OAc)_2$ catalyst and an Ag_2CO_3 oxidant. Various β-lactams could be obtained in excellent yield even on gram scale (Scheme 8.1).

Ge and coworkers developed $Cu(OAc)_2$-catalyzed intramolecular amidation of N-(quinolin-8-yl)pivalamide derivatives to mono-, spiro-, and bicyclic β-lactam

Transition Metal-Catalyzed Heterocycle Synthesis via C—H Activation, First Edition. Edited by Xiao-Feng Wu.
© 2016 Wiley-VCH Verlag GmbH & Co. KGaA. Published 2016 by Wiley-VCH Verlag GmbH & Co. KGaA.

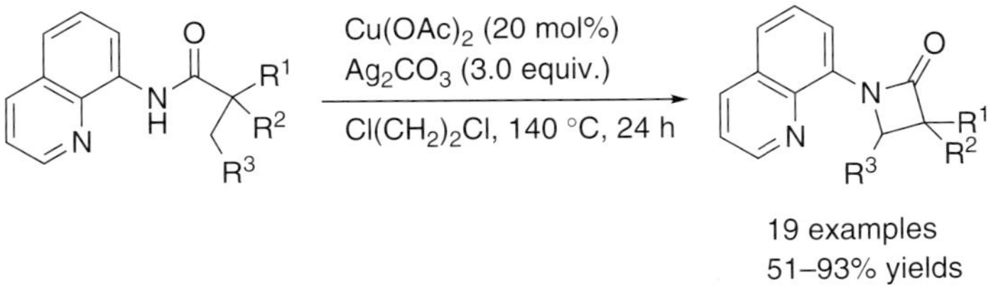

Scheme 8.1 $Cu(OAc)_2$ catalyzed lactams formation using 5-methoxyquinolyl as directing group.

derivatives with an 8-aminoquinolinyl group as the directing group (Scheme 8.2). In this process, benzoquinone (BQ) was chosen as the oxidant which is different from the previous reaction system. The C(sp^3)–H bonds of β-methyl groups are preferential to be activated via a five-membered-ring intermediate in the cyclometalation step. Moreover, the reactivity order of secondary β-C(sp^3) atoms among methylene groups is benzylic > ring > linear.

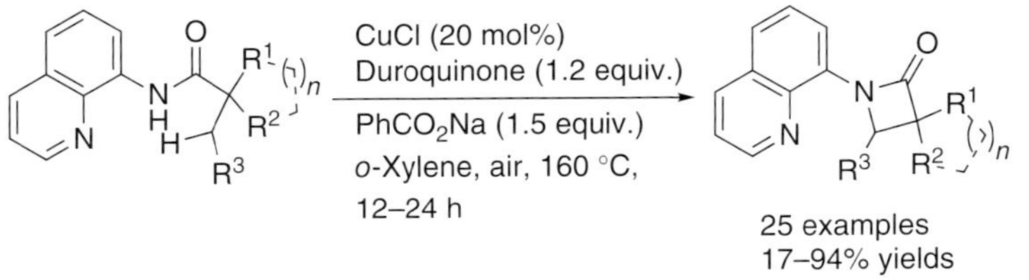

Scheme 8.2 CuCl catalyzed lactams formation using 5-methoxyquinolyl as directing group.

8.3
Five-Membered-Ring Formation

8.3.1
Copper-Catalyzed Synthesis of Pyrroles

An efficient Cu-catalyzed cycloisomerization of alkynyl imines into the 2-monosubstituted and 2,5-disubstituted pyrroles was reported by Gevorgyan and coworkers (Scheme 8.3). This methodology was efficient to the synthesis of various types of fused heteroaromatic compounds and even synthesis of (±) monomorine. A C(sp^2)–H of the substrate was activated during this copper-catalyzed pyrrole formation [12]. Moreover, they also realized the

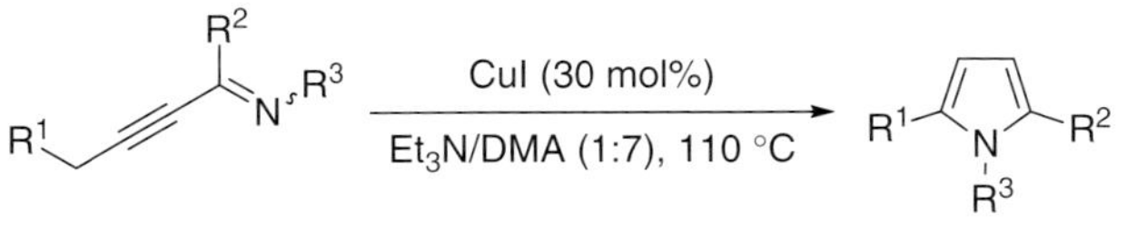

Scheme 8.3 CuI catalyzed synthesis of 2-monosubstituted and 2,5-disubstituted pyrroles.

highly diastereoselective total synthesis of (±)-tetraponerine T6 using this copper-catalyzed pyrrole synthesis process [13].

3-Azabicyclo[3.1.0] hex-2-enes and 4-carbonylpyrroles could be obtained respectively via copper-mediated/catalyzed reactions of *N*-allyl/propargyl enamine carboxylates under an O_2 atmosphere with different copper catalysts. Intramolecular cyclopropanation and carbooxygenation were involved during this reaction (Scheme 8.4). Three C(sp^2)–H bonds and one C(sp^3)–H bond were active for this transformation [14].

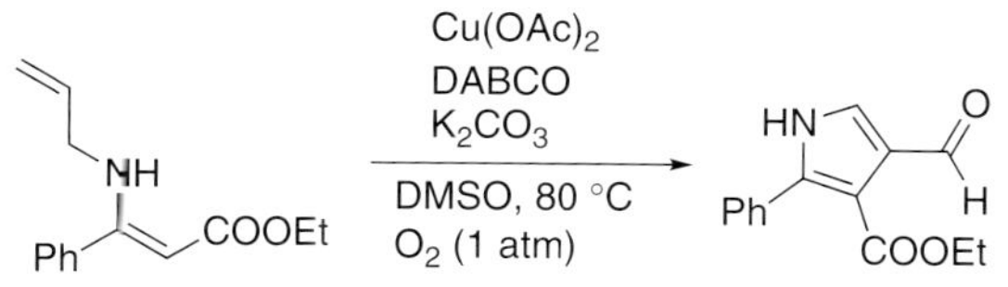

Scheme 8.4 $Cu(OAc)_2$ catalyzed synthesis of 3-Azabicyclo[3.1.0] hex-2-enes and 4-carbonylpyrroles.

Huang, Liang *et al.* developed a direct CuI-catalyzed synthesis of polysubstituted pyrroles from dialkyl ethylenedicarboxylates and β-enamino ketones or esters in moderate to good yields with molecular oxygen (1 atm) (Scheme 8.5). Variation of N-substituents, aromatic ring, alkyl, and ester could be obtained smoothly. In this procedure, molecular oxygen was used as the oxidant and the C(sp^2)–H bond of the β-enamino ketones was cleaved [15].

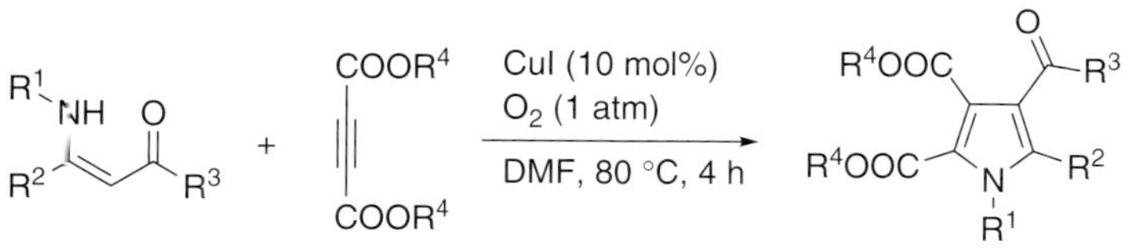

Scheme 8.5 CuI catalyzed synthesis of polysubstituted pyrroles.

The cycloaddition of α-metalated methyl isocyanides onto the triple bond of electron-deficient acetylenes to 2,3,4-trisubstituted pyrroles in 17–97% yield was realized in de Meijere group (Scheme 8.6). Several copper could catalyze the polysubstituted pyrrole formation. Moreover, they also developed a new copper catalyst system for the synthesis of 2,3-disubstituted pyrroles by the reaction of copper acetylides derived from unactivated terminal alkynes with substituted methyl isocyanides in 5–88% yield [16, 17].

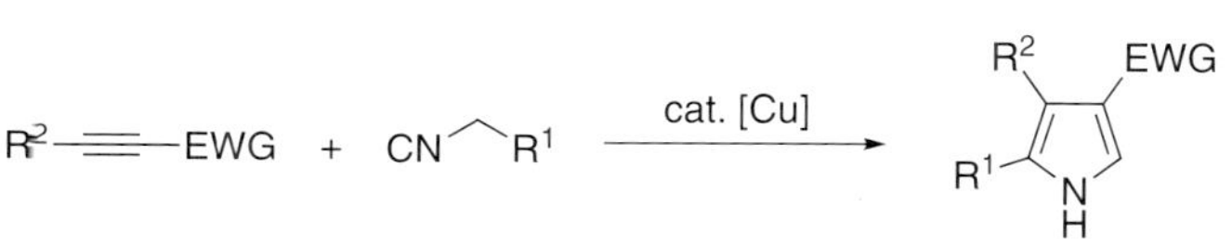

Scheme 8.6 Copper catalyzed synthesis of 2,3,4-trisubstituted pyrroles.

The regiocontrolled formation of pyrroles via a formal [3+2] cycloaddition of isocyanides and electron-deficient alkynes could be realized in different catalytic system. The Cu_2O/phen catalytic system or a phosphine catalyst could lead to two distinct regioisomeric pyrroles. The reaction under a catalytic amount of Cu_2O and phen gives 2,4-di-EWG-substituted pyrroles (electron withdrawing group), whereas the reaction under dppp (1,3-Bis(diphenylphosphino)propane) catalyst affords 2,3-di-EWG-substituted pyrroles selectively. This interesting switching of the regioselectivity is due to umpolung of the reactivity of the alkynes with different catalytic conditions [18] (Scheme 8.7).

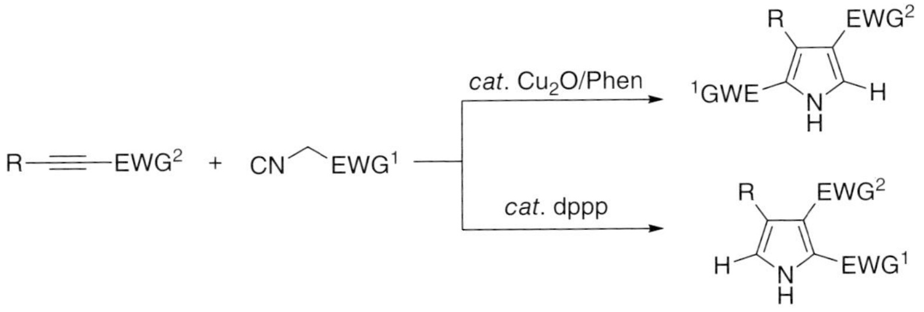

Scheme 8.7 The regiocontrolled formation of pyrroles via a formal [3+2] cycloaddition of isocyanides and electron-deficient alkynes.

Chiba and coworkers developed a $Cu(NTf_2)_2$-catalyzed synthesis of pyrroles from α-ethoxycarbonyl vinyl azides and ethyl acetoacetate through the 1,4-addition reaction of the acetoacetate to the vinyl azides [19]. Jiao and coworkers reported a copper- or nickel-catalyzed highly selective denitrogenative annulation of vinyl azides with acetaldehydes to 2,4- and 3,4-diaryl-substituted pyrroles. $Cu(OAc)_2$ could catalyze the formation of 2,4-diaryl-substituted pyrroles. This selective polysubstituted pyrrole synthesis could proceed under mild conditions without any acidic or basic conditions [20] (Scheme 8.8).

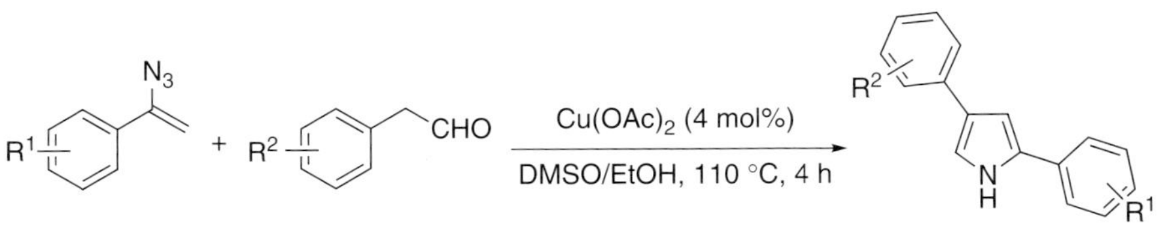

Scheme 8.8 $Cu(NTf_2)_2$-catalyzed synthesis of pyrroles from ethoxycarbonyl vinyl azides and ethyl acetoacetate.

CuOTf-catalyzed synthesis of polysubstituted pyrroles from α-diazoketones, nitroalkenes, and amines was reported by Lu, Wang, and coworkers. The corresponding polysubstituted pyrroles could be obtained in moderate yields using air as the oxidant. This cascade process of the polysubstituted pyrrole formation involves an NH insertion of carbene, a copper-catalyzed oxidative dehydrogenation of amine, and a [3+2] cycloaddition of azomethine ylide [21] (Scheme 8.9).

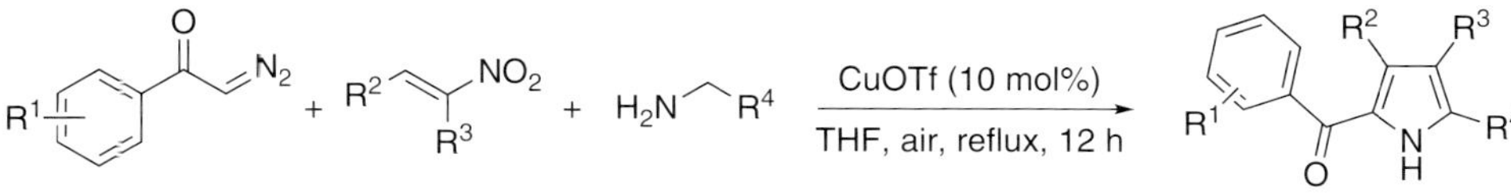

Scheme 8.9 CuOTf-catalyzed synthesis of polysubstituted pyrroles from diazoketones, nitroalkenes, and amines.

Recently, Guan *et al.* developed a novel $Cu(OAc)_2$-promoted oxidative coupling of enamides with electron-deficient alkynes for the synthesis of multisubstituted NH pyrroles. This reaction tolerates a wide range of functional groups and is a reliable procedure for the rapid elaboration of readily available enamides into a variety of diester-substituted NH pyrroles. The reaction proceeded through C–H and N–H bond functionalization of enamides Cu/O_2 system [22]. They also developed an efficient CuBr-catalyzed homocoupling of ketoxime carboxylates for the synthesis of symmetrical pyrroles [23] (Scheme 8.10).

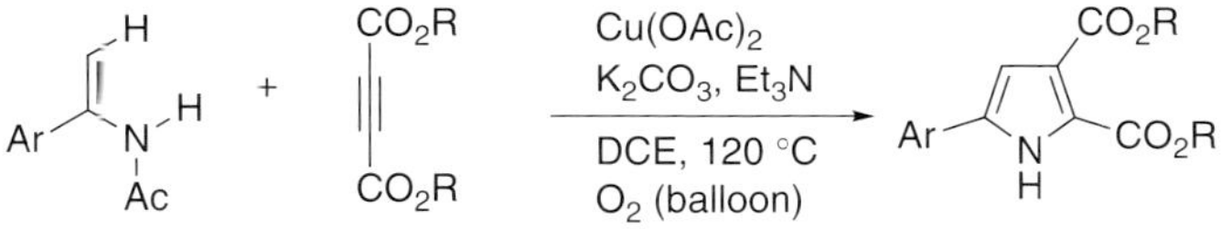

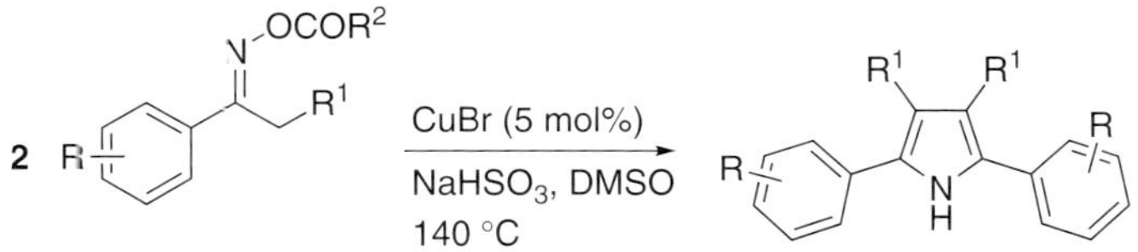

Scheme 8.10 $Cu(OAc)_2$-promoted oxidative coupling of enamides with electron-deficient alkynes for the synthesis of multisubstituted NH pyrroles.

8.3.2 Copper-Catalyzed Synthesis of Pyrrolidines

In 2002, Noack and Göttlich presented the first example of $CuPF_6$-catalyzed radical aminohydroxylation of double bonds of unsaturated *N*-benzoyloxyamines to generate pyrrolidine products. The N–O bond of *N*-benzoyloxyamines could be activated by adding an oxophilic (Lewis) acid that might coordinate to the benzoyloxy carbonyl oxygen and withdraw electron density. The cheap $BF_3{\cdot}Et_2O$ shows high activity and could give the best yields when used in equimolar amounts in toluene at 100 °C [24] (Scheme 8.11).

A Cu–Xantphos system [Cu(O–t-Bu)–Xantphos, 10–15 mol%]-catalyzed synthesis of pyrrolidine and piperidine derivatives was reported. Some pyrrolidine and piperidine derivatives could be obtained in excellent yields in alcoholic solvents via intramolecular hydroamination of unactivated terminal alkenes bearing an unprotected aminoalkyl substituent. Both primary and secondary amines with

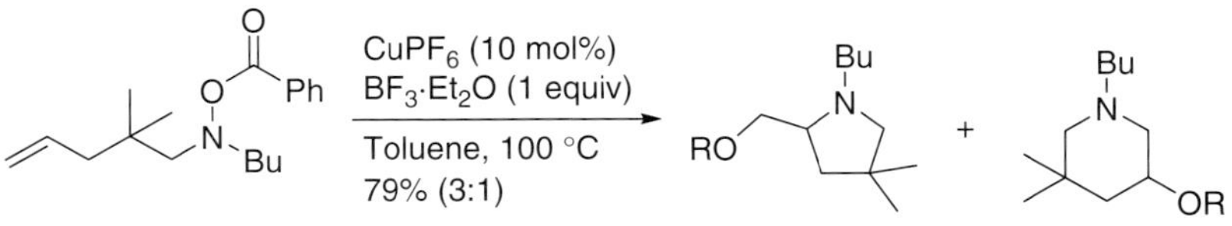

Scheme 8.11 $CuPF_6$-catalyzed pyrrolidines formation via radical aminohydroxylation of double bonds of unsaturated *N*-benzoyloxyamines.

a variety of functional groups are tolerated in this copper catalytic system [25] (Scheme 8.12).

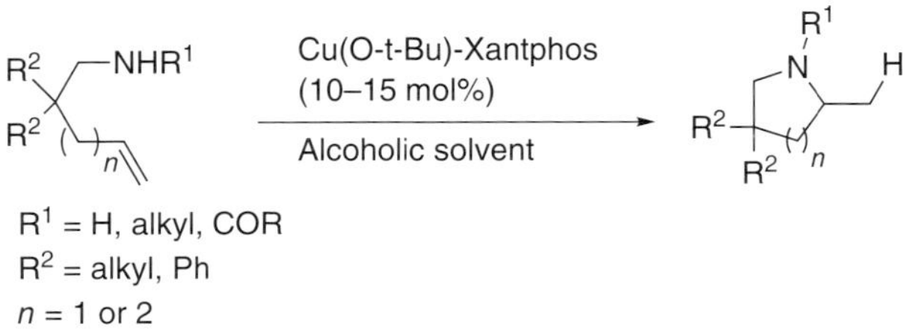

Scheme 8.12 A Cu-Xantphos system for the synthesis of pyrrolidine and piperidine derivatives.

Paderes and Chemler developed an intramolecular diastereoselective aminooxygenation of unactivated alkenes to pyrrolidines. $Cu(EH)_2/O_2$ catalytic system could lead to 2,5-*cis*- and 2,5-*trans*-pyrrolidines formation from α-substituted 4-pentenylsulfonamides in good to excellent yields and >20 : 1 selectivity. For β- and γ-substituted 4-pentenylsulfonamides, $Cu(OTf)_2$·bis(oxazoline) with O_2 catalytic system could give higher diastereoselectivities. Furthermore, enantioselective desymmetrization could be achieved (up to 98% ee) with β-allyl-4-pentenylsulfonamide [26] (Scheme 8.13).

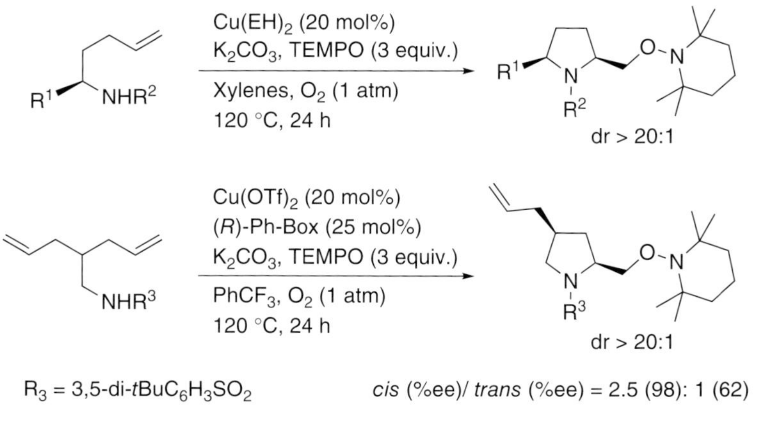

Scheme 8.13 Intramolecular diastereoselective aminooxygenation of unactivated alkenes to pyrrolidines.

The N-substituted pent-4-en-1-amines or N-substituted hex-5-en-1-amine could be converted to 2-chloromethylpyrrolidines or 2-chloromethylpiperidine in good isolated yields using 1 equiv. of $CuCl_2·2H_2O$. This intramolecular chloroamination of unfunctionalized olefins also could be realized using $MgCl_2$ as chlorine source, and reactions can be carried out in the presence of a catalytic amount of $CuCl_2$ (5 mol%). Pure oxygen atmosphere is better than air and it could accelerate conversion of the substrates [27] (Scheme 8.14).

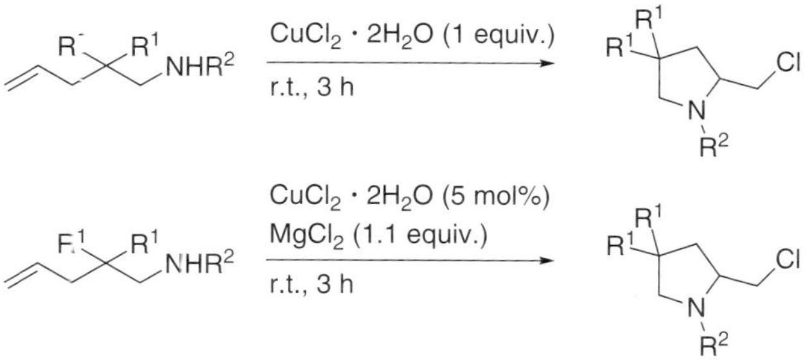

Scheme 8.14 Copper catalyzed synthesis of 2-chloromethylpyrrolidines.

The 1,3-dipolar cycloaddition of azomethine ylides to electron-deficient alkenes is one of the most powerful methods for the construction of highly substituted pyrrolidines with highly enantio- and endo-selectivity. Various chiral metal–ligand catalysts have been well developed for pyrrolidine synthesis from glycine ester aldimines and electron-deficient alkenes. Copper is a useful metal for this transformation and several catalytic systems have been reported. Some ligands such as BINAP (2,2′-bis(diphenylphosphino)-1,1′-binaphthyl), BINAP derivatives [28], P,N-ferrocene [29, 30], phosphoramide and selenophosphoramide [31], BiphamPhos [32], phosphoramidite [33], Segphos or DTBM-Segphos [34, 35], Fesulphos [36–38], Taniaphos [39] have been well developed for the copper catalysts in Carretero group (Scheme 8.15).

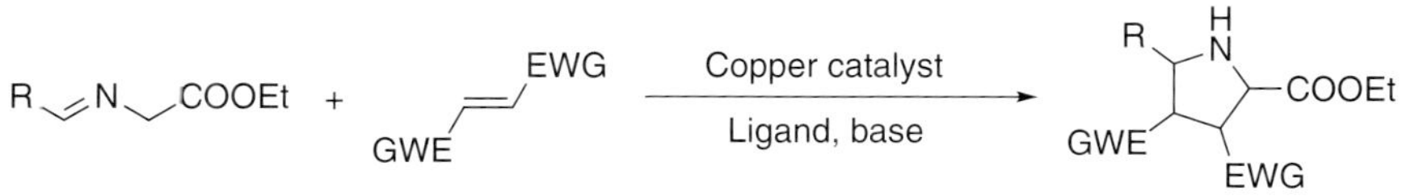

Scheme 8.15 1,3-dipolar cycloaddition of azomethine ylides to electron-deficient alkenes for the construction of pyrrolidines.

Scheidt and coworkers reported a $[Cu(OTf)]_2·C_6H_6$ catalyze the substituted pyrrolidine synthesis with excellent to good diastereoselectivities in high yields through three-component reaction of α-diazoester, imine, and various alkenes or alkynes. This 1,3-dipole is generated in situ and undergoes diastereoselective cycloadditions with activated dipolarophiles to afford highly substituted pyrrolidines. This reaction could generate four contiguous stereogenic centers in one operation by employing this copper catalytic system with C–H bond activation [40] (Scheme 8.16).

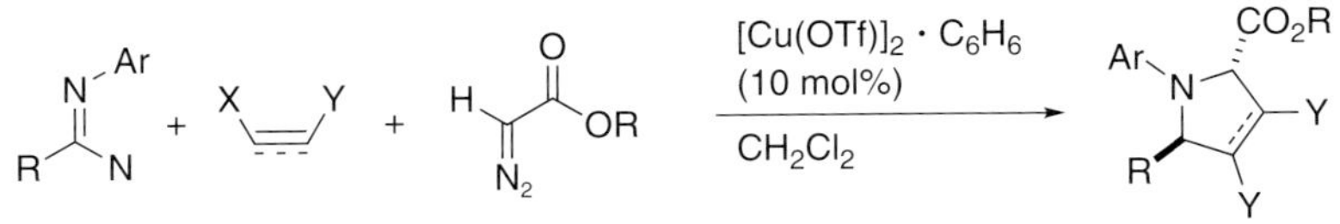

Scheme 8.16 $[Cu(OTf)]_2 \cdot C_6H_6$ catalyzed the substituted pyrrolidine synthesis through three-component reaction.

A range of 3-pyrrolines could be synthesized from conjugated dienes with common nitrene precursors catalyzed by $[Cu(hfacac)_2]$ via intermolecular [4+1] cycloaddition. Good to excellent *cis*/*trans* selectivity could be obtained when 1,4-disubstituted dienes are used as substrates. The *cis* or *trans* preference depends on the nature of the substituents of the substrates. Mechanistic studies disclose that this transformation proceeds through diene aziridination and subsequent ring expansion. The $[Cu(hfacac)_2]$ catalyst shows higher activity than other copper catalysts [41] (Scheme 8.17).

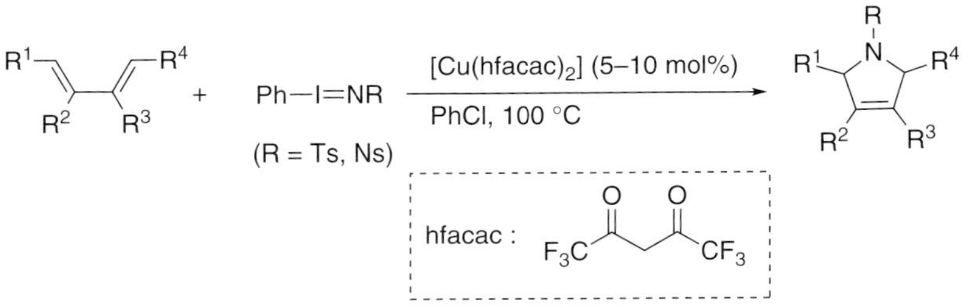

Scheme 8.17 $[Cu(hfacac)_2]$ catalyzed 3-pyrrolines synthesis via intermolecular [4+1] cycloaddition.

8.3.3 Copper-Catalyzed Synthesis of Indoles

The indole nucleus is present in various bioactive molecules and many selective protocols for its construction have been developed. Classical methods for the indole synthesis include the Fischer indole synthesis, the Batcho–Leimgruber synthesis from *o*-nitrotoluenes and dimethylformamide acetals, the Gassman synthesis from N-haloanilines, the reductive cyclization of *o*-nitrobenzyl ketones, and the Madelung cyclization of *N*-acyl-*o*-toluidines [42, 43].

Cacchi and coworkers developed an efficient CuI-catalyzed synthesis of a multisubstituted indole skeleton from readily available *N*-aryl enaminones with wide functional group tolerance including the whole range of halogen substituents. For instance, N-(2-bromophenyl)enaminone favors C–H activation process in comparison to C–Br activation process leading to 7-bromoindole with high selectivity. The halogen-substituted indoles are key intermediates for some biologically active compound synthesis via Suzuki–Miyaura cross-coupling reactions. Furthermore, indoles also could be obtained from α,β-ynones and primary amines directly by a sequential process and a wide variety of indole derivatives could be synthesized easily [44] (Scheme 8.18).

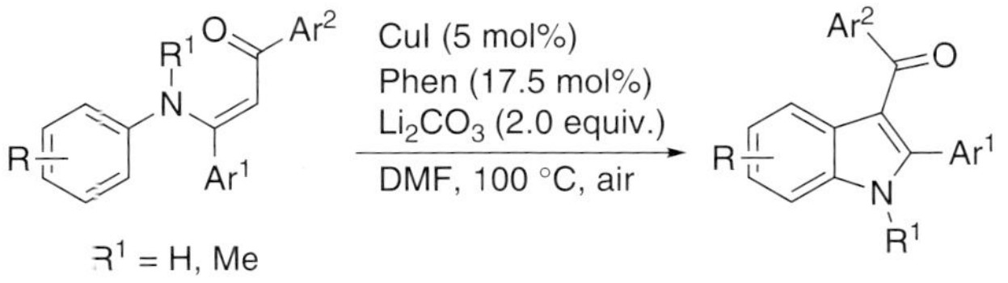

Scheme 8.18 CuI-catalyzed synthesis of a multisubstituted indole skeleton from *N*-aryl enaminones.

A novel three-component coupling reaction for the synthesis of 2-(aminomethyl) indoles and polycyclic indole derivatives from readily available N-protected ethynylanilines, amines, and aldehydes was reported by Ohno, Fujii, and coworkers. This is the first copper catalytic multicomponent construction of an indole ring that produces water as the only by-product. Two C–N bonds and one C–C bond are formed, whereas C(sp)–H bonds of alkynes and C(sp^2)–H bonds of aldehydes were activated in this reaction [45–48] (Scheme 8.19).

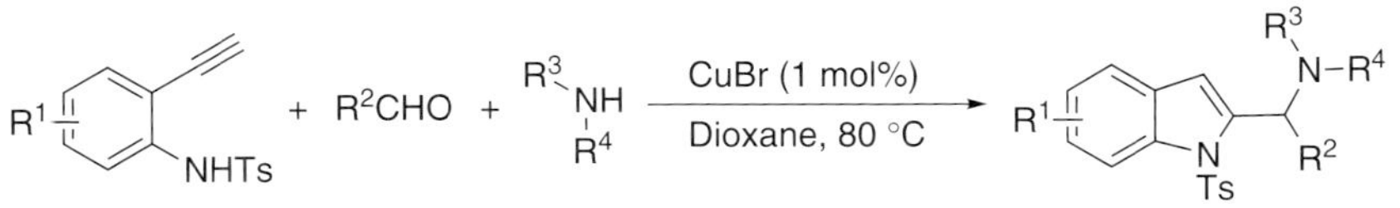

Scheme 8.19 CuBr catalyzed synthesis of 2-(aminomethyl) indoles and polycyclic indole derivatives from N-protected ethynylanilines, amines, and aldehydes.

Three-component reaction of 2-ethynylaniline, sulfonyl azide, and nitroolefin also could lead to polysubstituted indole derivatives in moderate to good yields with CuBr catalyst. The reaction conditions are extremely mild and various polysubstituted indoles were obtained successfully. Initially, a triazole intermediate was formed from alkyne and sulfonyl azide in the presence of CuBr and triethylamine. It could convert into reactive ketenimine by the ring-opening rearrangement. Subsequently intramolecular nucleophilic addition, intermolecular Michael addition, and tautomerization could lead to the desired polysubstituted indoles [49] (Scheme 8.20).

Patel *et al.* reported that CuBr catalyzed the synthesis of 3-aroylindoles from *o*-alkynylated *N*,*N*-dimethylamines through oxidative process using aqueous TBHP

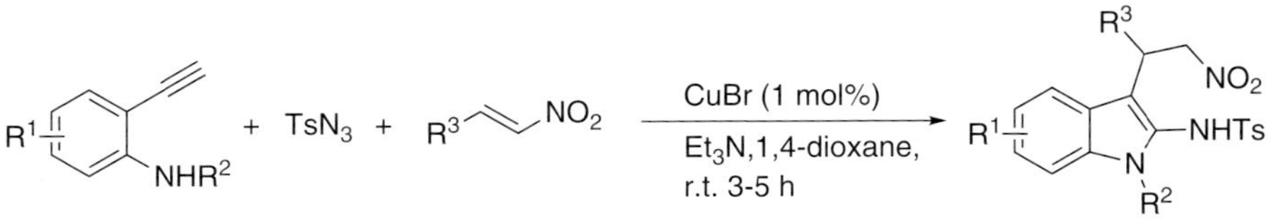

Scheme 8.20 Polysubstituted indole derivatives synthesis via three-component reaction of 2-ethynylaniline, sulfonyl azide, and nitroolefin.

(tert-butyl hydroperoxide) as the oxidant in DMSO (Dimethyl sulfoxide). The C–C and C–O bonds could be formed simultaneously through an intramolecular oxidative path involving the C(sp^3)–H bond activation. This reaction could proceed smoothly at a relatively lower temperature (80 °C) without any cocatalyst and additives [50] (Scheme 8.21).

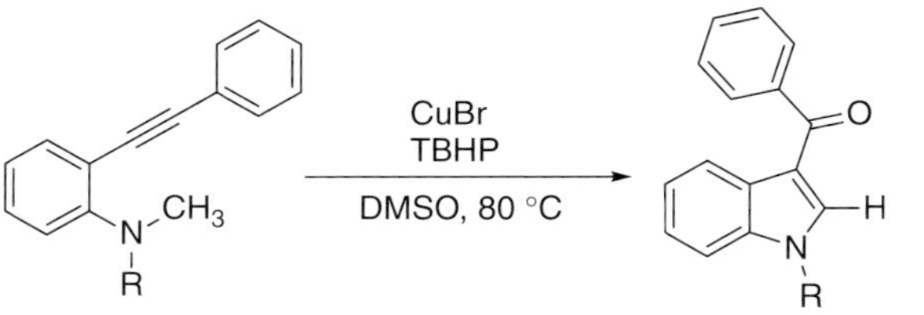

Scheme 8.21 CuBr catalyzed the synthesis of 3-aroylindoles from *o*-alkynylated *N*,*N*-dimethylamines.

8.3.4
Copper-Catalyzed Synthesis of Indolines

Chemler and coworkers developed a number of equivalent copper-promoted reactions that involve the intermolecular addition of amines to olefins to the synthesis of *N*-heterocycles via carboamination [51–54] and deamination [55, 56].

In 2008, they realized a copper-catalyzed enantioselective intramolecular aminooxygenation of olefins to give chiral indolines and pyrrolidines. The combination of Cu(OTf)$_2$/(4*R*,5*S*)-bis-Phbox ligand is crucial for this reaction and moderate to excellent enantioselectivity could be obtained. The primary carbon radical species generated from the intramolecular amination via C–H activation could be trapped with tetramethylaminopyridyl radical (TEMPO) to give the aminooxygenation product [57]. They also disclosed that the reaction is first order both in substrate and the [Cu(R,R)–Ph-bis(oxazoline)]OTf2 catalyst and zero order in TEMPO via kinetics studies. Furthermore, kinetic isotope effect studies support that the *cis*-aminocupration step is the rate-limiting step. The deuterium-labeling experiment shows that the C–O bond formation occurs through the direct carbon radical trapping with TEMPO. The C(4)-phenyl substitution on the bis(oxazoline) is crucial for high asymmetric induction of this reaction [58] (Scheme 8.22).

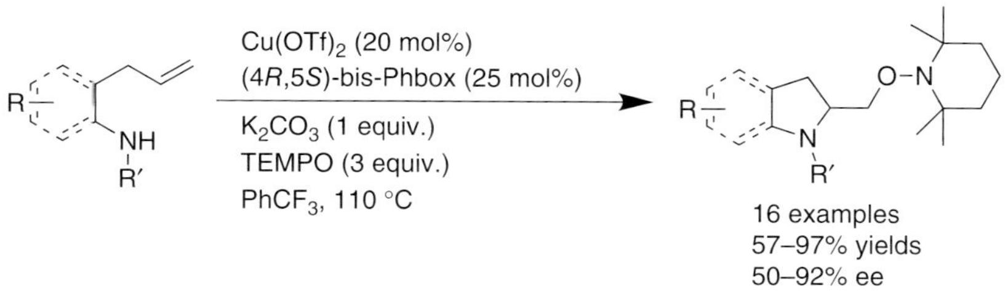

Scheme 8.22 Copper-catalyzed enantioselective intramolecular aminooxygenation of olefins.

They also reported the synthesis of the *N*-heterocycles 10*a*,11-dihydro-*10H*-indolo[1,2-*a*]indoles using $Cu(OTf)_2$/2,2′-bipyridyl system and MnO_2 as oxidant. When TEMPO is added to the reaction mixture, aminooxygenation product indolines could be obtained in good yields with O_2 as oxidant [59] (Scheme 8.23).

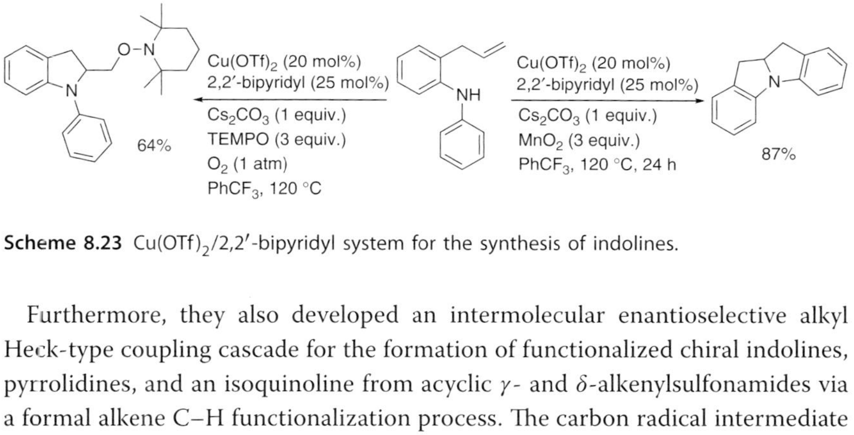

Scheme 8.23 $Cu(OTf)_2$/2,2′-bipyridyl system for the synthesis of indolines.

Furthermore, they also developed an intermolecular enantioselective alkyl Heck-type coupling cascade for the formation of functionalized chiral indolines, pyrrolidines, and an isoquinoline from acyclic γ- and δ-alkenylsulfonamides via a formal alkene C–H functionalization process. The carbon radical intermediate which is formed *in situ* under copper system could be oxidized to an alkene leading to the desired products [60] (Scheme 8.24).

The enantioselective alkene aminohalogenation reaction using 2-iodopropane as iodine source was also realized in this group. A range of chiral 2-iodomethyl indolines and pyrrolidines could be formed in good to excellent yield and enantioselectivity. In addition, 1,1-dichloroethene and (2,2-dibromo-1-methylcyclopropyl)benzene also could lead to aminochloride and aminobromide products in moderate yield and good enantioselectivity [61] (Scheme 8.25).

They also developed the $Cu(OTf)_2$-catalyzed enantioselective alkene hydroamination/cyclization to enantioenriched 2-methylindolines using 1,4-cyclohexadiene as hydrogen atom source. 1,4-Cyclohexadiene is relatively

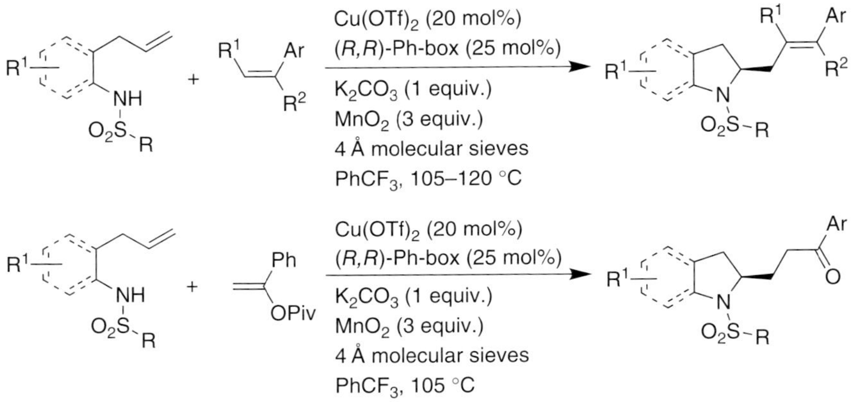

Scheme 8.24 Intermolecular enantioselective alkyl Heck-type coupling cascade for the formation of functionalized chiral indolines, pyrrolidines.

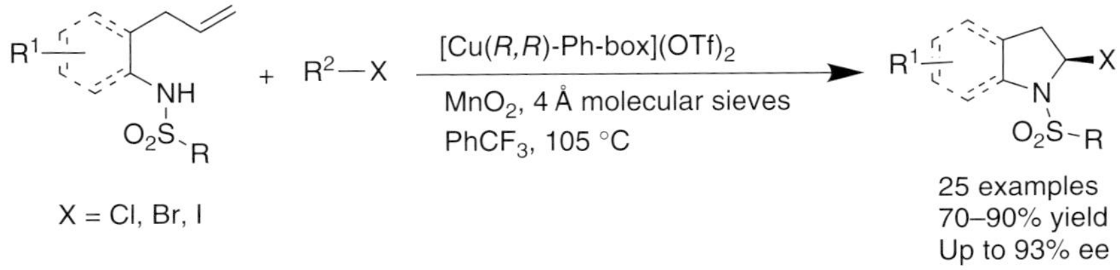

Scheme 8.25 The enantioselective alkene aminohalogenation reaction.

inexpensive and could produce a volatile by-product. 2,6-Di-*tert*-butyl-4-methylpyridine was a better base than K_2CO_3 for this cyclization reaction. The addition of flame-dried 4 Å molecular sieves to the reaction system is crucial for the selectivity and reproducibility [62] (Scheme 8.26).

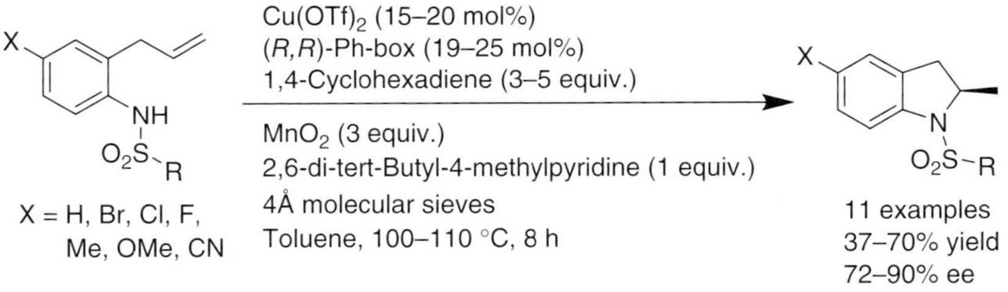

Scheme 8.26 $Cu(OTf)_2$-catalyzed enantioselective alkene hydroamination/cyclization to enantioenriched 2-methylindolines.

The fused indolines could be synthesized from 2-ethynylarylmethylenecyclopropane with sulfonyl azide using CuI as catalyst and Et_3N as base. A ketenimine species and 6π-electrocyclization could be involved for this transformation. The copper-catalyzed triazole intermediate formation from 2-ethynylaryl methylenecyclopropane with sulfonyl azide should be the key step for the generation of ketenimine species [63] (Scheme 8.27).

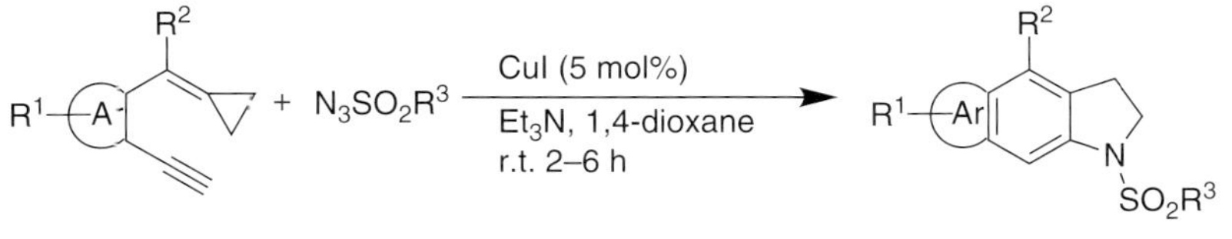

Scheme 8.27 Synthesis of fused indolines from 2-ethynylarylmethylenecyclopropane with sulfonyl azide.

Recently, Sodeoka *et al.* reported a CuI-catalyzed trifluoromethylation of N-protected allylaniline and homoallylaniline derivatives with Togni's reagent to trifluoromethylated heterocycles such as indolines in good to high yields. Both the $C(sp^2)$–H bonds of alkene and aryl rings were activated in the transformation [64] (Scheme 8.28).

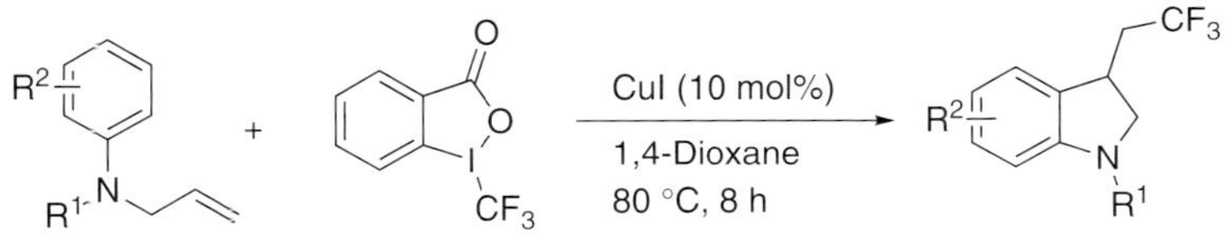

Scheme 8.28 CuI-catalyzed trifluoromethylation of N-protected allylaniline and homoallylaniline derivatives with Togni's reagent.

Recently, Hirano, Miura, and coworkers developed a $Cu(OAc)_2$-mediated indolines synthesis via intramolecular aromatic C–H amination using a picolinamide-type bidentate coordination group. The reaction also could proceed smoothly with catalytic $Cu(OAc)_2$ using MnO_2 as terminal oxidant. Some electron-rich thiophene- and indole-fused indoline analogs could be formed under the mild oxidative conditions. The deuterium-labeling experiments suggest that the picolinamide-type bidentate coordination group could direct this $C(sp^2)$–H bond activation process [65] (Scheme 8.29).

8.3.5 Copper-Catalyzed Synthesis of Oxindoles

The intramolecular synthesis of 3,3-disubstituted oxindoles through $C(sp^3)$–H and Ar–H coupling of anilides using potassium t-butoxide as base and stoichiometric $Cu(OAc)_2{\cdot}H_2O$ in DMF (N,N-Dimethylformamide) was reported by Taylor *et al.*[66]. This reaction also could be conducted under catalytic $Cu(OAc)_2{\cdot}H_2O$ in toluene or mesitylene with atmospheric oxygen as the reoxidant without an additional base. A range of substrates including several

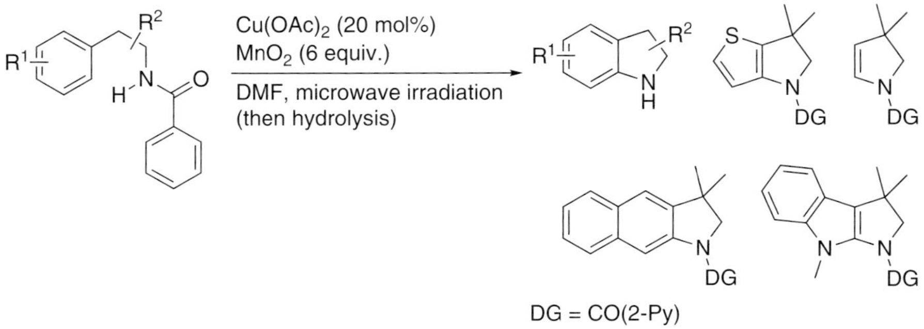

Scheme 8.29 $Cu(OAc)_2$-mediated indolines synthesis via intramolecular aromatic C–H amination.

containing acid-labile groups also could convert into corresponding products using mesitylene as solvent [67] (Scheme 8.30).

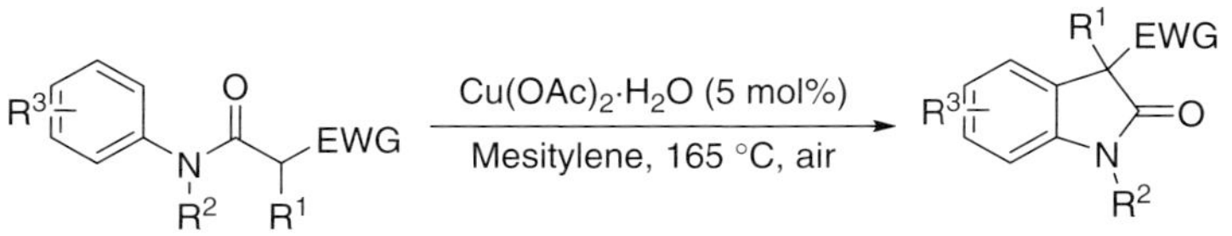

Scheme 8.30 The intramolecular synthesis of 3,3-disubstituted oxindoles through $C(sp^3)$–H and Ar–H coupling of anilides.

A $Cu(MeCN)_4PF_6$-catalyzed one-pot synthesis of trifluoromethylated oxindoles via cascade trifluoromethylation/1,4-aryl migration/desulfonylation/$C(sp^2)$–N bond formation of conjugated *N*-tosyl amides was developed by Nevado and coworkers. Togni's reagent is able to trifluoromethylate the alkene with the copper catalyst, and an amidyl radical intermediate could be generated. The amidyl radical can be engaged in a subsequent cyclization to give trifluoromethylated oxindoles. Radical inhibition experiments show that the reaction may occur via a radical mechanism and a crossover experiment suggests that the aryl migration is an intramolecular process in this transformation [68] (Scheme 8.31).

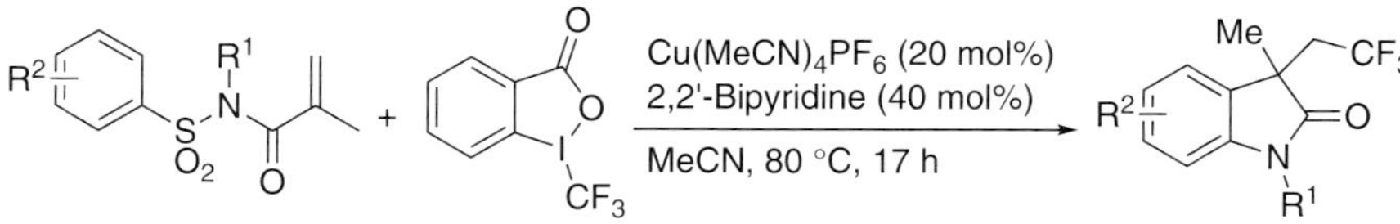

Scheme 8.31 $Cu(MeCN)_4PF_6$-catalyzed one-pot synthesis of trifluoromethylated oxindoles.

Recently, another novel reaction for the synthesis of trifluoromethylated oxindoles via $Cu(NO_3)_2{\cdot}2.5H_2O$-catalyzed aryl $C(sp^2)$–H functionalization was described by Liang, Lipshutz, and coworkers. The trifluoromethylated oxindoles could be obtained from N-arylacrylamides in good yields using the inexpensive and stable Langlois' reagent (CF_3SO_2Na) as the source of CF_3 radical. The reaction was performed using water as the solvent at room temperature. Notably, this aqueous medium using water-soluble copper catalyst can be recycled easily [69] (Scheme 8.32).

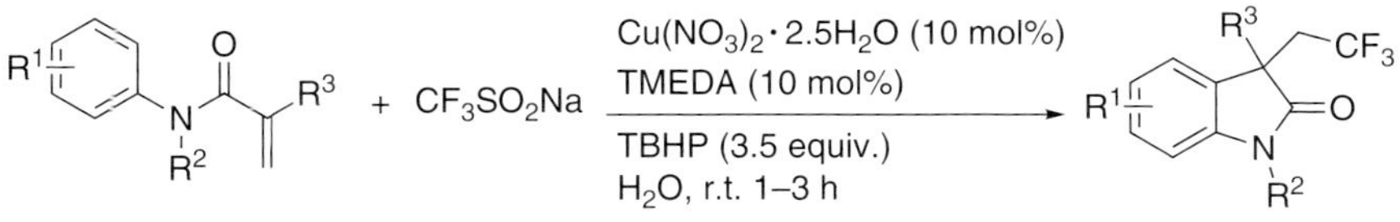

Scheme 8.32 $Cu(NO_3)_2 \cdot 2.5H_2O$-catalyzed synthesis of trifluoromethylated oxindoles.

Cu_2O could catalyze oxidative alkylarylation of acrylamides with simple alkanes to alkyl-substituted oxindoles using dicumyl peroxide (DCP) as radical initiator. Alkyl radical could be formed via hydrogen abstraction reaction with cumyloxyl radical which is generated from the homolysis of DCP. Then alkyl radical reacts with C=C bond followed by cyclization to benzene core and oxidation to give the alkyl-substituted oxindoles. Selective activation of unactivated C(sp^3)–H and C(sp^2)–H bonds by one single step is achieved in this transformation [70] (Scheme 8.33).

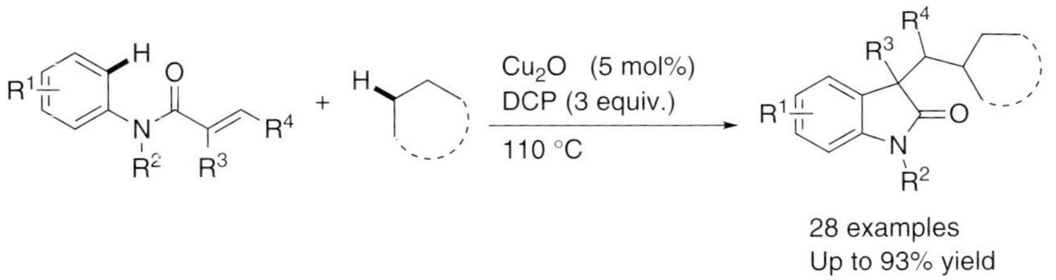

Scheme 8.33 Cu_2O catalyze oxidative alkylarylation of acrylamides with simple alkanes to alkyl-substituted oxindoles.

Synthesis of 3,3-disubstituted oxindoles from arylation– and vinylation–carbocyclization of electron-deficient alkenes with diaryliodonium salts catalyzed by CuCl catalyst was reported by Zhou, Li, and coworkers. A wide range of diaryl and vinyl(aryl)iodonium triflates and various substitutions on phenylacrylamides are compatible for this reaction. Both (±)-physostigmine and (±)-physovenine have been synthesized successfully in 41 and 36% overall yield using this vinyl addition/cyclization method [71]. Fu and coworkers developed a CuI-catalyzed synthesis of oxindole derivatives from readily available N-alkyl-N-phenylacrylamides and diaryliodonium triflates [72] (Scheme 8.34).

N-alkyl-N-phenylacrylamides and aryl aldehydes undergo oxidative cascade coupling reaction to 3-(2-oxo-2-arylethyl)indolin-2-ones in good yields using $CuCl_2$/TBHP as catalyst and oxidant, respectively. Both the C(sp^2)–H of the N-alkyl-N-phenylacrylamides and C(sp^2)–H bonds of the aldehyde were activated in one step under this oxidative system. The acyl radical which is generated from

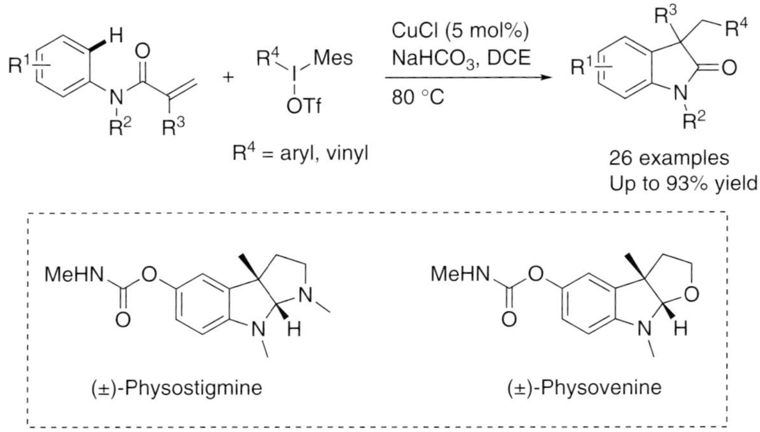

Scheme 8.34 Synthesis of 3,3-disubstituted oxindoles from arylation– and vinylation–carbocyclization of electron-deficient alkenes with diaryliodonium salts.

hydrogen abstraction of aldehyde in the Cu/TBHP system is a new radical source for this reaction [73] (Scheme 8.35).

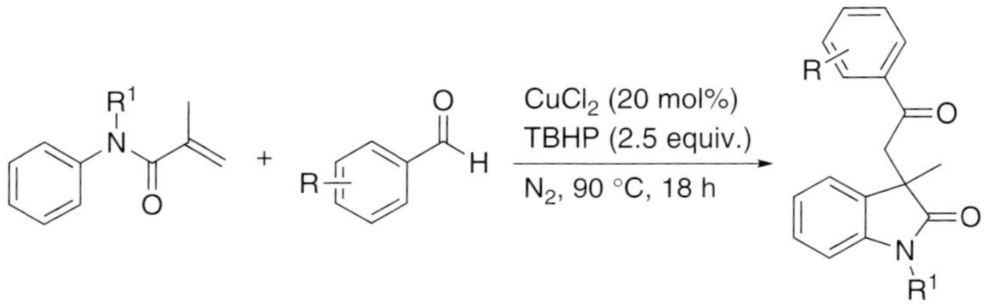

Scheme 8.35 Synthesis of 3-(2-oxo-2-arylethyl)indolin-2-ones from *N*-alkyl-*N*-phenylacrylamides and aryl aldehydes.

8.3.6
Copper-Catalyzed Synthesis of Indole-2,3-dione (Isatins)

A novel $CuCl_2$-catalyzed intramolecular C–H oxidation/acylation protocol using two C–H bonds as the reaction partners and O_2 as the terminal oxidant was reported by Li and coworkers (Scheme 8.36). Various substituted indoline-2,3-diones could be prepared in moderate to good yields with high tolerance of functional groups. The kinetic isotope effect ($k_H/k_D = 2:4$) was observed by FTIR (Fourier transform infrared spectroscopy) analysis, and this suggests that the C–H bond cleavage of aldehyde is the rate-limiting step for this transformation. A drawback of this transformation is the formation of inseparable regioisomers for *meta*-substituted substrates [74].

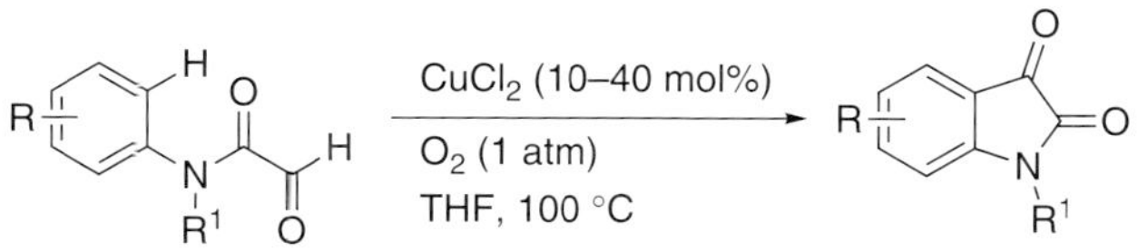

Scheme 8.36 $CuCl_2$-catalyzed intramolecular C–H oxidation/acylation protocol for the synthesis of indoline-2,3-diones.

Fu *et al.* developed a $Cu(OAc)_2$-catalyzed synthesis of indoline-2,3-dione derivatives from readily available secondary anilines and ethyl glyoxalate in moderate to good yields (Scheme 8.37). The picolinic acid was selected as the ligand and TBHP was selected as the oxidant for this reaction. The process including intermolecular copper-catalyzed selective oxidative *ortho*-site aromatic acylation of the N–H group in secondary anilines and intramolecular nucleophilic attack of the N–H group to the ester was proposed. Notably, this is the first dicarbonyl compound synthesis using ethyl glyoxalate as the dicarbonylating agent under copper catalyst [75].

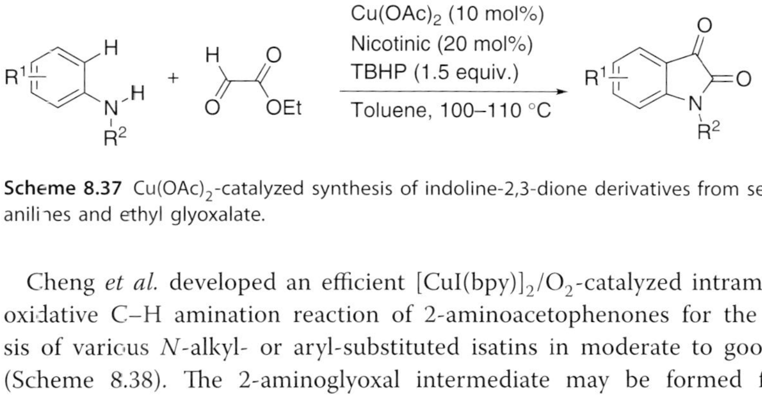

Scheme 8.37 $Cu(OAc)_2$-catalyzed synthesis of indoline-2,3-dione derivatives from secondary anilines and ethyl glyoxalate.

Cheng *et al.* developed an efficient $[CuI(bpy)]_2/O_2$-catalyzed intramolecular oxidative C–H amination reaction of 2-aminoacetophenones for the synthesis of various *N*-alkyl- or aryl-substituted isatins in moderate to good yields (Scheme 8.38). The 2-aminoglyoxal intermediate may be formed from 2-aminoaryl methyl ketone under oxidation conditions, and it could undergo intramolecular cyclization to give the desired isatins [76].

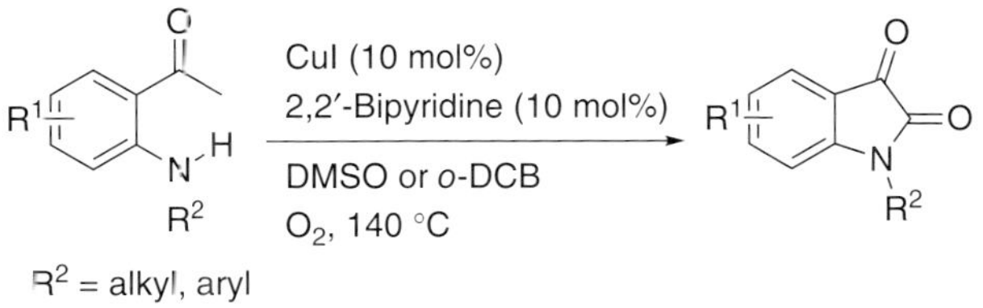

Scheme 8.38 $[CuI(bpy)]_2/O_2$-catalyzed intramolecular oxidative C–H amination reaction of 2-aminoacetophenones.

Ilangovan and Satish reported that tertiary amines can also be used for the synthesis of isatins under $Cu(OAc)_2{\cdot}H_2O$, NaOAc, and air system at relatively lower temperature (80 °C) (Scheme 8.39). The intramolecular C–N bond cleavage and C–N bond formation occur in a single step in this reaction [77].

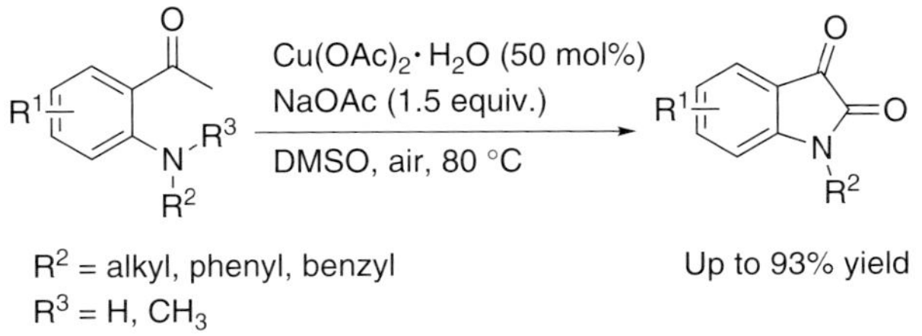

Scheme 8.39 $Cu(OAc)_2 \cdot H_2O$ catalyzed synthesis of isatins from tertiary amines.

8.3.7
Copper-Catalyzed Synthesis of Indolizines

A CuBr-catalyzed regioselective [3+2] cyclization of pyridines with alkenyldiazoacetates leading to functionalized indolizine derivatives is reported in 2010 (Scheme 8.40). A broad range of pyridine substrates with various substitutes is compatible under this mild copper-catalyzed cyclization system. Furthermore, this is the first successful example of metal-catalyzed cyclization of a π-deficient heterocyclic system with alkenyldiazo compounds. Initially, copper(I) alkenylcarbene species could be formed via decomposition of alkenyldiazoacetates. Then it would convert into the cuprate by Michael-type addition of the pyridine nitrogen. A cascade process including cyclization/reductive elimination/oxidative aromatization would result in the formation of the corresponding indolizidine [78].

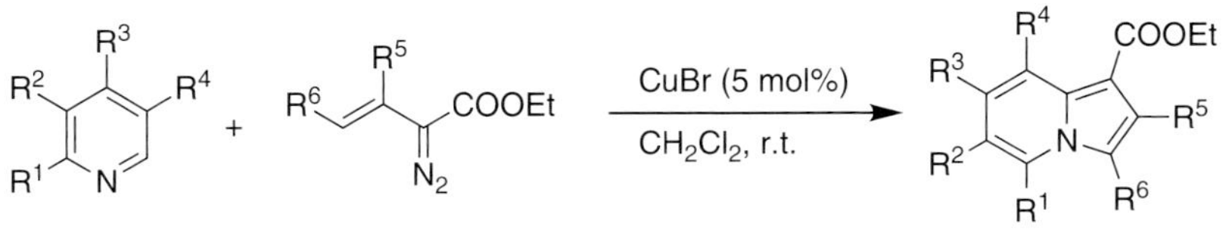

Scheme 8.40 CuBr-catalyzed regioselective [3+2] cyclization of pyridines with alkenyldiazoacetates.

8.3.8
Copper-Catalyzed Synthesis of Carbazoles

Chang *et al.* developed an intramolecular synthesis of carbazoles from N-substituted 2-amidobiphenyls via oxidative C–N bond formation (Scheme 8.41). This cyclization could take place readily under both Cu-catalyzed and metal-free conditions. The iodobenzene diacetate or bis(trifluoroacetoxy)iodobenzene undergoes this reaction to provide carbazole products in moderate to low yields.

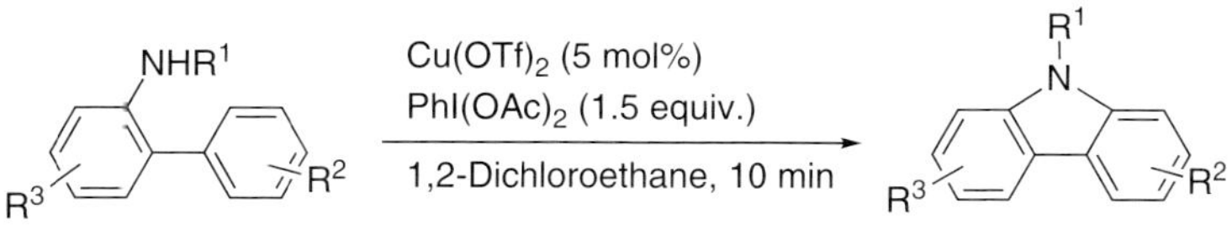

Scheme 8.41 Intramolecular synthesis of carbazoles from Nsubstituted 2-amidobiphenyls.

When catalytic $Cu(OTf)_2$ was added into the reaction mixture, the yield was improved efficiently and a range of products could be obtained in good to excellent yields. The results of kinetic profile, isotope effects, and radical inhibition experiments disclosed that the copper species is proposed to catalytically activate the hypervalent iodine(III) oxidants [79].

The copper-catalyzed synthesis of carbazoles from 2-aminobiphenyls via intramolecular C–H amination also could be realized using MnO_2 as terminal oxidant (Scheme 8.42). The picolinamide-based bidentate directing group which could be spontaneously removed after the coupling reaction was crucial for this reaction. This $Cu(OAc)_2$ catalytic system could proceed smoothly under I(III)-free conditions. Moreover, a few of heteroatom-containing carbazole core π-systems could be contained easily under the mild $Cu(OAc)_2/MnO_2$ catalytic conditions [80].

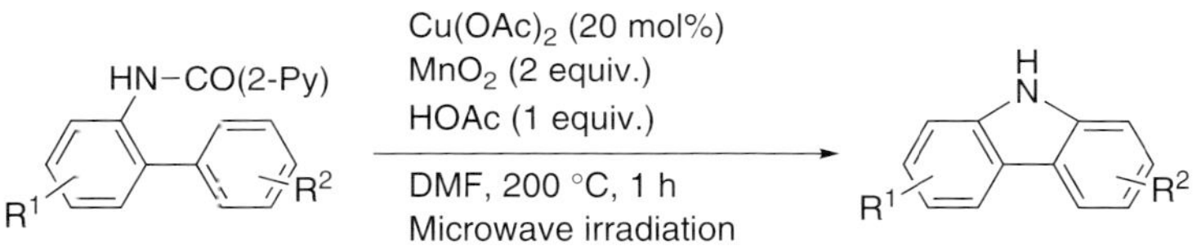

Scheme 8.42 Copper-catalyzed synthesis of carbazoles from 2-aminobiphenyls via intramolecular C–H amination using MnO_2 as terminal oxidant.

8.3.9
Copper-Catalyzed Synthesis of Imidazoles

Imidazoles are one of the most important classes of nitrogen -containing heterocyclic compounds and also important building blocks in biologically active compounds. Yamamoto *et al.* demonstrated a Cu_2O-catalyzed synthesis of 1,4-disubstituted imidazoles via cross-cycloaddition from alkylisocyanides and arylisocyanides using 1,10-phenanthroline as ligand (Scheme 8.43). The reaction

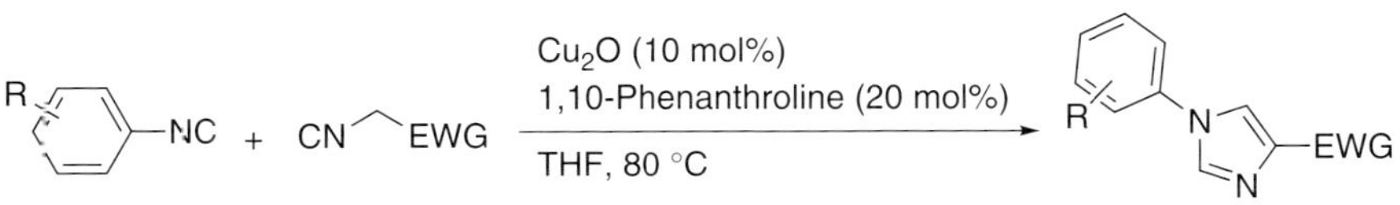

Scheme 8.43 Cu_2O-catalyzed synthesis of 1,4-disubstituted imidazoles from alkylisocyanides and arylisocyanidesa.

starts with the activation of a C–H bond of the alkylisocyanides to form the α-cuprioisocyanide or its tautomer through the extrusion of H_2O [81].

The trisubstituted imidazoles could be generated from ketones and benzylamines under CuI/BF_3·Et_2O catalytic system using O_2 as oxidant. The cocatalyst CuI/BF_3·Et_2O shows high reactivity for this transformation. Eight hydrogen atoms have been removed and four C(sp^3)–H bonds have been functionalized to form three new C–N bonds in this reaction (Scheme 8.44). This mild and efficient method provides a simple, easy-handling, and atom-economic way for the synthesis of trisubstituted imidazoles [82].

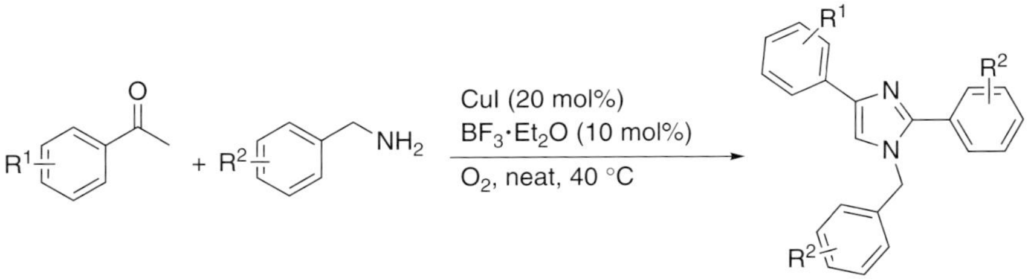

Scheme 8.44 CuI/BF_3 · Et_2O catalyzed trisubstituted imidazoles formation from ketones and benzylamines.

Chiba and coworkers developed a CuI-catalyzed intramolecular aerobic [3+2] annulation reaction of *N*-alkenyl amidines to bi- and tricyclic amidines under O_2 atmosphere (Scheme 8.45). A putative nitrene intermediate which might be generated under CuI/O_2 system is proposed for this transformation. And this nitrene intermediate could react with C=C double bond of the substrate as a 1,3-dipole with the activation of C(sp^2)–H bonds to give the desired products [83].

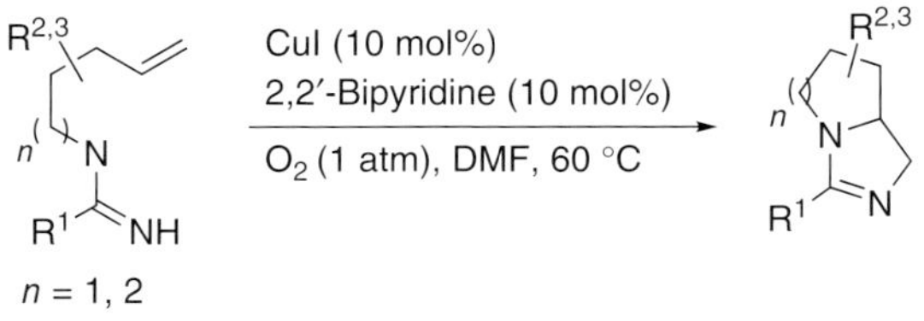

Scheme 8.45 CuI-catalyzed intramolecular aerobic [3+2] annulation reaction of *N*-alkenyl amidines to bi- and tricyclic amidines.

They also realized the Cu(OAC)$_2$-catalyzed aminoacetoxylation of *N*-alkenylamidines using PhI(OAc)$_2$ as an oxygen source for synthesis of 4-acetoxymethyl-4,5-dihydroimidazoles [84] (Scheme 8.46). The aliphatic C–H bonds of *N*-alkylamidines also could be activated to various dihydroimidazoles and tetrahydropyrimidines with Cu(OAc)$_2$/PhI(OAc)$_2$/K_3PO_4 catalytic system [85] (Scheme 8.46).

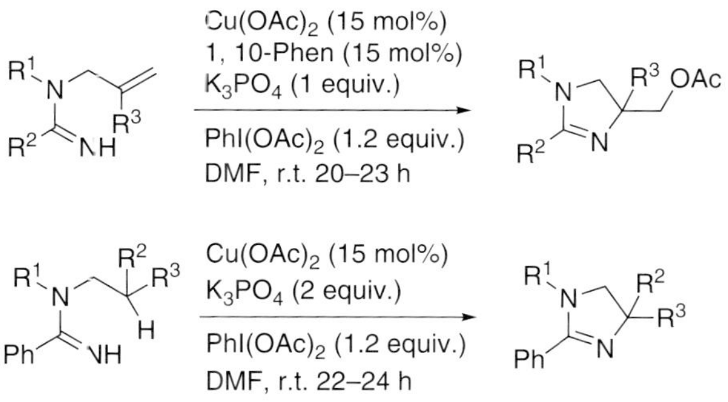

Scheme 8.46 Cu(OAC)$_2$-catalyzed synthesis of 4-acetoxymethyl-4,5-dihydroimidazoles and dihydroimidazoles.

However, the drawback of this reaction is the need of a stoichiometric PhI(OAc)$_2$ to maintain the catalytic turnover and it needs two-electron oxidation to realize the aliphatic C–H amination. They reported that the amidoximes such as readily available *N*-alkylamidoximes could be used as a precursor of the amidinyl radical to realize this C(sp^3)–H amination. A redox-neutral radical pathway including a Cu(I)–Cu(II) redox catalytic cycle is proposed for this reaction [86] (Scheme 8.47).

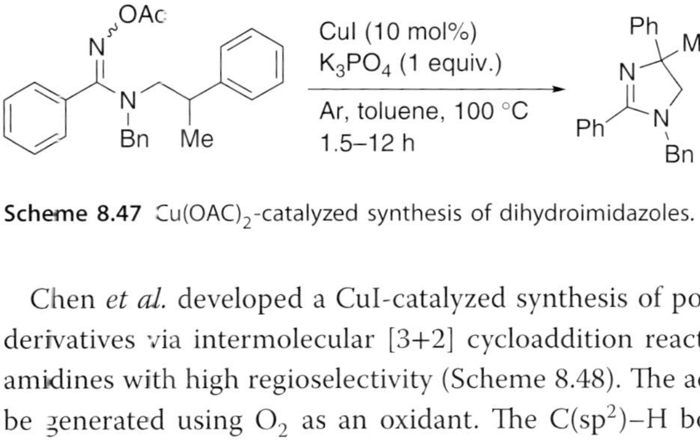

Scheme 8.47 Cu(OAC)$_2$-catalyzed synthesis of dihydroimidazoles.

Chen *et al.* developed a CuI-catalyzed synthesis of polysubstituted imidazole derivatives via intermolecular [3+2] cycloaddition reaction of nitroolefins and amidines with high regioselectivity (Scheme 8.48). The active Cu^{II} catalyst could be generated using O$_2$ as an oxidant. The C(sp^2)–H bond of nitroolefins was cleavaged and the NO$_2$ group was eliminated through leaving of one molecule of HNO$_2$ under this copper catalytic system [87].

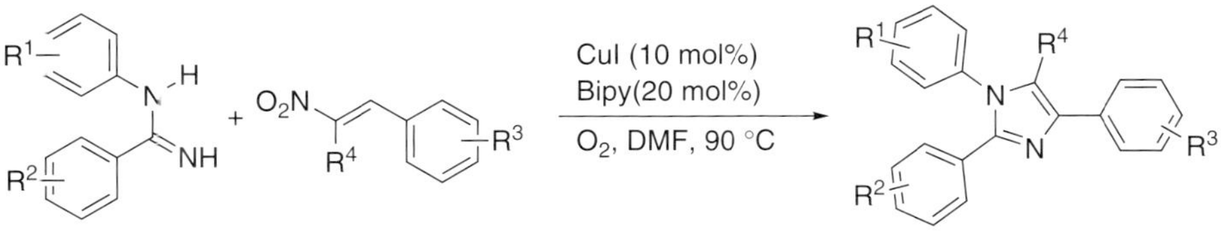

Scheme 8.48 CuI-catalyzed synthesis of polysubstituted imidazole derivatives via intermolecular [3+2] cycloaddition reaction of nitroolefins and amidines.

A series of 1,2,4-trisubstituted imidazoles could be synthesized efficiently from amidines and terminal alkynes with $CuCl_2·2H_2O$ catalyst under O_2 atmosphere. This oxidative reaction system involves catalytic amount of $CuCl_2·2H_2O$, Na_2CO_3, pyridine, and O_2 (1 atm) (Scheme 8.49). $CuCl_2·2H_2O$ catalyst could activate the C–H bond of alkyne and the N–H bond of amidine at the same time to generate an active species. The active Cu^{II} catalyst could be regenerated from Cu^{I} using O_2 [88].

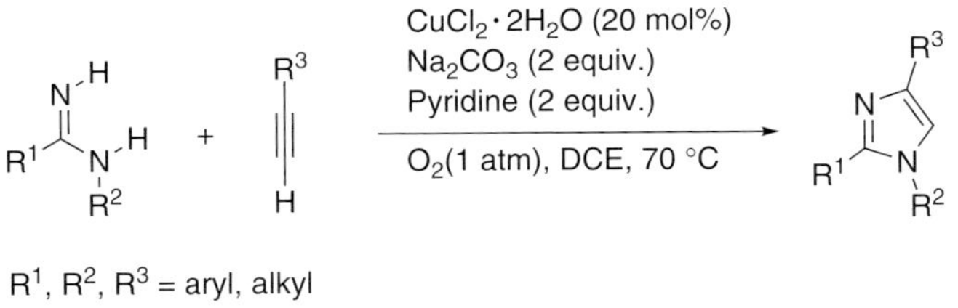

Scheme 8.49 Synthesis of 1,2,4-trisubstituted imidazoles from amidines and terminal alkynes.

8.3.10
Copper-Catalyzed Synthesis of Benzimidazoles

Brasche and Buchwald developed an efficient $Cu(OAc)_2$-catalyzed substituted benzimidazoles and substituted *N*-methyl-2-arylbenzimidazoles synthesis from readily available amidines in good to very good yields (Scheme 8.50). This transformation proceeds via direct C–H bond functionalization and C–N bond formation. The O_2 was used as the oxidant and water was generated as the only by-product. The starting material amidines were prepared easily from anilines with carbonitrile derivative in advance [89].

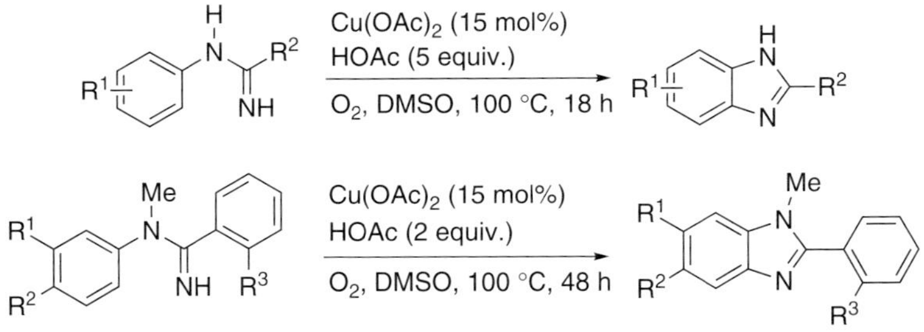

Scheme 8.50 $Cu(OAc)_2$-catalyzed synthesis of substituted benzimidazoles and substituted *N*-methyl-2-arylbenzimidazoles.

Benzimidazoles also could be obtained from benzamidines and boronic acids catalyzed by $Cu(OAc)_2$ under O_2 atmosphere (Scheme 8.51). This transformation might involve mono-*N*-arylation amidines synthesis via intermolecular C–N

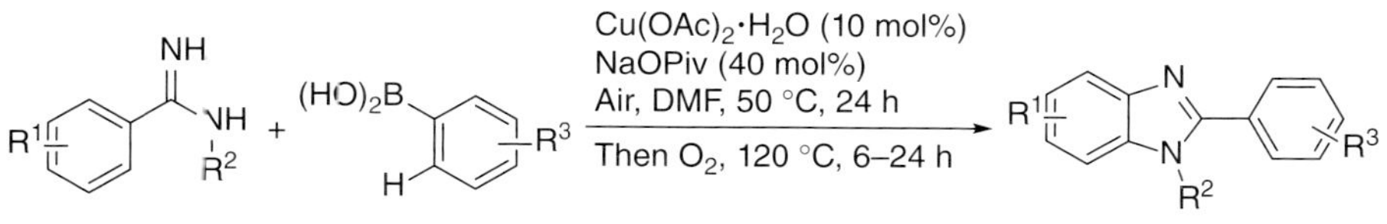

Scheme 8.51 Synthesis of benzimidazoles from benzamidines and boronic acids.

bond formation and then intramolecular C–H functionalization and C–N bond formation of the corresponding mono-*N*-arylation amidines [90].

A three-component reaction for the synthesis of functionalized benzimidazoles from terminal alkynes, *o*-aminoanilines, and *p*-tolylsulfonyl azide is developed by Wang and coworkers (Scheme 8.52). The C(sp)–H bond of alkyne was activated in this reaction via CuI-catalyzed azide–alkyne cycloaddtion (CuAAC) reaction which could lead to the formation of a ketenimine intermediate by the release of N_2. The benzimidazoles could be obtained by intramolecular nucleophilic addition and subsequent elimination of amide intermediate with 2% H_2SO_4 under reflux conditions [91]. They also developed another protocol for the synthesis of substituted benzimidazole derivatives via a sequential three-component cascade reaction from sulfonyl azides, alkynes, and 2-bromoanilines using CuI as catalyst. Similarly, the C–H bond of alkyne was activated at the process of copper-catalyzed azide–alkyne cycloaddtion (CuAAC) reaction [92].

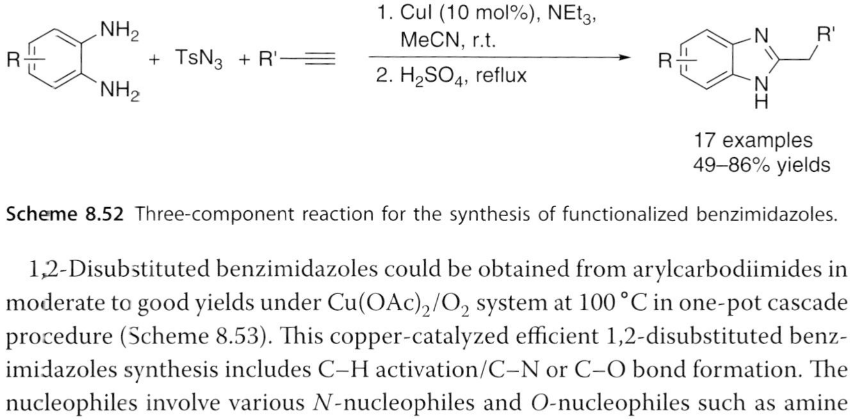

Scheme 8.52 Three-component reaction for the synthesis of functionalized benzimidazoles.

1,2-Disubstituted benzimidazoles could be obtained from arylcarbodiimides in moderate to good yields under $Cu(OAc)_2/O_2$ system at 100 °C in one-pot cascade procedure (Scheme 8.53). This copper-catalyzed efficient 1,2-disubstituted benzimidazoles synthesis includes C–H activation/C–N or C–O bond formation. The nucleophiles involve various *N*-nucleophiles and *O*-nucleophiles such as amine and alcohol. Electron-donating groups on the arylcarbodiimides accelerate this reaction in contrast to the electron-withdrawing groups [93].

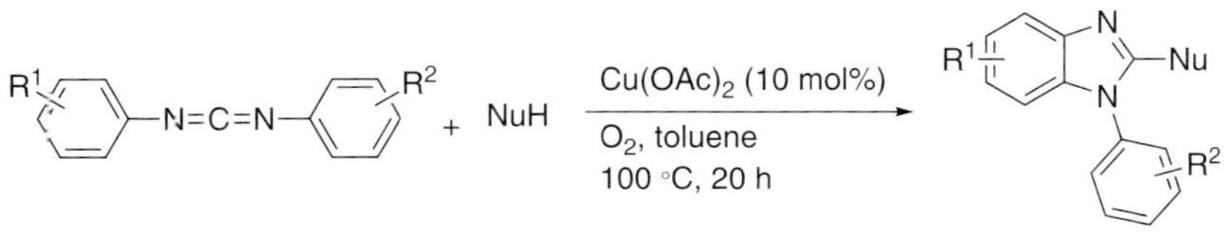

Scheme 8.53 Synthesis of 1,2-disubstituted benzimidazoles from arylcarbodiimides.

The use of primary amines as suitable substrates for the synthesis of 2-substituted benzimidazoles, benzoxazoles, and benzothiazoles with $CuBr_2$ catalyst was developed by Zhou and coworkers (Scheme 8.54). A key step in this reaction is the formation of an imine intermediate via oxidative dehydrogenation of benzylamine with the copper complex and molecular oxygen [94].

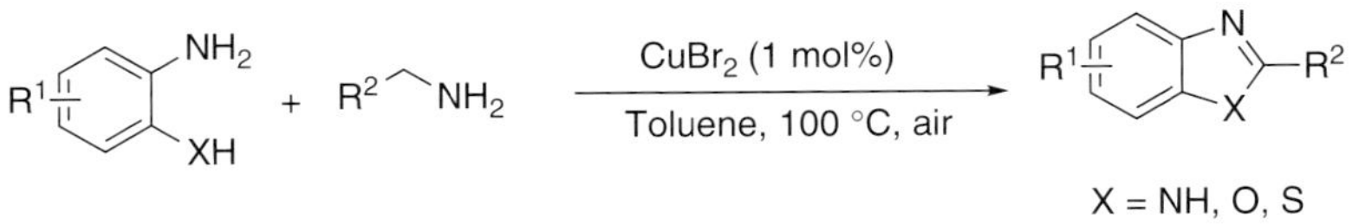

Scheme 8.54 Synthesis of 2-substituted benzimidazoles, benzoxazoles, and benzothiazoles from primary amines.

Fu *et al.* developed an efficient $Cu(OAc)_2$-catalyzed synthesis of imidazobenzimidazole derivatives via aerobic oxidative intramolecular $C(sp^2)$–H amination in excellent yields (Scheme 8.55). 1,10-Phenanthroline was chosen as the ligand and NaOAc was used as the base. Molecular oxygen acts as the oxidant to recycle the copper catalyst [95]. Kaliappan and coworkers reported an efficient, one-pot, copper-catalyzed $C(sp^2)$–H cascade amination to synthesize various imidazobenzimidazole derivatives from various azoles and aryl bromides or 2-bromopyridines [96].

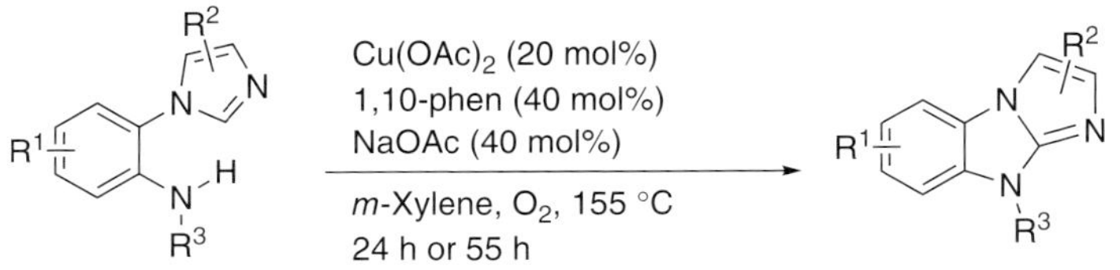

Scheme 8.55 $Cu(OAc)_2$-catalyzed synthesis of imidazobenzimidazole derivatives.

8.3.11
Copper-Catalyzed Synthesis of Imidazopyridines

Döring *et al.* developed a $CuCl_2$ -catalyzed synthesis of imidazo[1,5-*a*]pyridines, imidazo[1,5-*a*]imidazoles, and imidazo[5,1-*a*]isoquinolines starting from Schiff base which could be prepared readily from the appropriate aldehydes and primary amines [97] (Scheme 8.56).

A novel and efficient synthesis of pyrido[1,2-*a*]benzimidazoles from *N*-aryl-2-aminopyridines in moderate to excellent yields through direct intramolecular aromatic C–H amination was reported by Zhang, Zhu *et al.* (Scheme 8.57). The cocatalysts $Cu(OAc)_2$ and $Fe(NO_3)_3{\cdot}9H_2O$ are the best for this transformation in DMF under O_2 atmosphere. However, electron-withdrawing substituents in the meta position of the aniline ring and any position of the pyridine ring are

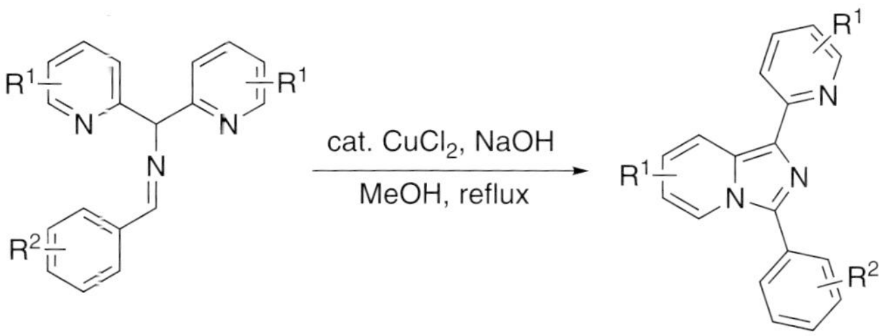

Scheme 8.56 $CuCl_2$ -catalyzed synthesis of imidazo[1,5-*a*]pyridines, imidazo[1,5-*a*]imidazoles, and imidazo[5,1-*a*]isoquinolines.

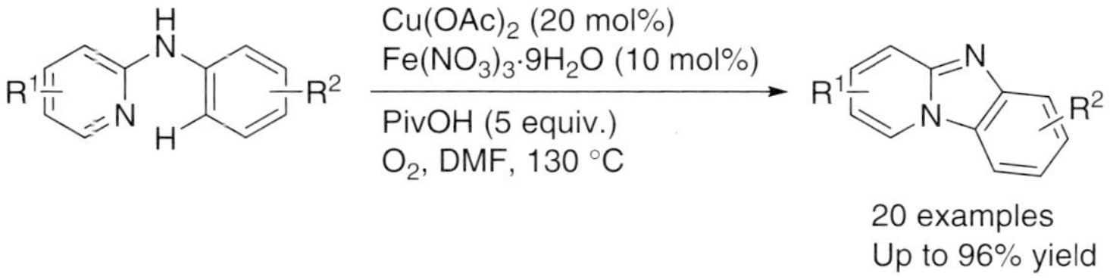

Scheme 8.57 synthesis of pyrido[1,2-*a*]benzimidazoles from *N*-aryl-2-aminopyridines.

unfavorable at present conditions. The mechanistic experiments suggest that a Cu(III)-catalyzed electrophilic aromatic substitution (S_EAr) pathway is operating for this reaction. Iron(III) is believed to help facilitate the formation of the more electrophilic Cu(III) species [98].

CuCl- and $Cu(OTf)_2$-catalyzed three-component coupling reaction of aryl, heteroaryl, and alkyl aldehydes with 2-aminopyridines and terminal alkynes to imidazopyridine derivatives was reported by Chernyak and Gevorgyan (Scheme 8.58). 2-Aminoquinoline and 2-aminoisoquinoline also could act as coupling partners to form imidazoquinoline and imidazoisoquinoline frameworks in good yields. Alpidem and zolpidem were synthesized using this novel reaction in a highly efficient one-pot procedure [99].

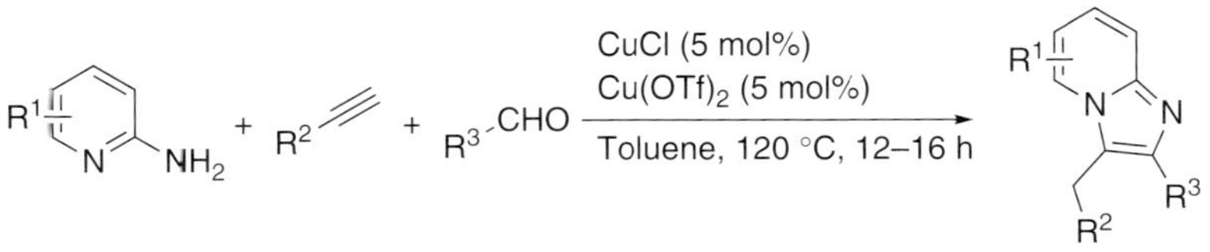

Scheme 8.58 Three-component coupling reaction of aryl, heteroaryl, and alkyl aldehydes with 2-aminopyridines and terminal alkynes.

The same group developed a copper-catalyzed unprecedented intramolecular dehydrogenative aminooxygenation (IDA) process that produces imidazo[1,2-a]pyridine-3-carbaldehydes and 1,2-disubstituted imidazole-4-carbaldehydes from readily available *N*-allyl-2-aminopyridines and substituted *N*-allylamidines using 20 mol% of Cu(hfacac)$_2$·xH$_2$O catalyst in DMF or DMA (N,N-Dimethylacetamide) under O_2 atmosphere. Substituted imidazo[1,2-a]pyridine-3-carbaldehydes and imidazole-4-carbaldehydes are obtained in moderate to good yields with a broad substrate scope and good functionality tolerance (Scheme 8.59). Mechanistic studies disclosed that the carbonyl oxygen in the aldehyde products is derived from O_2 [100]. Maes *et al.* reported a Cu(OAc)$_2$-catalyzed 6-, 7-, and 8-substituted pyrido [1,2-a]benzimidazoles synthesis process with additive such as 3,4,5-trifluorobenzoic acid (TFBA) or HOAc. The type of acid (pKa) proved to be crucial for the catalytic system [101].

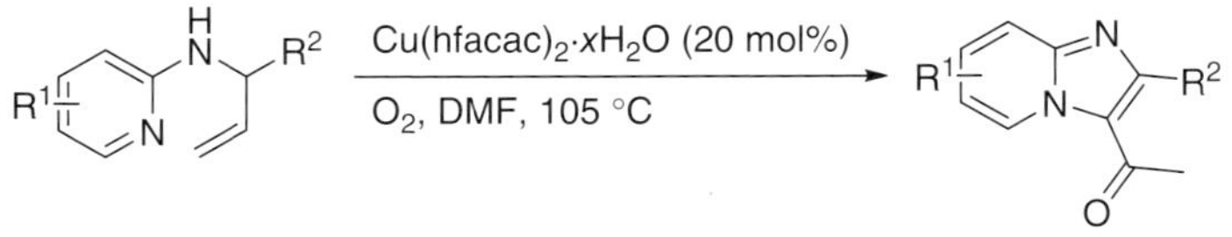

Scheme 8.59 Copper-catalyzed intramolecular dehydrogenative aminooxygenation.

Imidazo[1,2-a]pyridines also could be synthesized from aminopyridines and nitroolefins using CuBr as catalyst and air as oxidation agent in a one-pot procedure (Scheme 8.60). Various substitutes are tolerated in this reaction and many imidazo[1,2-*a*]pyridines could be obtained smoothly in moderate to good yields [102].

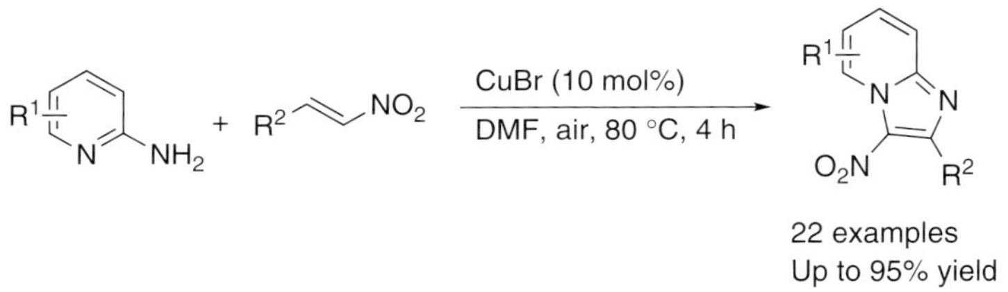

Scheme 8.60 Synthesis of imidazo[1,2-a]pyridines from aminopyridines and nitroolefins.

Wang, Ji, and coworkers developed a CuI/BF$_3$·Et$_2$O/O$_2$ system for the synthesis of heteroaromatic imidazo[1,2-*a*]pyridines from ketones with 2-aminopyridines (Scheme 8.61). The C(sp^3)–H bonds of ketones are activated under these oxidation conditions. A total of four hydrogen atoms are removed, and two new C–N bonds are formed in one step to construct the imidazo[1,2-*a*]pyridines [103]. Hajra *et al.* developed a Cu(OAc)$_2$·H$_2$O-catalyzed direct oxidative cyclization via C–H amination between 2-aminopyridines and methyl aryl/heteroaryl ketones

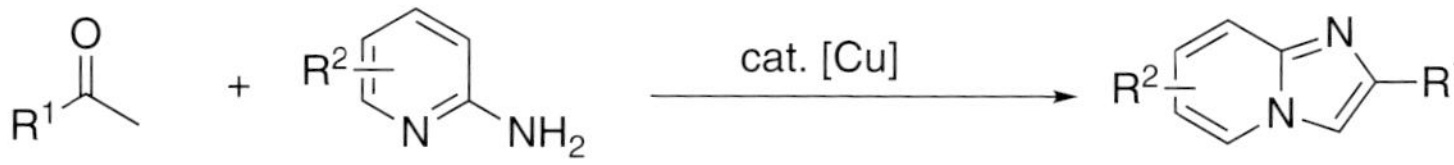

Scheme 8.61 Synthesis of heteroaromatic imidazo[1,2-*a*]pyridines.

under ambient air with 1,10-phenanthroline as ligand and ZnI_2 as additive [104]. $In(OTf)_3$ could be used as Lewis acids to accelerate the α-iodination step of this transformation [105]. Recently, Kumar *et al.* reported a CuI-catalyzed synthesis of imidazo[1,2-*a*]pyridines in dioxane without ligand and additive [106].

CuBr-catalyzed synthesis of multifunctional imidazo[1,5-a]pyridines through direct C(sp^3)–H amination of 2-acylpyridines with aliphatic amines using O_2 as the green oxidant under 80 °C was reported by Ye, Zeng, and coworkers (Scheme 8.62). The electron-rich or poor 2-acylpyridines and aliphatic amines are compatible in this transformation, and the desired products could be obtained in moderate to good yield [107].

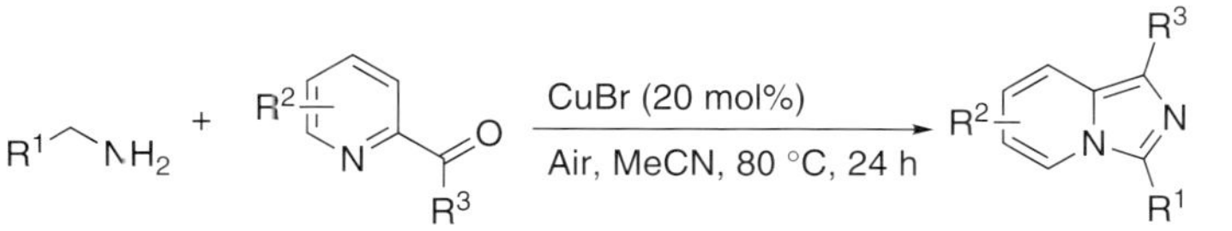

Scheme 8.62 CuBr-catalyzed synthesis of multifunctional imidazo[1,5-a]pyridines.

The synthesis of imidazo[1,2-*a*]pyridine derivatives from simple substituted pyridines is also developed recently. Fu *et al.* reported an efficient CuI-catalyzed aerobic oxidative C–H functionalization of substituted pyridines with *N*-(alkylidene)-4*H*-1,2,4-triazol-4-amines leading to imidazo[1,2-*a*]pyridine derivatives in moderate to good yields using O_2 (1 atm) as the oxidant (Scheme 8.63). The reaction proceeds by the cleavage of the N–N bond in the *N*-(alkylidene)-4*H*-1,2,4-triazol-4-amines and activation of an aryl C(sp^2)–H bond in the substituted pyridines [108].

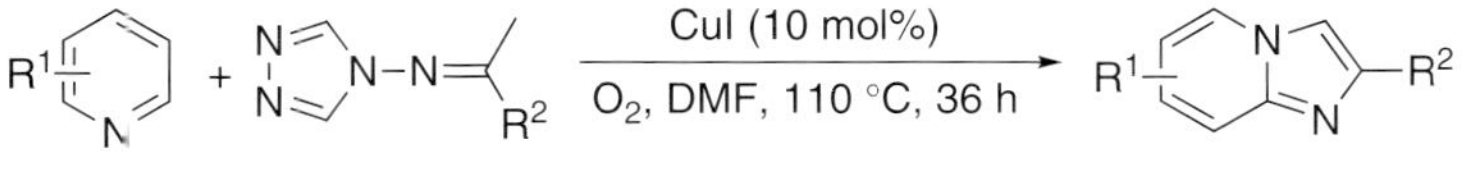

Scheme 8.63 Synthesis of imidazo[1,2-*a*]pyridine derivatives.

Jiang and coworkers developed a highly efficient synthesis of imidazo[1,2-a]pyridine derivatives from pyridine with ketone oxime esters using CuI catalyst (Scheme 8.64). The C(sp^3)–H bond of ketone oxime esters and C(sp^2)–H bond pyridinium were activated in this transformation. Air is used as the sole oxidant and HOAc and H_2O are the only green by-products [109].

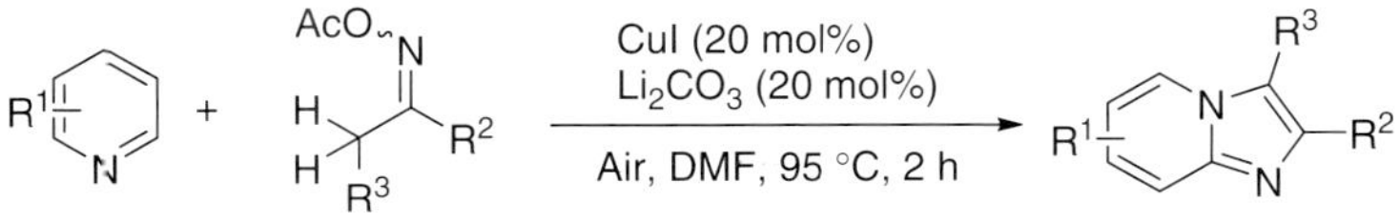

Scheme 8.64 Synthesis of imidazo[1,2-a]pyridine derivatives from pyridine with ketone oxime esters.

8.3.12 Copper-Catalyzed Synthesis of Pyrazoles and Indazoles

Pyrazoles and indazoles are important synthetic units in biologically active compounds and drugs. Glorius *et al.* reported a $Cu(OAc)_2$ mediates regiospecific preparation of tetrasubstituted pyrazoles from enamines and nitriles. The $C(sp^2)$–H bond of imine was cleavaged via elimination of HOAc molecule. $Cu(OAc)_2$ plays the role of the Lewis acid and oxidant. Moreover, an excess amount of nitrile was needed for this reaction [110]. To solve this problem, they report a modified synthetic approach using equivalent amounts of nitriles with an efficient $Cu(OAc)_2$ catalyst system and molecular oxygen as the oxidant [111].

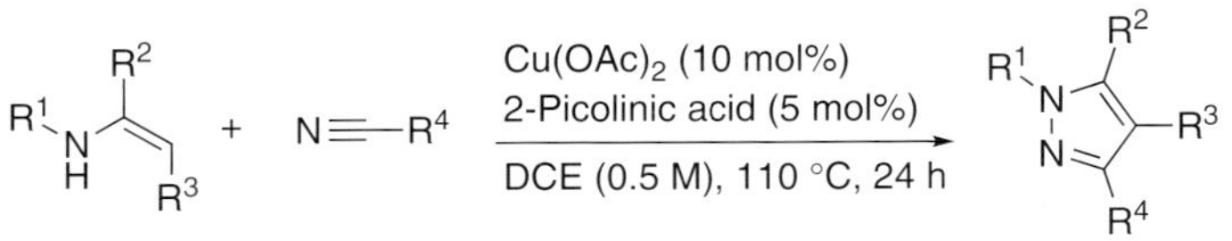

Scheme 8.65 Synthesis of tetrasubstituted pyrazoles from enamines and nitriles.

An intramolecular Cu(OAc)-catalyzed aerobic oxidative $C(sp^2)$–H amination of hydrazones to pyrazoles and indazoles using O_2 (1 atm) as the sole oxidant is reported by Jiang and coworkers (Scheme 8.66). A broad substrate scope and good functional group tolerance show that this aerobic oxidative transformation shows sustainable and practical application [112].

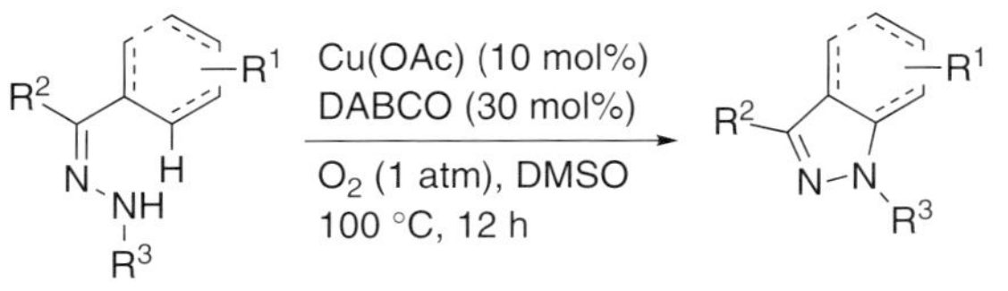

Scheme 8.66 Intramolecular synthesis of pyrazoles and indazoles via oxidative $C(sp^2)$–H amination.

Recently, an efficient CuBr-catalyzed aerobic intramolecular dehydrogenative cyclization reaction of *N*,*N*-disubstituted hydrazones to pyrazoles by a double $C(sp^3)$–H bond functionalization was developed in Ge's group (Scheme 8.67). This transformation includes $C(sp^3)$–H oxidation, cyclization, and aromatization for the formation of pyrazole products. This is the first example of an intramolecular copper-catalyzed dehydrogenative coupling reaction via an iminium ion intermediate by a $C(sp^3)$–H bond functionalization process [113].

Jiao *et al.* developed a CuI- and $AgCO_3$-cocatalyzed synthesis of pyrazolo[1,5-*a*]pyridine derivatives via dehydrogenative annulations of pyridinium ylide and terminal alkynes (Scheme 8.68). Ag_2CO_3 might assist the formation of copper acetylide from alkyne with Cu(I) catalyst. The C(sp)–H bond of alkyne

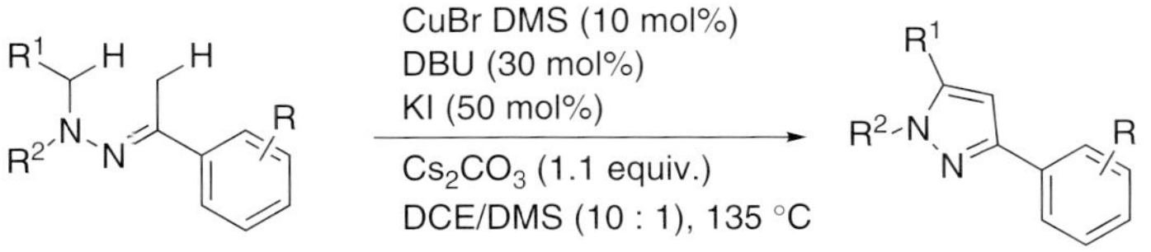

Scheme 8.67 Intramolecular dehydrogenative cyclization reaction of *N,N*-disubstituted hydrazones to pyrazoles.

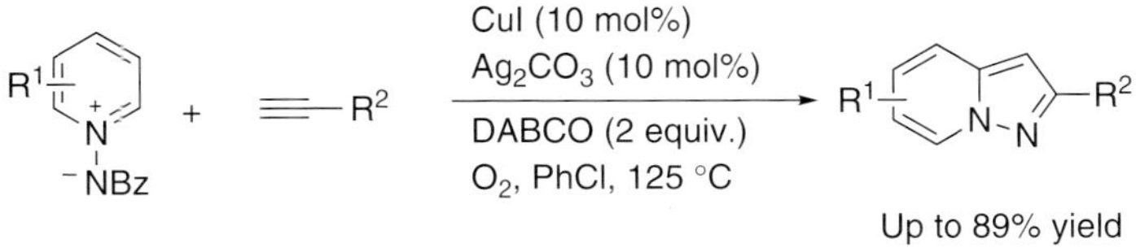

Scheme 8.68 Synthesis of pyrazolo[1,5-*a*]pyridine derivatives.

and C(sp^2)–H bond of pyridinium ylide were activated to generate the desired pyrazolo[1,5-*a*]pyridine product. 1,4-Diazabicyclo[2.2.2]octane (DABCO) promotes the expulsion of the benzoyl moiety or plays the role of ligand on Cu to facilitate this reaction [114].

8.3.13 Copper-Catalyzed Synthesis of Oxazoles

Wang *et al.* developed an efficient $Cu(OAc)_2$-catalyzed oxidative tandem cyclization of benzylamine and 1,3-diketone derivatives to polysubstituted oxazoles using TBHP as oxidant (Scheme 8.69). The condition of this reaction is mild and substrate scopes are wide. Four C(sp^3)–H bonds are cleavaged for this transformation. The iodointermediate which was formed via C–H bond activation of 1,3-diketone under TBHP/I_2 system is crucial for this reaction [115].

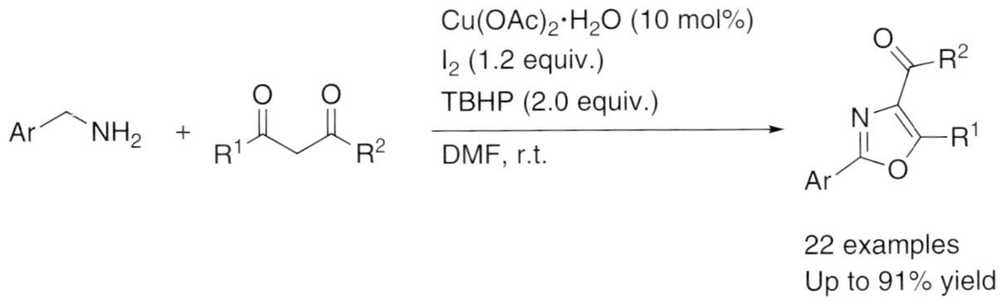

Scheme 8.69 $Cu(OAc)_2$-catalyzed oxidative tandem cyclization for the synthesis of polysubstituted oxazoles.

Recently, they also developed another one-pot synthesis of polyarylated oxazoles from benzylamine and benzil derivatives via oxidative dehydrogenation domino process using air as oxidant (Scheme 8.70). The reaction condition is mild and moderate to good yields could be obtained at room temperature. The C(sp^3)–H

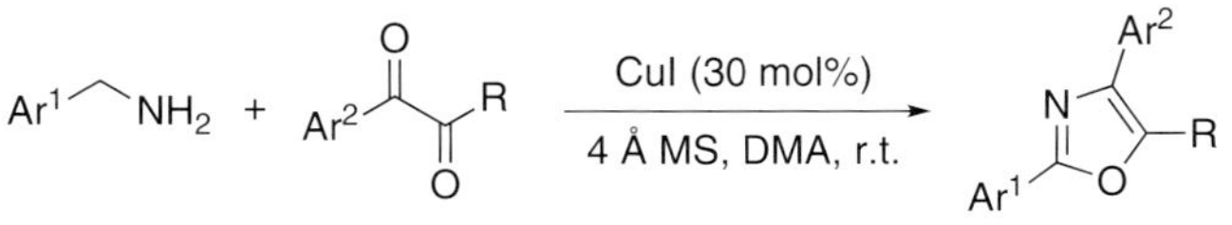

Scheme 8.70 Synthesis of polyarylated oxazoles from benzylamine and benzil derivatives.

bond of benzylamine was activated in this transformation [116]. Jiao *et al.* reported the synthesis of oxazoles from aldehydes, amines, and molecular oxygen through a $CuBr_2$-mediated aerobic oxidative dehydrogenative annulation [117].

A methodology of 2,5-disubstituted oxazole preparation from enamides was developed by Cheung and Buchwald using catalytic amounts of $CuBr_2$ in conjunction with $K_2S_2O_8$ as an oxidant and ethyl nicotinate as a ligand at room temperature (Scheme 8.71). One C(sp^2)–H bond of enamides was activated in the reaction. This reaction protocol can tolerate enamide substrates bearing aryl, heteroaryl, vinyl, and/or alkyl substituents to afford the corresponding oxazoles in moderate to high yields. The tetrabutylammonium bromide (TBAB) is an additive and it probably aids in the regeneration of catalyst $CuBr_2$ [118].

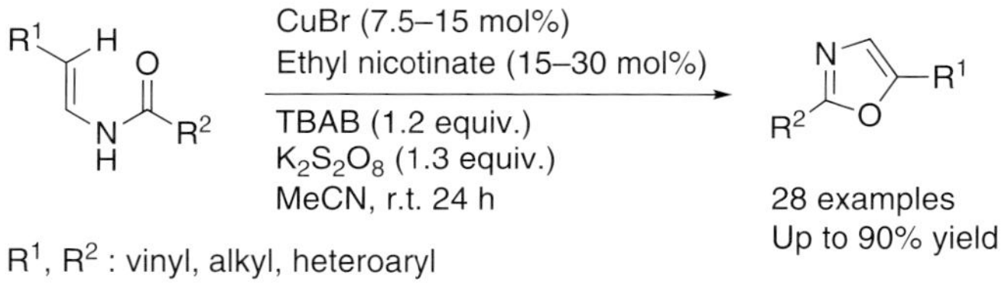

Scheme 8.71 2,5-Disubstituted oxazole preparation from enamides.

In the $CuBr{\cdot}SMe_2$/2,2′-bipyridine catalytic system under an O_2 atmosphere, dihydrooxazoles could be obtained through direct C–H oxygenation of C(sp^3)–H bonds of *N*-alkylamidines (Scheme 8.72). The proton could be abstracted *via* 1,5-H radical shift under this Cu/O_2-promoted radical process [119].

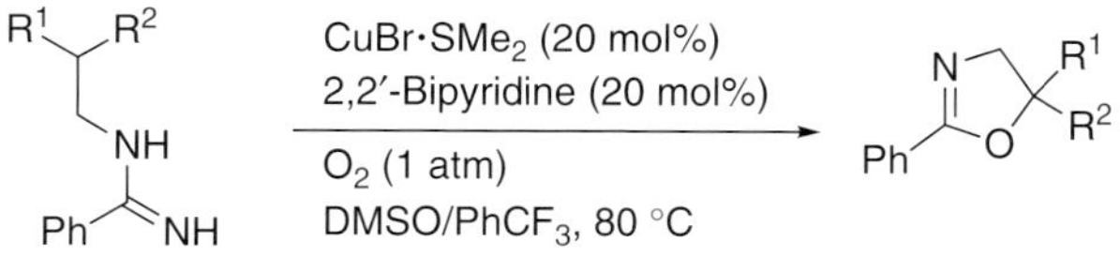

Scheme 8.72 Synthesis of dihydrooxazoles from *N*-alkylamidines through direct C–H oxygenation.

8.3.14
Copper-Catalyzed Synthesis of Benzoxazoles

A novel $Cu(OTf)_2$-catalyzed efficient synthesis of 2-arylbenzoxazoles through intramolecular oxidative C–O coupling reaction was developed by Ueda and

Nagasawa. This process for the benzoxazoles formation from benzanilides was realized with high functional group tolerance via regioselective C–H functionalization using O_2 as sole oxidant (Scheme 8.73). The C–O bond formation always takes place at the less sterically hindered position of the benzanilides [120]. This transformation also could conduct using air as the terminal oxidant in dichlorobenzene at 160 °C. Notably, 7-substituted benzoxazoles could be selectively obtained in moderate to high yields when the substrates have a directing group at the meta position. The initial coordination of benzanilide to copper catalyst would lead to directed *ortho*-metalation by C–H bond activation [121].

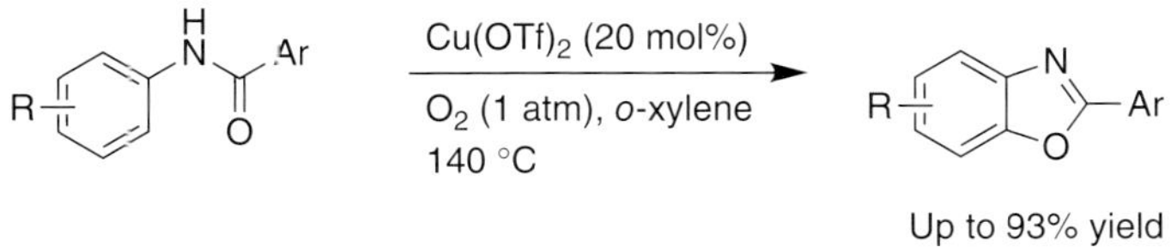

Scheme 8.73 Synthesis of 2-arylbenzoxazoles through intramolecular oxidative C–O coupling reaction.

Punniyamurthy *et al.* developed an intramolecular $Cu(OTf)_2$-catalyzed 2-arylbenzoxazole synthesis from *N*-benzyl bisaryloxime ethers via a cascade C–H functionalization/C–N/C–O bond formation (Scheme 8.74). The reactivity of the substrates with electron-withdrawing groups is higher than those bearing electron-donating groups. Furthermore, *N*-benzyl bisarylhydrazones also could convert into 2-aryl-*N*-benzylbenzimidazoles under this copper catalytic system. However, bisphenyloxime thioether substrate could not undergo this transformation under standard conditions [122, 123].

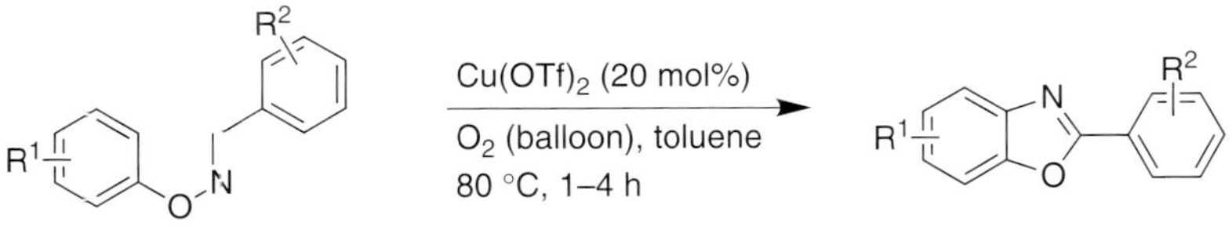

Scheme 8.74 Intramolecular synthesis of 2-arylbenzoxazole from *N*-benzyl bisaryloxime ethers.

8.3.15 Copper-Catalyzed Synthesis of 1,2,3-Triazoles

1,2,3-Triazoles are attractive compounds and widely used in materials, drugs, and bioconjugation chemistry [124–136]. 1,2,3-Triazoles could be synthesized by 1,3-dipolar cycloaddition of azides with alkynes under thermal conditions via the activation of C–H bond of alkyne [137–141]. The developments of "Click reaction," copper-catalyzed azide–alkyne cycloaddition (CuAAC), provide an efficient pathway for the synthesis of 1,2,3-triazoles [142, 135, 143, 144]. They have been well reviewed and we don't discuss it in detail in this chapter.

8.3.16
Copper-Catalyzed Synthesis of Tetrazoles

1,5-Disubstituted tetrazoles are ubiquitous structural units in organic synthesis. The functionalized starting materials are needed for the synthesis of 1,5-disubstituted tetrazoles in traditional strategies. Jiao and coworkers reported a novel CuI-promoted synthesis of 1,5-disubstituted tetrazoles by two C(sp^3)–H and one C–C bond cleavages using 2,3-dichloro-5,6-dicyano-1,4-benzoquinone (DDQ) as oxidant [145]. It is possible to functionalize simple and readily available hydrocarbons to valuable molecules via C–H and C–C bond cleavages using this unique strategy (Scheme 8.75).

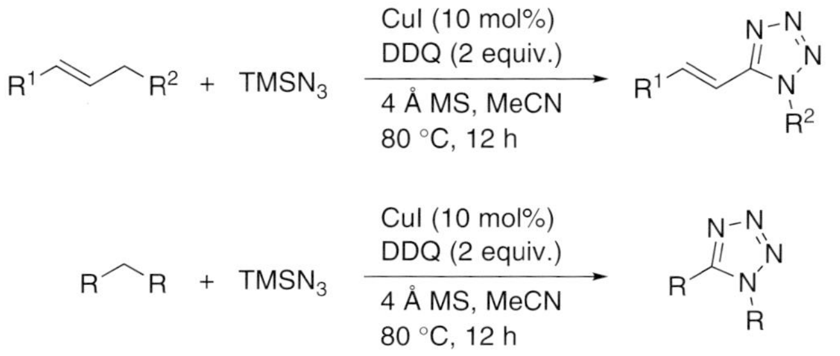

Scheme 8.75 Synthesis of 1,5-disubstituted tetrazoles by two C(sp^3)–H and one C–C bond cleavages.

8.3.17
Copper-Catalyzed Synthesis of Furans

Chemler *et al.* reported a copper-catalyzed doubly intramolecular alkene carboetherification of unactivated alkenols to form bridged-ring tetrahydrofurans using the 2,2′-(propane-2,2-diyl)bis(4,5-dihydrooxazole) as ligand and MnO_2 as oxidant (Scheme 8.76). The formation of C–C bond is thought to occur *via* carbon radical arene addition, and this step constitutes an efficient C–H functionalization [146].

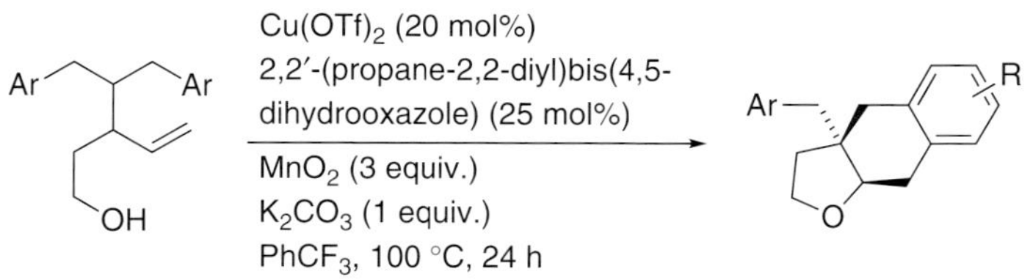

Scheme 8.76 Doubly intramolecular alkene carboetherification of unactivated alkenols to form tetrahydrofurans.

Recently, they report the enantioselective carboetherification reaction to corresponding tetrahydrobenzofuran (up to >95% ee) with the commercially available (S,S)-*t*Bu-box ligand. The intermolecular C–C bond could be formed by alkyl Heck-type couplings with vinyl arenes. DFT (Density functional theory) transition state calculations are consistent with a *cis*-oxycupration stereo-determining transition state, and there is good agreement between experimental and calculated levels of enantiomeric excess [147] (Scheme 8.77).

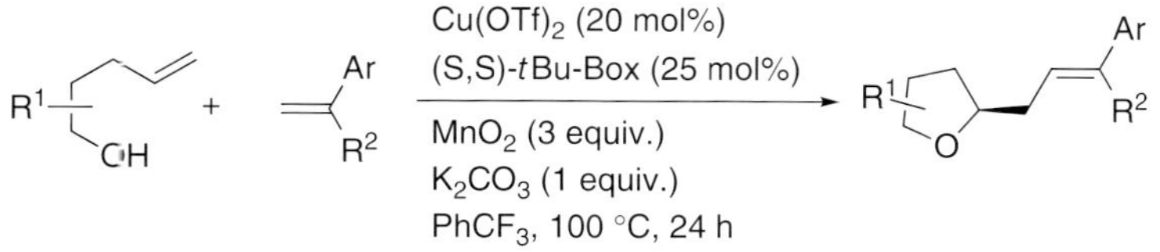

Scheme 8.77 Enantioselective carboetherification reaction for the synthesis of tetrahydrobenzofuran.

A Cu(OAc)$_2$-catalyzed aerobic C(sp^2)–H activation, followed by cycloetherification, has been developed by Zhu and coworkers to synthesize multisubstituted dibenzofurans from *o*-arylphenols. A catalytic amount of Cu (30 mol%) with 0.5 equiv. of Cs_2CO_3 and 1.0 equiv. of PivOH is employed under an air atmosphere to obtain the desired dibenzofurans in moderate to good yield (Scheme 8.78). The presence of a strong *para*-electron-withdrawing group such as NO_2, CN, and CHO on the phenol is essential for this transformation [148].

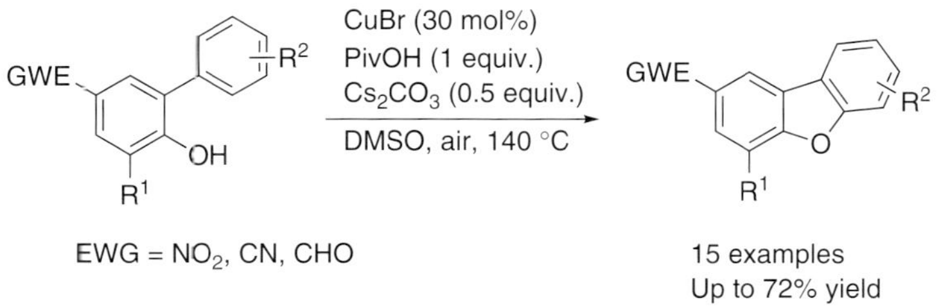

Scheme 8.78 Synthesis of multisubstituted dibenzofurans from *o*-arylphenols.

Regioselective synthesis of polysubstituted benzofurans from phenols and alkynes using a Cu(OTf)$_2$ catalyst and $ZnCl_2$ as Lewis acid in O_2 atmosphere in a one-pot procedure has been reported by Jiang *et al.* (Scheme 8.79). The transformation consists of a sequential nucleophilic addition of phenols to alkynes and oxidative cyclization. A wide variety of phenols and alkynes can be used in this reaction. The reaction might proceed via a Cu-catalyzed intermolecular nucleophilic addition and intramolecular Cu-catalyzed aryl C(sp^2)–H functionalization [149].

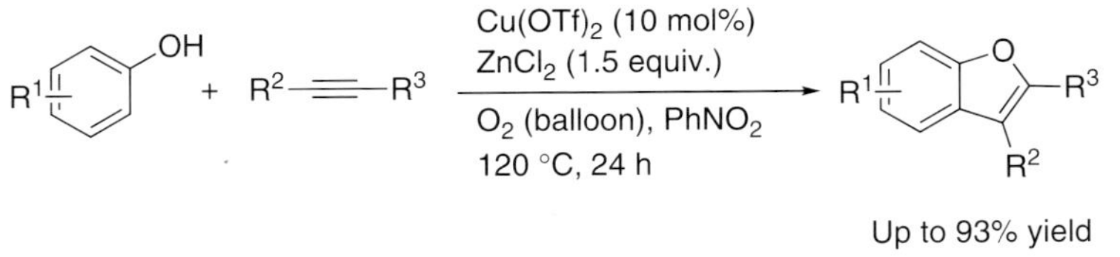

Scheme 8.79 Regioselective synthesis of polysubstituted benzofurans from phenols and alkynes.

8.4
Six-Membered-Ring Formation

8.4.1
Copper-Catalyzed Synthesis of Pyridines

An efficient $Cu(OTf)_2$-catalyzed 2,4,6-trisubstituted pyridine derivatives synthesis via C–N bond cleavage of aromatic methylamines and C(sp^3)–H bond activated of ketones was developed by Jiang's group (Scheme 8.80). A wide range of 2,4,6-trisubstituted pyridines could be obtained up to 95% yield at 100 °C under Cu/O_2 under this catalytic system with neat conditions. This process should be initiated by copper-catalyzed aerobic oxidative cleavage of C–N bond of aromatic methylamines [150].

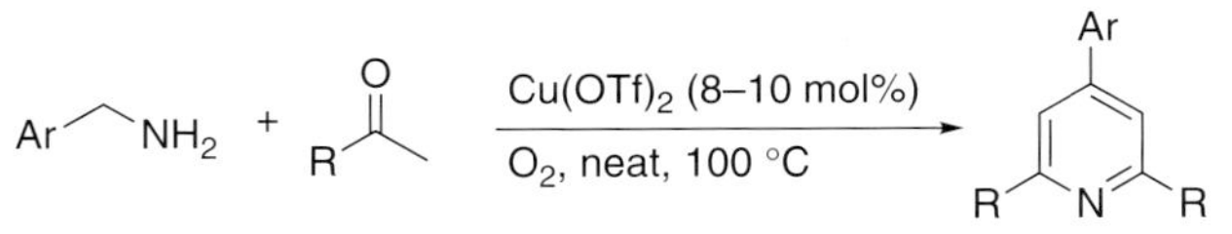

Scheme 8.80 Synthesis of 2,4,6-trisubstituted pyridine derivative from aromatic methylamines and ketones.

Jiao *et al.* developed copper-catalyzed diverse constructions of 3,5-diarylpyridines and 2-(1*H*)-pyridones through tandem Chichibabin-type cyclization, C(sp^3)–C(sp^3) bond cleavage, and aerobic oxidation with high selectivity. Azide, ceric ammonium nitrate (CAN), and 2-aminopyridine are efficient nitrogen donors in this Cu catalytic system for the selective synthesis of pyridines and pyridones [151] (Scheme 8.81).

Nakamura and coworkers have successfully developed an efficient and regioselective approach to multisubstituted pyridine N-oxide derivatives from (*E*)-*O*-propargylic α,β-unsaturated oximes via [2,3]-rearrangement to the *N*-allenyl nitrones followed by 6π-3-azatriene electrocyclization under $CuBr(PPh_3)_3/PPh_3$ system (Scheme 8.82). Since pyridine N-oxides have recently received much attention in catalytic C–H functionalization, the present methodology is useful to synthesize these substrates [152].

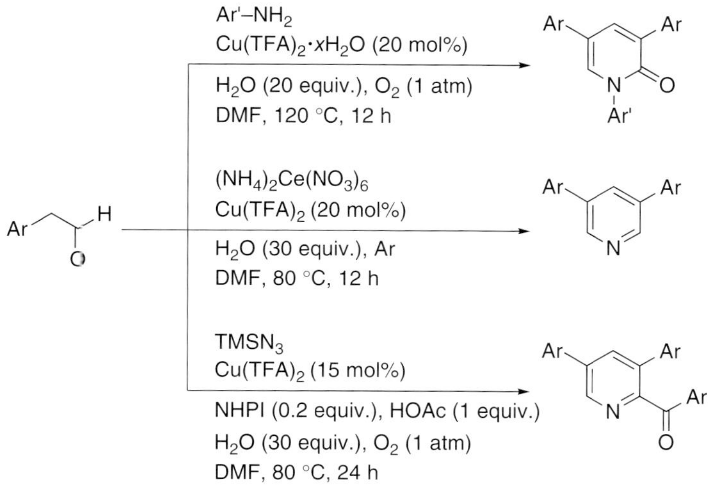

Scheme 8.81 Copper-catalyzed synthesis of 3,5-diarylpyridines and 2-(1*H*)-pyridones.

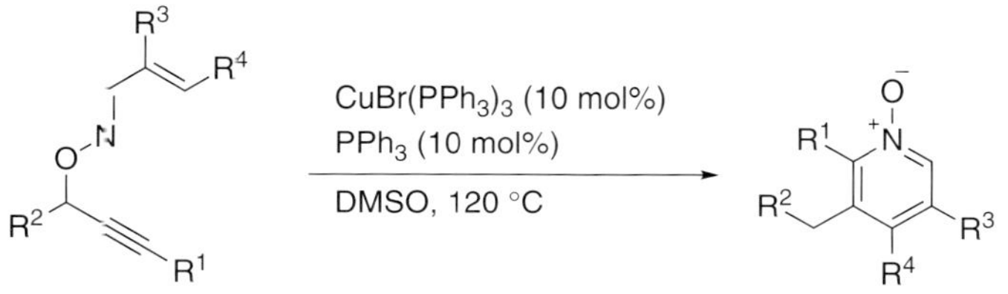

Scheme 8.82 Synthesis of multisubstituted pyridine N-oxide derivatives.

8.4.2 Copper-Catalyzed Synthesis of Quinolines

2-Aminobenzyl alcohol could be oxidatively cyclized with ketones catalyzed by a $CuCl_2$ catalyst with a KOH base in dioxane at 100 °C under O_2 atmosphere to afford quinolines in good yields. When aldehydes were used as the substrates, the desired quinolines also could be obtained in two steps under the similar conditions. Intramolecular aldol-type condensation and cyclodehydration to form quinolone are involved for this transformation [153] (Scheme 8.83).

A simple and practical $Cu(OTf)_2$-catalyzed synthesis of quinoline-2-carboxylates through tandem Grignard-type imine addition/Friedel–Crafts alkenylation of arenes with aryl or alkyl alkynes via activation of C–H bonds was reported in 2009 (Scheme 8.84). The quinoline-2-carboxylates products could be obtained in high yields from a number of readily available alkynes and imines at room temperature [154].

Chiba and coworkers developed a $CuBr{\cdot}SMe_2$-catalyzed synthesis of highly substituted quinolines from *N*-(2-alkynylaryl)enamine carboxylates under aerobic conditions via intramolecular carbooxygenation of alkynes (Scheme 8.85).

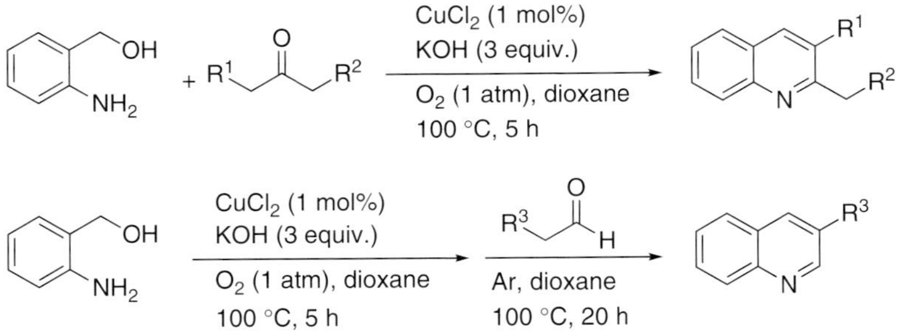

Scheme 8.83 Synthesis of quinolines from 2-aminobenzyl alcohol with ketones.

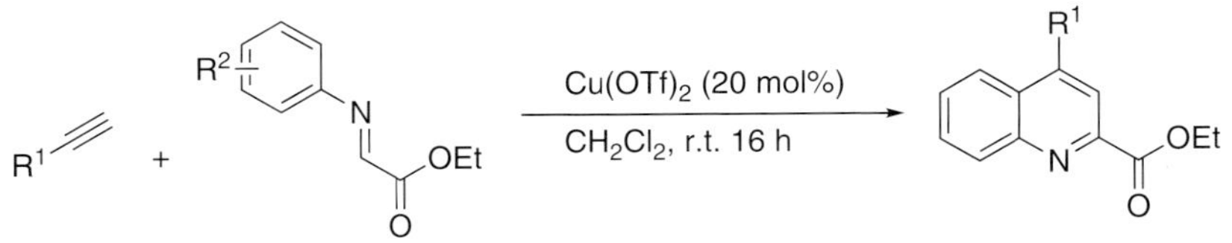

Scheme 8.84 $Cu(OTf)_2$-catalyzed synthesis of quinoline-2-carboxylates.

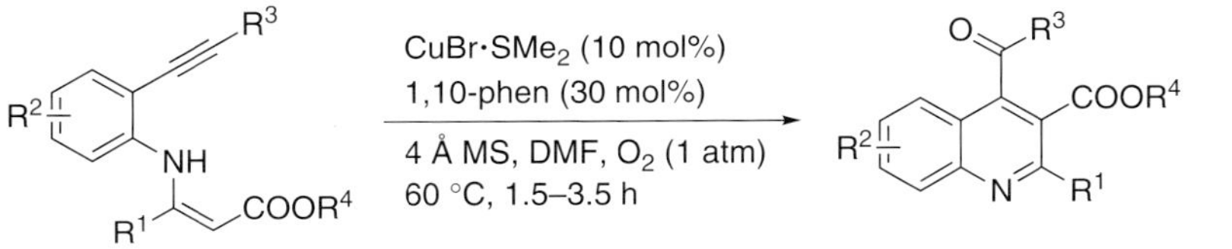

Scheme 8.85 Synthesis of highly substituted quinolines from *N*-(2-alkynylaryl)enamine carboxylates.

This strategy also could be applied for *N*-alkynylamidines for aminooxygenation of alkynes to imidazole and quinazoline derivatives. It is noted that other metals such as Fe(III), Co(II), and Pd(II) have no activation for this transformation [155].

Liang *et al.* demonstrated a novel $CuCl_2$-catalyzed synthesis of 4-carbonylquinolines through aerobic oxidative intramolecular cyclization of enynes using O_2 as an oxygen source. A Cu(III) and Cu(II) mechanism is involved for this transformation for the C–H bond activation [156] (Scheme 8.86).

A novel and efficient approach to 4-sulfonamidoquinolines catalyzed by CuI via a cascade reaction of sulfonyl azides with alkynyl imines has been developed by Cheng and Cui (Scheme 8.87). The reaction process includes 1,3-dipole cycloaddition/ketenimine formation/6π-electrocyclization/[1,3]-H shift. Various 4-sulfonamidoquinolines were afforded in moderate to good yields under this mild copper catalytic system [157].

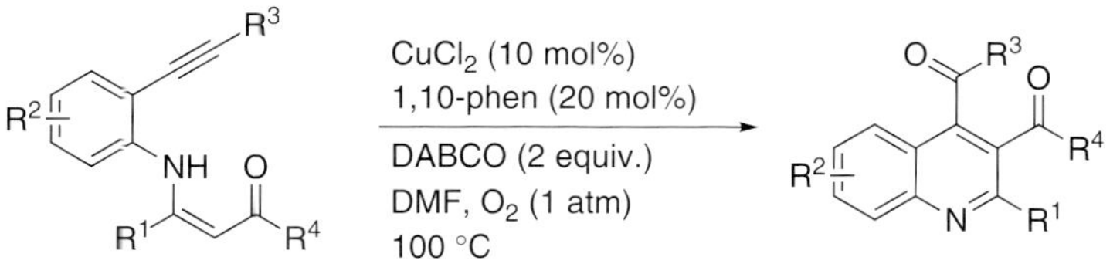

Scheme 8.86 Synthesis of 4-carbonylquinolines through aerobic oxidative intramolecular cyclization of enynes.

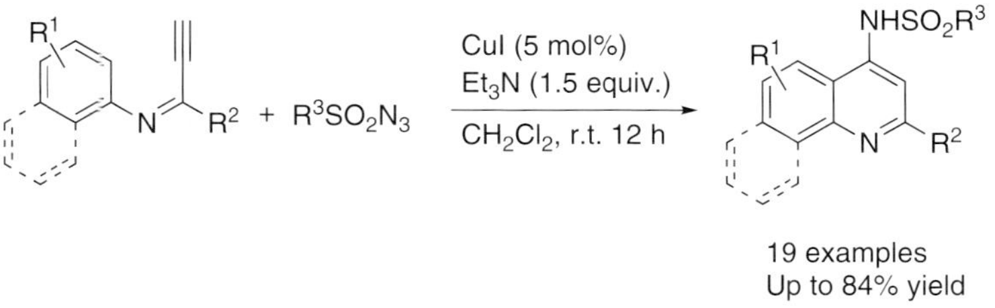

Scheme 8.87 Synthesis of 4-sulfonamidoquinolines from sulfonyl azides with alkynyl imines.

A novel and efficient method for the synthesis of polysubstituted quinolines catalyzed by CuBr through aerobic oxidative dehydrogenative annulation of anilines and aldehydes by C–H functionalization, C–C formation, and cleavage was reported recently. Various substituted quinolines could be synthesized in good yields under the standard conditions. It is noteworthy that air was used in this procedure as the oxidant and it makes the procedure very practical and economical [158] (Scheme 8.88).

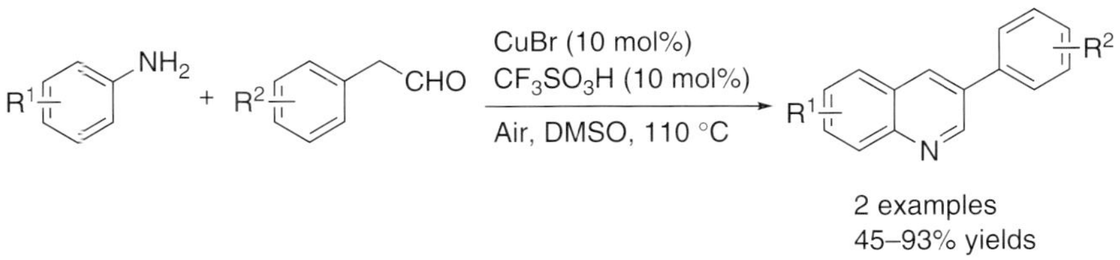

Scheme 8.88 Synthesis of polysubstituted quinolines through oxidative dehydrogenative annulation of anilines and aldehydes.

Swamy and coworkers developed a $Cu(OTf)_2$-catalyzed synthesis of highly conjugated cyclopenta[c]quinolines and indenylindoles via ring expansion/intramolecular electrophilic substitution of 3-dienylindoles using air as the oxidant (Scheme 8.89). The reaction proceeds by Brønsted acid-mediated Friedel–Crafts alkenylation and isomerization, followed by copper-catalyzed dehydrogenation/oxidative ring expansion. The $C(sp^3)$–H bonds of methyl group in propargyl alcohol were activated, and the carbon atom becomes a part of a six-membered ring in the final products [159].

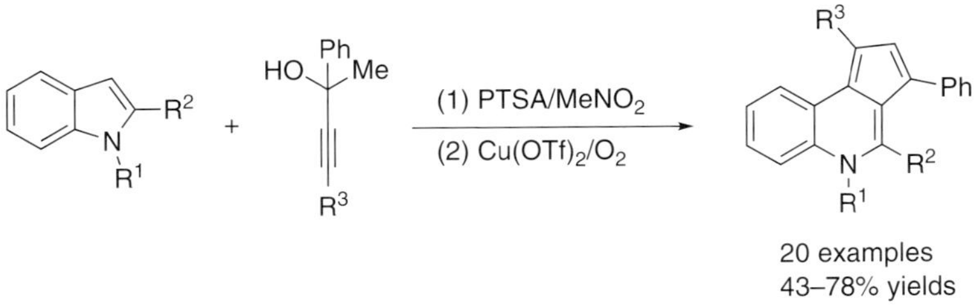

Scheme 8.89 Synthesis of highly conjugated cyclopenta[c]quinolines.

A simple $CuCl_2/O_2$ catalyst system for the oxidative cyclization of *N*-methylanilines with electron-deficient olefins to substituted tetrahydroquionoline was developed by Hirano, Miura, and coworkers (Scheme 8.90). One C(sp^3)–H bond of methyl group and one C(sp^2)–H bond of arene were cleaved in this reaction. The present efficient copper/oxygen catalytic system provides a direct access to tetrahydroquionoline frameworks easily [160]. Seidel *et al.* reported CuBr-catalyzed oxidative Povarov reactions of various *N*-aryl amines to synthesize tetrahydroazepine derivatives in moderate yields using TBHP as the terminal oxidant. Both the C(sp^3)–H and C(sp^2)–H bonds are active during the process [161] (Scheme 8.90).

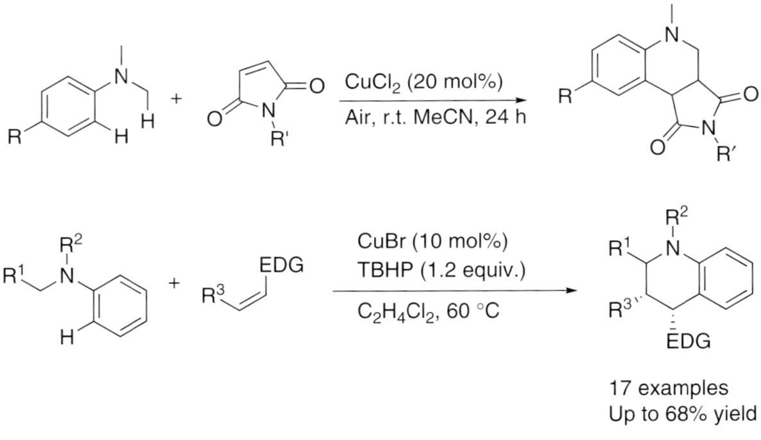

Scheme 8.90 Oxidative cyclization of N-methylanilines with electron-deficient olefins to substituted tetrahydroquionoline.

Taylor and coworkers report a copper(II) 2-ethylhexanoate catalyst for the synthesis of oxindoles, thio-oxindoles, 3,4-dihydro-1*H*-quinolin-2-ones, and 1,2,3,4-tetrahydroquinolines from linear precursors by direct C(sp^3)–H and Ar–H coupling using atmospheric oxygen as the oxidant (Scheme 8.91). The reaction procedure is open to the air and easy to handle [162].

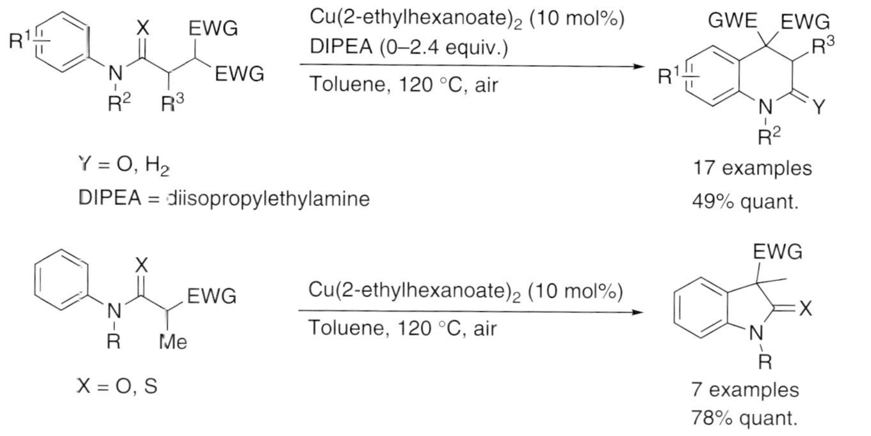

Scheme 8.91 Synthesis of oxindoles, thio-oxindoles, 3,4-dihydro-1*H*-quinolin-2-ones, and 1,2,3,4-tetrahydroquinolines.

8.4.3
Copper-Catalyzed Synthesis of Isoquinolines

Wu *et al.* described a $Cu(OTf)_2$-catalyzed reactions of N′-(2-alkynylbenzylidene) hydrazides with diethyl phosphite for the synthesis of isoquinolin-1-ylphosphonate. Some isoquinolin-1-ylphosphonates could be obtained in moderate yield (Scheme 8.92). The C(sp^2)–H bond of imine was cleaved and new C–P bond was formed [163]. They also reported a AgOTf- and $Cu(OAc)_2$-cocatalyzed cascade reaction of N′-(2-alkynylbenzylidene)-hydrazide with allenoate to 2-carbonyl *H*-pyrazolo[5,1-*a*]isoquinolines under mild conditions. The AgOTf/$Cu(OAc)_2$ cooperative catalysis is essential for this transformation. The reaction might proceed through a peroxy–copper(III) intermediate [164] (Scheme 8.92).

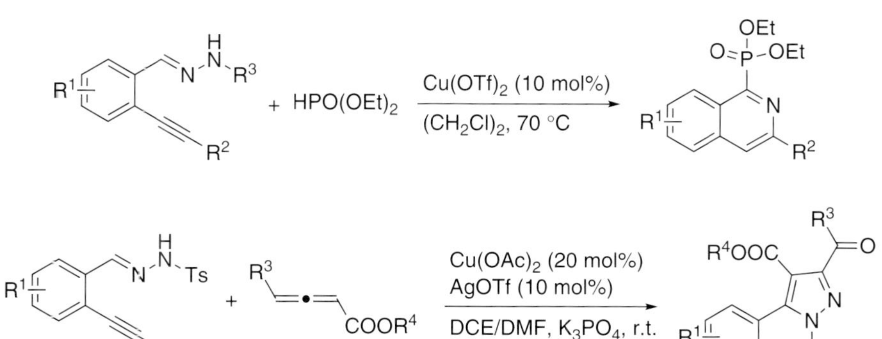

Scheme 8.92 Synthesis of isoquinolin-1-ylphosphonate from N′-(2-alkynylbenzylidene) hydrazides with diethyl phosphite.

A novel CuI-catalyzed domino four-component coupling/cyclization reaction for the synthesis of 3-(aminomethyl)isoquinolines from 2-ethynylbenzaldehydes, paraformaldehyde, secondary amine, and t-BuNH$_2$ in DMF was reported by Fujii, Ohno, and coworkers (Scheme 8.93). The C(sp)–H bond was activated and one C–C and three C–N bonds were formed in one procedure [165].

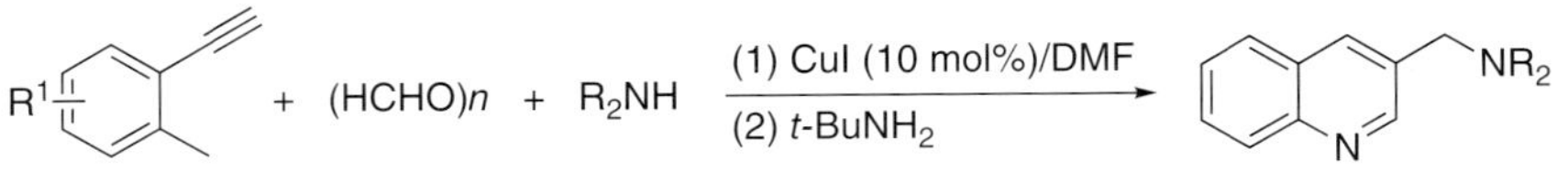

Scheme 8.93 Synthesis of 3-(aminomethyl)isoquinolines from 2-ethynylbenzaldehydes, paraformaldehyde, secondary amine, and t-BuNH$_2$.

8.4.4
Copper-Catalyzed Synthesis of Quinolinones

Li *et al.* developed a new route to quinolinone synthesis catalyzed by Cu(acac)$_2$ through oxidative (sp3)–H functionalization–carbocyclization–ketonization cascade (Scheme 8.94). This efficient protocol provides a new pathway to synthesize the quinolinone framework via implanting an oxygen atom from molecular oxygen. The desired quinolinones could be obtained in moderate to good yields under the mild conditions [166].

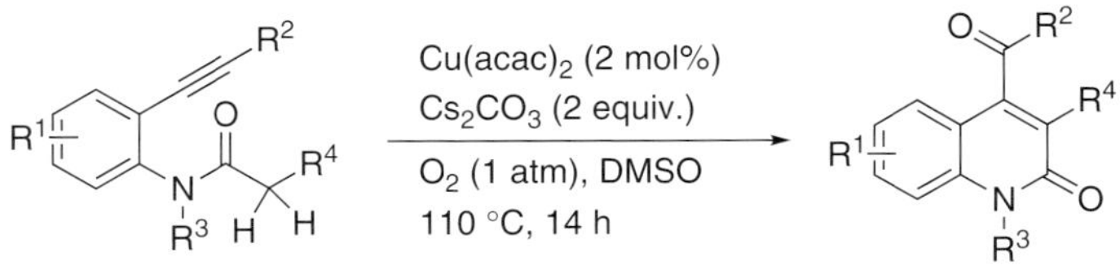

Scheme 8.94 Quinolinone synthesis through oxidative (sp3)–H functionalization–carbocyclization–ketonization cascade.

Cacchi *et al.* developed a novel CuI-catalyzed construction of the 2-quinolones from 3,3-diarylacrylamides via an intramolecular C–H functionalization/C–N bond-forming process (Scheme 8.95). High selectivity was observed for unsymmetrical 3,3-diarylacrylamides through using substrates where one aromatic ring bears *o*-methyl, chloro, or bromo substituents [167].

Duan *et al.* developed a Cu$_2$O-catalyzed synthesis of highly functionalized dihydroquinolinones through tandem oxidative cyclization of cinnamamides with benzyl hydrocarbons (Scheme 8.96). The reaction proceeds smoothly via the direct cross-dehydrogenative coupling of C(sp^3)–H and C(sp^2)–H bonds using tert-butylperoxy benzoate (TBPB) as oxidant. This is the first example

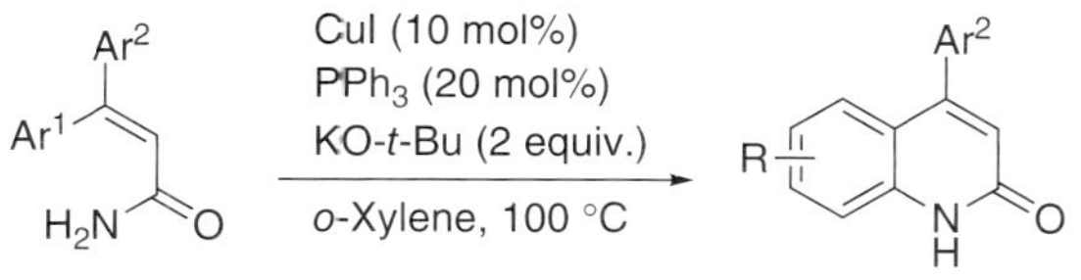

Scheme 8.95 Synthesis of 2-quinolones ring from 3,3-diarylacrylamides.

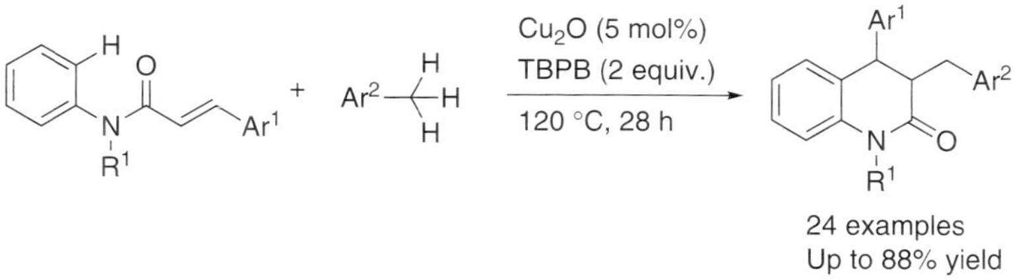

Scheme 8.96 Synthesis of highly functionalized dihydroquinolinones from cinnamamides with benzyl hydrocarbons.

of regioselective intramolecular 6-endo-trig cyclization of cinnamamides with reactants containing unactivated C(sp^3)–H bonds [168].

8.4.5
Copper-Catalyzed Synthesis of Acridones

A CuI-catalyzed synthesis of *N*-aryl acridones from 2-(aryllamino)benzophenones *via* intramolecular C(sp^2)–H bond amination using air as oxidant is disclosed by Deng *et al.* (Scheme 8.97). The electron-rich substrates were more active than electron-deficient ones. The catalyst Cu(I) could react with the substrate to form a reactive Cu(III) intermediate under O_2 via an electrophilic metalation or a C–H bond activation. Then reductive elimination could lead to the product and regenerate active Cu(I) catalyst [169].

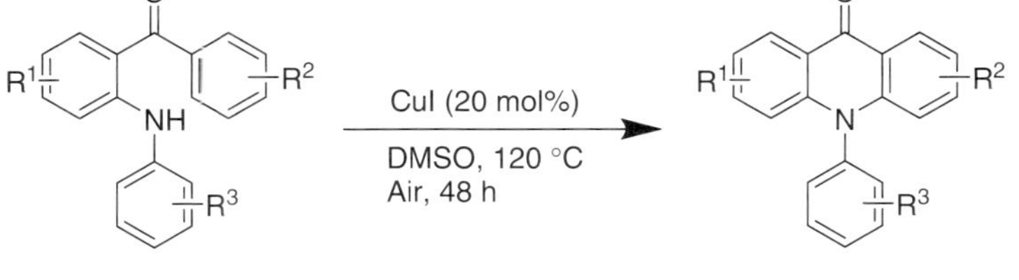

Scheme 8.97 Synthesis of *N*-aryl acridones from 2-(aryllamino)benzophenones.

A copper-catalyzed efficient procedure for the synthesis of 10-methylacridin-9(10*H*)-ones from 2-(methylamino)benzophenones has been developed by Zhu and coworkers (Scheme 8.98). The CuTc could catalyze intramolecular aromatic

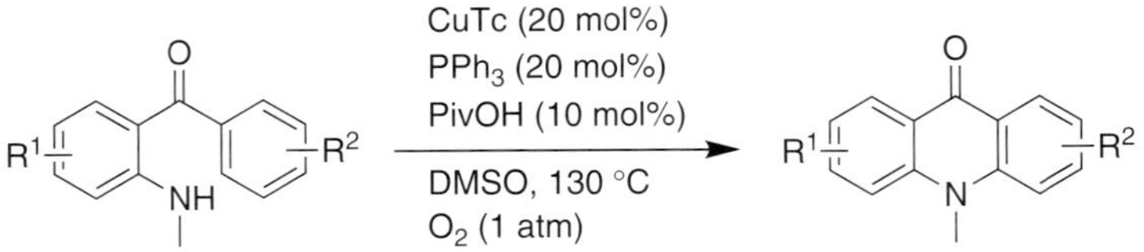

Scheme 8.98 Synthesis of 10-methylacridin-9(10*H*)-ones from 2-(methylamino) benzophenones.

C–H amination by using O_2 as the sole oxidant, PPh_3 as ligand, and PivOH as additive to provide 10-methylacridin-9(10*H*)-ones in moderate to good yields. In addition, 2-allylamino- and 2-(benzylamino)benzophenones as well as unprotected substrates can also undergo the C–H amination reaction to deliver the corresponding cyclization products successfully. The mechanistic studies suggest that C–H activation is involved in a rate-limiting step [170].

Cheng *et al.* have successfully developed a CuI/bpy catalytic system with very low catalyst loading (1.0 mol%) for the synthesis of acridone derivatives from 2-aminobenzophenones through C–H functionalization and C–N bond formation (Scheme 8.99). This protocol is successfully applied to the total synthesis of arborinine. It is noteworthy that the present catalytic system involves an unusual pseudo-1,2-migration of the substituent on the arene ring leading to the formation of two regioisomers [171].

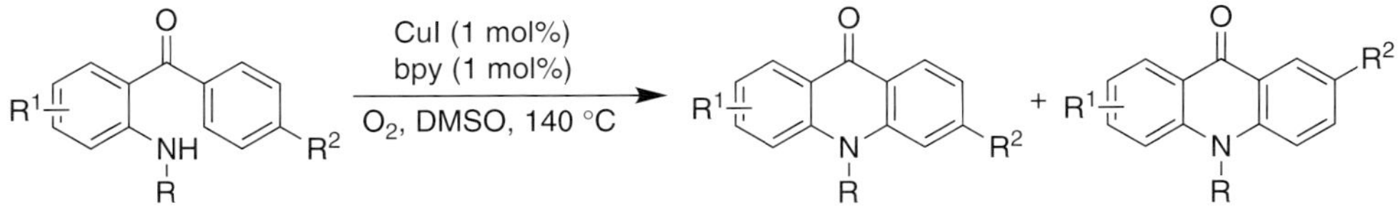

Scheme 8.99 Synthesis of acridone derivatives from 2-aminobenzophenones.

A CuI-catalyzed synthesis of acridones via intramolecular cyclization including C(sp^2)–H bond activation and C(O)–CH_3 bond cleavage process using air as the oxidant was reported in 2013 (Scheme 8.100). Many substituents on the aromatic rings are tolerable in the reaction, and the acridin-9(10*H*)-ones could be obtained in moderate to excellent yields. ^{13}C-labeling experiments show that only about 86% of carbon atom of carbonyl originates from the substrate. They proposed that a copper-catalyzed intramolecular Friedel–Crafts-type reaction pathway is disfavored [172]. Shortly after the previous report, Fu and coworkers developed a relevant aerobic synthesis of acridone derivatives from 1-[2-(arylamino)aryl]ethanones under $Cu(O_2CCF_3)_2$/pyridine/O_2 catalytic system [173].

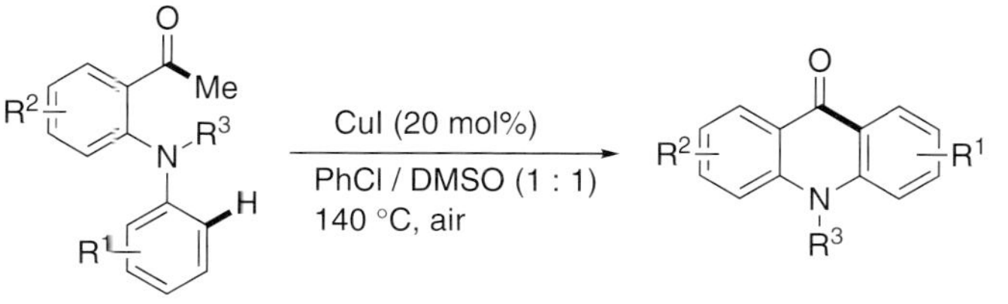

Scheme 8.100 Synthesis of acridones via intramolecular cyclization.

8.4.6
Copper-Catalyzed Synthesis of Phenanthridine

Chiba *et al.* developed an efficient copper -catalyzed synthesis of phenanthridine derivatives from biaryl-2-carbonitriles and Grignard reagents under an O_2 atmosphere (Scheme 8.101). This $Cu(OAc)_2$-catalyzed C–N bond formation involves an aromatic C–H bond activation process. This reaction proceeds via N–H imine formation by the nucleophilic addition of a Grignard reagent to biaryl-2-carbonitrile and intramolecular aryl C(sp^2)–H functionalization with copper catalyst [174].

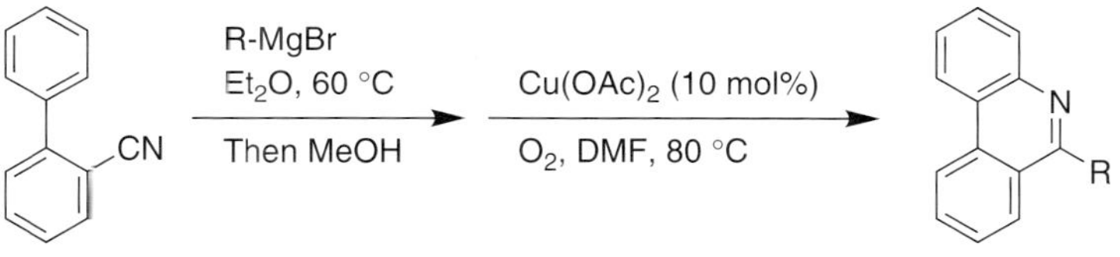

Scheme 8.101 Synthesis of phenanthridine derivatives from biaryl-2-carbonitriles and Grignard reagents.

Liu and coworkers developed a facile synthetic method of 6-alkyl-substituted phenanthridine derivatives catalyzed by CuF_2 or Cu_2O via free-radical addition/cyclization of isocyanides with simple alkanes and alcohols using DCP as oxidant (Scheme 8.102). This transformation includes selective functionalization of unactivated C(sp^3)–H and C(sp^2)–H bonds in one reaction. The cumyloxyl radical would be formed through homolysis of the O–O bond in DCP with the assistance of copper catalyst. Hydrogen abstraction of an alkane or α-hydroxy-C–H of the alcohol by the cumyloxyl radical would lead to an alkyl radical. Addition of the alkyl radical to the biphenyl isocyanide provides the imidoyl radical intermediate. It undergoes cyclization/oxidation process or oxidation/cyclization process to give the desired product [175]. Simultaneity, Huang and coworkers reported a cascade alkylarylation reaction of 2-isocyanobiphenyls with simple alkanes to synthesize 6-alkyl phenanthridines via dual C(sp^3)–H/C(sp^2)–H functionalizations using Cu_2O/DTBP (di-*tert*-butyl peroxide) system [176].

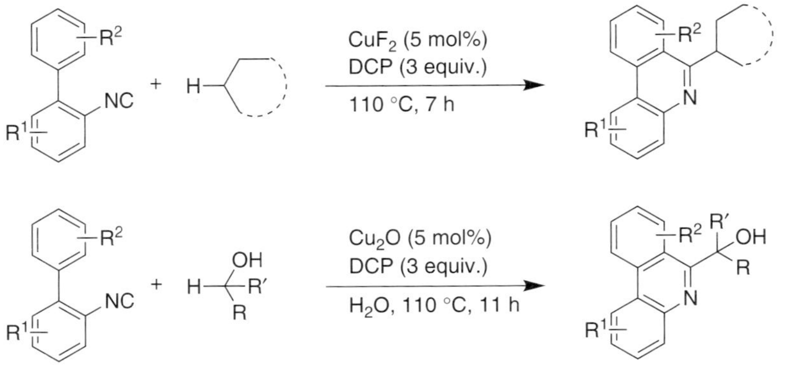

Scheme 8.102 Synthesis of 6-alkyl-substituted phenanthridine derivatives.

8.4.7
Copper-Catalyzed Synthesis of Quinazoline and Quinazolinones

An efficient $Cu(OTf)_2$-catalyzed synthesis of quinazolines from amidines and DMSO was reported by Xiong, Zhang, and coworkers. DMF, DMA, NMP (N-Methyl-2-pyrrolidone), or TMEDA (N,N,N′,N′-tetramethyl-ethane-1,2-diamine) was also a suitable carbon source for this transformation (Scheme 8.103). This process includes oxidative amination of N–H bonds and methyl C(sp^3)–H bonds of various carbon sources followed by intramolecular C–C bond formation with C(sp^2)–H bond activation. The quinazolines could be obtained with high selectivity and good yields via this domino reaction [177].

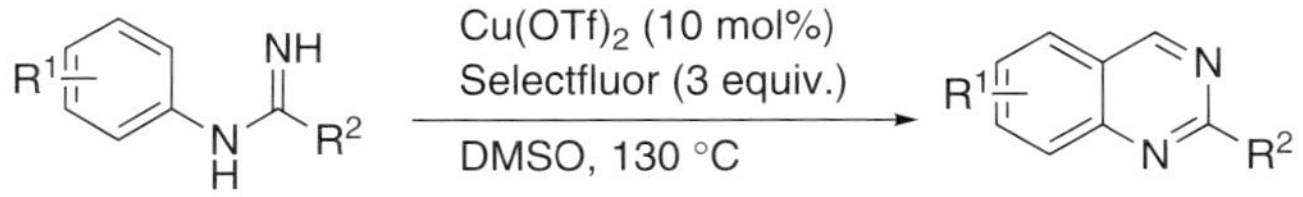

Scheme 8.103 Synthesis of quinazolines from amidines and DMSO.

Zou, Zhang *et al.* developed an efficient $Cu(OAc)_2$-catalyzed synthesis of indolo[1,2-*c*]quinazoline derivatives using environmentally friendly air as the oxidant (Scheme 8.104). Various indolo[1,2-*c*]quinazoline derivatives could be generated from readily available 2-(2-halophenyl)-1*H*-indoles and (aryl)methanamines in moderate to good yields via C(sp^3)–H bond activation [178].

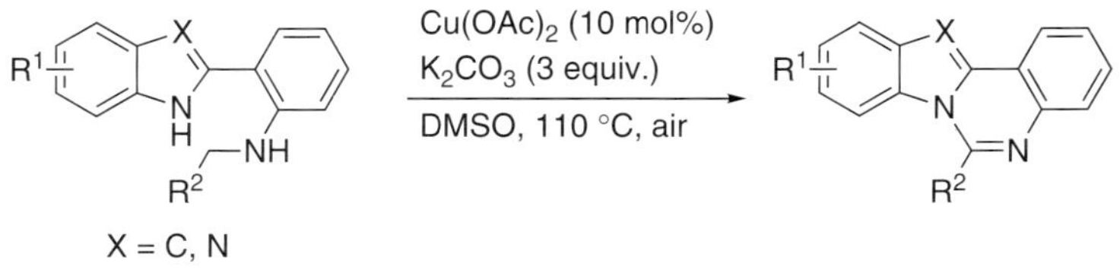

Scheme 8.104 Synthesis of indolo[1,2-*c*]quinazoline derivatives.

5-Substituted imidazo/benzimidazo[2,1-*b*]quinazolinones could be synthesized from (1*H*-imidazol-1-yl)[2-(alkylamino)phenyl]methanones and (1*H*-benzimidazol-1-yl)[2-(alkylamino)phenyl]-methanones catalyzed by CuCl in the absence of ligand using air as the oxidant (Scheme 8.105). The method is tolerant toward functional groups in the substrates, and a range of 5-substituted imidazo/benzimidazo[2,1-*b*]quinazolinones could be obtained in moderate to good yields. A C(sp^2)–H bond was activated to form a new C–N bond [179].

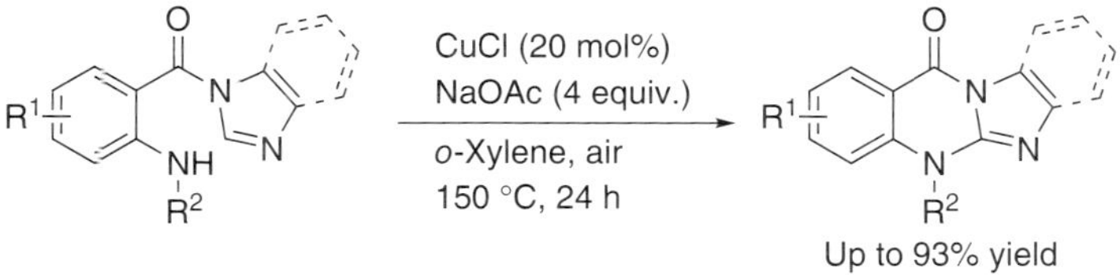

Scheme 8.105 Synthesis of 5-substituted imidazo/benzimidazo[2,1-*b*]quinazolinones.

Zhou, Yin *et al.* disclosed an efficient CuCl-catalyzed synthesis of 2-hetarylquinazolin-4(3*H*)-ones through aerobic oxidative C(sp^3)–H amination of (2-azaaryl)methanes with 2-aminobenzamides using oxygen as the sole oxidant under mild conditions (Scheme 8.106). Three C(sp^3)–H and three N–H bonds are removed in one reaction to produce the highly valuable *N*-heterocycles. This copper-catalyzed tandem oxidation–amination–cyclization transformation is easy to handle, and various prepared 2-hetaryl-substituted quinazolinones could be obtained in moderate to good yields [180].

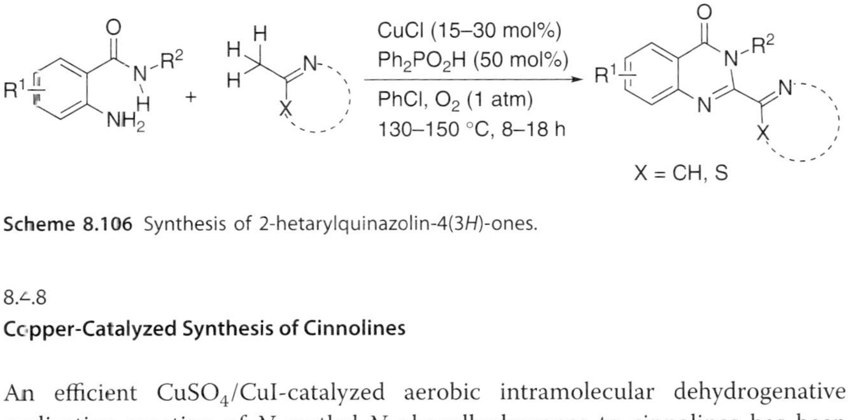

Scheme 8.106 Synthesis of 2-hetarylquinazolin-4(3*H*)-ones.

8.4.8
Copper-Catalyzed Synthesis of Cinnolines

An efficient $CuSO_4$/CuI-catalyzed aerobic intramolecular dehydrogenative cyclization reaction of *N*-methyl-*N*-phenylhydrazones to cinnolines has been developed by Ge and coworkers through sequential C(sp^3)–H oxidation, cyclization, and aromatization processes (Scheme 8.107). This transformation is the first example of copper-catalyzed coupling reactions of hydrazones through a C(sp^3)–H bond functionalization pathway. This transformation starts with the oxidation of *N*-methyl-*N*-phenylhydrazones into aldehyde intermediate through the activation of C(sp^3)–H under $CuSO_4/O_2$ catalytic system [181].

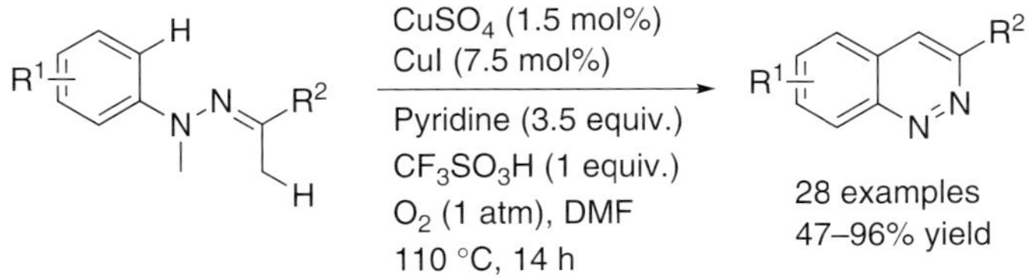

Scheme 8.107 Synthesis of cinnolines from *N*-methyl-*N*-phenylhydrazones.

8.4.9 Copper-Catalyzed Synthesis of Pyrimidinone

Lu, Wang *et al.* developed an efficient preparation of dihydropyrimidin-4-ones from readily available *N*-(prop-2-yn-1-yl)amides and sulfonyl azides using CuI as catalyst (Scheme 8.108). This cascade reaction includes copper-catalyzed alkyne–azide cycloaddition, formation and intramolecular nucleophilic addition of ketenimine, and subsequent rearrangement. The C(sp)–H bond was activated through copper-catalyzed alkyne–azide cycloaddition process in this reaction [182].

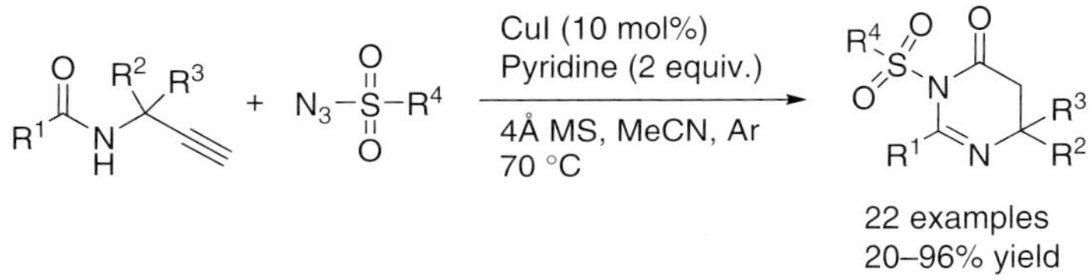

Scheme 8.108 Synthesis of dihydropyrimidin-4-ones .

8.4.10 Copper-Catalyzed Synthesis of 1,4-Dihydropyrazine Derivatives

An efficient $Cu(CF_3CO_2)_2$-catalyzed aerobic oxidative intramolecular alkene $C(sp^2)$–H amination leading to 1,4-dihydropyrazine derivatives from substituted 3-methyleneisoindolin-1-ones using air as the oxidant was developed by Fu and coworkers (Scheme 8.109). The corresponding *N*-heterocycles 1,4-dihydropyrazine derivatives were obtained in good to excellent yields with wide functional group tolerance [183].

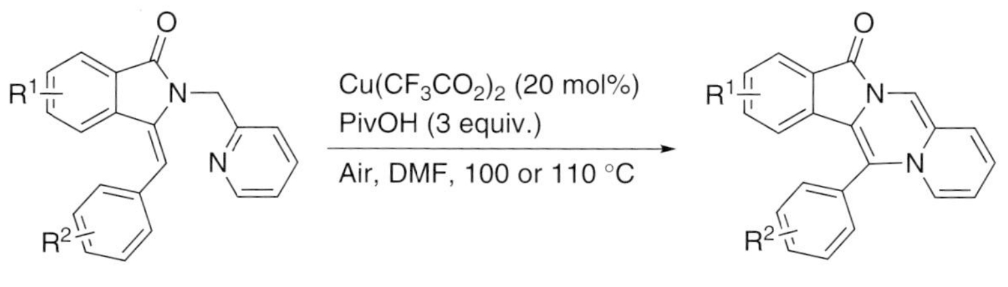

Scheme 8.109 Synthesis of 1,4-dihydropyrazine derivatives.

8.4.11 Copper-Catalyzed Synthesis of 1,3-Oxazines

Dihydro-1,3-oxazine derivatives are important heterocyclics in medicinal chemistry. Maycock *et al.* developed a $Cu(OAc)_2$-catalyzed intramolecular synthesis of benzo- or naphtho-2,3-dihydro-1,3-oxazines from tertiary amines through C–H bond oxidative activation and new C–O bond formation (Scheme 8.110). This reaction could be performed in an open vessel without the addition of external oxidants or additives. Moreover, naphthoxazines could be obtained with 100% diastereoselectivity [184].

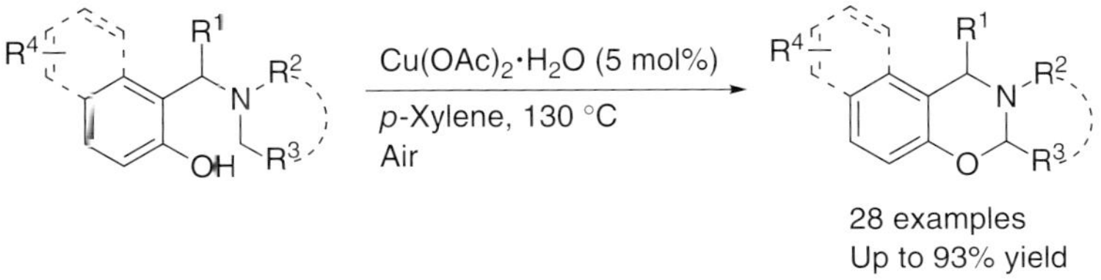

Scheme 8.110 Synthesis of benzo- or naphtho-2,3-dihydro-1,3-oxazines.

Bi. Zhang *et al.* reported a novel $Cu(OTf)_2$-catalyzed preparation of benzoxazine derivatives from readily available *N-para*-tolylamides in the presence of Selectfluor and water through the intermolecular C–H-activated dehydrogenative cross-coupling reaction (Scheme 8.111). The $C(sp^3)$–H bond of benzylic methyl and $C(sp^2)$–H bonds of arene were activated, and new C–C and C–O bonds were formed. The desired benzoxazine derivatives could be obtained in good to excellent yield. However, the substrates such as *N-para*-tolylacetamide, 3-methyl-*N-para*-tolylbutanamide, and 2-phenyl-*N-para*-tolylacetamide have

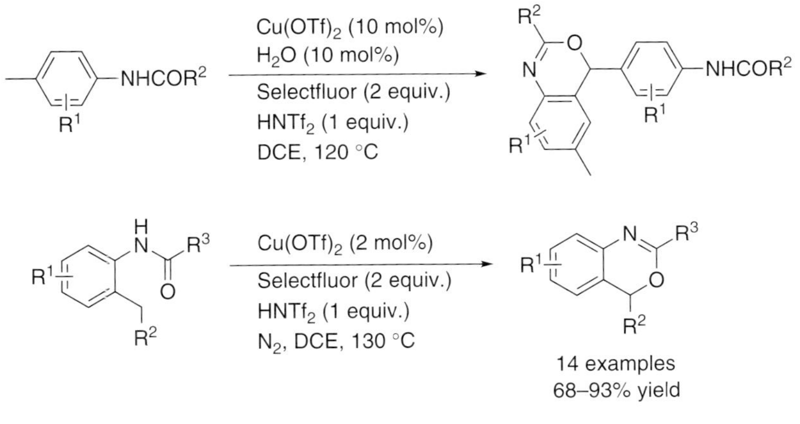

Scheme 8.111 Synthesis of benzoxazine derivatives.

no activation under standard conditions [185]. They also developed a $Cu(OTf)_2$-catalyzed intramolecular benzylic C–O cyclization of *N*-*o*-tolylbenzamide using Selectfluor as an oxidant for the efficient synthesis of 4*H*-3,1-benzoxazines. The exclusive selectivity for benzylic C–H over aromatic C–H bonds is realized [186] (Scheme 8.111).

8.4.12 Copper-Catalyzed Synthesis of Oxazinone Derivatives

Maiti *et al.* developed a $CuCl_2{\cdot}H_2O$-catalyzed dihydro-oxazinone derivative synthesis using O_2 as oxidant via cross-dehydrogenative coupling. One $C(sp^3)$–H bond was cleaved and a new $C(sp^3)$–O bond was formed (Scheme 8.112). Mechanistic studies including O-18 labeling, kinetic experiments, and deuterium labeling suggested that a Cu/O_2-based active species is responsible for this transformation. This method provides a concise and efficient preparation of an investigational new medicine CX-614 which might be applied to the treatment for Parkinson's and Alzheimer's diseases [187].

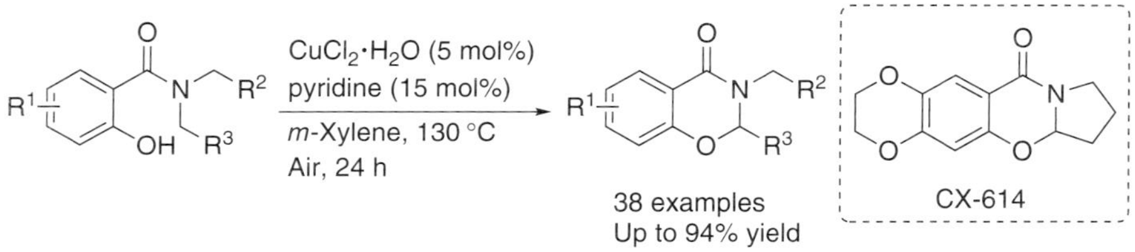

Scheme 8.112 Synthesis of dihydro-oxazinone derivatives.

8.4.13 Copper-Catalyzed Synthesis of Chroman Derivatives

A copper-catalyzed enantioselective aziridination of aryl cinnamyl ethers and intramolecular arylation of the *in situ* generated tethered aziridine was developed by Hajra and Sinha. This method provides an easy access of *N*-sulfonyl-protected *trans*-3-amino-4-arylchromans with high regio-, diastereo- (dr > 99 : 1), and enantioselectivity (ee up to 95%) (Scheme 8.113). Three $C(sp^2)$–H bonds were activated in this transformation under this $Cu(OTf)_2/Cu(ClO_4)_2{\cdot}6H_2O$ catalyst system. The major side reaction is the C–H insertion of nitrene at the $O–CH_2$ unit followed by oxidative cleavage [188].

Adrio and Hii reported that an air- and moisture-stable $Cu(OTf)_2$–bipy catalyzes the addition of phenols to 1,3-dienes under aerobic conditions in a tandem hydroalkoxylation–rearrangement–hydroalkylation sequence, furnishing benzopyrans in moderate to good yields, and can be recycled without any loss in catalytic activity [189] (Scheme 8.114).

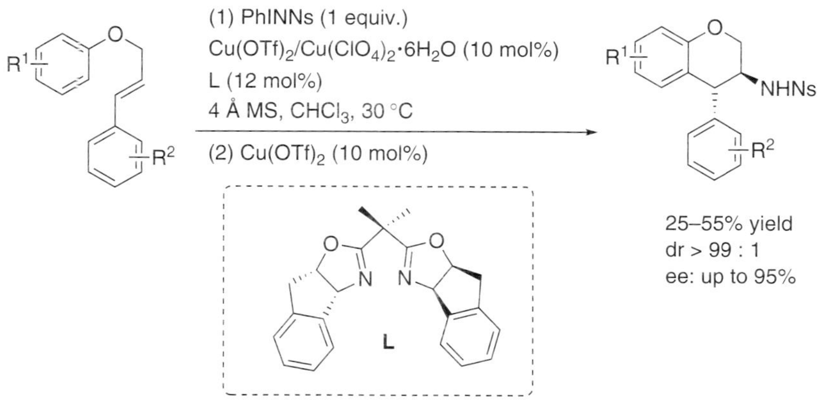

Scheme 8.113 Enantioselective synthesis of *N*-sulfonyl-protected *trans*-3-amino-4-arylchromans.

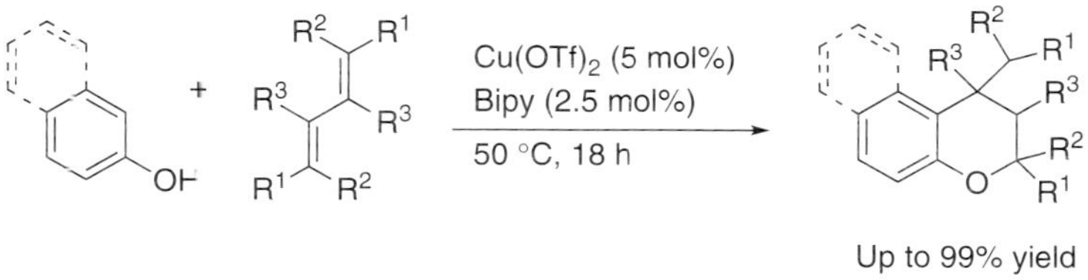

Scheme 8.114 Synthesis of benzopyrans from phenols and 1,3-dienes.

8.4.14
Copper-Catalyzed Synthesis of Benzolactone Derivatives

Benzolactones are important structural motifs in natural products and biologically active compounds. Gallardo-Donaire and Martin reported an efficient $Cu(OAc)_2$-catalyzed synthesis of benzolactones under mild conditions (Scheme 8.115). The intramolecular C–H hydroxylation of 2-arylbenzoic acids afforded benzolactones in moderate to good yields using $(PhCO_2)_2$ as the oxidant [190].

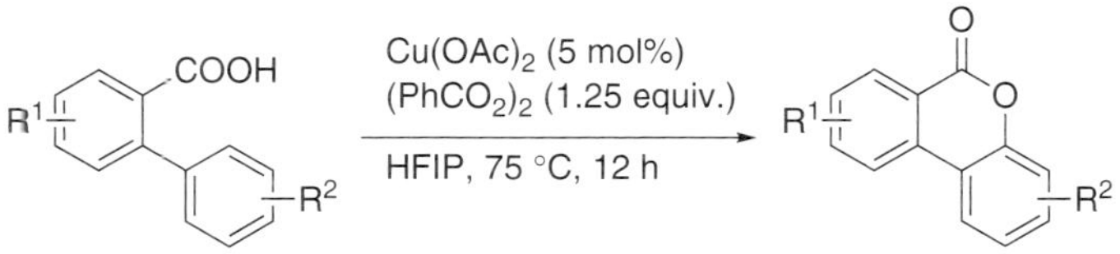

Scheme 8.115 Synthesis of benzolactones from 2-arylbenzoic acids.

8.4.15
Copper-Catalyzed Synthesis of Coumarin Derivatives

Coumarins are common and useful structures and they are widely present in natural products and biological and therapeutic activities. Lu and Ding *et al.* reported a $Cu(OAc)_2$-catalyzed synthesis of trifluoromethylated coumarins in moderate yields *through* direct trifluoromethylation of activated alkynes with Togni's reagent (Scheme 8.116). Preliminary mechanistic studies indicate that the reaction proceeds through a CF_3 radical addition to activated alkynes, followed by sequential oxidation cyclization to desired trifluoromethylated coumarins [191].

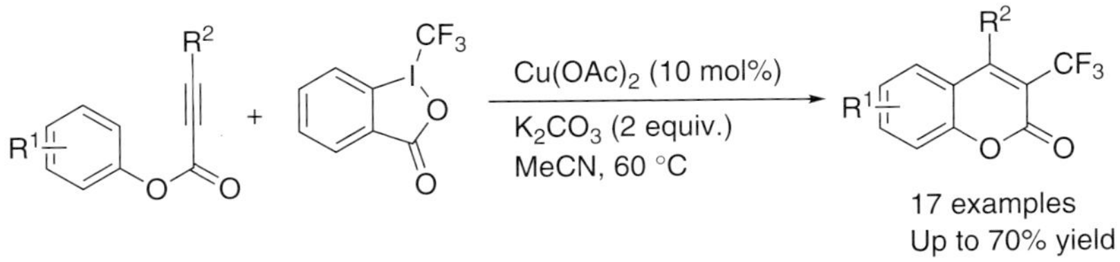

Scheme 8.116 Synthesis of trifluoromethylated coumarins.

Wang, Liu, and coworkers developed a $CuBr_2$-catalyzed synthesis of coumarins using a wide range of α-EWG-substituted ketene S,S-acetals, including a-aroyl, acetyl, cyano, ethoxycarbonyl groups with salicylaldehydes under very mild conditions [192] (Scheme 8.117).

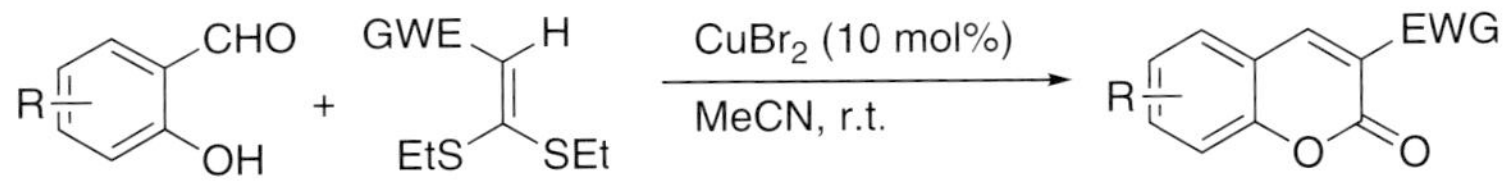

Scheme 8.117 Synthesis of coumarins from substituted ketene S,S-acetals.

Wang *et al.* developed a mild and efficient procedure for the synthesis of substituted iminocoumarin derivatives through a multicomponent reaction of sulfonyl azides, terminal alkynes, and salicylaldehydes (Scheme 8.118). The CuI/Et_3N catalytic system could give the desired products iminocoumarin in good to excellent yields. The C(sp)–H could be activated via the formation of ketenimine species in this transformation [193].

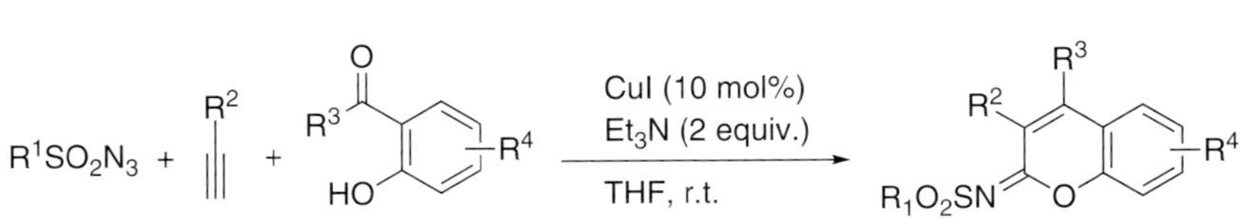

Scheme 8.118 Synthesis of substituted iminocoumarin derivatives.

A number of 3-triazolyl-2-iminochromenes were synthesized with CuI catalyst through three-component condensation. Initially, a CuI-catalyzed cycloaddition between 2-azidoacetonitrile and an acetylene could lead to a triazole intermediate. Then the neighboring methylene group would be activated and react with a salicylaldehyde inducing an aldol cyclization–dehydration sequence to give the corresponding 3-triazolyl-2-iminochromenes. The C(sp)–H, C(sp^2)–H, and C(sp^3)–H bonds were activated in one pot for this transformation [194] (Scheme 8.119).

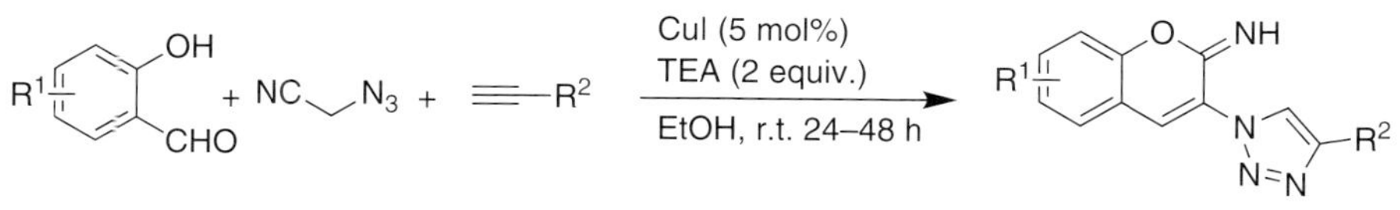

Scheme 8.119 Synthesis of 3-triazolyl-2-iminochromenes.

Murugavel and Punniyamurthy developed a CuI-catalyzed one-pot synthesis of iminocoumarin aryl methyl ethers from ynal, phenol, and sulfonyl azide via a cascade [3+2]-cycloaddition, 1,3-pseudopericyclic ketenimine rearrangement, 1,4-conjugate addition, and aldol-type condensation at room temperature under air [195] (Scheme 8.120).

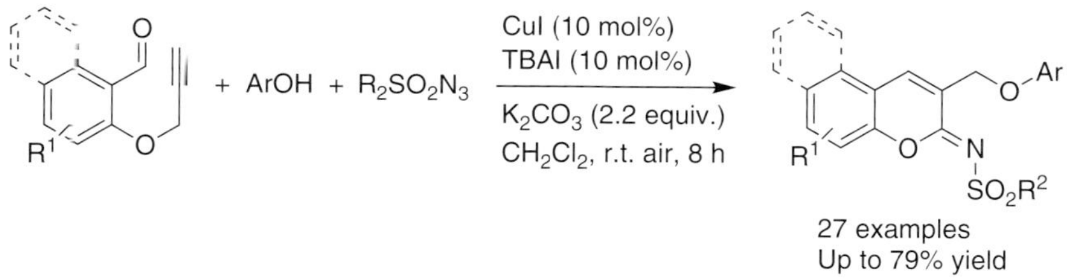

Scheme 8.120 Synthesis of iminocoumarin aryl methyl ethers.

8.4.16
Copper-Catalyzed Synthesis of Xanthone Derivatives

Wang *et al.* demonstrated an intermolecular $CuCl_2$-catalyzed *ortho*-acylation of phenols with various aryl aldehydes in the presence of triphenylphosphine using air as an oxidant. Furthermore, when 2-substituted aryl aldehydes (OMe, Br, NO_2) reacted with phenols, xanthones were obtained in high yield without the preactivation of aldehydes. This is a concise and straightforward strategy to construct xanthones compared to conventional methods (Scheme 8.121). The xanthone products are achieved via the *ortho*-acylation of phenols with 2-substituted aryl aldehydes and then the ring-closed reaction under basic conditions using OMe, Br, or NO_2 as leaving groups [196].

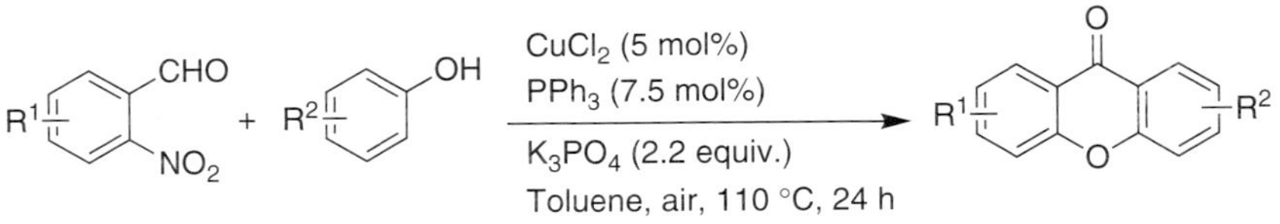

Scheme 8.121 Synthesis of xanthone derivatives via *ortho*-acylation of phenols with aryl aldehydes.

8.4.17 Copper-Catalyzed Synthesis of N,S-Heterocycles

Zeng and Chemler developed a copper-catalyzed preparation of six-membered N,S-heterocycle sultam derivatives using 3 equiv. of MnO_2 as the oxidant (Scheme 8.122). The enantiomerically enriched N,S-heterocycle sultams were prepared by an asymmetric intramolecular carboamination of alkenes with moderate to good yield and high enantioselectivity. An intramolecular Cu-catalyzed aryl C(sp^2)–H functionalization process is included for this reaction [54].

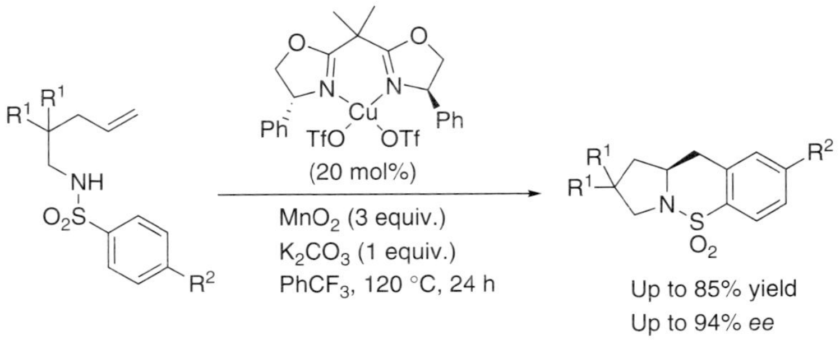

Scheme 8.122 Synthesis of N,S-heterocycle sultam derivatives.

Matsunaga and Kanai *et al.* reported a $Cu(CH_3CN)_4BF_4$-catalyzed intermolecular aminoarylation of aliphatic alkenes that provides straightforward access to sultams in moderate to excellent yields. Both terminal and internal alkenes could react with N-fluorobenzenesulfonimide (NFSI) under 2.5 mol% of $Cu(CH_3CN)_4BF_4$ at 60 °C (Scheme 8.123). The C(sp^2)–H bond of aryl rings of NFSI was activated and new C–C, C–S, C–N bonds were formed in this reaction [197].

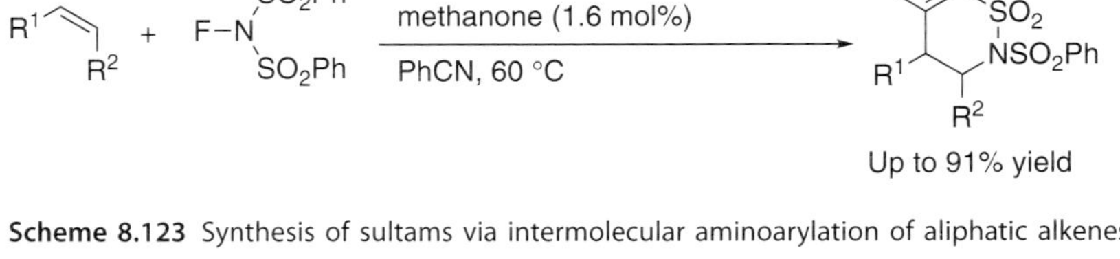

Scheme 8.123 Synthesis of sultams via intermolecular aminoarylation of aliphatic alkenes.

Benzo[*b*][1,4]thiazine-4-carbonitrile derivatives could be synthesized by a CuI-catalyzed coupling of 2-aminobenzothiazoles and terminal alkynes under ambient air (Scheme 8.124). Various benzo[*b*][1,4]thiazine-4-carbonitriles which are not readily accessible by conventional methods could be obtained in moderate to good yields. Preliminary mechanistic studies indicate that the reaction possibly proceeds through the ring opening of thiazole moiety followed by sequential oxidative coupling with alkyne and intramolecular cyclization to afford the products [198].

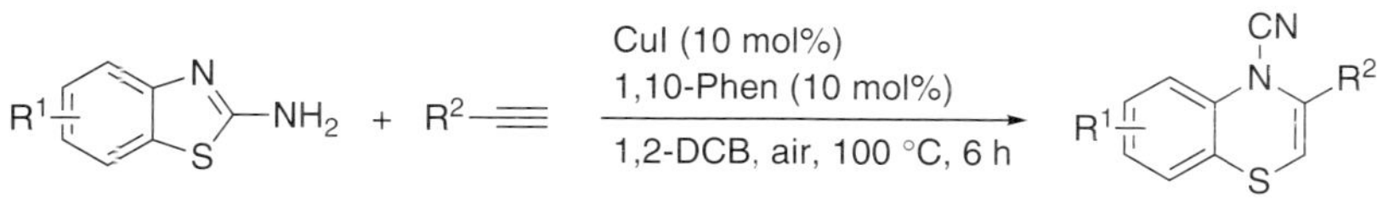

Scheme 8.124 Synthesis of Benzo[*b*][1,4]thiazine-4-carbonitriles.

8.5 Summary

Synthesis of N-, N,O-, O-, and N,S-heterocycles including four-, five-, and six-membered rings via copper-catalyzed C–H functionalization has been well developed. Many heterocycles could be obtained easily and efficiently using copper catalysts. However, it is still a challenge for chemists to develop some new copper catalytic systems for the heterocycle synthesis. Firstly, environmentally friendly oxidants such as O_2 are greatly suitable, and they are about to get more attention. Secondly, the development of copper-catalyzed asymmetric synthesis of chiral heterocycles is rarely reported, and new catalytic systems are still urgent. Thirdly, the direct C(sp^3)–H functionalization using inactivated hydrocarbons is required while they could provide some valuable molecules from simple and readily available substrates. Finally, the detailed mechanistic studies are also desired to understand these transformations and develop new catalytic reactions in this area.

References

1. Hashimoto, M., Obora, Y., Sakaguchi, S., and Ishii, Y. (2008) *J. Org. Chem.*, **73**, 2894–2897.
2. Hashimoto, M., Obora, Y., and Ishii, Y. (2009) *Org. Process Res. Dev.*, **13**, 411–414.
3. Krow, G.R. (1981) *Tetrahedron*, **37**, 1283–1307.
4. Aube, J. and Milligan, G.L. (1991) *J. Am. Chem. Soc.*, **113**, 8965–8966.
5. Milligan, G.L., Mossman, C.J., and Aube, J. (1995) *J. Am. Chem. Soc.*, **117**, 10449–10459.
6. Kunieda, T., Nagamatsu, T., Higuchi, T., and Hirobe, M. (1988) *Tetrahedron Lett.*, **29**, 2203–2205.
7. Bonache, M.A., Gerona-Navarro, G., Martín-Martínez, M., García-López, M.T., López, P., Cativiela, C., and González-Muñiz, R. (2003) *Synlett*, 1007–1011.
8. Kanwar, S. and Sharma, S.D. (2006) *Bull. Chem. Soc. Jpn.*, **79**, 1748–1752.
9. Knapp, S. and Gibson, F.S. (1992) *Org. Synth.*, **70**, 101–110.

10. Wang, Z., Ni, J., Kuninobu, Y., and Kanai, M. (2014) *Angew. Chem. Int. Ed.*, **53**, 3496–3499.
11. Wu, X., Zhao, Y., Zhang, G., and Ge, H. (2014) *Angew. Chem. Int. Ed.*, **53**, 3706–3710.
12. Kel'in, A.V., Sromek, A.W., and Gevorgyan, V. (2001) *J. Am. Chem. Soc.*, **123**, 2074–2075.
13. Kim, J.T. and Gevorgyan, V. (2002) *Org. Lett.*, **4**, 4697–4699.
14. Toh, K.K., Wang, Y.-F., Ng, E.P.J., and Chiba, S. (2011) *J. Am. Chem. Soc.*, **133**, 13942–13945.
15. Yan, R.-L., Luo, J., Wang, C.-X., Ma, C.-W., Huang, G.-S., and Liang, Y.-M. (2010) *J. Org. Chem.*, **75**, 5395–5397.
16. Larionov, O.V. and de Meijere, A. (2005) *Angew. Chem. Int. Ed.*, **44**, 5664–5667.
17. Lygin, A.V., Larionov, O.V., Korotkov, V.S., and de Meijere, A. (2009) *Chem. Eur. J.*, **15**, 227–236.
18. Kamijo, S., Kanazawa, C., and Yamamoto, Y. (2005) *J. Am. Chem. Soc.*, **127**, 9260–9266.
19. Chiba, S., Wang, Y.-F., Lapointe, G., and Narasaka, K. (2008) *Org. Lett.*, **10**, 313–316.
20. Chen, F., Shen, T., Cui, Y., and Jiao, N. (2012) *Org. Lett.*, **14**, 4926–4929.
21. Hong, D., Zhu, Y., Li, Y., Lin, X., Lu, P., and Wang, Y. (2011) *Org. Lett.*, **13**, 4668–4671.
22. Zhao, M.-N., Ren, Z.-H., Wang, Y.-Y., and Guan, Z.-H. (2014) *Chem. Eur. J.*, **20**, 1839–1842.
23. Ran, L., Ren, Z.-H., Wang, Y.-Y., and Guan, Z.-H. (2014) *Green Chem.*, **16**, 112–115.
24. Noack, M. and Göttlich, R. (2002) *Chem. Commun.*, 536–537.
25. Ohmiya, H., Moriya, T., and Sawamura, M. (2009) *Org. Lett.*, **11**, 2145–2147.
26. Paderes, M.C. and Chemler, S.R. (2011) *Eur. J. Org. Chem.*, **2011**, 3679–3684.
27. Liu, G.-Q., Li, W., and Li, Y.-M. (2013) *Adv. Synth. Catal.*, **355**, 395–402.
28. Oderaotoshi, Y., Cheng, W., Fujitomi, S., Kasano, Y., Minakata, S., and Komatsu, M. (2003) *Org. Lett.*, **5**, 5043–5046.
29. Gao, W., Zhang, X., and Raghunath, M. (2005) *Org. Lett.*, **7**, 4241–4244.
30. Yan, X.-X., Peng, Q., Zhang, Y., Zhang, K., Hong, W., Hou, X.-L., and Wu, Y.-D. (2006) *Angew. Chem. Int. Ed.*, **45**, 1979–1983.
31. Shi, M. and Shi, J.-W. (2007) *Tetrahedron: Asymmetry*, **18**, 645–650.
32. Wang, C.-J., Liang, G., Xue, Z.-Y., and Gao, F. (2008) *J. Am. Chem. Soc.*, **130**, 17250–17251.
33. Castelló, L.M., Nájera, C., Sansano, J.M., Larrãnaga, O., de Cózar, A., and Cossío, F.P. (2013) *Org. Lett.*, **15**, 2902–2905.
34. Robles-Machín, R., González-Esguevillas, M., Adrio, J., and Carretero, J.C. (2010) *J. Org. Chem.*, **75**, 233–236.
35. López-Pérez, A., Adrio, J., and Carretero, J.C. (2009) *Angew. Chem. Int. Ed.*, **48**, 340–343.
36. Martín-Matute, B., Pereira, S.I., Peña-Cabrera, E., Adrio, J., Silva, A.M.S., and Carretero, J.C. (2007) *Adv. Synth. Catal.*, **349**, 1714–1724.
37. Cabrera, S., Arrayás, R.G., Martín-Matute, B., Cossío, F.P., and Carretero, J.C. (2007) *Tetrahedron*, **63**, 6587–6602.
38. Cabrera, S., Arrayás, R.G., and Carretero, J.C. (2005) *J. Am. Chem. Soc.*, **127**, 16394–16395.
39. Llamas, T., Arrayás, R.G., and Carretero, J.C. (2006) *Org. Lett.*, **8**, 1795–1798.
40. Galliford, C.V., Beenen, M.A., Nguyen, S.T., and Scheidt, K.A. (2003) *Org. Lett.*, **5**, 3487–3490.
41. Wu, Q., Hu, J., Ren, X., and Zhou, J. (2011) *Chem. Eur. J.*, **17**, 11553–11558.
42. Sundberg, R.J. (ed) (1996) *Indoles*, Academic Press, London.
43. Sundberg, R.J. (1984) in *Comprehensive Heterocyclic Chemistry*, vol. **4** (eds A.R. Katritzky and C.W. Rees), Pergamon Press, Oxford, pp. 313–376.
44. Bernini, R., Fabrizi, G., Sferrazza, A., and Cacchi, S. (2009) *Angew. Chem. Int. Ed.*, **48**, 8078–8081.
45. Ohno, H., Ohta, Y., Oishi, S., and Fujii, N. (2007) *Angew. Chem. Int. Ed.*, **46**, 2295–2298.
46. Ohta, Y., Chiba, H., Oishi, S., Fujii, N., and Ohno, H. (2009) *J. Org. Chem.*, **74**, 7052–7058.

47. Suzuki, Y., Ohta, Y., Oishi, S., Fujii, N., and Ohno, H. (2009) *J. Org. Chem.*, **74**, 4246–4251.
48. Ohta, Y., Chiba, H., Oishi, S., Fujii, N., and Ohno, H. (2008) *Org. Lett.*, **10**, 3535–3538.
49. Chen, Z., Zheng, D., and Wu, J. (2011) *Org. Lett.*, **13**, 848–851.
50. Gogoi, A., Guin, S., Rout, S.K., and Patel, B.K. (2013) *Org. Lett.*, **15**, 1802–1805.
51. Sherman, E.S., Chemler, S.R., Tan, T.B., and Gerlits, O. (2004) *Org. Lett.*, **6**, 1573–1575.
52. Sherman, E.S., Fuller, P.H., Kasi, D., and Chemler, S.R. (2007) *J. Org. Chem.*, **72**, 3896–3905.
53. Fuller, P.H. and Chemler, S.R. (2007) *Org. Lett.*, **9**, 5477–5480.
54. Zeng, W. and Chemler, S.R. (2007) *J. Am. Chem. Soc.*, **129**, 12948–12949.
55. Zabawa, T.P., Kasi, D., and Chemler, S.R. (2005) *J. Am. Chem. Soc.*, **127**, 11250–11251.
56. Zabawa, T.P. and Chemler, S.R. (2007) *Org. Lett.*, **9**, 2035–2038.
57. Fuller, P.H., Kim, J.-W., and Chemler, S.R. (2008) *J. Am. Chem. Soc.*, **130**, 17638–17639.
58. Paderes, M.C., Keister, J.B., and Chemler, S.R. (2013) *J. Org. Chem.*, **78**, 506–515.
59. Sherman, E.S. and Chemler, S.R. (2009) *Adv. Synth. Catal.*, **351**, 467–471.
60. Liwosz, T.W. and Chemler, S.R. (2012) *J. Am. Chem. Soc.*, **134**, 2020–2023.
61. Bovino, M.T. and Chemler, S.R. (2012) *Angew. Chem. Int. Ed.*, **51**, 3923–3927.
62. Turnpenny, B.W., Hyman, K.L., and Chemler, S.R. (2012) *Organometallics*, **31**, 7819–7822.
63. Li, S., Luo, Y., and Wu, J. (2011) *Org. Lett.*, **13**, 3190–3193.
64. Egami, H., Shimizu, R., Kawamura, S., and Sodeoka, M. (2013) *Angew. Chem. Int. Ed.*, **52**, 4000–4003.
65. Takamatsu, K., Hirano, K., Satoh, T., and Miura, M. (2015) *J. Org. Chem.*, **80**, 3242–3249.
66. Perry, A. and Taylor, R.J.K. (2009) *Chem. Commun.*, 3249–3251.
67. Klein, J.E.M.N., Perry, A., Pugh, D.S., and Taylor, R.J.K. (2010) *Org. Lett.*, **12**, 3446–3449.
68. Kong, W., Casimiro, M., Merino, E., and Nevado, C. (2013) *J. Am. Chem. Soc.*, **135**, 14480–14483.
69. Yang, F., Klumphu, P., Liang, Y.-M., and Lipshutz, B.H. (2014) *Chem. Commun.*, **50**, 936–938.
70. Li, Z., Zhang, Y., Zhang, L., and Liu, Z.-Q. (2014) *Org. Lett.*, **16**, 382–385.
71. Zhou, B., Hou, W., Yang, Y., Feng, H., and Li, Y. (2014) *Org. Lett.*, **16**, 1322–1325.
72. Shi, L., Wang, Y., Yang, H., and Fu, H. (2014) *Org. Biomol. Chem.*, **12**, 4070–4073.
73. Gong, W., Xu, L., Ji, T., Xie, P., Qi, X., Pittman, C.U., and Zhou, A. (2014) *RSC Adv.*, **4**, 6854–6857.
74. Tang, B.-X., Song, R.-J., Wu, C.-Y., Liu, Y., Zhou, M.-B., Wei, W.-T., Deng, G.-B., Yin, D.-L., and Li, J.-H. (2010) *J. Am. Chem. Soc.*, **132**, 8900–8902.
75. Liu, T., Yang, H., Jiang, Y., and Fu, H. (2013) *Adv. Synth. Catal.*, **355**, 1169–1176.
76. Huang, P.-C., Gandeepan, P., and Cheng, C.-H. (2013) *Chem. Commun.*, **49**, 8540–8542.
77. Ilangovan, A. and Satish, G. (2013) *Org. Lett.*, **15**, 5726–5729.
78. Barluenga, J., Lonzi, G., Riesgo, L., López, L.A., and Tomás, M. (2010) *J. Am. Chem. Soc.*, **132**, 13200–13202.
79. Cho, S.H., Yoon, J., and Chang, S. (2011) *J. Am. Chem. Soc.*, **133**, 5996–6005.
80. Takamatsu, K., Hirano, K., Satoh, T., and Miura, M. (2014) *Org. Lett.*, **16**, 2892–2895.
81. Kanazawa, C., Kamijo, S., and Yamamoto, Y. (2006) *J. Am. Chem. Soc.*, **128**, 10662–10663.
82. Cai, Z.-J., Wang, S.-Y., and Ji, S.-J. (2012) *Org. Lett.*, **14**, 6068–6071.
83. Wang, Y.-F., Zhu, X., and Chiba, S. (2012) *J. Am. Chem. Soc.*, **134**, 3679–3682.
84. Sanjaya, S. and Chiba, S. (2012) *Org. Lett.*, **14**, 5342–5345.
85. Chen, H., Sanjaya, S., Wang, Y.-F., and Chiba, S. (2013) *Org. Lett.*, **15**, 212–215.
86. Chen, H. and Chiba, S. (2014) *Org. Biomol. Chem.*, **12**, 42–46.

87. Tang, D., Wu, P., Liu, X., Chen, Y.-X., Guo, S.-B., Chen, W.-L., Li, J.-G., and Chen, B.-H. (2013) *J. Org. Chem.*, **78**, 2746–2750.
88. Li, J. and Neuville, L. (2013) *Org. Lett.*, **15**, 1752–1755.
89. Brasche, G. and Buchwald, S.L. (2008) *Angew. Chem. Int. Ed.*, **47**, 1932–1934.
90. Li, J., Bénard, S., Neuville, L., and Zhu, J. (2012) *Org. Lett.*, **14**, 5980–5983.
91. She, J., Jiang, Z., and Wang, Y.G. (2009) *Synlett*, 2023–2027.
92. Jin, H., Xu, X., Gao, J., Zhong, J., and Wang, Y. (2010) *Adv. Synth. Catal.*, **352**, 347–350.
93. He, H.-F., Wang, Z.-J., and Bao, W. (2010) *Adv. Synth. Catal.*, **352**, 2905–2912.
94. Xiao, T., Xiong, S., Xie, Y., Dong, X., and Zhou, L. (2013) *RSC Adv.*, **3**, 15592–15595.
95. Wang, X., Jin, Y., Zhao, Y., Zhu, L., and Fu, H. (2012) *Org. Lett.*, **14**, 452–455.
96. Subramanian, P. and Kaliappan, K.P. (2014) *Eur. J. Org. Chem.*, **2014**, 5986–5997.
97. Bluhm, M.E., Ciesielski, M., Görls, H., and Döring, M. (2002) *Angew. Chem. Int. Ed.*, **41**, 2962–2965.
98. Wang, H., Wang, Y., Peng, C., Zhang, J., and Zhu, Q. (2010) *J. Am. Chem. Soc.*, **132**, 13217–13219.
99. Chernyak, N. and Gevorgyan, V. (2010) *Angew. Chem. Int. Ed.*, **49**, 2743–2746.
100. Wang, H., Wang, Y., Liang, D., Liu, L., Zhang, J., and Zhu, Q. (2011) *Angew. Chem. Int. Ed.*, **50**, 5678–5681.
101. Masters, K.-S., Rauws, T.R.M., Yadav, A.K., Herrebout, W.A., Van der Veken, B., and Maes, B.U.W. (2011) *Chem. Eur. J.*, **17**, 6315–6320.
102. Yan, R.-L., Yan, H., Ma, C., Ren, Z.-Y., Gao, X.-A., Huang, G.-S., and Liang, Y.-M. (2012) *J. Org. Chem.*, **77**, 2024–2028.
103. Cai, Z.-J., Wang, S.-Y., and Ji, S.-J. (2013) *Adv. Synth. Catal.*, **355**, 2686–2692.
104. Bagdi, A.K., Rahman, M., Santra, S., Majee, A., and Hajra, A. (2013) *Adv. Synth. Catal.*, **355**, 1741–1747.
105. Zhang, Y., Chen, Z., Wu, W., Zhang, Y., and Su, W. (2013) *J. Org. Chem.*, **78**, 12494–12504.
106. Pericherla, K., Kaswan, P., Khedar, P., Khungar, B., Parang, K., and Kumar, A. (2013) *RSC Adv.*, **3**, 18923–18930.
107. Li, M., Xie, Y., Ye, Y., Zou, Y., Jiang, H., and Zeng, W. (2014) *Org. Lett.*, **16**, 6232–6235.
108. Yu, J., Jin, Y., Zhang, H., Yang, X., and Fu, H. (2013) *Chem. Eur. J.*, **19**, 16804–16808.
109. Huang, H., Ji, X., Tang, X., Zhang, M., Li, X., and Jiang, H. (2013) *Org. Lett.*, **15**, 6254–6257.
110. Neumann, J.J., Suri, M., and Glorius, F. (2010) *Angew. Chem. Int. Ed.*, **49**, 7790–7794.
111. Suri, M., Jousseaume, T., Neumann, J.J., and Glorius, F. (2012) *Green Chem.*, **14**, 2193–2196.
112. Li, X., He, L., Chen, H., Wu, W., and Jiang, H. (2013) *J. Org. Chem.*, **78**, 3636–3646.
113. Zhang, G., Zhao, Y., and Ge, H. (2013) *Angew. Chem. Int. Ed.*, **52**, 2559–2563.
114. Ding, S., Yan, Y., and Jiao, N. (2013) *Chem. Commun.*, **49**, 4250–4252.
115. Wan, C., Zhang, J., Wang, S., Fan, J., and Wang, Z. (2010) *Org. Lett.*, **12**, 2338–2341.
116. Hu, P., Wang, Q., Yan, Y., Zhang, S., Zhang, B., and Wang, Z. (2013) *Org. Biomol. Chem.*, **11**, 4304–4307.
117. Xu, Z., Zhang, C., and Jiao, N. (2012) *Angew. Chem. Int. Ed.*, **51**, 11367–11370.
118. Cheung, C.W. and Buchwald, S.L. (2012) *J. Org. Chem.*, **77**, 7526–7537.
119. Wang, Y.-F., Chen, H., Zhu, X., and Chiba, S. (2012) *J. Am. Chem. Soc.*, **134**, 11980–11983.
120. Ueda, S. and Nagasawa, H. (2008) *Angew. Chem. Int. Ed.*, **47**, 6411–6413.
121. Ueda, S. and Nagasawa, H. (2009) *J. Org. Chem.*, **74**, 4272–4277.
122. Guru, M.M., Ali, M.A., and Punniyamurthy, T. (2011) *J. Org. Chem.*, **76**, 5295–5308.
123. Guru, M.M., Ali, M.A., and Punniyamurthy, T. (2011) *Org. Lett.*, **13**, 1194–1197.
124. Nair, D.P., Podgórski, M., Chatani, S., Gong, T., Xi, W., Fenoli, C.R., and Bowman, C.N. (2014) *Chem. Mater.*, **26**, 724–744.

125. Yim, C.-B., Dijkgraaf, I., Merkx, R., Versluis, C., Eek, A., Mulder, G.E., Rijkers, D.T.S., Boerman, O.C., and Liskamp, R.M.J. (2010) *J. Med. Chem.*, **53**, 3944–3953.

126. Tripathi, R.P., Yadav, A.K., Ajay, A., Bisht, S.S., Chaturvedi, V., and Sinha, S.K. (2010) *Eur. J. Med. Chem.*, **45**, 142–148.

127. Meldal, M. and Tornoe, C.W. (2008) *Chem. Rev.*, **108**, 2952–3015.

128. Chen, H., Taylor, J.L., and Abrams, S.R. (2007) *Bioorg. Med. Chem. Lett.*, **17**, 1979–1983.

129. Bock, V.D., Speijer, D., Hiemstra, H., and Van Maarseveen, J.H. (2007) *Org. Biomol. Chem.*, **5**, 971–975.

130. Moses, J.E. and Moorhouse, A.D. (2007) *Chem. Soc. Rev.*, **36**, 1249–1262.

131. Wang, J., Sui, G., Mocharla, V.P., Lin, R.J., Phelps, M.E., Kolb, H.C., and Tseng, H.-R. (2006) *Angew. Chem. Int. Ed.*, **45**, 5276–5281.

132. Manetsch, R., Krasiski, A., Radi, Z., Raushel, J., Taylor, P., Sharpless, K.B., and Kolb, H.C. (2004) *J. Am. Chem. Soc.*, **126**, 12809–12818.

133. Agard, N.J., Prescher, J.A., and Bertozzi, C.R. (2004) *J. Am. Chem. Soc.*, **126**, 15046–15047.

134. Kolb, H.C. and Sharpless, K.B. (2003) *Drug Discovery Today*, **8**, 1128–1137.

135. Rostovtsev, V.V., Green, L.G., Fokin, V.V., and Sharpless, K.B. (2002) *Angew. Chem. Int. Ed.*, **41**, 2596–2599.

136. Maliakal, A., Lem, G., Turro, N.J., Ravichandran, R., Suhadolnik, J.C., DeBellis, A.D., Wood, M.G., and Lau, J. (2002) *J. Phys. Chem. A*, **106**, 7680–7689.

137. Huisgen, R. (1961) *Proc. Chem. Soc.*, 357–396.

138. Huisgen, R. (1963) *Angew. Chem. Int. Ed. Engl.*, **2**, 565–598.

139. Huisgen, R. (1963) *Angew. Chem. Int. Ed. Engl.*, **2**, 633–645.

140. Yang, C.-H., Sherf, H.-J., Wang, R.-H., and Wang, J.-C. (2002) *J. Chin. Chem. Soc.*, **49**, 95–102.

141. Shriteha, T., Kodlaaa, A., Abu-Orabib, S.T., Madwara, R., and Atfeh, A. (2011) *Jordan J. Chem.*, **6**, 81–89.

142. Kolb, H.C., Finn, M.G., and Sharpless, K.B. (2001) *Angew. Chem. Int. Ed.*, **40**, 2004–2021.

143. Worrell, B.T., Malik, J.A., and Fokin, V.V. (2013) *Science*, **340**, 457–460.

144. Hein, J.E. and Fokin, V.V. (2010) *Chem. Soc. Rev.*, **39**, 1302–1315.

145. Chen, F., Qin, C., Cui, Y., and Jiao, N. (2011) *Angew. Chem. Int. Ed.*, **50**, 11487–11491.

146. Miller, Y., Miao, L., Hosseini, A.S., and Chemler, S.R. (2012) *J. Am. Chem. Soc.*, **134**, 12149–12156.

147. Bovino, M.T., Liwosz, T.W., Kendel, N.E., Miller, Y., Tyminska, N., Zurek, E., and Chemler, S.R. (2014) *Angew. Chem. Int. Ed.*, **53**, 6383–6387.

148. Zhao, J., Wang, Y., He, Y., Liu, L., and Zhu, Q. (2012) *Org. Lett.*, **14**, 1078–1081.

149. Zeng, W., Wu, W., Jiang, H., Huang, L., Sun, Y., Chen, Z., and Li, X. (2013) *Chem. Commun.*, **49**, 6611–6613.

150. Huang, H., Ji, X., Wu, W., Huang, L., and Jiang, H. (2013) *J. Org. Chem.*, **78**, 3774–3782.

151. Li, Z., Huang, X., Chen, F., Zhang, C., Wang, X., and Jiao, N. (2015) *Org. Lett.*, **17**, 584–587.

152. Nakamura, I., Zhang, D., and Terada, M. (2010) *J. Am. Chem. Soc.*, **132**, 7884–7886.

153. Cho, C.S., Ren, W.X., and Shim, S.C. (2006) *Tetrahedron Lett.*, **47**, 6781–6785.

154. Huang, H., Jiang, H., Chen, K., and Liu, H. (2009) *J. Org. Chem.*, **74**, 5476–5480.

155. Toh, K.K., Sanjaya, S., Sahnoun, S., Chong, S.Y., and Chiba, S. (2012) *Org. Lett.*, **14**, 2290–2292.

156. Xia, X.-F., Zhang, L.-L., Song, X.-R., Liu, X.-Y., and Liang, Y.-M. (2012) *Org. Lett.*, **14**, 2480–2483.

157. Cheng, G. and Cui, X. (2013) *Org. Lett.*, **15**, 1480–1483.

158. Yan, R., Liu, X., Pan, C., Zhou, X., Li, X., Kang, X., and Huang, G. (2013) *Org. Lett.*, **15**, 4876–4879.

159. Gangadhararao, G., Uruvakilli, A., and Swamy, K.C.K. (2014) *Org. Lett.*, **16**, 6060–6063.

160. Nishino, M., Hirano, K., Satoh, T., and Miura, M. (2011) *J. Org. Chem.*, **76**, 6447–6451.
161. Min, C., Sanchawala, A., and Seidel, D. (2014) *Org. Lett.*, **16**, 2756–2759.
162. Hurst, T.E., Gorman, R.M., Drouhin, P., Perry, A., and Taylor, R.J.K. (2014) *Chem. Eur. J.*, **20**, 14063–14073.
163. Yu, X., Ding, Q., Chen, Z., and Wu, J. (2009) *Tetrahedron Lett.*, **50**, 4279–4282.
164. Chen, Z., Gao, L., Ye, S., Ding, Q., and Wu, J. (2012) *Chem. Commun.*, **48**, 3975–3977.
165. Ohta, Y., Oishi, S., Fujii, N., and Ohno, H. (2008) *Chem. Commun.*, 835–837.
166. Xie, P., Wang, Z.-Q., Deng, G.-B., Song, R.-J., Xia, J.-D., Hu, M., and Li, J.-H. (2013) *Adv. Synth. Catal.*, **355**, 2257–2262.
167. Berrino, R., Cacchi, S., Fabrizi, G., and Goggiamani, A. (2012) *J. Org. Chem.*, **77**, 2537–2542.
168. Zhou, S.-L., Guo, L.-N., Wang, S., and Duan, X.-H. (2014) *Chem. Commun.*, **50**, 3589–3591.
169. Zhou, W., Liu, Y., Yang, Y., and Deng, G.-J. (2012) *Chem. Commun.*, **48**, 10678–10680.
170. Huang, J., Wan, C., Xu, M.-F., and Zhu, Q. (2013) *Eur. J. Org. Chem.*, **2013**, 1876–1880.
171. Huang, P.-C., Parthasarathy, K., and Cheng, C.-H. (2013) *Chem. Eur. J.*, **19**, 460–464.
172. Zhou, W., Yang, Y., Liu, Y., and Deng, G.-J. (2013) *Green Chem.*, **15**, 76–80.
173. Yu, J., Yang, H., Jiang, Y., and Fu, H. (2013) *Chem. Eur. J.*, **19**, 4271–4277.
174. Zhang, L., Ang, G.Y., and Chiba, S. (2010) *Org. Lett.*, **12**, 3682–3685.
175. Li, Z., Fan, F., Yang, J., and Liu, Z.-Q. (2014) *Org. Lett.*, **16**, 3396–3399.
176. Zhu, Z.-Q., Wang, T.-T., Bai, P., and Huang, Z.-Z. (2014) *Org. Biomol. Chem.*, **12**, 5839–5842.
177. Lv, Y., Li, Y., Xiong, T., Pu, W., Zhang, H., Sun, K., Liu, Q., and Zhang, Q. (2013) *Chem. Commun.*, **49**, 6439–6441.
178. Sang, P., Xie, Y., Zou, J., and Zhang, Y. (2012) *Org. Lett.*, **14**, 3894–3897.
179. Chen, L., Li, C., Bi, X., Liu, H., and Qiao, R. (2012) *Adv. Synth. Catal.*, **354**, 1773–1779.
180. Li, Q., Huang, Y., Chen, T., Zhou, Y., Xu, Q., Yin, S.-F., and Han, L.-B. (2014) *Org. Lett.*, **16**, 3672–3675.
181. Zhang, G., Miao, J., Zhao, Y., and Ge, H. (2012) *Angew. Chem. Int. Ed.*, **51**, 8318–8321.
182. Wang, J., Wang, J., Lu, P., and Wang, Y. (2013) *J. Org. Chem.*, **78**, 8816–8820.
183. Lu, J., Jin, Y., Liu, H., Jiang, Y., and Fu, H. (2011) *Org. Lett.*, **13**, 3694–3697.
184. Deb, M.L., Dey, S.S., Bento, I., Barros, M.T., and Maycock, C.D. (2013) *Angew. Chem. Int. Ed.*, **52**, 9791–9795.
185. Xiong, T., Li, Y., Bi, X., Lv, Y., and Zhang, Q. (2011) *Angew. Chem. Int. Ed.*, **50**, 7140–7143.
186. Li, Y., Li, Z., Xiong, T., Zhang, Q., and Zhang, X. (2012) *Org. Lett.*, **14**, 3522–3525.
187. Modak, A., Dutta, U., Kancherla, R., Maity, S., Bhadra, M., Mobin, S.M., and Maiti, D. (2014) *Org. Lett.*, **16**, 2602–2605.
188. Hajra, S. and Sinha, D. (2011) *J. Org. Chem.*, **76**, 7334–7340.
189. Adrio, L.A. and Hii, K.K. (2008) *Chem. Commun.*, 2325–2327.
190. Gallardo-Donaire, J. and Martin, R. (2013) *J. Am. Chem. Soc.*, **135**, 9350–9353.
191. Li, Y., Lu, Y., Qiu, G., and Ding, Q. (2014) *Org. Lett.*, **16**, 4240–4243.
192. Yuan, H.-J., Wang, M., Liu, Y.-J., and Liu, Q. (2009) *Adv. Synth. Catal.*, **351**, 112–116.
193. Cui, S.-L., Lin, X.-F., and Wang, Y.-G. (2006) *Org. Lett.*, **8**, 4517–4520.
194. Qian, W., Amegadzie, A., Winternheimer, D., and Allen, J. (2013) *Org. Lett.*, **15**, 2986–2989.
195. Murugavel, G. and Punniyamurthy, T. (2013) *Org. Lett.*, **15**, 3828–3831.
196. Hu, J., Adogla, E.A., Ju, Y., Fan, D., and Wang, Q. (2012) *Chem. Commun.*, **48**, 11256–11258.
197. Kaneko, K., Yoshino, T., Matsunaga, S., and Kanai, M. (2013) *Org. Lett.*, **15**, 2502–2505.
198. Mitra, S., Chakraborty, A., Mishra, S., Majee, A., and Hajra, A. (2014) *Org. Lett.*, **16**, 5652–5655.

9
Fe- and Ag-Catalyzed Synthesis of Heterocycles

Jin-Heng Li and Ren-Jie Song

9.1
Introduction

Transition metal-catalyzed activation of C–H bonds has made significant progress, the applications of which have received growing efforts to the construction of various heterocycle units relevant to natural products, pharmaceuticals, and materials. However, the majority of these currently available C–H bond activation transformations require noble transition metal catalysts, typically including palladium, rhodium, or ruthenium. Therefore, the development of green, sustainable, and economical transition metal catalysts for the catalytic C–H activation is one of the major challenges.

In the past decade, inexpensive iron and silver have proven to be efficient catalysts for the synthesis of heterocycles through the C–H functionalization. Particularly, iron, as an economical and environmentally benign catalyst, has attracted much attention in the activation of unreactive C–H bonds. This chapter will cover recent advances in the iron- or silver-catalyzed C–H activation methods for the synthesis of heterocycles through the formation of new C–C, C–N, C–O, and C–P bonds. Moreover, progress in the iron- or silver-mediated synthesis of heterocycles by the C–H oxidative coupling strategy is also briefly introduced.

9.2
Iron-Catalyzed Synthesis of Heterocycles

Iron is an abundant, inexpensive, and environmentally benign metal in the first transition series with numerous important applications in a wide range of organic transformations and enzyme chemistry. In recent years, researchers have made impressive advances in the iron-catalyzed C–H bond activation [1]. These methods are successful through the formation of various C–C, C–N, and C–O bonds and have become prevalent and efficient tools to assemble valuable nitrogen- and oxygen-containing heterocycles.

Transition Metal-Catalyzed Heterocycle Synthesis via C—H Activation, First Edition. Edited by Xiao-Feng Wu.
© 2016 Wiley-VCH Verlag GmbH & Co. KGaA. Published 2016 by Wiley-VCH Verlag GmbH & Co. KGaA.

9.2.1
Iron-Catalyzed Synthesis of Nitrogen-Containing Heterocycles

The synthesis of 2-arylindoles via noble Rh-catalyzed intramolecular imine C(sp^2)–H amination reaction of 2-azideostilbenes has been reported by Driver and coworkers [2]. After that, they expanded the C–H amination strategy to synthesize benzimidazoles using iron(II) bromide as the catalyst, which proceeded via the intramolecular amination between an azide and an imine C(sp^2)–H bond (Scheme 9.1). The results showed that the mode of reactivity of the aryl azide relied on the composition of the linker between the aryl azide and the pendant C–H bond [3].

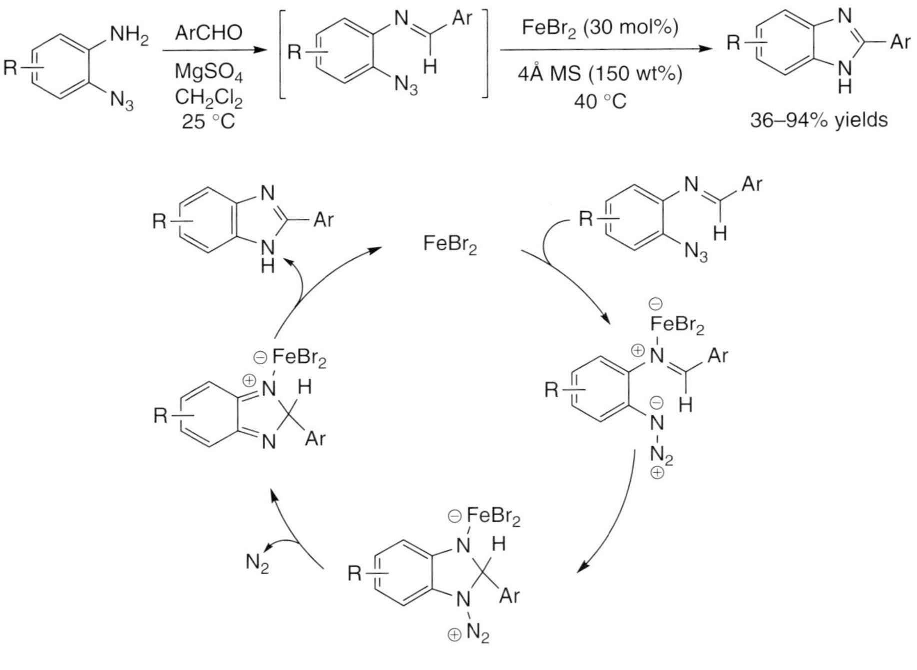

Scheme 9.1 Synthesis of Benzimidazoles from Aryl Azides.

Although the exact mechanism remains unclear, they have proposed a Lewis acid-catalyzed mechanism to elucidate the C–H amination process according to the reactivity trends of electron-deficient imines and electron-rich azides with higher conversion (Scheme 9.1).

An unexpected observation during the iron(II) bromide-catalyzed amination of alkyl C(sp^3)–H bonds with azides has been illustrated by Driver and coworkers (Scheme 9.2) [4]: sequential ethereal C(sp^3)–H bond amination and 1,2-shift reaction of *ortho*-substituted aryl azides occurred to afford 2,3-disubstituted indoles in the presence of iron(II) bromide. The results demonstrated that the preference for the 1,2-shift component of the tandem reaction was established to be Me < 1° <2° < Ph.

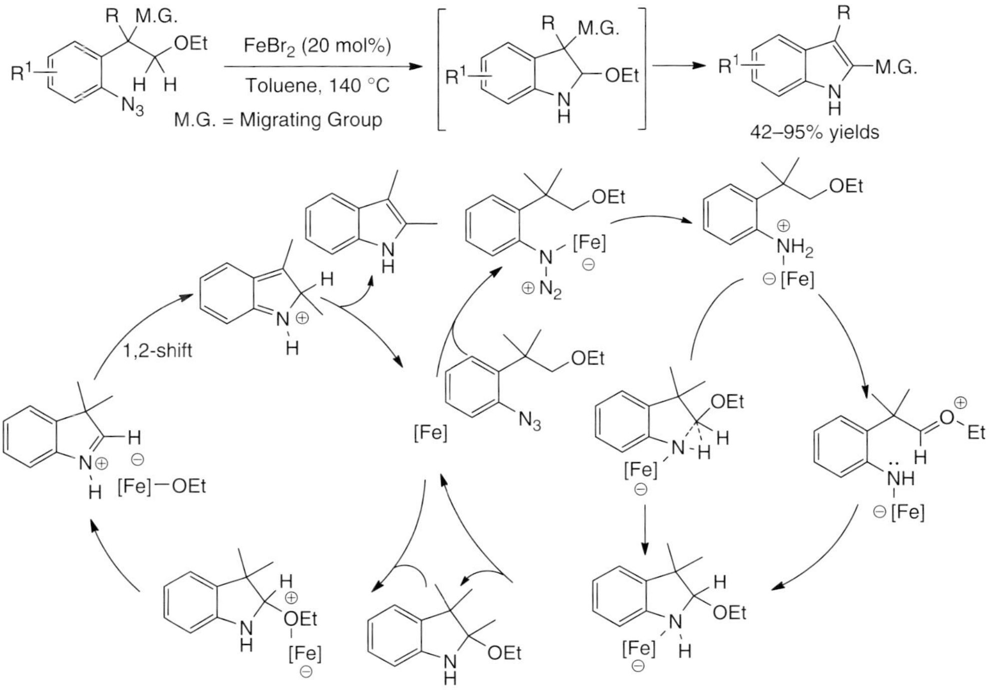

Scheme 9.2 Synthesis of indoles via Amination of Alkyl C(sp^3)–H Bonds with Azides.

Bolm and coworkers have developed a new route to the synthesis of indole derivatives by Fe(OTf)$_2$-catalyzed intramolecular C(sp^2)–H amination reaction of azidoacrylates (Scheme 9.3) [5]. This method proceeds through the amination of a vinyl azide with an aryl C(sp^2)–H bond, and the iron catalytic system is efficient

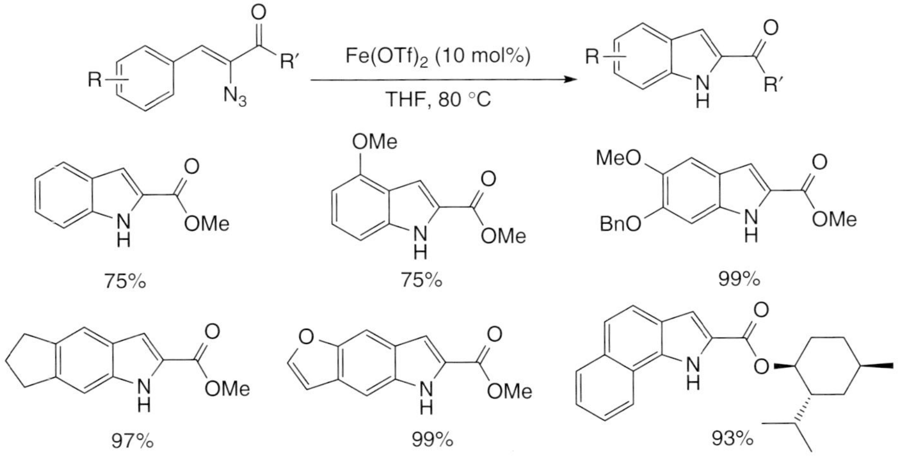

Scheme 9.3 Synthesis of Indoles via Intramolecular C–H Amination.

to tolerate a broad range of functional groups, such as MeO, Me, F, Cl, CF_3, and Ph groups. Moreover, substrates with *meta*- and *para*-disubstituted aryl groups afforded only single isomers in excellent yields.

The transition metal-catalyzed direct C–H aminations with the free N–H bonds are also powerful tools for heterocycle synthesis. For example, Bao and Zhang have established a new $FeBr_3/O_2$ catalytic system for the synthesis of substituted 1,3-diaryl-substituted indazoles and trisubstituted pyrazoles (Scheme 9.4) [6]. This transformation proceeded via the C(sp^2)–H activation/C–N bond formation process with excellent functional group toleration: various substituted phenylhydrazones were converted into the corresponding N-containing heterocycles in moderate to excellent yields, and substrates containing electron-donating groups were more active than substrates containing electron-withdrawing groups. Furthermore, the reaction time was shortened when using substrates derived from α,β-unsaturated ketones and phenylhydrazine.

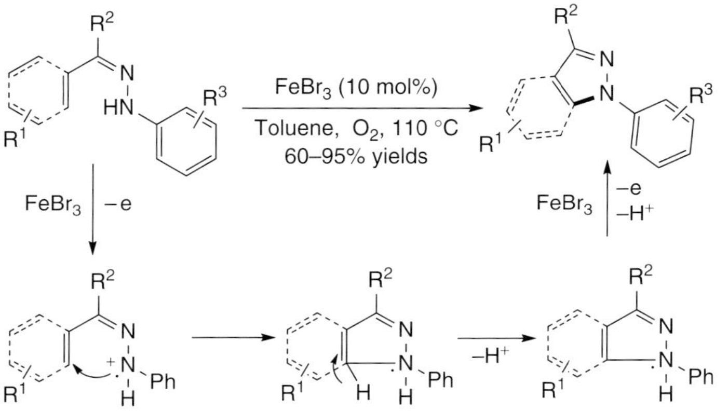

Scheme 9.4 Synthesis of 1*H*-Indazoles and 1*H*-Pyrazoles via Intramolecular C(sp^2)–H Amination.

Fullerene derivatives occupy a privileged position in chemistry as they have many potential applications in materials and biomolecules. Recently, Wang and You reported a $FeCl_3$-catalyzed synthesis of oxazolidino[4,5:1,2] [60]fullerenes via the intermolecular tandem reaction of [60]fullerene with *tert*-butyl *N*-substituted carbamates to access through the direct amination between the C(sp^2)–H bonds and the N–H bonds (Scheme 9.5) [7].

Imidazopyridines are important nitrogen-containing heterocycles that show a wide range of bioactivities. Hajra and coworkers have developed an iron-catalyzed oxidative diamination of nitroalkene with 2-aminopyridine to access 2-nitro-3-arylimidazo[1,2-*a*]pyridines under mild and aerobic reaction conditions (Scheme 9.6) [8]. In particular, the method allows the formation of two new C–N bonds through sequential Michael addition and vinyl C(sp^2)–H/N–H functionalizations and represents the first example of employing molecular oxygen as the terminal oxidant for the selective C–H amination.

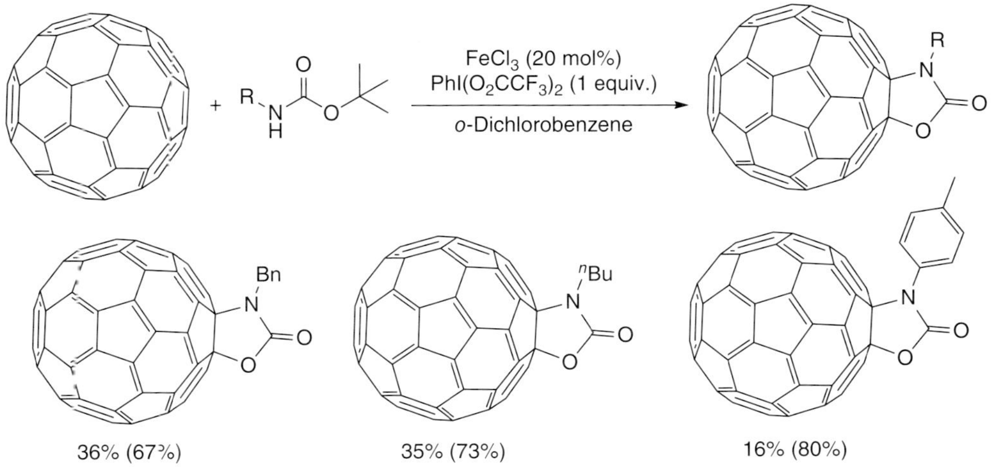

Scheme 9.5 Synthesis of Oxazolidino[4,5:1,2][60]fullerenes.

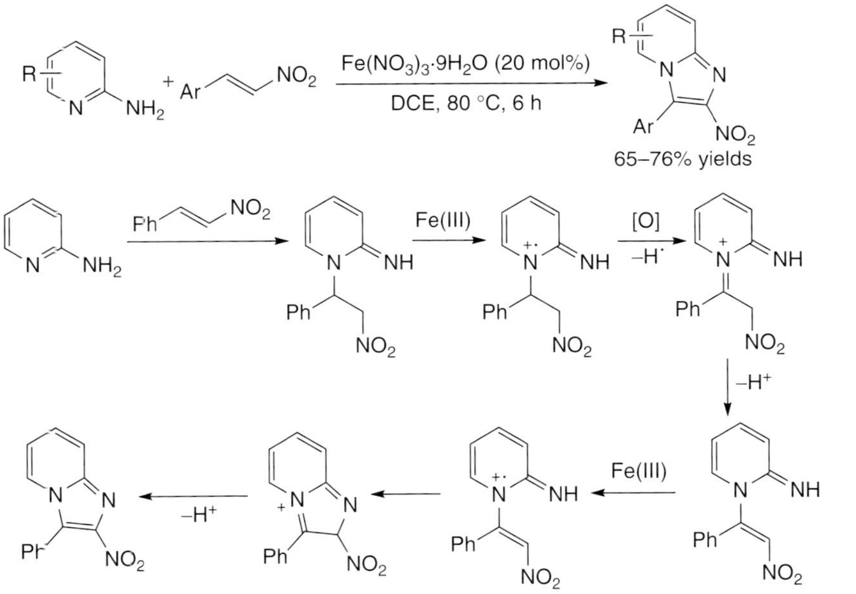

Scheme 9.6 Synthesis of Imidazopyridines by Oxidative Diamination of Nitroalkenes with 2-Aminopyridines

The mechanism for the iron-catalyzed oxidative diamination of nitroalkene with 2-aminopyridine was proposed (Scheme 9.6). The authors suggest that the first step proceeds via the Michael addition, followed by intramolecular cyclization and deprotonation that afford the desired imidazopyridines.

Functionalization of an unactivated C(sp^3)–H bond is one of the biggest challenges for chemists and continues to attract the attention of the synthetic

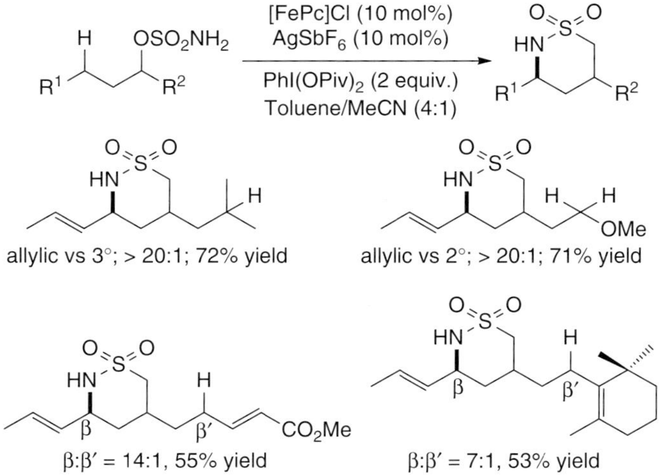

Scheme 9.7 Synthesis of 1,2-Thiazinane 1,1-Dioxides via Intramolecular Allylic C(sp^3)–H Amination.

community. In 2012, White and Paradine reported a highly selective allylic C(sp^3)–H amination reaction under inexpensive, nontoxic iron catalysis (Scheme 9.7) [9]. Importantly, in polyolefinic substrates, the site selectivity could be controlled by the electronic and steric character of the allylic C–H bond. Within this process, allylic C–H amination is strongly preferred over aziridination as well as over amination of stronger C–H bonds.

Meanwhile, an iron-promoted tandem reaction of anilines with styrene oxides via C–C cleavage and C–H activation for the synthesis of 3-arylquinolines has been developed (Scheme 9.8) [10]. In the presence of $FeCl_3$, a variety of

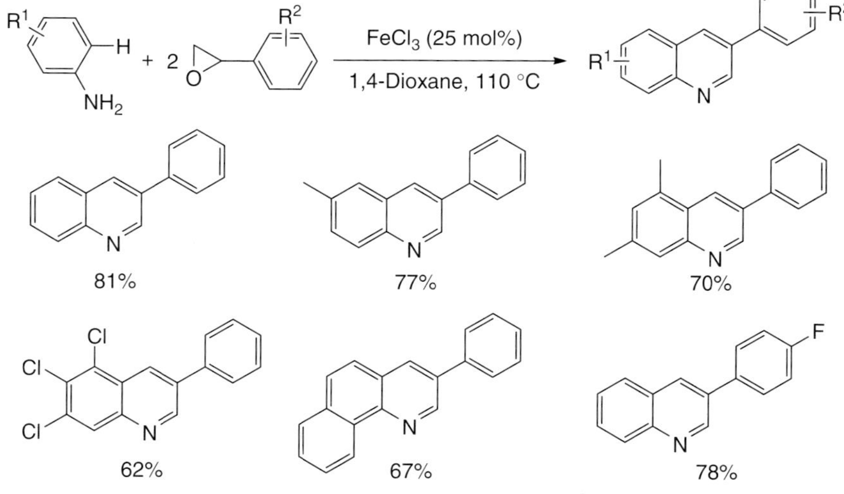

Scheme 9.8 Synthesis of Quinolines by Tandem Reaction of Anilines with 2-Aryloxiranes.

3-arylquinolines were obtained in 50–82% yields. It was noted that this is an inexpensive and convenient approach to 3-arylquinolines from simple and readily available starting materials and reagents.

Very recently, Gu, Li, and coworkers described an iron-catalyzed tandem oxidative process for the synthesis of benzoxazoles from easily available substrates, in which an alkyl C(sp^3)–H bond adjacent to an oxygen atom was directly aminated with a N–H bond (Scheme 9.9) [11]. In addition, excellent functional group tolerance under relatively mild conditions and great regioselectivity were exhibited.

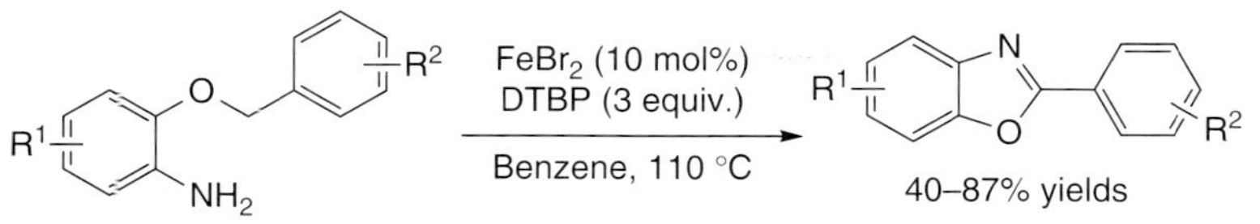

Scheme 9.9 Synthesis of Benzoxazoles by Tandem C(sp^3)–H Oxidative Amination.

A straightforward iron-catalyzed divergent oxidative tandem synthesis of dihydroquinazolines and quinolines from N-alkylanilines using a TEMPO oxoammonium salt as a mild and nontoxic oxidant has been described by García Mancheño and coworkers (Scheme 9.10) [12]. In this reaction, Fe(OTf)$_2$ was the Lewis acid

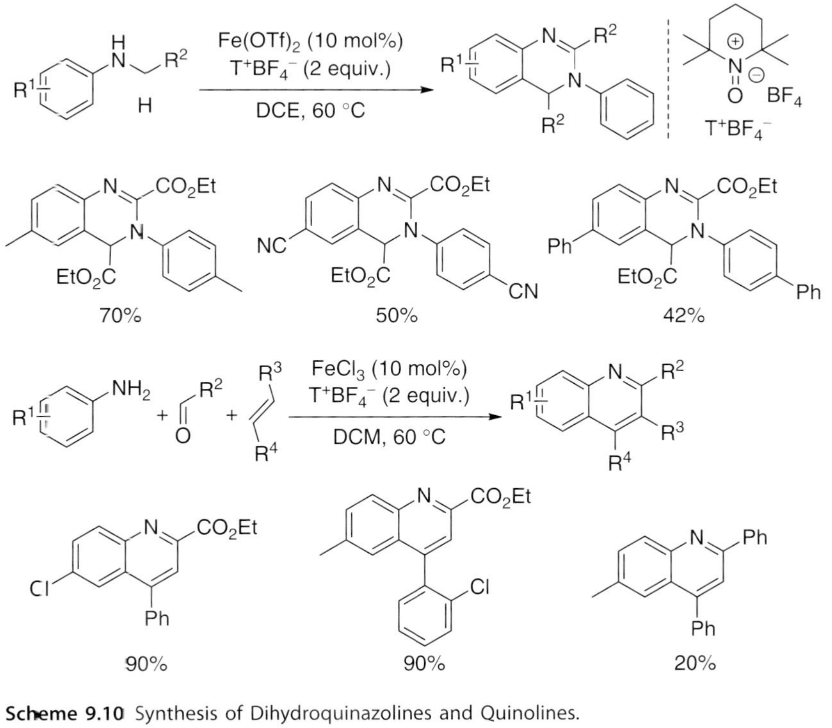

Scheme 9.10 Synthesis of Dihydroquinazolines and Quinolines.

catalyst of choice for the formation of dihydroquinazolines, whereas $FeCl_3$ led to better results for the synthesis of quinolines. This transformation implies that, for both syntheses, direct oxidative functionalization of a α-C(sp^3)–H bond of the *N*-alkylanilines occurs, leading to C–N bond formation or C–C bond formation upon homocondensation or reaction with simple olefins, respectively. Cyclization followed by a final oxidation generates these classes of interesting bioactive heterocycles in one synthetic transformation. In addition, the one-pot multicomponent synthesis of quinolines from anilines, aldehydes, and olefins has also been successfully developed under these mild oxidative conditions.

In 2010, Zhu and coworkers first reported the direct intramolecular C(sp^2)–H amination of *N*-arylpyridine-2-amines using a combination of $Cu(OAc)_2$ and $Fe(NO_3)_3$ as a bimetallic catalytic system to furnish pyrido[1,2-*a*]benzimidazoles in satisfactory yields [13]. The authors believe that a Cu(III)-catalyzed electrophilic aromatic substitution (SEAr) pathway is operating in this process according to the results of mechanistic studies, wherein iron(III) acts a unique role to facilitate the formation of the more electrophilic Cu(III) species because in the absence of iron(III), a much less efficient and reversible Cu(II)-mediated SEAr process takes place.

Encouraged by these results, Maes and coworkers have extended this similar strategy to the direct intramolecular C(sp^2)–H amination of 5-(pyridine-2-ylamino)pyrimidine-2,4-(1*H*,3*H*)-diones for the synthesis of pyrido[1,2-*e*]purines only using $FeCl_2$ as the catalyst combined with O_2 as the oxidant (Scheme 9.11) [14]. Interestingly, iron has proved to be superior to copper catalysis in this C(sp^2)–H amination process. The method has excellent functional group compatibility, and the chemoselectivity toward halogens will allow postfunctionalization in the annulated ring.

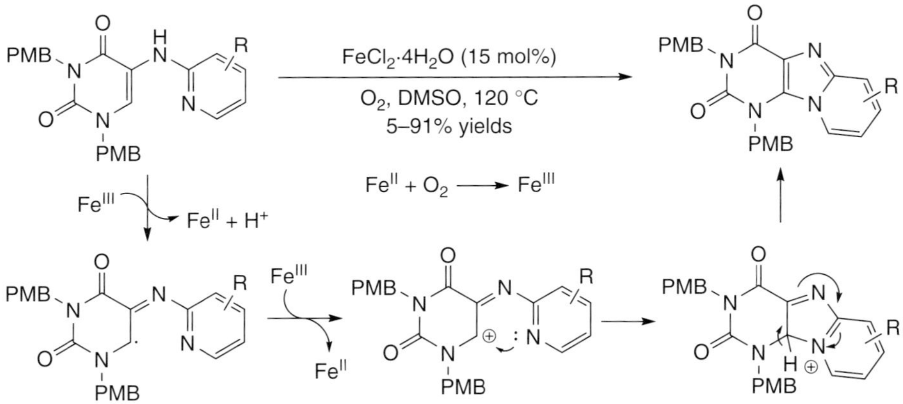

Scheme 9.11 Synthesis of Pyrido [1,2-*e*]purines via Direct Intramolecular C(sp^2)–H Amination.

The iron-catalyzed direct intramolecular C(sp^2)–H amination of 2-benzhydrylpyridines for the synthesis of pyrido[1,2-*a*]indoles has also been developed by Sekar and Karthikeyan (Scheme 9.12) [15]. In the presence of $FeCl_2 \cdot 4H_2O$, $FeCl_2 \cdot 6H_2O$, and PivOH (10 equiv.), 2-benzhydrylpyridine derivatives underwent the intramolecular C–H amination leading to the corresponding pyrido[1,2-*a*]indoles in 58–77% yields under nitrogen atmosphere. However, only triarylmethanols were obtained when the reaction underwent under oxygen atmosphere.

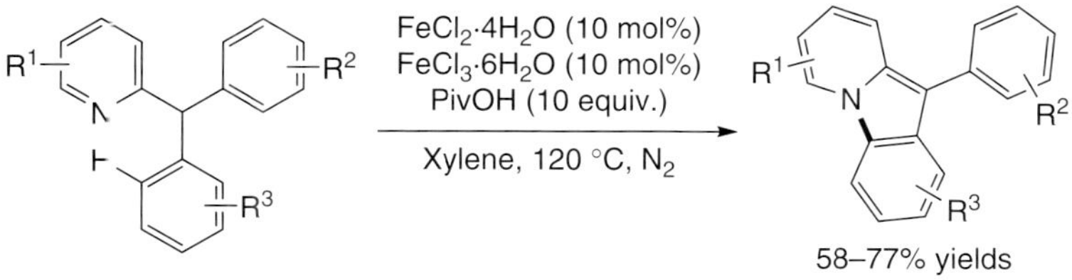

Scheme 9.12 Synthesis of Pyrido[1,2-*a*]indoles via Intramolecular C(sp^2)–H Amination of 2-Benzhydrylpyridines.

Over the past few years, transition metal-catalyzed C–H functionalization methods for the formation of the C–C bonds have been well explored, many of which have included a handful of impressive strategies and synthetically valuable heterocycles when the starting materials contain heteroatoms. Among them, iron salts showed particular fascination for such purposes. In 2010, Liang and coworkers reported a new iron-catalyzed dual C–H activation between aryl C(sp^2)–H and vinyl C(sp^2)–H bond activation to build substituted indoles in the presence of $Cu(OAc)_2 \cdot CuCl_2$ (Scheme 9.13) [16]. This reaction showed highly functional group tolerance and exhibited different properties from the palladium-catalyzed reactions of this type.

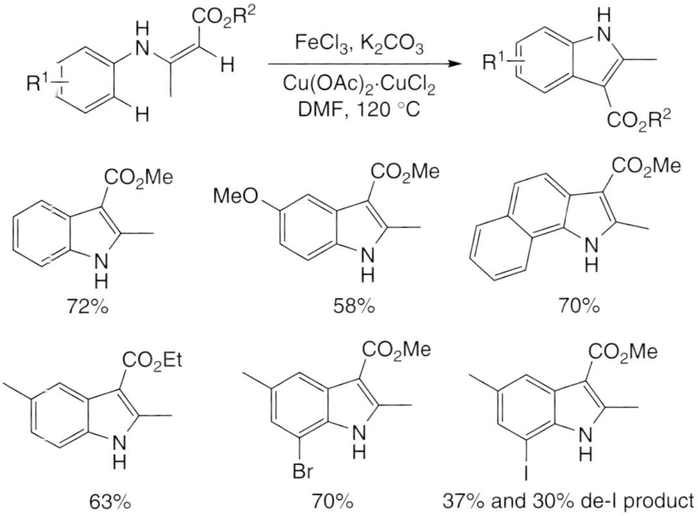

Scheme 9.13 Direct C(sp^2)–H Oxidative Coupling to Build Indoles.

Li and coworkers chose ethers, compounds containing the C(sp^3)–H bond adjacent to the oxygen atom, to achieve the 1,2-difunctionalization of alkenes in the presence of peroxides (Scheme 9.14) [17]. In the presence of $FeCl_3$, *tert*-butyl hydroperoxide (TBHP), and DBU, a variety of activated alkenes were successfully reacted with an aryl C(sp^2)–H bond and a C(sp^3)–H bond adjacent to a heteroatom, giving functionalized 3-(2-oxoethyl)indolin-2-ones in moderate to high yields. This oxidative 1,2-alkylarylation reaction is catalyzed by economical and environmentally benign iron and represents the first example of metal-catalyzed 1,2-difunctionalization of an alkene with an aryl C(sp^2)–H bond and a C(sp^3)–H bond adjacent to a heteroatom for the simultaneous formation of two new C–C bonds. Interestingly, this C–H oxidative coupling protocol was applicable to thioether and tertiary alkylamine. Although an iron-catalyzed radical-mediated mechanism was proposed for the C–H oxidative coupling protocol, the real role of iron catalyst needs to be further studied.

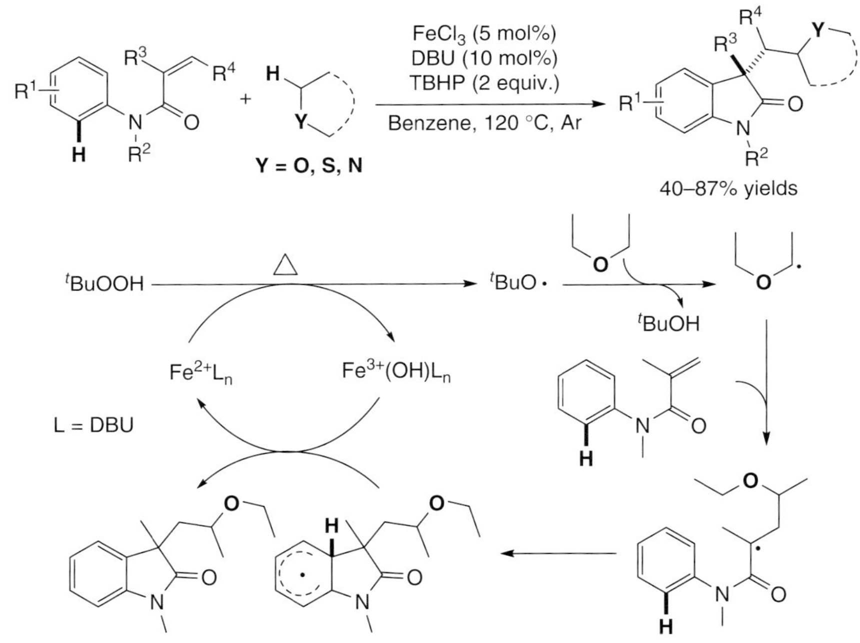

Scheme 9.14 Synthesis of 3-(2-Oxoethyl)indolin-2-ones by 1,2-Difunctionalization of Alkenes.

Very recently, Loh and Lu found that the C(sp^3)–H bonds adjacent to the chloride atom were highly reactive to react with activated alkenes recently (Scheme 9.15) [18]. In the presence of $FeCl_2$ and PhI_2OTf, the cascade carbochloromethylation of activated alkenes with the C(sp^3)–H bonds adjacent to the chloride atom and aryl C(sp^2)–H bonds was successfully performed to access chloro-containing oxindoles.

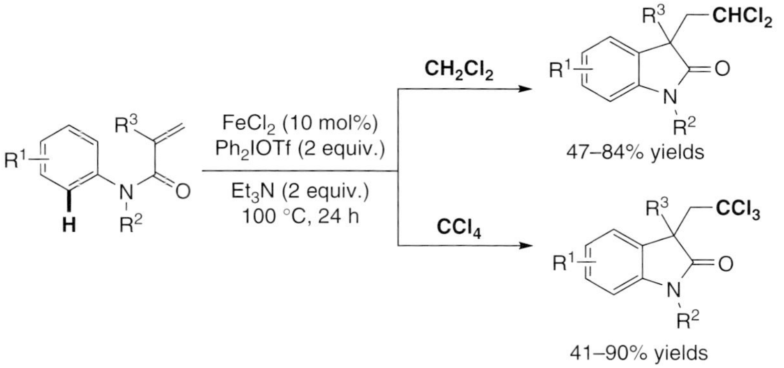

Scheme 9.15 Synthesis of Chloro-Containing Oxindoles via 1,2-Carbochloromethylation of Alkenes.

The Li group [19a] and Cheng group [19b] have independently reported that peroxides as alkyl resource underwent the 1,2-alkylarylation with activated alkenes (Scheme 9.16). The method allows the simultaneous formation of two C–C bonds for the assembly of quaternary oxindoles.

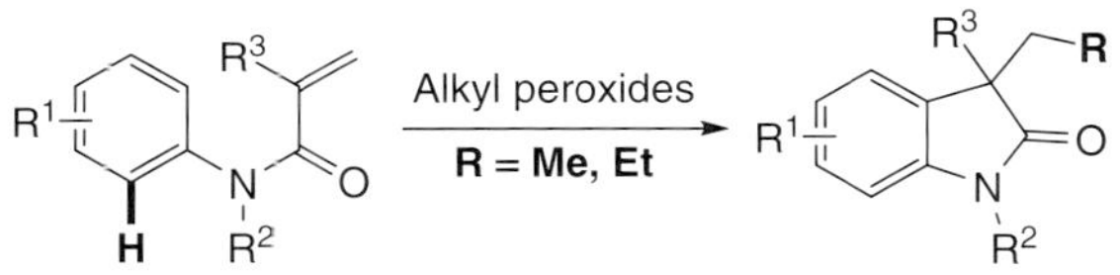

Li's conditions:
$Fe(OAc)_2$, DABCO, DMSO, 120 °C 46–72% yields
Cheng's conditions:
$FeCl_2$, benzene, 135 °C 49–84% yields

Scheme 9.16 1,2-Alkylarylation with Activated Alkenes Using Peroxides as Alkyl Resource.

The C–H oxidative coupling strategy was expanded to the 1,2-carbonylarylation of *N*-arylacrylamides with alcohols for the assembly of carbonyl-containing oxindoles using the $Fe(OAc)_2$ and TBHP catalytic system (Scheme 9.17) [20]. Notably, the desired oxindole could be constructed in 28% yield even without Fe catalyst.

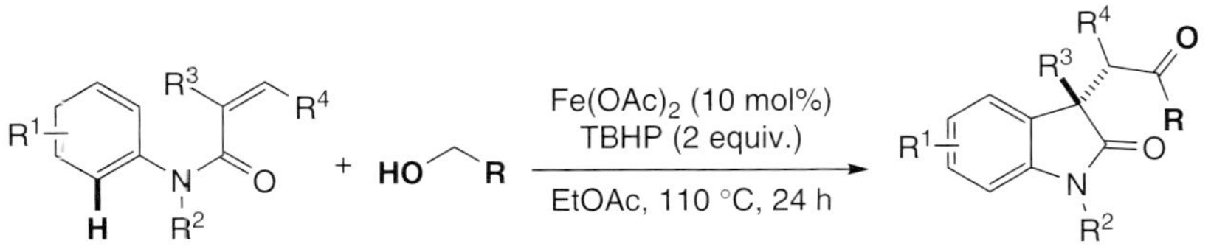

Scheme 9.17 1,2-Carbonylarylation of *N*-Arylacrylamides with Alcohols.

The carbonyl radicals were found to be readily formed by the oxidative cleavage of the C–N bonds of carbazates. Very recently, carbazates as easily accessible and safe alkoxycarbonyl radical precursors for the 1,2-alkoxycarbonylarylation of *N*-aryl acrylamides with aryl C(sp^2)–H bonds were developed (Scheme 9.18) [21]. The method provided a general way for the construction of alkoxycarbonylated oxindoles. Moreover, the use of economical and environmentally benign iron as the catalyst makes this transformation more sustainable and practical.

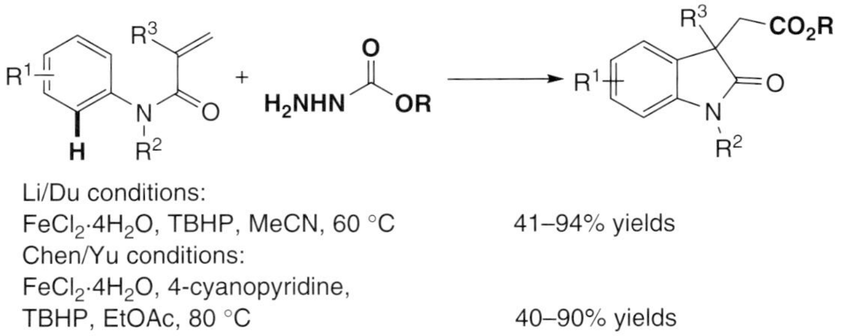

Scheme 9.18 1,2-Alkoxycarbonylarylation of *N*-aryl acrylamides for the Construction of Alkoxycarbonylated Oxindoles.

The iron-catalyzed aryldifluoromethylation of activated alkenes with $PhSO_2CF_2I$ and aryl C(sp^2)–H bonds to difluoromethylated oxindoles has also been established by Wang and coworkers (Scheme 9.19) [22]. This method provides a useful route to fluorine-containing oxindoles.

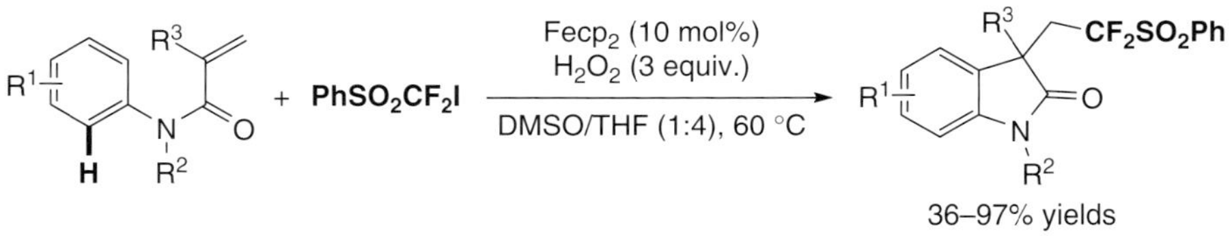

Scheme 9.19 1,2-Aryldifluoromethylation of Activated Alkenes with $PhSO_2CF_2I$.

Readily accessible 2-isocyanobiphenyls are important synthetic blocks in organic synthesis. Studer and coworkers have disclosed a novel method for the synthesis of 6-aroylated phenanthridines starting with readily prepared 2-isocyanobiphenyls and aryl aldehydes using small amounts of $FeCl_3$ (0.4 mol%) and the commercially available and cheap ^{t}BuOOH oxidant (Scheme 9.20) [23]. The reaction achieves the addition of aryl aldehyde to the isonitrile functionality and subsequent intramolecular C–H cyclization. The iron-catalyzed strategy using 2-isocyanobiphenyls as the starting materials was employed to assemble 6-carboxylated phenanthridines (Scheme 9.20) [24].

The cycloaddition reaction has become one of the most important methodologies in synthesis to build diverse carbo- and heterocyclic ring systems. Among them transition metal-catalyzed aromatic C(sp^2)–H functionalization strategy, in

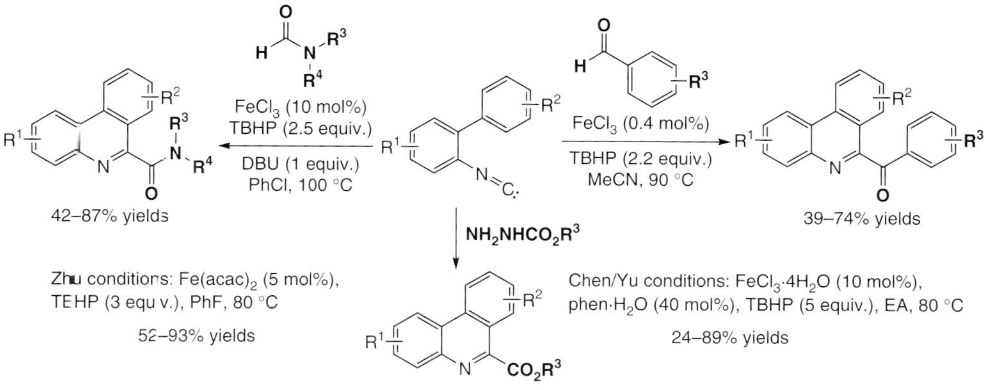

Scheme 9.20 Synthesized of 6-Carboxylated Phenanthridines.

combination with its subsequent annulation with unsaturated compounds, has become particularly attractive because it avoids the prior preparation of activated substrates with reducing chemical waste generated. However, the majority of these transformations are restricted to the requirement of noble transition metal catalytic systems, examples of which using inexpensive iron catalysts are scarce [25]. In 2012, Liu and coworkers illustrated a [4+2] annulation method for the synthesis of quinolines by iron-catalyzed C–H functionalization/oxidative coupling tandem reaction of *N*-alkyl anilines with alkynes or alkenes through dual C–H functionalization (Scheme 9.21) [26].

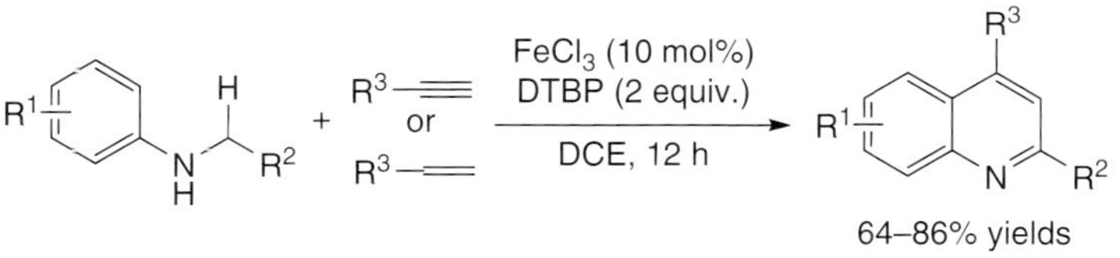

Scheme 9.21 Synthesis of Quinolines by C–H functionalization/oxidative coupling tandem reaction.

The iron-catalyzed [4+2] annulation of *ortho*-nitroanilines with phenethylamines was also developed by Nguyen and coworkers (Scheme 9.22), in which

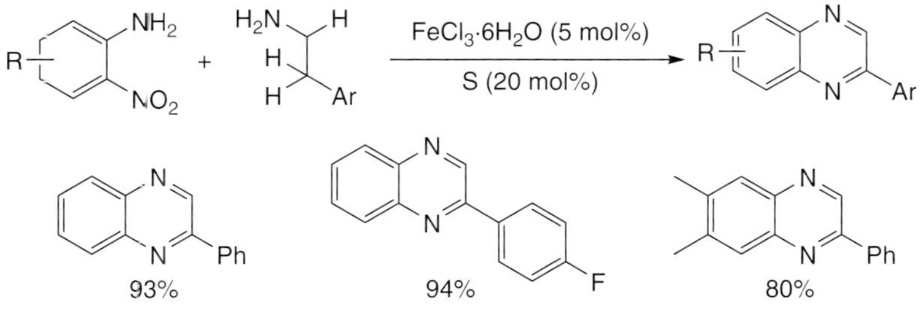

Scheme 9.22 [4+2] Annulation of *ortho*-Nitroanilines with Phenethylamines.

the redox condensation and C–H functionalization cascades occurred utilizing $FeCl_3{\cdot}6H_2O$ (5 mol%) and elemental sulfur (20 mol%) as the catalyst partner [27].

9.2.2
Iron-Catalyzed Synthesis of Oxygen-Containing Heterocycles

Oxygen-containing heterocycles represent a ubiquitous structural motif in numerous natural products, pharmaceuticals, and organic materials. Therefore, considerable efforts have been devoted to the development of efficient methods for oxygen-containing heterocycle synthesis. One typical class of these transformations for such purposes is the annulation strategy involving the C–H oxidative coupling.

As shown in Scheme 9.23, Li and coworkers demonstrated that the combination of $FeCl_3{\cdot}6H_2O$ and *di-tert*-butyl peroxide offers a novel and efficient method for the construction of polysubstituted benzofurans from the annulation reaction of simple phenols and β-ketoesters, which are expected to give coumarins in the well-known Pechmann condensation [28]. A variety of phenols reacted with β-ketoesters to provide a range of benzofuran products in moderate to excellent yields.

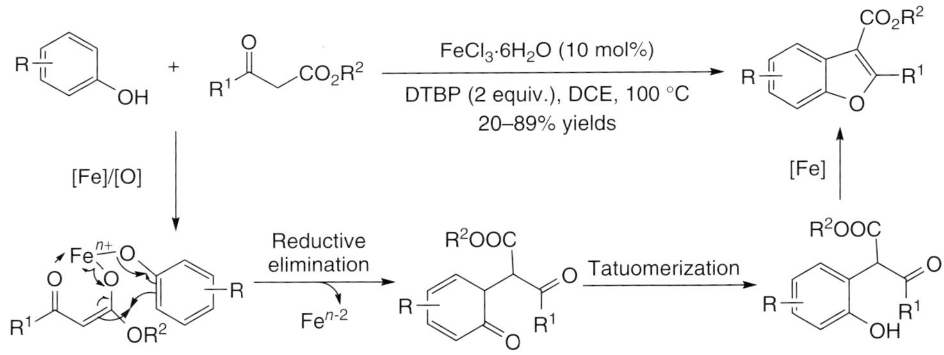

Scheme 9.23 Construction of Polysubstituted Benzofurans through Annulation of Simple Phenols and β-Keto Esters.

In 2012, Pappo and coworkers also reported a chemo-, regio-, and stereoselective $FeCl_3$/1,10-phenanthroline-catalyzed cross-dehydrogenative-coupling reaction between phenols and α-substituted β-ketoesters (Scheme 9.24) [29]. The reaction creates a new quaternary carbon center within a polycyclic hemiacetal or polycyclic spirolactone architecture.

The iron-catalyzed oxidative cross-coupling of phenols with alkenes was developed for the preparation of the pharmacologically important 2,3-dihydrobenzofurans (Scheme 9.25) [30]. The reaction was applicable to a variety of alkenes including styrene, α-alkyl- and α-arylstyrenes, *β*-alkyl styrenes, and stilbenes.

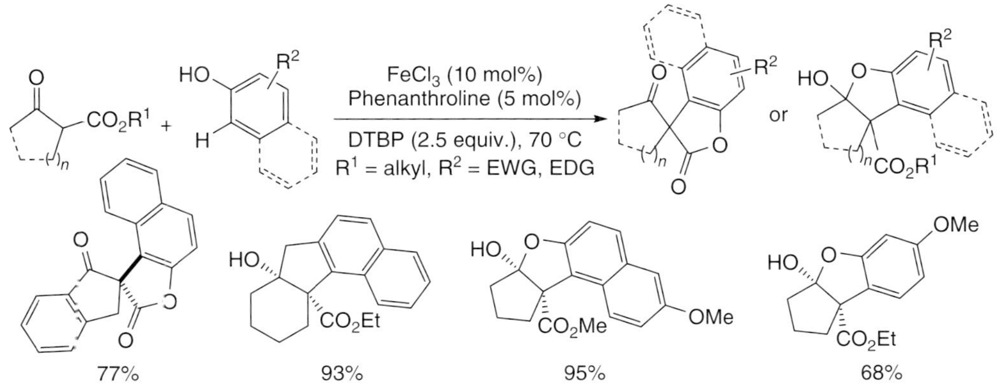

Scheme 9.24 Cross-Dehydrogenative-Coupling between Phenols and α-Substituted β-Ketoesters.

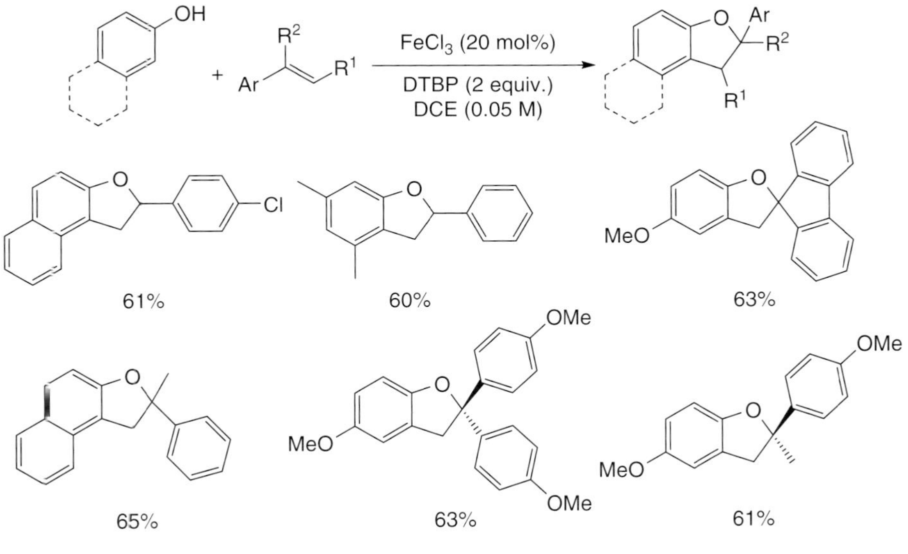

Scheme 9.25 Synthesis of 2,3-Dihydrobenzofurans by Oxidative Cross-Coupling of Phenols with Alkenes.

Generally, the formation of homocoupling by-products from simple phenols during the oxidative coupling process takes place. To overcome these disadvantages, Lei and coworkers employed the $FeCl_3$/DDQ catalytic system to realize the oxidative cross-coupling/cyclization of simple phenols with alkenes at room temperature (Scheme 9.26) [31]. Mechanistic investigations revealed that the reaction proceeds through a radical pathway, and the high selectivity is due to the coordination of the iron Lewis acid.

The microwave irradiation tool has been widely utilized to promote organic transformations. An iron-catalyzed, microwave-promoted method reported by Li and coworkers for the synthesis of 9-substituted xanthenes has been developed

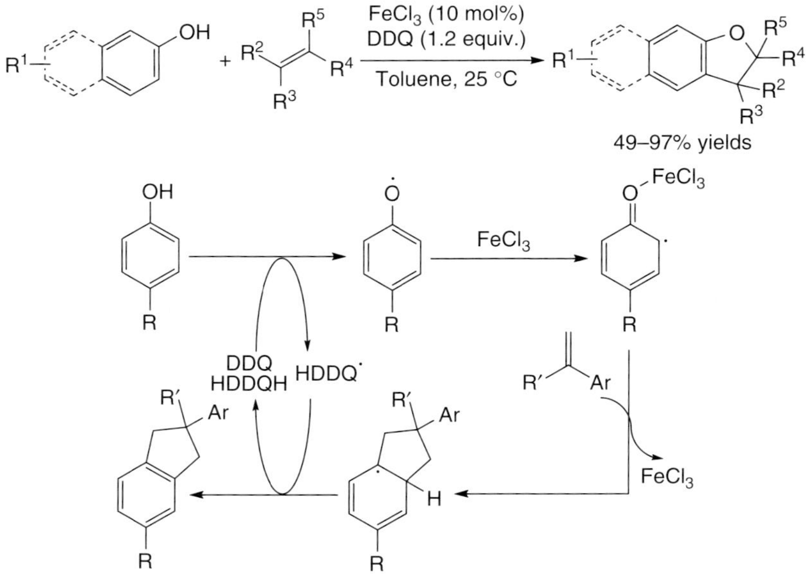

Scheme 9.26 Oxidative Cross-Coupling/Cyclization of Simple Phenols with Alkenes.

(Scheme 9.27) [32]. The method proceeds via a cascade benzylation–cyclization process, in which benzyl acetates, benzyl bromides, and benzyl carbonates are suitable benzylating reagents.

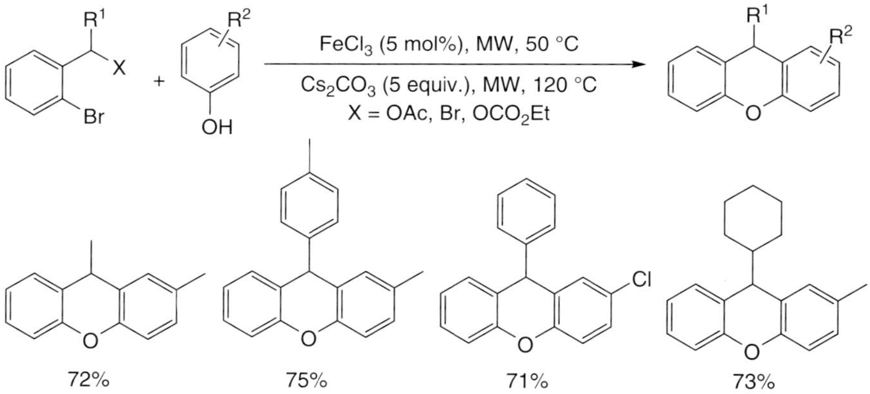

Scheme 9.27 Microwave-Promoted Synthesis of 9-Substituted Xanthenes.

Recently, Studer and coworkers reported an iron-catalyzed intramolecular cyclization between an aldehyde C(sp^2)–H bond and an aryl C(sp^2)–H bond (Scheme 9.28) [33]. In the presence of $FeCp_2$ and TBHP, a variety of xanthones were prepared from *ortho*-formyl biphenylethers in moderate to good yields.

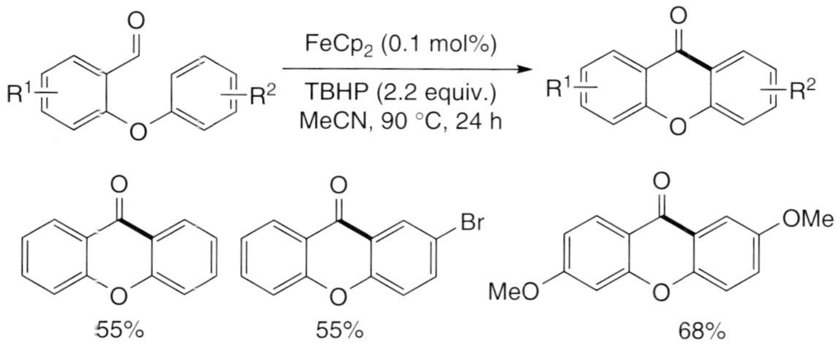

Scheme 9.28 Intramolecular Cyclization for the Construction of Xanthones.

A directing effect on metal (oxo)-promoted C(sp^3)–H hydroxylations for selectively synthesizing dihydrofuran-2(3*H*)-ones has been developed (Scheme 9.29) [34]. A range of electronic, steric, and stereoelectronic carboxylic acid substrates could achieve the C–H oxidation reaction using the nonheme iron complex Fe(PDP) and H_2O_2 to enable installation of C-2 directing oxidation in the naturally occurring oxidation state and stereoconfiguration, thus making the method useful in the natural product synthesis.

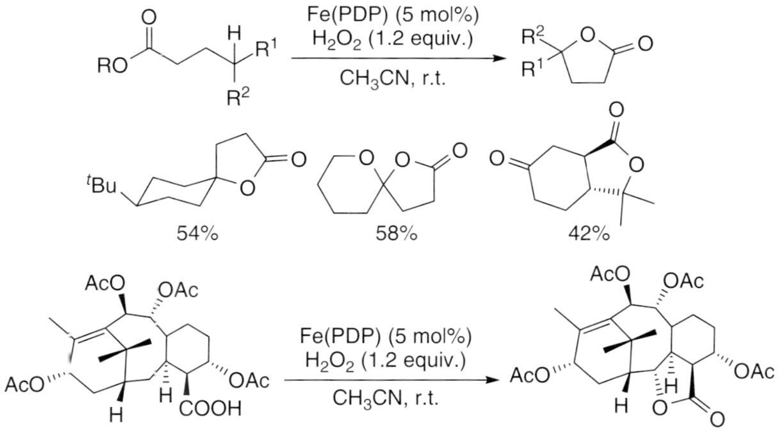

Scheme 9.29 Synthesis of Dihydrofuran-2(3*H*)-ones through C(sp^3)–H Hydyoxylations.

9.3 Silver-Catalyzed Synthesis of Heterocycles

In contrast with other transition metals, silver salts have long been believed to have low catalytic efficiency. Thus, they are most commonly used as cocatalysts or Lewis acids to promote the organic transformations. Recent reports have showed that silver salts exhibited interesting activity, thus inspiring considerable effort to develop new expedient silver-catalyzed approaches. Undoubtedly, silver-catalyzed reactions have emerged as important synthetic methods for the synthesis of various heterocyclic compounds because the use of silver salts can offer the economic and mild reaction conditions in comparison to the analogous noble transition metal-catalyzed processes [35].

9.3.1
Silver-Catalyzed Synthesis of Nitrogen-Containing Heterocycles

The use of silver salts as catalysts for the direct C–H amination has been rarely reported. In 2004, the He group first reported a silver-catalyzed intramolecular cyclization of alkyl $C(sp^3)$–H bonds with the free N–H bonds, which is stereospecific and practical for constructing amine-containing organic molecules (Scheme 9.30) [36]. Within this C–H amination reaction, $[Ag(^tBu_3tpy)_2NO_3]^+$ intermediate was detected by electrospray mass spectrometry.

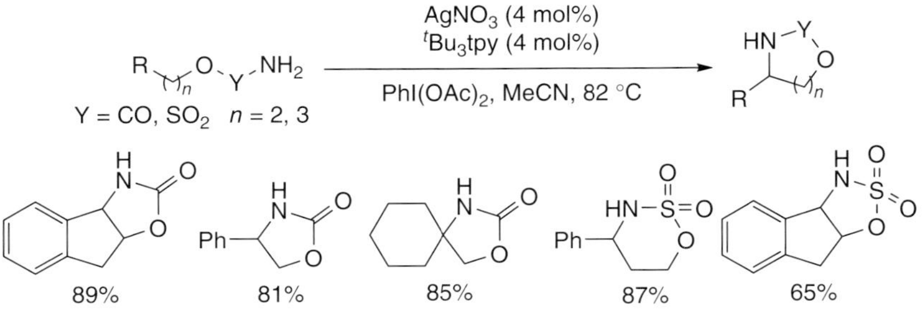

Scheme 9.30 Intramolecular Cyclization of Alkyl $C(sp^3)$–H Bonds with the Free N–H Bonds.

Subsequently, Schomaker and coworkers have developed the first example of ligand-controlled and tunable silver-catalyzed $C(sp^3)$–H amination method (Scheme 9.31) [37]. Simple silver catalysts supported by common nitrogenated ligands were used to tune a nitrene transfer reaction between two different types of C–H bonds: while silver catalysts supported by ^{t}Bubipy appear to prefer amination of the most electron-rich C–H bond, silver supported by a tpa ligand is more sensitive to the steric environment around the C–H bond, as well as the bond dissociation energy.

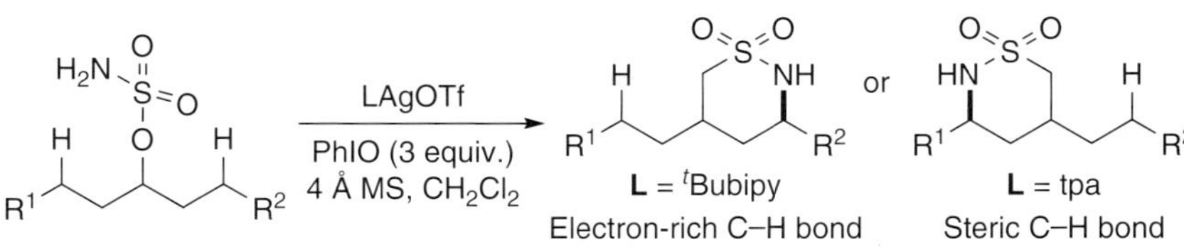

Scheme 9.31 $C(sp^3)$-H Amination between Two Different Types of C–H Bonds.

In the past decades, transition metal-catalyzed C–H oxidative coupling reactions have provided one of the most straightforward and atom-economical strategies for the synthesis of heterocyclic compounds through the formation of the new C–C bonds, and the powerful transition metal catalysts included silver catalysts. In 2013, Yang and coworkers reported a silver-catalyzed arylphosphorylation of *N*-arylacrylamides with $Ph_2P(O)H$ and aryl $C(sp^2)$–H bonds for the synthesis of phosphorylated oxindoles (Scheme 9.32) [38]. In the presence of $AgNO_3$ and $Mg(NO_3)_2{\cdot}6H_2O$, a range of diverse phosphorylated oxindoles were synthesized in good to excellent yields.

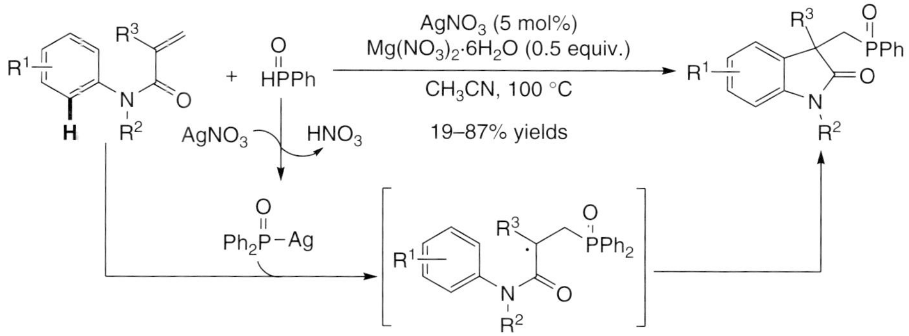

Scheme 9.32 Arylphosphorylation of *N*-Arylacrylamides with $Ph_2P(O)H$ for the Synthesis of Phosphorylated Oxindoles.

Mechanistically, they proposed that Ag^+ first reacted with $Ph_2P(O)H$ to form the crucial active intermediate, which promotes the reaction. Worthy of note that this is the first time for the hydrophosphorylation of alkenes via silver salt to be catalyzed.

Prompted by the success of silver-catalyzed carbophosphorylation of activated alkenes, the Yang group [39a] and Jiao group [39b] have independently documented a new C–H oxidative coupling strategy for the formation of the azidyl radical by carboazidation of arylacrylamides with $TMSN_3$ using $AgNO_3$ catalyst and oxidizing reagents (e.g., $Zr(NO_3)_4$ or $Ce(SO_4)_2$) $Zr(NO_3)_4{\cdot}5H_2O$ (Scheme 9.33). Through their methods, various N_3-containing oxindoles were prepared smoothly in moderate to good yields.

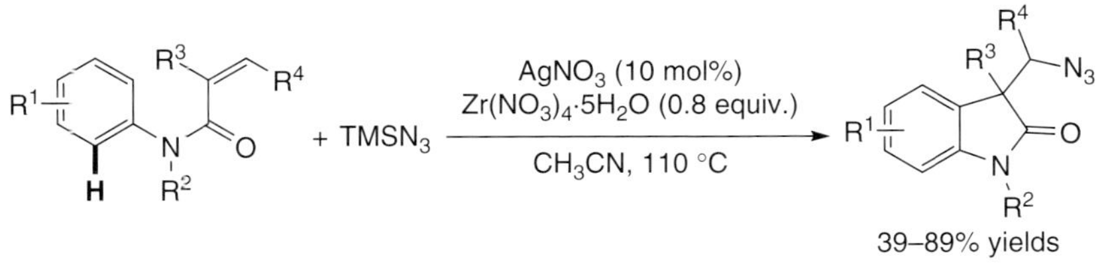

Scheme 9.33 1,2-Carboazidation of Arylacrylamides with $TMSN_3$.

In 2013, Guo/Duan group employed a new $AgNO_3/K_2S_2O_8$ oxidative system to achieve the 1,2-alkylarylation of activated alkenes with α-$C(sp^3)$–H bonds of 1,3-dicarbonyl compounds, in which two new C–C bonds were formed (Scheme 9.34)

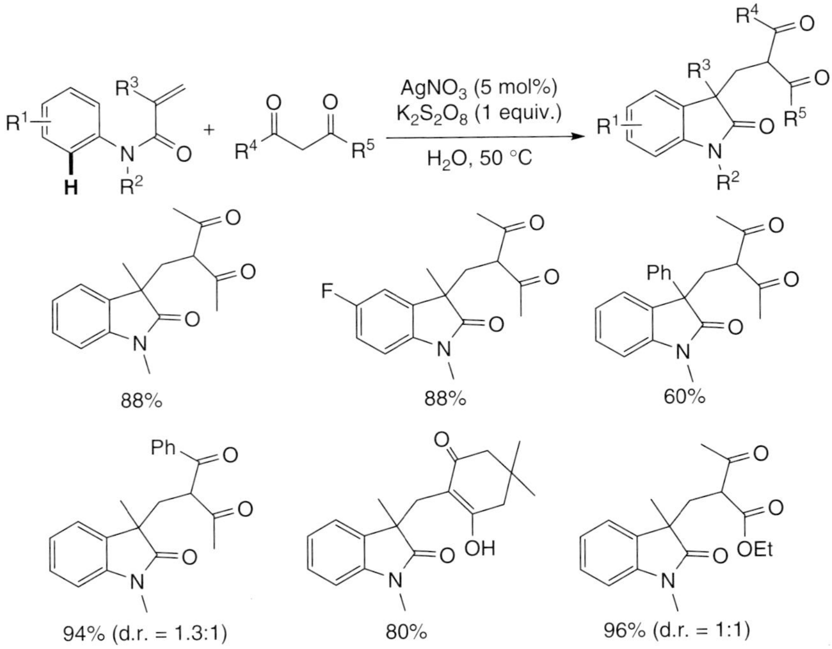

Scheme 9.34 1,2-Alkylarylation of Activated Alkenes with α-C(sp^3)–H Bonds of 1,3-Dicarbonyl compounds.

[40]. Gratifyingly, the same conditions could be extended to the reaction of simple ketones. They proposed that the Ag-catalyzed reaction proceeded via a sequential intermolecular radical addition and intramolecular radical substitution.

The $AgNO_3/K_2S_2O_8$ oxidative system was applicable to acylarylation of activated alkenes with easily available α-oxocarboxylic acids through sequential oxidative decarboxylation/C–H functionalization cyclization. This transformation provides a facile and mild route to carboxylated oxindoles in good to excellent yields with excellent functional group tolerance (Scheme 9.35) [41].

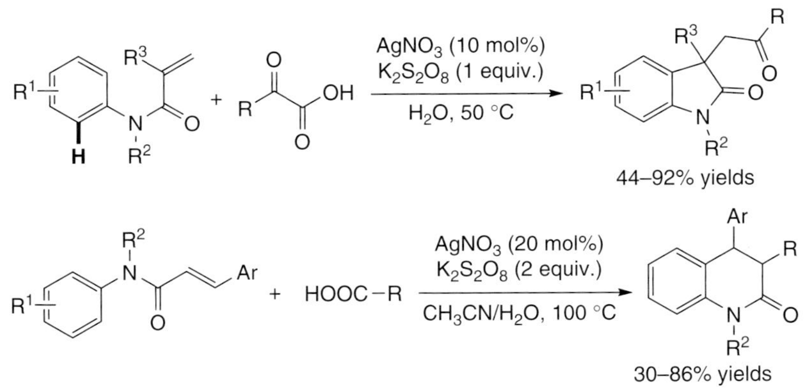

Scheme 9.35 1,2-Acylarylation of Activated Alkenes.

Interestingly, the chemoselectivity was shifted toward 3,4-disubstituted dihydroquinolin-2(1*H*)-ones simply by changing substrates from α-oxocarboxylic acids to aryl acids (Scheme 9.35).

9.3.2 Silver-Catalyzed Synthesis of Oxygen- or Phosphorus-Containing Heterocycles

The C–H functionalization and C–O cyclization strategy represents a shortcut to access oxygen-containing heterocycles. Very recently, Xu and coworkers reported a new silver-catalyzed C–H functionalization/C–O cyclization for the synthesis of benzo[*c*]chromen-6-ones (Scheme 9.36) [42]. Using a combination of $AgNO_3$ catalyst, $(NH_4)_2S_2O_8$ oxidant, and KOAc base in CH_2Cl_2/H_2O at room temperature enabled 2-aryl carboxylic acids to deliver lactones in good to excellent yields.

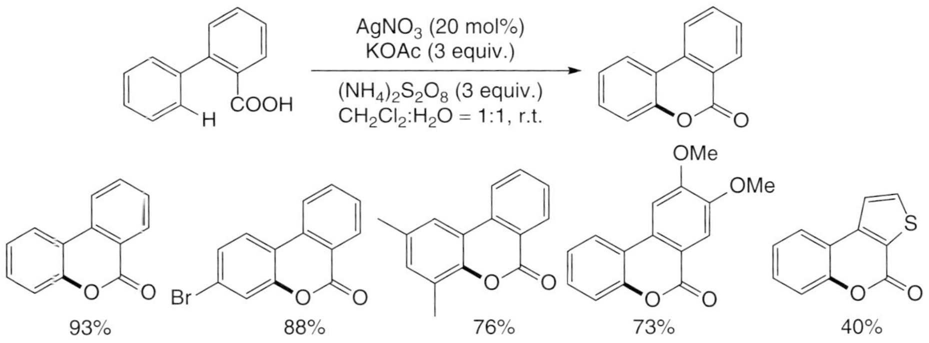

Scheme 9.36 Synthesis of Benzo[c]chromen-6-ones by C–H functionalization/C–O Cyclization.

Silver-catalyzed 1,2-arylphosphorylation of activated alkynes has been developed for assembling 3-phosphonated coumarins (Scheme 9.37) [43]. This method employs a cheap and nontoxic silver salt to catalyze the domino C–P and C–C bond formation of alkynoates for the first time.

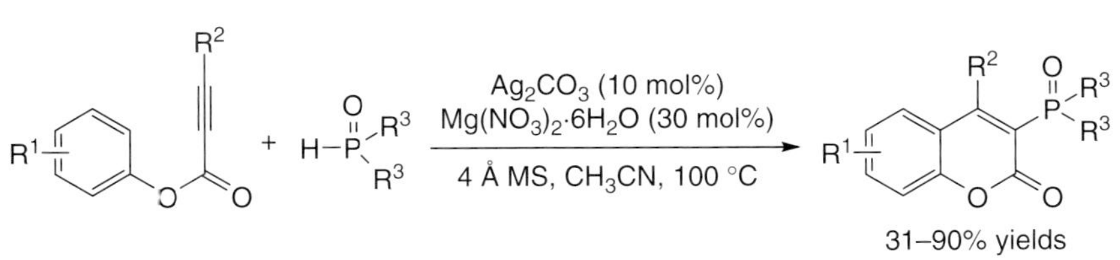

Scheme 9.37 1,2-Arylphosphorylation of Activated Alkynes to access 3-Phosphonated coumarins.

Phosphorus-containing heterocycles have diverse biological, medicinal, optical, and electronic properties. In 2013, the Duan group [44a] and Miura group [44b] independently reported a silver-mediated C–H activation strategy for the

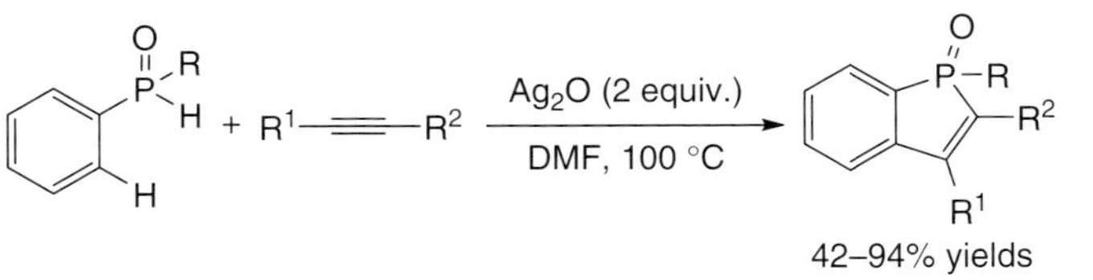

Scheme 9.38 Synthesis of Benzo[*b*]phosphole Oxides through Silver-Mediated C–H Activation.

synthesis of benzo[*b*]phosphole oxides (Scheme 9.38). The catalytic system is very simple: stoichiometric amounts of Ag_2O (2 equiv.) and AgOAc (1 equiv.) or catalytic amounts of Ag_2O (5 mol%) combined with $Zn(NO_3)_2 \cdot 6H_2O$ (1 equiv.) oxidant. Under the optimal conditions, diphenylphosphine oxides could react with disubstituted acetylenes to form the corresponding benzo[*b*]-phosphole oxides in good yield.

Mechanistically, the Duan group suggests that diphenylphosphine oxide is first reacted with diphenylacetylene under the catalysis of transition metals. The initial assumption is to utilize the coordination ability of the P-atom toward transition metals for the formation and functionalization of possible metallacycle intermediates via *ortho*-C–H bond activation. An unusual aryl migration on the P-atom derived from a C–P bond cleavage and a new C–P bond formation was also observed and demonstrated to proceed via the radical process (Scheme 9.39).

9.4 Conclusion and Outlook

Iron and silver, the common elements on Earth, have been widely used in chemistry and industry. Though their catalytic efficiency in synthesis is usually lower than the noble transition metals, particularly palladium, rhodium, or ruthenium, both iron and silver are abundant, inexpensive, and environmentally benign metals that exhibited interesting catalytic activity under the viable reaction conditions. For these reasons, Fe- and Ag-catalyzed synthesis of heterocycles has recently received a growing interest of synthetic chemists. This chapter summarized recent efforts on the discovery of the Fe- and Ag-catalyzed activation of various C–H bonds for the assembly of diverse heterocycles. These new Fe- and Ag-catalyzed methods are successful for heterocycle synthesis through sequential C–H functionalization and annulation and allow the formation of C–N, C–O, or C–O bonds by the functionalization of the C–H bonds, including aryl $C(sp^2)$–H bonds, vinyl $C(sp^2)$–H bonds, and alkyl $C(sp^3)$–H bonds. Particularly, many methods require oxidants to trigger the reactions. One typical class of these transformations is the C–H oxidative coupling with many functional groups, such as alkyl, hydroxyl, carbonyl, trifluoromethyl, azidyl, and phosphoryl groups, thus making the heterocycle framework more diverse. Owing to their nontoxic, high step and atom economy, excellent functional group tolerance, and broad substrate scope, the Fe- and Ag-catalyzed activation of C–H bonds will have rising usage in the pharmaceutical, agrochemical, and material fields.

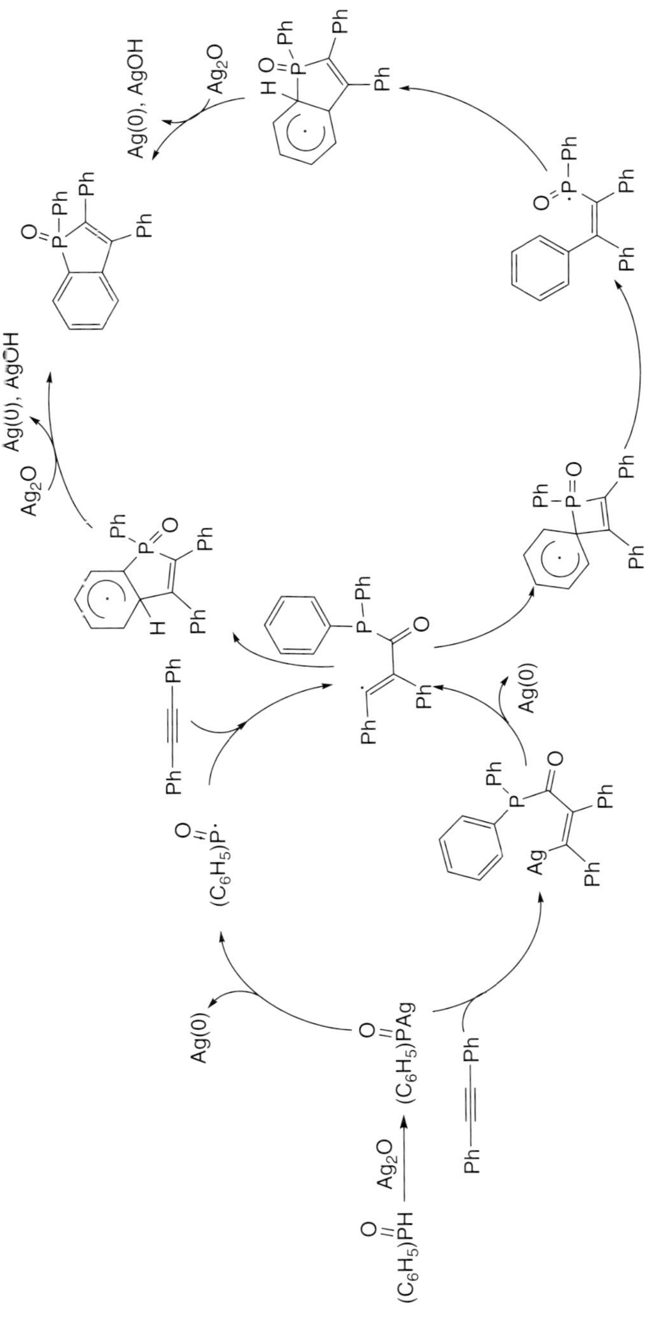

Scheme 9.39 Possible Mechanism of Silver-Mediated Synthesis of Benzo[*b*]phosphole Oxides.

Despite impressive progress in the field, approaches of the Fe- and Ag-catalyzed activation of C–H bonds for the synthesis of heterocycles are not abundant in comparison to the analogous noble transition metal (e.g., palladium, rhodium, and ruthenium)-catalyzed transformations. In addition, the obtained heterocycles are limited to some special nitrogen- and oxygen-containing heterocyclic framework, Thus, the development of new efficient and general Fe- and Ag-catalyzed routes for the activation of C–H bonds to access a wide range of heterocycles, especially embedded in the synthetically interesting, is one of the most challenging research areas in modern organic synthesis.

References

1. (a) Bauer, I. and Knölker, H.-J. (2015) *Chem. Rev.*, **115**, 3170–3387. (b) Song, R.-J., Liu, Y., Xie, Y.-X., and Li, J.-H. (2015) *Synthesis*, **47**, 1195–1209. (c) Sun, C.-L., Li, B.-J., and Shi, Z.-J. (2011) *Chem. Rev.*, **111**, 1293–1314; (d) Sarhan, A.A.O. and Bolm, C. (2009) *Chem. Soc. Rev.*, **38**, 2730–2744; (e) Correa, A., Garcia Mancheno, O., and Bolm, C. (2008) *Chem. Soc. Rev.*, **37**, 1108–1117.
2. Shen, M., Leslie, B.E., and Driver, T.G. (2008) *Angew. Chem. Int. Ed.*, **47**, 5056–5059.
3. Shen, M. and Driver, T.G. (2008) *Org. Lett.*, **10**, 3367–3370.
4. Nguyen, Q., Nguyen, T., and Driver, T.G. (2013) *J. Am. Chem. Soc.*, **135**, 620–623.
5. Bonnamour, J. and Bolm, C. (2011) *Org. Lett.*, **13**, 2012–2014.
6. Zhang, T. and Bao, W. (2013) *J. Org. Chem.*, **78**, 1317–1322.
7. You, X. and Wang, G.-W. (2014) *J. Org. Chem.*, **79**, 117–121.
8. Monir, K., Bagdi, A.K., Ghosh, M., and Hajra, A. (2014) *Org. Lett.*, **16**, 4630–4633.
9. Paradine, S.M. and White, M.C. (2012) *J. Am. Chem. Soc.*, **134**, 2036–2039.
10. Zhang, Y., Wang, M., Li, P., and Wang, L. (2012) *Org. Lett.*, **14**, 2206–2209.
11. Gu, L., Wang, W., Xiong, Y., Huang, X., and Li, G. (2014) *Eur. J. Org. Chem.*, 319–322.
12. Rohlmann, R., Stopka, T., Richter, H., and García Mancheño, O. (2013) *J. Org. Chem.*, **78**, 6050–6064.
13. Wang, H., Wang, Y., Peng, C., Zhang, J., and Zhu, Q. (2010) *J. Am. Chem. Soc.*, **132**, 13217–13219.
14. Maes, J., Rauws, T.R.M., and Maes, B.U.W. (2013) *Chem. Eur. J.*, **19**, 9137–9141.
15. Karthikeyan, I. and Sekar, G. (2014) *Eur. J. Org. Chem.*, **2014**, 8055–8063.
16. Guan, Z.-H., Yan, Z.-Y., Ren, Z.-H., Liu, X.-Y., and Liang, Y.-M. (2010) *Chem. Commun.*, **46**, 2823–2825.
17. Wei, W.-T., Zhou, M.-B., Fan, J.-H., Liu, W., Song, R.-J., Liu, Y., Hu, M., Xie, P., and Li, J.-H. (2013) *Angew. Chem. Int. Ed.*, **52**, 3638–3641.
18. Lu, M.-Z. and Loh, T.-P. (2014) *Org. Lett.*, **16**, 4698–4701.
19. (a) Fan, J.-H., Zhou, M.-B., Liu, Y., Wei, W.-T., Ouyang, X.-H., Song, R.-J., and Li, J.-H. (2014) *Synlett*, **25**, 657–660; (b) Dai, Q., Yu, J., Jiang, Y., Guo, S., Yang, H., and Cheng, J. (2014) *Chem. Commun.*, **50**, 3865–3867.
20. Ouyang, X.-H., Song, R.-J., and Li, J.-H. (2014) *Eur. J. Org. Chem.*, 3395–3401.
21. (a) Xu, X., Tang, Y., Li, X., Hong, G., Fang, M., and Du, X. (2014) *J. Org. Chem.*, **79**, 446–451; (b) Wang, G., Wang, S., Wang, J., Chen, S.-Y., and Yu, X.-Q. (2014) *Tetrahedron*, **70**, 3466–3470.
22. Wang, J.-Y., Zhang, X., Bao, Y., Xu, Y.-M., Cheng, X.-F., and Wang, X.-S. (2014) *Org. Biomol. Chem.*, **12**, 5582–5585.
23. Leifert, D., Daniliuc, C.G., and Studer, A. (2013) *Org. Lett.*, **15**, 6286–6289.
24. (a) Tu, H.-Y., Liu, Y.-R., Chu, J.-J., Hu, B.-L., and Zhang, X.-G. (2014) *J. Org. Chem.*, **79**, 9907–9912; (b) Pan, C., Han, J., Zhang, H., and Zhu, C. (2014) *J. Org. Chem.*, **79**, 5374–5378; (c) Wang,

G., Chen, S.-Y., and Yu, X.-Q. (2014) *Tetrahedron Lett.*, **55**, 5338–5341.

25. (a) Thansandote, P. and Lautens, M. (2009) *Chem. Eur. J.*, **15**, 5874–5883; (b) Satoh, T. and Miura, M. (2010) *Synthesis*, **2010**, 3395–3401; (c) Colby, D.A., Bergman, R.G., and Ellman, J.A. (2010) *Chem. Rev.*, **110**, 624–655; (d) Satoh, T. and Miura, M. (2010) *Chem. Eur. J.* **16**, 11212–11222;(e) Arockiam, P.B., Bruneau, C., and Dixneuf, P.H. (2012) *Chem. Rev.*, **112**, 5879–5918; (f) Simmons, E.M. and Hartwig, J.F. (2012) *Angew. Chem. Int. Ed.*, **51**, 3066–3072; (g) Song, G., Wang, F., and Li, X. (2012) *Chem. Soc. Rev.*, **41**, 3651–3678.
26. Liu, P., Li, Y., Wang, H., Wang, Z., and Hu, X. (2012) *Tetrahedron Lett.*, **53**, 6654–6656.
27. Nguyen, T.B., Retailleau, P., and Al-Mourabit, A. (2013) *Org. Lett.*, **15**, 5238–5241.
28. Guo, X., Yu, R., Li, H., and Li, Z. (2009) *J. Am. Chem. Soc.*, **131**, 17387–17393.
29. Parnes, R., Kshirsagar, U.A., Werbeloff, A. Regev, C., and Pappo, D. (2012) *Org. Lett.*, **14**, 3324–3327.
30. Kshirsagar, U.A., Regev, C., Parnes, R., and Pappo, D. (2013) *Org. Lett.*, **15**, 3174–3177.
31. Huang, Z., Jin, L., Feng, Y., Peng, P., Yi, H., and Lei, A. (2013) *Angew. Chem. Int. Ed.*, **52**, 7151–7155.
32. Xu, X., Xu, X., Li, H., Xie, X., and Li, Y. (2010) *Org. Lett.*, **12**, 100–103.
33. Wertz, S., Leifert, D., and Studer, A. (2013) *Org. Lett.*, **15**, 928–931.
34. Bigi, M.A., Reed, S.A., and White, M.C. (2012) *J. Am. Chem. Soc.*, **134**, 9721–9726.
35. (a) Munöz, M.P. (2014) *Chem. Soc. Rev.*, **43**, 3164–3183; (b) Álvarez-Corral, M., Muñoz-Dorado, M., and Rodríguez-García, I. (2008) *Chem. Rev.*, **108**, 3174–3198; (c) Yamamoto, Y. (2008) *Chem. Rev.*, **108**, 3199–3222.
36. Cui, Y. and He, C. (2004) *Angew. Chem. Int. Ed.*, **43**, 4210–4212.
37. Alderson, J.M., Phelps, A.M., Scamp, R.J., Dolan, N.S., and Schomaker, J.M. (2014) *J. Am. Chem. Soc.*, **136**, 16720–16723.
38. Li, Y.-M., Sun, M., Wang, H.-L., Tian, Q.-P., and Yang, S.-D. (2013) *Angew. Chem. Int. Ed.*, **52**, 3972–3976.
39. (a) Wei, X.-H., Li, Y.-M., Zhou, A.-X., Yang, T.-T., and Yang, S.-D. (2013) *Org. Lett.*, **15**, 4158–4161; (b) Yuan, Y., Shen, T., Wang, K., and Jiao, N. (2013) *Chem. Asian J.*, **8**, 2932–2935.
40. Wang, H., Guo, L.-N., and Duan, X.-H. (2013) *Chem. Commun.*, **49**, 10370–10372.
41. (a) Wang, H., Guo, L.-N., and Duan, X.-H. (2013) *Adv. Synth. Catal.*, **355**, 2222; (b) Mai, W.-P., Wang, J.-T., Yang, L.-R., Yuan, J.-W., Xiao, Y.-M., Mao, P., and Qu, L.-B. (2014) *Org. Lett.*, **16**, 204–207.
42. Dai, J.-J., Xu, W.-T., Wu, Y.-D., Zhang, W.-M., Gong, Y., He, X.-P., Zhang, X.-Q., and Xu, H.-J. (2015) *J. Org. Chem.*, **80**, 911–919.
43. Mi, X., Wang, C., Huang, M., Zhang, J., Wu, Y., and Wu, Y. (2014) *Org. Lett.*, **16**, 3356–3359.
44. (a) Chen, Y.-R. and Duan, W.-L. (2013) *J. Am. Chem. Soc.*, **135**, 16754–16757; (b) Unoh, Y., Hirano, K., Satoh, T., and Miura, M. (2013) *Angew. Chem. Int. Ed.*, **52**, 12975–12979.

10
Heterocycles Synthesis via Co-Catalyzed C–H Bond Functionalization

Naohiko Yoshikai

10.1
Introduction

The ability of cobalt complexes to mediate cycloaddition reactions has been known for a long time. Representative examples include [2+2+2] cycloaddition of three alkyne units, [2+2+2] cycloaddition of two alkyne units and a nitrile molecule, and [2+2+1] cycloaddition of alkyne, alkene, and carbon monoxide (Pauson–Khand reaction), which have been extensively explored and practiced by the synthetic community [1, 2]. However, cobalt has not been recognized as a versatile metal for C–H activation for a long time, and the first use of a cobalt complex in a C–H functionalization reaction dates back to the 1950s. Murahashi reported on a $Co_2(CO)_8$-catalyzed *ortho*-carbonylation reaction of benzaldimine under high-pressure carbon monoxide at a high temperature, which affords an isoindolinone derivative (Scheme 10.1a) [3]. Under similar reaction conditions, azobenzene also undergoes *ortho*-carbonylation to afford an indazolone derivative or a quinazoline-2,4-dione derivative depending on the reaction temperature (Scheme 10.1b) [4]. These reactions most likely involve cobalt-mediated activation (cyclometalation) of the *ortho* C–H bonds and represent notable early examples of heterocycle synthesis via C–H activation.

Despite the seminal findings of Murahashi, cobalt had been rarely used in catalytic C–H functionalization. With respect to heterocycle synthesis, a notable example known before the late 2000s would be the reaction of an azobenzene derivative with diphenylacetylene reported by Kisch and coworkers in 1994 (Scheme 10.2) [5]. In the presence of a cobalt complex $CoH(N_2)(PPh_3)_3$ or $Co(H)_3(PPh_3)_3$, both of the *ortho* C–H bonds of the azobenzene derivative undergo insertion of a diphenylacetylene molecule, affording a simple dialkenylated product or its annulated product, that is, a dihydrocinnoline derivative, depending on the substituents.

Another notable early example of heterocycle synthesis via cobalt-mediated C–H activation would be the transformation of *N*,*N*-diallylanilines into quinoline derivatives reported by Jones and coworkers [6]. The reaction of *N*,*N*-diallylaniline in the presence of a catalytic amount of $Co_2(CO)_8$ under

Transition Metal-Catalyzed Heterocycle Synthesis via C—H Activation, First Edition. Edited by Xiao-Feng Wu.
© 2016 Wiley-VCH Verlag GmbH & Co. KGaA. Published 2016 by Wiley-VCH Verlag GmbH & Co. KGaA.

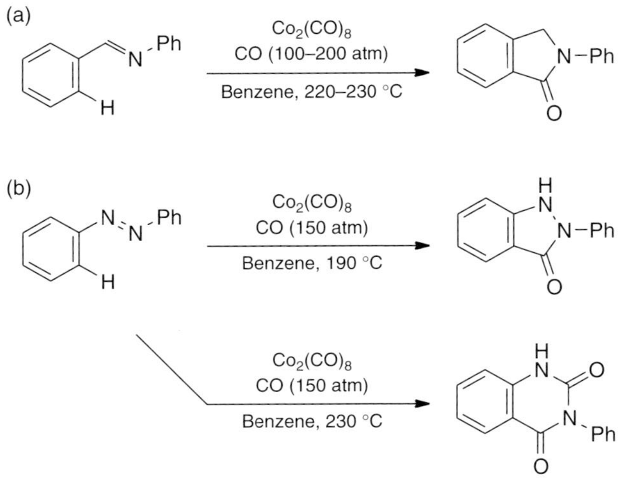

Scheme 10.1 $Co_2(CO)_8$-catalyzed ortho-carbonylation of benzaldimine (a) and azobenzene (b) leading to *N*-heterocycles.

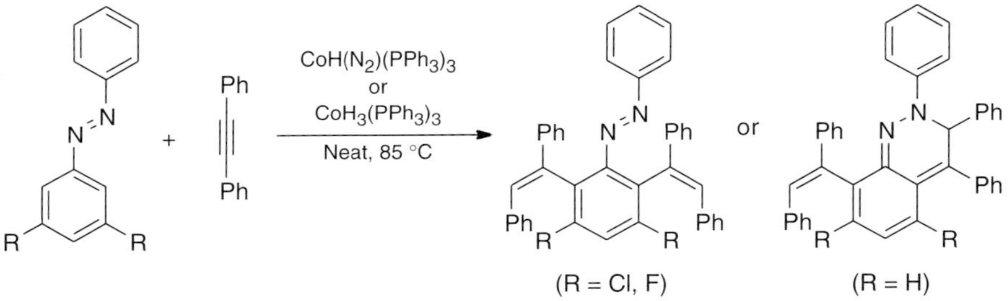

Scheme 10.2 Reaction of azobenzene derivative and diphenylacetylene.

CO atmosphere affords 2-ethyl-3-methylquinoline through dehydrogenative cyclization (Scheme 10.3a). Furthermore, the same catalytic system also promotes a cross-coupling of *N*-benzylideneaniline and *N*,*N*-diallylaniline to afford 2-phenyl-3-methylquinoline, where the latter reactant serves as a donor

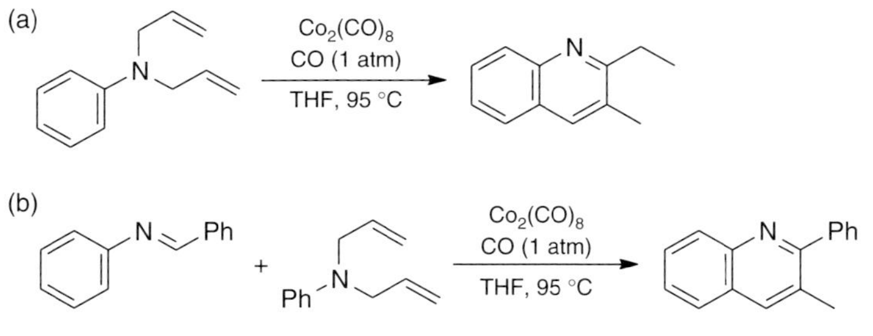

Scheme 10.3 $Co_2(CO)_8$-catalyzed synthesis of quinoline derivatives from *N*,*N*-diallylaniline (a) and from *N*-benzylideneaniline and *N*,*N*-diallylaniline (b).

of two allyl groups and ends up in aniline (Scheme 10.3b). A complex reaction pathway initiated by nitrogen-assisted *ortho* C–H metalation with a $Co_2(CO)_7$ fragment has been proposed on the basis of mechanistic experiments.

While major advances in the area of C–H functionalization have been made with catalysts based on rare and expensive transition metals such as rhodium, palladium, ruthenium, and iridium [7], increasing interest in the sustainability aspect of catalysis has stimulated researchers toward the development of alternative catalysts based on naturally abundant first-row transition metals including cobalt [8]. As such, a growing number of cobalt-catalyzed C–H functionalization reactions, including those for heterocycle synthesis, have been reported over the last several years to date (early 2015) [9]. The purpose of this chapter is to provide an overview of such recent advancements with classification according to the nature of the catalytically active cobalt species involved in the C–H activation event. Besides inner-sphere C–H activation reactions catalyzed by low-valent and high-valent cobalt complexes, nitrene and carbene C–H insertion reactions promoted by cobalt(II)–porphyrin metalloradical catalysts are also discussed.

10.2
Heterocycle Synthesis via Low-Valent Cobalt-Catalyzed C–H Activation

The ability of low-valent cobalt complexes to undergo chelation-assisted aromatic C–H activation has been demonstrated by the aforementioned catalytic examples (Schemes 10.1–10.3) as well as by a series of stoichiometric cyclometalation reactions using $MeCo(PMe_3)_4$ [10, 11]. Recently, the scope of low-valent cobalt-catalyzed directed C–H functionalization has been substantially expanded, as summarized in review articles by Yoshikai and Ackermann [9a–c]. In particular, a catalyst generated from an appropriate combination of a cobalt salt, a ligand, and a Grignard reagent proved effective for promoting alkenylation or alkylation of a $C(sp^2)$–H bond with an alkyne or an alkene, respectively, via chelation-assisted C–H activation [12]. Thus far, two examples of heterocycle synthesis based on such C–H alkenylation or alkylation have been reported, as described below.

One is an intermolecular annulation reaction of an α,β-unsaturated imine bearing an *N-para*-methoxyphenyl (PMP) group with an internal alkyne that leads to a dihydropyridine derivative (Scheme 10.4) [13]. The reaction is achieved at a mild temperature (40 °C) using a catalytic system consisting of $CoBr_2$, $P(3\text{-}ClC_6H_4)_3$, and isopropylmagnesium bromide, tolerating functional groups such as chloro, bromo, and cyano groups. The reaction is proposed to consist of cobalt-catalyzed and uncatalyzed processes. The former process involves imine-assisted oxidative addition of the vinylic C–H bond to cobalt, migratory insertion of the alkyne into the Co–H bond, and reductive elimination of the dialkenylcobalt species. The resulting azatriene intermediate then undergoes uncatalyzed 6π-electrocyclization to afford the dihydropyridine derivative as the final product. The oxidation state of the active cobalt catalyst is assumed to be

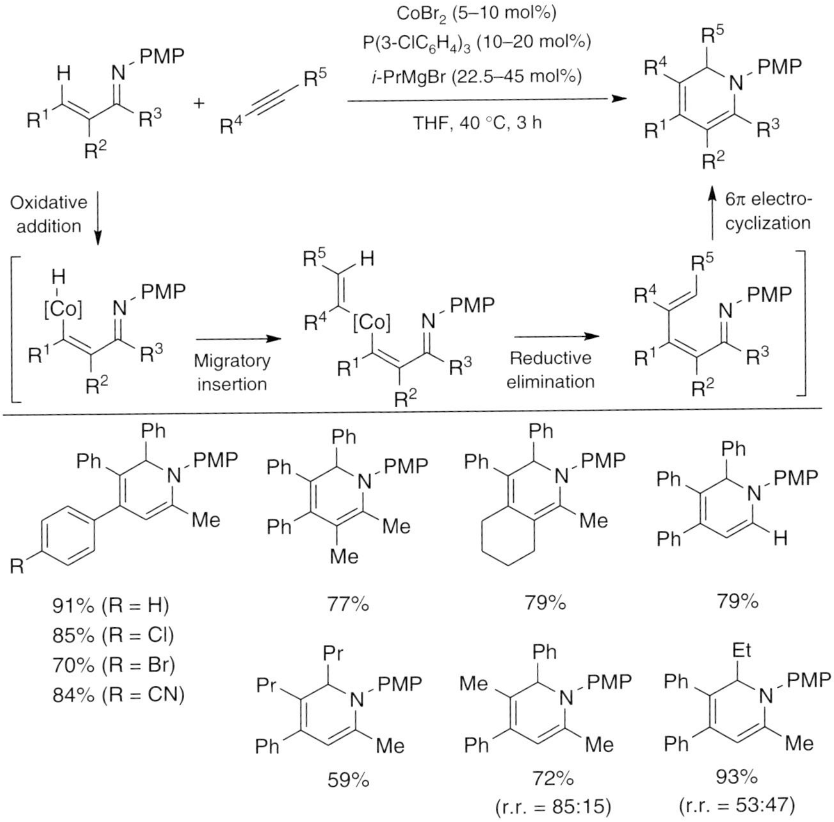

Scheme 10.4 Annulation of α,β-unsaturated imine and internal alkyne. r.r. refers to regioisomer ratio (major regioisomers are shown).

zero [14]. Note that the same type of reaction was previously achieved by Bergman and Ellman using a rhodium(I) catalyst and an imine bearing an *N*-benzyl group [15], who also demonstrated the utility of the corresponding dihydropyridine product as a precursor to pyridine, tetrahydropyridine, piperidine, and tropane derivatives [16]. Thus, the products of the cobalt-catalyzed reaction would also serve as valuable synthetic intermediates for such *N*-heterocycles.

Another example is an intramolecular cyclization reaction of an indole substrate bearing an olefin tether and an aldimine directing group on the N atom and the C3 position, respectively (Scheme 10.5) [17]. With a homoallyl group on the N atom, intramolecular directed C2-alkylation of the indole substrate is achieved with the aid of a cobalt–N-heterocyclic carbene (NHC) catalyst generated from $CoBr_2$, an imidazoli(ni)um salt, and Me_3SiCH_2MgCl. Interestingly, the regiochemistry of the cyclization reaction significantly depends on the NHC ligand used. Thus, Co–SIMes and Co–IPr catalysts afford dihydropyrroloindole

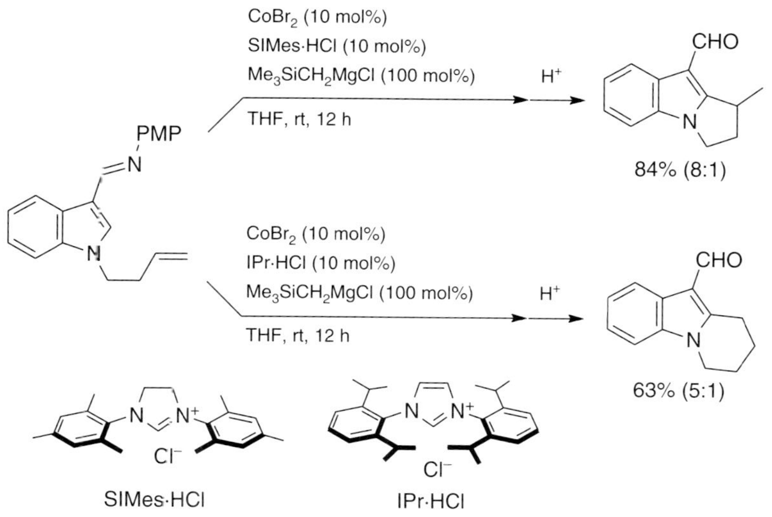

Scheme 10.5 Intramolecular C2-alkylation of indole leading to dihydropyrroloindole or tetrahydropyridoindole derivative.

and tetrahydropyridoindole derivatives, respectively, with decent regioselectivities. Deuterium-labeling experiments suggest that the regioselectivity of the reaction is determined in the olefin insertion step prior to the C–C bond-forming reductive elimination step.

The addition of an aldehydic C–H bond across an unsaturated bond (e.g., C=C, C=O, C≡C), hydroacylation, represents an atom-economical route to carbonyl compounds [18]. When performed in intramolecular settings, such reactions could serve as attractive methods for the synthesis of carbo- and heterocyclic compounds bearing carbonyl moieties. Indeed, rhodium(I) – (chiral) phosphine complexes have proved to be versatile catalysts for the (enantioselective) intramolecular hydroacylation of olefins and ketones, affording cyclic products such as cyclopentanones, indanones, phthalides, and other carbo- and heterocycles. On the other hand, cobalt complexes have rarely been used as catalysts for intramolecular hydroacylation. Isolated early examples include $(Ph_3P)_2Co(dppe)$-catalyzed cyclization of 4-pentenal to cyclopentanone by Vinogradov *et al.* and $Cp^*Co(CH_2{=}CHSiMe_3)_2$-catalyzed ($Cp^*$ = 1,2,3,4,5-pentamethylcyclopentadienyl) cyclization of a benzaldehyde derivative bearing a dimethylvinylsiloxy group on the ortho position by Lenges and Brookhart (Scheme 10.6) [19, 20].

In 2014, Yang and Yoshikai reported enantioselective intramolecular hydroacylation reactions of ketones and olefins using cobalt–chiral diphosphine catalysts (Scheme 10.7) [21]. A cobalt catalyst generated by the reduction of $CoBr_2$ with indium powder in the presence of 1,2-bis((2R,5R)-2,5-diphenylphospholano)ethane ((R,R)-Ph-BPE) promotes cyclization of a variety of 2-acylbenzaldehydes to afford phthalide derivatives in moderate

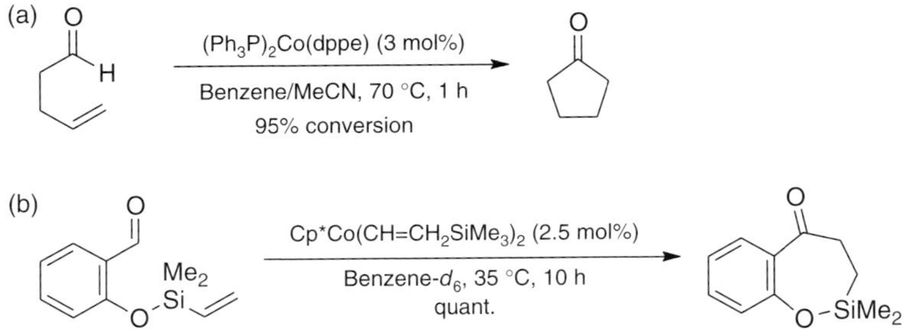

Scheme 10.6 Early examples of cobalt-catalyzed intramolecular hydroacylation using cobalt(0)-phosphine (a) and Cp*-cobalt(I) (b) catalysts.

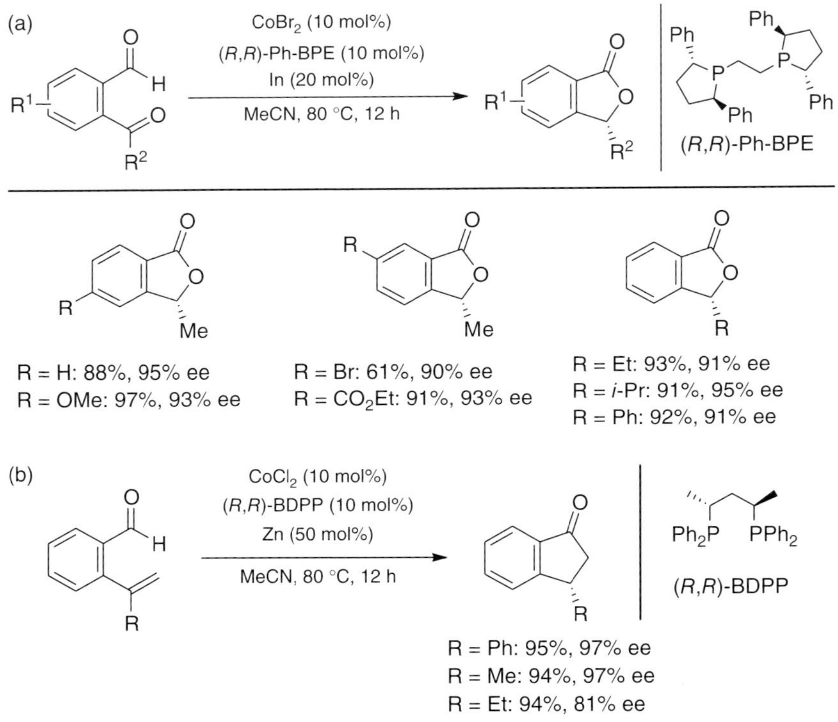

Scheme 10.7 Enantioselective intramolecular hydroacylation of ketones (a) and olefins (b) using cobalt–chiral diphosphine catalysts.

to good yields with enantioselectivities as high as those achieved with a chiral rhodium catalyst [22]. Another cobalt catalyst generated from $CoCl_2$, (2R,4R)-2,4-bis(diphenylphosphino)pentane ((R,R)-BDPP), and zinc dust allows enantioselective cyclization of 2-alkenylbenzaldehydes to indanone derivatives, but the scope of this reaction is somewhat limited compared with that of a rhodium-catalyzed variant [23]. Deuterium-labeling experiments on the ketone hydroacylation suggested a catalytic cycle consisting of aldehyde C–H oxidative addition to cobalt(I) [24], insertion of the C=O moiety into the Co–H bond, and C–O reductive elimination of the acyl(alkoxy)cobalt species, the last step being rate-determining.

While the heterocycle-forming reactions discussed above commonly involve C–H oxidative addition, migratory insertion, and reductive elimination steps, Yoshikai and coworkers also developed a different approach to heterocycles through cobalt-catalyzed C–H activation. They found an addition reaction of an arylzinc reagent to an internal alkyne that is uniquely catalyzed by a cobalt–Xantphos complex to afford an ortho-alkenylarylzinc reagent (Scheme 10.8) [25]. The reaction is considered to proceed through insertion of the alkyne into an arylcobalt species, vinyl-to-aryl 1,4-cobalt migration, and transmetalation of the resulting *ortho*-alkenylarylcobalt species with the arylzinc reagent. In light of an analogous 1,4-rhodium migration process [26, 27], the 1,4-cobalt migration is assumed to proceed through oxidative addition of the ortho C–H bond to the alkenylcobalt(I) species followed by reductive elimination of the alkenyl and the hydride ligand on the cobalt(III) intermediate. Capitalizing on this "migratory arylzincation" reaction, they have devised new methods for the synthesis of some members of the benzo[*b*]heterole family of heterocycles as described below.

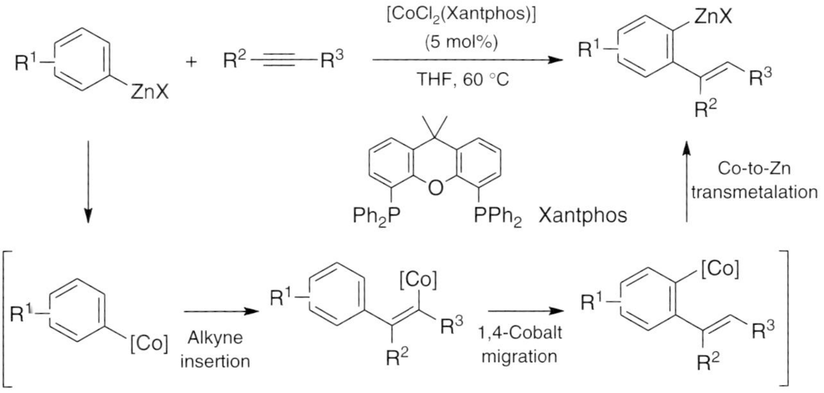

Scheme 10.8 Addition of arylzinc reagent to alkyne involving 1,4-cobalt migration.

The *ortho*-alkenylarylzinc reagent obtained by the migratory arylzincation serves as a precursor to benzo[*b*]chalcogenophenes such as benzo[*b*]thiophenes and benzo[*b*]selenophenes (Scheme 10.9) [28]. Treatment of the zinc intermediate with sulfur powder in the presence of CuI allows one-pot synthesis of a benzo[*b*]thiophene derivative (route A). Alternatively, the zinc intermediate is quenched with iodine to give ortho-alkenylaryl iodide, which is then subjected to copper-catalyzed reaction with sulfur or selenium powder to furnish the corresponding benzo[*b*]thiophene or benzo[*b*]selenophene derivative (route B). Collectively, these approaches allow for expedient synthesis of functionalized benzo[*b*]thiophenes and benzo[*b*]selenophenes, many of which are difficult to

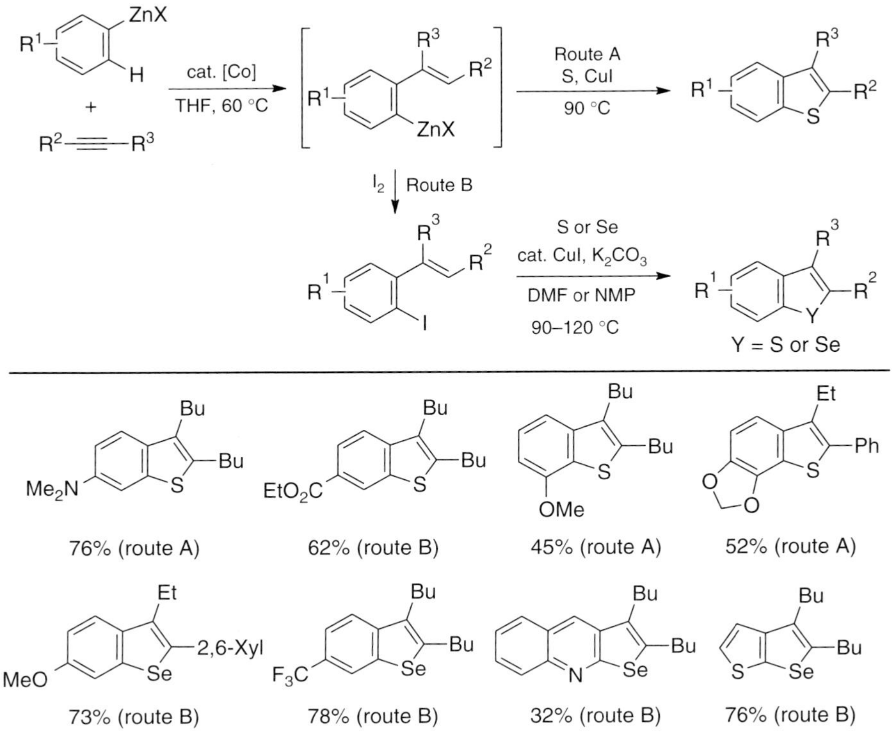

Scheme 10.9 Synthesis of benzo[*b*]thiophenes and benzo[*b*]selenophenes through cobalt-catalyzed migratory arylzincation and copper-mediated/copper-catalyzed chalcogenative cyclization.

be accessed by existing methods. While the exact mechanism for the formation of the benzo[*b*]chalcogenophene ring is not clear, these cyclization reactions would involve conversion of the C–Zn or C–I bond into a C–chalcogenide anion or radical moiety, attack of the chalcogenide moiety to the pendant olefin, and oxidative aromatization.

The migratory arylzincation reaction can also be utilized for the construction of benzo[*b*]phosphole derivatives through sequential one-pot multicomponent coupling (Scheme 10.10) [29]. Thus, the *ortho*-alkenylarylzinc reagent is subjected to copper-catalyzed electrophilic trapping with dichlorophenylphosphine and then to oxidation with hydrogen peroxide or sulfur powder, affording a benzo[*b*]phosphole oxide or sulfide derivative (route A). A benzo[*b*]phosphole derivative may also be constructed by sequential exposure of the zinc reagent to phosphorus trichloride, a Grignard reagent, and the oxidant (route B). The latter route is particularly attractive in terms of modular and diversity-oriented synthesis, because it allows flexible modification of the "benzo" moiety, the C2–C3 moiety, and the P-substituents of the benzo[*b*]phosphole core at the same time and hence enables systematic evaluation of substituent effects on the photophysical properties of benzo[*b*]phosphole derivatives.

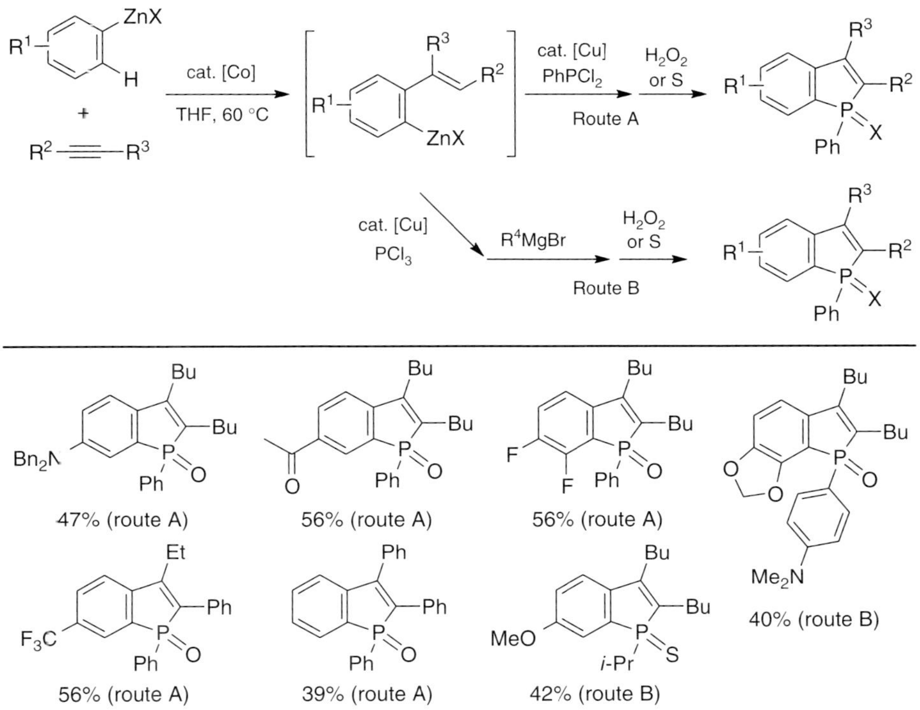

Scheme 10.10 Synthesis of benzo[*b*]phospholes through one-pot multicomponent coupling.

10.3 Heterocycle Synthesis via High-Valent Cobalt-Catalyzed C–H Activation

Since the late 2000s, high-valent Cp*Rh^{III} complexes have emerged as versatile catalysts for C–C and C–heteroatom bond-forming reactions via chelation-assisted C(sp^2)–H activation [30]. In particular, a number of annulation reactions of directing group-appended arenes and olefins with unsaturated substrates such as alkynes and olefins have been achieved under oxidative or redox-neutral conditions, which allow for the atom- and step-economical synthesis of various heterocycles. In 2013, Matsunaga, Kanai, and coworkers demonstrated the competence of an analogous Cp*Co^{III} complex as a catalyst for directed C–H activation by the development of direct addition reactions of 2-arylpyridines to *N*-sulfonylimines and α,β-unsaturated ketones (Scheme 10.11) [31]. Since then, the repertoire of Cp*Co^{III}-catalyzed C–H functionalization has steadily expanded to date [9d], which now includes a few examples of heterocycle-forming reactions as described below.

Matsunaga, Kanai, and coworkers developed a Cp*Co^{III}-catalyzed redox-neutral annulation reaction of an *N*-carbamoyl-indole bearing a morpholinyl group and an internal alkyne to afford a pyrroloindolone derivative (Scheme 10.12) [32]. The annulation reaction employs a catalytic system consisting of a cationic

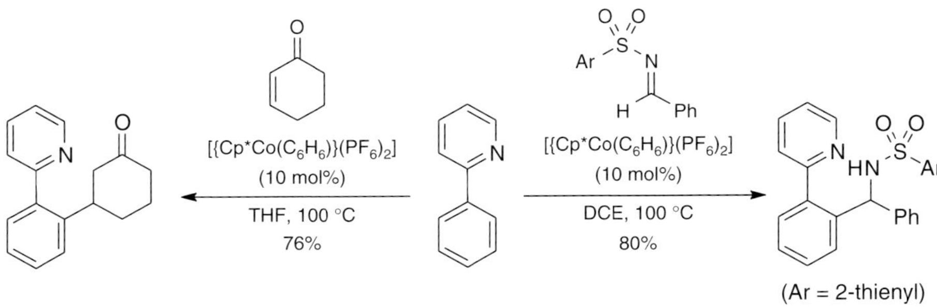

Scheme 10.11 The first examples of Cp*Co^{III}-catalyzed C–H functionalization.

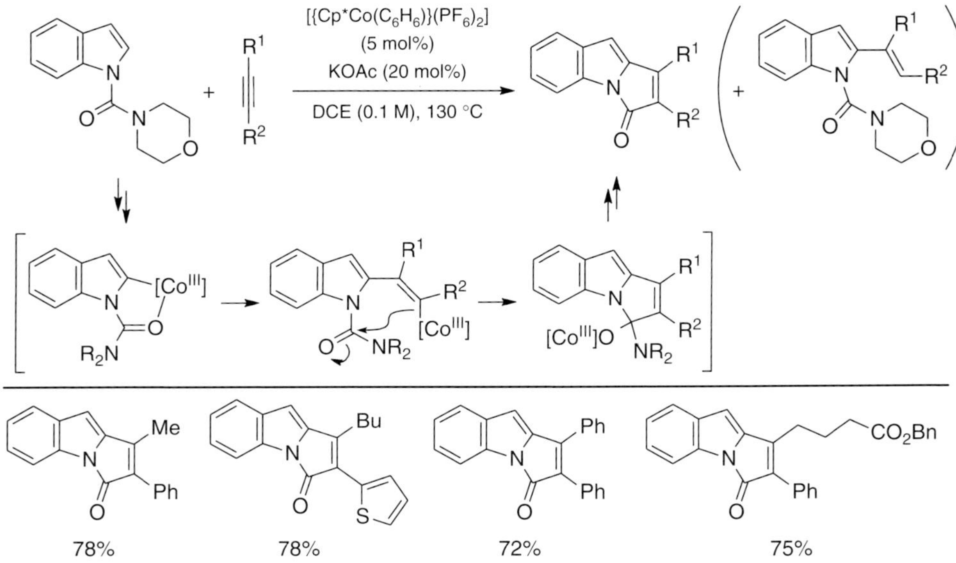

Scheme 10.12 Pyrroloindolone synthesis through Cp*Co^{III}-catalyzed annulation of *N*-carbamoyl-indole and internal alkyne.

complex [{Cp*Co(C_6H_6)}(PF_6)$_2$] and an additive KOAc and requires careful choice of the carbamoyl group, the concentration (0.1 M), and the reaction temperature (130 °C) to minimize the formation of a C2-alkenyled indole as a minor by-product. With a dimethylcarbamoyl group and at a higher concentration (0.5 M) and a lower temperature (80 °C), the C2-alkenylation becomes the dominant reaction pathway. The mechanism of the annulation reaction is proposed to involve carbamoyl-directed C2-metalation with assistance of the acetate anion [33], insertion of the alkyne into the Co–indolyl bond, and nucleophilic attack of the resulting alkenylcobalt species to the carbamoyl moiety. Interestingly, irrespective of the carbamoyl group, the use of a Cp*Rh^{III} catalyst instead of the Cp*Co^{III} catalyst exclusively affords the C2-alkenylated indole [34]. The drastic difference of the reactivities of the Cp*Co^{III} and Cp*Rh^{III} catalysts is ascribed to higher nucleophilicity of the alkenylcobalt(III) species than that of the alkenylrhodium(III) species.

The nucleophilicity of an organocobalt(III) species was utilized by Hummel and Ellman to achieve condensation of an azobenzene and an aldehyde to afford an *N*-aryl-2*H*-indazole derivative (Scheme 10.13) [35]. A combination of a cationic Cp*Co^{III} complex bearing a noncoordinating $B(C_6F_5)_4$ anion and acetic acid promotes direct addition of the azobenzene to the aldehyde through azo-directed ortho-metalation and subsequent insertion of the C=O bond into the aryl–Co^{III} bond. While these catalytic steps are likely reversible, intramolecular dehydrative condensation of the azo and the alcohol moieties in the aldehyde adduct serves as a driving force for the formation of the indazole product. The reaction shows a substrate scope comparable to that of the same reaction catalyzed by a Cp*Rh^{III} catalyst [36] and is applicable to both symmetrical and unsymmetrical azobenzene derivatives and to aromatic, heteroaromatic, and aliphatic aldehydes.

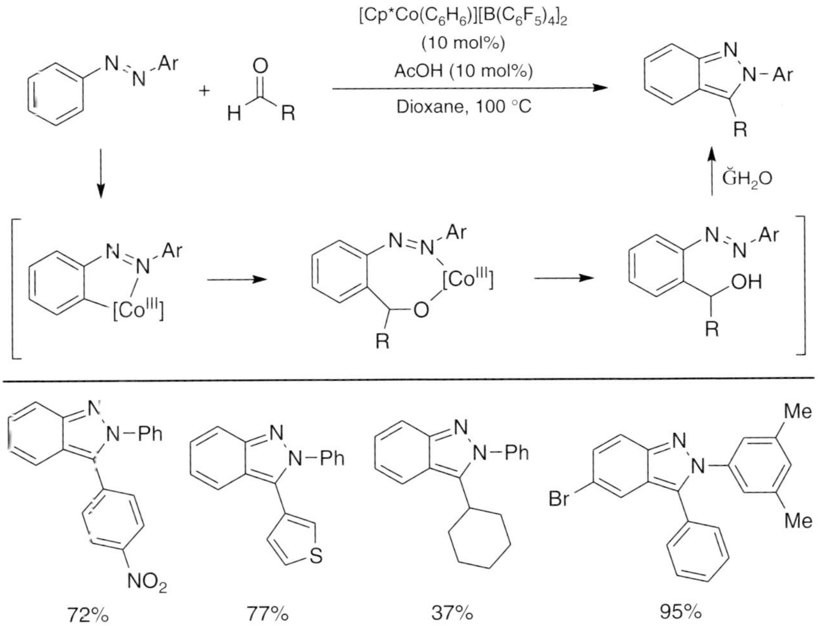

Scheme 10.13 Indazole synthesis through Cp*Co^{III}-catalyzed C–H functionalization/dehydrative cyclization of azobenzene and aldehyde.

As was demonstrated earlier with Cp*Rh^{III} catalysis [37], the above Cp*Co^{III} catalytic system also proved effective for the condensation of an α,β-unsaturated oxime with an aldehyde to afford a furan derivative (Scheme 10.14) [35]. The reaction likely involves oxime-directed vinylic C–H metalation with the Cp*Co^{III} catalyst and subsequent insertion of the aldehyde into the alkenylcobalt(III)

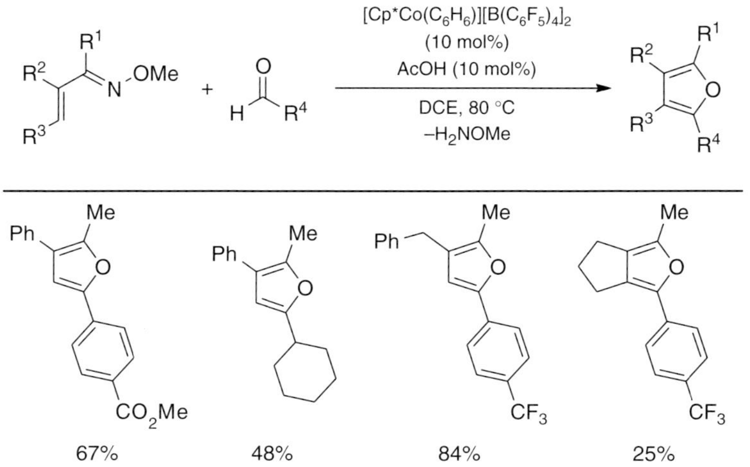

Scheme 10.14 Furan synthesis through Cp*Co^{III}-catalyzed C–H functionalization/deaminative cyclization of α,β-unsaturated oxime and aldehyde.

species. The resulting aldehyde adduct would undergo intramolecular condensation of the oxime and the alcohol moiety to give the furan product while liberating *O*-methylhydroxylamine.

On the way to further extension of the scope of Cp*Co^{III}-catalyzed C–H functionalization, Glorius and coworkers developed a condensation reaction of 2-arylpyridine derivatives and diazoesters to form unique polycyclic heteroaromatics having 6*H*-pyrido[2,1-*a*]isoquinolin-6-one skeletons (Scheme 10.15) [38]. The reaction is achieved by the combination of a bench-stable Co^{III} precatalyst [Cp*Co(CO)I_2] [39], a silver salt ($AgSbF_6$), and an acetate source (KOAc) in trifluoroethanol. The in situ-formed Cp*Co^{III} catalyst is proposed to play a dual role in this condensation reaction. First, it promotes formal carbene insertion into the ortho C–H bond through pyridine-directed C–H metalation, cobalt–carbene formation, carbene insertion into the aryl–Co^{III} bond, and protodemetalation. Second, it acts as a Lewis acid to facilitate nucleophilic attack of the pyridine moiety to the ester group, which eventually leads to the product through aromatization and elimination of methanol. The thus-synthesized polycyclic heteroarenes exhibit bright and color-tunable fluorescence in solution and in the solid state.

While a number of transition metal-catalyzed directed C–H functionalization reactions, including many of the reactions discussed above, employ monodentate directing groups, recent years have witnessed an emergence of "bidentate" directing groups that allow for tight binding of the catalyst and the substrate and thus facilitate otherwise difficult C–H activation/functionalization reactions. Since the seminal work of Daugulis and coworkers in 2005 on palladium-catalyzed direct arylation of C(sp^3)–H and C(sp^2)–H bonds in the β-position of an 8-aminoquinoline-derived amide group [40], the scope of such bidentate directing group-assisted C–H functionalization has been substantially expanded

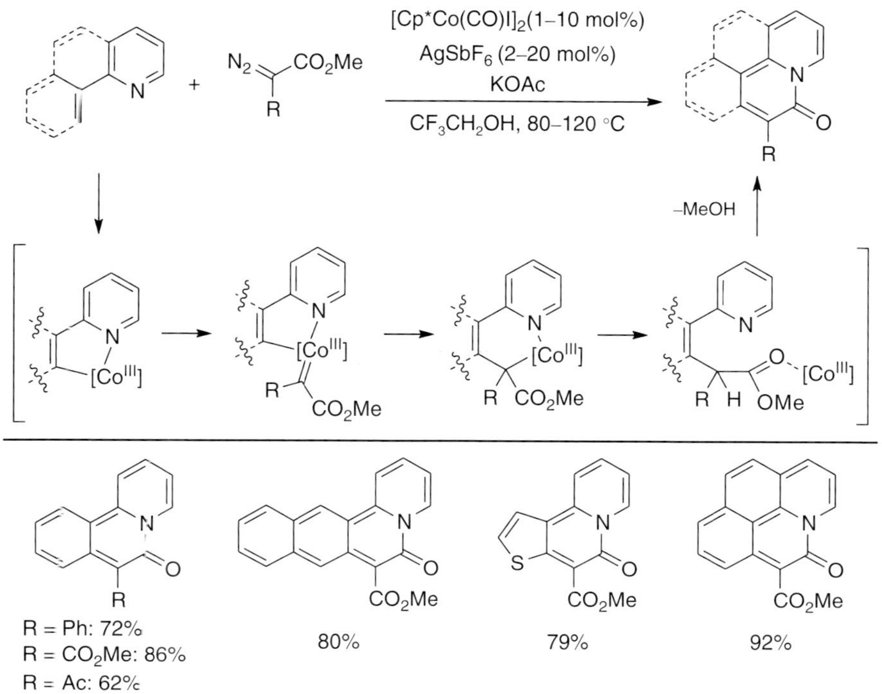

Scheme 10.15 Cp*Co^{III}-catalyzed condensation of 2-arylpyridine and diazoester.

with respect to the type of reaction (e.g., arylation, alkenylation, alkylation, alkynylation, C–heteroatom bond formation), the structure of the directing group, and the transition metal catalyst used (e.g., palladium, ruthenium, nickel, copper, iron) [41].

In 2014, Grigorjeva and Daugulis reported a series of heterocycle-forming reactions featuring cobalt-catalyzed C–H activation assisted by a bidentate directing group. A reaction system comprising a simple cobalt salt $Co(OAc)_2{\cdot}4H_2O$ as a precatalyst and sodium pivalate and manganese(III) acetate as additives promotes oxidative annulation of 8-aminoquinoline-bearing benzamides and alkynes under air, affording isoquinolone derivatives (Scheme 10.16) [42]. The reaction tolerates a variety of functional groups on both the benzamide and the alkyne substrates. Furthermore, the reaction is applicable not only to internal alkynes but also to terminal alkynes bearing electronically diverse substituents. A propose reaction mechanism involves aminoquinoline-assisted cyclometalation with concomitant oxidation of the cobalt(II) precatalyst, insertion of the alkyne into the resulting cobalt(III)cycle, and C–N reductive elimination.

With a slight modification of the catalytic system, olefins can also be used as reaction partners for the cobalt-catalyzed oxidative annulation with the benzamide substrates (Scheme 10.17) [43]. The reaction features mild room-temperature conditions and is applicable to terminal olefins including styrene,

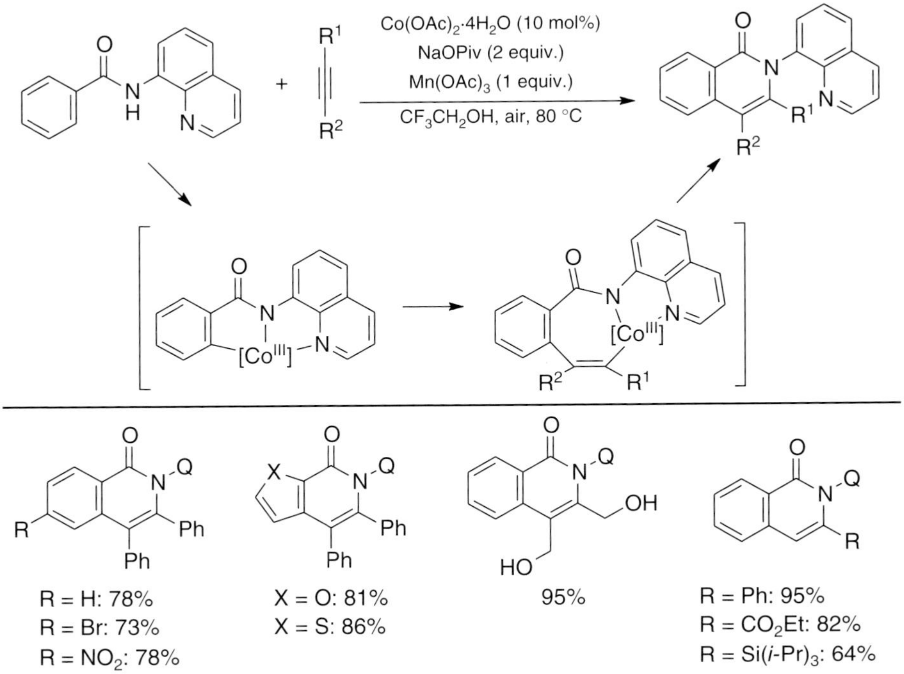

Scheme 10.16 Cobalt-catalyzed oxidative annulation of 8-aminoquinoline-bearing benzamide and alkyne (Q = 8-quinolinyl).

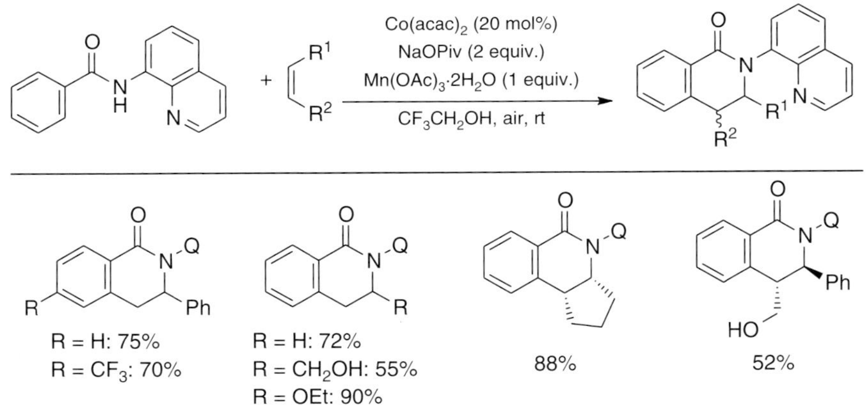

Scheme 10.17 Cobalt-catalyzed oxidative annulation of 8-aminoquinoline-bearing benzamide and olefin (Q = 8-quinolinyl).

ethylene, alkyl olefin, and vinyl ether and internal olefins such as cycloalkene and cinnamyl alcohol. Interestingly, both *E*- and *Z*-isomers of cinnamyl alcohol afford the annulation product with *trans* configuration.

Grigorjeva and Daugulis further extended the scope of the cobalt catalysis of 8-aminoquinoline benzamide to achieve oxidative carbonylation

(Scheme 10.18) [44]. The use of the catalytic system used in the annulation with olefins (Scheme 10.17) under an atmosphere of carbon monoxide and air effects *ortho*-carbonylation of benzamides and concomitant C–N bond formation, thus affording phthalimide derivatives in moderate to excellent yields. As is the case with the annulation reactions of alkynes and olefins, the carbonylation reaction exhibits tolerance to a broad range of functional groups. The quinolinyl group on the product can be removed by treatment with ammonia.

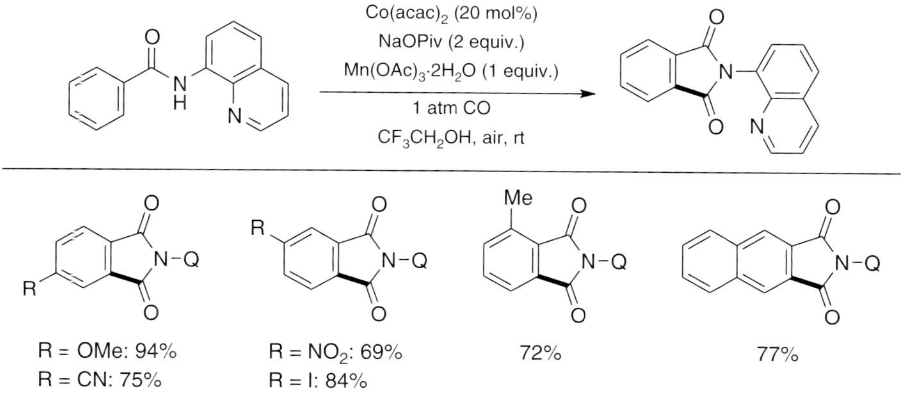

Scheme 10.18 Cobalt-catalyzed oxidative carbonylation of 8-aminoquinoline-bearing benzamide (Q = 8-quinolinyl).

10.4 Heterocycle Synthesis via C–H Functionalization under Co(II)-Based Metalloradical Catalysis

C–H alkylation and amination reactions involving metal–carbenoid and metal–nitrenoid species have been developed for many years, most extensively with (chiral) dirhodium(II) carboxylate and carboxamidate complexes as catalysts [45]. When performed in intramolecular settings, such reactions offer versatile methods for the (enantioselective) synthesis of hetero- and carbocycles. In the past decade, Zhang and coworkers had explored the catalysis of cobalt(II)–porphyrin complexes for carbene- and nitrene-transfer reactions [46] and revealed a radical nature of such processes as a distinct mechanistic feature compared with typical metal (e.g., rhodium)-catalyzed carbenoid and nitrenoid reactions [47]. Described below are examples of heterocycle synthesis via cobalt(II)–porphyrin-catalyzed intramolecular C–H amination or C–H alkylation.

In 2007, cobalt(II)–tetraphenylporphyrin (TPP) complex was reported to be a competent catalyst for intramolecular amination of benzylic C–H bond through decomposition of sulfonyl azide (Scheme 10.19) [48]. Thus, arylsulfonyl azides

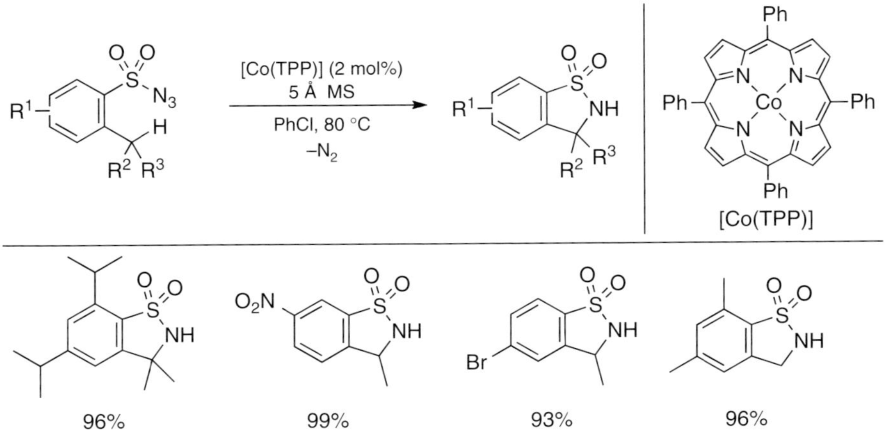

Scheme 10.19 Intramolecular benzylic C–H amination with arylsulfonyl azide catalyzed by [Co(TPP)].

bearing *ortho*-isopropyl, ethyl, or methyl groups undergo denitrogenative cyclization to afford five-membered benzosultam derivatives in excellent yields. Note that, with a longer alkyl group (e.g., propyl, butyl) on the *ortho* position, the reaction affords a mixture of five- and six-membered benzosultams due to competitive amination of the benzylic and homobenzylic C–H bonds.

The scope of the cobalt-catalyzed C–H amination was extended to phosphoryl azide derivatives (Scheme 10.20) [49]. Amination of the benzylic C–H bond of an alkyl group *ortho* to the phosphoryl azide is efficiently promoted by a cobalt(II) complex [Co(**P1**)] with a D_{2h}-symmetric tetraarylporphyrin ligand, which features proximal amide functionalities, thus affording a six-membered cyclic product. The amide functionalities of the catalyst appear to play critical roles in the amination reaction presumably by facilitating catalyst–substrate

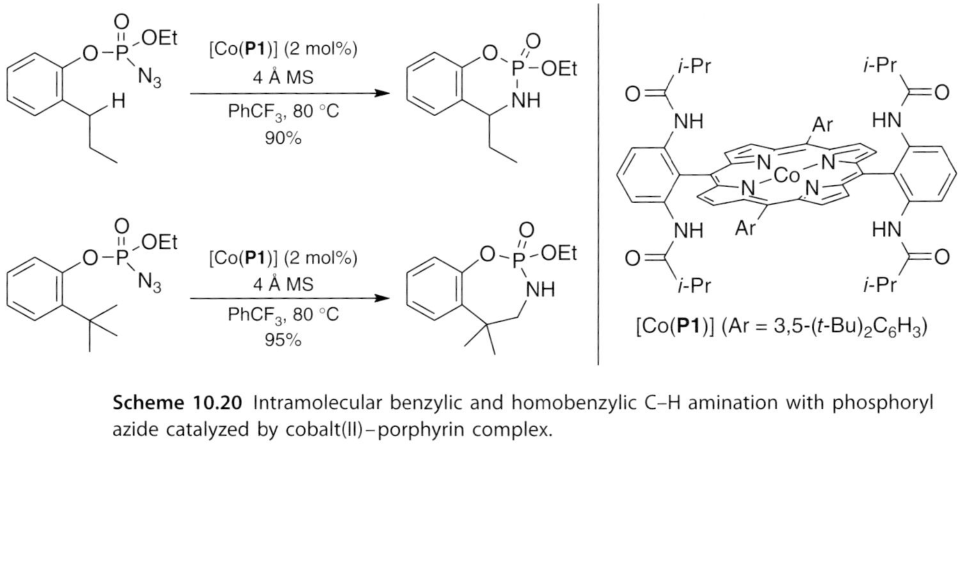

Scheme 10.20 Intramolecular benzylic and homobenzylic C–H amination with phosphoryl azide catalyzed by cobalt(II)–porphyrin complex.

association through hydrogen bonding, because no desired product is obtained with the simple [Co(TPP)] catalyst. When the substrate bears a *tert*-butyl group on the *ortho* position, the reaction affords a seven-membered cyclic product through amination of the homobenzylic C–H bond.

The elaborate cobalt(II)–porphyrin catalyst [Co(**P1**)] further enabled intramolecular amination of C(sp^3)–H bonds in simple alkyl chains, adjacent to electron-withdrawing groups, or at the allylic or propargylic positions through decomposition of sulfamoyl azide (Scheme 10.21) [50]. Thus, the catalyst promotes nitrene insertion into such C(sp^3)–H bonds located at the γ-position of the sulfamoyl group to afford six-membered cyclic sulfamides in excellent yields, which serve as versatile 1,3-diamine scaffolds. On the basis of mechanistic experiments, a mechanism that is distinct from related Rh^{II}-catalyzed C–H amination was proposed. Thus, the 15-electron metalloradical catalyst [Co(**P1**)] decomposes the sulfamoyl azide to form a "Co^{III}-radical nitrene" species, which

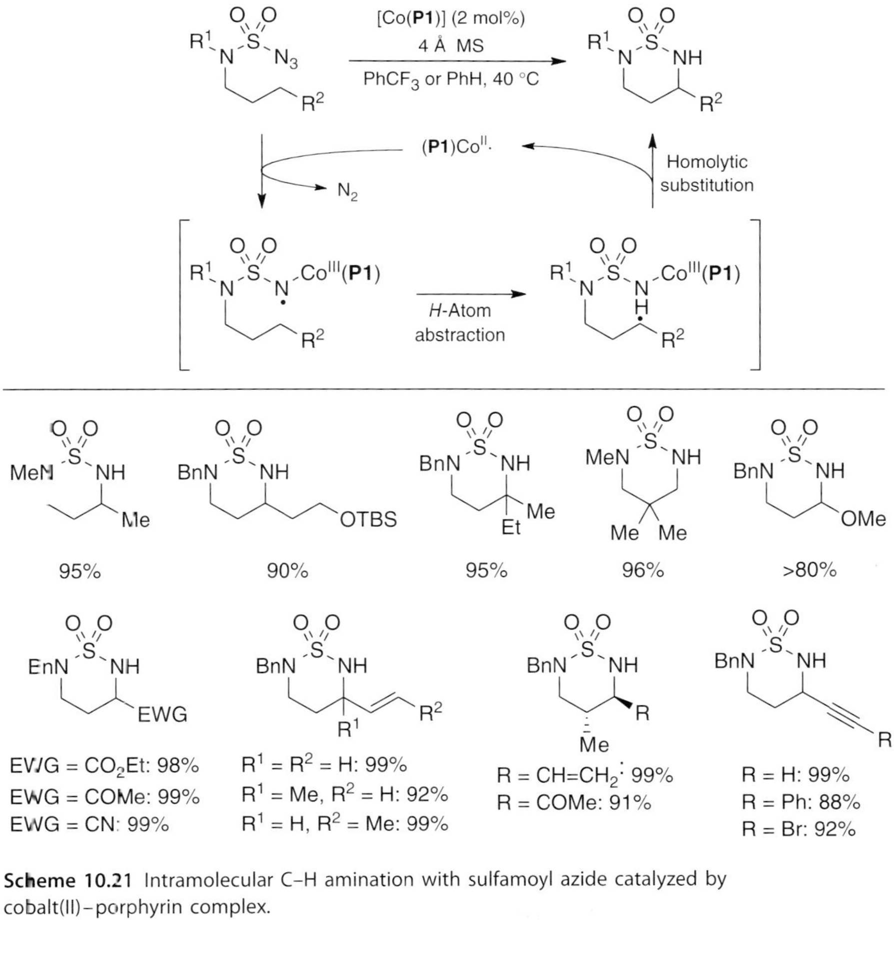

Scheme 10.21 Intramolecular C–H amination with sulfamoyl azide catalyzed by cobalt(II)–porphyrin complex.

abstracts the hydrogen atom in a 1,6-fashion. The resulting alkyl radical undergoes homolytic substitution of the cobalt(III)–amide moiety to afford the cyclic product and regenerate the Co^{II} catalyst.

Zhang and coworkers also achieved cobalt(II)-catalyzed enantioselective intramolecular alkylation of a benzylic or allylic C–H bond with an acceptor/acceptor-substituted diazo compound as a carbenoid precursor (Scheme 10.22) [51]. Carbene transfer reactions of such diazo compounds are known to be challenging, because (i) they are electron-poor and hence less reactive toward typical Lewis acidic metal catalysts and (ii) once formed, the corresponding metal carbenoids would be highly electrophilic and thus difficult to be tamed. With a cobalt(II)–tetraarylporphyrin complex [Co(**P2**)] bearing chiral amide functionalities as a catalyst, a diazo substrate bearing sulfonyl and ester groups undergoes stereoselective intramolecular C–H alkylation to afford a five-membered cyclic sulfone with high diastereo- and enantioselectivities. The reaction is proposed to go through α-Co^{III}–alkyl radical (Co^{III}–carbene radical), which undergoes 1,5-hydrogen abstraction followed by intramolecular radical substitution. The amide N–H bonds are assumed to be hydrogen-bonded to the sulfonyl and the ester groups, thus controlling the stereochemical course of the reaction.

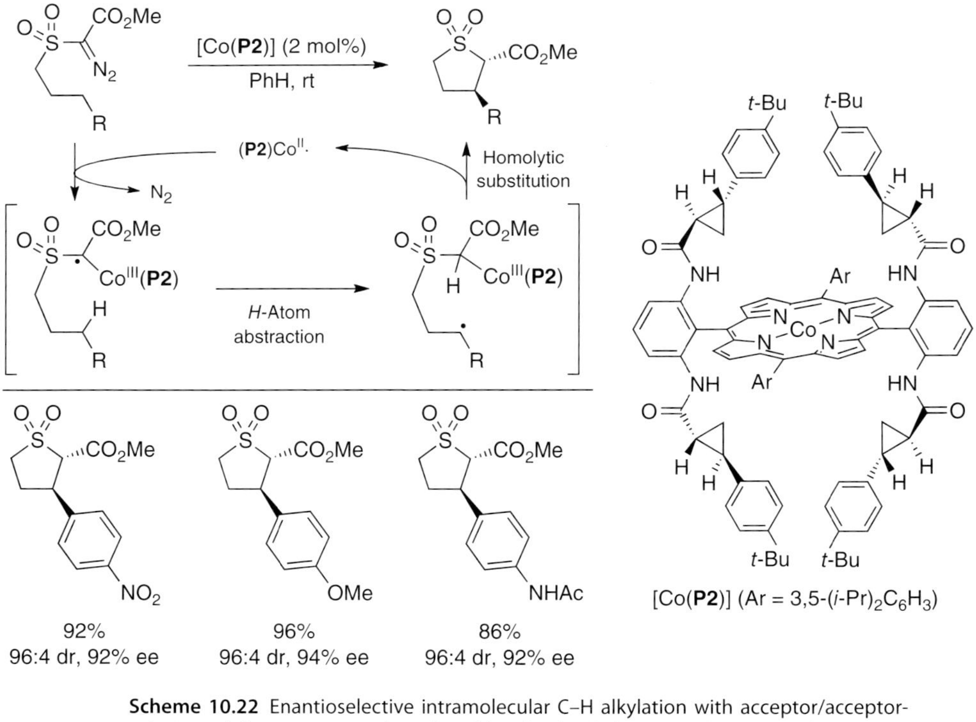

Scheme 10.22 Enantioselective intramolecular C–H alkylation with acceptor/acceptor-substituted diazo compound catalyzed by chiral cobalt(II)–porphyrin complex.

10.5
Summary and Conclusions

The last decade has seen significant advancements in cobalt-catalyzed C–H functionalization through both inner-sphere and outer-sphere C–H activation mechanisms. Inner-sphere C–H activation can now be achieved in various settings with both low-valent and high-valent cobalt catalysts. With careful choice of the starting materials, the catalyst, and the reaction conditions, such C–H activation processes offer valuable direct and indirect routes to various heterocycles, heteroaromatics in particular. On the other hand, cobalt(II)–porphyrin complexes have proved to be efficient catalysts for intramolecular C–H amination and alkylation reactions through outer-sphere C–H cleavage with cobalt(III)–nitrene and cobalt(III)–carbene radical species. With this distinct mechanistic feature, the cobalt(II) metalloradical catalysis offers unique methods for the synthesis of saturated heterocycles. Collectively, the heterocycle-forming reactions discussed in this chapter illustrate that cobalt catalysis may not only offer viable and affordable alternative to analogous rhodium catalysis but also enable unique and unprecedented synthetic transformations, although its overall synthetic scope is still narrower than that of rhodium and other noble metal catalyses. Further efforts of chemists in this field are desired to enhance the versatility and practicality of cobalt-catalyzed C–H activation for heterocycle synthesis.

References

1. (a) Varela, J.A. and Saá, C. (2003) *Chem. Rev.*, **103**, 3787–3801; (b) Kotha, S., Brahmachary, E., and Lahiri, K. (2005) *Eur. J. Org. Chem.*, 4741–4767; (c) Chopade, P.R. and Louie, J. (2006) *Adv. Synth. Catal.*, **348**, 2307–2327; (d) Heller, B. and Hapke, M. (2007) *Chem. Soc. Rev.*, **36**, 1085–1094; (e) Domínguez, G. and Pérez-Castells, J. (2011) *Chem. Soc. Rev.*, **40**, 3430–3444.
2. (a) Brummond, K.M. and Kent, J.L. (2000) *Tetrahedron*, **56**, 3263–3283; (b) Gibson, S.E. and Stevenazzi, A. (2003) *Angew. Chem. Int. Ed.*, **42**, 1800–1810; (c) Blanco-Urgoiti, J., Añorbe, L., Pérez-Serrano, L., Domínguez, G., and Pérez-Castells, J. (2004) *Chem. Soc. Rev.*, **33**, 32–42; (d) Gibson, S.E. and Mainolfi, N. (2005) *Angew. Chem. Int. Ed.*, **44**, 3022–3037; (e) Shibata, T. (2006) *Adv. Synth. Catal.*, **348**, 2328–2336; (f) Lee, H.-W. and Kwong, F.-Y. (2010) *Eur. J. Org. Chem.*, 789–811.
3. Murahashi, S. (1955) *J. Am. Chem. Soc.*, **77**, 6403–6404.
4. Murahashi, S. and Horiie, S. (1956) *J. Am. Chem. Soc.*, **78**, 4816–4817.
5. Halbritter, G., Knoch, F., Wolski, A., and Kisch, H. (1994) *Angew. Chem. Int. Ed.*, **33**, 1603–1605.
6. (a) Jacob, J., Cavalier, C.M., Jones, W.D., Godleski, S.A., and Valente, R.R. (2002) *J. Mol. Catal. A*, **182**, 565–570; (b) Jacob, J. and Jones, W.D. (2003) *J. Org. Chem.*, **68**, 3563–3568; (c) Li, L. and Jones, W.D. (2007) *J. Am. Chem. Soc.*, **129**, 10707–10713.
7. (a) Kakiuchi, F. and Kochi, T. (2008) *Synthesis*, 3013–3039; (b) Colby, D.A., Bergman, R.G., and Ellman, J.A. (2010) *Chem. Rev.*, **110**, 624–655; (c) Lyons, T.W. and Sanford, M.S. (2010) *Chem. Rev.*, **110**, 1147–1169; (d) Engle, K.M., Mei, T.-S., Wasa, M., and Yu, J.-Q. (2012) *Acc. Chem. Res.*, **45**, 788–802; (e) Arockiam, P.B., Bruneau, C., and Dixneuf, P.H. (2012) *Chem. Rev.*, **112**, 5879–5918.
8. Kulkarni, A.A. and Daugulis, O. (2009) *Synthesis*, 4087–4109.

9. (a) Gao, K. and Yoshikai, N. (2014) *Acc. Chem. Res.*, **47**, 1208–1219; (b) Yoshikai, N. (2014) *Bull. Chem. Soc. Jpn.*, **87**, 843–857; (c) Ackermann, L. (2014) *J. Org. Chem.*, **79**, 8948–8954; (d) Yoshikai, N. (2015) *ChemCatChem*, **7**, 732–734.
10. (a) Klein, H.-F., Helwig, M., Koch, U., Flörke, U., and Haupt, H.J. (1993) *Z. Naturforsch. B.*, **48**, 778–784; (b) Klein, H.-F., Camadanli, S., Beck, R., Leukel, D., and Flörke, U. (2005) *Angew. Chem. Int. Ed.*, **44**, 975–977.
11. Albrecht, M. (2010) *Chem. Rev.*, **110**, 576–623.
12. Early reports: (a) Gao, K., Lee, P.-S., Fujita, T., and Yoshikai, N. (2010) *J. Am. Chem. Soc.*, **132**, 12249–12251; (b) Gao, K. and Yoshikai, N. (2011) *J. Am. Chem. Soc.*, **133**, 400–402; (c) Ilies, L., Chen, Q., Zeng, X., and Nakamura, E. (2011) *J. Am. Chem. Soc.*, **133**, 5221–5223.
13. Yamakawa, T. and Yoshikai, N. (2013) *Org. Lett.*, **15**, 196–199.
14. (a) Lee, P.-S., Fujita, T., and Yoshikai, N. (2011) *J. Am. Chem. Soc.*, **133**, 17283–17295; (b) Fallon, B.J., Derat, E., Amatore, M., Aubert, C., Chemla, F., Ferreira, F., Perez-Luna, A., and Petit, M. (2015) *J. Am. Chem. Soc.*, **137**, 2448–2451.
15. Colby, D.A., Bergman, R.G., and Ellman, J.A. (2008) *J. Am. Chem. Soc.*, **130**, 3645–3651.
16. (a) Duttwyler, S., Lu, C., Rheingold, A.L., Bergman, R.G., and Ellman, J.A. (2012) *J. Am. Chem. Soc.*, **134**, 4064–4067; (b) Ischay, M.A., Takase, M.K., Bergman, R.G., and Ellman, J.A. (2013) *J. Am. Chem. Soc.*, **135**, 2478–2481; (c) Duttwyler, S., Chen, S., Lu, C., Mercado, B.Q., Bergman, R.G., and Ellman, J.A. (2014) *Angew. Chem. Int. Ed.*, **53**, 3877–3880; (d) Duttwyler, S., Chen, S., Takase, M.K., Wiberg, K.B., Bergman, R.G., and Ellman, J.A. (2013) *Science*, **339**, 678–682.
17. Ding, Z. and Yoshikai, N. (2013) *Angew. Chem. Int. Ed.*, **52**, 8574–8578.
18. (a) Jun, C.-H., Jo, E.-A., and Park, J.-W. (2007) *Eur. J. Org. Chem.*, 1869–1881; (b) Willis, M.C. (2010) *Chem. Rev.*, **110**, 725–748; (c) Leung, J.C. and Krische, M.J. (2012) *Chem. Sci.*, **3**, 2202–2209; (d) Murphy, S.K. and Dong, V.M. (2014) *Chem. Commun.*, **50**, 13645–13649.
19. Vinogradov, M.G., Tuzikov, A.B., Nikishin, G.I., Shelimov, B.N., and Kazansky, V.B. (1988) *J. Organomet. Chem.*, **348**, 123–134.
20. Lenges, C.P. and Brookhart, M. (1997) *J. Am. Chem. Soc.*, **119**, 3165–3166.
21. Yang, J. and Yoshikai, N. (2014) *J. Am. Chem. Soc.*, **136**, 16748–16751.
22. Phan, D.H.T., Kim, B., and Dong, V.M. (2009) *J. Am. Chem. Soc.*, **131**, 15608–15609.
23. Kundu, K., McCullagh, J.V., and Morehead, A.T. Jr., (2005) *J. Am. Chem. Soc.*, **127**, 16042–16043.
24. Fiebig, L., Kuttner, J., Hilt, G., Schwarzer, M.C., Frenking, G., Schmalz, H.-G., and Schäfer, M. (2013) *J. Org. Chem.*, **78**, 10485–10493.
25. Tan, B.-H., Dong, J., and Yoshikai, N. (2012) *Angew. Chem. Int. Ed.*, **51**, 9610–9614.
26. (a) Ma, S. and Gu, Z. (2005) *Angew. Chem. Int. Ed.*, **44**, 7512–7517; (b) Shi, F. and Larock, R.C. (2010) *Top. Curr. Chem.*, **292**, 123–164.
27. Sasaki, K., Nishimura, T., Shintani, R., Kantchev, E.A.B., and Hayashi, T. (2012) *Chem. Sci.*, **3**, 1278–1283.
28. Wu, B. and Yoshikai, N. (2013) *Angew. Chem. Int. Ed.*, **52**, 10496–10499.
29. Wu, B., Santra, M., and Yoshikai, N. (2014) *Angew. Chem. Int. Ed.*, **53**, 7543–7546.
30. (a) Satoh, T. and Miura, M. (2010) *Chem. Eur. J.*, **16**, 11212–11222; (b) Patureau, F.W., Wencel-Delord, J., and Glorius, F. (2012) *Aldrichimica Acta*, **45**, 31–41; (c) Song, G., Wang, F., and Li, X. (2012) *Chem. Soc. Rev.*, **41**, 3651–3678; (d) Kuhl, N., Schröder, N., and Glorius, F. (2014) *Adv. Synth. Catal.*, **356**, 1443–1460.
31. Yoshino, T., Ikemoto, H., Matsunaga, S., and Kanai, M. (2013) *Angew. Chem. Int. Ed.*, **52**, 2207–2211.
32. Ikemoto, H., Yoshino, T., Sakata, K., Matsunaga, S., and Kanai, M. (2014) *J. Am. Chem. Soc.*, **136**, 5424–5431.
33. Ackermann, L. (2011) *Chem. Rev.*, **111**, 1315–1345.

34. Schipper, D.J., Hutchinson, M., and Fagnou, K. (2010) *J. Am. Chem. Soc.*, **132**, 6910–6911.
35. Hummel, J.R. and Ellman, J.A. (2015) *J. Am. Chem. Soc.*, **137**, 490–498.
36. Lian, Y., Bergman, R.G., Lavis, L.D., and Ellman, J.A. (2013) *J. Am. Chem. Soc.*, **135**, 7122–7125.
37. Lian, Y., Huber, T., Hesp, K.D., Bergman, R.G., and Ellman, J.A. (2013) *Angew. Chem. Int. Ed.*, **52**, 629–633.
38. Zhao, D., Kim, J.H., Stegemann, L., Strassert, C.A., and Glorius, F. (2015) *Angew. Chem. Int. Ed.* doi: 10.1002/anie.201411994
39. Sun, B., Yoshino, T., Matsunaga, S., and Kanai, M. (2014) *Adv. Synth. Catal.*, **356**, 1491–1495.
40. Zaitsev, V.G., Shabashov, D., and Daugulis, O. (2005) *J. Am. Chem. Soc.*, **127**, 13154–13155.
41. Rouquet, G. and Chatani, N. (2013) *Angew. Chem. Int. Ed.*, **52**, 11726–11743.
42. Grigorjeva, L. and Daugulis, O. (2014) *Angew. Chem. Int. Ed.*, **53**, 10209–10212.
43. Grigorjeva, L. and Daugulis, O. (2014) *Org. Lett.*, **16**, 4684–4687.
44. Grigorjeva, L. and Daugulis, O. (2014) *Org. Lett.*, **16**, 4688–4690.
45. (a) Doyle, M.P., Yu, L., and Ratnikov, M.O. (2013) *Org. React.*, **80**, 1–131; (b) Davies, H.M.L. and Lian, Y. (2012) *Acc. Chem. Res.*, **45**, 923–935; (c) Roizen, J.L., Harvey, M.E., and Bois, J.D. (2012) *Acc. Chem. Res.*, **45**, 911–922; (d) Collet, F., Lescot, C., and Dauban, P. (2011) *Chem. Soc. Rev.*, **40**, 1926–1936; (e) Davies, H.M.L. and Morton, D. (2011) *Chem. Soc. Rev.*, **40**, 1857–1869; (f) Doyle, M.P., Duffy, R., Ratnikov, M., and Zhou, L. (2010) *Chem. Rev.*, **110**, 704–724; (g) Davies, H.M.L. and Manning, J.R. (2008) *Nature*, **451**, 417–424.
46. (a) Lu, H. and Zhang, X.P. (2011) *Chem. Soc. Rev.*, **40**, 1899–1909; (b) Che, C.-M., Lo, V.K.-Y., Zhou, C.-Y., and Huang, J.-S. (2011) *Chem. Soc. Rev.*, **40**, 1950–1975.
47. (a) Dzik, W.I., Xu, X., Zhang, X.P., Reek, J.N.H., and de Bruin, B. (2010) *J. Am. Chem. Soc.*, **132**, 10891–10902; (b) Lu, H., Dzik, W.I., Xu, X., Wojtas, L., de Bruin, B., and Zhang, X.P. (2011) *J. Am. Chem. Soc.*, **133**, 8518–8521; (c) Lyaskovskyy, V., Suarez, A.I.O., Lu, H., Jiang, H., Zhang, X.P., and de Bruin, B. (2011) *J. Am. Chem. Soc.*, **133**, 12264–12273.
48. Ruppel, J.V., Kamble, R.M., and Zhang, X.P. (2007) *Org. Lett.*, **9**, 4889–4892.
49. Lu, H., Tao, J., Jones, J.E., Wojtas, L., and Zhang, X.P. (2010) *Org. Lett.*, **12**, 1248–1251.
50. (a) Lu, H., Jiang, H., Wojtas, L., and Zhang, X.P. (2010) *Angew. Chem. Int. Ed.*, **49**, 10192–10196; (b) Lu, H., Jiang, H., Hu, Y., Wojtas, L., and Zhang, X.P. (2011) *Chem. Sci.*, **2**, 2361–2366; (c) Lu, H., Hu, Y., Jiang, H., Wojtas, L., and Zhang, X.P. (2012) *Org. Lett.*, **14**, 5158–5161; (d) Lu, H., Li, C., Jiang, H., Lizardi, C.L., and Zhang, X.P. (2014) *Angew. Chem. Int. Ed.*, **53**, 7028–7032.
51. Cui, X., Xu, X., Jin, L.-M., Wojtas, L., and Zhang, X.P. (2015) *Chem. Sci.*, **6**, 1219–1224.

11
Ir-Catalyzed Heterocycles Synthesis

Yasushi Obora

11.1
Introduction

Organic transformations which employ waste-minimizing protocols are important targets in the search for environmentally benign synthetic methods. To achieve this, considerable attention has been paid to the development of these synthetic methodologies. In particular, reactions involving direct C–H bond activation have been the focus of many investigations.

As an alternative, iridium complexes show exciting catalytic activities in various organic transformations for C–C bond formation. Iridium complexes have been known to be effective catalysts for hydrogenation [1–5] and hydrogen transfers [6–27], including in enantioselective synthesis [28–47]. The catalytic activity of iridium complexes also covers a wide range for dehydrogenation [48–54], metathesis [55], hydroamination [56–61], hydrosilylation [62], and hydroalkoxylation reactions [63] and has been employed in alkyne–alkyne and alkyne–alkene cyclizations and allylic substitution reactions [64–114]. In addition, Ir-catalyzed asymmetric 1,3-dipolar cycloaddition of α,β-unsaturated nitriles with nitrone was reported [115].

Therefore, direct functionalization by iridium-catalyzed reactions involving C–H bond activation would provide an alternative protocol to the existing multistep organic synthesis [116–130]. To date, iridium-catalyzed C–H activation of aromatic rings for reactions such as borylation [131–136], alkylation/alkenylation [137–143] and silylation [144–146], and cross-coupling [147, 148] has been investigated.

This chapter will summarize the recent progress on the iridium-catalyzed heterocyclization reactions involving C–H bond activation.

Transition Metal-Catalyzed Heterocycle Synthesis via C—H Activation, First Edition. Edited by Xiao-Feng Wu.
© 2016 Wiley-VCH Verlag GmbH & Co. KGaA. Published 2016 by Wiley-VCH Verlag GmbH & Co. KGaA.

11.2
Ir-Catalyzed Heterocyclization by *ortho*-Aryl C–H Activation

11.2.1
Ir-Catalyzed [3+2] Cyclization of Ketimines with 1,3-Dienes/Alkynes

Nishimura and coworkers reported the Ir-catalyzed cyclization of *N*-sulfonyl ketimines with 1,3-dienes using a cationic Ir complex derived from $[IrCl(cod)]_2$ (cod = 1,5-cyclooctadiene), which provided spiroaminoindane derivatives [149]. In a typical example, *N*-sulfonyl ketimine was reacted with isoprene (2 equiv.) and a catalytic amount of $[IrCl(cod)]_2$ (5 mol%) and 1,4-diazabicyclo[2.2.2]octane (DABCO) (5 mol%) and $NaBAr^F{}_4$ ($Ar^F = 3,5\text{-}(CF_3)_2C_6H_3$) (10 mol%) in toluene for 20 h at 80 °C and led to the corresponding spiroaminoindane derivatives in 94% yield (Scheme 11.1). The reaction proceeded in high regio- and diastereoselectivity with only 5% of a regioisomer being produced (Scheme 11.1).

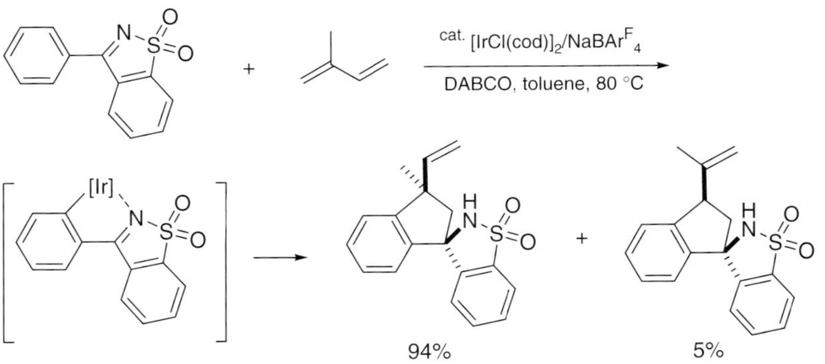

Scheme 11.1

The use of a cationic iridium complex efficiently catalyzed this transformation through the formation of an aryliridium(I) intermediate by *ortho*-aryl C–H bond activation, followed by reaction with the 1,3-diene.

The enantioselective [3+2] cyclization of *N*-acyl ketimines and 1,3-dienes involving C–H activation has been achieved by the use of a cationic iridium catalyst with chiral diene ligands and was reported by the same authors [150]. In a typical example, 3-phenyl-3-hydroxyisoindolin-1-one, which *in situ* formed a *N*-acyl ketimine by dehydration, reacted with isoprene in the presence of a catalytic amount (5 mol%) of iridium complex including a dimethyl-substituted chiral tetrafluorobenzobarrelene ligand $[IrCl((S,S)\text{-Me-tfb*})]_2$, with DABCO (5 mol%) and $NaBAr^F{}_4$ (10 mol%) in toluene for 20 h at 80 °C which led to the chiral spiroisoindolin-1-one in 84% yield with 91% enantiomeric excess (Scheme 11.2 and Figure 11.1). The use of a chiral diene ligand is also highly effective in achieving the cyclization with high regio- and enantioselectivity. The

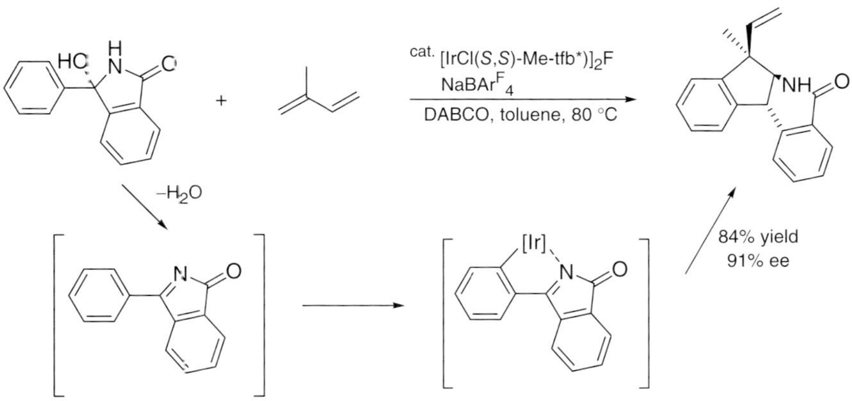

Scheme 11.2

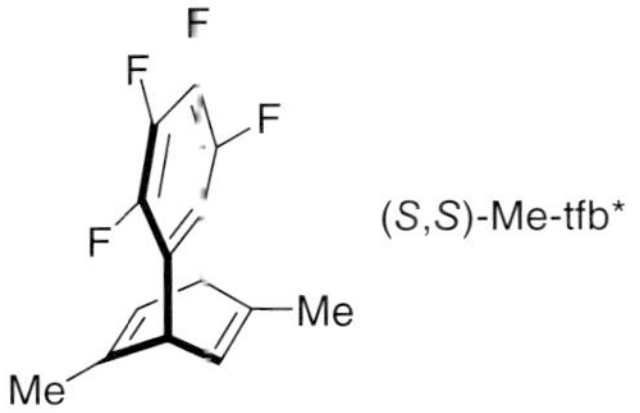

Figure 11.1 Structure of (*S*,*S*)-Me-tfb*

absolute configuration of the resulting heteroannulation product was determined to be (1*R*, 3*R*).

Additionally, the same group reported the cyclization of *N*-acyl ketimines with alkynes using iridium-catalyzed *ortho*-aryl C–H activation [151]. For instance, the reaction of 3-hydroxy-3-phenylisoindolin-1-one with diphenylacetylene with a catalytic amount of $[IrCl(cod)]_2$ (5 mol%) combined with $NaBAr^F{}_4$ (10 mol%) in toluene for 20 h at 80 °C led to the corresponding aminoindane derivative in 97% yield (Scheme 11.3). In contrast to the aforementioned catalytic [3+2] annulations of ketimines with 1,3-dienes, the use of DABCO as an additive totally

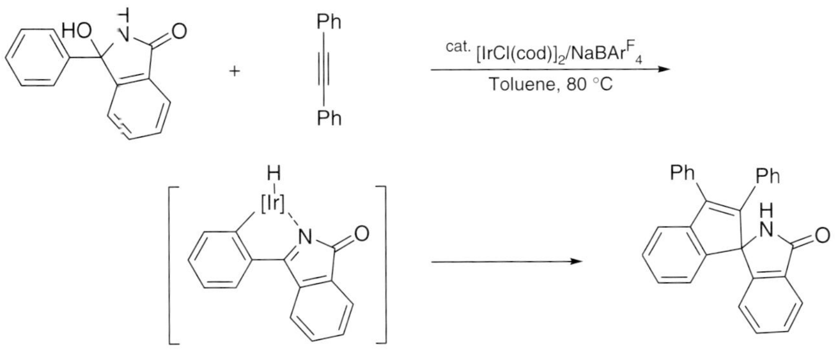

Scheme 11.3

suppressed this reaction. It is proposed that this reaction proceeds by the formation of an aryliridium–hydridoiridium(III) intermediate by oxidative addition to the *ortho*-aryl C–H bond. Subsequent alkyne insertion into the Ir–H bond gave an alkenyliridium(III) intermediate followed by cyclization, producing the desired aminoindane derivatives.

11.2.2
Ir-Catalyzed Cyclization of Benzoic Acid to Give 2-Hydroxy-6*H*-benzo[*c*]chromen-6-ones

Ison and coworkers reported that the formation of 2-hydroxy-6*H*-benzo[*c*]chromen-6-ones can be achieved by C–H bond functionalization of benzoic acid [152]. In a typical example, the benzoic acid (1 equiv.) was reacted with benzoquinone (2 equiv.) in the presence of NaOAc (2 equiv.) and a catalytic amount (10 mol%) of $[Cp^*IrCl_2]_2$ in toluene at 120 °C for 24 h and gave 2-hydroxy-6*H*-benzo[*c*]chromen-6-one in 83% yield (Scheme 11.4).

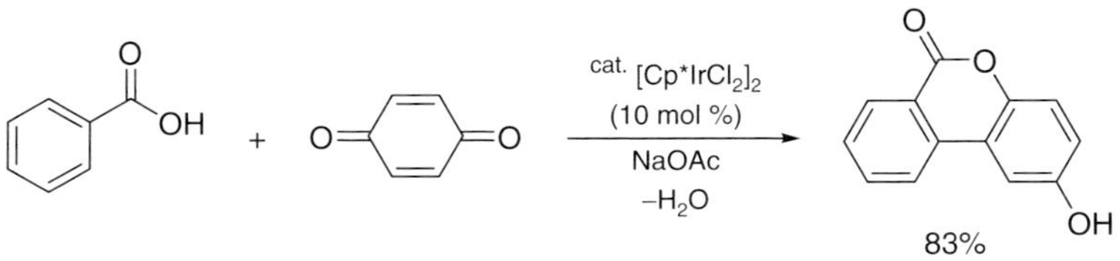

Scheme 11.4

In this reaction, the use of NaOAc is crucial to the reaction because Cp*Ir(OAc), derived from $[Cp^*IrCl_2]_2$ and NaOAc, is the active catalyst. It has been proposed that the reaction proceeds by initial C–H bond activation at the *ortho*-aryl position of benzoic acid to form an oxoirida cycle followed by insertion of the benzoquinone into the Ir–C bond of the metallacycle, giving the desired benzo[*c*]chromen-6-one.

11.2.3
Ir-Catalyzed Cyclization of *N*-Arylcarbamoyl Chlorides with Alkynes

Iridium complex-catalyzed cyclization of an *N*-arylcarbamoyl chloride with an alkyne has been reported by Tsuji and coworkers [153]. In a typical example, *N*-methyl-*N*-phenylcarbamoyl was reacted with 5-decyne and a catalytic amount of $[IrCl(cod)]_2$ (2.5 mol%) and additional cod (30 mol%) in refluxing *o*-xylene for 20 h to give 3,4-dimethyl-1-methyl-2-quinolone in 92% yield (Scheme 11.5). During this reaction, no indole product formed by decarbonylation was observed. This reaction is proposed to proceed by oxidative addition of *N*-arylcarbamoyl chloride to Ir(I), giving a carbamoyl chloroiridium(III) species. Subsequently, the formation of a five-membered iridacycle by *ortho*-aryl C–H activation followed by insertion of the alkene and reductive elimination produces the 2-quinolone derivative.

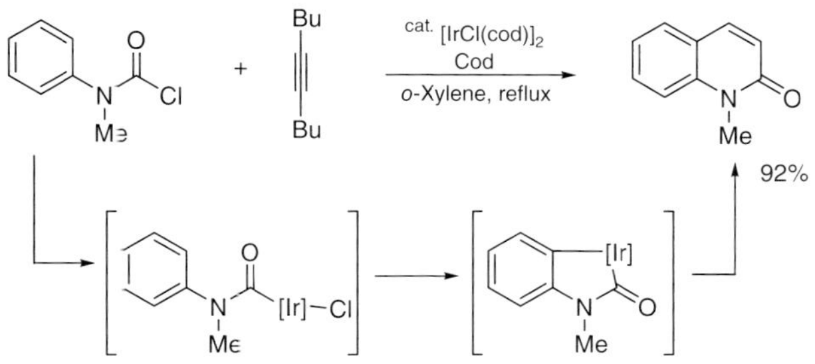

Scheme 11.5

11.3 Ir-Catalyzed Heterocyclization by Benzylic C–H Activation

11.3.1 Ir-Catalyzed *N*-Cyclization of Aryl Azides

Driver and coworkers reported on iridium-catalyzed *N*-heterocyclization by benzylic C–H bond activation [154]. In this case, treatment of an *ortho*-substituted aryl azide with a catalytic amount of $[Ir(OMe)(cod)]_2$ (2 mol%) at 25 °C gave a mixture of the corresponding aniline, indole, and indoline in 19, 13, and 58% yield, respectively, for a total yield of 90% (Scheme 11.6). The introduction of the electron-withdrawing CF_3 group on the aryl substituent led to indoline derivatives in high yield with high selectivity.

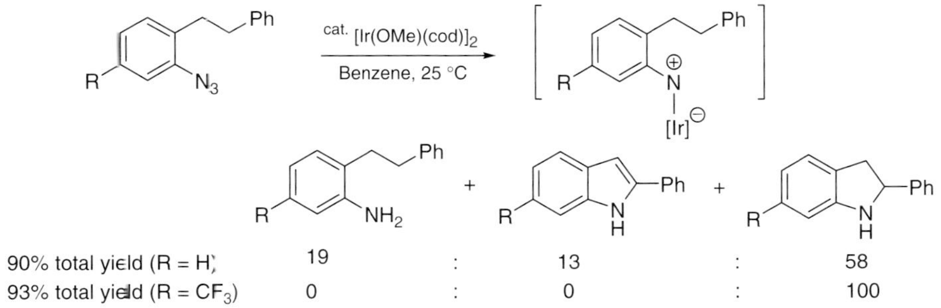

Scheme 11.6

The reaction is proposed to proceed through the formation of an electrophilic iridium nitrenoid.

11.3.2 Ir-Catalyzed Silylation of Benzylic Amines and 2,*N*-Dialkylanilines *via* Aryl C–H Bond Activation

Hartwig reported that the iridium-catalyzed silylation of benzylic amines to afford azasilolane compounds [155].

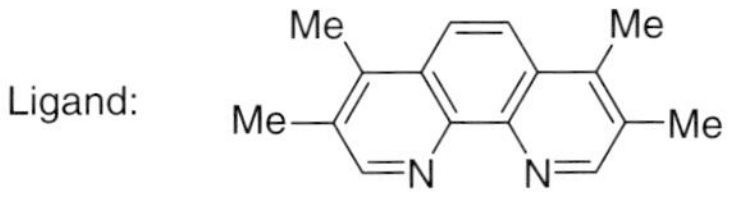

Figure 11.2 Structure of Ligand

In a typical example, the reaction of *N*-methylbenzylamine with diethylsilane was carried out in the presence of $[Ir(cod)(OMe)]_2$ afforded hydridosilylamine. Subsequently, the reaction of the resulting hydridosilylamine in the presence of $[Ir(cod)(OMe)]_2$ combined with 3,4,7,8-tetramethyl-1,10-phenanthroline as a ligand and norbornene as hydrogen acceptor at 80 °C led to the corresponding azasilolane derivatives in 95% yield (Scheme 11.7, Figure 11.2).

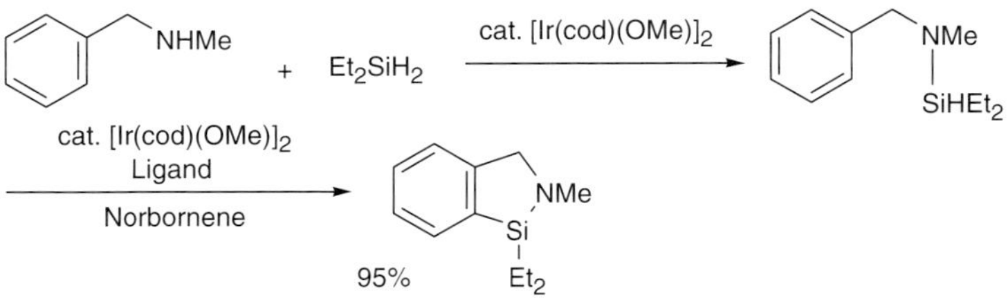

Scheme 11.7

The methodology also applied to the silylative cyclization reaction of 2,*N*-dialkylamines. Thus the reaction of 2,5-dimethyl-*N*-methylaniline with diethylsilane subjected to dehydrogenative coupling and provide hydridosilylaniline. Subsequently, selective silylation of the hydridosilylamine led to the corresponding arylamine derivatives in 91% yield (Scheme 11.8).

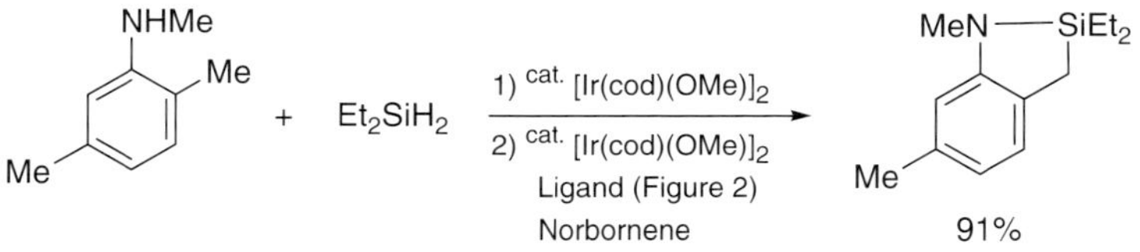

Scheme 11.8

This reaction consists of dehydrogenative silylation of *N*-methylbenzylamine and 2,*N*-dialkylamine by hydrosilane involving *ortho*-aryl C–H or benzyl C–H bond activation.

11.4 Ir-Catalyzed Heterocyclization by sp^3 C–H Activation

11.4.1 Ir-Catalyzed *N*-Cyclization of Aryl Azides

Sames and coworkers reported that the catalytic oxidative cyclization of amide can be achieved by an iridium carbene complex, which consists of $[Ir(coe)_2Cl]_2$,

IPr [*N*,*N*′-bis-(2,6-diisopropylphenyl)imidazolyl] carbine ligand, and norbornene (4 equiv.) [156]. In a typical example, the treatment of an amide that has an alkene moiety with the aforementioned Ir catalyst gave cyclization of the alkene–amide moiety through coupling of the sp^3 C–H bonds. The reaction is proposed to proceed by the formation of an amide-coordinated iridium–IPr–Cl complex as the key intermediate, which reacts by C–H bond activation of the double bond followed by C–C formation to give the cyclized product (Scheme 11.9).

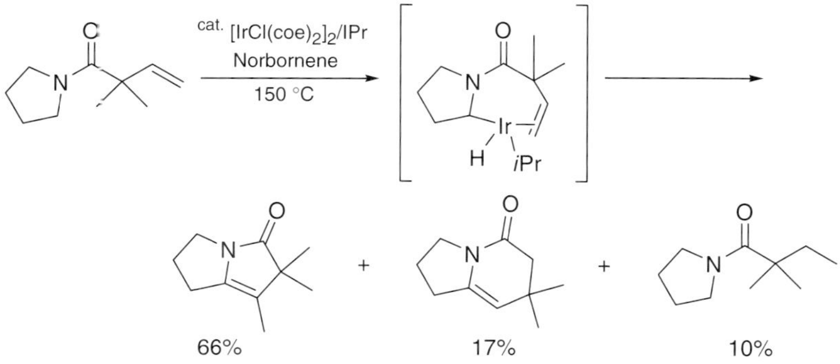

Scheme 11.9

11.5 Heterocyclization by Ir Catalyst as Lewis Acid

Shimizu and coworkers reported that quinolone can be synthesized by an Ir-catalyzed three-component coupling reaction of an amine, an aldehyde, and a ketone [157, 158]. Thus, a mixture of aniline and benzaldehyde was allowed to react with a catalytic amount (5 mol%) of $[IrHCl_2(cod))]_2$ at room temperature for 1 h. Butanal was added to the mixture and heated to 90 °C for 12 h under oxygen, giving 3-ethyl-2-phenylquinoline in 79% yield along with the a small amount of 3-ethyl-2-propylquinoline in 8% yield (Scheme 11.10). When the reaction was carried out under an argon atmosphere, *N*-benzylideneaniline was derived by hydrogen transfer from the Ir catalyst in 20–30% as a by-product.

In this reaction, *N*-benzylideneimine was formed initially by the reaction of aniline with the aldehyde, and the iridium complex catalyst serves as a Lewis acid.

11.6 Ir-Catalyzed Heterocyclization by C–H Bond Activation through Transfer Hydrogenation

11.6.1 Ir-Catalyzed *N*-Heterocyclization of Naphthylamines with Diols

Iridium-catalyzed transfer hydrogenation of alcohol substrates also provides a useful method of C–C bond formation.

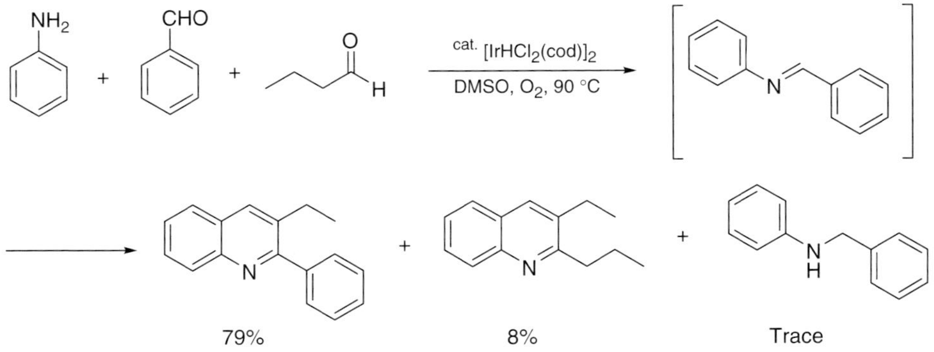

Scheme 11.10

Ishii and coworkers reported that the *N*-heterocyclization of naphthylamines with diols can be achieved with an iridium catalyst. In a typical example, the reaction of 1-naphthylamine with 1,3-propanediol was carried out with a catalytic amount of $IrCl_3$ (5 mol%), *rac*-2,2′-bis(diphenylphosphino)-1,1′-binaphthyl (BINAP) (10 mol%), and the corresponding 7,8-benzoquinoline was obtained in 96% yield (Scheme 11.11) [159]. The proposed reaction mechanism indicates that the imine intermediate is formed by the reaction of the amine and aldehyde by Ir-catalyzed dehydrogenation. Subsequent hydrogenation by the *in situ* generated Ir hydride leads to an aminoalcohol followed by cyclization to the desired quinoline products.

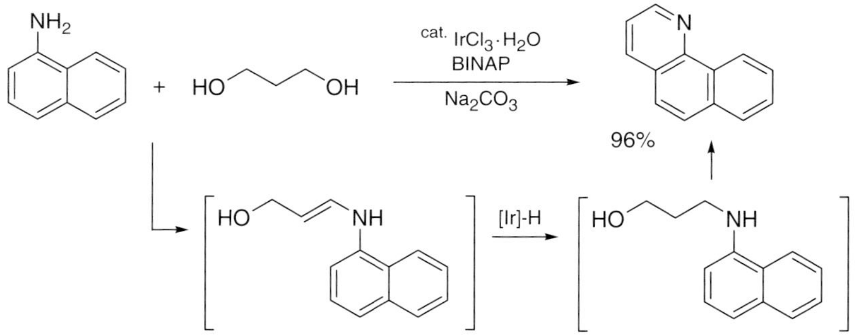

Scheme 11.11

11.6.2
Ir-Catalyzed Reaction of Anilines with Diols to Give 2,3-Disubstituted Indoles

Madsen and coworkers reported Ir-catalyzed synthesis of 2,3,-disubstituted indoles from anilines and diols. A typical reaction involved the treatment of aniline with 2,3-butanediol in the presence of a catalytic amount of $[Cp^{*}IrCl_2]_2$ (1 mol%) combined with methanesulfonic acid (5 mol%) at 170 °C and gave 2,3-dimethylindole in 76% yield [160]. The reaction is proposed to proceed by the formation of an α-hydroxyimine followed by isomerization to an α-aminoketone intermediate. Subsequently, the cyclization of the aminoketone gives the indole product (Scheme 11.12)

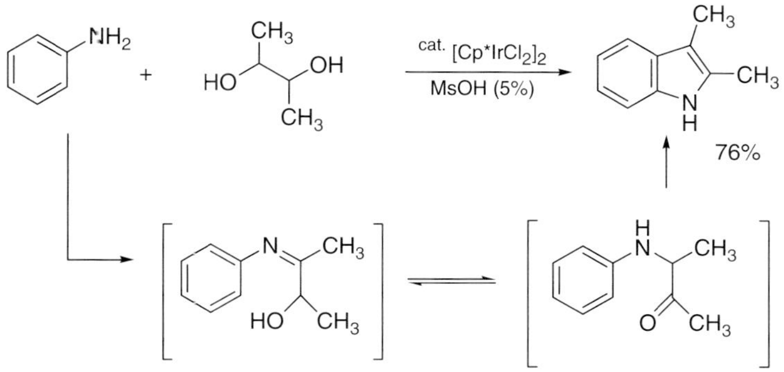

Scheme 11.12

11.6.3
Ir-Catalyzed Synthesis of Indole from 2-Aminoaryl Ethyl Alcohol

Grigg and coworkers reported that the Ir-catalyzed synthesis of indoles from 2-aminoaryl ethyl alcohols [20]. In a typical example, 2-aminophenyl ethyl alcohol and a catalytic amount of $[Cp^*IrCl_2]_2$ (2.5 mol%) were combined with KOH (2 equiv.) at 110 °C for 24 h to give 3-benzylindole in 78% yield with high selectivity (Scheme 11.13).

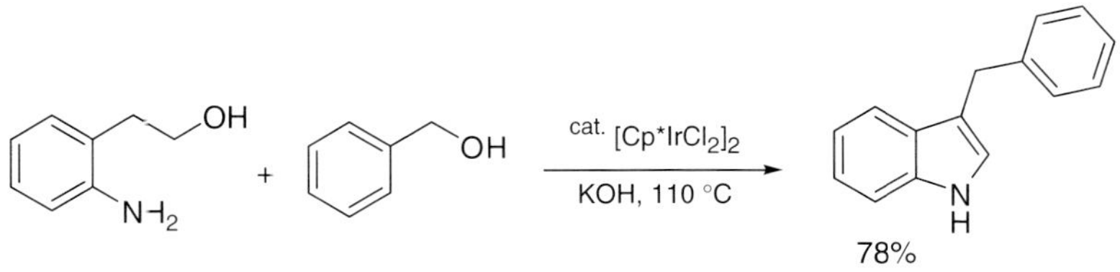

Scheme 11.13

11.6.4
Ir Catalysts with Pyrazoyl and Pyrazoyl-1,2,3-bidentate (N–N) Ligands for the Synthesis of Tricyclic Indoles

Messerle and coworkers reported that a cationic iridium(III) complex with a methyl-bridged bipyrazoyl or pyrazoyl-1,2,3-triazole (N–N) ligand $[Ir(N\text{–}N)Cp^*Cl]BAr^F{}_4$ ($BAr^F{}_4$ = tetrakis[3,5-bis(trifluoromethyl)phenyl]borate) had catalytic activity for the synthesis of tricyclic indoles [161]. The process consists of two incremental steps as shown in Schemes 11.12 and 11.13. First, the treatment of 2-(hydroxyalk-1-ynyl)aniline with a catalytic amount of the iridium catalyst [Ir(N–N)Cp*Cl]BAr$^F{}_4$ (5 mol%) in toluene-d_6 at 110 °C affords 2-(4-hydroxybutyl)indole (Scheme 11.14). This reaction proceeds by the Ir complex-catalyzed intramolecular hydroamination of the alkynylamine.

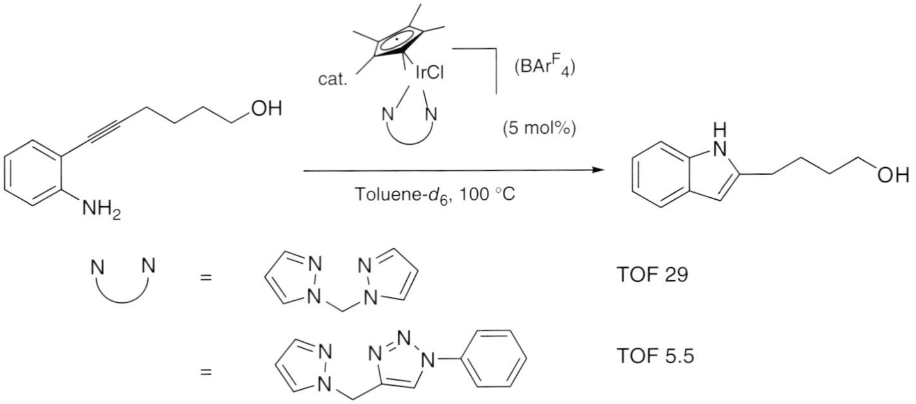

Scheme 11.14

Then, the resulting 2-(4-hydroxybutyl)indole was reacted in the same cationic Ir(III) catalyst system with KOH at 110 °C to afford 1,2,3,4-tetrahydrocarbazole as the product (Scheme 11.15). Here, the iridium catalyst bearing a methyl-bridged bipyrazoyl or a pyrazoyl-1,2,3-triazole (N–N) ligand [Ir(N–N)Cp*Cl]BAr$^F{}_4$ is much more effective than $[Cp^*IrCl_2]_2$ (TOF = 0.2 h^{-1}). This reaction has been proposed to proceed by a C3-alkylation reaction.

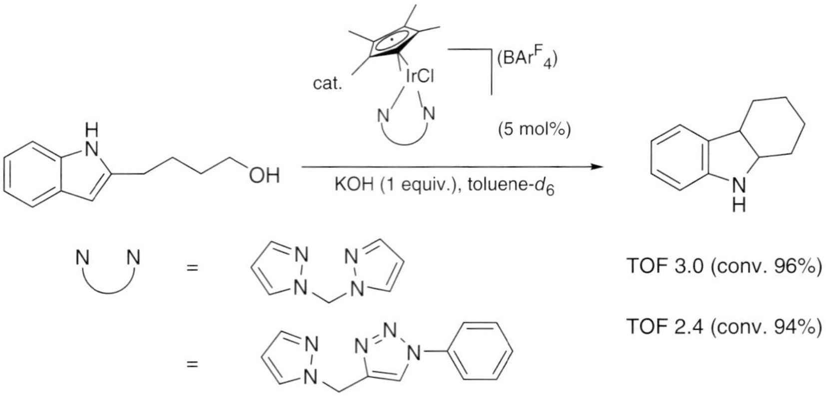

Scheme 11.15

As shown in Schemes 11.14 and 11.15, the reaction proceeded by a tandem C–N and C–C bond formation, which includes the hydroamination reaction followed by alkylation at the C3 position of the resulting 2-(hydroxyalkyl)indole to give tricyclic indoles. Thus the one-pot synthesis of 1,2,3,4-tetrahydrocarbazole from 2-(hydroxyalk-1-ynyl)aniline could also be achieved using the same Ir catalyst $[Ir(N{-}N)Cp^{*}Cl]BAr^{F}{}_{4}$.

11.7 Miscellaneous Reactions

11.7.1 Ir-Catalyzed Arylative Cyclization of Alkynones with Arylboronic Acid

Lam and coworkers reported that the Ir-catalyzed arylative cyclization of alkynone with arylboronic acid afforded cyclic products [162]. Here, 1 equiv. of alkynone was reacted with phenylboronic acid (1.5 equiv.) with a catalytic amount of $[IrCl_2(cod)_2]$ (1.5 mol%) and KF (1.5 equiv.) (to act as a base) and *tert*-butanol (1.5 equiv.) in toluene at 65 °C and provided the tricyclic compound through arylative cyclization in 72% yield along with the 2 : 1 adduct in 27% yield (Scheme 11.16).

In this reaction, 1,4-metal migration by the Ir and aryl-H induced the C–H activation during the reaction. Organic transformations exploiting 1,4-metal migrations are also known for other metals such as Pd, Rh, Pt, Ni, and Co.

This reaction is proposed to proceed the formation of an Ir–Ph species generated by the reaction of the iridium complex with the arylboronic acid, followed by insertion of the Ir–Ph to alkyne group, giving an alkenyliridium intermediate. This then undergoes a 1,4-migration by the Ir and aryl-H (Scheme 11.16). Subsequently, the nucleophilic attack of the aryl-Ir species to the carbonyl group in the alkynone followed by protonation by *tert*-butanol leads to the product (Scheme 11.14). When the reaction of the additional alkynone with the aryl-Ir intermediate takes place, the 2 : 1 adduct was obtained by a sequential 1,4-migration. The 1,4-migration preferentially goes to the sterically less hindered position of the phenyl group.

11.7.2 *N*-Heterocyclization of Aminoalcohol by Ir Catalyst with a Triazolyl-diylidene Ligand

Peris and coworkers reported that an Ir-triazolyl-diylidene complex showed catalytic activity in the cyclization of aminoalcohols [163]. Thus, the reaction of 2-aminophenyl ethyl alcohol with benzyl alcohol can be carried out with a catalytic amount of the Ir-triazolyl-diylidene complex (1 mol%) combined with KOH (2 equiv.) in toluene at 110 °C to give a mixture of bisindoylmethane and 3-benzylindole in 87 : 13 ratio (Scheme 11.17). The reaction is proposed to proceed by the initial formation of a 3-(phenylmethylene)-3*H*-indole intermediate followed by a Michael addition, giving bisindoylmethanes. Transfer

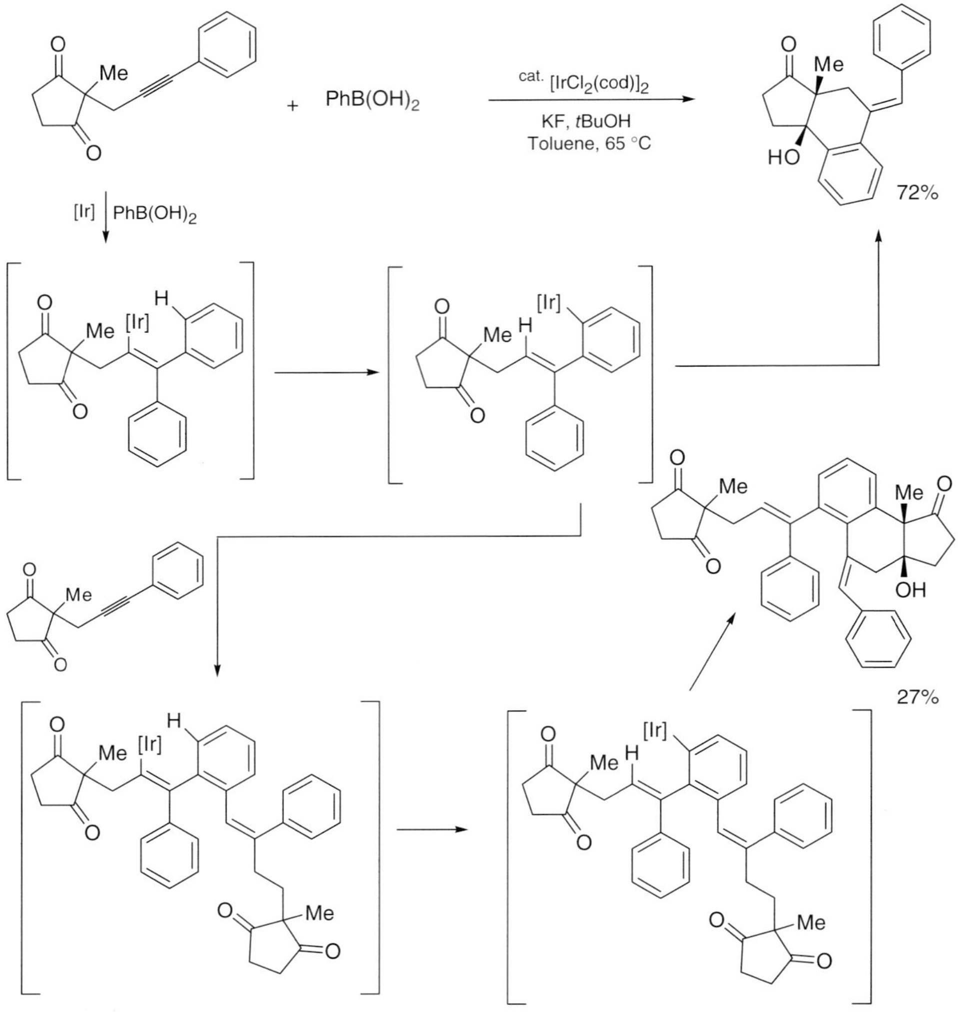

Scheme 11.16

hydrogenation of methylene-3*H*-indole by benzyl alcohol can also lead to the monobenzylated product (Scheme 11.17).

11.7.3
Synthesis of Indoles from Aminoalcohol and Alkynyl Alcohols by Ir–Pt Catalyst

Peris and coworkers reported on a Ir/P-bimetallic complex having a 1,2,4-trimethyltriazolylidene ligand act as an efficient catalyst for the tandem oxidative cyclization of 2-(*ortho*-aminopheny)ethanol to afford indole derivatives [164]. In a typical example, 2-(*ortho*-aminophenyl)ethanol (1 equiv.) was reacted in the

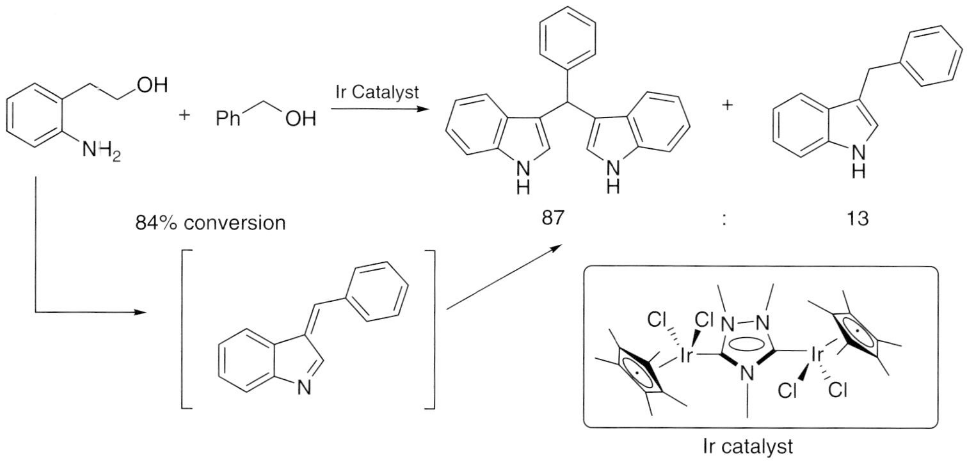

Scheme 11.17

presence of the Ir–Pt complex (5 mol%) and silver triflate (10 mol%) in toluene at 110 °C. After 12 h, 4-pentyn-1-ol (1.2 equiv.) was added to the mixture and stirred at the same reaction temperature for 15 h, giving the indole in 81% (Scheme 11.18).

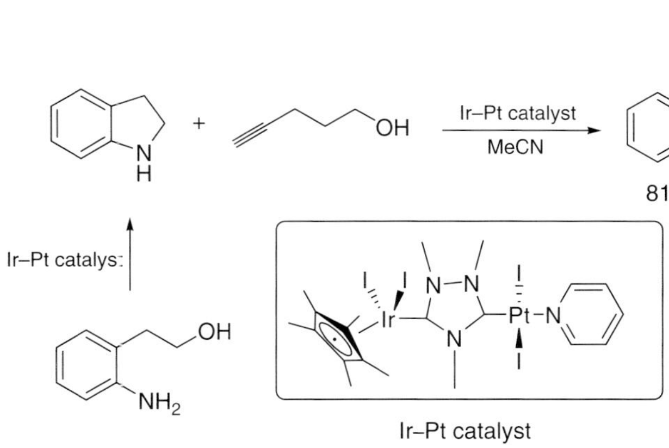

Scheme 11.18

11.7.4
Synthesis of Pyrrolo[1,2-*a*]quinoxalines by Iridium Complex-Catalyzed Annulation of 2-Alkylquinoxalkines

Johannes and coworkers reported on the Ir-catalyzed cyclization of 2-alkylquinoxalines to pyrrolo[1,2-*a*]quinoxalines [165]. Thus the reaction of 2-methyl-3-phenylquinoxaline was carried out in the presence of Ir(acac)$_3$

(10 mol%) with glycerol to give 1-methyl-4-phenylpyrrolo[1,2-*a*]quinoxaline (Scheme 11.19).

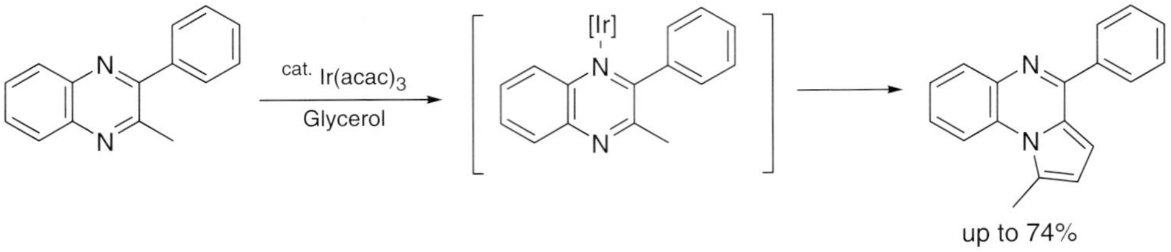

Scheme 11.19

In this reaction, glycerol serves as the carbon source for the formation of the pyrrole ring.

11.7.5 Ir-MOF-Catalyzed Hydrosilylation/*Ortho*-Silylation to Benzoxasiloles

Lin and coworkers reported that an iridium complex with bipyridyl and phenanthroline ligands in a metal-organic framework (mBPV-MOF-Ir and mPT-MOF-Ir) has shown catalytic activity in heterocyclization reactions via C–H bond activation such as tandem hydrosilylation/*ortho*-silylation [166]. Thus the treatment of benzaldehyde with diethylsilane in the presence of a catalytic amount of the Ir-MOFs (0.5 mol%) in *n*-heptane at 25 °C for 24 h gives a benzylic silyl ether by hydrosilylation. The reaction mixture was continued to be heated at 115 °C for 3–4.5 h inducing dehydrogenative *ortho*-silylation, giving benzoxasilole in almost quantitative yield (Scheme 11.20).

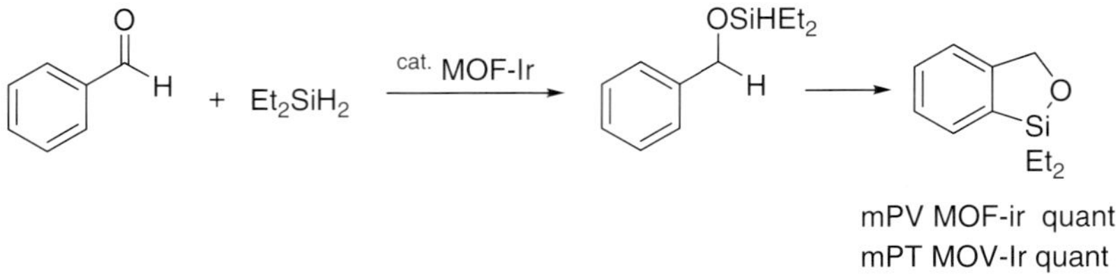

Scheme 11.20

The Ir-MOF catalyst could be reused at least 15 times in this reaction without loss of catalytic activity.

The Ir-MOF catalyst also serves as a catalyst for the tandem dehydrocoupling of *N*-methylbenzylamine with diethylsilane, giving a hydrido-silylamine, followed by intramolecular dehydrogenative silylation via *ortho*-aryl C–H bond which leads to the azasilolane in excellent yield.

11.7.6 Synthesis of Furanes and Pyrroles Involving Alkylation of 1,3-Dicarbonyl Compounds by Iridium–Tin Bimetallic Catalyst

Roy and coworkers reported that Ir–Sn bimetallic complex serves as efficient catalyst for the synthesis of furans and pyrroles from 1,3-dicarbonyl compounds. This reaction proceeds initial alkylation of 1,3-dicarbonyl compounds with propargylic alcohol to form propargylated 1,3-dicarbonyl intermediate, followed by ring closure which led to the cyclization products [167].

In a typical example for the preparation of furans, the reaction of propargylic alcohol with dibenzoylmethane was carried out in the presence of a catalytic amount of $[Ir(cod)(SnCl_3)Cl(\mu\text{-}Cl)]_2$ combined with Cs_2CO_3 in 1,2-dichloroethane at 80 °C which gave the corresponding functionalized furanes in 67% yield (Scheme 11.21).

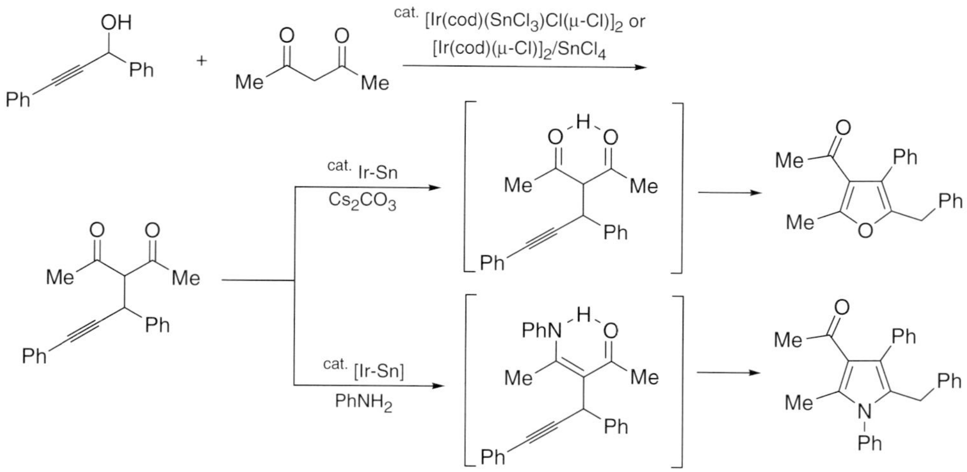

Scheme 11.21

On the other hand, the reaction of the propargylic alcohol with dibenzoylmethane was performed in the presence of a catalytic amount of $[Ir(cod)(\mu\text{-}Cl)]_2$ and $SnCl_4$ followed by addition of aniline, and the reaction was performed at 80 °C which led to the formation of pyrrole derivatives in 78% yield (Scheme 11.21).

11.8 Summary and Conclusions

In conclusion, this chapter described some representative examples of the iridium-catalyzed synthesis of heterocycles involving C–H bond activation. The reactions include oxidative *ortho*-aryl C–H, benzyl, and sp^3–CH activation steps. Additionally, the design of ligands and the framework of catalytically active Ir complex suitable for heterocyclization via C–H bond activation have also been described. Compared with the more traditional use of palladium and rhodium as catalysts, the chemistry of Ir is still young. However, because these reactions are rapidly growing in impact as efficient catalysts for various organic transformations, the use of Ir catalysts in organic transformations is increasing.

Therefore, we hope that this chapter will inspire the reader to use iridium complexes as an alternative and inherently active metal catalyst for innovative organic transformations involving C–H activation.

References

1. Dobereiner, G.E., Nova, A., Schley, N.D., Hazari, N., Miller, S.J., Eisenstein, O., and Craabtree, R.H. (2011) *J. Am. Chem. Soc.*, **133**, 7547–7562.
2. Church, T.L. and Andersson, P.G. (2008) *Coord. Chem. Rev.*, **252**, 513–531.
3. Roseblade, S.J. and Pfaltz, A. (2007) *Acc. Chem. Res.*, **40**, 1402–1411.
4. Källström, K., Munslow, I., and Andersson, P.G. (2006) *Chem. Eur. J.*, **12**, 3194–3200.
5. Cui, X. and Burgess, K. (2005) *Chem. Rev.*, **105**, 3272–3296.
6. Obora, Y. (2014) *ACS Catal.*, **4**, 3972–3981.
7. Imm, S., Báhn, S., Zhang, M., Neubert, L., Neumann, H., Klasovsky, F., Pfeffer, J., Haas, T., and Beller, M. (2011) *Angew. Chem. Int. Ed.*, **50**, 7599–7603.
8. Michlik, S., Hille, T., and Kempe, R. (2012) *Adv. Synth. Catal.*, **354**, 847–862.
9. Kawahara, R., Fujita, K., and Yamaguchi, R. (2011) *Adv. Synth. Catal.*, **353**, 1161–1168.
10. Debereiner, G.E. and Crabtree, R.H. (2010) *Chem. Rev.*, **110**, 681–703.
11. Choi, J., MacArther, H.R., Brookhart, M., and Goldman, A.S. (2011) *Chem. Rev.*, **111**, 1761–1779.
12. Pan, S. and Shibata, T. (2013) *ACS Catal.*, **3**, 704–712.
13. Krische, M.J. (2009) *Angew. Chem. Int. Ed.*, **48**, 34–46.
14. Suzuki, T. (2011) *Chem. Rev.*, **111**, 1825–1845.
15. Guillena, G., Ramon, D.J., and Yus, M. (2010) *Chem. Rev.*, **110**, 1611–1641.
16. Nixon, T.D., Whittlesey, M., and Williams, J.M.J. (2009) *Dalton Trans.*, 753–762.
17. Zhou, J. and Fang, J. (2011) *J. Org. Chem.*, **76**, 7730–7736.
18. Li, J.-Q. and Andersson, P.G. (2013) *Chem. Commun.*, **49**, 6131–6133.
19. Fang, J., Zhou, J., and Fang, Z. (2013) *RSC Adv.*, **3**, 334–336.
20. Whitney, S., Grigg, R., Derrick, A., and Keep, A. (2007) *Org. Lett.*, **9**, 3299–3302.
21. Blacker, A.J., Farah, M.M., Hall, M.I., Marsden, S.P., Saidi, O., and Williams, J.M.J. (2009) *Org. Lett.*, **11**, 2039–2042.
22. Eary, C.T. and Clausen, D. (2006) *Tetrahedron Lett.*, **47**, 6899–6902.
23. Yang, S.C., Shue, Y.J., and Liu, P.C. (2002) *Organometallics*, **21**, 2013–2016.
24. Fujita, K. and Yamaguchi, R. (2005) *Synlett*, 560–571.
25. Fujita, K., Yamamoto, K., and Yamaguchi, R. (2002) *Org. Lett.*, **16**, 2691–2694.
26. Fujita, K., Takahashi, Y., Owaki, M., Yamamoto, K., and Yamaguchi, R. (2004) *Org. Lett.*, **6**, 2785–2788.
27. Fujita, K., Enoki, Y., and Yamaguchi, R. (2006) *Org. Synth.*, **83**, 217–221.
28. Verendel, J.J., Zhou, T., Li, J.-Q., Paptchikhine, A., Lebedev, O., and Andersson, P.G. (2010) *J. Am. Chem. Soc.*, **132**, 8880–8881.
29. Song, S., Zhu, S.-F., Pu, L.-Y., and Zhou, Q.-L. (2013) *Angew. Chem. Int. Ed.*, **52**, 6072–6075.
30. Tang, W., Sun, Y., Xu, L., Wang, T., Fan, Q., Lam, K.-H., and Chaan,

A.S.C (2010) *Org. Biomol. Chem.*, **8**, 3464–3471.

31. Wang W.B., Lu, S.M., Yang, P.Y., Han, X.W., and Zhou, Y.G. (2003) *J. Am. Chem Soc.*, **125**, 10536–10537.
32. Yang, P.Y. and Zhou, Y. (2004) *Tetrahedron: Asymmetry*, **15**, 1145–1149.
33. Lu, S.M., Han, X.W., and Zhou, Y.G. (2004) *Adv. Synth. Catal.*, **346**, 909–912.
34. Lu, S.M., Wang, Y.Q., Han, X.W., and Zhou Y.G. (2006) *Angew. Chem. Int. Ed.*, **45**, 2260–2263.
35. Wang, X.-B. and Zhou, Y.-G. (2008) *J. Org. Chem.*, **73**, 5640–5642.
36. Wang, D.W., Wang, X.B., Wang, D.S., Lu, S M., Zhou, Y.G., and Li, Y.X. (2009) *J. Org. Chem.*, **74**, 2780–2787.
37. Xu, L., Lam, K., Ji, J., Wu, J., Fan, Q.H., Lo, W.H., and Chan, A.S.C. (2005) *Chem. Commun.*, 1390–1392.
38. Lam, K., Xu, L., Feng, L., Fan, Q., Lam, F., Lo, W., and Chan, A.S.C. (2005) *Adv. Synth. Catal.*, **347**, 1755–1758.
39. Qiu, L., Kwong, F., Wu, J., Lam, W., Chan, S., Yu, W., Li, Y., Guo, R., Zhou, Z., and Chan, A.S.C. (2006) *J. Am. Chem. Soc.*, **128**, 5955–5965.
40. Tang, W.J., Zhu, S.F., Xu, L.J., Zhou, Q.L., Fan, Q.H., Zhou, H.F., Lam, K., and Chan, A.S.C. (2007) *Chem. Commun.*, 613–615.
41. Wang, Z.J., Deng, G.J., Li, Y., He, Y.M., Tang, W.J., and Fan, Q.H. (2007) *Org. Lett.* **9**, 1243–1246.
42. Chan, S.H., Lam, K.H., Li, Y.M., Xu, L.J., Tang, W.J., Lam, F., Lo, W.H., Yu, W.Y., Fan, Q.H., and Chan, A.S.C. (2007) *Tetrahedron: Asymmetry*, **18**, 2625–2631.
43. Reetz, M. and Li, X. (2006) *Chem. Commun.*, 2159–2160.
44. Tadaoka, H., Cartigny, D., Nagano, T., Gosavi, T., Ayad, T., Genet, J.P., Ohshima, T., Ratovelomanana-Vidal, V., and Mashima, K. (2009) *Chem. Eur. J.*, **15**, 9990–9994.
45. Wang, D., Zhou, J., Wang, D., Guo, Y., and Zhou, Y. (2010) *Tetrahedron Lett.*, **51**, 525–528.
46. Tang, W.J., Xu, L.J., Fan, Q.H., Wang, J., Fan, B.M., Lam, K.H., and Chan, A.S.C. (2009) *Angew. Chem. Int. Ed.*, **48**, 9135–9138.
47. Wang, X.B., Zeng, W., and Zhou, Y.G. (2008) *Tetrahedron Lett.*, **49**, 4922–4924.
48. Yao, W., Zhang, Y., Jia, X., and Huang, Z. (2014) *Angew. Chem. Int. Ed.*, **53**, 1390–1394.
49. Kuklin, S.A., Sheloumov, A.M., Dolgushin, F.M., Ezernitskaya, M.G., Peregudov, A.S., Petrovskii, A.A., and Koridza, A.A. (2006) *Organometallics*, **25**, 5466–5475.
50. Adams, J.J., Arulsamy, N., and Roddick, D.M. (2012) *Organometallics*, **31**, 1439–1447.
51. Gruver, B.C., Adams, J.J., Warner, S.J., Arulsamy, N., and Roddick, D.M. (2011) *Organometallics*, **30**, 5133–5140.
52. Chianese, A.R., Mo, A., Lampland, N.L., Swartz, R.L., and Bremer, P.T. (2010) *Organometallics*, **29**, 3019–3026.
53. Zuo, W. and Braunstein, P. (2012) *Organometallics*, **31**, 2606–2615.
54. Ahuja, R., Punji, B., Findlater, M., Supplee, C., Schinski, W., Brookhart, M., and Goldman, A.S. (2011) *Nat. Chem.*, **3**, 167–172.
55. Brawn, R.A., Guimarães, C.R.W., McClure, K.F., and Lilas, S. (2012) *Org. Lett.*, **14**, 4802–4805.
56. Field, L.D., Messele, B.A., Vuong, K.Q., and Turner, P. (2005) *Organometallics*, **24**, 4241–4250.
57. Li, X., Chianese, A.R., Vogel, T., and Crabtree, R.H. (2005) *Org. Lett.*, **7**, 5437–5440.
58. Kashiwame, Y., Kuwta, S., and Ikariya, T. (2012) *Organometallics*, **31**, 8444–8455.
59. Burling, S., Field, L.D., Messerle, B.A., and Turner, P. (2004) *Organometallics*, **23**, 1714–1721.
60. Casalnuovo, A.L., Calabrese, J.C., and Milstein, D. (1988) *J. Am. Chem. Soc.*, **110**, 6738–6744.
61. Dorta, R., Egli, P., Zurcher, F., and Togni, A. (1997) *J. Am. Chem. Soc.*, **119**, 10857–10858.
62. Iglessias, M., Perez-Nicolas, M., Miguel, P.J.S., Polo, V., Fernandez-Alvarez, F.J., Perez-Torrente, J., and Oro, L.A. (2012) *Chem. Commun.*, **48**, 9480–9482.
63. Ho, J.H.H., Choy, S.W.S., Macgregor, S.A., and Messerle, B.A. (2011) *Organometallics*, **30**, 5978–5984.

64. Liu, W.B., He, H., Dai, L.-X., and You, S.-L. (2008) *Org. Lett.*, **10**, 1815–1818.
65. Liu, W.B., He, H., Dai, L.-X., and You, S.-L. (2009) *Synthesis*, 2076–2082.
66. Stanley, L.M. and Hartwig, J.F. (2009) *Angew. Chem. Int. Ed.*, **48**, 7841–7844.
67. Takeuchi, R. and Kezuka, S. (2006) *Synthesis*, 3349–3366.
68. Helmchen, G., Dahnz, A., Dübon, P., Schelwies, M., and Weihofen, R. (2007) *Chem. Commun.*, 675–684.
69. Janssen, J.P. and Halmchen, G. (1997) *Tetrahedron Lett.*, **38**, 8025–8026.
70. Zhao, D., Fananas-Mastral, M., Chang, M.-C., Otten, E., and Feringa, B.L. (2014) *Chem. Sci.*, **5**, 4216–4220.
71. Teichert, J.F., Fananas-Mastral, M., and Feringa, B.L. (2011) *Angew. Chem. Int. Ed.*, **50**, 688–691.
72. Crabtree, R.H., Felkin, H., and Morris, G.E. (1977) *J. Organomet. Chem.*, **141**, 205–210.
73. Itoh, J., Han, S.B., and Krische, M.J. (2009) *Angew. Chem. Int. Ed.*, **48**, 6313–6316.
74. Spiess, S., Welter, C., Frack, G., Taquet, J.-P., and Helmchen, G. (2008) *Angew. Chem. Int. Ed.*, **47**, 7652–7655.
75. Bondzic, B.P., Farwick, A., Liebich, J., and Eilbracht, P. (2008) *Org. Biomol. Chem.*, **6**, 3723–3731.
76. Pouy, M.J., Leitner, A., Weix, D.J., Ueno, S., and Hartwig, J.F. (2007) *Org. Lett.*, **9**, 3949–3952.
77. Markovic, D. and Hartwig, J.F. (2007) *J. Am. Chem. Soc.*, **129**, 11680–11681.
78. Lee, J.H., Shin, S., Kang, J., and Lee, S. (2007) *J. Org. Chem.*, **72**, 7443–7446.
79. Spiess, S., Berthold, C., Weinhofen, R., and Helmchen, G. (2007) *Org. Biomol. Chem.*, **5**, 2357–2360.
80. Yamashita, Y., Gopalarathnam, A., and Hartwig, J.F. (2007) *J. Am. Chem. Soc.*, **129**, 7508–7509.
81. Defieber, C., Ariger, M.A., Moriel, P., and Carreira, E.M. (2007) *Angew. Chem. Int. Ed.*, **46**, 3139–3143.
82. Singh, O.V. and Han, H. (2007) *J. Am. Chem. Soc.*, **129**, 774–775.
83. Nemoto, T., Sakamoto, T., Matsumoto, T., and Hamada, Y. (2006) *Tetrahedron Lett.*, **47**, 8737–8740.
84. Vazquez-Serreno, L.D., Owens, B.T., and Buriak, J.M. (2006) *Inorg. Chim. Acta*, **339**, 2786–2797.
85. Shekhar, S., Trantow, B., Leitner, A., and Hartwig, J.F. (2006) *J. Am. Chem. Soc.*, **128**, 11770–11771.
86. Polet, D., Alexakis, A., Tissot-Croset, K., Corminboeuf, C., and Ditrich, K. (2006) *Chem. Eur. J.*, **12**, 3596–3609.
87. Leitner, A., Shekhar, S., Pouy, M.J., and Hartwig, J.F. (2005) *J. Am. Chem. Soc.*, **127**, 15506–15514.
88. Welter, C., Moreno, R.M., Streiff, S., and Helmchen, G. (2005) *Org. Biomol. Chem.*, **3**, 3266–3268.
89. Weihofen, R., Dahnz, A., Tverskoy, O., and Helmchen, G. (2005) *Chem. Commun.*, 3541–3543.
90. Polet, D. and Alexakis, A. (2005) *Org. Lett.*, **7**, 1621–1624.
91. Leitner, A., Shu, C., and Hartwig, J.F. (2005) *Org. Lett.*, **7**, 1093–1096.
92. Miyabe, H., Matsumura, A., Moriyama, K., and Takemoto, Y. (2004) *Org. Lett.*, **6**, 4631–4634.
93. Shu, C., Leitner, A., and Hartwig, J.F. (2004) *Angew. Chem. Int. Ed.*, **45**, 4797–4800.
94. Tissot-Croset, K., Prolet, D., and Alexakis, A. (2004) *Angew. Chem. Int. Ed.*, **43**, 2426–2428.
95. Welter, C., Koch, O., Lipowsky, G., and Helmchem, G. (2004) *Chem. Commun.*, 896–897.
96. Kiener, C.A., Shu, C., Incarvito, C., and Hartwig, J.F. (2003) *J. Am. Chem. Soc.*, **125**, 14272–14273.
97. Gao, N., Guo, X.-W., Zheng, S.-C., Yang, W.-K., and Zhao, X.-M. (2012) *Tetrahedron*, **68**, 9413–9418.
98. Lipowsky, G. and Helmchen, G. (2004) *Chem. Commun.*, 116–117.
99. Ohmura, T. and Hartwig, J.F. (2002) *J. Am. Chem. Soc.*, **124**, 15164–15165.
100. Wu, Q.-F., He, H., Liu, W.-B., and You, S.-L. (2010) *J. Am. Chem. Soc.*, **132**, 11418–11419.
101. Stanley, L.M. and Hartwig, J.F. (2009) *J. Am. Chem. Soc.*, **131**, 8971–8983.
102. Onodera, G., Shimizu, Y., Kimura, J., Kobayashi, J., Ebihara, Y., Kondo, K., Sakata, K., and Takeeuchi, R. (2012) *J. Am. Chem. Soc.*, **134**, 10515–10531.

103. Onodera, G., Suto, M., and Takeuchi, R. (2012) *J. Org. Chem.*, **77**, 908–920.
104. Takeuchi, R., Tanaka, S., and Nakaya, Y. (2001) *Tetrahedron Lett.*, **42**, 2991–2994.
105. Takeuchi, R. (2002) *Synlett*, 1954–1965.
106. Takeuchi, R. and Nakaya, Y. (2003) *Org. Lett.*, **5**, 3659–3662.
107. Kezuka, S., Okado, T., Niou, E., and Takeuchi, R. (2005) *Org. Lett.*, **7**, 1711–1714.
108. Kezuka, S., Tanaka, S., Ohe, T., Nakaya, Y., and Takeuchi, R. (2006) *J. Org. Chem.*, **71**, 543–552.
109. Onodera, G., Matsuzawa, M., Aizawa, T., Kitahara, T., Shimizu, Y., Kezuka, S., and Takeuchi, R. (2008) *Synlett*, 755–758.
110. Takeuchi, R. and Kashio, M. (1997) *Angew. Chem. Int. Ed. Engl.*, **36**, 263–265.
111. Takeuchi, R. and Kashio, M. (1998) *J. Am. Chem. Soc.*, **120**, 8647–8655.
112. Takeuchi, R. and Tanabe, K. (2000) *Angew. Chem. Int. Ed.*, **39**, 1975–1978.
113. Onodera, G., Watabe, K., Matsubara, M., Oda, K., and Kezuka, S. (2008) *Adv. Synth. Catal.*, **350**, 2725–2732.
114. Onodera, G., Kato, M., Kawano, R., Kometani, Y., and Takeuchi, R. (2009) *Org. Lett.*, **11**, 5038–5041.
115. Carmona, D., Lamata, M.P., Viguri, F., Rodríguez, R., Lahoz, F.J., Fabra, M.J., and Oro, L.A. (2009) *Tetrahedron: Asymmetry*, **20**, 1197–1205.
116. Lin, Y., Ma, D., and Lu, X. (1987) *Tetrahedron Lett.*, **28**, 3249–3252.
117. Nishinaka, Y., Satoh, T., Miura, M., Morisaka, H., Nomura, M., Matsui, H., and Yamaguchi, C. (2001) *Bull. Chem. Soc. Jpn.*, **74**, 1727–1736.
118. Zhang, Y.J., Skucas, E., and Krische, M.J. (2009) *Org. Lett.*, **11**, 4248–4250.
119. Wiley, J.S. and Heinekey, D.M. (2002) *Inorg. Chem.*, **41**, 4961–4966.
120. Clot, E., Chen, J.Y., Lee, D.H., Sung, S.Y., Appelhans, L.N., Faller, J.W., Crabtree, R.H., and Eisenstein, O. (2004) *J. Am. Chem. Soc.*, **126**, 8795–8804.
121. Yung, C.M., Skadden, M.B., and Bergman, R.G. (2004) *J. Am. Chem. Soc.*, **126**, 13033–13043.
122. Parisien, M., Valette, D., and Fagnou, K. (2005) *J. Org. Chem.*, **70**, 7578–7584.
123. Bhalla, G. and Periana, R.A. (2005) *Angew. Chem. Int. Ed.*, **44**, 1540–1543.
124. Bhalla, G., Liu, X.Y., Oxgaard, J., Goddard, W.A., and Periana, R.A. (2005) *J. Am. Chem. Soc.*, **127**, 11372–11389.
125. Leung, D.H., Bergman, R.G., and Raymond, K.N. (2006) *J. Am. Chem. Soc.*, **128**, 9781–9797.
126. Davies, D.L., Donald, S.M.A., Al-Duaij, O., Macgregor, S.A., and Poelleth, M. (2006) *J. Am. Chem. Soc.*, **128**, 4210–4211.
127. Voutchkova, A.M. and Crabtree, R.H. (2009) *J. Mol. Catal. A: Chem.*, **312**, 1–6.
128. Boutadla, Y., Al-Duaij, O., Davies, D.L., Griffith, G.A., and Singh, K. (2009) *Organometallics*, **28**, 433–445.
129. Bischof, S.M., Eas, D.H., Meier, S.K., Oxgaard, J., Nielsen, R.J., Bhalla, G., Goddard, W.A., and Periana, R.A. (2010) *Organometallics*, **29**, 742–756.
130. He, H., Liu, L.-X., and You, S.-L. (2009) *J. Am. Chem. Soc.*, **131**, 8346–8347.
131. Cho, J.-Y., Tse, M.K., Holmes, D., Maleczka, R.E. Jr.,, and Smith, M.R. III, (2002) *Science*, **295**, 305–308.
132. Ishiyama, T., Takagi, J., Ishida, K., Miyaura, N., Anastasi, N.R., and Hartwig, J.F. (2002) *J. Am. Chem. Soc.*, **124**, 390–391.
133. Lu, B. and Falck, J.R. (2008) *Angew. Chem. Int. Ed.*, **47**, 7508–7510.
134. Lyall, C.L., Shotton, C.C., Perez-Salvia, M., Pantos, G.D., and Lewis, S.E. (2014) *Chem. Commun.*, **50**, 13837–13840.
135. Mkhalid, I.A.I., Barnard, J.H., Marder, T.B., Murphy, J.M., and Hartwig, J.F. (2010) *Chem. Rev.*, **110**, 890–931.
136. Teraoka, T., Hiroto, S., and Shinokubo, H. (2011) *Org. Lett.*, **13**, 3012–3015.
137. Wu, Y., Zhou, Y.Y., and Li, Y. (2015) *J. Org. Chem.*, **80**, 1946–1951.
138. Satoh, T., Nishinaka, Y., Miura, M., and Nomura, M. (1999) *Chem. Lett.*, 615–618.
139. Matsumoto, T., Taube, D.J., Periana, R.A., Taube, H., and Yoshida, H. (2000) *J. Am. Chem. Soc.*, **122**, 7414–7415.
140. Dorta, R. and Togni, A. (2003) *Chem. Commun.*, 760–762.

141. Tsuchikama, K., Kasagawa, M., Hashimoto, Y., Endo, K., and Shibata, T. (2008) *J. Organomet. Chem.*, **693**, 3939–3942.
142. Takebayashi, S. and Shibata, T. (2012) *Organometallics*, **31**, 4114–4117.
143. Pan, S., Ryu, N., and Shibata, T. (2012) *J. Am. Chem. Soc.*, **134**, 17474–17477.
144. Cheng, C. and Hartwig, J.F. (2015) *Chem. Rev.*, **115**, 8946–8975.
145. Cheng, C. and Hartwig, J.F. (2015) *J. Am. Chem. Soc.*, **137**, 592–595.
146. Ishiyama, T., Saili, T., Kishida, E., Sasaki, I., Ito, H., and Miyaura, N. (2013) *Org. Biomol. Chem.*, **11**, 8162–8165.
147. Fujita, K., Nonogaw, M., and Yamaguchi, R. (2004) *Chem. Commun.*, 1926–1927.
148. Join, B., Yamamoto, T., and Itami, K. (2009) *Angew. Chem. Int. Ed.*, **48**, 3644–3647.
149. Nishimura, T., Ebe, Y., and Hayashi, T. (2013) *J. Am. Chem. Soc.*, **135**, 2092–2095.
150. Nishimura, T., Nagatomo, M., and Hayashi, T. (2013) *Chem. Sci.*, **4**, 4499–4504.
151. Nagatomo, M. and Nishimura, T. (2014) *Chem. Commun.*, **50**, 6274–6277.
152. Engelman, K.L., Feng, Y., and Ison, E.A. (2011) *Organometallics*, **30**, 4572–4577.
153. Iwai, T., Fujihara, T., Terao, J., and Tsuji, Y. (2010) *J. Am. Chem. Soc.*, **132**, 9602–9603.
154. Sun, K., Sachwani, R., Richert, K.J., and Driver, T.G. (2009) *Org. Lett.*, **11**, 3598–3601.
155. Li, Q., Driess, M., and Hartwig, J.F. (2014) *Angew. Chem. Int. Ed.*, **53**, 8471–8474.
156. DeBoef, B., Pastine, S.J., and Sames, D. (2004) *J. Am. Chem. Soc.*, **126**, 6556–6557.
157. Igarashi, T., Inada, T., Sekioka, T., Nakajima, T., and Shimizu, I. (2005) *Chem. Lett.*, **34**, 106–107.
158. Nakajima, T., Inada, T., Igarashi, T., Sekioka, T., and Shimizu, I. (2006) *Bull. Chem. Soc. Jpn.*, **79**, 1941–1949.
159. Aramoto, H., Obora, Y., and Ishii, Y. (2009) *J. Org. Chem.*, **74**, 628–633.
160. Tursky, M., Lorentz-Petersen, L.L.R., Olsen, L.B., and Madsen, R. (2010) *Org. Biomol. Chem.*, **8**, 5576–5582.
161. Wong, C.M., Vuong, K.Q., Gatus, M.R.D., Hua, C., Bhadbhade, M., and Messerle, B.A. (2012) *Organometallics*, **31**, 7500–7510.
162. Partridge, B.M., Gonzalez, J.S., and Lam, H.W. (2014) *Angew. Chem. Int. Ed.*, **53**, 6523–6527.
163. Zanardi, A., Corberan, R., Mata, J.A., and Peris, E. (2008) *Organometallics*, **27**, 3570–3576.
164. Zanardi, A., Mata, J.A., and Peris, E. (2010) *Chem. Eur. J.*, **16**, 13109–13115.
165. Ammermann, S., Hrib, C., Jones, P.G., du Mont, W.-W., Kowalsky, W., and Johannes, H.-H. (2012) *Org. Lett.*, **14**, 5090–5093.
166. Manna, K., Zhang, T., Greene, F.X., and Lin, W. (2015) *J. Am. Chem. Soc.*, **137**, 2665–2673.
167. Chatterjee, P.N. and Roy, S. (2011) *Tetrahedron*, **67**, 4569–4577.

12
Au- and Pt-Catalyzed C–H Activation/Functionalizations for the Synthesis of Heterocycles

Yuanjing Xiao and Junliang Zhang

12.1
Introduction

Divergent properties of gold and platinum complexes, such as their mild carbophilic π Lewis acid and their ability to insert into C–H bonds as well as their ability to undergo redox transformations, have been gradually and intensively disclosed by chemists in the past years. A tremendous growth has been witnessed in the exploitation of these divergent properties to affect a range of transformations involving cycloisomerization, C–H bond activation/functionalization, cross-coupling reactions, and so on. This growth has been reflected in a number of excellent reviews [1]. Among those many different products accessible via these transformations, heterocycles are probably the most important class of compounds in terms of their significant impact on the synthesis of agrochemicals and pharmaceuticals. C–H bond activation is commonly followed by functionalization, the overall transformation therefore being a C–H functionalization, involving a distinct C–H bond activation step. In this chapter we will mainly discuss recent significant advances in the synthesis of heterocycles via Au and Pt-catalyzed C–H bond activation/functionalization process. Those processes directly decorating heterocycles via C–H activation and C–H functionalization processes will not be included here. These results are organized according to the different types of heterocycles, that is, *O*-heterocycles, *N*-heterocycles, *S*-heterocycles, *O*-heterocycles, *N*-heterocycles, and fused polycyclic heterocycles. Each section is divided into smaller subsections according to the size of the heterocyclic ring, that is, five- and six-membered and other ring-sized heterocycles. Furthermore, each subsection is further classified according to C(sp)–H, C(sp^2)–H, or C(sp^3)–H functionalizations. For some typical reactions, the detailed mechanism will be also included for better understanding.

Transition Metal-Catalyzed Heterocycle Synthesis via C—H Activation, First Edition. Edited by Xiao-Feng Wu.
© 2016 Wiley-VCH Verlag GmbH & Co. KGaA. Published 2016 by Wiley-VCH Verlag GmbH & Co. KGaA.

12.2
Synthesis of *O*-Heterocycles

O-Containing heterocycles are important substructures found in numerous bioactive natural products and important pharmaceuticals. The diversity of the structures encountered, as well as their biological and pharmaceutical relevance, has motivated research aimed at the development of new economical, efficient, and selective synthetic strategies to access these compounds [2, 3]. For the past dozen years, spectacular achievements have been made toward the synthesis of oxygen-containing heterocycles via Au- and Pt-catalyzed C–H activation/functionalization process.

12.2.1
Synthesis of Five-Membered *O*-Heterocycles

12.2.1.1 Via Au-Catalyzed C(sp)–H Functionalization

The interaction of gold species with alkynes is prolific. In terms of the formation of heterocycles, there are two modes between the interactions of gold species with alkynes. The general mode is that gold seeks π-electron density increasing the electrophilic character of the alkyne upon coordination, thus enabling the attack of heteroatom nucleophiles in an intra- or intermolecular manner due to its soft Lewis acidity [4]. The other mode is the direct formation of gold acetylide species from terminal alkynes and gold complex via C(sp)–H bond activation. Gold acetylide species can be used as organometallics in nucleophilic addition and substitution reactions as well as in catalytic redox process for heterocycle synthesis.

In 2010, Gouverneur and coworkers reported a novel cascade cyclization–intermolecular alkynylation reactions toward the synthesis of five-membered *O*-heterocycles. β-Alkynyl-γ-butenolides **2** could be efficiently prepared from allenoates **1** and terminal alkynes in the presence of a gold catalyst with the use of Selectfluor as an external oxidant (Scheme 12.1) [5]. This process involves a direct C(sp)–H functionalization of terminal alkynes. This novel gold-catalyzed cascade cyclization oxidative cross-coupling process provides a basis for the development of novel cascade reactions combining traditional gold catalysis and intermolecular oxidative alkynylation.

12.2.1.2 Via Au-Catalyzed Aryl C(sp^2)–H Functionalization

In 2006, Li and coworkers reported a Au-catalyzed aryl C(sp^2)–H functionalization of phenols and naphthols **3** with dienes toward the synthesis of benzofuran **4** (Scheme 12.2) [6]. This reaction proceeds via a Au(III)-catalyzed intermolecular addition of aryl C–H bond to dienes followed by the intramolecular addition of the phenol O–H bond to afford benzofuran **4**.

A similar process leading to dihydrobenzofurans **6** through a $Ph_3PAuOTf$-catalyzed Claisen rearrangement of aryl allyl ethers **5** followed by a tandem phenol addition was reported by He *et al.* Controlled experiments revealed that

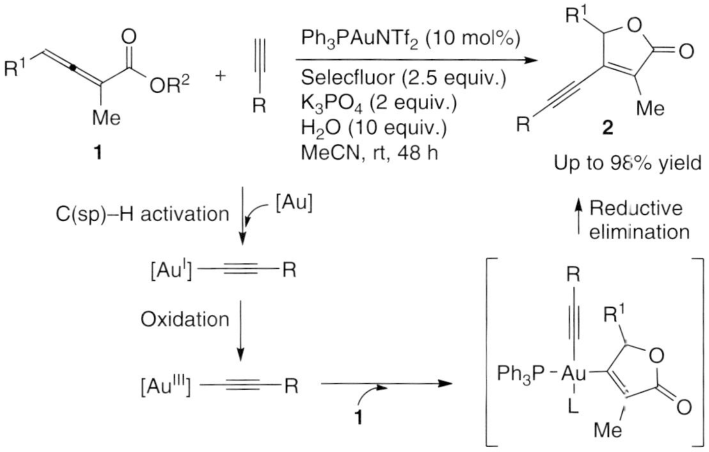

Scheme 12.1 Au-catalyzed synthesis of β-alkynyl-γ-butenolides.

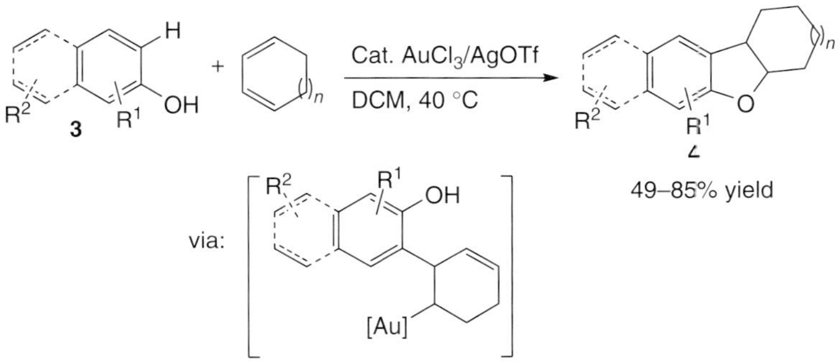

Scheme 12.2 Au-catalyzed synthesis of benzofurans.

gold(III) works better for mediating the Claisen rearrangement, while gold(I) is more efficient to catalyze the phenol addition step (Scheme 12.3) [7].

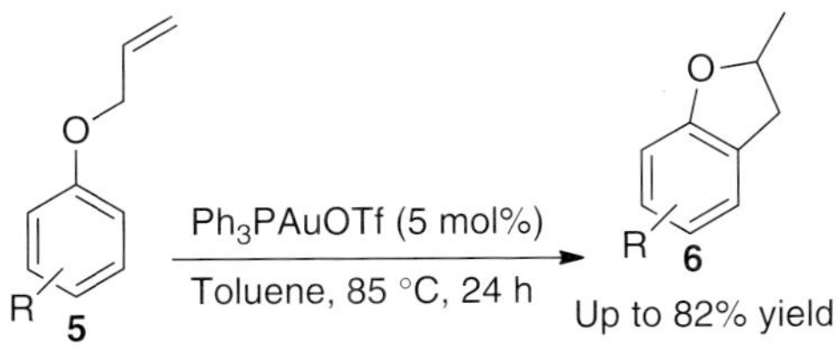

Scheme 12.3 Au-catalyzed synthesis of dihydrobenzofurans.

Very recently, a two-step synthetic approach to aryl benzofuranone skeleton **8**, a prominent structural motif in natural products, was developed by Liu, Zhang, and coworkers (Scheme 12.4) [8]. This approach consists of a Au-catalyzed intermolecular reaction of unprotected phenols with α-aryl α-diazoacetates **7** followed

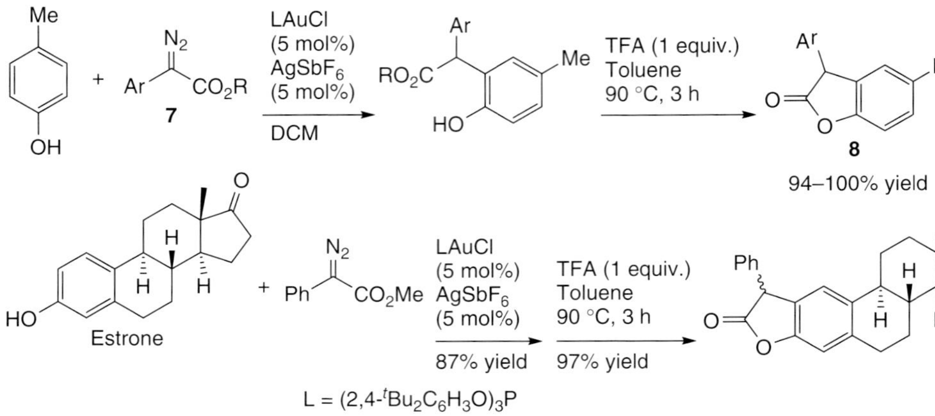

Scheme 12.4 A two-step synthetic approach to construct aryl benzofuranone skeleton via Au-catalyzed aryl C–H functionalization followed by lactonization.

by TFA-mediated intramolecular lactonization. The first step of reaction proceeds via gold-catalyzed direct *ortho*-aryl C–H bond functionalization of unprotected phenols with α-aryl α-diazoacetates **7** rather than O–H insertion. This procedure can be used to *late-stage modification* of natural products or pharmaceutical compounds such as estrone because of its operational convenience and mild reaction conditions.

12.2.1.3 **Via Au-Catalyzed C(sp^3)–H Functionalization**

In 2010, Malacria and coworkers developed a novel Au-catalyzed cyclization of 1,6-enyne **9**, furnishing allene-substituted tetrahydrofuran **11** in excellent yield. The proposed mechanism involves an initial enyne cyclization to generate the cationic gold–vinyl complex **10** and subsequent unexpected [1,5]-hydride shift. Deuteration experiments support the [1,5]-hydride shift and rule out the intermolecular scrambling. This cascade reaction is unusual because the cyclization precedes followed by an uncommon [1,5]-hydride shift (Scheme 12.5) [9].

12.2.1.4 **Via Pt-Catalyzed C(sp^3)–H Functionalization**

A Pt-catalyzed C(sp^3)–H functionalization with *ortho*-alkyl-substituted aromatic carboxylic acids **12** for the synthesis of aryl lactones **13** was explored by Chang and coworkers in 2006 (Scheme 12.6) [10]. Kinetic isotope studies reveal that the reaction takes place presumably by the chelation assistance of metal catalyst to the carboxylic group followed by the C–H bond activation.

12.2.1.5 **Via Au-Catalyzed C(sp)–H and C(sp^3)-H Functionalization**

In 2014, by taking advantage of the unique redox property and carbophilic π-acidity of gold, You and coworkers accomplished an efficient access to 3-alkynyl polysubstituted furans **15** from readily available 1,3-dicarbonyl compounds **14** and terminal alkynes under mild reaction conditions (Scheme 12.7)

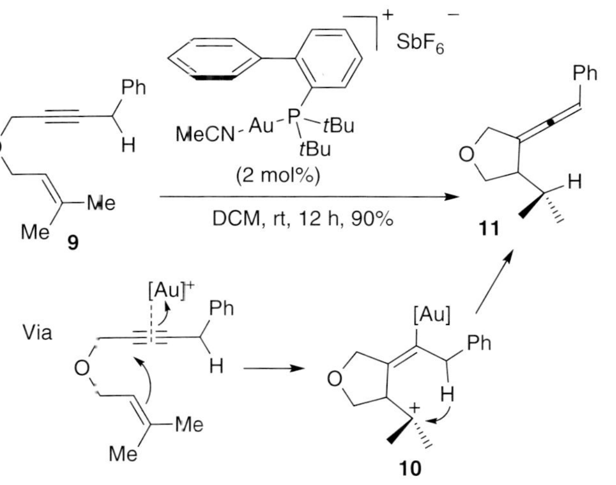

Scheme 12.5 Au-catalyzed synthesis of allene-substituted tetrahydrofuran.

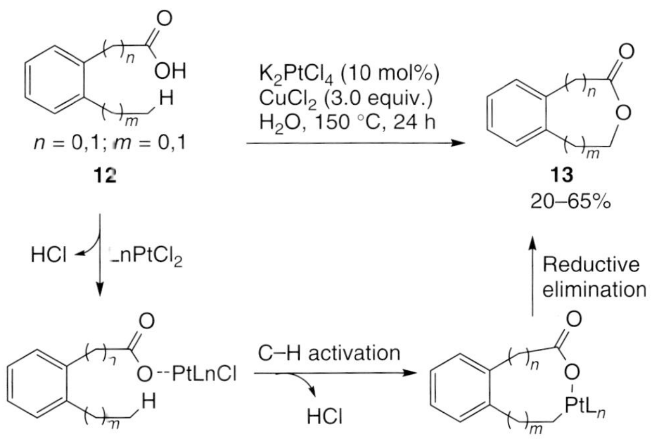

Scheme 12.6 Au-catalyzed synthesis of aryl lactones.

[11]. This reaction was believed to proceed via a tandem sequence consisting of gold-catalyzed C(sp^3)–H/C(sp)–H cross-coupling/cyclization/oxidative alkynylation.

12.2.2
Synthesis of Six-Membered *O*-Heterocycles

12.2.2.1 Via Au-Catalyzed C(sp)–H Functionalization

In 2006, Li and Yao demonstrated a gold-catalyzed reaction of *ortho*-alkynylaryl aldehyde **16** with terminal alkynes for the synthesis of 1-alkynyl-1*H*-isochromenes **17** (Scheme 12.8) [12]. They proposed that an *in situ*-generated

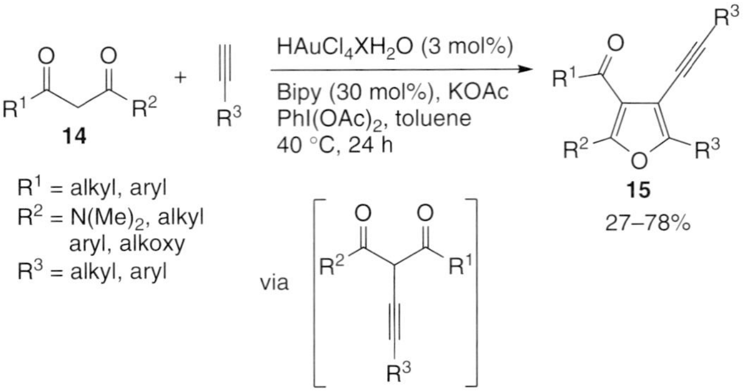

Scheme 12.7 Au-catalyzed synthesis of 3-alkynyl polysubstituted furans.

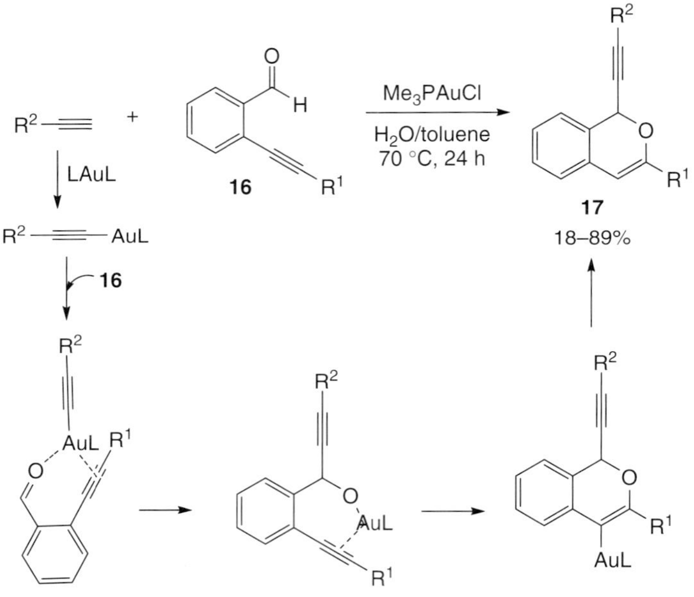

Scheme 12.8 Au-catalyzed synthesis of 1-alkynyl-1*H*-isochromenes.

gold acetylide would react with *ortho*-alkynylaryl aldehyde **16**, followed by the gold-mediated heterocyclization.

12.2.2.2 Via Au-Catalyzed Formyl C(sp^2)–H Functionalization

Using a similar procedure for the synthesis of 1-alkynyl-1*H*-isochromenes **17**, the group of Li realized a Au-catalyzed synthesis of isoflavanones **20** by using salicylaldehyde **18** instead of *ortho*-alkynylaryl aldehyde **16** as substrates (Scheme 12.9) [13]. The initial complexation of the gold(I) catalyst with salicylaldehyde **18**

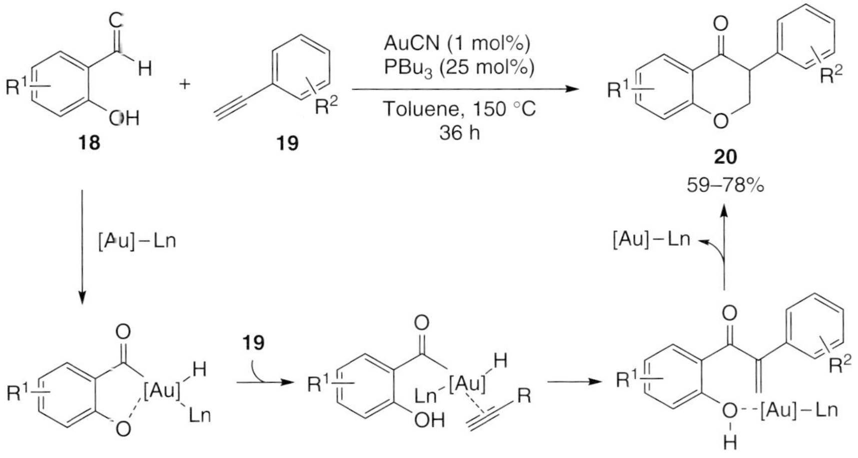

Scheme 12.9 Au-catalyzed synthesis of isoflavanones.

would then undergo oxidative addition of the aldehyde C–H bond to generate an acyl gold(III) hydride intermediate. These intermediate complexes would react with phenylacetylene via hydroacylation. A subsequent conjugate addition of the hydroxy group to the α,β-unsaturated ketone would give the desired isoflavanone derivative **20** and regenerate the gold catalyst.

12.2.2.3 Via Au-Catalyzed Aryl C(sp^2)–H Functionalization

The groups of Reetz and He independently reported a gold(III)-catalyzed hydroarylation of normal alkynes and electron-deficient alkynes **21**, respectively. This reaction could be well extended to the intramolecular version to give coumarin derivatives **22** as products (Scheme 12.10) [14].

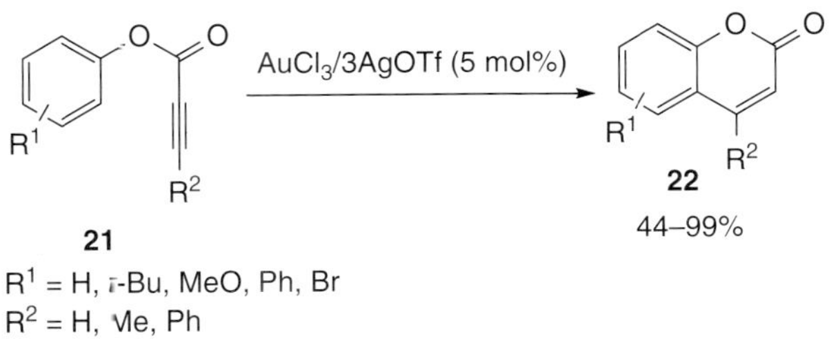

Scheme 12.10 Au-catalyzed synthesis of coumarin derivatives.

Later, He and coworkers developed a new gold(III)-catalyzed stereospecific cycloalkylation of electron-rich arenes with tethered epoxides **23**, which provided a rapid access to 3-chromanols **24** (Scheme 12.11) [15]. The proposed reaction mechanism proceeds through C–H activation of the arene, followed by addition to the tethered epoxide (path A); however, a second possible mechanism in which

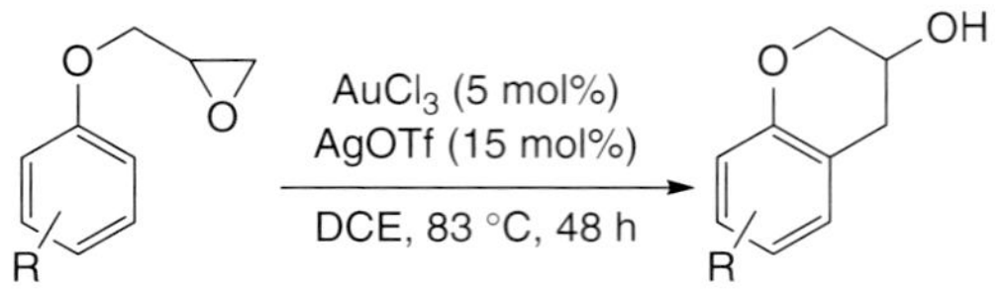

23 24 58–83%

Scheme 12.11 Au-catalyzed synthesis of 3-chromanol.

the gold catalyst acts purely as a Lewis acid to activate the epoxide cannot be ruled out (path B) (Scheme 12.12).

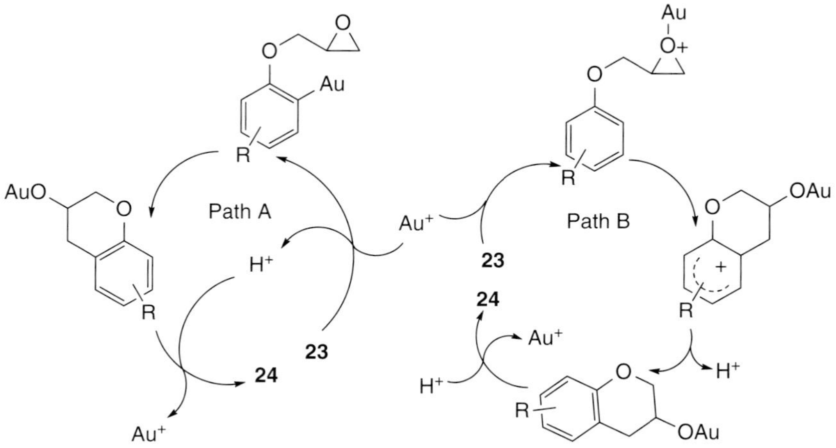

Scheme 12.12 Two possible mechanisms for Au-catalyzed synthesis of 3-chromanol from epoxides 23.

In 2007, a gold(I)-catalyzed cascade reaction of diynes and electron-rich aromatics and heteroaromatics for the synthesis of complex α-pyrones was successfully realized by Schreiber and Luo (Scheme 12.13) [16]. This reaction involves the [3,3]-sigmatropic rearrangement of the diynes **25** to an yne-allene intermediate **26**, which is supported by the observed racemization of enantioenriched substrates. Gold(I)-induced a 6-*endo*-dig cyclization of intermediate **26** would give the oxocarbenium intermediate **27**, which could be trapped by electron-rich aromatics and heteroaromatics to afford complex α-pyrones **28**.

Recently, gold-catalyzed intermolecular oxidative functionalization of alkynes using pyridine/quinoline *N*-oxides, nitrones, or nitrosobenzenes as oxidants has received much attention, which was pioneered by Zhang and Liu [17]. In 2012, Zhang and coworkers successfully applied this strategy for the efficient synthesis of chroman-3-ones **32** using propargyl aryl ethers **29** as substrates (Scheme 12.14) [18]. The common intermediate α-oxo gold carbenes **31** were most likely formed in the reaction. They found that $Me_4tBuXPhosAuNTf_2$ with bulky ligand was

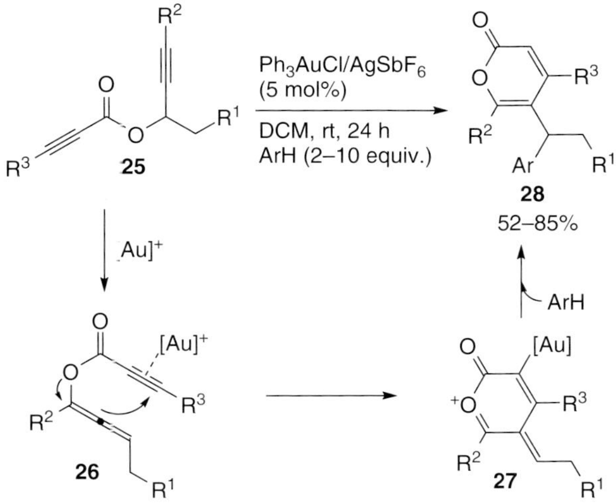

Scheme 12.13 Au-catalyzed synthesis of complex α-pyrones.

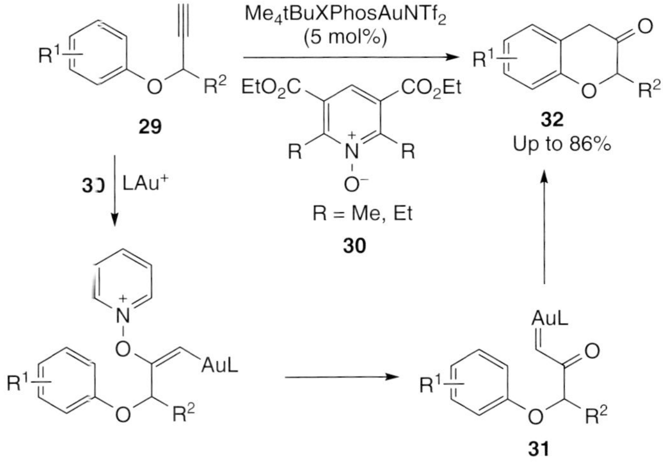

Scheme 12.14 Au-catalyzed synthesis of chroman-3-ones.

the best catalyst for promoting 6-*endo-dig* cyclizations selectively over typically favored 5-*exo-dig* ones. Two easily accessible pyridine N-oxides **30** derived from Hantzsch esters were found to be highly effective oxidants.

12.2.2.4 Via Au-Catalyzed C(sp^3)–H Functionalization

A gold-catalyzed benzylic C–H functionalization for the synthesis of 1*H*-isochromene derivatives **34** was disclosed by Hashmi and coworkers in their study of the intramolecular hydroalkoxylation of alkynes (Scheme 12.15) [19]. A proposed mechanism for this reaction is shown in Scheme 12.16. The reaction

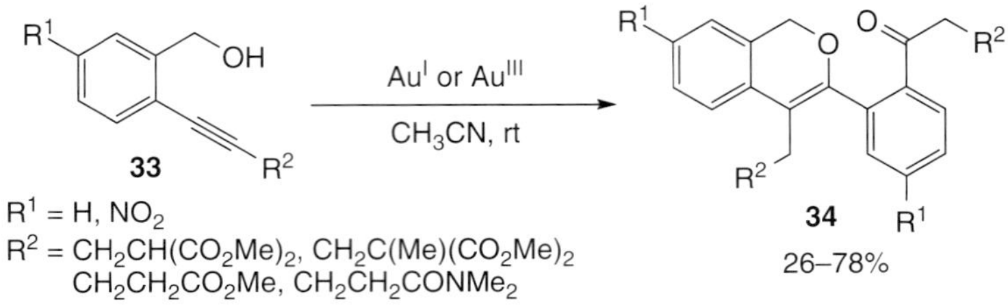

Scheme 12.15 Au-catalyzed synthesis of 1*H*-isochromene derivatives.

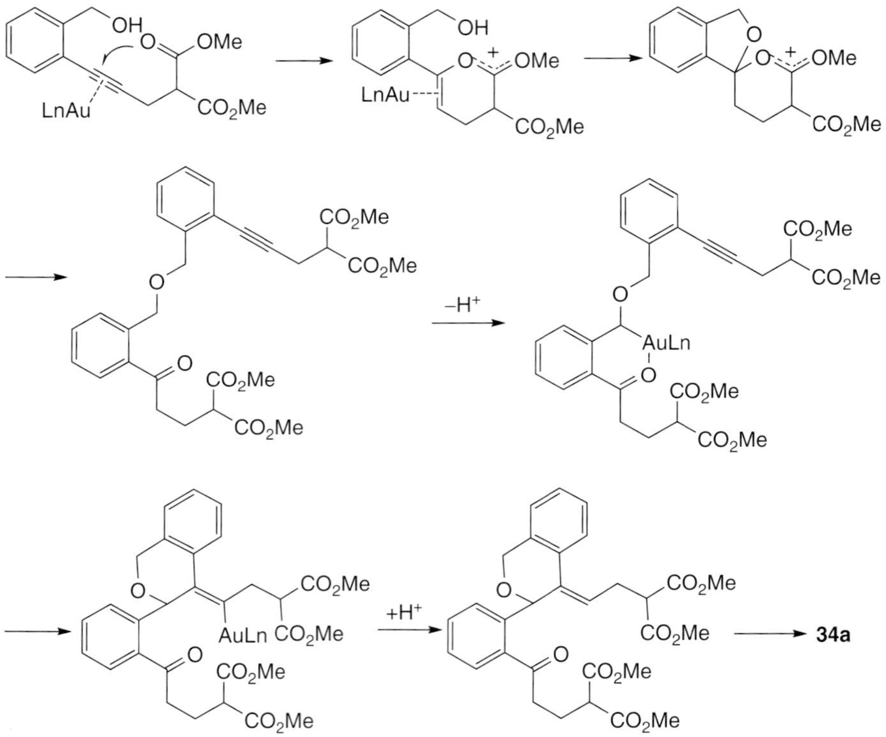

Scheme 12.16 Proposed mechanism for the unprecedented dimerization formation via Au-catalyzed benzylic C–H activation.

might start by the nucleophilic addition of the carbonyl ester group onto the activated alkyne, followed by the intramolecular addition of the hydroxy moiety affording tricyclic intermediate. The hydroxy group of a second molecule of substrate performs the ring opening of this system to afford the corresponding ether, which then undergoes electrophilic C(sp^3)–H activation at the benzylic position forming a C–Au bond. The C–Au bond is stabilized by coordination/chelation with the adjacent carbonyl group. Upon insertion of the alkyne, protonolysis and isomerization of the double bond can afford the corresponding dimer product 1*H*-isochromene derivatives (Scheme 12.16).

12.3 Synthesis of *N*-Heterocycles

Nitrogen-containing heterocycles show a variety of biological activities and represent privileged structures compared to other types of heterocycles. Thus the preparation of the *N*-heterocycles is the core of organic synthesis throughout the history of synthesis chemistry [20]. Most investigations involving the synthesis of heterocycles via Au- and Pt-catalyzed C–H functionalization reactions have been directed toward the synthesis of nitrogen-containing heterocycles.

12.3.1 Synthesis of Five-Membered *N*-Heterocycles

12.3.1.1 Via Au-Catalyzed C(sp)–H Functionalization

Indolizines, featured by an *N*-bridgehead bicyclic ring system fused with both π-excessive pyrrole and π-deficient pyridine, are frequently found in a variety of naturally occurring and biologically active compounds. Therefore, a direct and efficient synthesis of indolizines is highly desirable. In 2007, Liu and Yan reported a gold-catalyzed multicomponent coupling/cycloisomerization reaction of heteroaryl aldehydes, amines, and alkynes under solvent-free conditions or in water, furnishing a rapid access to aminoindolizines **35** (Scheme 12.17) [21]. The addition of *in situ*-formed gold acetylide to iminium and subsequent gold-mediated heterocyclization by the activation of alkyne would produce the final product.

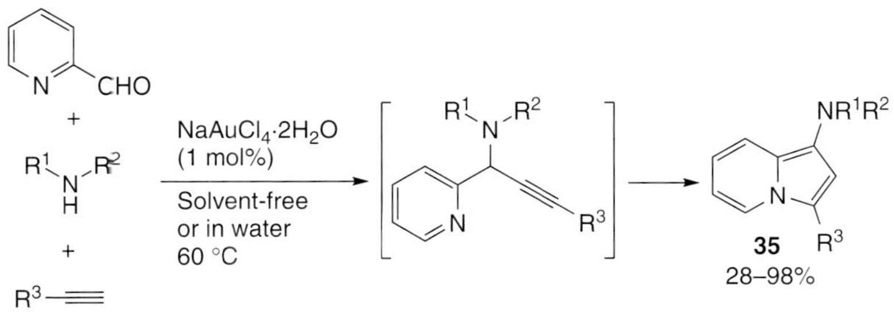

Scheme 12.17 Au-catalyzed synthesis of aminoindolizines.

12.3.1.2 Via Au-Catalyzed C(sp)–H and Alkenyl C(sp^2)–H Functionalization

Similarly, a three-component annulation of bisalkynes, hydrazines, and aldehydes for the synthesis of fused tricyclic dihydropyrazoles **36** was explored by Fujii and Ohno (Scheme 12.18) [22]. This process consists of Au-catalyzed C(sp)–H functionalization and alkenyl C(sp^2)–H functionalization.

12.3.1.3 Via Au-Catalyzed C(sp)–H, C(sp^3)–H, or Aryl C(sp^2)–H Functionalization

Very recently, Hashmi, Ohno, and coworkers reported a novel method for the preparation of bicyclic and tricyclic pyrroles via Au-catalyzed sequential

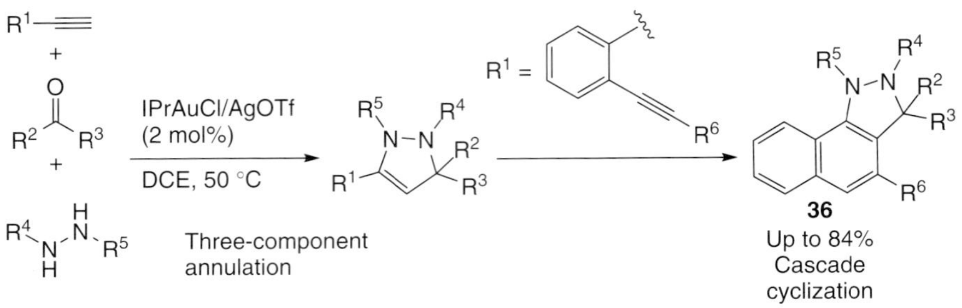

Scheme 12.18 Au-catalyzed synthesis of dihydropyrazoles.

C(sp)–H/C(sp^3)–H or C(sp)–H/C(sp^2)–H functionalization (Scheme 12.19) [23]. A cationic dual gold-catalyzed intramolecular cyclization of *N*-propargyl ynamides **37** provided bicyclic and tricyclic pyrroles **39** in moderate to good yields. A key step in this reaction was the formation of a gold vinylidene intermediate **38** via the nucleophilic addition of a gold acetylide onto an ynamide triple bond at the *β*-position of the nitrogen. Both aryl and alkyl C–H bonds can be used for the second cyclization step, and the corresponding pyrroles are obtained in moderate to good yields.

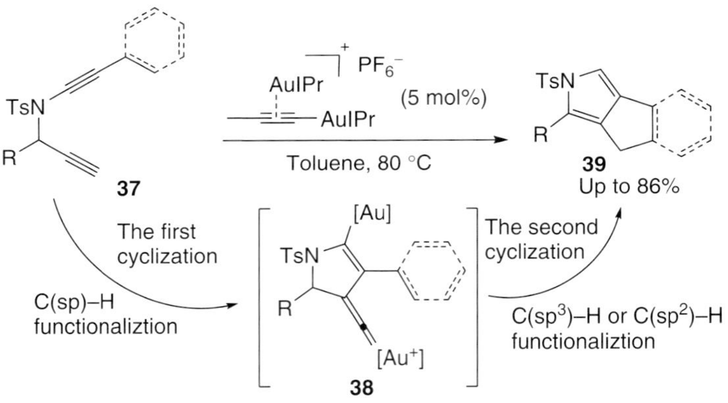

Scheme 12.19 Au-catalyzed synthesis of bicyclic and tricyclic pyrroles.

12.3.1.4 Via Au-Catalyzed Aryl C(sp^2)–H Functionalization

In 2011, Zhang and coworkers developed a gold-catalyzed intramolecular tandem hydroamination/oxidative coupling reaction of *N*,*N*-allyl alkyl-*N*′-aryl urea **40** using Selecfluor as an oxidant, which provide a facile access to tricyclic indolines **41**. This is the first oxidative cross coupling reaction between an aryl C(sp^2)–H and an *in situ*-generated alkyl gold intermediate (Scheme 12.20) [24]. Deuterium labeling and kinetic isotope effect studies along with the isolation of alkyl gold intermediates strongly support an electrophilic aromatic substitution for the C–H functionalization and a subsequent inner-sphere concerted reductive elimination for the C(sp^2)–C(sp^3) bond formation.

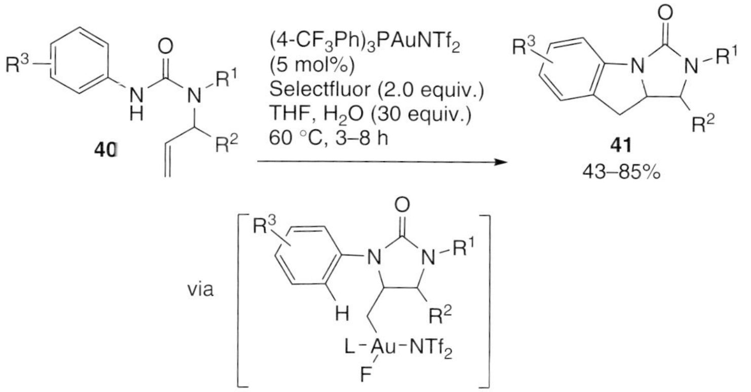

Scheme 12.20 Au-catalyzed synthesis of indolines.

In 2011, a gold-catalyzed one-pot reaction of *ortho*-azidophenylalkyne **42** with electron-rich aromatics and heteroaromatics was independently developed by Zhang and Gagosz (Scheme 12.21) [25], which gave functional indoles **43a** in good to excellent yields. This one-pot reaction consists of an intramolecular C–N formation and intermolecular aryl C(sp^2)–H functionalization/C–C bond-forming procedure. This new approach achieves umpolung reactivity of indole at the 3-position through gold catalysis. The reaction proceeds via a gold carbene intermediate which is formed by an intramolecular nitrene transfer from the azido group. Interestingly, when an allylic alcohol was used as the nucleophile, indolin-3-ones **43b** were obtained via nucleophilic addition with a-imino gold carbine intermediate, followed by a Claisen rearrangement sequence [25b].

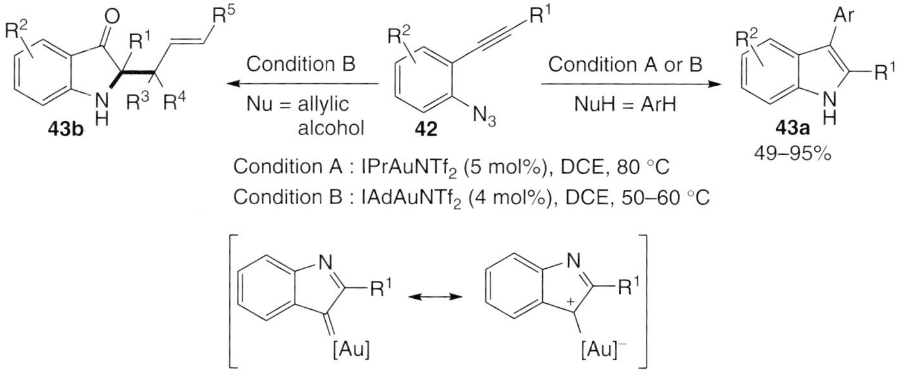

Scheme 12.21 Au-catalyzed synthesis of functional indoles and indolin-3-ones.

In the same year, Zhang also reported a gold-catalyzed synthesis of 2-alkylindoles **46** from *N*-arylhydroxylamines **44** and aliphatic terminal alkynes (Scheme 12.22) [26]. The addition of *N*-arylhydroxylamines to aliphatic terminal alkynes in the presence of gold catalyst affords *O*-alkenyl-*N*-arylhydroxylamines **45**, which undergo facile *in situ* sequential 3,3-rearrangement

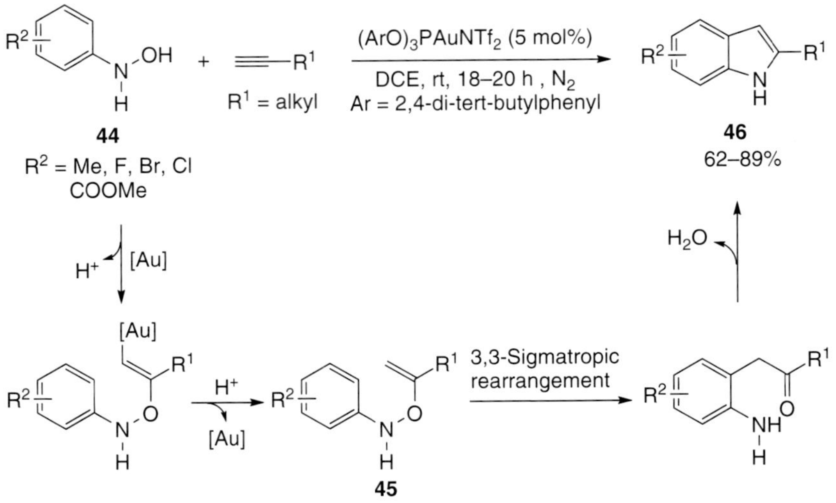

Scheme 12.22 Au-catalyzed synthesis of 2-alkylindoles.

and cyclodehydration to afford 2-alkylindoles **46** with regiospecificity under exceptionally mild reaction conditions.

In order to extend this chemistry for the synthesis of other types of indoles besides 2-alkylindoles, the same group disclosed an efficient synthesis of *N*-protected indoles **48** from *N*-arylhydroxamic acids/*N*-aryl-*N*-hydroxycarbamates **47** and a variety of alkynes via gold and zinc cooperative catalysis (Scheme 12.23) [27]. They found that catalytic $Zn(OTf)_2$ enhances the nucleophilicity of these hydroxylamine derivatives via the formation of deprotonated chelates, which is similar to the metal ion catalysis in metalloenzymes. This chemistry is a rare example of cooperative dual catalysis involving gold catalyst.

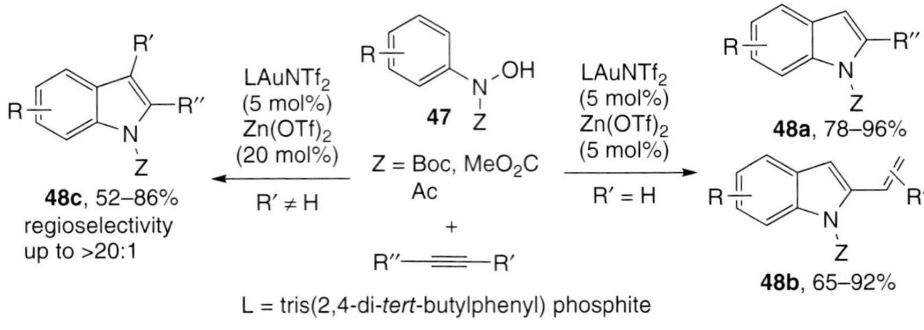

Scheme 12.23 Au-catalyzed synthesis of *N*-protected indoles.

An efficient catalytic C–H functionalization by means of gold carbenoids for the one-step synthesis of 3-acyloxindole derivatives **50** has been explored by Zhang and Qian in 2012 (Scheme 12.24) [28]. 3-Acyloxindole derivatives were prepared in good yields by a gold-catalyzed tandem oxidation/C–H functionalization reaction of electron-deficient *N*-arylpropiolamides **49** using 2-bromopydine

N-oxidant as an external oxidant. The reaction proceeds efficiently with extremely good substrate scope and significant opportunities for structural diversification. This method provides a diazo compound-free access to this type of products [29].

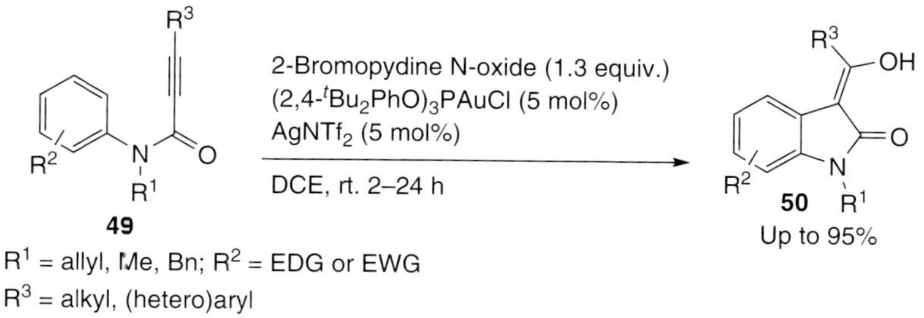

Scheme 12.24 Au-catalyzed synthesis of 3-acyloxindole derivatives.

A gold(I)-catalyzed hydroarylation of an allene with indole was developed by Toste and Zeldin (Scheme 12.25) [30]. This reaction could be applied toward the total synthesis of flinderoles B **54** and flinderoles C **55**, members of a new class of antimalarial bisindole alkaloids isolated from plants of the *Flindersia* genus. The key intermediate pyrrolidine **53** in the total synthesis of flinderoles B and C could be efficiently constructed as a single diastereomer via Au-catalyzed intramolecular hydroarylation of a pendant allene **52**, which was prepared from commercially available tryptophol **51**.

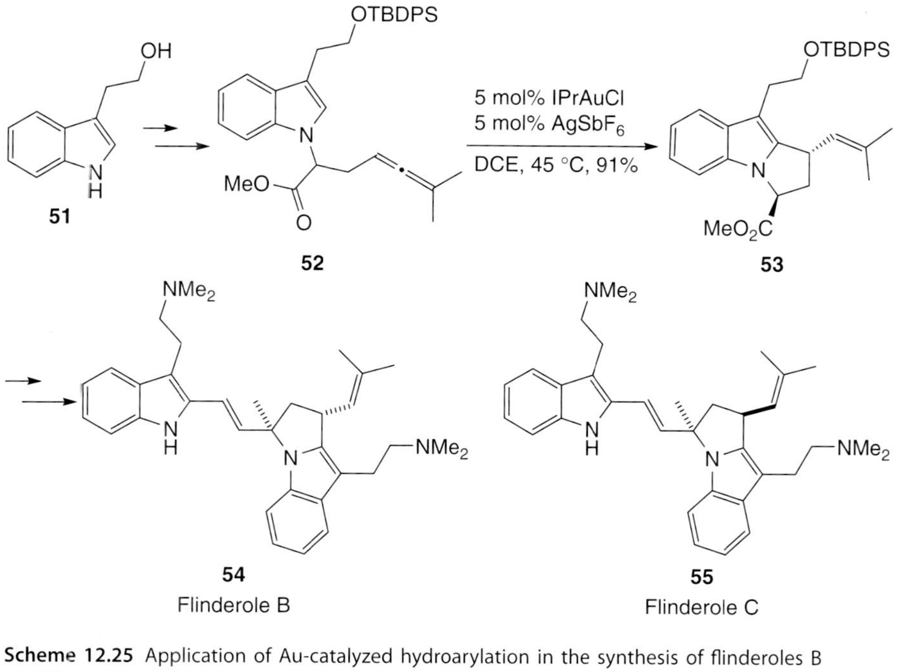

Scheme 12.25 Application of Au-catalyzed hydroarylation in the synthesis of flinderoles B and C.

12.3.1.5 Via Au-Catalyzed C(sp^3)–H Functionalization

In 2007, Che and coworkers successfully applied the Au-catalyzed C(sp^3)–H functionalization to the synthesis of five-membered *N*-heterocycles such as lactams (Scheme 12.26) [31]. In the presence of Au(I) complexes, the intramolecular addition of β-ketoamide **56** to unactivated alkenes can efficiently afford highly substituted lactams **57**, which are commonly found in natural products and biologically active molecules.

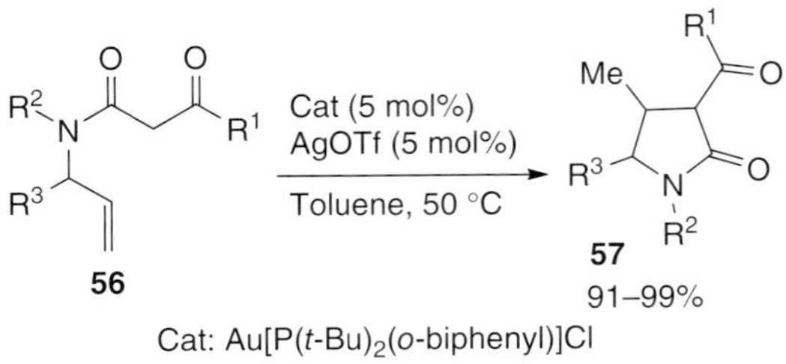

Scheme 12.26 Au-catalyzed synthesis of lactam.

12.3.1.6 Via Au-Catalyzed Miscellaneous Reactions

Using azides as nitrene precursors and electrocyclizations of destabilized 1-azapentadienium intermediates, Zhang and coworkers have realized a gold-catalyzed synthesis of 2,3-dihydro-1*H*-pyrrolizines **59** having electron-withdrawing groups in the 5-position (Scheme 12.27) [32]. A key step in this reaction was the formation of a reactive α-imino gold carbene intermediate via Au-catalyzed 5-*exo-dig* cyclization of linear azidoenynes **58**. The synthetical significance of this transformation is demonstrated by an elegant formal synthesis of 7-methoxymitosene (Eq. (12.1)).

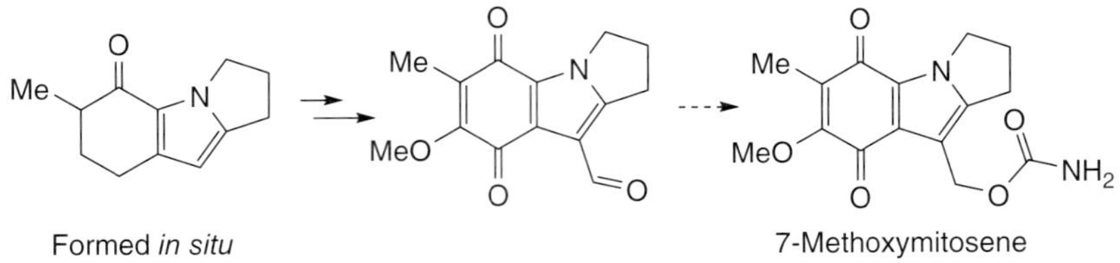

(12.1)

A facile two-step, one-pot method toward the synthesis of a range of fused pyrroles **64** from *N*-alkynylhydroxamines salt **61** and readily enolizable ketones **60** was lately developed by Zhang and coworkers (Scheme 12.28) [33]. The reaction starts firstly by condensing an *N*-alkynylhydroxammonium salt **61** with a readily enolizable ketone **60** under mild basic conditions, which produce an *N*-hydroxyenamine intermediate **62**. This intermediate undergoes a gold-catalyzed cyclization to give an *N*,O-dialkenylhydroxamine **63**, followed by a rapid [3,3]-sigmatropic rearrangement to give cyclic imino ketone. The final

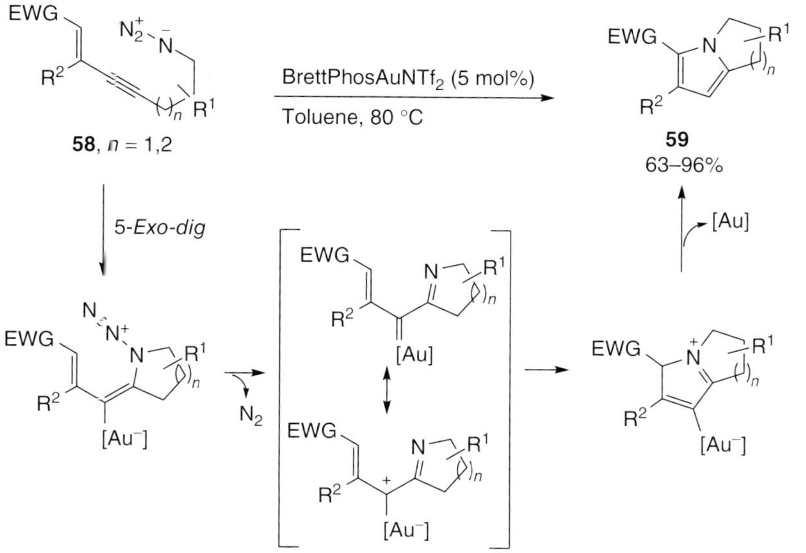

Scheme 12.27 Au-catalyzed synthesis of 2,3-dihydro-1*H*-pyrrolizines.

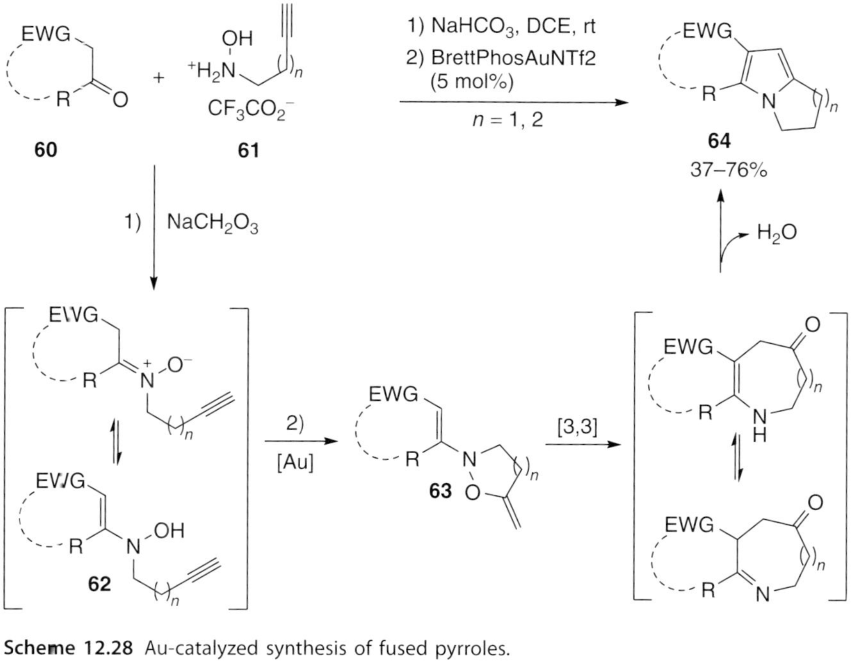

Scheme 12.28 Au-catalyzed synthesis of fused pyrroles.

product fused pyrroles were obtained after tautomerization of cyclic imino ketone to enamine and subsequent transannular dehydrative cyclization.

12.3.2 Synthesis of Six-Membered *N*-Heterocycles

12.3.2.1 Via Au-Catalyzed C(sp)–H and Aryl C(sp^2)–H Functionalization

A simple and effective catalytic method to form quinoline derivatives **65** by using a $AuCl_3$-catalyzed reaction of propargyl amines or by a $AuCl_3$/CuBr-catalyzed three-component reaction of aldehydes, amines, and alkynes was reported by Wang and coworkers (Scheme 12.29) [34]. The reaction proceeds via a sequential Au-catalyzed C(sp)–H and aryl C(sp^2)–H functionalization. These processes can provide a diverse range of quinoline derivatives in moderate to good yields from simple starting materials.

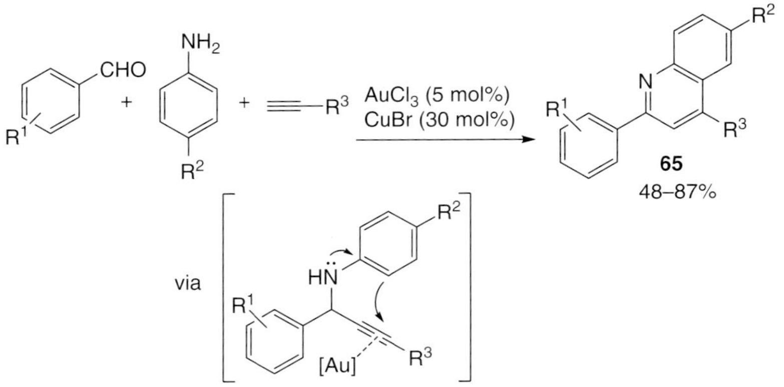

Scheme 12.29 Au-catalyzed synthesis of quinoline derivatives.

12.3.2.2 Via Au-Catalyzed Formyl C(sp^2)–H Functionalization

During the course of their extensive study on Au-catalyzed synthesis of *O*-heterocycles[12-13], Li and coworkers found that the cascade reaction consisting of Au-catalyzed formyl C_{sp2}–H functionalization could be applied for the synthesis of the six-membered *N*-heterocycles (Scheme 12.30) [35]. 2,3-Dihydro-1-tosylquinolin-4(1*H*)-one derivatives, azaflavanones **67**, could be efficiently

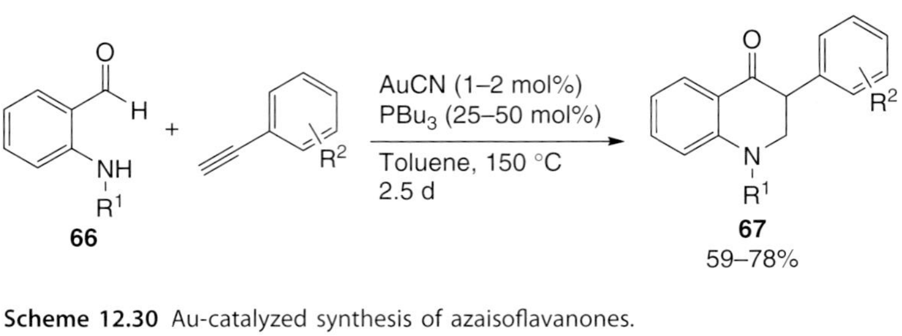

Scheme 12.30 Au-catalyzed synthesis of azaisoflavanones.

constructed via a gold-catalyzed annulation of 2-tosylaminobenzaldehyde **66** and alkynes.

12.3.2.3 Via Au-Catalyzed C(sp)–H and C(sp^3)–H Functionalization

By taking advantage of Ugi four-component reaction (U-4CR) and inspired by Zhang's and Hashmi's work on gold-catalyzed dual $\sigma-\pi$ activation chemistry, Van der Eycken and coworkers realized a gold-catalyzed regioselective tandem cyclization of *N*-propynylbutynamide **68**, which provides a distinctive methodology for the architecture of (spiro)cyclopentapyridinones **69** (Scheme 12.31) [36].

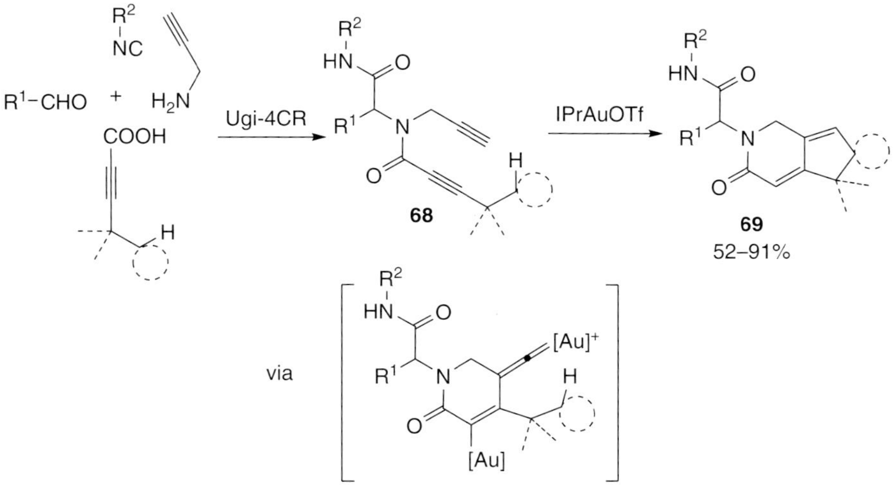

Scheme 12.31 Au-catalyzed synthesis of (spiro)cyclopentapyridinones.

12.3.2.4 Via Au-Catalyzed Aryl C(sp^2)–H Functionalization

Fujii, Ohno, and coworkers developed a gold-catalyzed intramolecular aryl C(sp^2)–H functionalization with allenic anilines for the synthesis of dihydroquinoline (Scheme 12.32) [37]. This intramolecular gold-catalyzed 6-*endo* hydroarylation of allenic anilines **70** could afford the dihydroquinolines **71** in

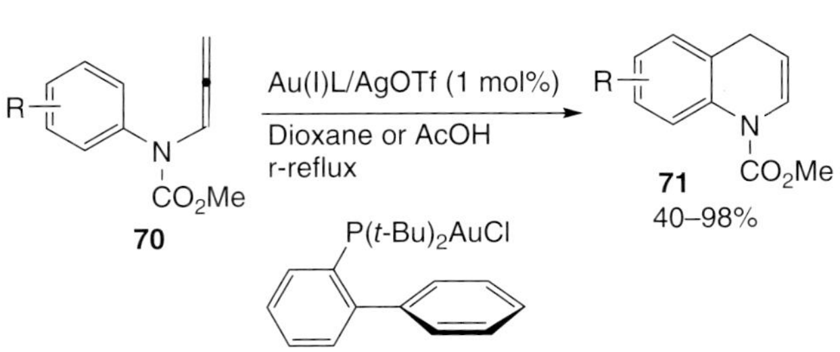

Scheme 12.32 Au-catalyzed synthesis of dihydroquinoline.

moderate to good yields. It was found that electron-donating substitutions on the arene are required for the reaction which preferentially produces six-membered rings.

Pyrroles are electron-rich heteroaromatic systems which easily undergo electrophilic substitution at the 2-position. Nelson, Liu, and coworkers utilized allenes of type **72** bearing pyrrole skeleton as intramolecular electrophiles, PPh_3AuOTf as catalyst, and successfully constructed a six-membered heterocyclic ring in the product **73** with a highly diastereoselectivity (dr = 97 : 3), which was a key intermediate in an enantioselective total synthesis of (−)-rhazinilam **74** (Scheme 12.33) [38].

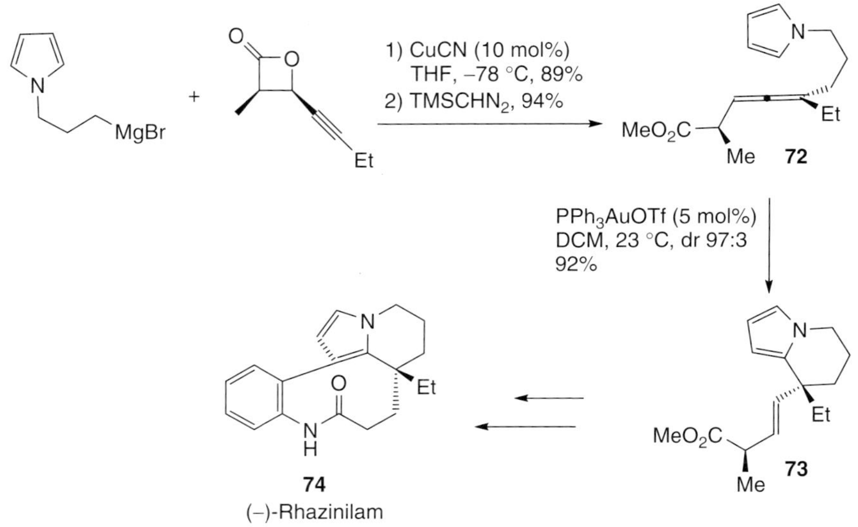

Scheme 12.33 Au-catalyzed hydroarylation in the synthesis of (−)-rhazinilam.

The synthesis of dihydroquinoline derivatives **75** and **76** by a gold-catalyzed aryl $C(sp^2)$–H functionalization of aromatic amines with alkynes has also been realized under microwave irradiation by Che and coworkers (Scheme 12.34) [39].

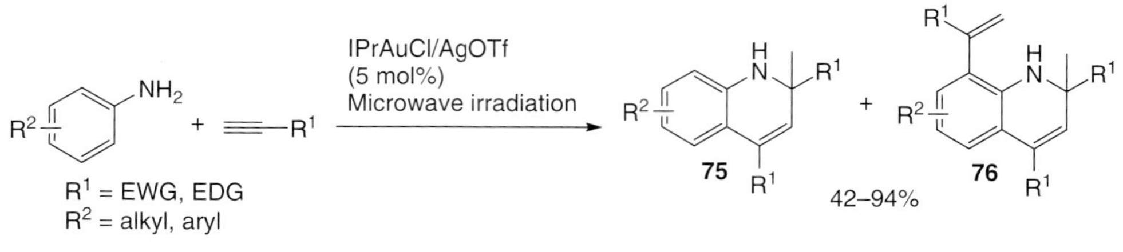

Scheme 12.34 Au-catalyzed synthesis of 1,2-dihydroquinolines.

A series of substituted 1,2-dihydroquinolines were obtained in moderate to good yields. This method features a short reaction time and a broad substrate scope.

12.3.2.5 Via Pt-Catalyzed Aryl C(sp^2)–H Functionalization

A $PtBr_2$-catalyzed oxidative dehydrogenation of alkylamines with anilines toward the synthesis of quinoline derivatives **77** was accomplished by Brunet and coworkers (Scheme 12.35) [40]. The plausible mechanism of this reaction might consist of Pt-catalyzed aliphatic α-C–H activation of alkylamines and aromatic *ortho*-C–H activation of anilines.

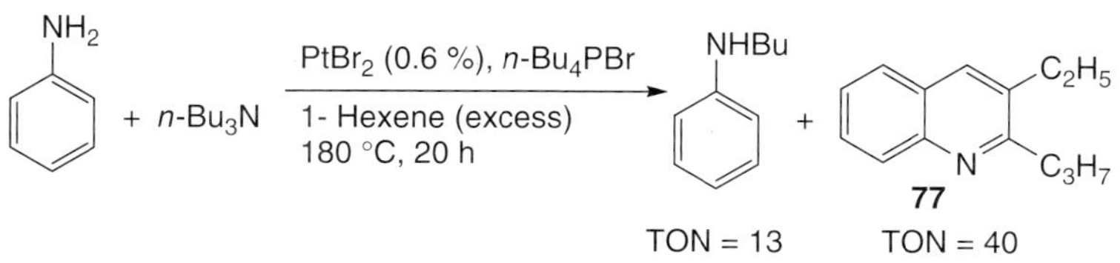

Scheme 12.35 Pt-catalyzed synthesis of quinoline derivatives.

Using a similar Pt-catalyzed intermolecular oxidative dehydrogenation hydroamination followed by intramolecular aryl C(sp^2)–H fuctionalization process, Poli and coworkers developed a simple protocol for the synthesis of quinaldine **78** from readily available aniline and ethylene (Scheme 12.36) [41]. It was found that the selectivity of the catalytic reaction between aniline and ethylene in the presence of the Brunet catalyst ($PtBr_2/Br^-$) shifts from the hydroamination product *N*-ethylaniline to the heterocyclization product

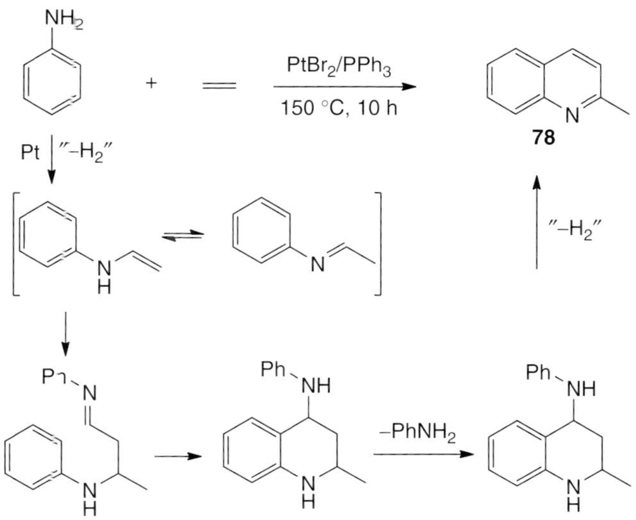

Scheme 12.36 Pt-catalyzed synthesis of quinaldine derivatives.

2-methylquinoline (quinaldine) **78** when the reactions were conducted in the presence of PPh_3 (1 equiv. per Pt atom).

12.3.2.6 Via Au-Catalyzed C(sp^3)–H Functionalization

Building on their previous discovery of the generation of α-oxo gold carbenes through the oxidation by tethered *N*-oxides, Zhang and coworkers designed the hydride shift-triggered (C(sp^3)–H functionalization) construction of piperidinones **80** starting from a range of secondary amines **79** in a two-step synthetic approach (Scheme 12.37) [42a]. The synthetical utility of this transformation is demonstrated by an elegant total synthesis of (±)-cermizine C (Eq. (12.2)). Later, this methodology was also applied to the stereoselective total synthesis of the natural product (+)-lentiginosine (Scheme 12.38) [42b].

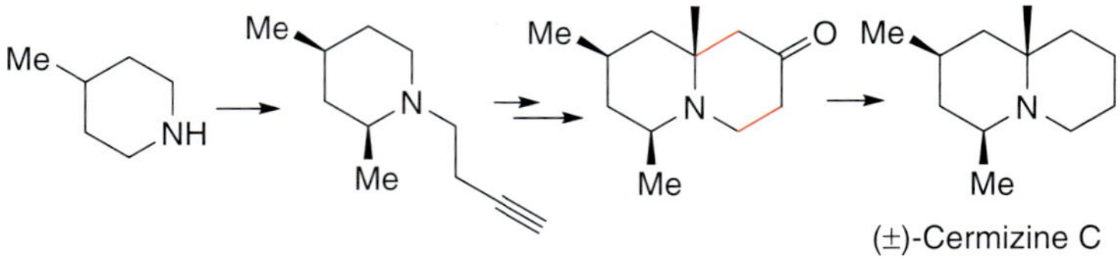

(12.2)

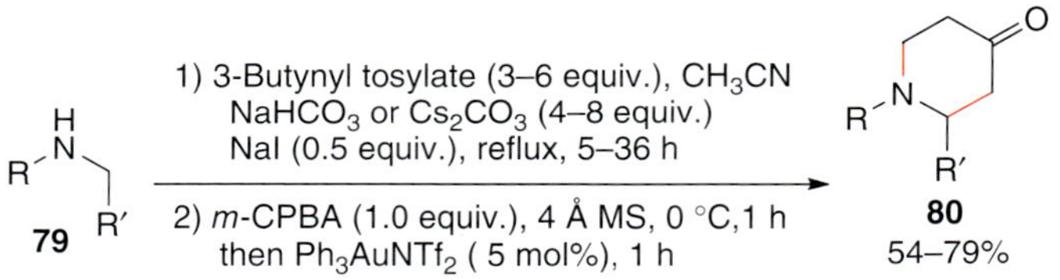

Scheme 12.37 Au-catalyzed synthesis of piperidinones.

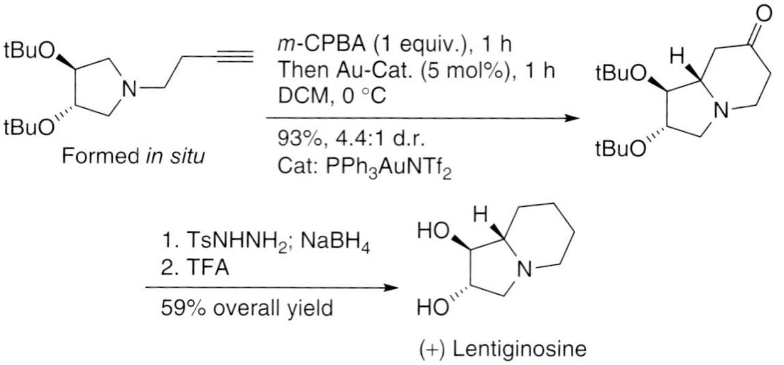

Scheme 12.38 Total synthesis of (+)-lentiginosine via a key Au catalysis.

Initially, the authors proposed the mechanism involving α-oxo gold carbine formation. Lately, Zhang and Houk presented a revised mechanism for the Au-catalyzed system, which involves a concerted heteroretroene reaction based on both DFT calculations and deuterium-labeling studies. Accordingly, the internal redox step occurs by the unusual heteroretroene transition state (Scheme 12.39) [42c].

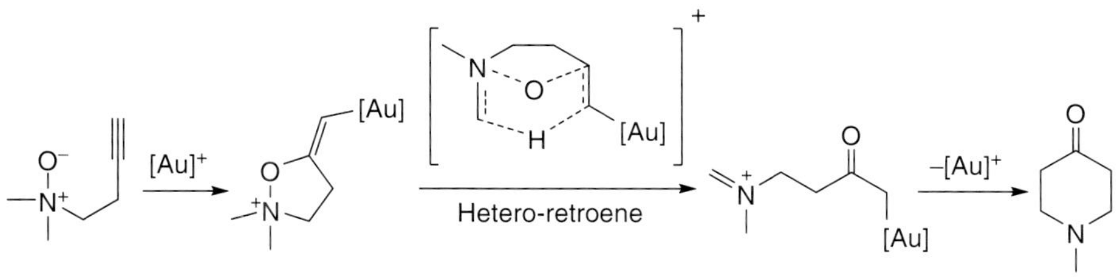

Scheme 12.39 Proposed mechanism for Au-catalyzed synthesis of piperidinones.

A relay catalytic hydroamination/redox reaction for the synthesis of cyclic aminals **83** from tertiary amine-substituted 3-en-1-yne derivatives **81** and various aniline derivatives **82** was recently developed by Gong and coworkers (Scheme 12.40) [43]. The gold carbene complex [(IPr)AuNTf$_2$] and TfOH were found to be optimal cocatalysts for this cascade reaction. In their preliminary studies, the authors found that this reaction could be carried out in a highly enantioselective manner, despite that 2 equiv. of an expensive chiral phosphoric acid are required. Interestingly, the diastereoselectivity for a particular substrate appears to be independent of the catalyst or ee. This suggests that the enantioselectivity may result from the ring-closure step, rather than the hydride shift step.

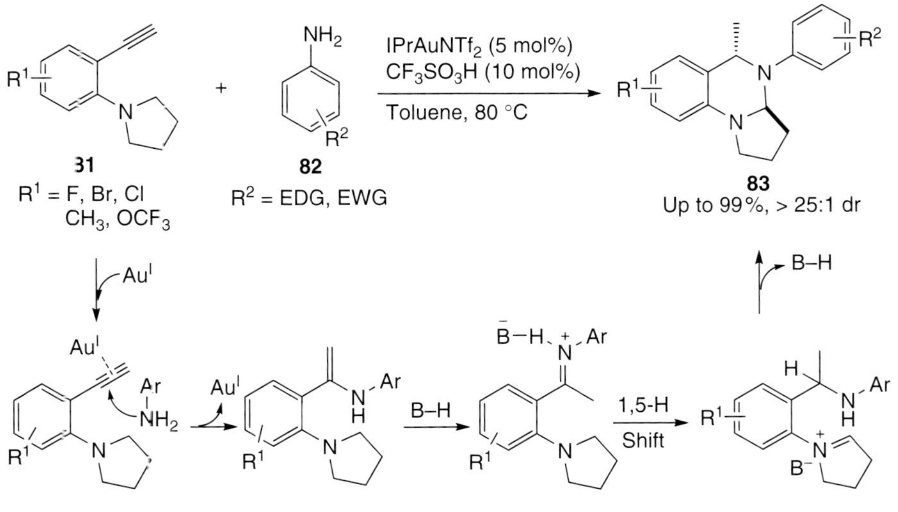

Scheme 12.40 Gold/Brønsted acid cocatalyzed synthesis of cyclic aminals.

12.3.3
Synthesis of Seven-Membered *N*-Heterocycles

12.3.3.1 Via Au-Catalyzed C(sp^2)–H Functionalization

In 2009, Zhang and coworkers described an efficient Au-catalyzed one-pot synthesis of seven-membered *N*-heterocycles tetrahydrobenz[*b*]azepin-4-ones **85** from aniline substrates with tethered terminal alkynes **84**. Compound **85** were efficiently constructed by Au-catalyzed intramolecular cyclization of the tertiary aniline N-oxides *in situ* formed by oxidation of the corresponding aniline substrates with *m*-CPBA (Scheme 12.41) [44].

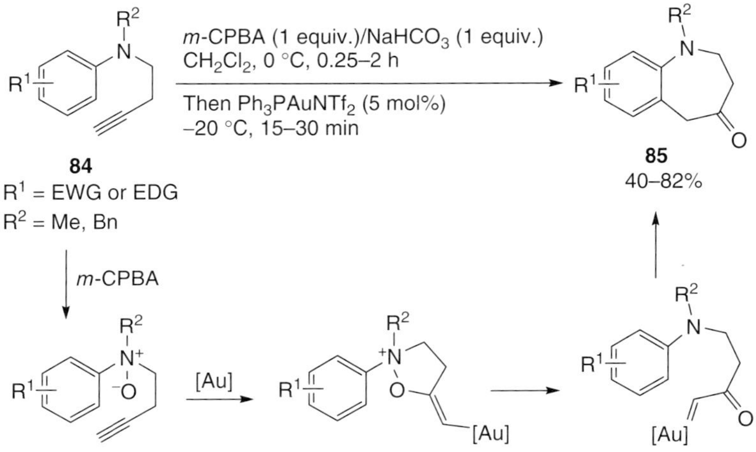

Scheme 12.41 Au-catalyzed synthesis of tetrahydrobenz[*b*]azepin-4-ones.

12.3.3.2 Via Au-Catalyzed C(sp^3)–H Functionalization

In a subsequent publication, Zhang and coworkers also found that homologous alkynyl amines **86** are also viable substrates for the same transformation, allowing for the generation of larger-ring analogs such as azepan-4-ones **87** (Scheme 12.42) [45].

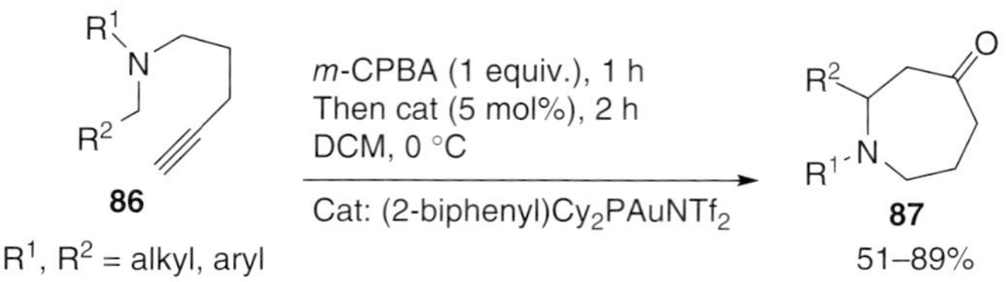

Scheme 12.42 Au-catalyzed synthesis of azepan-4-ones.

12.4
Synthesis of S-Heterocycles

12.4.1
Synthesis of Seven-Membered S-Heterocycles via Au-Catalyzed Aryl C(sp^2)–H Functionalization

Toste and Zhang independently developed a series of Au(I) complex catalyzed rearrangement of alkynyl sulfoxides **88** to benzothiepinones **89** (Scheme 12.43) [46a,b]. Initially, a common mechanism via an α-oxo gold carbene intermediate was proposed. Recently, a combination of experiments and density functional theory studies by the Zhang group disclosed that the involvement of gold carbene intermediates in the previous formation of dihydrobenzothiepinones is highly unlikely. Instead, a [3,3]-sigmatropic rearrangement of the initial cyclization intermediate was proposed. With internal alkyne substrates, the intermediacy of a gold carbene species becomes possible, and the generation of a gold carbene species becomes competitive with the [3,3]-sigmatropic rearrangement [46c].

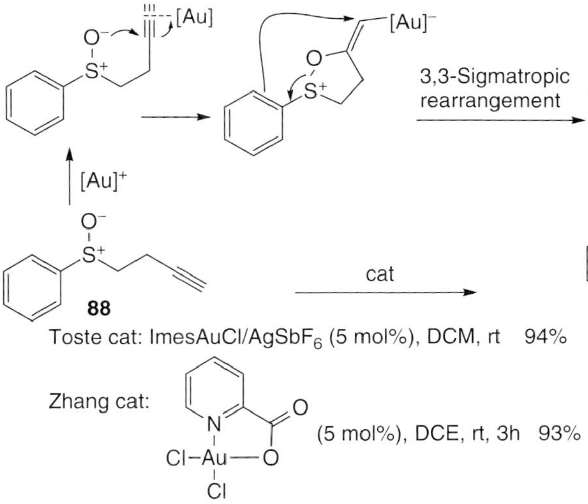

Scheme 12.43 Au-catalyzed synthesis of benzothiepinones.

12.5
Synthesis of O-Heterocycles and N-Heterocycles

12.5.1
Synthesis of Five-Membered O-Heterocycles and N-Heterocycles

12.5.1.1 Via Au-Catalyzed C(sp)–H Functionalization

In 2012, Zhang and Hashmi independently described a novel Au-catalyzed cycloisomerization of (2-ethynylphenyl)alkynes **90** [47]. The group of Zhang found that using of a pyridine N-oxide as a weak base can facilitate the formation of alkynylgold intermediate **91**. Another molecule of gold can then activate the other C–C triple bond of **91** via the intermediacy of **92**, which undergoes a 5-*endo-dig* cyclization to afford gold vinylidene intermediate **93**. This novel gold species is highly reactive and undergoes facile intramolecular O–H and N–H insertions as well as C(sp^3)–H insertions leading to highly efficient formation of tricyclic furan, pyrrole derivatives **94**, and indene structures (Scheme 12.44) [47a].

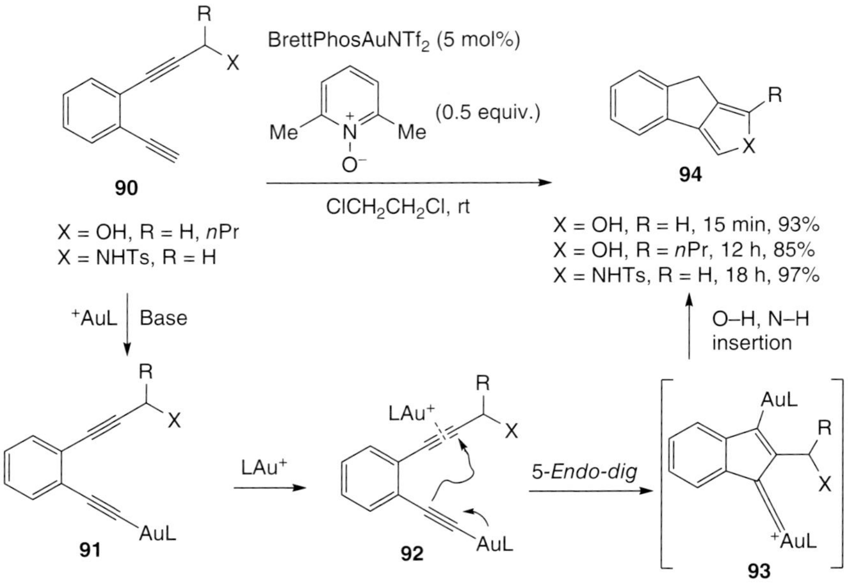

Scheme 12.44 Au-catalyzed synthesis of tricyclic furan and pyrrole.

Interestingly, when the substrate was switched from benzene-1,2-dialkyne **90** to *cis*-enediyne **95**, 6-*endo-dig* cyclization reaction pathway was observed. Two regioisomeric *ortho*-auroaryl cation intermediates **96** and **98** were proposed via dual gold-catalyzed 6-*endo-dig* cyclization model, which can be mutual isomerization via aryne-coordinated gold complex **97**. *Ortho*-aurophenyl cation intermediates **98** could be trapped by a tethered hydroxy or sulfonamide group to deliver the seven-membered *O*- or *N*-heterocycles **99** (Scheme 12.45) [48].

Using a similar dual gold catalysis of diynes model, Hashmi, Hansmann, and coworkers explored a dual gold-catalyzed cyclization of 3,4-diethynylthiophenes to generate pentaleno[*c*]thiophenes through gold–vinylidenes and C–H bond activation. In these cases, 5-*endo*-dig cyclization pathway was observed. This protocol can be used to construct thiophene/furan or thiophene/pyrrole **101** system skeleton from 3,4-diethynylthiophene derivatives **100** (Scheme 12.46) [49].

Inspired by their proposed novel reaction mode involving σ, π-type dual activation of diynes substrates [50], Hashmi and coworkers developed an alternative

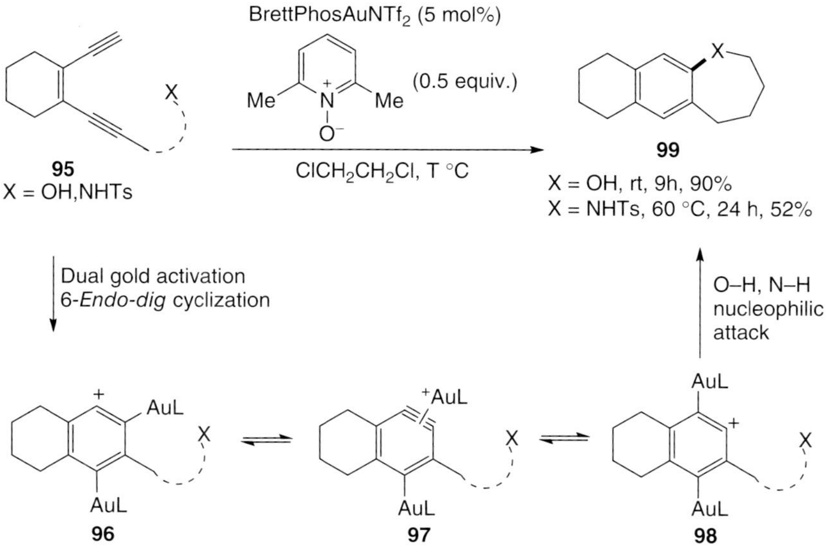

Scheme 12.45 Au-catalyzed synthesis of tetrahydrooxepin and tetrahydroazepin.

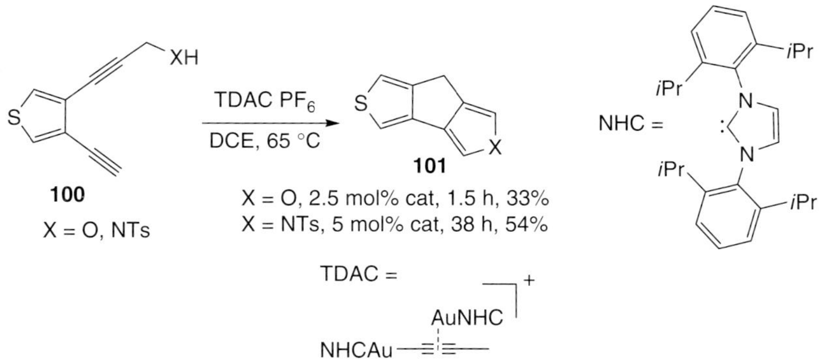

Scheme 12.46 Au-catalyzed synthesis of tricyclic thiophene/furan or thiophene/pyrrole.

approach for gold vinylidene intermediate formation by using a leaving group replacing the π-activated alkyne. They demonstrated that the formation of gold vinylidene intermediates **103** is also possible by the cyclization of gold acetylides alone and therefore not strictly dependent on a dual activation mode. The protocol can be utilized to synthesize tetrahydrofuran or pyrrolidine derivatives bearing a vinyl sulfonate moiety **104** from readily available substituted alkynyl sulfonates **102** (Scheme 12.47) [51].

12.5.1.2 **Via Au-Catalyzed Aryl C(sp^2)–H Functionalization**

A gold(I)-catalyzed tandem heterocyclization/Friedel–Crafts addition (aryl C(sp^2)–H functionalization) between enynes **105** and electron-rich arenes **106**

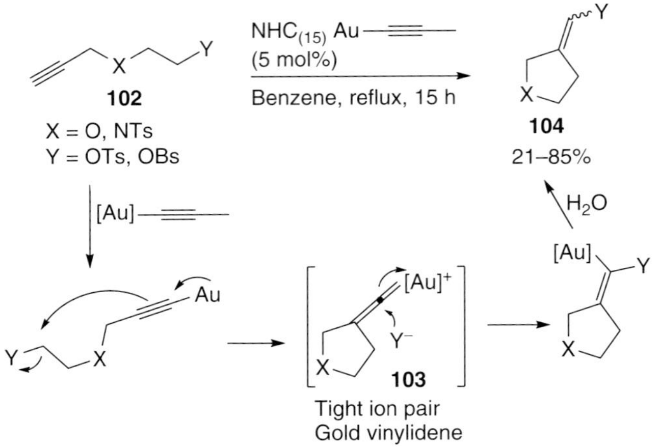

Scheme 12.47 Au-catalyzed synthesis of tetrahydrofuran or pyrrolidine derivatives.

for 3-methylenetetrahydrofuran or 3-methylenepyrrolidine derivatives **108** synthesis was reported by Genêt, Michelet, and coworkers (Scheme 12.48) [52a]. They believed that the enyne cyclization occurs first to give a transient carbenic intermediate **107**, which is subsequently attacked by the electron-rich arene. A similar system was also reported almost simultaneously by the Echavarren research group [52b].

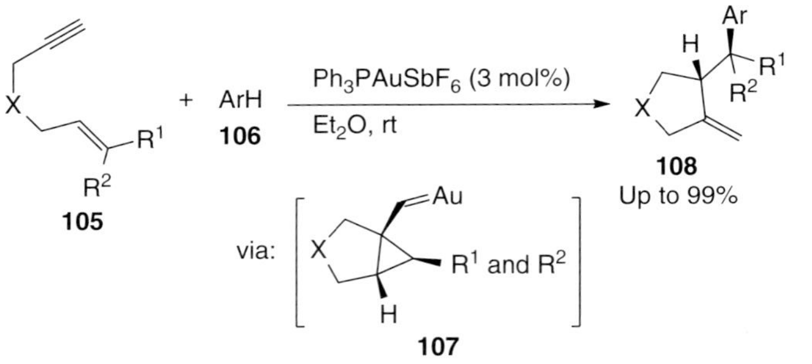

Scheme 12.48 Au-catalyzed synthesis of 3-methylenetetrahydrofuran or 3-methylenepyrrolidine derivatives.

12.5.2 Synthesis of Six-Membered *O*-Heterocycles and *N*-Heterocycles

12.5.2.1 Via Pt or Au-Catalyzed Aryl C(sp^2)–H Functionalization

In 2003, the group of Echavarren intensively studied the intramolecular hydroarylation (aryl C(sp^2)–H functionalization) of *N*-propargyl-*N*-tosyl anilines **109** and resorcinol derivatives **111** under the catalysis of Pt^{II} or Au^{I} complexes. The reaction proceeds via 6-*endo-dig* pathways to afford 1,2-dihydroquinolines **110** or chromene derivatives **112**, respectively. A DFT study on the cyclization mechanism shows a kinetic and thermodynamic preference for 6-*endo-dig* versus 5-*exo-dig* cyclizations in Pt^{II}-catalyzed processes (Scheme 12.49) [53].

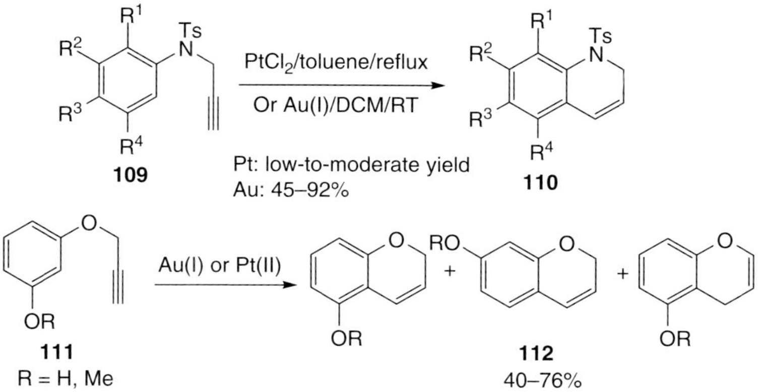

Scheme 12.49 Pt- and Au-catalyzed synthesis of 1,2-dihydroquinolines and chromenes.

In the same year, the research group of Sames found that $PtCl_4$ proved to be a better hydroarylation catalyst for arene–yne substrates **113** with diverse structural features, including propargyl ethers, propargylamines, and alkynoate esters, providing good to excellent yields of the 6-*endo* products **114** (chromenes, dihydroquinolines, and coumarins). In contrast, Pt(II), Pd(II), and Ga(III) salts were shown to be sensitive to the substitution on the alkyne moiety (Scheme 12.50) [54].

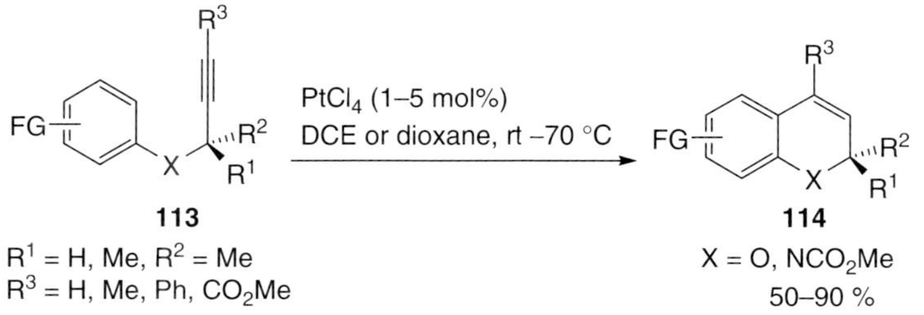

Scheme 12.50 Pt- and Au-catalyzed synthesis of chromenes, dihydroquinolines, and coumarins.

This Pt-catalyzed hydroarylation was further applied to the total synthesis of deguelin **117**, a rotenoid with promising chemopreventive properties (Scheme 12.51) [55]. Although $PtCl_4$ proved to be superior to $PtCl_2$ as catalyst in model systems for the 6-*endo* hydroarylation of alkynones and alkynoates, the cyclization of **115** with $PtCl_4$ led only to poor yields of **116**. However, the reaction of **115** with $PtCl_2$ as catalyst provided **116** in excellent yield. The synthesis of racemic deguelin **117** was completed by selective demethylation with boron trichloride, followed by a base-catalyzed intramolecular oxa-Michael addition.

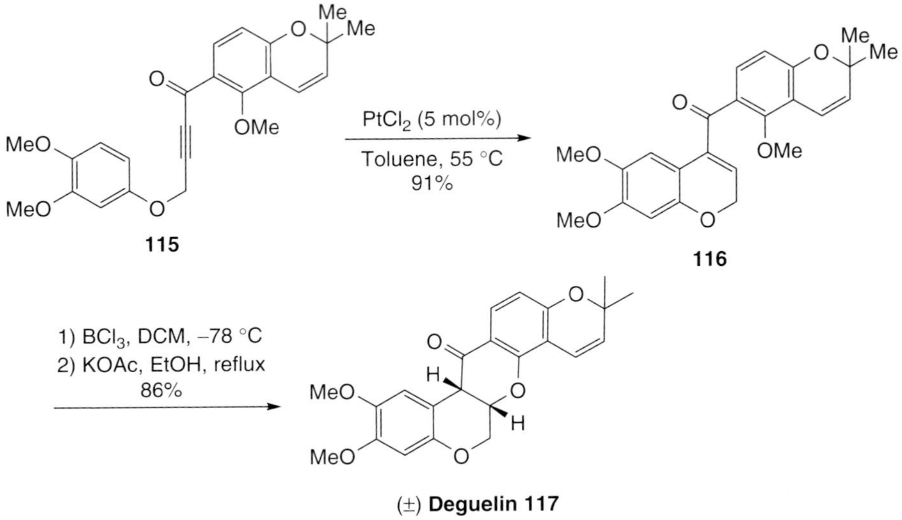

Scheme 12.51 Total synthesis of deguelin via a key Pt(II) catalysis.

In 2006, Liu and coworkers reported a gold-catalyzed intramolecular cycloaddition of diynes with tethered arenes **118** to synthesize 1,3-dihydroindeno[2,1-c]pyran or 2,3-dihydro-1*H*-indeno[2,1-*c*]pyridine derivatives **120** (Scheme 12.52) [56]. On the basis of deuterium labeling, they believed that the first reaction step is the intramolecular arylation of one alkyne to form vinylgold(I) intermediate **119**,

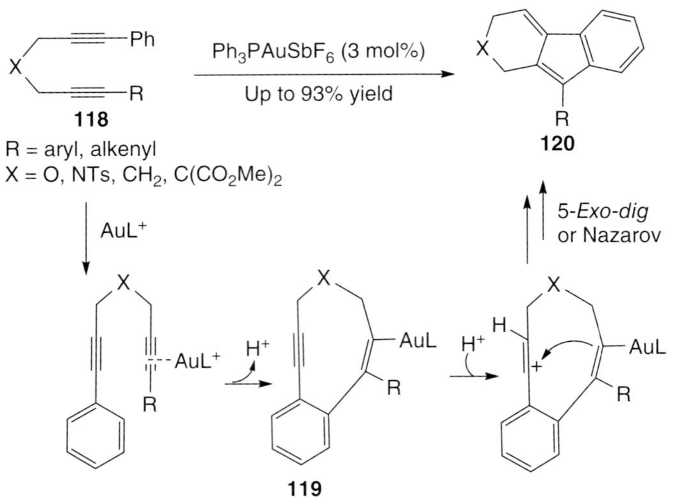

Scheme 12.52 Au-catalyzed synthesis of 1,3-dihydroindeno[2,1-*c*]pyran and 2,3-dihydro-1*H*-indeno[2,1-*c*]pyridine derivatives.

which then undergoes either a 5-*exo-dig* addition or Nazarov cyclization to give the final product.

Using a combination of $AuCl_3$/AgOTf as the catalytic system, Jean and Weghe realized a gold-catalyzed intramolecular hydroarylation of unactivated olefins, which provided a facile access to dihydrobenzopyrans, tetralins, and tetrahydroquinolines **122** (Scheme 12.53) [57]. A variety of homoallyl aryl ethers **121** with

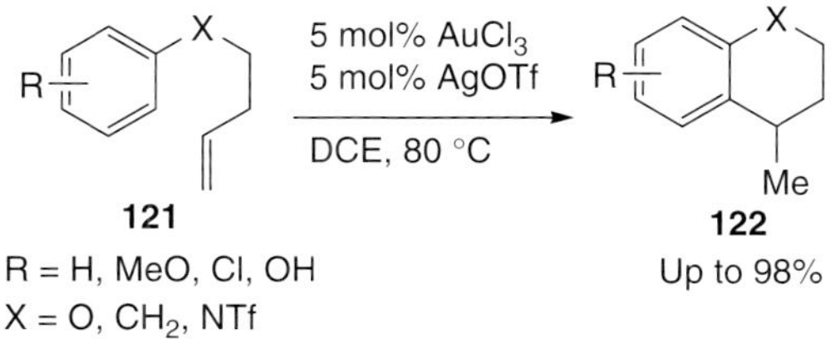

Scheme 12.53 Au-catalyzed synthesis of dihydrobenzopyrans and tetrahydroquinolines.

various functional groups such as methoxy, chloride, and free phenol worked well under standard optimal conditions. The nature of the substituent on the aromatic ring has an important effect on the reaction. The presence of a strong electron-withdrawing group such as a nitro group prevented the hydroarylation reaction. Mechanistic studies support that this catalytic hydroarylation is consistent with an electrophilic aromatic substitution process.

12.6 Synthesis of Fused Polycyclic Polyheterocycles

12.6.1 Via Au-Catalyzed Aryl C(sp^2)–H Functionalization

In 2006, Echavarren and Ferrer reported a facile annulation of six- to eight-membered rings on indoles by gold-catalyzed cyclization with alkynes to afford azepino[4,5-*b*]indole derivatives **124** or indoloazocines **125**. Regiochemical selectivity of the annulation is controlled by the oxidation state of the gold catalyst. Au(I) complex is the best catalyst for the formation of six- and seven-membered rings through 6-*endo-dig*, 6-*exo-dig*, and 7-*exo-dig* cyclizations. Indoloazocines are obtained with $AuCl_3$ as catalyst through a rare 8-*endo-dig* process (Scheme 12.54) [58].

Hashmi and coworkers found that intramolecular β-alkenylated furan product, polyheterocycle 4,5,6,7-tetrahydrofuro[2,3-*c*]pyridine derivatives **127** rather than phenolic products [59], could be efficiently constructed via gold(I)-catalyzed *exo*-selective intramolecular hydroarylation of silylated γ-alkynylfurans **126** (Scheme 12.55) [60]. Substrates bearing different silylated groups have a different reactivity. They also found that gold(I) catalyst efficiently serves as a hydroarylation catalyst for the silylated γ-alkynylfurans, while the platinum(II)

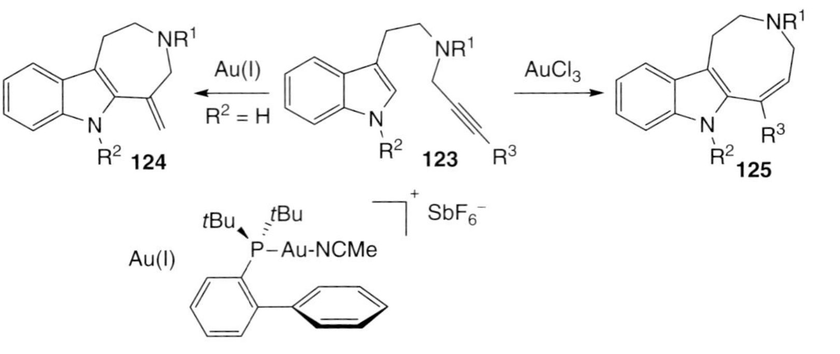

Scheme 12.54 Au-catalyzed synthesis of azepino[4,5-*b*]indole and indoloazocines.

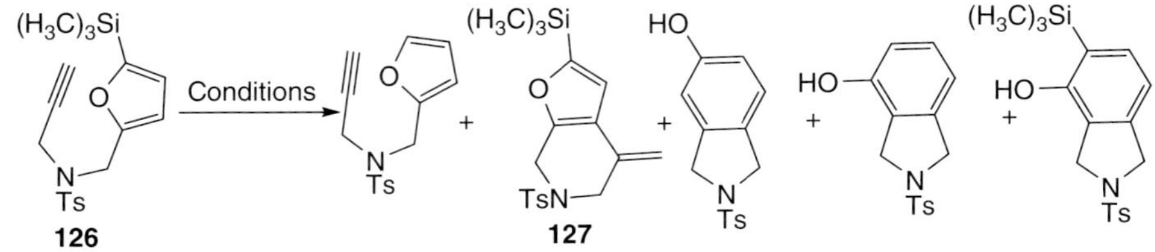

$[(Ph_3PAu)_2Cl][BF_4]$ (5 mol%) CH_3CN, rt, 2 min	--	Quantitative yield	--	--	--
$Na[AuCl_4]$ (5 mol%) CH_3CN, 70 °C	22%	--	Σ 26% Ratio 60:40		--
$PtCl_2(MeCN)_2$ (5 mol%) CH_3CN, 70 °C	--	14%	--	27%	26%

Scheme 12.55 Au-catalyzed synthesis of 4,5,6,7-tetrahydrofuro[2,3-*c*]pyridine derivatives.

catalyst shows a high selectivity toward the phenolic products. Gold(III) catalysts show a significant potential to serve as a desilylation catalyst, which ultimately leads to a mixture of hydroarylation and desilylated phenolic products.

Interestingly, Yu and coworkers found that seven-membered ring-fused furans and thiophene **129** could be efficiently constructed via gold(I)-catalyzed *endo*-selective intramolecular α-alkenylation of β-yne-furans **128**. Preliminary DFT calculations have been carried out to understand the experimentally observed regioselectivity (Scheme 12.56) [61].

A gold catalysis-initiated cascade process toward the synthesis of electron-rich arene-fused hexahydroquinolizinones **131** was developed by Zhang and Liu in 2012 (Scheme 12.57) [62]. In this cascade reaction, a gold-catalyzed amide cyclization to a tethered C–C triple bond initiates a subsequent Friedel–Crafts type cyclization, followed by a Ferrier rearrangement. Synthetic utility of this

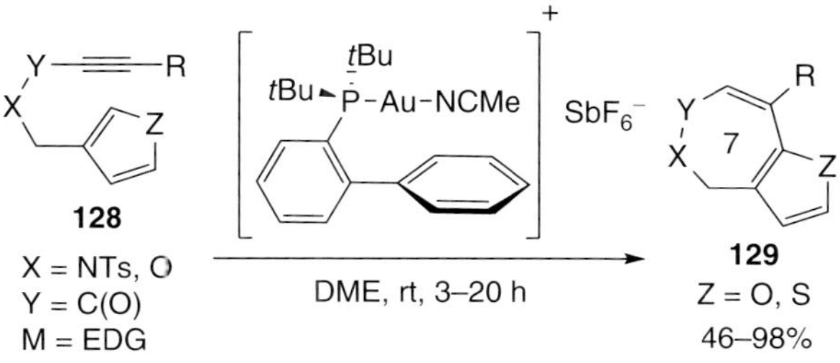

Scheme 12.56 Synthesis of seven-membered ring-fused furans and thiophene.

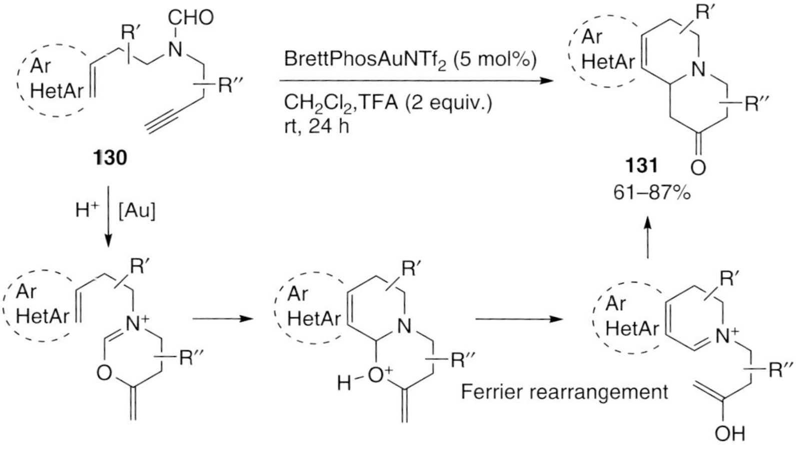

Scheme 12.57 Au-catalyzed synthesis of hexahydroquinolizinones.

new method is illustrated by a succinct and stereoselective synthesis of dihydrocorynantheol **132** without the use of any protecting group and a formal synthesis of yohimbine **133a** and β-yohimbine **133b** (Eq. (12.3)).

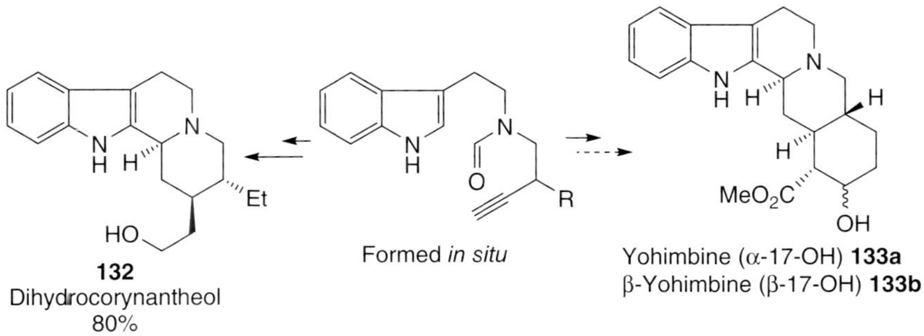

(12.3)

Under the cocatalysis of Au(I)/chiral Brønsted acid, a cascade hydroamination/hydroarylation of alkynols **134** was developed by Patil and coworkers, which provided an efficient route for the synthesis of enantiomerically enriched pyrrole-embedded aza-heterocyclic scaffolds **136** bearing a quaternary carbon stereocenter (Scheme 12.58) [63]. The method is very general and works well for a range of pyrrole-based aromatic amines **135**. The presence of a hydroxyl group in the alkyne tether turned out to be very crucial for obtaining products in high yields and enantioselectivities (Scheme 12.59).

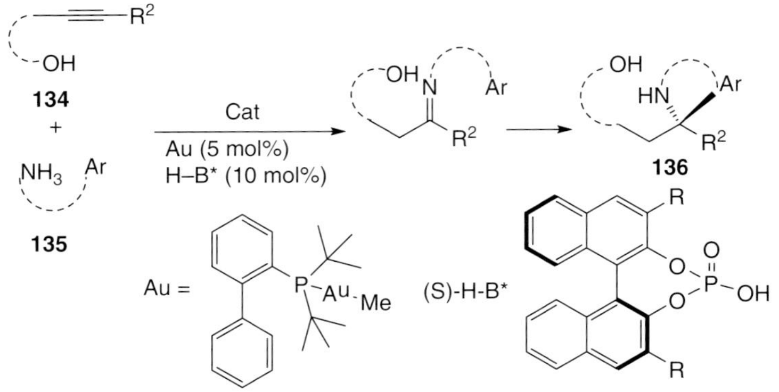

Scheme 12.58 Au(I)/chiral Brønsted acid cocatalyzed strategy for the synthesis of enantiomerically enriched aza-heterocyclic scaffolds.

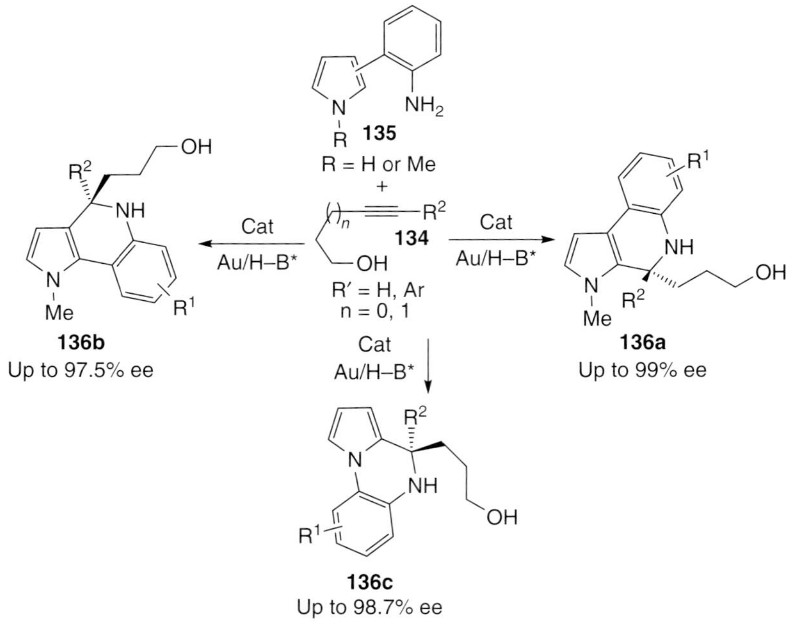

Scheme 12.59 Au(I)/chiral Brønsted acid cocatalyzed synthesis of pyrrole-embedded aza-heterocycles.

12.6.2 Via Au- or Pt-Catalyzed Aryl C(sp^2)–H Functionalization

In 2012, Van der Eycken and coworkers realized a regioselective synthesis of pyrrolopyridinones **137** and pyrroloazepinones **138** (Scheme 12.60) [64]. They employed an Ugi reaction, which is followed by a gold(I)- or platinum(II)-catalyzed intramolecular hydroarylation. The method allows the selective formation of six- and seven-membered pyrrole-fused heterocycles from easily available starting materials under mild reaction conditions.

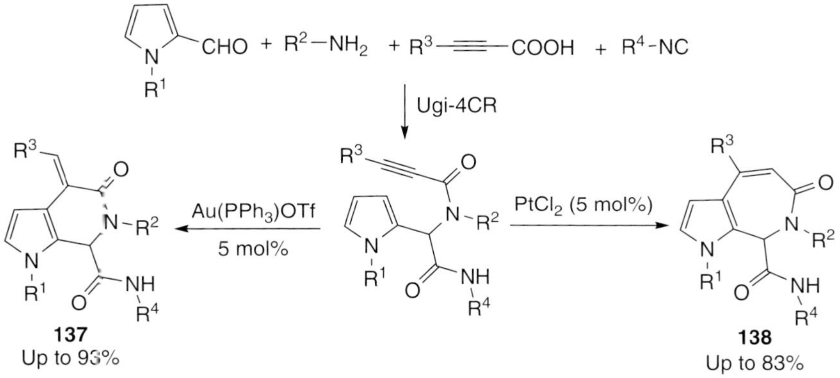

Scheme 12.60 Au- or Pt-catalyzed synthesis of pyrrolopyridinones and pyrroloazepinones.

Using a similar procedure, they successfully extend this methodology to synthesize thiophenoazepinone, azepinoindole, and azepinobenzothiophene derivatives **140** by using readily available 2-formylthiophene, 2-formylindole, or 2-formylbenzothiophene instead of 2-formyl-*N*-methylpyrrole in U-4CR reaction. Interestingly, only *endo*-selectivities were observed in the subsequent Au- or Pt-catalyzed intramolecular hydroarylation (Scheme 12.61) [65].

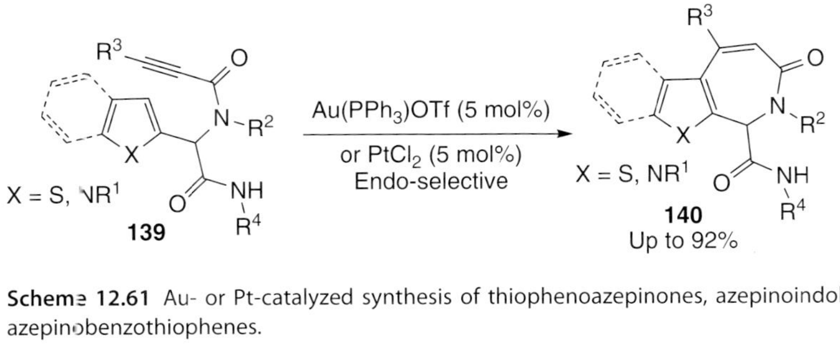

Scheme 12.61 Au- or Pt-catalyzed synthesis of thiophenoazepinones, azepinoindoles, and azepinobenzothiophenes.

12.6.3
Via Pt-Catalyzed Aryl C(sp^2)–H Functionalization

A Pt-catalyzed intramolecular aryl C(sp^2)–H functionalization was developed by Beller and coworkers with pyrrole-2-carboxamide alkyne **141** in the presence of H_2PtCl_6 $6H_2O$, which was applied to construct poly-*N*-heterocycles (Scheme 12.62) [66]. Pyrrolo[3,2-*c*]azepin-4-one derivatives **142** were obtained in moderate to good yields. Interestingly, this 7-*endo-dig* cyclization process involved a concomitant rearrangement of the amidocarbonyl group from the 2- to the 3-position of the pyrrole ring.

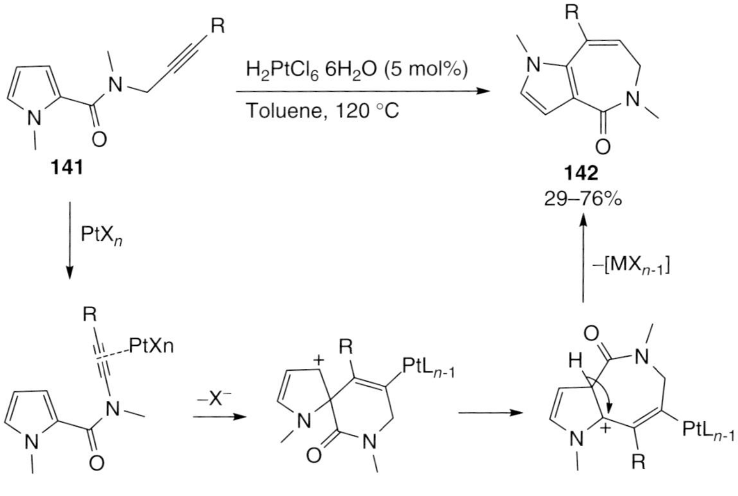

Scheme 12.62 Pt-catalyzed synthesis of pyrrolo[3,2-c]azepin-4-one derivatives.

Using a similar procedure, they extend this methodology to the preparation of indoloazepinones **144**. Interestingly, contrary to pyrrole-substituted alkynes **141**, no rearrangement of the amide substituent was observed on indole-substituted alkynes **143** (Scheme 12.63) [67].

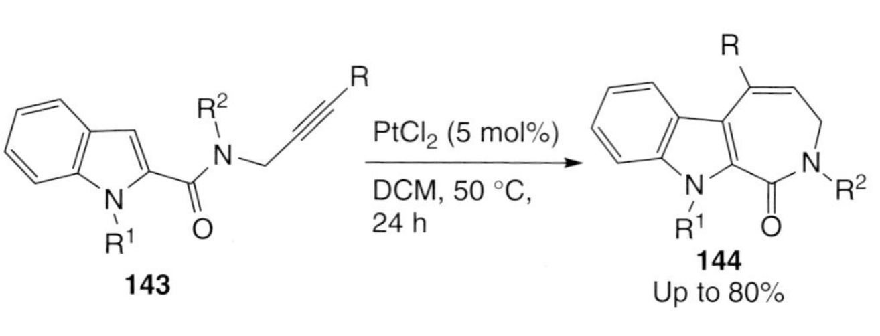

Scheme 12.63 Pt-catalyzed synthesis of indoloazepinones.

12.6.4 Via Au-Catalyzed C(sp³)–H Functionalization

Gagosz extensively investigated on the gold-catalyzed cyclizations of ether–alkyne systems **145** and **147** bearing both terminal and internal alkynes in 2010. A wide range of structurally important spiro or fused dihydrofurans **146** and dihydropyrans **148** were dexterously constructed via a 1,5-hydride shift/cyclization sequence using alkyne as hydride acceptors (Scheme 12.64) [68]. This hydroalkylation process, which could be applied to the terminal as well as ester-substituted alkynes, allows the efficient conversion of secondary or tertiary C(sp^3)–H bonds into new C–C bonds under practical conditions. The stereoselectivity of the cycloisomerization process toward the formation of a new five- or six-membered cycle appears to be dependent on steric factors and alkyne substitution.

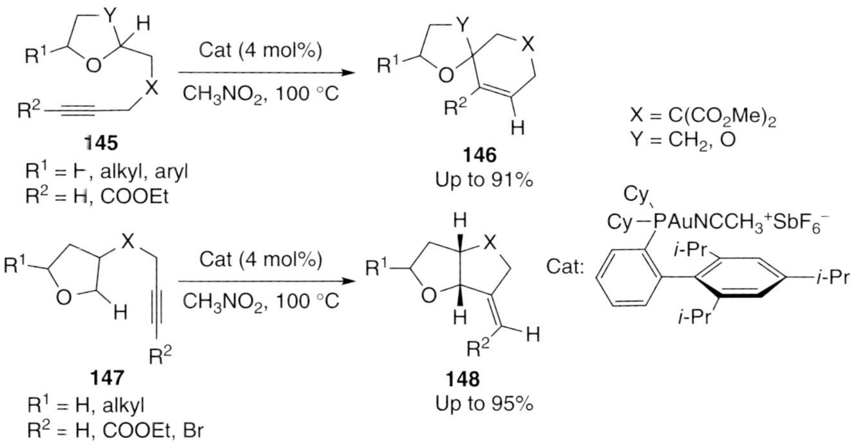

Scheme 12.64 Au-catalyzed synthesis of spiro or fused dihydrofurans and dihydropyrans.

Later, the same group found that allenes could also be used as hydride acceptors in the cyclization of related ether **149** catalyzed by cationic gold(I). Various fused tetrahydrofurans and tetrahydropyrans **150** could be obtained in high yields (Scheme 12.65) [69].

Zhang and coworkers reported two divergent redoxisomerization pathways for yne-enones **151** to afford ring-fused tetrahydroquinolines and tetrahydroazepines. Product selectivity depends on whether a carbophilic or oxophilic Lewis acid was used. Ring-fused tetrahydroazepines **152** could be obtained in moderate to excellent yields via the carbophilic gold(I)-catalyzed domino reaction under the mild conditions (Scheme 12.66) [70].

An enantioselective variant of this process was later accomplished by employing a catalytic amount of a chiral gold–bisphosphine complex. The resulting furan-fused azepines **154** could be obtained in good to excellent yields with satisfactory enantioselectivity (Scheme 12.67) [71].

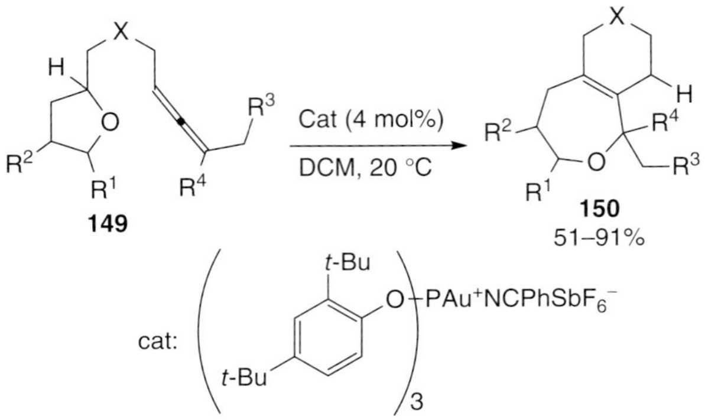

Scheme 12.65 Au-catalyzed synthesis of fused tetrahydrofurans and tetrahydropyrans.

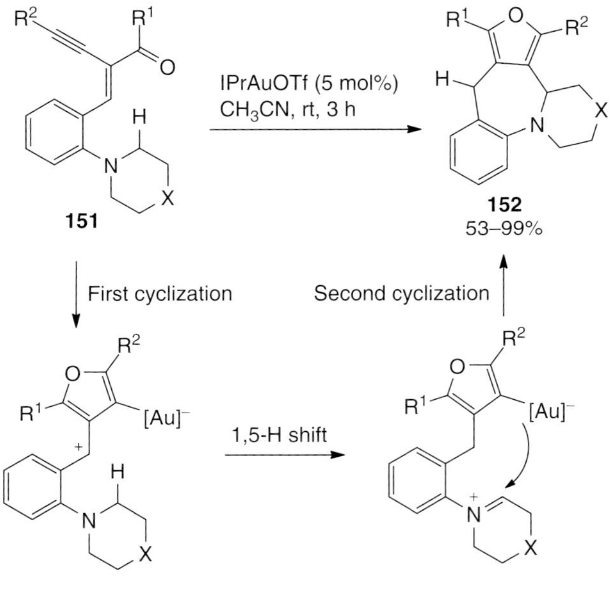

Scheme 12.66 Au-catalyzed synthesis of ring-fused tetrahydroazepines.

12.6.5
Via Pt-Catalyzed C(sp^3)–H Functionalization

Oh and coworkers successively developed a series of highly efficient methods for the construction of various types of fused polycyclic heterocycles from enynals with a pendant unsaturated bond [72]. Enynals with a benzyloxy substituent at the propargylic position **155** were successfully cyclized via platinum-catalyzed Huisgen-type cycloaddition to form polycyclic Pt–carbene complex intermediate **156**, which undergo insertion into a C–H bond to afford various types of fused polycyclic heterocycles **157** (Scheme 12.68) [73].

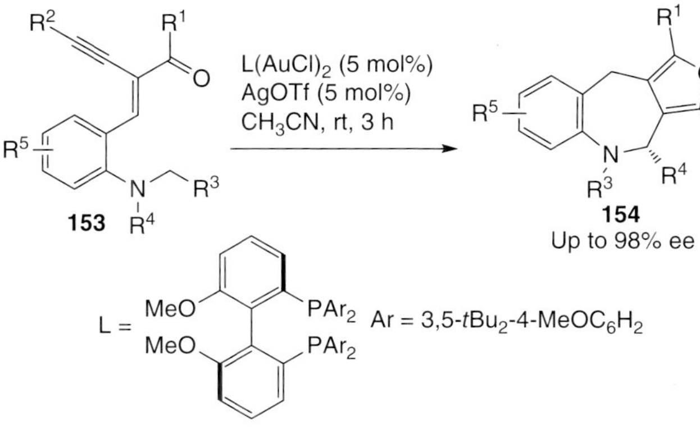

Scheme 12.67 Au-catalyzed asymmetric synthesis of ring-fused tetrahydroazepines.

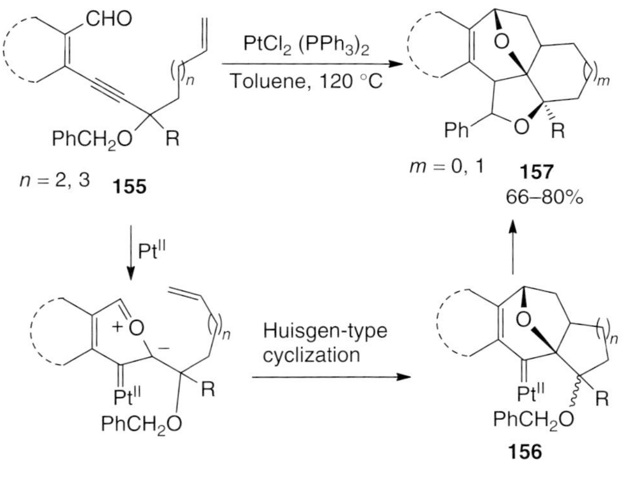

Scheme 12.68 Pt-catalyzed synthesis of fused polycyclic heterocycles.

12.7 Conclusions

Much progress in gold- and platinum-catalyzed heterocycle synthesis via C–H bond activation/functionalization processes has been made in the past years. This new strategy for the construction of heterocycle scaffolds includes intramolecular cyclizations and intermolecular cyclizations covering C(sp)–H, C(sp^2)–H, or C(sp^3)–H functionalizations, which achieve the formation of multiple bonds in one operation. The strategy exhibits high transformation efficiency, good functional group tolerance, and good chemo-, regio-, and stereochemistry selectivity and has become a highly versatile tool for the construction of a broad variety of complex heterocyclic compounds. However, the potential of Au and Pt catalysis for C–H bond activation/functionalization is still at their infancies and will undoubtedly continue to gain momentum over the coming years. In particular, enantioselective gold and platinum-catalyzed C–H activation/functionalizations for the synthesis of chiral heterocycles will receive much more attention in the future.

References

1. Very recently published reviews on Au or Pt-catalyzed chemical transformation: (a) Fürstner, A. (2009) *Chem. Soc. Rev.*, **38**, 3208–3221; (b) Sohel, S.M.A. and Liu, R.S. (2009) *Chem. Soc. Rev.*, **38**, 2269–2281; (c) Hashmi, A.S.K. (2010) *Pure Appl. Chem.*, **82**, 657–668; (d) Hashmi, A.S.K. and Hubbert, C. (2010) *Angew. Chem. Int. Ed.*, **49**, 1010–1012; (e) Garcia, P., Malacria, M., Aubert, C., Gandon, V., and Fensterbank, L. (2010) *ChemCatChem*, **2**, 493–497; (f) Sengupta, S. and Shi, X. (2010) *ChemCatChem*, **2**, 609–619; (g) Das, A., Sohel, S.M.A., and Liu, R.S. (2010) *Org. Biomol. Chem.*, **8**, 960–979; (h) Bandini, M. (2011) *Chem. Soc. Rev.*, **40**, 1358–1367; (i) Krause, N. and Winter, C. (2011) *Chem. Rev.*, **111**, 1994–2009; (j) De Haro, T. and Nevado, C. (2011) *Synthesis*, **11**, 2530–2539; (k) Wegner, H.A. and Auzias, M. (2011) *Angew. Chem. Int. Ed.*, **50**, 8236–8247; (l) Boorman, T.C. and Larrosa, I. (2011) *Chem. Soc. Rev.*, **40**, 1910–1925; (m) Rudolph, M. and Hashmi, A.S.K. (2011) *Chem. Commun.*, **47**, 6536–6544; (n) Gaillard, S., Cazin, C.S.J., and Nolan, S.P. (2012) *Acc. Chem. Res.*, **45**, 778–787; (o) Rudolph, M. and Hashmi, A.S.K. (2012) *Chem. Soc. Rev.*, **41**, 2448–2462; (p) Zhang, D.-H., Zhang, Z., and Shi, M. (2012) *Chem. Commun.*, **48**, 10271–10279; (q) Haibach, M.C. and Seidel, D. (2014) *Angew. Chem. Int. Ed.*, **53**, 5010–5036; (r) Zhang, L. (2014) *Acc. Chem. Res.*, **47**, 877–888; (s) Xie, J., Pan, C., Abdukade, A., and Zhu, C.J. (2014) *Chem. Soc. Rev.*, **43**, 5245–5256; (t) Obradors, C. and Echavarren, A.M. (2014) *Chem. Commun.*, **50**, 16–28; (u) Yamamoto, Y. (2014) *Chem. Soc. Rev.*, **43**, 1575–1600.
2. (a) Sundberg, R.J. (1984) in *Comprehensive Heterocyclic Chemistry*, vol. **5** (eds A.R. Katrizky and C.W. Rees), Pergamon Press, New York, p. 313; (b) Lipshutz, B.H. (1986) *Chem. Rev.*, **86**, 795–819; (c) Heaney, H. (1974) in *Natural Products Chemistry* (ed K. Nakanishi), Kodansha, Toyko, p. 297; (d) Brown, T.H., Armitage, M.A., Blakemore, R.C., Blurton, P., Durant, G.J., Ganellin, C.R., Ife, R.G., Parsons, M.E., Rawlings, D.A., and Slingsby, B.P. (1990) *Eur. J. Med. Chem.*, **25**, 217–226.
3. (a) Hou, X.L., Cheung, H.Y., Hon, T.Y., Kwan, P.L., Lo, T.H., Tong, S.Y., and Wong, H.N.C. (1998) *Tetrahedron*, **54**, 1955–2020; (b) Keay, B.A. (1999) *Chem. Soc. Rev.*, **28**, 209–215; (c) Gilchrist, T.L. (1999) *J. Chem. Soc., Perkin Trans. 1*, 2849–2866; (d) Hou, X.L., Yang, Z., and Wong, H.N.C. (2002) in *Progress in Heterocyclic Chemistry*, vol. **14** (eds G.W. Gribble and T.L. Gilchrist), Pergamon Press, Oxford, p. 139;(e) Arundale, E. and Mikeska, L.A. (1952) *Chem. Rev.*, **51**, 505–555.
4. Fürstner, A. and Davies, P.W. (2007) *Angew. Chem. Int. Ed.*, **46**, 3410–3449.
5. Hopkinson, M.N., Ross, J.E., Giuffredi, G.T., Gee, A.D., and Gouverneur, V. (2010) *Org. Lett.*, **12**, 4904–4907.
6. Nguyen, R.V., Yao, X.-Q., and Li, C.-J. (2006) *Org. Lett.*, **8**, 2397–2399.
7. Reich, N.W., Yang, C.-G., Shi, Z., and He, C. (2006) *Synlett*, 1278–1280.
8. Yu, Z., Ma, B., Chen, M., Wu, H., Liu, L., and Zhang, J. (2014) *J. Am. Chem. Soc.*, **136**, 6904–6907.
9. Harrak, Y., Simonneau, A., Malacria, M., Gandon, V., and Fensterbank, L. (2010) *Chem. Commun.*, **46**, 865–867.
10. Lee, J.M. and Chang, S. (2006) *Tetrahedron Lett.*, **47**, 1375–1379.

11. Ma, Y., Zhang, S., Yang, S., Song, F., and You, J. (2014) *Angew. Chem. Int. Ed.*, **53**, 7870–7834.
12. Yao, X.-Q. and Li, C.-J. (2006) *Org. Lett.*, **8**, 1953–1955.
13. Skouta, R. and Li, C.-J. (2007) *Angew. Chem. Int. Ed.*, **46**, 1117–1119.
14. (a) Reetz, M.T. and Sommer, K. (2003) *Eur. J. Org. Chem.*, **2003**, 3485–3496; (b) Shi, Z. and He, C. (2004) *J. Org. Chem.*, **69**, 3669–3671.
15. Shi, Z. and He, C. (2004) *J. Am. Chem. Soc.*, **126**, 5964–5965.
16. Luo, T. and Schreiber, S.L. (2007) *Angew. Chem. Int. Ed.*, **46**, 8250–8253.
17. (a) Ye, L., He, W., and Zhang, L. (2011) *Angew. Chem. Int. Ed.*, **50**, 3236–3239; (b) Ye, L., He, W., and Zhang, L. (2010) *J. Am. Chem. Soc.*, **132**, 8550–8551; (c) Ye, L., Cui, L., Zhang, G., and Zhang, L. (2010) *J. Am. Chem. Soc.*, **132**, 3258–3259; (d) Lu, B., Li, C., and Zhang, L. (2010) *J. Am. Chem. Soc.*, **132**, 14070–14072; (e) He, W., Li, C., and Zhang, L. (2011) *J. Am. Chem. Soc.*, **133**, 8482–8485; (f) Vasu, D., Hung, H.-H., Bhunia, S., Gawade, S.A., Das, A., and Liu, R.-S. (2011) *Angew. Chem. Int. Ed.*, **50**, 6911–6914; (g) Mukherjee, A., Dateer, R.B., Chaudhuri, R., Bhunia, S., Karad, S.N., and Liu, R.-S. (2011) *J. Am. Chem. Soc.*, **133**, 15372–15375; (h) Bhunia, S., Chorpade, S., Huple, D.B., and Liu, R.-S. (2012) *Angew. Chem. Int. Ed.*, **51**, 2939–2942; (i) Chorpade, S., Su, M.-D., and Liu, R.-S. (2013) *Angew. Chem. Int. Ed.*, **52**, 4229–4234; (j) Chen, C.-H., Tsai, Y.-C., and Liu, R.-S. (2013) *Angew. Chem. Int. Ed.*, **52**, 4599–4603; (k) Karad, S.N. and Liu, R.-S. (2014) *Angew. Chem. Int. Ed.*, **53**, 5444–5448; (l) Henrion, G., Chavas, T.E.J., Goff, X.L., and Gagosz, F. (2013) *Angew. Chem. Int. Ed.*, **52**, 6277–6282; (m) Qian, D. and Zhang, J. (2011) *Chem. Commun.*, **47**, 11152–11154; (n) Qian, D., Hu, H., Liu, F., Tang, B., Ye, W., Wang, Y., and Zhang, J. (2014) *Angew. Chem. Int. Ed.*, **53**, 13751–13755.
18. Wang, Y., Ji, K., Lan, S., and Zhang, L. (2012) *Angew. Chem. Int. Ed.*, **51**, 1915–1918.
19. Hashmi, A.S.K., Schäöfer, S., Wölfle, M., Diez Gil, C., Fischer, P., Laguna, A., Blanco, M.C., and Gimeno, M.C. (2007) *Angew. Chem. Int. Ed.*, **46**, 6184–6187.
20. For recent reviews on the synthesis of *N*-heterocycles: (a) Wu, X.-F., Neumann, H., and Beller, M. (2013) *Chem. Rev.*, **113**, 1–35; (b) Dhakshinammoorthy, A. and Garcia, H. (2014) *Chem. Soc. Rev.*, **43**, 5750–5765; (c) Otto, N. and Opatz, T. (2014) *Chem. Eur. J.*, **20**, 13064–13077; (d) He, L., Nie, H., Qiu, G., Gao, Y., and Wu, J. (2014) *Org. Biomol. Chem.*, **12**, 9045–9053; (e) Ishoey, M. and Nielsen, T.E. (2014) *Chem. Eur. J.*, **20**, 8832–8840; (f) Majumdar, K.C., De, N., Ghosh, T., and Roy, B. (2014) *Tetrahedron*, **70**, 4827–4868; (g) He, L., Li, H., Chen, J., and Wu, X.-F. (2014) *RSC Adv.*, **4**, 12065–12077; (h) Hu, Y., Xin, X., and Wan, B. (2015) *Tetrahedron Lett.*, **56**, 32–52; (i) Cornil, J., Gonnard, L., Bensoussan, C., Serra-Muns, A., Gnamm, C., Commandeur, C., Commandeur, M., Reymond, S., Guérinot, A., and Cossy, J. (2015) *Acc. Chem. Res.*, **48**, 761–773; (j) Yuan, J., Liu, C., and Lei, A. (2015) *Chem. Commun.*, **51**, 1394–1409; (k) Guo, X., Gu, D., Wu, Z., and Zhang, W. (2015) *Chem. Rev.*, **115**, 1622–1651.
21. Yan, B. and Liu, Y. (2007) *Org. Lett.*, **9**, 4323–4326.
22. Suzuki, Y., Naoe, S., Oishi, S., Fujii, N., and Ohno, H. (2012) *Org. Lett.*, **14**, 326–329.
23. Tokimizu, Y., Wieteck, M., Rudolph, M., Oishi, S., Fujii, N., Hashmi, A.S.K., and Ohno, H. (2015) *Org. Lett.*, **17**, 604–607.
24. Zhang, G., Luo, Y., Wang, Y., and Zhang, L. (2011) *Angew. Chem. Int. Ed.*, **50**, 4450–4454.
25. (a) Lu, B., Luo, Y., Liu, L., Ye, L., Wang, Y., and Zhang, L. (2011) *Angew. Chem. Int. Ed.*, **50**, 8358–8362; (b) Wetzel, A. and Gagosz, F. (2011) *Angew. Chem. Int. Ed.*, **50**, 7354–7358.
26. Wang, Y., Ye, L., and Zhang, L. (2011) *Chem. Commun.*, **47**, 7815–7817.
27. Wang, Y., Liu, L., and Zhang, L. (2013) *Chem. Sci.*, **4**, 739–746.
28. Qian, D. and Zhang, J. (2012) *Chem. Commun.*, **48**, 7802–7804.

29. Wang, H.-L., Li, Z., Wang, G.-W., and Yang, S.-D. (2011) *Chem. Commun.*, **47**, 11336–11338.
30. Zeldin, R.M. and Toste, F.D. (2011) *Chem. Sci.*, **2**, 1706–1709.
31. Zhou, C.Y. and Che, C.-M. (2007) *J. Am. Chem. Soc.*, **129**, 5828–5829.
32. Yan, Z., Xiao, Y., and Zhang, L. (2012) *Angew. Chem. Int. Ed.*, **51**, 8624–8627.
33. Zheng, Z., Tu, H., and Zhang, L. (2014) *Chem. Eur. J.*, **20**, 2445–2448.
34. Xiao, F., Chen, Y., Liu, Y., and Wang, J. (2008) *Tetrahedron*, **64**, 2755–2761.
35. Skouta, R. and Li, C.-J. (2007) *Synlett*, 1759–1762.
36. Vachhani, D.D., Galli, M., Jacobs, J., Van Meervelt, L., and Van der Eycken, E.V. (2013) *Chem. Commun.*, **49**, 7171–7173.
37. Watanabe, T., Oishi, S., Fujii, N., and Ohno, H. (2007) *Org. Lett.*, **9**, 4821–4824.
38. Liu, Z., Wasmuth, A.S., and Nelson, S.G. (2006) *J. Am. Chem. Soc.*, **128**, 10352–10353.
39. Liu, X.-Y., Ding, P., Huang, J.-S., and Che, C.-M. (2007) *Org. Lett.*, **9**, 2645–2648.
40. Anguille, S., Brunet, J.-J., Chu, N.C., Diallo, O., Pages, C., and Vincendeau, S. (2006) *Organometallics*, **25**, 2943–2948.
41. Béthegnies, A., Dub, P.A., and Poli, R. (2013) *Organometallics*, **32**, 1882–1891.
42. (a) Cui, L., Peng, Y., and Zhang, L. (2009) *J. Am. Chem. Soc.*, **131**, 8394–8395; (b) Cui, L. and Zhang, L. (2010) *Sci. China, Ser. B Chem.*, **53**, 113; (c) Noey, E.L., Luo, Y., Zhang, L., and Houk, K.N. (2012) *J. Am. Chem. Soc.*, **134**, 1078–1084.
43. He, Y.-P., Wu, H., Chen, D.-F., Yu, J., and Gong, L.-Z. (2013) *Chem. Eur. J.*, **19**, 5232–5237.
44. Cui, L., Zhang, G., Peng, Y., and Zhang, L. (2009) *Org. Lett.*, **11**, 1225–1228.
45. Cui, L., Ye, L.W., and Zhang, L. (2010) *Chem. Commun.*, **46**, 3351–3353.
46. (a) Shapiro, N.D. and Toste, F.D. (2007) *J. Am. Chem. Soc.*, **129**, 4160–4161; (b) Li, G. and Zhang, L. (2007) *Angew. Chem. Int. Ed.*, **46**, 5156–5159; (c) Lu, B., Li, Y., Wang, Y., Aue, D.H., Luo, Y., and Zhang, L. (2013) *J. Am. Chem. Soc.*, **135**, 8512–8524.
47. (a) Ye, L., Wang, Y., Aue, D.H., and Zhang, L. (2012) *J. Am. Chem. Soc.*, **134**, 31–34; (b) Hashmi, A.S.K., Braun, I., Nösel, P., Schädlich, J., Wieteck, M., Rudolph, M., and Rominger, F. (2012) *Angew. Chem. Int. Ed.*, **51**, 4456–4460; (c) Hashmi, A.S.K., Wieteck, M., Braun, I., Nöel, P., Jongbloed, L., Rudolph, M., and Rominger, F. (2012) *Adv. Synth. Catal.*, **354**, 555–562; (d) Hashmi, A.S.K., Braun, I., Rudolph, M., and Rominger, F. (2012) *Organometallics*, **31**, 644–661.
48. Wang, Y., Yepremyan, A., Ghorai, S., Todd, R., Aue, D.H., and Zhang, L. (2013) *Angew. Chem. Int. Ed.*, **52**, 7795–7799.
49. Hansmann, M.M., Tŝupova, S., Rudolph, M., Rominger, F., and Hashmi, A.S.K. (2014) *Chem. Eur. J.*, **20**, 2215–2223.
50. (a) Hashmi, A.S.K., Wieteck, M., Braun, I., Rudolph, M., and Rominger, F. (2012) *Angew. Chem. Int. Ed.*, **51**, 10633–10637; (b) Hansmann, M.M., Rudolph, M., Rominger, F., and Hashmi, A.S.K. (2013) *Angew. Chem. Int. Ed.*, **52**, 2593–2598; (c) Braun, I., Asiri, A.M., and Hashmi, A.S.K. (2013) *ACS Catal.*, **3**, 1902–1907.
51. Bucher, J., Wurm, T., Nalivela, K.S., Rudolph, M., Rominger, F., and Hashmi, A.S.K. (2014) *Angew. Chem. Int. Ed.*, **53**, 3854–3858.
52. (a) Toullec, P.Y., Genin, E., Leseurre, L., Genêt, J.-P., and Michelet, V. (2006) *Angew. Chem. Int. Ed.*, **45**, 7427–7430; (b) Amijs, C.H.M., Ferrer, C., and Echavarren, A.M. (2007) *Chem. Commun.*, 698–700.
53. (a) Martín-Matute, B., Nevado, C., Cárdenas, D.J., and Echavarren, A.M. (2003) *J. Am. Chem. Soc.*, **125**, 5757–5766; (b) Nevado, C. and Echavarren, A.M. (2005) *Chem. Eur. J.*, **11**, 3155–3164.
54. (a) Pastine, S.J., Youn, S.W., and Sames, D. (2003) *Org. Lett.*, **5**, 1055–1058; (b) Pastine, S.J., Youn, S.W., and Sames, D. (2003) *Tetrahedron*, **59**, 8859–8868.
55. Pastine, S.J. and Sames, D. (2003) *Org. Lett.*, **5**, 4053–4055.
56. Lian, J.-J., Chen, P.-C., Lin, Y.-P., Ting, H.-C., and Liu, R.-S. (2006) *J. Am. Chem. Soc.*, **128**, 11372–11373.

57. Jean, M. and Weghe, P. (2011) *Tetrahedron Lett.*, **52**, 3509–3513.
58. Ferrer, C. and Echavarren, A.M. (2006) *Angew. Chem. Int. Ed.*, **45**, 1105–1109.
59. Hashmi, A.S.K., Frost, T.M., and Bats, J.W. (2000) *J. Am. Chem. Soc.*, **122**, 11553–11554.
60. Hashmi, A.S.K., Kurpejoviæc, E., Frey, W., and Bats, J.W. (2007) *Tetrahedron*, **63**, 5879–5885.
61. Dong, Z., Liu, C.-H., Wang, Y., Lin, M., and Yu, Z.-X. (2013) *Angew. Chem. Int. Ed.*, **52**, 14157–14161.
62. Liu, L. and Zhang, L. (2012) *Angew. Chem. Int. Ed.*, **51**, 7301–7304.
63. Shinde, V.S., Mane, M.V., Vanka, K., Mallick, A., and Patil, N.T. (2015) *Chem. Eur. J.*, **21**, 975–979.
64. Modha, S.G., Kumar, A., Vachhani, D.D., Sharma, S.K., Parmar, V.S., and Van der Eycken, E.V. (2012) *Chem. Commun.*, **48**, 10916–10918.
65. Kumar, A., Vachhani, D.D., Modha, S.G., Sharma, S.K., Parmar, V.S., and Van der Eycken, E.V. (2013) *Synthesis*, **45**, 2571–2582.
66. Gruit, M., Michalik, D., Tillack, A., and Beller, M. (2009) *Angew. Chem. Int. Ed.*, **48**, 7212–7216.
67. Gruit, M., Pews-Davtyan, A., and Beller, M. (2011) *Org. Biomol. Chem.*, **9**, 1148–1159.
68. Jurberg, I.D., Odabachian, Y., and Gagosz, F. (2010) *J. Am. Chem. Soc.*, **132**, 3543–3552.
69. Bolte, B. and Gagosz, F. (2011) *J. Am. Chem. Soc.*, **133**, 7696–7699.
70. Zhou, G.H. and Zhang, J. (2010) *Chem. Commun.*, **46**, 6593–6595.
71. Zhou, G.H., Liu, F., and Zhang, J. (2011) *Chem. Eur. J.*, **17**, 3101–3104.
72. (a) Kim, N., Kim, Y., Park, W., Sung, D., Gupta, A.K., and Oh, C.H. (2005) *Org. Lett.*, **7**, 5289–5291; (b) Oh, C.H., Ryu, J.H., and Lee, H.I. (2007) *Synlett*, 2337–2342; (c) Oh, C.H., Lee, J., Lee, S., Yi, H., and Hong, C. (2009) *Chem. Eur. J.*, **15**, 71–74.
73. Oh, C.H., Lee, J., Lee, S., Kim, J., and Hong, C. (2008) *Angew. Chem. Int. Ed.*, **47**, 7505–7507.

13
Heterocycle Synthesis Based on Visible-Light-Induced Photocatalytic C–H Functionalization

Wei Ding†, Wei Guo†, Ting-Ting Zeng, Liang-Qiu Lu, and Wen-Jing Xiao

13.1
Introduction

The development of new reactions with high efficiencies and selectivities using mild/green conditions is an important goal for synthetic organic chemists. A century ago, Giacomo Ciamician postulated that clean, cost-effective, and environmentally benign photochemical transformations had the capacity to replace high-energy synthetic processes [1]. However, the inability of most organic compounds to absorb visible light has limited the use of photochemical synthesis. Hence, utilizing visible-light-absorbing inorganic and organic compounds as photocatalysts (**PCs**) that sensitize organic molecules may allow scientists to overcome this barrier. The most commonly employed visible-light photocatalysts are the polypyridyl complexes of ruthenium and iridium, such as $Ru(bpy)_3Cl_2$, (bpy: 2,2′-bipyridine) *fac*-$Ir(ppy)_3$, (ppy: 2-phenylpyridine) and organic dyes (i.e., Eosin Y and Rose Bengal). Photocatalysts absorb visible light in order to generate a photoexcited state, and they usually have long lifetimes and often remarkable redox properties. Three distinctive pathways, namely, reductive quenching, oxidative quenching, and energy transfer, are frequently used to describe photocatalytic mechanisms enabled by visible light (Scheme 13.1) [2]. Reductive quenching requires that the reductive quencher, such as a tertiary amine, reduces the photoexcited catalyst to its low-valence state. In oxidative quenching, an oxidative quencher, such as methyl viologen, oxidizes the catalyst in the photoexcited state, yielding a high-valence state. When the energy of the photoexcited catalyst is similar to the triplet energy of the substrate, energy transfer from **PC*** to the substrate may afford an excited state capable of undergoing additional organic transformations.

Heterocyclic moieties are known to occur in a wide range of biologically active compounds, including both natural products and pharmaceutical compounds [3]. Over the last decade, efficient heterocycle synthesis via C–H functionalization

† These two authors contributed equally to this chapter.

Transition Metal-Catalyzed Heterocycle Synthesis via C—H Activation, First Edition. Edited by Xiao-Feng Wu.
© 2016 Wiley-VCH Verlag GmbH & Co. KGaA. Published 2016 by Wiley-VCH Verlag GmbH & Co. KGaA.

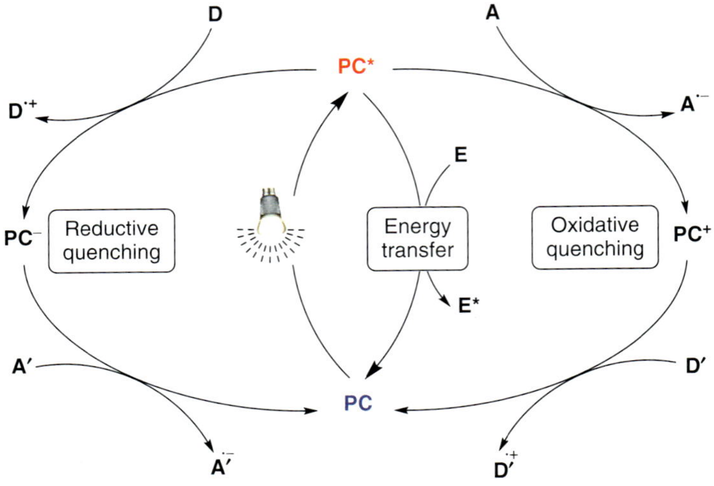

Scheme 13.1 Mechanism profile of visible-light-induced photocatalysis.

has attracted increasing attention from both the academic and industrial communities, yielding materials used in applications in both medicinal and process chemistry [4]. Owing to its energy-conservant and environmentally benign features, visible-light-induced photoredox catalysis has undergone substantial and rapid development over the past decade. Recently, the use of visible-light-induced photocatalytic C–H functionalization [5] for the *de novo* construction of heterocyclic systems and direct C–H functionalization of heterocycles has emerged as growing fields in synthetic organic chemistry. In this chapter, we will highlight achievements in this field.

13.2
de novo Synthesis of Heterocycles

13.2.1
Photocatalytic sp^3 C–H Functionalization for Heterocycle Synthesis

Since the initial report from Stephenson [6] in 2010, visible-light-induced α-amino C–H functionalization has developed rapidly [5, 7]. In these reactions, amines are good electron donors and can be easily oxidized by the excited photocatalyst into an amino radical cation intermediate **I** (Scheme 13.2). The lower bond dissociation energy of the α-amino C–H bond enables easy conversion of intermediate **I** to the corresponding iminium ion **II** though hydrogen atom abstraction. On the other hand, the formation of **I** significantly decreases the pK_a of the proton at α-position of the amino radical cation. Deprotonation of intermediate **I** yields a

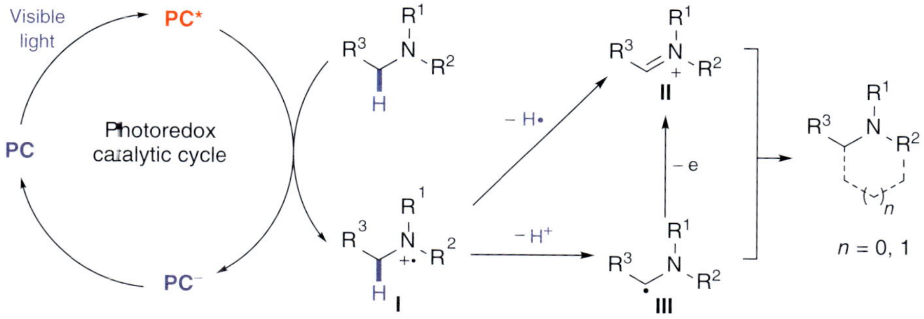

Scheme 13.2 Main pathways for visible-light-induced photocatalytic α-amino C–H functionalization.

nucleophilic α-aminoalkyl radical **III**. A subsequent one-electron oxidation of **III** produces the iminium ion **II** in the presence of oxidants. These two types of reactive intermediates are generally formed during the visible-light-mediated α-amino C–H functionalization of amines, yielding nitrogen-containing heterocycles.

In 2011, Xiao and coworkers [8a] utilized a visible-light-mediated oxidation/[3+2] cycloaddition/aromatization sequence for the construction of pyrrolo[2,1-*a*]isoquinolines **3** (Eq. (13.1)). Not long thereafter, Rueping's group [8b] independently reported a similar [3+2] cycloaddition reaction (Eq. (13.2)). More recently, Zhao and coworkers developed two types of novel photocatalysts including C_{60}-Bodipy dyad triplet photosensitizers [9] and Porous material-immobilized iodo-Bodipy [10], which were successfully applied to the same transformation.

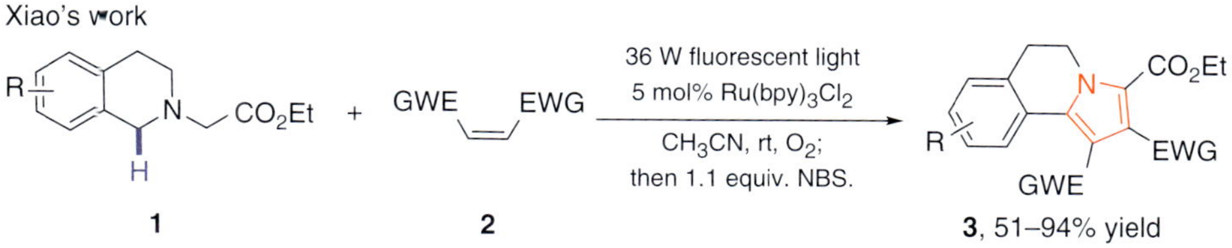

(13.1)

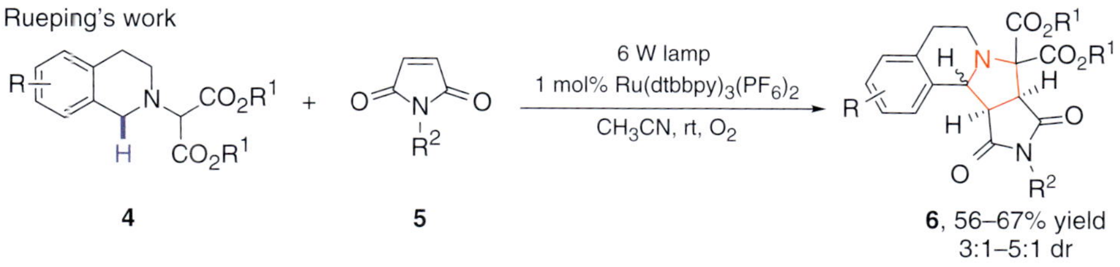

(13.2)

A potential mechanism featuring an azomethine ylide as the key intermediate is depicted in Scheme 13.3. Initially, the excited photocatalyst (**PC***) is quenched

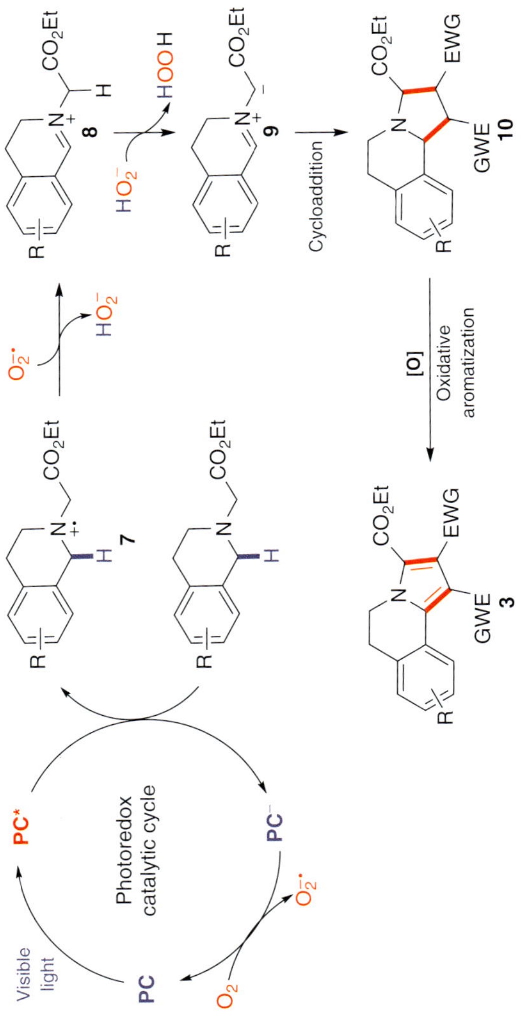

Scheme 13.3 Possible mechanism for the [3+2] cycloaddition reactions.

by a tetrahydroisoquinoline ester **1** to form the low-oxidation ground-state PC^-, releasing the amino radical cation **7**. Molecular oxygen then oxidizes $\mathbf{PC^-}$ to **PC**, completing the photoredox catalytic cycle. The oxygen radical anion from the previous step abstracts the reactive benzyl hydrogen from the radical cation **7**, yielding the iminium ion **8**. At this point, the generated HO_2^- may deprotonate the α-position of the ester group to afford the azomethine ylide **9**, a 1,3-dipole. Cycloaddition with a variety of dipolarophiles, followed by aromatization, delivers the desired product **3**.

As a continuation of their previous work on visible-light-induced syntheses of heterocycles, Xiao and coworkers disclosed a series of intramolecular oxidative cyclization reactions of chiral diamines, which afford highly substituted tetrahydroimidazoles **12** with good diastereoselectivity [11]. Under irradiation with visible light, the benzylic C–H bonds in the diamines **11**, which were prepared from natural amino acids, are oxidized to give iminium ions **13**. Subsequent intramolecular trapping of the iminium fragment by a nitrogen anion occurs in the presence of strong base. Due to steric repulsion, intramolecular nucleophilic attack at the *Re* face was thought to be favorable in comparison to attack at the *Si* face (Scheme 13.4).

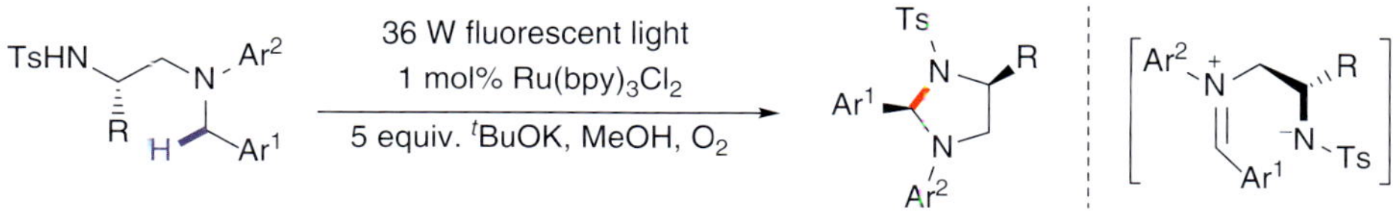

11

12, 45–94% yields
2:1–>19:1 dr

13
Re-face attack

Scheme 13.4 Visible-light-mediated intramolecular oxidative cyclization of diamines.

In 2012, Xiao's group [12] applied the same strategy in the construction of isoquino[2,1-*a*][3,1]oxazine and isoquino[2,1-*a*]pyrimidine derivatives (Eq. (13.3)). Under the optimized reaction conditions, different kinds of tetrahydroisoquinolines, bearing oxygen or nitrogen nucleophiles, underwent oxidative cyclization to deliver heterocycles in good to excellent yields. Soon after, Marvin and coworkers extended this strategy to unreactive amines [13]. Such intramolecular reactions were also realized with a palladium(II)-porphyrin complex (**PdF$_{20}$TPP**), which was developed by Che and coworkers [14].

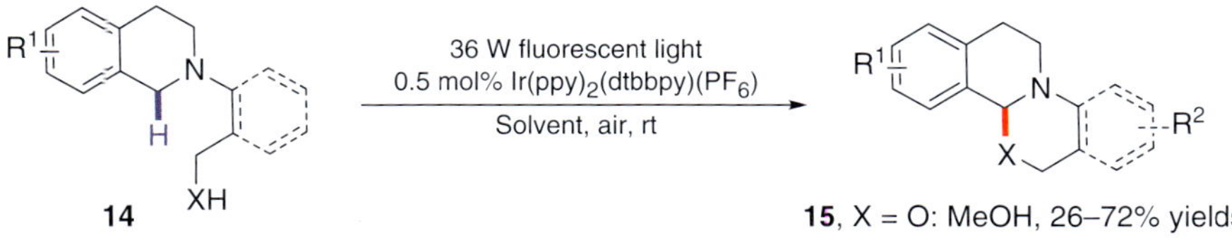

14

15, X = O: MeOH, 26–72% yields
X = NTs: MeOH/CH_2Cl_2, 70–86% yields

(13.3)

Sulfonylmethyl isocyanide is a useful synthon as a 1,3-dipole. In early 2014, Xiao's group [15] made a further contribution to the syntheses of nitrogen-containing heterocycles with their report of visible-light photocatalysis using secondary amines and isocyanides as starting materials (Eq. (13.4)). In this photocatalytic cascade, readily accessible glycine derivatives **16** were oxidized to imines and then underwent a formal [3+2] cycloaddition/proton transfer/aromatization process to yield various biologically significant 1,5-disubstituted imidazoles **18**. Notably, nitrogen-protected imidazo-[1,5-*a*]quinoxalin-4-one **20**, which may be transformed to a novel tyrosine kinase (PTK) inhibitor candidate **21** according to a literature procedure, was obtained in an excellent yield (Eq. (13.5)).

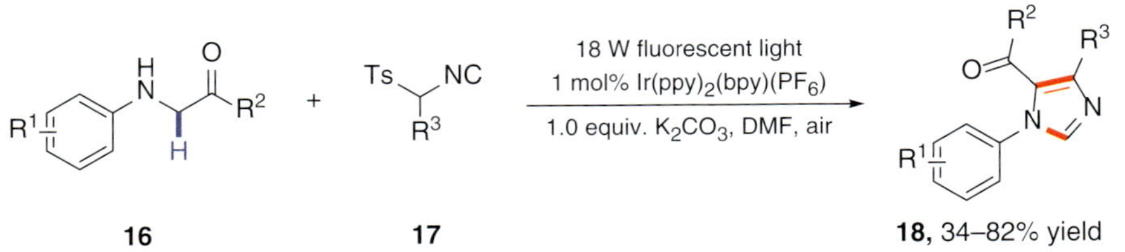

(13.4)

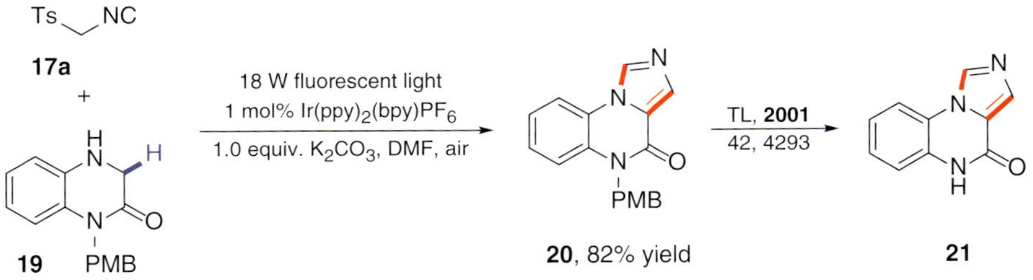

(13.5)

In 2013, Zhu and coworkers [16] discovered a visible-light-mediated tandem C–H activation/retro-aza-Michael/oxidation/[3+2]-cycloaddition sequence using α-ketoesters **23** as coupling reagents with tetrahydroisoquinolines (**22**, Scheme 13.5). Biologically important, functionalized isoxazolidines **24** were synthesized under optimized reaction conditions in moderate yields with excellent diastereoselectivities. According to their experimental results and literature reports, the authors proposed a possible mechanism, as depicted in Scheme 13.5. Initially, a reactive iminium ion **26** was formed through a visible-light-induced photocatalytic cycle. Next, with the aid of TfOH, the iminium ion was trapped by an ester **23a**, resulting in isolation of the Mannich reaction product **27**. Due to the special characteristics of α-ketoesters, this intermediate **26** favors the enol conformation **28** and undergoes a retro-aza-Michael reaction to give

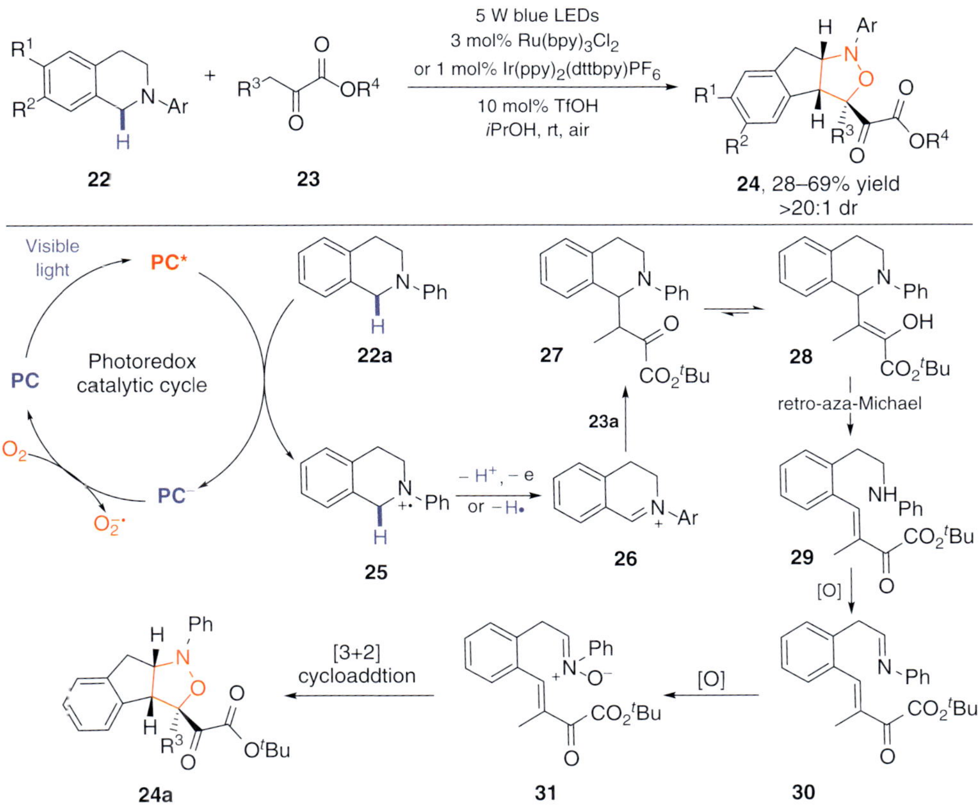

Scheme 13.5 Visible-light-mediated tandem aerobic C–H/C–N cleavage.

the secondary amine **29**. The resulting intermediate **29** is quickly oxidized to a nitrone **31** that undergoes an intramolecular [3+2] cycloaddition reaction to deliver the bicyclic isoxazolidine **24a**.

In a manner different from Zhu's, Rueping and coworkers [17] utilized *N*-substituted hydroxylamines **32** as starting materials for the generation of the key intermediate nitrones **35** under visible-light-induced aerobic oxidation conditions (Scheme 13.6). Following intermolecular cycloaddition with **33**, intermediate **35** was transformed into a range of isoxazolidines. Three possible pathways for the formation of the nitrone intermediate were proposed, and none could be ruled out.

Aside from the trapping of iminium ion intermediates, visible-light-photocatalytic *α*-amine C–H functionalization can be achieved through the use of highly reactive *α*-aminoalkyl radicals. In 2012, Reiser and coworkers [18] developed a visible-light-induced coupling of *α*-aminoalkyl radicals and *α*,*β*-unsaturated carbonyl compounds. Through intramolecular reactions of this kind, radical conjugate addition followed by dehydrogenation led to the

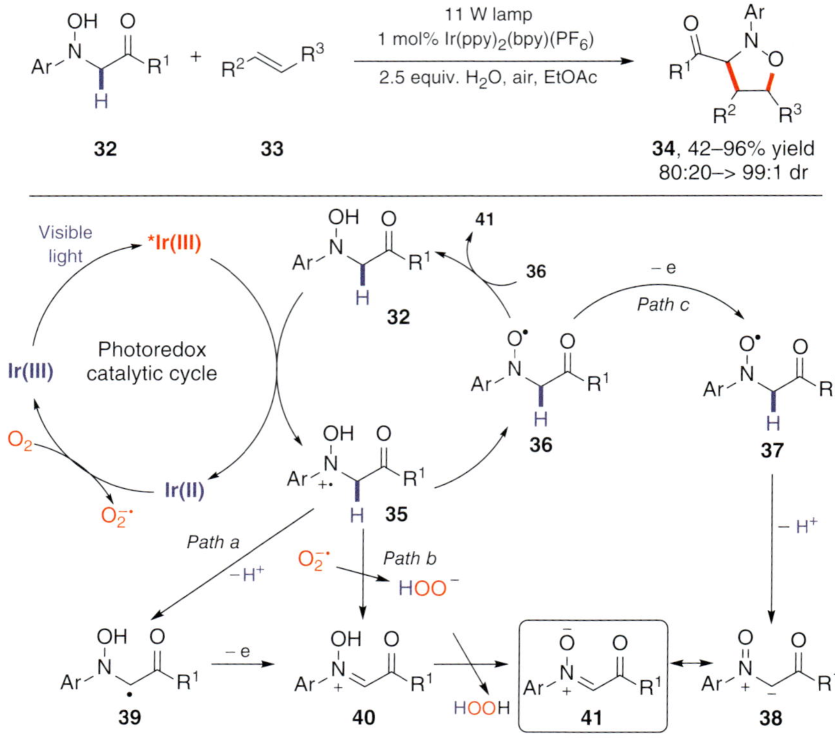

Scheme 13.6 [3+2] cycloaddition reactions of *N*-substituted hydroxylamines with alkenes.

formation of 5,6-dihydroindolo[2,1-*a*]tetrahydroisoquinolines, albeit in low yields (Scheme 13.7). In this work, the nucleophilic α-aminoalkyl radical was produced from an amino radical cation via a deprotonation process. Then, intramolecular radical conjugate addition, followed by a reduction/protonation and dehydrogenation sequence, gave the final heterocycle **45**.

After Reiser's report, Rueping *et al.* used the radical conjugate addition strategy in the syntheses of various 3-formyl indole derivatives (Scheme 13.8) [19]. To gain an understanding of the mechanism, reactions involving aminoenone **50a** were conducted. As expected, this reaction gave a mixture of **51a**, in 36% yield, and the Michael addition/dehydrogen product **52a**, in 21% yield, under standard conditions. Moreover, **52a** could be subsequently transformed into **51a** under the same conditions. This strongly supported the idea that the Michael addition/dehydrogen product (**52**) might be an intermediate in the generation of 3-formyl indole products. ^{18}O-labeling experiments revealed molecular oxygen to be the source of the formyl oxygen in the final product.

Using *ortho*-alkynylated *N*,*N*-dialkylamines as substrates, Zhou and coworkers demonstrated radical addition to alkynes followed by C–O bond formation [20]

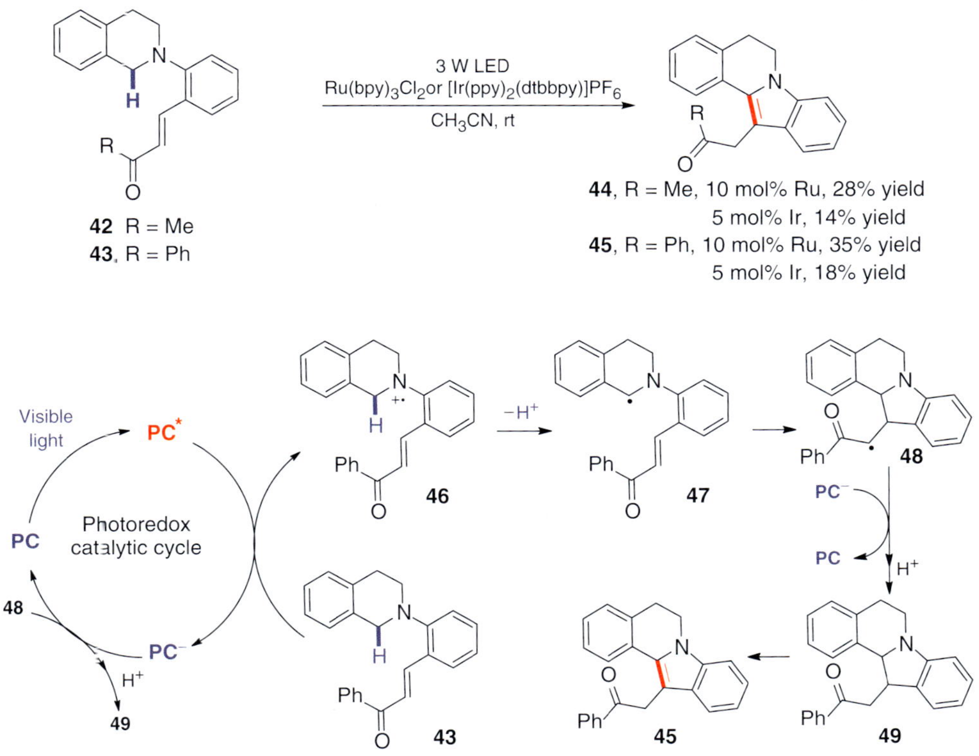

Scheme 13.7 Visible-light-induced intramolecular trapping of α-amino radicals.

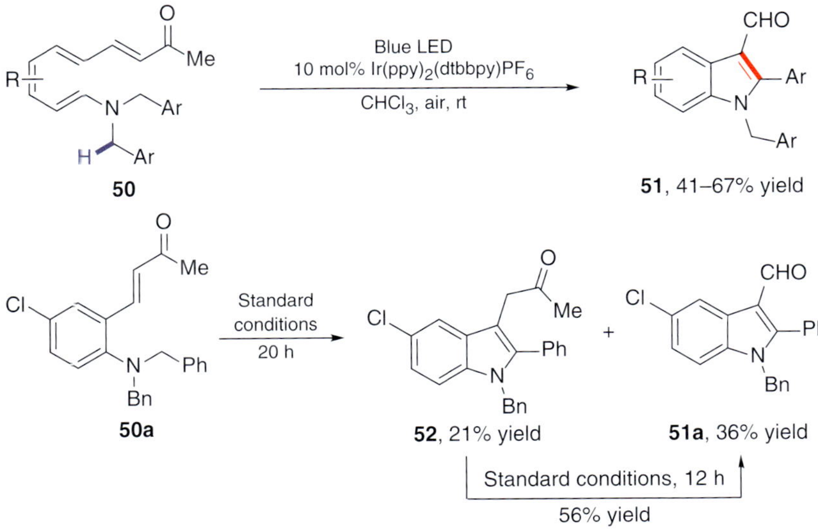

Scheme 13.8 Visible-light-induced synthesis of 3-formyl indole derivatives.

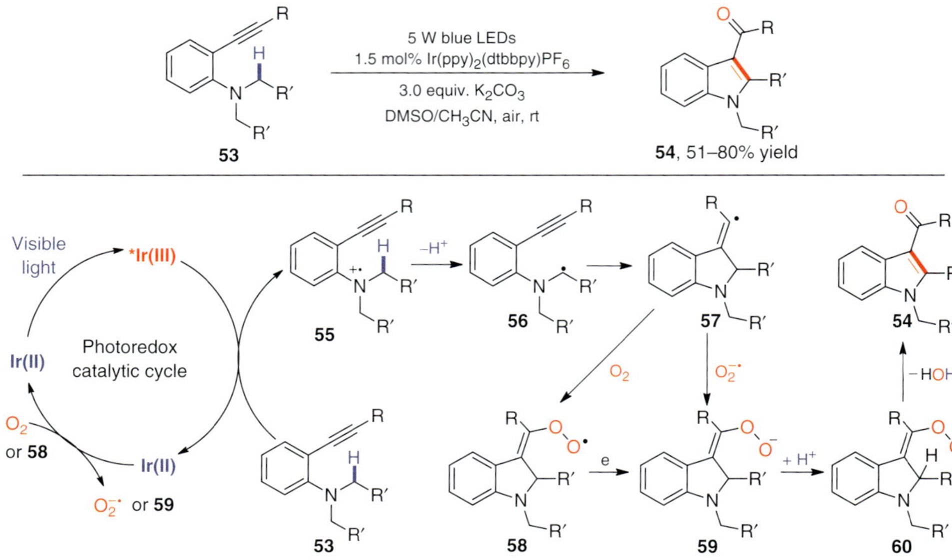

Scheme 13.9 Visible-light-induced intramolecular oxidative cyclization.

(Scheme 13.9). Various 2-aryl-3-acylindoles were assembled through this visible-light-induced, α-amino, sp^3 C–H functionalization reaction. Initially, α-aminoalkyl is a radical generated from tertiary amine **53**. This reactive radical then undergoes an intramolecular radical addition to the alkyne skeleton, giving vinyl radicals that are captured by molecular oxygen or superoxide anion radicals ($O_2^{-\bullet}$) to provide vinyl peroxides. Intramolecular hydrogen atom abstraction and dehydration deliver the final products.

In 2012, Bian, Yu, and coworkers [21] discovered an intermolecular visible-light-mediated radical addition/cyclization process useful for coupling tertiary anilines with maleimides (Scheme 13.10). In a previous report on α-aminoalkyl radical-involved reactions, anaerobic conditions were important for the inhibition of further oxidation of the reactive radical [18, 22]. However, Bian and Yu *et al.* found air to be necessary as a green external oxidant. In the first step, the α-aminoalkyl radical **65** is formed from a tertiary amine through reductive quenching (Scheme 13.10, path a) and added to maleimides to generate **66**. Then, **66** undergoes cyclization to form radical **67**, which is then rapidly oxidized to tetrahydroquinoline **63** by O_2. A hydrogen atom transfer (HAT) process between HOO$^\cdot$ and the reaction substrate **61** to form radical intermediate **65** (Scheme 13.10, path b) is not ruled out by the data in this study.

Recently, Rueping's group [19] reported a similar reaction involving α-amino radicals (Scheme 13.11). Instead of maleimides, 2-benzylidenemalononitriles were used as electron-deficient alkenes in reactions with anilines that are

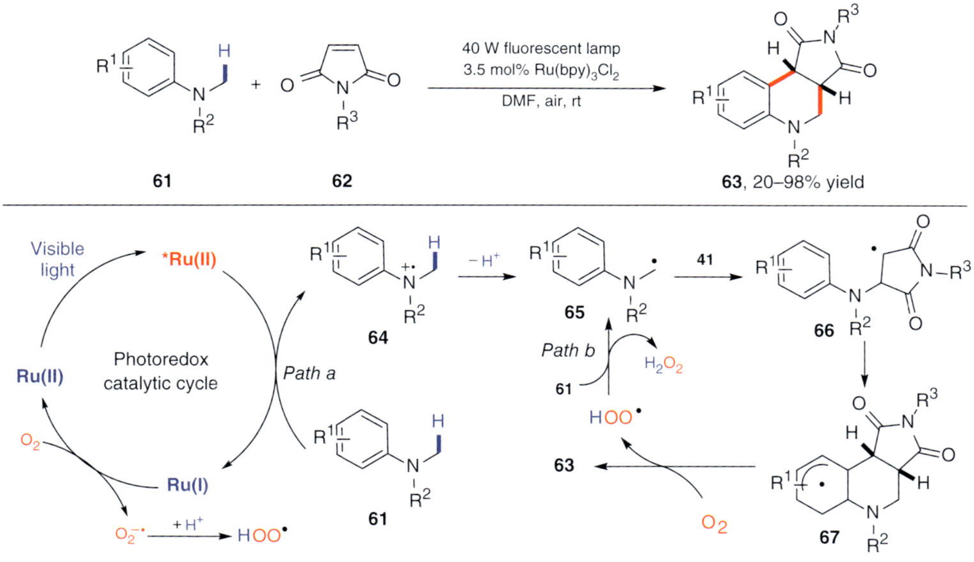

Scheme 13.10 Visible-light-induced radical cyclization of tertiary amines and maleimides.

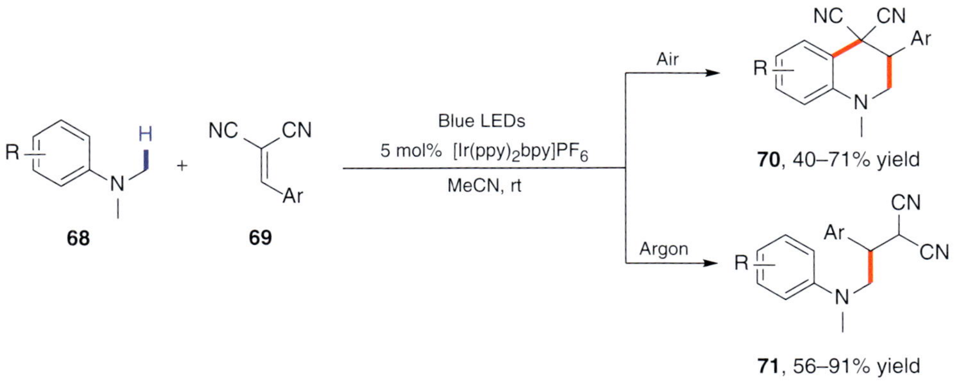

Scheme 13.11 Radical cyclization of tertiary amines and 2-benzylidenemalononitriles.

controlled by the presence, or absence, of oxygen. In the absence of oxygen, visible-light-induced intermolecular additions/cyclizations of *N*,*N*-dimethylanilines and alkenes **68** provided tetrahydroquinoline derivatives in good to excellent yields. When O_2 was excluded, a one-step radical addition took place smoothly and exclusively.

The methodology discussed previously for the syntheses of tetrahydroquinoline derivatives has been further developed by Tan [23] and Bach [24]. In Tan's work (Eq. (13.6)), a wide range of *N*-aryl glycines were used instead of

N,*N*-dimethylanilines, showing decarboxylation and subsequent reactions with maleimides to afford nitrogen-unprotected tetrahydroquinoline derivatives. Such compounds were difficult to obtain in satisfactory yields by Bian and Rueping's method. In this context, *N*-aryl glycines seem to be useful candidates as the α-amino radical source. In 2014, Bach and coworker extended the range of the electron-deficient alkene starting materials to include various five- or six-membered α,β-unsaturated lactones and lactams using *N*-methyl-*N*-((trimethylsilyl)methyl)aniline as the other reaction component in a similar visible-light-induced radical addition/cyclization pathway (Eq. (13.7)). Although these two reactions were not direct C–H functionalizations of amines to heterocycles, the sp^2 C–H functionalization of the arene ring was a key step in the construction of nitrogen-containing heterocycles.

Tan's work

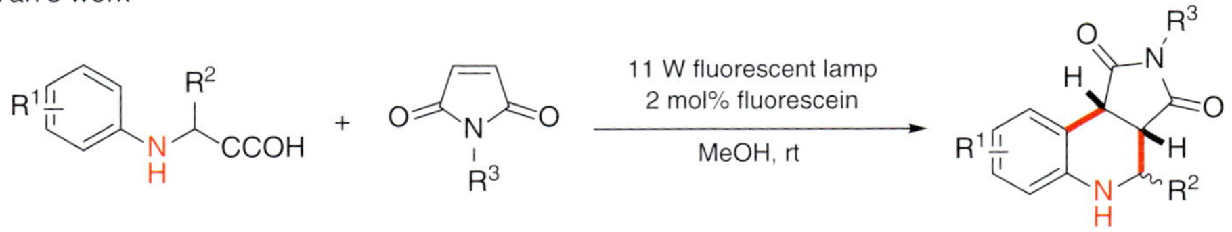

72 **73** **74**, 20–98% yield

(13.6)

Bach's work

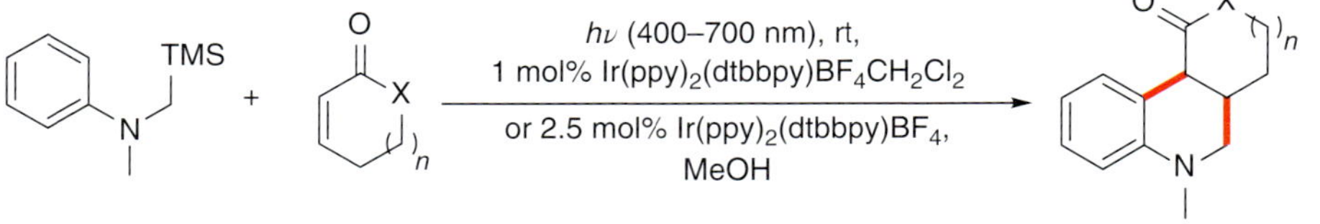

75 **76** n = 0, 1: X = O, CH$_2$, NBoc, NTs **77**, 25–49% yield

(13.7)

In 2013, Pandey *et al.* [25] reported a cycloetherification reaction via direct benzylic C–H activation under photoredox catalysis conditions (Scheme 13.12). Various cyclic ethers were obtained in moderate to good yields using DCN (1,4-dicyanonaphthalene) as the photoredox catalyst and molecular oxygen as the oxidant. It should be noted that the 450 W Hanovia medium pressure lamp, which was housed in a Pyrex glass immersion well (>300 nm), was used rather than common visible-light sources in this cycloetherification reaction. As shown in the following text, substrates **78** were oxidized to benzyl cations through two sequential photoredox cycles, followed by intramolecular etherification to produce the desired product **79**.

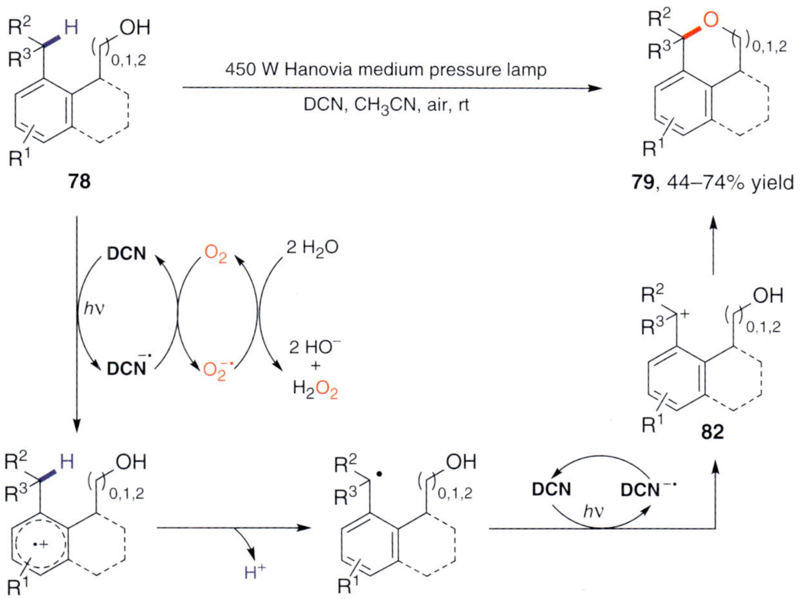

Scheme 13.12 Benzylic C–H activation by photoredox catalysis.

13.2.2
Photocatalytic sp^2 C–H Functionalization for Heterocycle Synthesis

Generally, sp^2 C–H bonds are much less reactive than α-amino sp^3 C–H units. However, sp^2 C–H activation is substantially easier after radical addition to heteroarenes due to the strong driving force of rearomatization.

In 2009, Stephenson and coworkers [26] developed a reductive dehalogenation of activated carbon–halogen bonds via visible-light photoredox catalysis. Further investigation [27] on the reactive alkyl radical intermediate reported the selective sp^2 C–H intermolecular bond alkylation of indoles and pyrroles at the C2 position. Organic halides could be reduced by a Ru(I) species, which was generated through reductive quenching of an excited Ru(II) complex, to give the reactive radical intermediate **85** (Scheme 13.13). Radical addition to the heterocyclic indoles or pyrroles, followed by two possible aromatization pathways, produced the desired heterocycle **84**.

In Stephenson's work, the double cyclization product **90** was also obtained in good yields as a single diastereoisomer (Eq. (13.8)). This cascade radical cyclization strategy was also used by other groups for the efficient syntheses of oxindoles (Eq. (13.9)) [28] and 6-trifuoromethyl phenanthridines (Eq. (13.10)) [29].

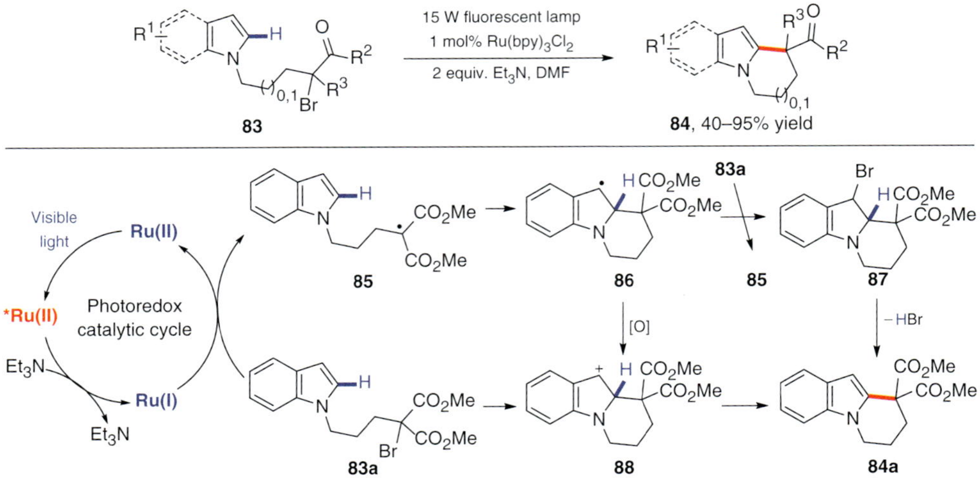

Scheme 13.13 Visible-light-induced intramolecular alkylation of indoles and pyrroles.

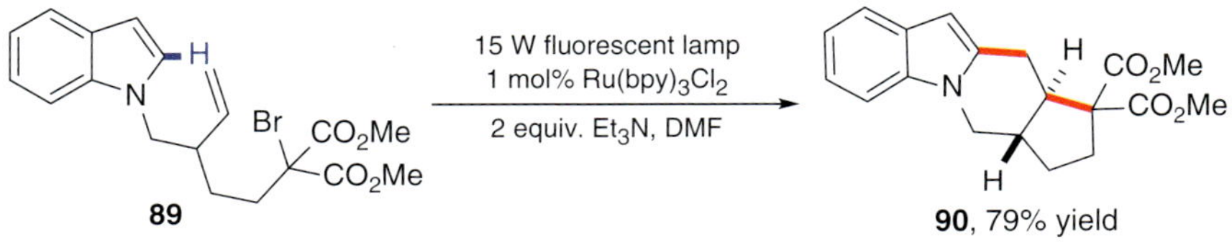

(13.8)

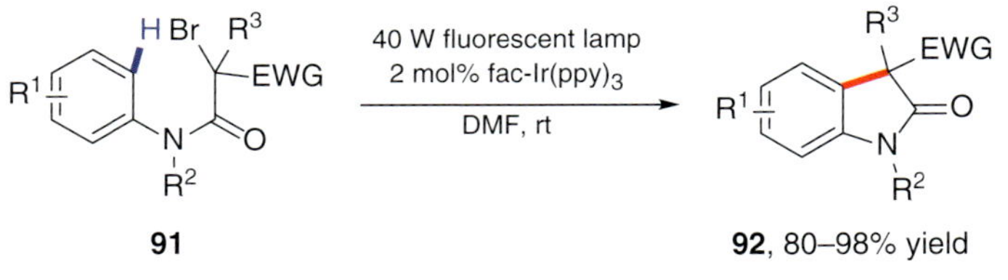

(13.9)

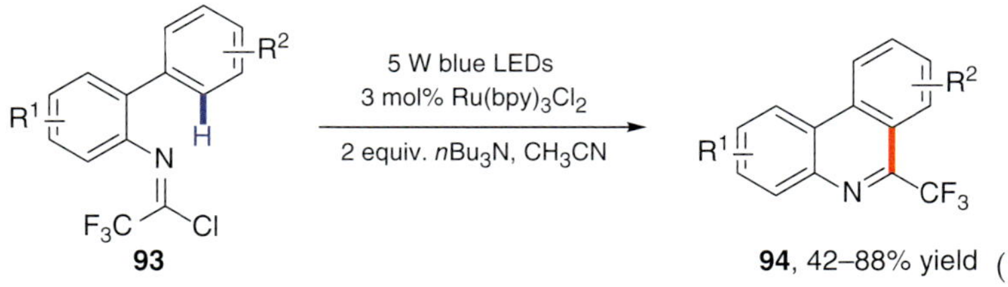

(13.10)

In 2011, Stephenson's group developed a tandem radical cyclization/sigmatropic rearrangement reaction [30] that was used to prepare tricyclic pyrrolidinone **96** from organic halide **95** through photoredox catalysis (Scheme 13.14). The possible intermediate **97** could be isolated and further transformed to the desired

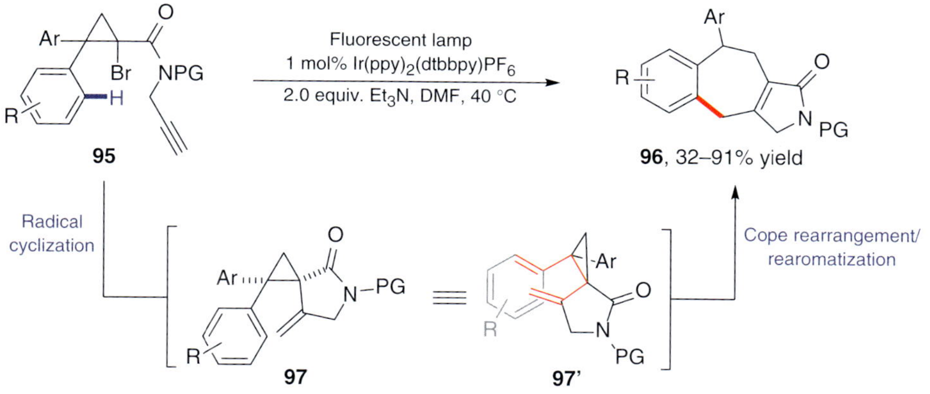

Scheme 13.14 Visible-light-induced intramolecular radical cyclization.

product in the absence of any redox reaction conditions, indicating that the tandem radical cyclization/sigmatropic rearrangement process was the most likely pathway.

After the report on the visible-light-induced [4+2] benzannulation of biaryldiazonium salts and alkynes [31], Zhou and coworkers [32] also demonstrated the radical cyclization of trifluoroacetimidoyl chlorides with alkynes, yielding various 2-trifluoromethyl quinolones (Scheme 13.15). It is worth noting that this is the first example where the activation of sp^2 C–Cl bonds by a photoredox catalyst

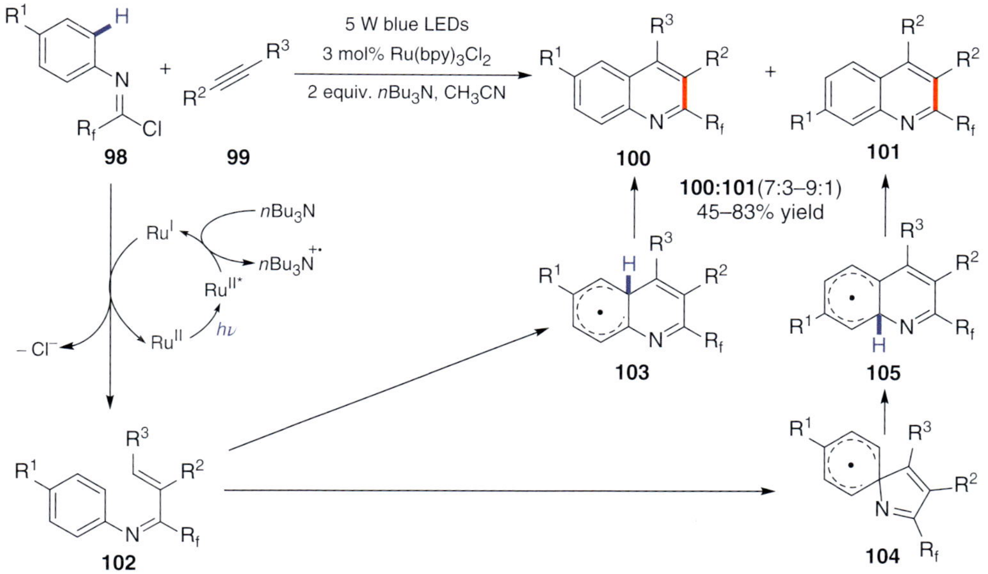

Scheme 13.15 Visible-light-induced cyclization of trifluoroacetimidoyl chlorides with alkynes.

is used to generate imidoyl radicals. To better understand the reaction selectivity, two possible pathways, which diverge after the reductive quenching cycle of photocatalyst, were proposed.

In 2012, Li and coworkers [33] reported an aerobic visible-light-mediated protocol for the construction of 2-substituted benzothiazoles (Scheme 13.16). The authors found that the reaction mixture turned green early in the reaction, as is often considered an indication of the presence of $Ru(bpy)^{3+}$. As a result, a pathway involving oxidative quenching by molecular oxygen was proposed. Single-electron transfer between the deprotonated thioamide and the Ru(III) species produces the radical **109**, which facilitates radical cyclization followed by an aromatization cascade to yield the final products.

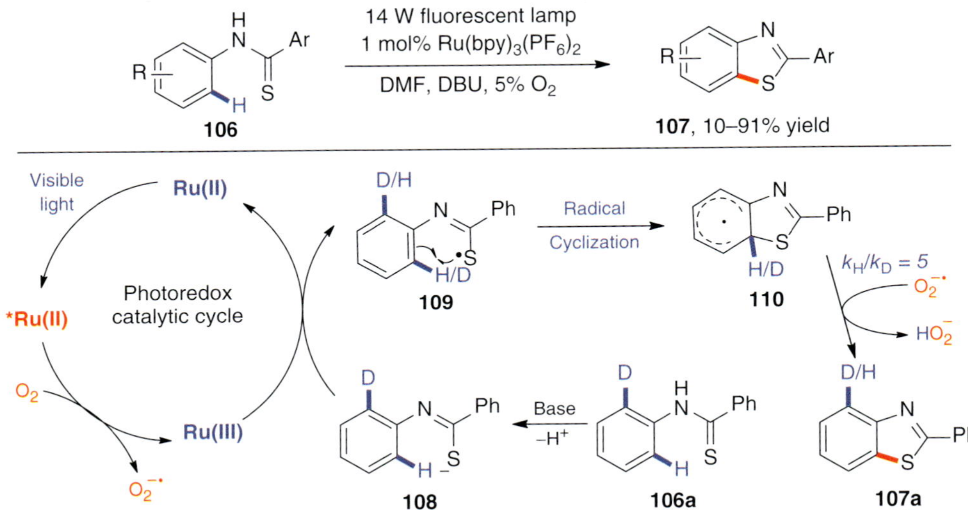

Scheme 13.16 Visible-light-mediated synthesis of 2-substituted benzothiazoles.

A successful transformation based on the reactivity of nitrogen-centered radical cations was achieved by Zheng *et al.* in 2012 [34]. In Zheng's process, a nitrogen-centered radical cation undergoes electrophilic addition to a tethered alkene instead of transformation into an iminium ion α-amino radical (see Section 13.2.1 for details) or carbon-centered radical via a ring-opening process [35]. In this reaction, the addition of silica gel significantly accelerates the process, perhaps through the absorption of O_2, facilitating single-electron oxidation and acting as a source of protons. It is worth mentioning that *gem*-di-substituted styryl anilines such as **113** and **115** were transformed into 2,3-disubstituted indoles **114** and **116**. This indicates the formation of a benzylic cation intermediate, which undergoes a 1,2-carbon shift. A plausible mechanism for this visible-light-induced indole synthesis is shown in Scheme 13.17. The C–N bond formation by electrophilic addition of **113**, generated through reductive

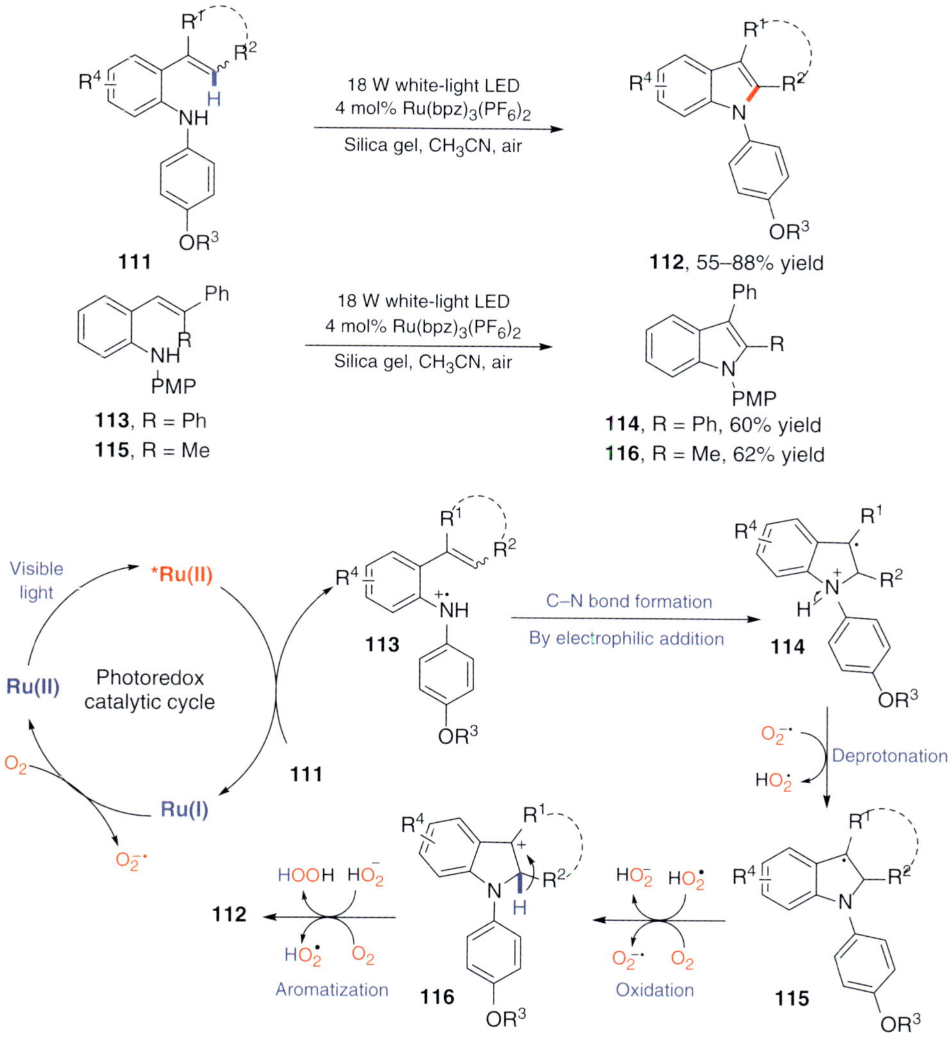

Scheme 13.17 Visible-light-induced *N*-aryl indole synthesis.

quenching cycles, gave **114**. This intermediate is then transformed to the desired indole via a deprotonation/oxidation/aromatization cascade.

Another possible pathway for reactivity via nitrogen-centered radical cation intermediates involves delocalization of the radical the aromatic substituents. In 2013, Collins and coworkers [36] reported the visible-light-mediated photoredox synthesis of carbazoles from arylamines **117** through C–H activation and C–C bond formation. A possible mechanism for this conversion is depicted in Scheme 13.18. The excited sensitizer $[Cu(Xantphos)(dmp)]^{+*}$ is reduced by I_2, generating a radical dication $[Cu(Xantphos)(dmp)]^{2+}$ (dmp: 2,9-dimethyl-1,

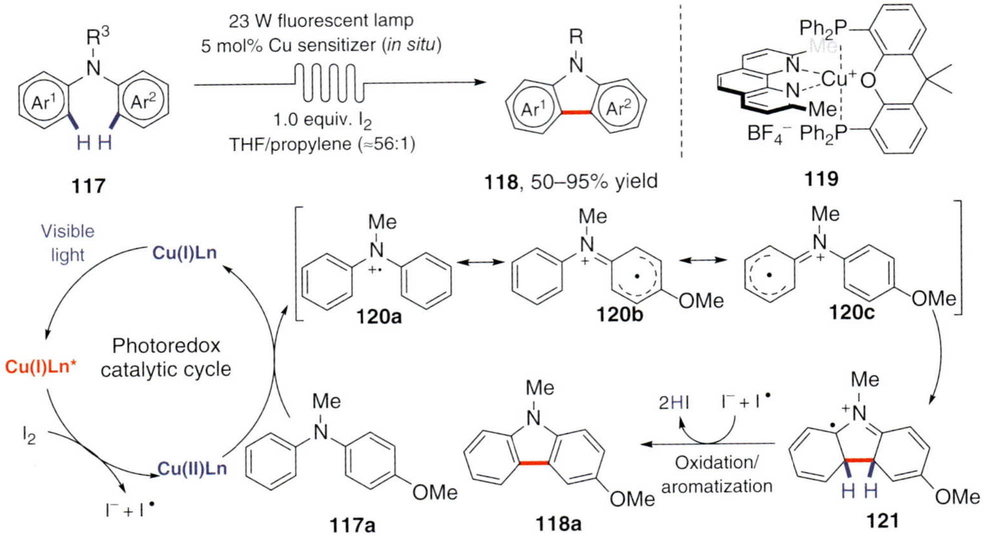

Scheme 13.18 Visible-light-induced carbazoles synthesis.

10-phenanthroline) that participates in the oxidation of the di- or triarylamine to afford the radical cation **120b**. This radical cation **120b**, which was delocalized across one of the aromatic substituents, induces intramolecular C–C bond formation, a second oxidation and aromatization to deliver the desired carbazoles. This transformation was also highlighted by the greatly accelerated continuous-flow conditions and the use of a Cu-based sensitizer, which could be rapidly screened.

The phenoxy radical may also undergo delocalization. Utilizing mesoporous graphitic carbon nitride (mpg-C_3N_4) as a photocatalyst, Liu and Wang [37] achieved the oxidative coupling of resveratrol and analogous compounds (Scheme 13.19). Under irradiation with visible light, electrons in mpg-C_3N_4 are excited from the valence band into the conduction band, forming electron/hole pairs. The formation of the phenoxy radical occurred though either direct hydrogen abstraction or proton transfer following electron transfer. The coupling of M5 and M10, derived from the delocalization of radical **122**, followed by tautomeric rearrangement and intramolecular nucleophilic attack, gave δ-viniferin.

A recent report from Yoon's group [38] described the oxidative [3+2] cycloadditions of phenols and electron-rich styrenes (Eq. (13.11)). A large family of dihydrobenzofurans, including two kinds of natural products **128a** and **129**, was synthesized in good yields (Eqs. (13.12) and (13.13)). An essential factor in this reaction is the transformation of phenols to resonance-stabilized phenoxonium cations **131** via a two-step oxidation. The cation **131** is then trapped by the electron-rich styrenes to afford the dihydrobenzofurans (Scheme 13.20). The authentic active catalyst in this transformation is the insoluble $Ru(bpz)_3^{2+}$ (bpz: 2,2′-bipyrazine) persulfate complex, as confirmed by the fact that $Ru(bpz)_3(S_2O_8)$, prepared from $Ru(bpz)_3(PF_6)_2$ with $(Bu_4N)_2(S_2O_8)$ in the dark, proved to be an effective catalyst with no induction period.

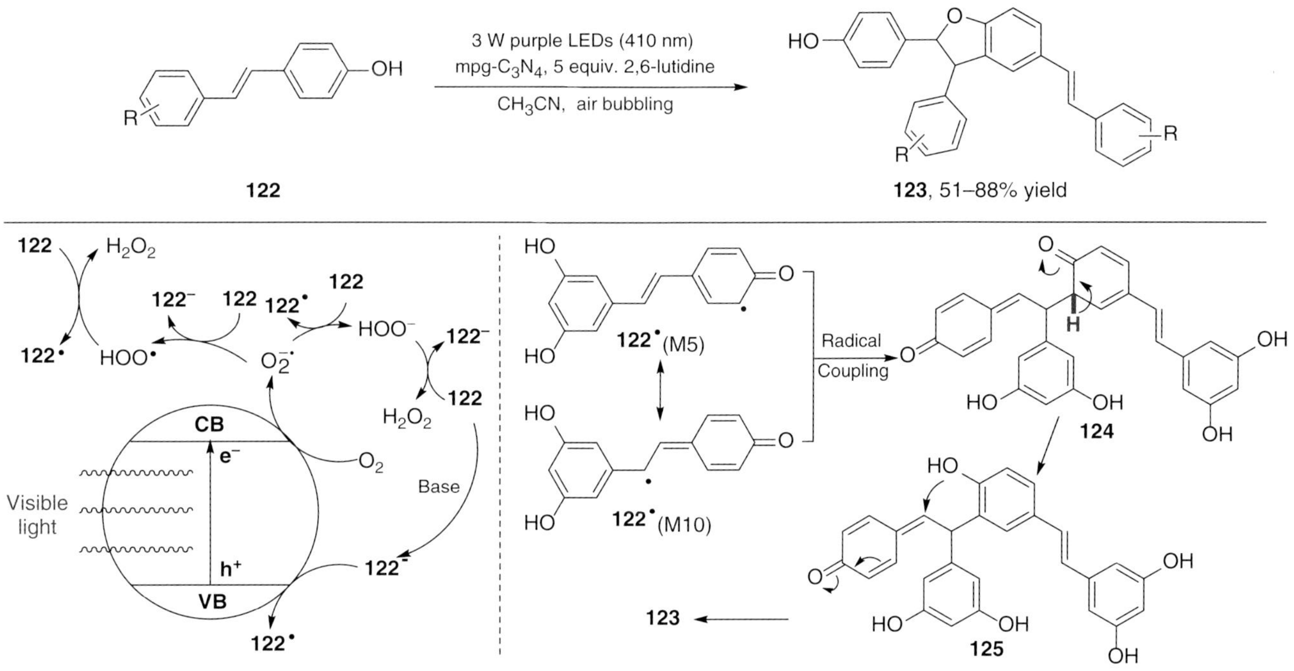

Scheme 13.19 Visible-light-induced δ-viniferin synthesis.

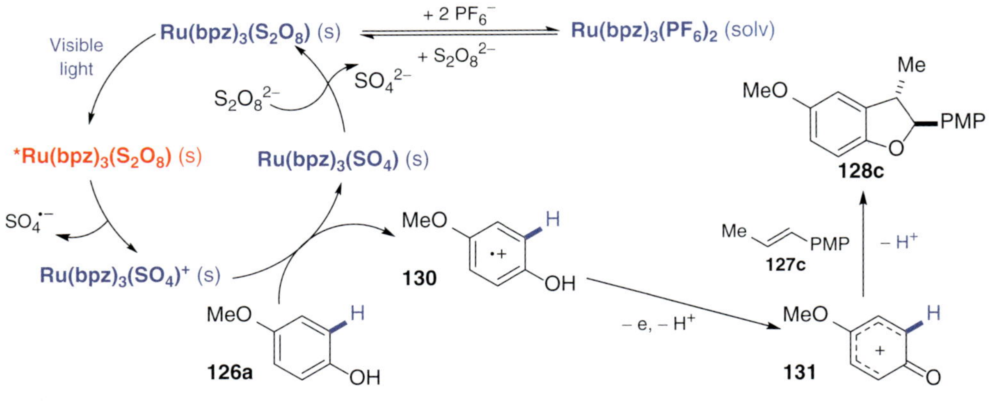

Scheme 13.20 Visible-light-induced oxidative [3+2] cycloadditions of phenols and styrenes.

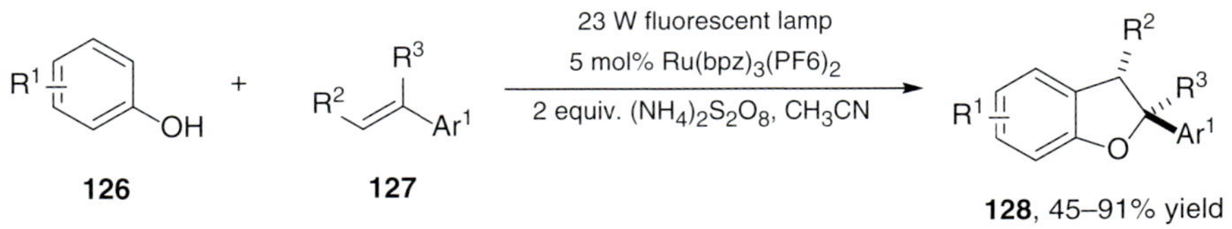

(13.11)

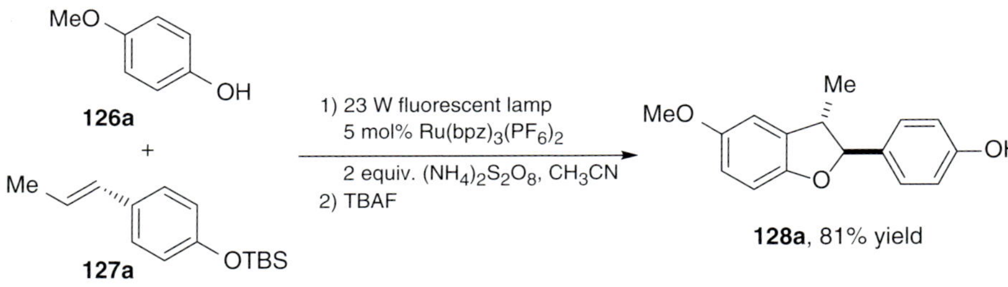

(13.12)

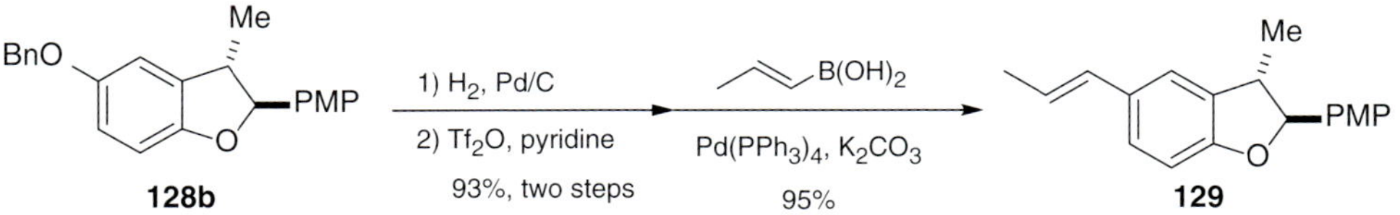

(13.13)

One of the most important routes to diverse 3,3-disubstituted oxindoles is the tandem radical cyclization of *N*-arylacrylamides [39]. Visible-light photoredox catalysis has proved to be a powerful tool in providing reactive radical intermediates. The development of photocatalytic radical cyclizations of *N*-arylacrylamides has been ongoing for some time, with the first example provided by Zhu and coworkers [40]. After the initial discovery of the decarboxylation/radical C–H functionalization reaction, Togni's reagent [41], aryl diazonium salts [42], fluoroalkylsulfonyl chlorides [43], acetonitrile [44], acetone [44], and alkyl halides [45] were used in reactions with *N*-arylacrylamides to yield different kinds of oxindoles. The general mechanism for this kind of reaction includes radical production through oxidative quenching cycles, radical addition to *N*-arylacrylamides, radical cyclization and rearomatization with the release of the photocatalyst (Scheme 13.21).

Isocyanides, which are isoelectronic with carbon monoxide, can undergo insertion reactions to provide *N*-containing heterocycles. In 2013, Yu and coworkers [46] developed a photoredox catalytic, somophilic isocyanide insertion (HAS) yielding 6-alkylated phenanthridine derivatives from biphenyl isocyanide and electron-deficient bromides (Eq. (13.14)). Interestingly, when ethyl bromofluoroacetate or ethyl bromodifluoroacetate were used as the substrate [47], a subsequent decarboxylation efficiently produced mono- and difluoromethylated phenanthridine derivatives (Eq. (13.15)).

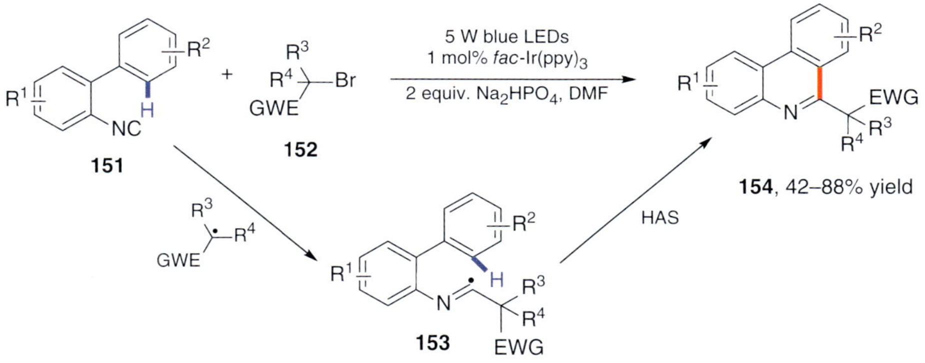

(13.14)

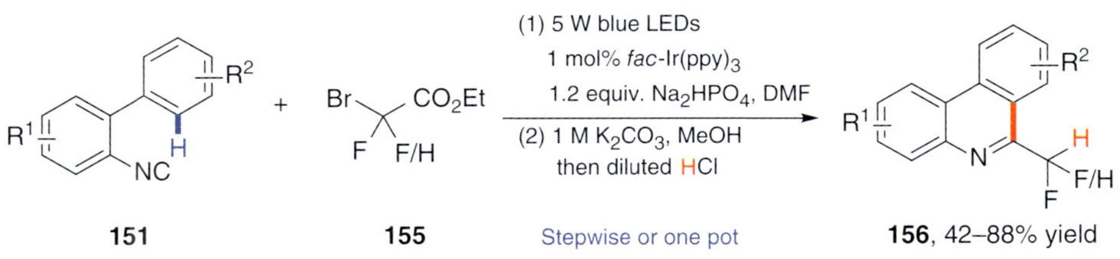

(13.15)

This somophilic isocyanide insertion reaction is also suitable for use with hydrazines [48], arylsulfonyl chlorides [49], diaryliodonium salts [50], and Umemoto's reagent [50] (Scheme 13.22). Polycyclic quinoxalines can also be synthesized by the visible-light-induced decarboxylative radical cyclization of

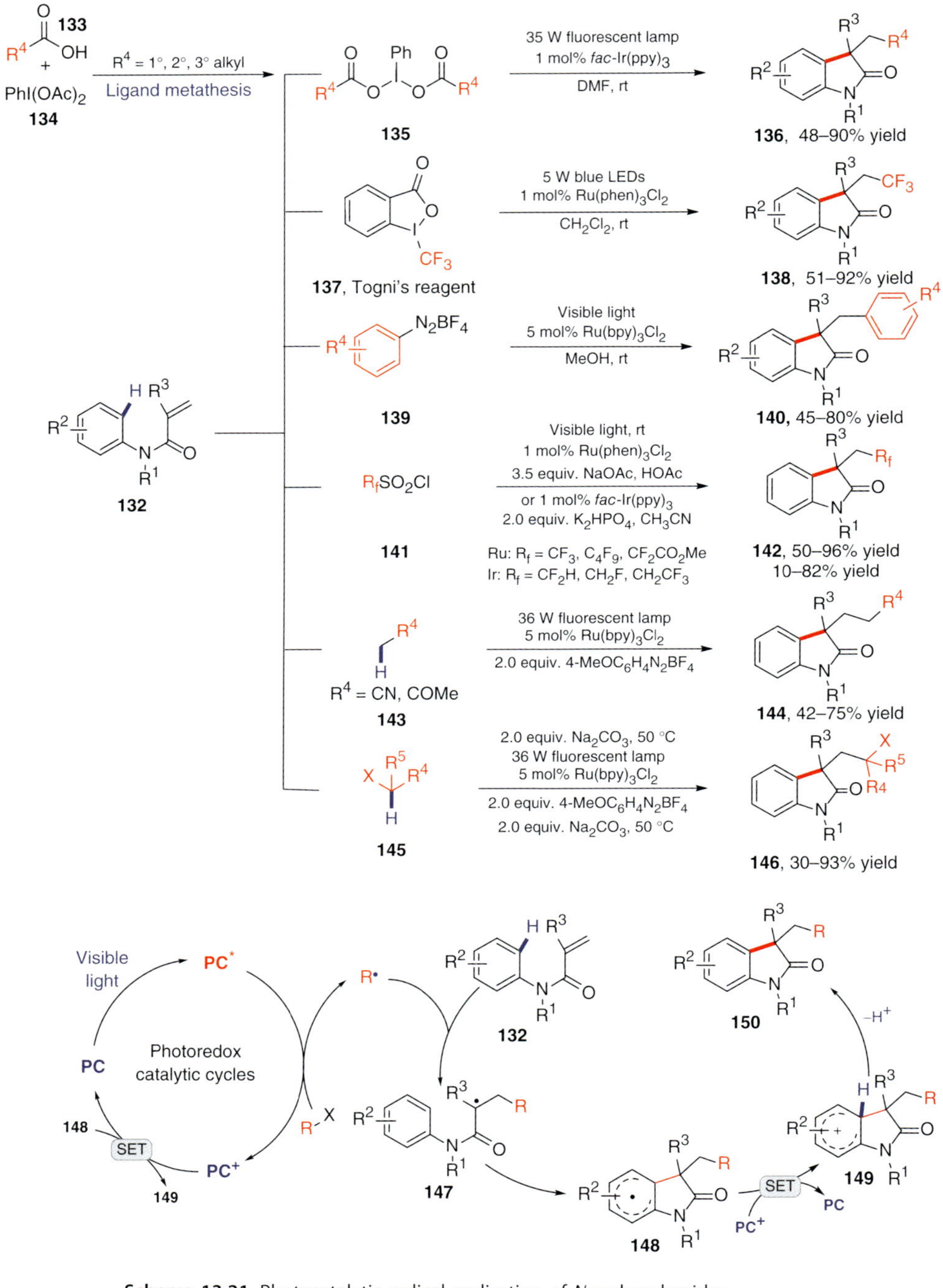

Scheme 13.21 Photocatalytic radical cyclization of *N*-arylacrylamides.

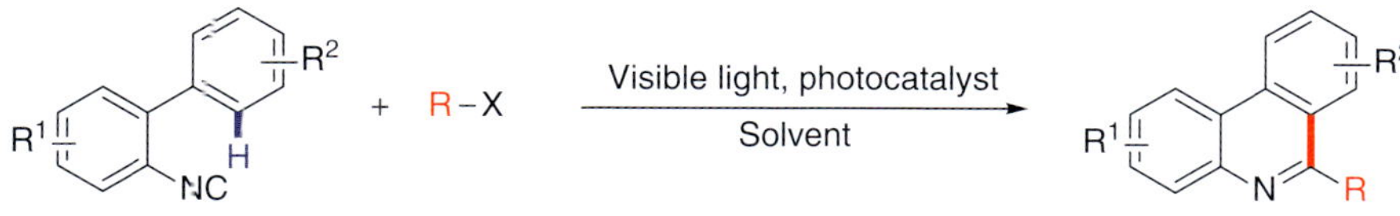

R–X	Condition	Yield of products
157, RNHNH$_2$ or RNHNH$_2$·HCl	5 W blue LEDs, 5 mol% Eosin B 1.5 eq. K_2CO_3, DMSO	**158**, 56–93% yield
159, **Ar**SO_2Cl	5 W blue LEDs, 5 mol% Eosin Y 1.5 eq. K_2HPO_4, CH_3CN	**160**, 51–82% yield
161, Ar_2IBF_4	3 W blue LED 2.5 mol% Ir(ppy)$_2$(dtbbpy)PF_6 2.0 eq. Na_2CO_3, MeOH	**162**, 52–77% yield
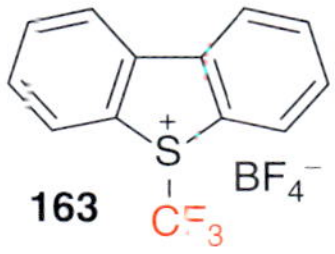 **163**	3 W blue LED, 1 mol% Ru(bpy)$_3$Cl$_2$ DMF	**164**, 30–88% yield

Scheme 13.22 Visible-light-mediated somophilic isocyanide insertion reactions.

ortho-heterocycle-substituted arylisocyanides [51] (Eq. (13.16)).

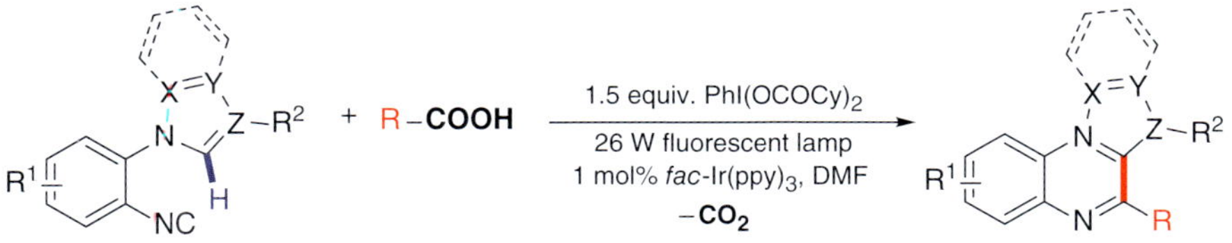

165 **166** **167**, 30–84% yield

(13.16)

As an alternative to the reductive quenching and oxidative quenching pathways, direct energy transfer [52] between the excited state of photocatalyst and substrate is an important pathway in photoredox catalysis. In 2014, Yoon [53] and Xiao [54] reported visible-light-induced transformation of azide. An energy transfer between vinyl or styryl azides and the excited catalyst gives the nitrene intermediate, followed by nitrene insertion into the sp^2 C–H bond. However, a ring-closure/rearrangement pathway was thought to be more reasonable in Yoon's work and is strongly supported by the observation of aziridine **175** (Eq. (13.16)).

Xiao's work

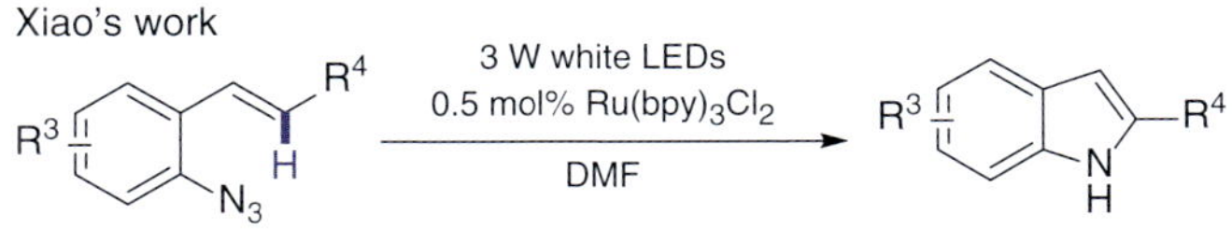

168 **169**, 39–99% yield (13.17)

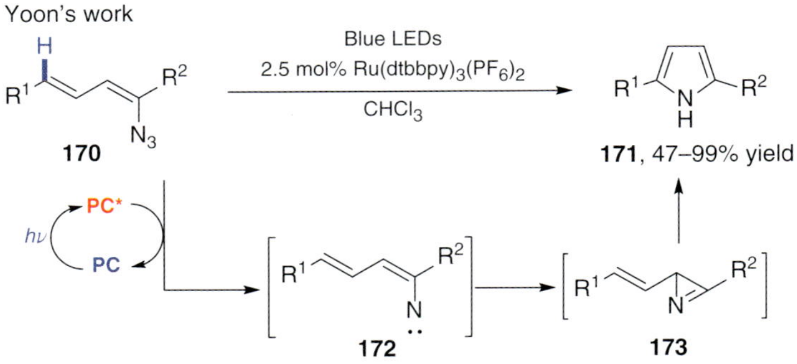

(13.18)

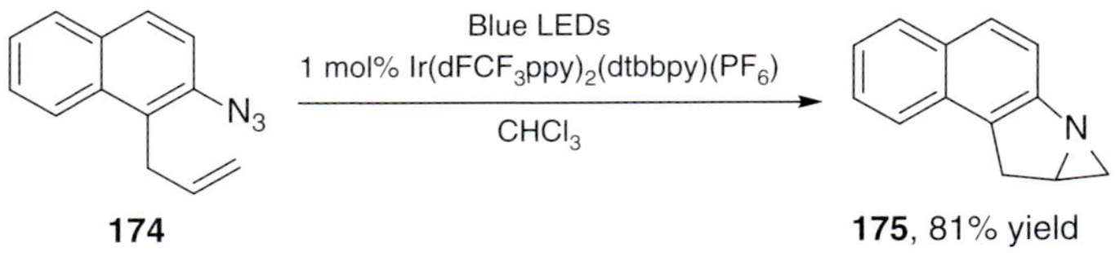

(13.19)

The Xiao group [55] has recently made great strides in applying the energy transfer strategy to heterocycle synthesis via C–H functionalization, demonstrating the visible-light-induced intramolecular cyclization and H-transfer of diazoamides results in the generation of 3-ester-2-oxindole intermediates via a carbene intermediate resulting from the release of N_2. Subsequent conversion of the 3-ester-2-oxindole into more important 3-ester-3-hydroxy-2-oxindole derivatives by visible-light-induced phototandem catalysis, using a single iridium photocatalyst, was likewise shown. Energy transfer between an excited photocatalyst and substrates (**176**) give **176** in an excited state, furnishing the carbene intermediate by N_2 release.

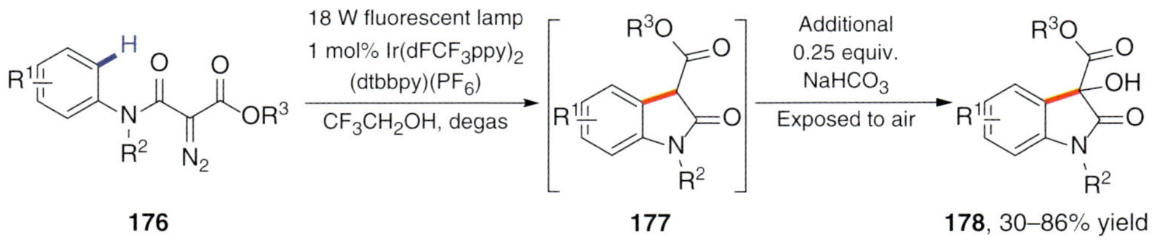

(13.20)

Transition metal-catalyzed C–H activation has proven to be a reliable method for the construction of complex products. However, the requirement of stoichiometric quantities of external oxidants has limited the sustainability of this method. In 2014, Rueping and coworkers [56] developed an oxidative Heck reaction that included recycling of the metal catalyst through a photoredox process. As a continuation of their work on photoredox catalysis, they recently disclosed a new combined photoredox- and palladium-catalyzed system useful for indole synthesis involving C–H activation [57]. The mechanism is depicted in Scheme 13.23. The traditional addition of stoichiometric quantities of an external oxidant was replaced by catalytic photoredox cycles utilizing O_2 as the external oxidant.

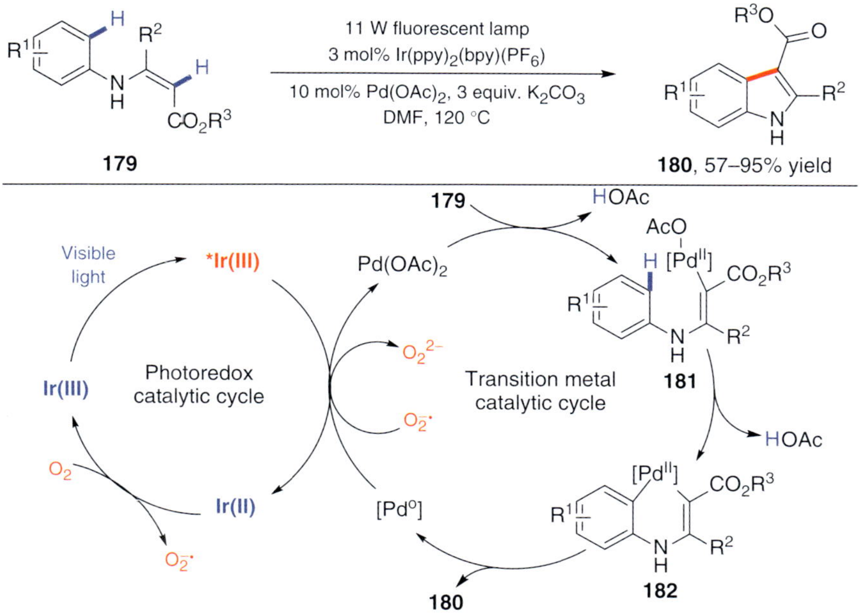

Scheme 13.23 Visible-light-mediated oxidative Heck reactions.

13.3 Direct C–H Functionalization of Heteroarenes

13.3.1 The Photocatalytic Alkylation of Heteroarenes

The direct alkylation of heteroarenes, one of the most useful C(sp^2)–C(sp^3) bond-forming reactions in all of organic synthesis, plays a pivotal role in the syntheses of medicinally and biologically relevant molecules. Great effort has been invested in the search for more efficient and general methods than those currently employed; however, methods for the direct C–H alkylation of electron-rich heteroaromatic compounds remain comparatively few in number. In 2010, Stephenson and coworkers developed a protocol for the intermolecular C–H alkylation of electron-rich heteroarenes (substituted indoles, pyrroles, and furans) with diethyl bromomalonate, through visible-light-induced photoredox catalysis using $Ru(bpy)_3Cl_2$ as photocatalyst (Scheme 13.24) [58]. It is worth

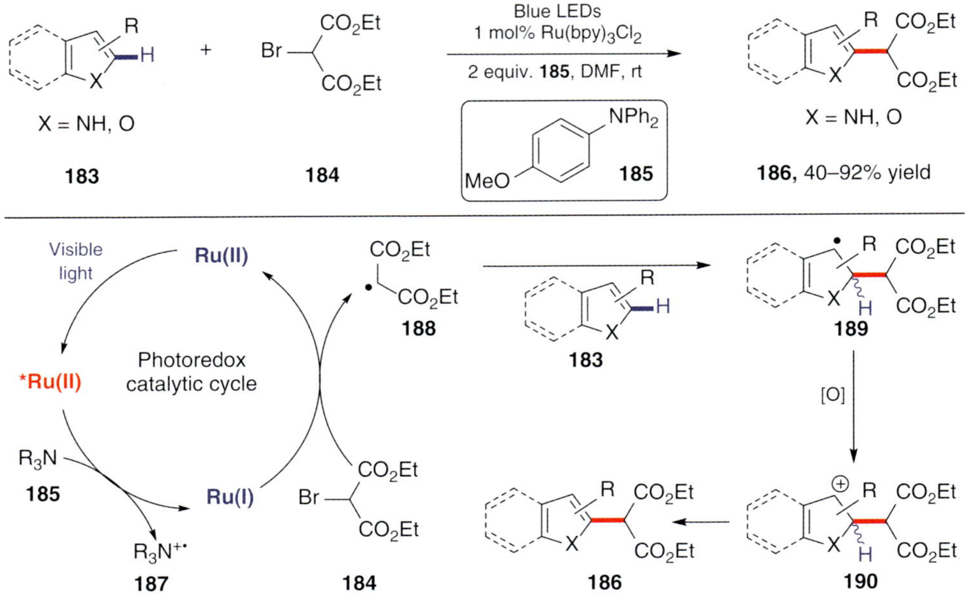

Scheme 13.24 Visible-light-mediated direct C–H alkylation of electron-rich heterocycles.

noting that the choice of tertiary amines, which were particularly effective reductive quenchers in the photoredox catalytic cycle, was the key to this reaction. Proper use of trialkylamines led to the reductive dehalogenation of bromomalonate as a result of their capacities to act as hydrogen atom donors. However, the use of aromatic amines eliminated this problem, allowing intermolecular transformation. 4-Methoxy-*N*,*N*-diphenylaniline **185** was found to be the best electron donor.

A plausible mechanism for the alkylation, involving reductive quenching cycles, has been proposed by Stephenson and coworkers (Scheme 13.24). First, the Ru(II) photocatalyst was excited by visible light, generating excited Ru(II)*, which was reductively quenched by **185** to afford Ru(I) and the radical ammonium cation **187**. The Ru(I) species reduced diethyl bromomalonate **184** to the malonyl radical **188**, regenerating Ru(II). The selective coupling of the malonyl radical **188** with indoles at the 2-position afforded a benzylic radical adduct **189**. Single-electron oxidation and subsequent deprotonation of the resulting cation generated the 2-alkylated indole products.

In 2011, Stephenson applied this visible-light photocatalytic indole alkylation reaction to the total synthesis of the natural product gliocladin **C**, which contains a C3–C3′ linked indole/pyrroloindoline core (Scheme 13.25) [59]. Through effective blocking of the 2-position of the indole **191** with a formyl group, a visible-light-mediated intermolecular alkylation reaction with bromopyrroloindoline **192** led to formation of the desired C3–C3′ coupling product **193** in an

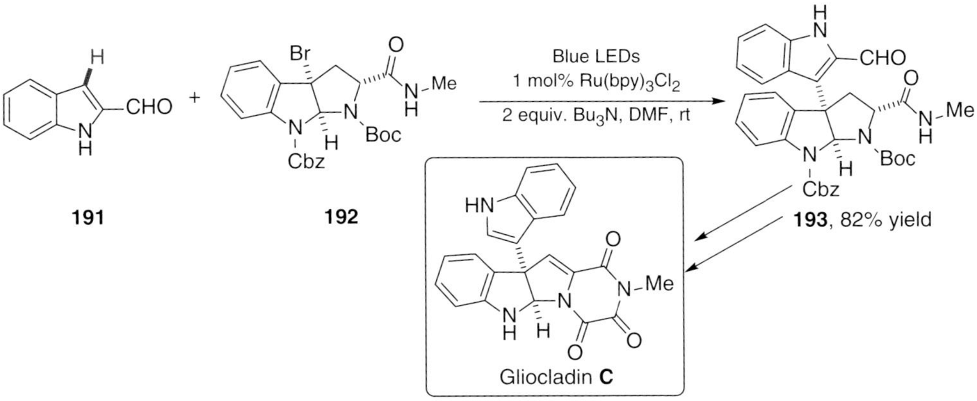

Scheme 13.25 Total synthesis of gliocladin **C** enabled by visible-light photoredox catalysis.

excellent yield. In total, starting with commercially available materials, gliocladin **C** was synthesized in 10 steps in 30% overall yield. This work showed that visible-light photoredox catalysis is not only a viable method for the synthesis of simple organic small molecules but may also be employed in the construction of complex natural products.

Recently, another example of visible-light-induced direct C–H alkylation of indoles was achieved by Melchiorre and coworkers in the absence of photocatalyst (Scheme 13.26) [60]. This metal-free photochemical strategy was key in the direct benzylation and phenacylation of 2- and 3-substituted 1*H*-indoles with electron-accepting benzyl and phenacyl bromides, respectively. The 3-substituted 1*H*-indoles were selectively alkylated at the C2 position. However, when C2 position is already substituted, alkylation proceeds efficiently at the C3 position. In this work, the conjunction of substituted 1*H*-indoles **194** with electron-accepting benzyl bromides **195** generated an electron donor–acceptor (EDA) complex **197**. The EDA complex was isolated, characterized by X-ray single-crystal spectroscopy, and shown to be capable of visible-light absorption. Under visible-light irradiation, the EDA complex **197** underwent an electron transfer from indole to benzyl bromide.

The introduction of small alkyl groups, such as methyl units, into pharmaceuticals may address problems associated with on- and off-target activity, metabolism, and pharmacokinetic profile. However, installing these groups onto complex heterocycles via direct C–H functionalization remains a challenging task. In 2014, DiRocco's group described a visible-light-mediated catalytic photoredox strategy to be effective in the efficient installation of alkyl groups (methyl, ethyl, and cyclopropyl) into a variety of complex heteroarenes (Scheme 13.27) [61]. This mild method was applied to the late-stage functionalization of biologically active heterocycles and proved to be an important tool for drug discovery.

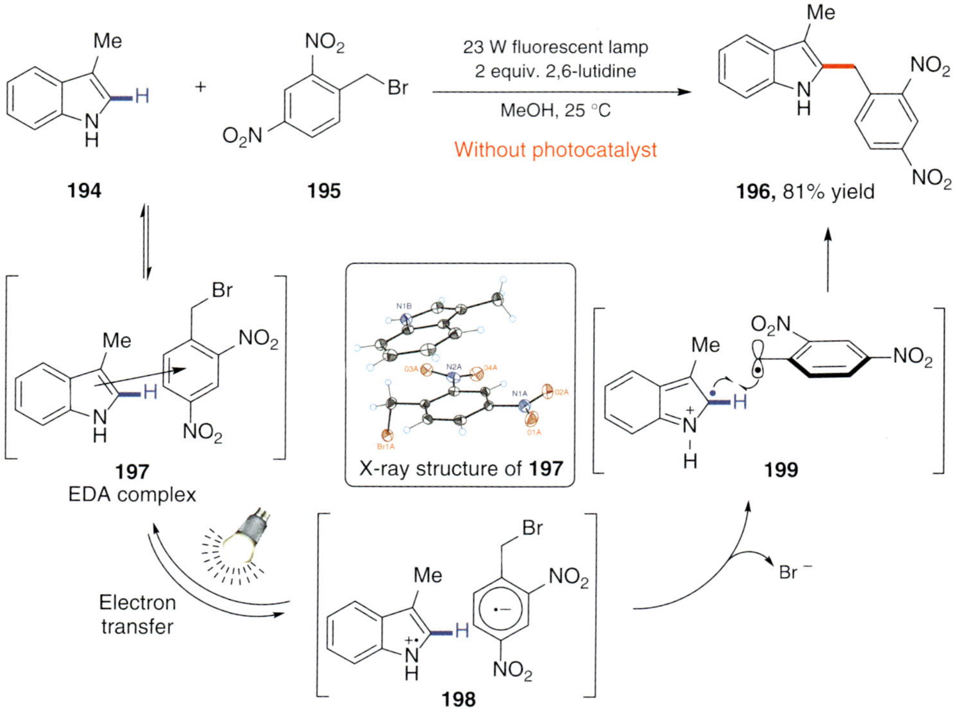

Scheme 13.26 Visible-light-induced C–H alkylation of indoles without photocatalyst.

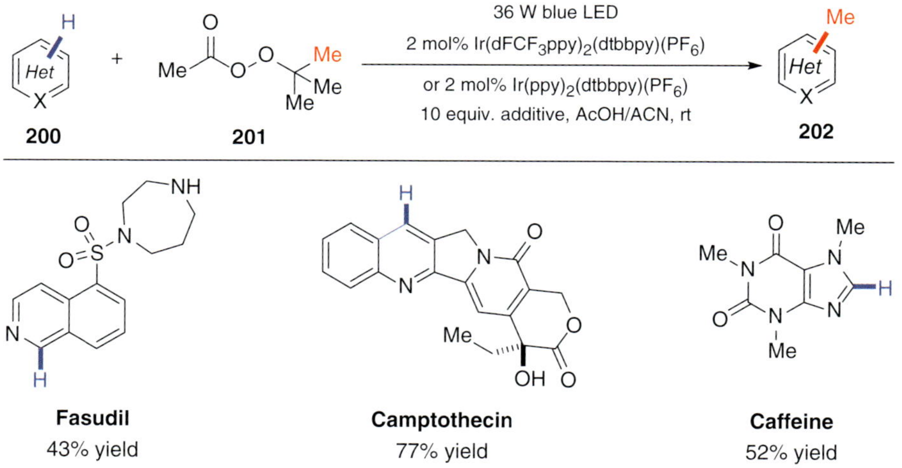

Scheme 13.27 Visible-light-induced direct C–H methylation of biologically active heterocycles.

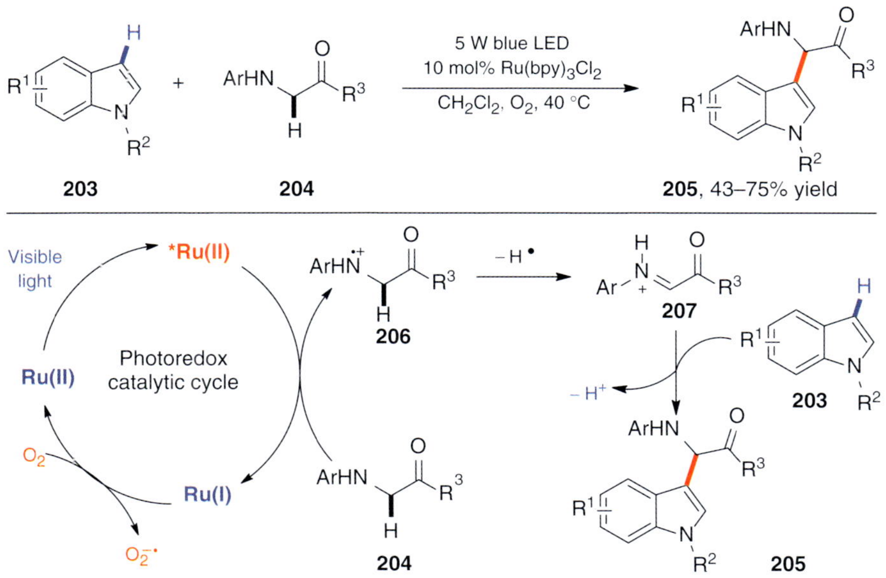

Scheme 13.28 Visible-light-mediated α-arylation of α-amino carbonyl compounds.

In 2012, a new method toward the functionalization of secondary amines through visible-light-induced aerobic oxidation was developed by Li and coworkers (Scheme 13.28) [62]. In this work, they established a general methodology for the construction of a number of functionalized 2-(1*H*-indol-3-yl)-2-aminocarbonyl compounds from α-amino carbonyls and indoles, using $Ru(bpy)_3Cl_2$ as the photocatalyst and O_2 as the oxidant, under visible-light irradiation. The authors proposed a pathway for this reaction that included a reductive quenching cycle. It is likely that in the mechanism at play in the visible-light-induced photocatalytic α-amino C–H functionalization (Scheme 13.2), the iminium ion **207** is generated through a deprotonation/oxidation sequence. After this deprotonation, nucleophilic addition of the indole **203** to the iminium ion **207** afforded the desired product **205**.

At the same time, Rueping and coworkers reported a similar functionalization of peptides and glycine esters through a combination of visible-light photoredox catalysis and Lewis acid catalysis (Eqs. (13.21) and (13.22)) [63]. The addition of the Lewis acid $Zn(OAc)_2$ was essential for C–H arylation due to its capacity to activate the electrophilic imine intermediate.

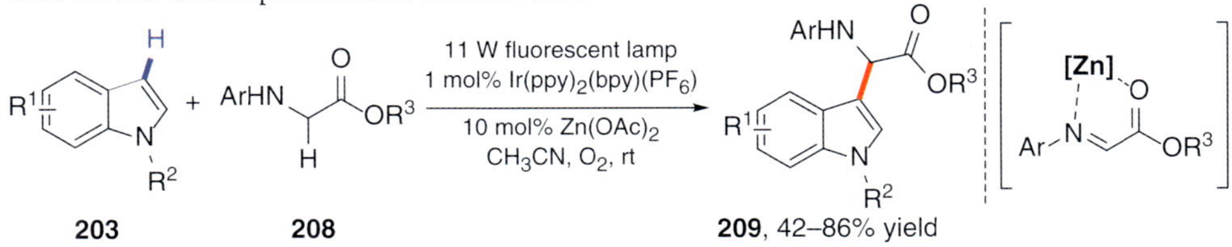

(13.21)

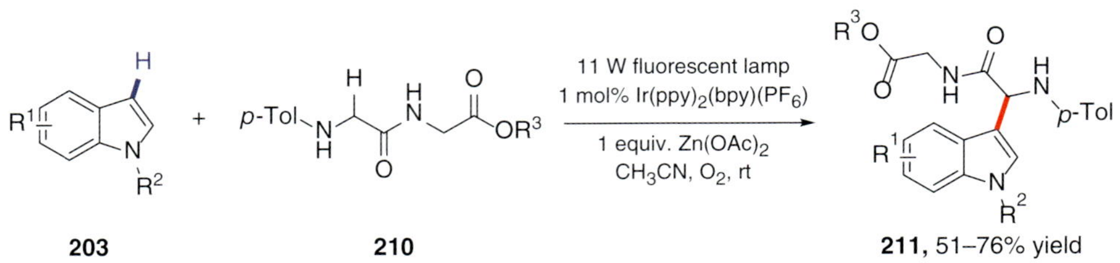

(13.22)

Very recently, Stephenson used a visible-light-mediated photocatalytic C–H functionalization method to rapidly introduce a morpholinomethyl unit into pyridazine **212** in one step (Scheme 13.29) [64]. By doing so, an important intermediate of the pharmaceutical agent **LY2784544**, a selective inhibitor of JAK2-V617F, which has undergone clinical trials for the treatment of several myeloproliferative disorders, was directly provided in moderate yields. This methodology was extended in the syntheses of a range of analogs utilizing commercially available *N*-methyl-substituted tertiary amines.

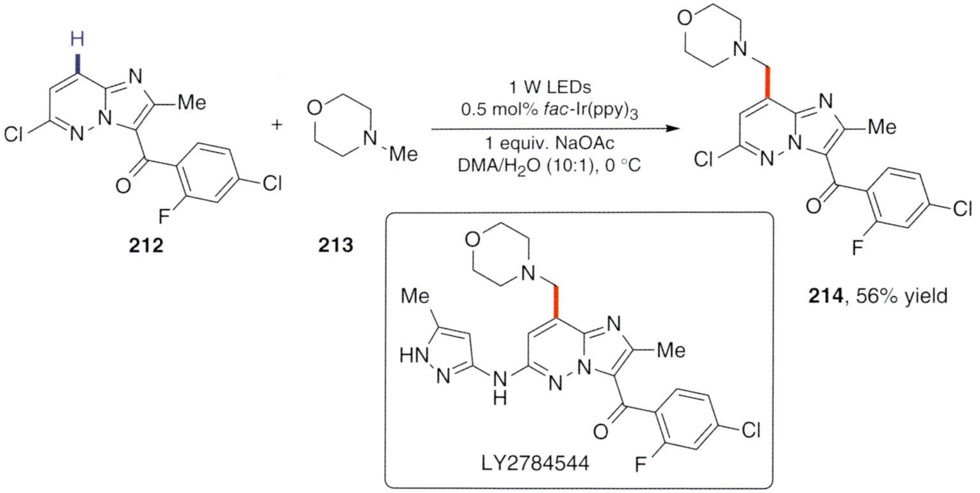

Scheme 13.29 Preparation of LY2784544 enabled by visible-light photoredox catalysis.

In 2014, MacMillan's group revealed reported the light-induced generation of α-oxyalkyl radicals from widely available ethers via a HAT pathway. This strategy was used for the radical coupling of ethers to a range of electron-deficient heteroarenes through a Minisci-type mechanism, affording high-value α-oxyalkylated heteroarenes (Scheme 13.30) [65]. A plausible mechanism for the reaction was proposed by MacMillan and coworkers. Oxidative quenching of the photoexcited Ir(III)* catalyst by the persulfate anion afforded Ir(IV), the sulfate dianion, and the sulfate radical anion, which abstracts a hydrogen atom from the

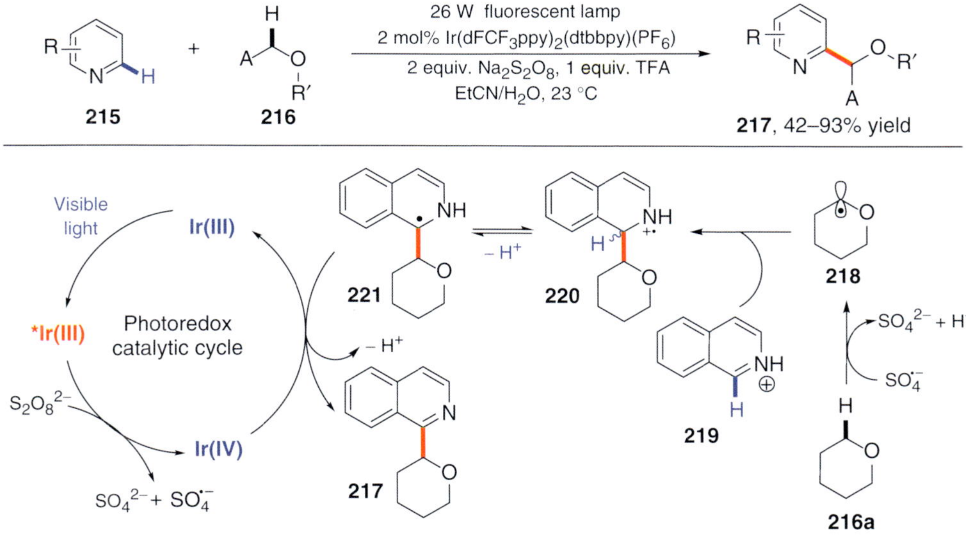

Scheme 13.30 Visible-light-mediated C–H oxyalkylation of electron-deficient heteroarenes.

ether **216a** to give the α-oxyalkyl radical **218**. Subsequent Minisci-type radical addition and oxidation affords the desired product.

The trifluoromethyl (CF_3) group is an important structural motif in many pharmaceutical molecules due to its unique chemical properties including lipophilicity, metabolic stability, and binding selectivity. The incorporation of the trifluoromethyl group into organic molecules can augment their biological activities. In 2011, MacMillan and coworkers described the direct trifluoromethylation of arenes and heteroarenes via a visible-light-mediated photoredox catalytic sequence (Eq. (13.23)) [66a]. It is worth noting that trifluoromethanesulfonyl chloride (CF_3SO_2Cl) **224**, a replacement for CF_3I, is employed as an effective precursor to the trifluoromethyl radical in order to diminish competition from aryl iodination reactions. A wide range of five- or six-atom heteroarenes were capable of performing this photocatalytic trifluoromethylation process, smoothly affording the corresponding coupling products in good to excellent yields.

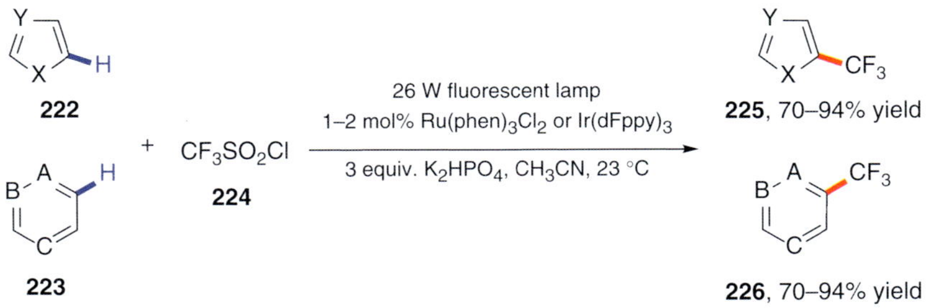

(13.23)

The authors have proposed a plausible mechanism for the trifluoromethylation reaction (Scheme 13.31). They assumed that a SET reduction of CF_3SO_2Cl **224** by the photoexcited catalyst gives the oxidatively active catalyst and a CF_3SO_2Cl radical anion **227**, which rapidly collapses to generate the CF_3 radical by the release of SO_2 and chloride. The CF_3 radical then adds to a heteroarene to form the radical **228**. Single-electron oxidation and subsequent deprotonation of the resulting cation **229** furnishes the trifluoromethyl-substituted heteroarene products.

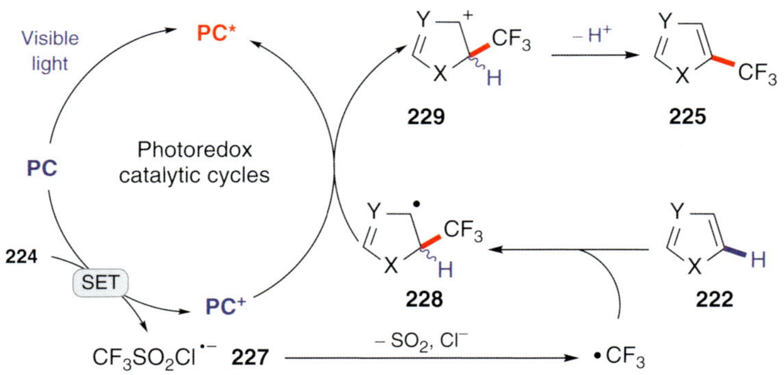

Scheme 13.31 Mechanism of C–H trifluoromethylation of heteroarenes.

Around the same time, the photocatalytic trifluoromethylation of electron-rich heteroarenes such as indole, furan, pyrrole, and thiophene using CF_3I as the trifluoromethane source was reported by Cho and coworkers (Eq. (13.24)) [66b].

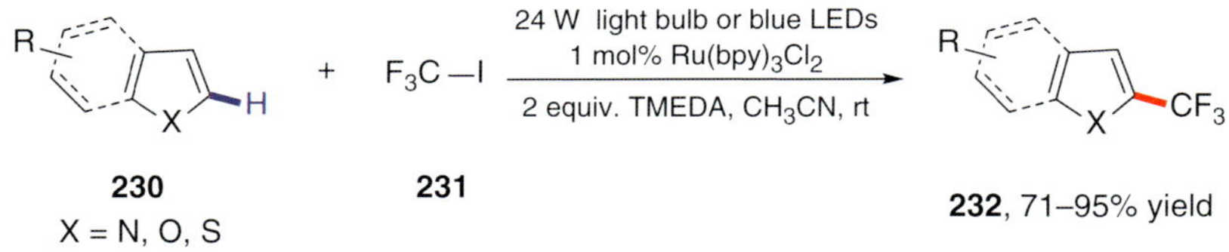

(13.24)

Recently, Blechert's group used a heterogeneous photocatalyst, mesoporous graphitic carbon nitride (mpg-CN), to achieve the trifluoromethylation of heteroarenes with CF_3SO_2Cl under visible-light irradiation (Scheme 13.32) [67].

The difluoromethyl group CF_2H is also an important structural motif in many pharmaceutical molecules. In 2014, Wang and coworkers developed a catalytic visible-light photoredox process for the difluoro(phenylsulfonyl)methylation of electron-rich N-, O-, and S-containing heteroarenes via an electrophilic radical-type pathway (Scheme 13.33) [68]. $PhSO_2CF_2I$ **236** was utilized as the difluoromethyl source to give a variety of substituted heteroarenes **237** in good to excellent yields. Desulfonylation of **237** proceeds smoothly, mediated by Mg at room temperature, to afford the corresponding difluoromethyl heteroarenes **238**.

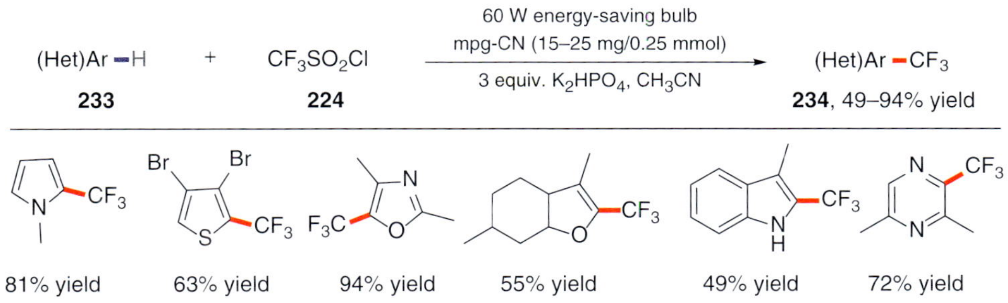

Scheme 13.32 Heterogeneous catalytic approaches for the C–H trifluoromethylation of heteroarenes.

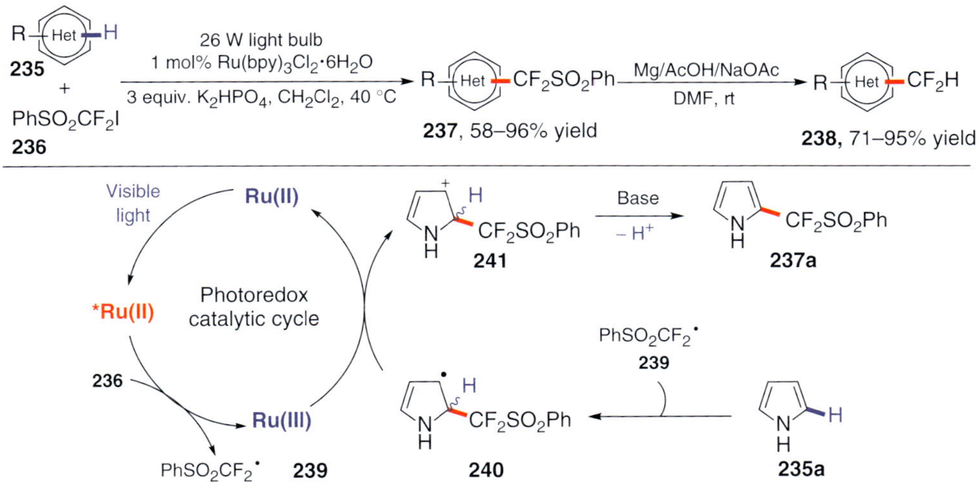

Scheme 13.33 Direct C–H difluoromethylation of electron-rich heteroarenes.

A possible pathway was proposed to explain the difluoromethylation reaction involving the key initial reduction of $PhSO_2CF_2I$ **236** by Ru(II)* followed by a single-electron oxidation to generate the difluoromethyl-substituted products.

Shortly after, Cho also demonstrated the direct C–H difluoromethylation of heteroarenes through a mild, efficient, and convenient method using visible-light photoredox catalysis (Scheme 13.34a) [69a]. A range of heteroarenes were difluoromethylated using $BrCF_2CO_2Et$ as the difluoromethylation reagent and *fac*-$Ir(ppy)_3$ as the photocatalyst with visible-light irradiation at room temperature. The same transformation was also achieved by Qing and coworkers (Scheme 13.34b) [69b] using $Ru(phen)_3Cl_2 \cdot xH_2O$ as the photocatalyst and *p*-methyl-C_6H_4-NMe_2 as reductive quencher.

Recently, Liu's group described the direct C–H difluoroacetamidation and difluoromethylenephosphonation of heteroarenes via visible-light photoredox

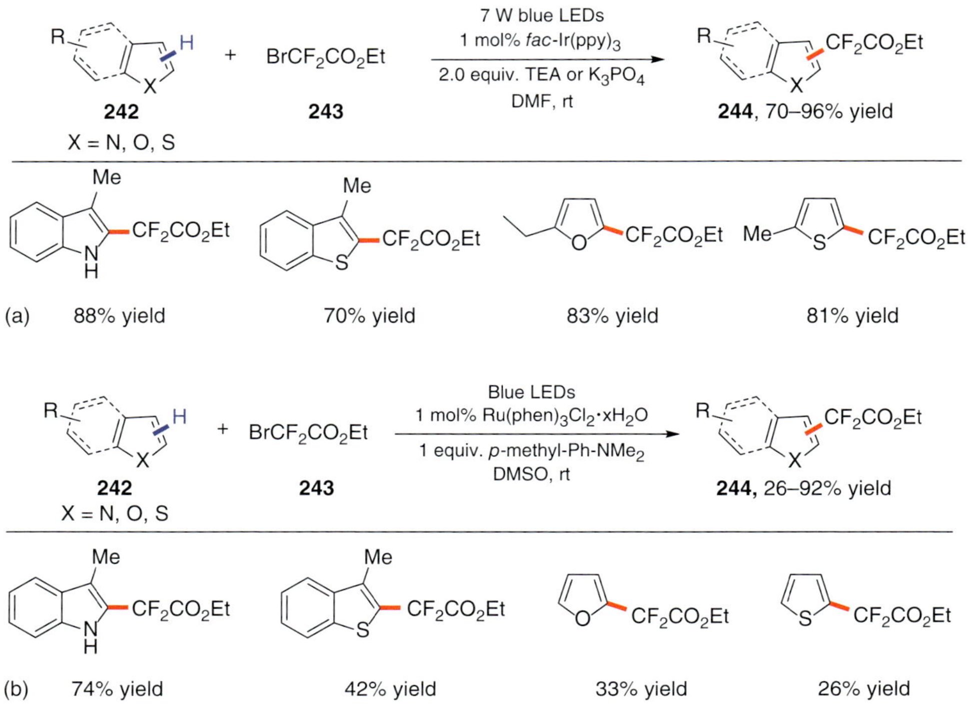

Scheme 13.34 Visible-light-mediated direct C–H difluoromethylation of heteroarenes. (a) Cho's work and (b) Qing's work.

catalysis (Eqs. (13.25) and (13.26)) [70]. Using commercially available bromodifluoroacetamides **246** and diethyl bromodifluoromethyl phosphonate **248** as precursors to the difluoromethyl radical, *fac*-Ir(ppy)$_3$ as photocatalyst, and a 3 W 450 nm LED light source, a variety of difluoromethyl heteroarenes were generated in moderate to excellent yields at room temperature.

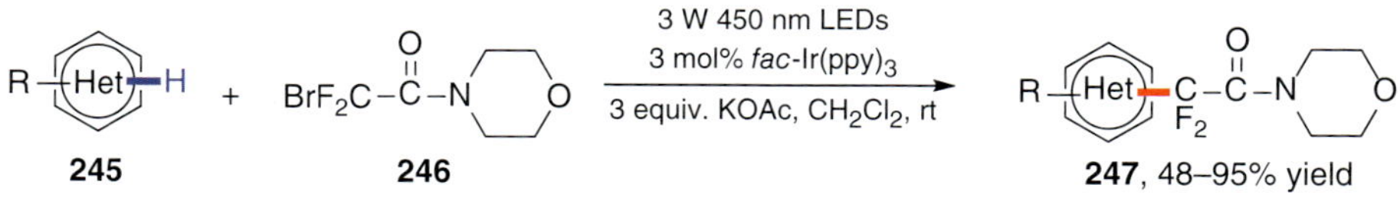

(13.25)

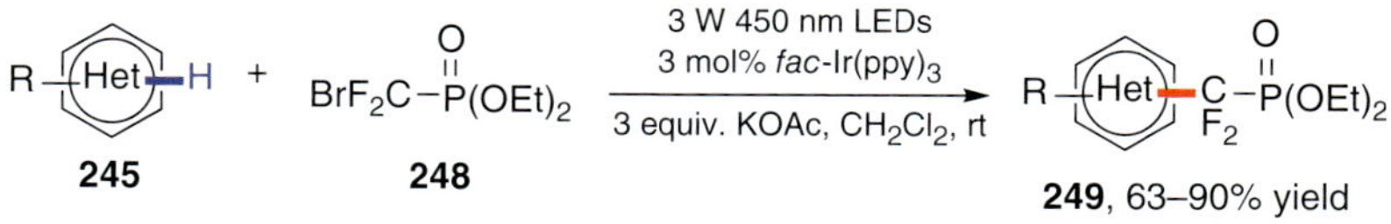

(13.26)

13.3.2
The Photocatalytic Arylation of Heteroarenes

Arylated heteroarenes are ubiquitous in pharmaceuticals, agrochemicals, natural products, and organic materials. Recently, direct C–H arylations of heteroarenes via visible-light photoredox catalysis have proven to be the most efficient method for the construction of aryl–heteroaryl bonds. In 2012, König and coworkers developed a metal-free, C–H arylation of a range of electron-rich heteroarenes (i.e., furans, pyrroles, and thiophenes) under mild conditions using readily available aryldiazonium salts as precursors for aryl radicals and the Eosin Y organic dye as a photocatalyst (Scheme 13.35) [71]. A plausible mechanism for the arylation involving oxidative quenching cycles has been proposed by König and coworkers. Two possible pathways for the generation of the carbocation intermediate **255** from radical adduct **254** were suggested: (a) oxidation by the Eosin Y radical cation and (b) oxidation by the aryl diazonium salts **251** in a radical chain transfer mechanism.

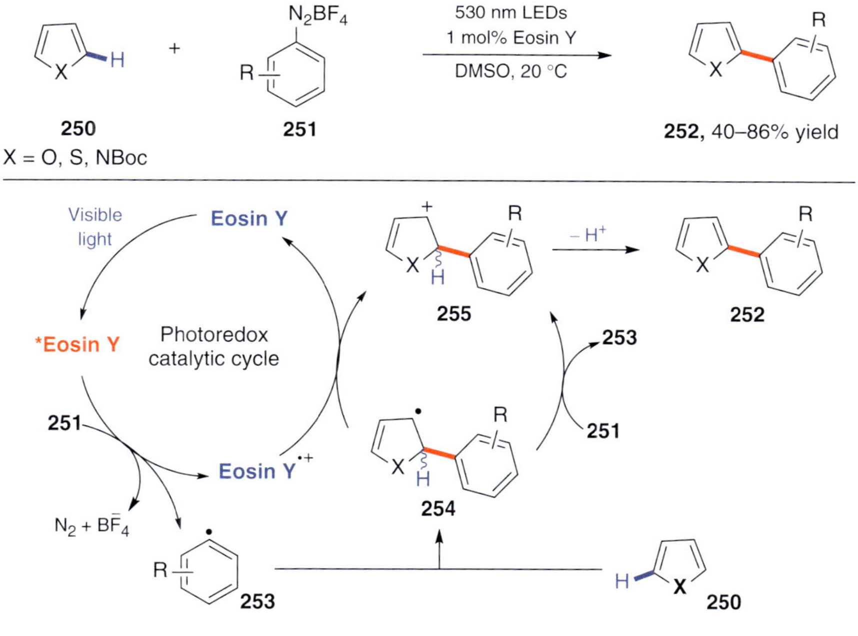

Scheme 13.35 Visible-light-mediated direct C–H arylation of electron-rich heterocycles.

Compared to processes with electron-rich heteroarenes, the direct C–H arylation of electron-deficient *N*-heteroarenes is often difficult. Recently, an efficient visible-light-induced arylation of electron-deficient *N*-heteroarenes with aryl diazonium salts has been developed by Xue (Scheme 13.36) [72]. Using $Ru(bpy)_3Cl_2 \cdot 6H_2O$ as photocatalyst, the reaction proceeds smoothly

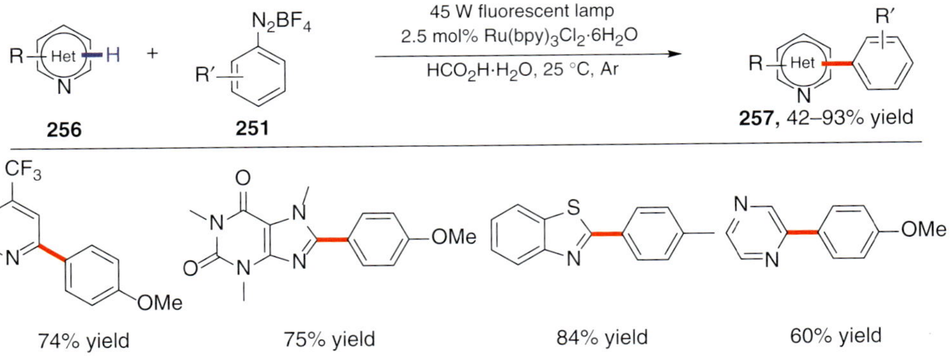

Scheme 13.36 Visible-light-mediated direct C–H arylation of *N*-heteroarenes.

at room temperature under irradiation provided by a commercial household lightbulb in water. The mild reaction conditions and use of water as the reaction medium make this transformation an environmentally friendly strategy for the construction of aryl–heteroaryl motifs.

Recently, Lei and coworkers described a visible-light-induced direct C–H arylation of isoquinoline with an aryldiazonium salt (Scheme 13.37) [73]. The arylation reaction was applied to a range of substituted isoquinoline and aryldiazonium salts with good functional group tolerance. The method was further applied to the total synthesis of the isoquinoline alkaloids menisporphine and daurioxoisoporphine C in four and five steps, respectively, from readily available starting materials.

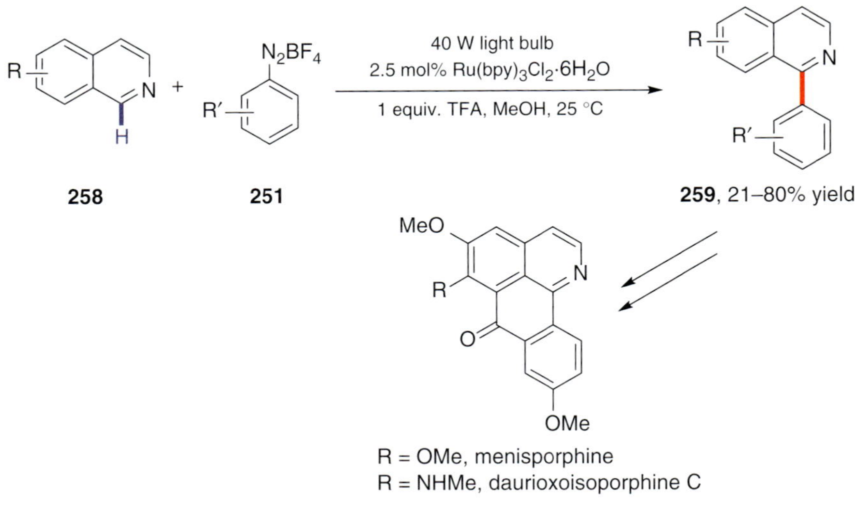

Scheme 13.37 Visible-light-mediated direct C–H arylation of isoquinoline.

13.3.3
The Photocatalytic Amination and Sulfuration of Heteroarenes

Construction of C–X bonds via a visible-light photocatalytic strategy is a growing field in the synthetic organic chemistry. Direct C–H amination of arenes and heteroarenes provides straightforward and streamlined access to aromatic amine moieties, which are widely found in pharmaceuticals, agrochemicals, and organic materials. In 2014, Sanford described a mild and efficient visible-light-induced photocatalytic method for the direct C–H amination of a variety of arene and heteroarene substrates (Scheme 13.38) [74]. The design of *N*-acyloxyphthalimides, which can be readily prepared from *N*-hydroxyphthalimide and carboxylic acid derivatives, as nitrogen-based radical sources is a key step toward enabling the advancement of this chemistry. Screening of a series of electronically varied *N*-acyloxyphthalimides, trifluoromethylacyloxyphthalimide **261** was found to be the best choice for the amination. This unique methodology provides an alternative mode of C–H activation for the syntheses of aromatic amines.

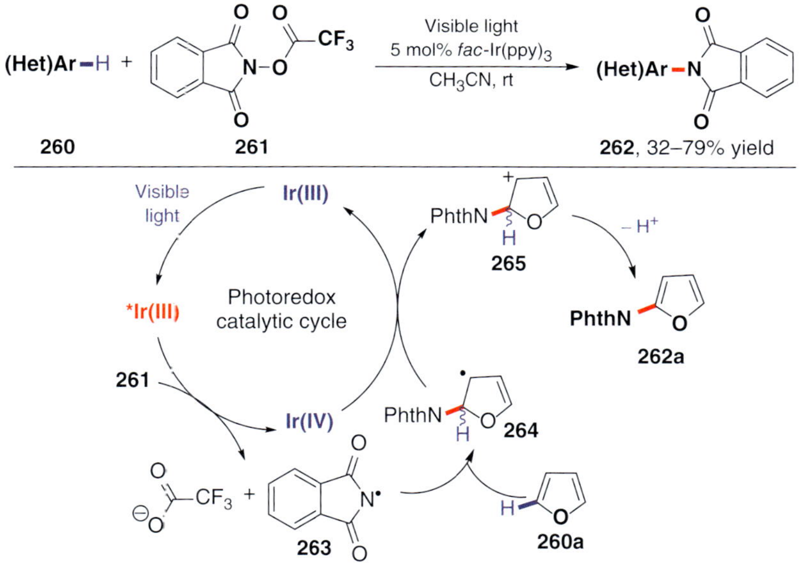

Scheme 13.38 Visible-light-mediated direct C–H amination of heteroarenes.

A possible catalytic cycle for this transformation, involving an oxidative quenching pathway, was proposed by the authors (Scheme 13.38). Single-electron transfer from photoexcited Ir(III)* to trifluoromethylacyloxyphthalimide **261** generates the *N*-centered radical (PhthN) **263**, trifluoroacetate anion, and Ir(IV). The addition of radical **263** to a heteroarene forms the carbon radical intermediate **264**, which is then oxidized by Ir(IV) with subsequent deprotonation providing the aminated product **262a**.

Recently, Lee and coworkers found that nitrogen-centered radicals could be generated from commercially available *N*-chlorophthalimide with low photocatalyst loadings. By this method, the direct C–H imidation of pyridines was achieved through visible-light-mediated photoredox catalysis (Scheme 13.39) [75].

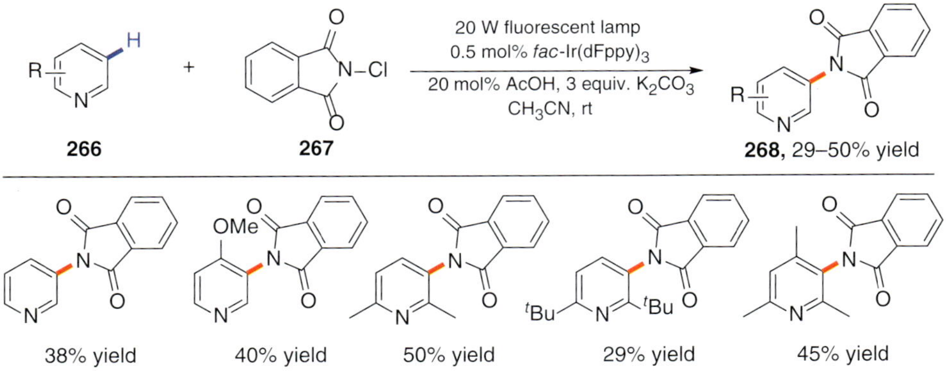

Scheme 13.39 Visible-light-mediated direct C–H amination of pyridines.

Soon after, Xue's group also demonstrated a visible-light photocatalytic transformation of chloroamines to nitrogen-centered radicals. This strategy was utilized for the direct C–H amination of benzoxazoles in a one-pot, two-step process (Eq. (13.27)) [76]. This one-pot process circumvents the needs for purification of the chloroamines, showing the potential of this amination for synthetic applications.

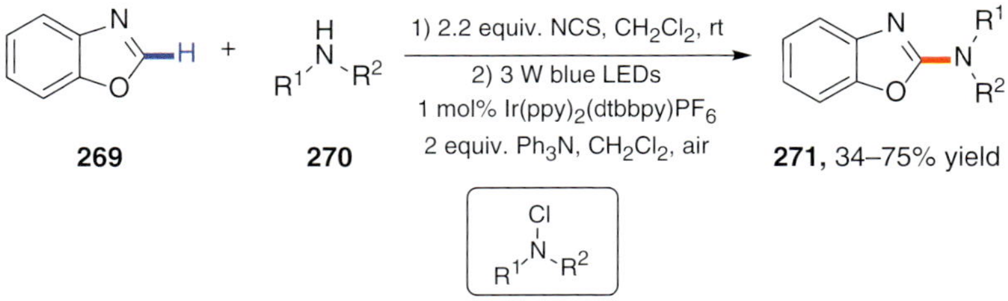

(13.27)

In 2014, Yu and coworkers used the hydroxylamine derivative *N*,*O*-ditosyl-*N*-methylhydroxylamine **273** as a tunable nitrogen source for the direct C–H amidation of heteroarenes through a visible-light-mediated photocatalytic method (Eq. (13.28)) [77a]. A variety of heteroarenes, such as indoles, furans, and pyrroles, smoothly underwent this amidation reaction at room temperature. Very recently, Studer successfully introduced *N*-aminopyridinium salts, which are readily prepared from cheap starting materials, as efficient *N*-centered radical precursors. These novel *N*-radical precursors can be used for the direct C–H amidation of heteroarenes via visible-light-induced photocatalysis employing $Ru(bpy)_3Cl_2$ as the photocatalyst (Eq. (13.29)) [77b].

Yu's work

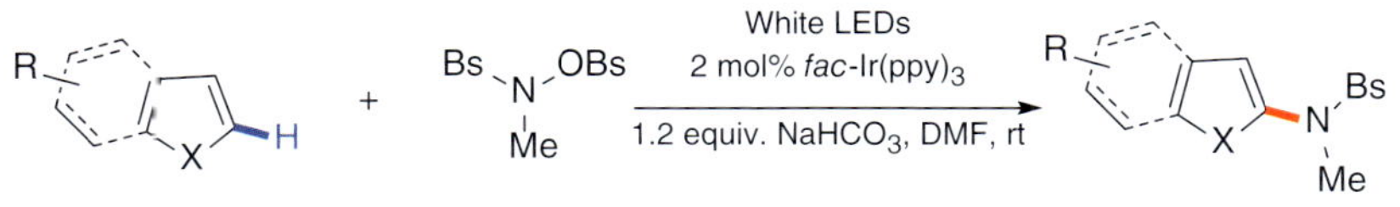

272 X = N, O **273** **274,** 47–98% yield

(13.28)

Studer's work

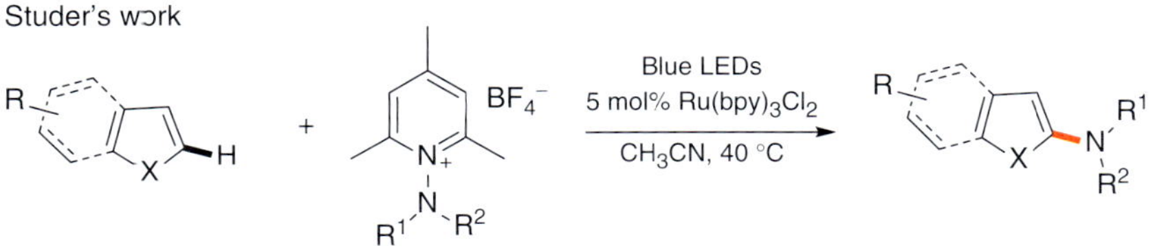

275 X = N, O, S **276** **277,** 46–96% yield

(13.29)

In addition, Luo's group has used a visible-light photocatalytic approach for the C–H imidation of a number of heteroarenes with bromosaccharin **279** as the nitrogen source (Scheme 13.40) [78]. This reaction proceeds via a radical chain pathway with termination by electron transfer–proton transfer (ET–PT).

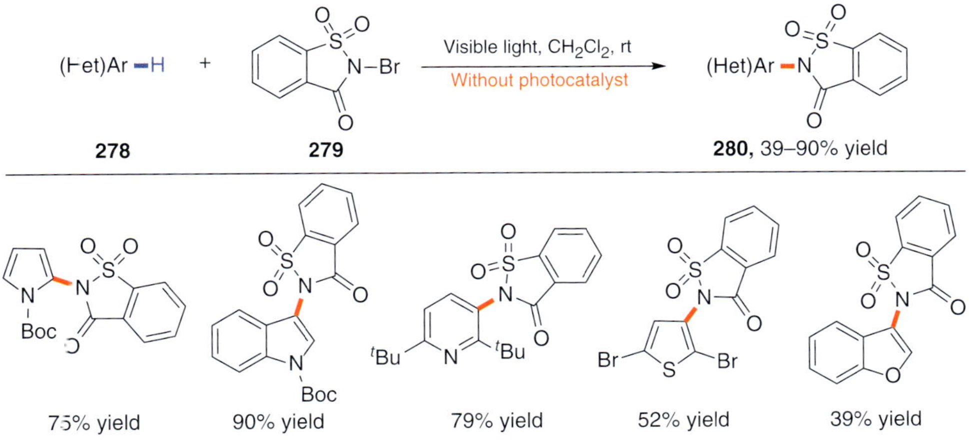

Scheme 13.40 Visible-light-mediated direct C–H imidation of heteroarenes without catalyst.

Recently, König reported the visible-light-mediated direct C–H amidation of electron-rich heteroarenes with benzoyl azides as aminating agents in the presence of phosphoric acid, affording a variety of heteroaromatic amides with good regioselectivities and yields (Scheme 13.41) [79]. This reaction proceeds smoothly using a $Ru(bpy)_3Cl_2 \cdot 6H_2O$ photocatalyst under visible-light irradiation with dinitrogen as the only by-product. A possible mechanism for the amidation reaction has been proposed involving initial irradiation of the Ru(II) photocatalyst with visible light to produce excited Ru(II)*, followed by energy transfer from Ru(II)* to benzoyl azides **282** to give triplet **284**. The loss of dinitrogen from **284** generates the benzoyl nitrene **286**, which converts into the corresponding isocyanate **285** via the Curtius rearrangement. The protonation of the benzoyl nitrene **286** under strongly acidic conditions affords the electrophilic nitrenium ion **287**. The reaction of the nitrenium ion **287** with heteroarene **281a** gives the carbenium ion intermediate **288**. Finally, the deprotonation and rearomatization of the carbenium ion intermediate **288** furnishes the amidation product.

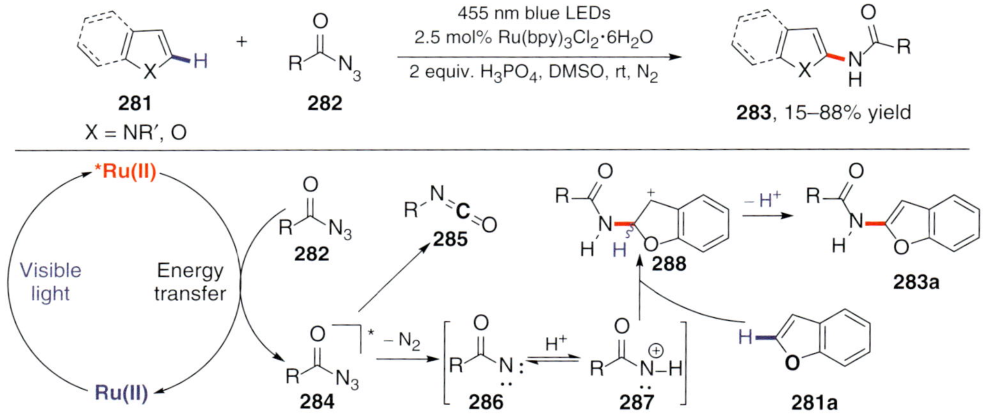

Scheme 13.41 Visible-light-mediated direct C–H amidation of electron-rich heteroarenes.

In 2012, Zheng and coworkers developed a mild and convenient approach for the construction of C–S bonds via visible-light photocatalysis (Scheme 13.42) [80]. This method allows for the synthesis of a range of 1-methyl-3-(arylthio)-1*H*-indoles **291** through the direct C–H sulfenylation of *N*-methylindoles **289** in moderate yields using readily available arylsulfonyl chlorides.

In 2014, Li's group developed an efficient visible-light-promoted aerobic method for the direct C3 thiocyanation of indoles (Scheme 13.43) [81]. This transformation occurs smoothly under metal-free conditions using the readily available Rose Bengal organic dye as the photocatalyst with environmentally benign air as the terminal oxidant. A plausible mechanism for this thiocyanation reaction starts with single-electron transfer between ammonium thiocyanate NH_4SCN **293** and photoexcited RB* affording radical **295** and an RB radical

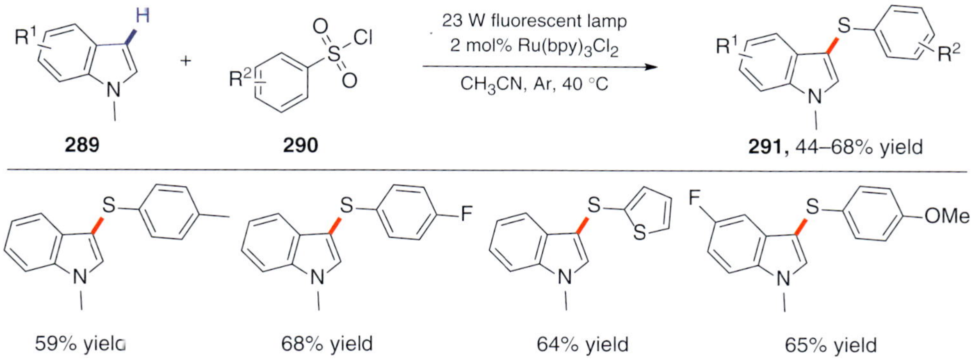

Scheme 13.42 Visible-light-mediated 3-sulfenylation of *N*-methylindoles.

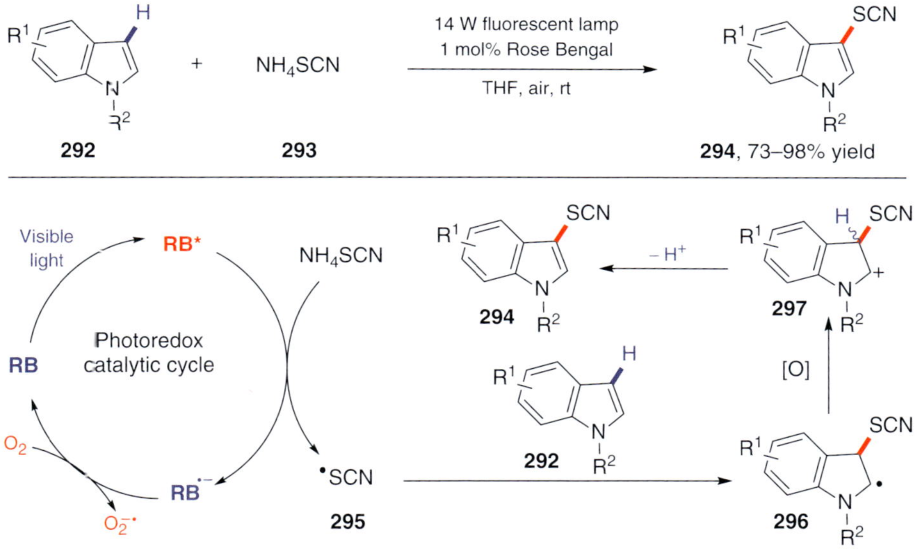

Scheme 13.43 Visible-light-mediated direct C–H thiocyanation of indoles.

anion. Subsequent oxidation of the RB radical anion by molecular oxygen regenerates the ground-state Rose Bengal. Electrophilic addition of radical **295** to an indole **292** gives a carbon radical intermediate **296**. The oxidation and subsequent deprotonation of the intermediate **296** affords the final 3-thiocyanoindole **294**.

13.4 Summary and Outlook

Since 2008, visible-light-induced photocatalysis has experienced vigorous development in hopes of its utilization as a "green chemistry" technique. Through this strategy, heterocycle syntheses involving C–H functionalization became widely studied and applied. In this chapter, we have briefly introduced these advancements in the *de novo* synthesis and functionalization of heterocycles via visible-light-induced photocatalytic C–H functionalization. Compared with traditional C–H bond functionalization, photoredox catalysis avoids harsh conditions such as high temperatures and catalyst loadings. As noted, almost all the described reactions take place at room temperature under irradiation from low-energy light sources. The development of such approaches toward heterocycle synthesis will undoubtedly make tremendous contributions to both researches in synthetic methodology and in industrial application.

Despite these tremendous advances, several challenges still remain to be solved. *First*, the construction of heterocycles is mainly based on the functionalization of the C–H bond adjacent to nitrogen atom due to its high-activity. Expanding the substrate diversity is a pressing need. *Second*, very few optically pure heterocycles have been synthesized through this strategy. *Third*, more theoretical and spectroscopic methods need to be incorporated into this research. Such work would provide insight into reaction mechanisms and allow for the rational design of new reactions. *Last, but not least*, further application of this reaction methodology to the syntheses of natural products and pharmaceuticals are required to show their full potential.

References

1. Ciamician, G. (1912) *Science*, **36**, 385–394.
2. (a) Zeitler, K. (2009) *Angew. Chem. Int. Ed.*, **48**, 9785–9789; (b) Yoon, T.P., Ischay, M.A., and Du, J. (2010) *Nat. Chem.*, **2**, 527–532; (c) Narayanam, J.M. and Stephenson, C.R.J. (2011) *Chem. Soc. Rev.*, **40**, 102–113; (d) Teplý, F. (2011) *Collect. Czech. Chem. Commun.*, **76**, 859–917; (e) Xuan, J. and Xiao, W.-J. (2012) *Angew. Chem. Int. Ed.*, **51**, 6828–6838; (f) Hari, D.P. and König, B. (2013) *Angew. Chem. Int. Ed.*, **52**, 4734–4743; (g) Prier, C.K., Rankic, D.A., and MacMillan, D.W. (2013) *Chem. Rev.*, **113**, 5322–5363; (h) Xi, Y., Yi, H., and Lei, A.-W. (2013) *Org. Biomol. Chem.*, **11**, 2387–2403; (i) Xuan, J., Lu, L.-Q., Chen, J.-R., and Xiao, W.-J. (2013) *Eur. J. Org. Chem.*, **2013**, 6755–6770; (j) Hari, D.P. and König, B. (2014) *Chem. Commun.*, **50**, 6688–6699; (k) Hopkinson, M.N., Sahoo, B., Li, J.-L., and Glorius, F. (2014) *Chem. Eur. J.*, **20**, 3874–3886; (l) Schultz, D.M. and Yoon, T.P. (2014) *Science*, **343**, 985–993.
3. Lednicer, D. (2008) *Strategies for Organic Drug Synthesis and Design*, John Wiley & Sons, Inc., Hoboken, NJ.
4. (a) Ackermann, L., Vicente, R., and Kapdi, A.R. (2009) *Angew. Chem. Int. Ed.*, **48**, 9792–9826; (b) Alberico, D., Scott, M.E., and Lautens, M. (2007) *Chem. Rev.*, **107**, 174–238; (c) Bellina, F. and Rossi, R. (2009) *Tetrahedron*, **65**, 10269–10310; (d) Seregin, I.V. and Gevorgyan, V. (2007) *Chem. Soc. Rev.*, **36**, 1173–1193.
5. (a) Xie, J., Jin, H., Xu, P., and Zhu, C. (2014) *Tetrahedron Lett.*, **55**, 36–48; (b) Shi, L. and Xia, W. (2012) *Chem. Soc. Rev.*, **41**, 7687–7697.
6. Condie, A.G., Gonzalez-Gomez, J.C., and Stephenson, C.R.J. (2010) *J. Am. Chem. Soc.*, **132**, 1464–1465.
7. (a) Hu, J., Wang, J., Nguyen, T.H., and Zheng, N. (2013) *Beilstein J. Org. Chem.*, **9**, 1977–2001; (b) Maity, S. and Zheng, N. (2012) *Synlett*, **23**, 1851–1856; (c) Rüping, M., König, R.M., and Atodiresei, I. (2013) in *Chemical Photocatalysis* (ed

B. König), Walter de Gruyter, Berlin, pp. 169–184.

8. (a) Zou, Y.-Q., Lu, L.-Q., Fu, L., Chang, N.-J., Chen, J.-R., and Xiao, W.-J. (2011) *Angew. Chem. Int. Ed.*, **50**, 7171–7175; (b) Rueping, M., Leonori, D., and Poisson, T. (2011) *Chem. Commun.*, **47**, 9615–9617.
9. Huang, L. and Zhao, J. (2013) *Chem. Commun.*, **49**, 3751–3753.
10. Guo, S., Zhang, H., Huang, L., Guo, Z., Xiong, G., and Zhao, J. (2013) *Chem. Commun.*, **49**, 8689–8691.
11. Xuan, J., Cheng, Y., An, J., Lu, L.Q., Zhang, X.X., and Xiao, W.J. (2011) *Chem. Commun.*, **47**, 8337–8339.
12. Xuan, J., Feng, Z.-J., Duan, S.-W., and Xiao, W.-J. (2012) *RSC Adv.*, **2**, 4065–4068.
13. Mathis, C.L., Gist, B.M., Frederickson, C.K., Midkiff, K.M., and Marvin, C.C. (2013) *Tetrahedron Lett.*, **54**, 2101–2104.
14. To, W.P., Liu, Y., Lau, T.C., and Che, C.M. (2013) *Chem. Eur. J.*, **19**, 5654–5664.
15. Deng, Q.-H., Zou, Y.-Q., Lu, L.-Q., Tang, Z.-L., Chen, J.-R., and Xiao, W.-J. (2014) *Chem. Asian J.*, **9**, 2432–2435.
16. Xie, J., Xue, Q., Jin, H., Li, H., Cheng, Y., and Zhu, C. (2013) *Chem. Sci.*, **4**, 1281–1285.
17. Hou, H., Zhu, S., Pan, F., and Rueping, M. (2014) *Org. Lett.*, **16**, 2872–2875.
18. Kohls, P., Jadhav, D., Pandey, G., and Reiser, O. (2012) *Org. Lett.*, **14**, 672–675.
19. Zhu, S., Das, A., Bui, L., Zhou, H., Curran, D.P., and Rueping, M. (2013) *J. Am. Chem. Soc.*, **135**, 1823–1829.
20. Zhang, P., Xiao, T., Xiong, S., Dong, X., and Zhou, L. (2014) *Org. Lett.*, **16**, 3264–3267.
21. Ju, X., Li, D., Li, W., Yu, W., and Bian, F. (2012) *Adv. Synth. Catal.*, **354**, 3561–3567.
22. Miyake, Y., Nakajima, K., and Nishibayashi, Y. (2012) *J. Am. Chem. Soc.*, **134**, 3338–3341.
23. Chen, L., Chao, C.S., Pan, Y., Dong, S., Teo, Y.C., Wang, J., and Tan, C.H. (2013) *Org. Biomol. Chem.*, **11**, 5922–5925.
24. Lenhart, D. and Bach, T. (2014) *Beilstein J. Org. Chem.*, **10**, 890–896.
25. Pandey, G., Pal, S., and Laha, R. (2013) *Angew. Chem. Int. Ed.*, **52**, 5146–5149.
26. Narayanam, J.M.R., Tucker, J.W., and Stephenson, C.R.J. (2009) *J. Am. Chem. Soc.*, **131**, 8756–8757.
27. Tucker, J.W., Narayanam, J.M., Krabbe, S.W., and Stephenson, C.R.J. (2010) *Org. Lett.*, **12**, 368–371.
28. Ju, X., Liang, Y., Jia, P., Li, W., and Yu, W. (2012) *Org. Biomol. Chem.*, **10**, 498–501.
29. Fu, W., Zhu, M., Xu, F., Fu, Y., Xu, C., and Zou, D. (2014) *RSC Adv.*, **4**, 17226.
30. Tucker, J.W. and Stephenson, C.R.J. (2011) *Org. Lett.*, **13**, 5468–5471.
31. Xiao, T., Dong, X., Tang, Y., and Zhou, L. (2012) *Adv. Synth. Catal.*, **354**, 3195–3199.
32. Dong, X., Xu, Y., Liu, J.J., Hu, Y., Xiao, T., and Zhou, L. (2013) *Chem. Eur. J.*, **19**, 16928–16933.
33. Cheng, Y., Yang, J., Qu, Y., and Li, P. (2012) *Org. Lett.*, **14**, 98–101.
34. Maity, S. and Zheng, N. (2012) *Angew. Chem. Int. Ed.*, **51**, 9562–9566.
35. Maity, S., Zhu, M., Shinabery, R.S., and Zheng, N. (2012) *Angew. Chem. Int. Ed.*, **51**, 222–226.
36. Hernandez-Perez, A.C. and Collins, S.K. (2013) *Angew. Chem. Int. Ed.*, **52**, 12696–12700.
37. Song, T., Zhou, B., Peng, G.W., Zhang, Q.B., Wu, L.Z., Liu, Q., and Wang, Y. (2014) *Chem. Eur. J.*, **20**, 678–682.
38. Blum, T.R., Zhu, Y., Nordeen, S.A., and Yoon, T.P. (2014) *Angew. Chem. Int. Ed.*, **53**, 11056–11059.
39. (a) Chen, J.-R., Yu, X.-Y., and Xiao, W.-J. (2015) *Synthesis*, **47**, 604–629; (b) Mai, W., Wang, J., Yang, L., Yuan, J., Mao, P., Xiao, Y., and Qu, L. (2014) *Chin. J. Org. Chem.*, **34**, 1958.
40. Xie, J., Xu, P., Li, H., Xue, Q., Jin, H., Cheng, Y., and Zhu, C. (2013) *Chem. Commun.*, **49**, 5672–5674.
41. Xu, P., Xie, J., Xue, Q., Pan, C., Cheng, Y., and Zhu, C. (2013) *Chem. Eur. J.*, **19**, 14039–14042.
42. Fu, W., Xu, F., Fu, Y., Zhu, M., Yu, J., Xu, C., and Zou, D. (2013) *J. Org. Chem.*, **78**, 12202–12206.
43. Tang, X.J., Thomoson, C.S., and Dolbier, W.R. Jr., (2014) *Org. Lett.*, **16**, 4594–4597.

44. Li, J.-H., Jiang, G.-F., Zhang, J.-L., Liu, Y., and Song, R.-J. (2014) *Synlett*, **25**, 1031–1035.
45. Liu, Y., Zhang, J.-L., Song, R.-J., and Li, J.-H. (2014) *Org. Chem. Front.*, **1**, 1289–1294.
46. Jiang, H., Cheng, Y., Wang, R., Zheng, M., Zhang, Y., and Yu, S. (2013) *Angew. Chem. Int. Ed.*, **52**, 13289–13292.
47. Sun, X. and Yu, S. (2014) *Org. Lett.*, **16**, 2938–2941.
48. Xiao, T., Li, L., Lin, G., Wang, Q., Zhang, P., Mao, Z.-W., and Zhou, L. (2014) *Green Chem.*, **16**, 2418.
49. Gu, L., Jin, C., Liu, J., Ding, H., and Fan, B. (2014) *Chem. Commun.*, **50**, 4643–4645.
50. He, Z., Bae, M., Wu, J., and Jamison, T.F. (2014) *Angew. Chem. Int. Ed.*, **53**, 14451–14455.
51. Yu, S., Zhang, Y., Wang, R., Jiang, H., Cheng, Y., Kadi, A., and Fun, H.-K. (2014) *Synthesis*, **46**, 2711–2726.
52. For selective examples, see: (a) Zou, Y.-Q., Duan, S.-W., Meng, X.-G., Hu, X.-Q., Gao, S., Chen, J.-R., and Xiao, W.-J. (2012) *Tetrahedron*, **68**, 6914–6919; (b) Lu, Z. and Yoon, T.P. (2012) *Angew. Chem. Int. Ed.*, **51**, 10329–10332.
53. Farney, E.P. and Yoon, T.P. (2014) *Angew. Chem. Int. Ed.*, **53**, 793–797.
54. Xia, X.-D., Xuan, J., Wang, Q., Lu, L.-Q., Chen, J.-R., and Xiao, W.-J. (2014) *Adv. Synth. Catal.*, **356**, 2807–2812.
55. Xia, X.D., Ren, Y.L., Chen, J.R., Yu, X.L., Lu, L.Q., Zou, Y.Q., Wan, J., and Xiao, W.-J. (2015) *Chem. Asian J.*, **10**, 124–128.
56. Fabry, D.C., Zoller, J., Raja, S., and Rueping, M. (2014) *Angew. Chem. Int. Ed.*, **53**, 10228–10231.
57. Zoller, J., Fabry, D.C., Ronge, M.A., and Rueping, M. (2014) *Angew. Chem. Int. Ed.*, **53**, 13264–13268.
58. Furst, L., Matsuura, B.S., Narayanam, J.M., Tucker, J.W., and Stephenson, C.R.J. (2010) *Org. Lett.*, **12**, 3104–3107.
59. Furst, L., Narayanam, J.M., and Stephenson, C.R.J. (2011) *Angew. Chem. Int. Ed.*, **50**, 9655–9659.
60. Kandukuri, S.R., Bahamonde, A., Chatterjee, I., Jurberg, I.D., Escudero-Adan, E.C., and Melchiorre, P. (2015) *Angew. Chem. Int. Ed.*, **54**, 1485–1489.
61. DiRocco, D.A., Dykstra, K., Krska, S., Vachal, P., Conway, D.V., and Tudge, M. (2014) *Angew. Chem. Int. Ed.*, **53**, 4802–4806.
62. Wang, Z.Q., Hu, M., Huang, X.C., Gong, L.B., Xie, Y.X., and Li, J.H. (2012) *J. Org. Chem.*, **77**, 8705–8711.
63. Zhu, S. and Rueping, M. (2012) *Chem. Commun.*, **48**, 11960–11962.
64. Douglas, J.J., Cole, K.P., and Stephenson, C.R.J. (2014) *J. Org. Chem.*, **79**, 11631–11643.
65. Jin, J. and MacMillan, D.W.C. (2015) *Angew. Chem. Int. Ed.*, **54**, 1565–1569.
66. (a) Nagib, D.A. and MacMillan, D.W. (2011) *Nature*, **480**, 224–228; (b) Iqbal, N., Choi, S., Ko, E., and Cho, E.J. (2012) *Tetrahedron Lett.*, **53**, 2005–2008.
67. Baar, M. and Blechert, S. (2015) *Chem. Eur. J.*, **21**, 526–530.
68. Su, Y.-M., Hou, Y., Yin, F., Xu, Y.-M., Li, Y., Zheng, X., and Wang, X.-S. (2014) *Org. Lett.*, **16**, 2958–2961.
69. (a) Jung, J., Kim, E., You, Y., and Cho, E.J. (2014) *Adv. Synth. Catal.*, **356**, 2741–2748; (b) Lin, Q., Chu, L., and Qing, F.-L. (2013) *Chin. J. Chem.*, **31**, 885–891.
70. (a) Wang, L., Wei, X.-J., Jia, W.-L., Zhong, J.-J., Wu, L.-Z., and Liu, Q. (2014) *Org. Lett.*, **16**, 5842–5845; (b) Wang, L., Wei, X.-J., Lei, W.-L., Chen, H., Wu, L.-Z., and Liu, Q. (2014) *Chem. Commun.*, **50**, 15916–15919.
71. Hari, D.P., Schroll, P., and König, B. (2012) *J. Am. Chem. Soc.*, **134**, 2958–2961.
72. Xue, D., Jia, Z.H., Zhao, C.J., Zhang, Y.Y., Wang, C., and Xiao, J. (2014) *Chem. Eur. J.*, **20**, 2960–2965.
73. Zhang, J., Chen, J., Zhang, X., and Lei, X. (2014) *J. Org. Chem.*, **79**, 10682–10688.
74. Allen, L.J., Cabrera, P.J., Lee, M., and Sanford, M.S. (2014) *J. Am. Chem. Soc.*, **136**, 5607–5610.
75. Kim, H., Kim, T., Lee, D.G., Roh, S.W., and Lee, C. (2014) *Chem. Commun.*, **50**, 9273–9276.
76. Xue, D., Wang, J.-D., Liu, Y.-X., Wang, C., and Xiao, J. (2014) *Synlett*, **25**, 2013–2018.

77. (a) Qin, Q. and Yu, S. (2014) *Org. Lett.*, **16**, 3504–3507; (b) Greulich, T.W., Daniliuc, C.G., and Studer, A. (2015) *Org. Lett.*, **17**, 254–257.

78. Song, L., Zhang, L., Luo, S., and Cheng, J.P. (2014) *Chem. Eur. J.*, **20**, 14231–14234.

79. Brachet, E., Ghosh, T., Ghosh, I., and König, B. (2015) *Chem. Sci.*, **6**, 987–992.

80. Chen, M., Huang, Z.T., and Zheng, Q.Y. (2012) *Chem. Commun.*, **48**, 11686–11688.

81. Fan, W., Yang, Q., Xu, F., and Li, P. (2014) *J. Org. Chem.*, **79**, 10588–10592.

14 Heterogeneous C–H Activation for the Heterocycle Synthesis

Lin He and Matthias Beller

14.1 Introduction

The design of green and economically feasible catalytic system for the production of fine chemicals is one of the priorities in the modern chemistry [1]. Advances in C–H activation contribute to this goal by offering a wealth of improved protocols for the construction of elaborated architectures [2]. In this context, various transition metal complexes have been developed for the intramolecular version of C–H activation, providing an atom-efficient method for the synthesis of heterocycles [3]. But the practical limitations with regard to catalyst handling, recyclability, and separation of the catalyst from products impede the transfer of these methodologies into large-scale industrial use. Economic and environmental incentives are driving forth the development of heterogeneous catalytic processes [4].

The ubiquity and inherent stability of C–H bonds have rendered the use of heterogeneous systems rather difficult. Normally, supported catalysts require more drastic reaction conditions than homogeneous counterparts. By using higher temperatures, the somewhat lower activities can be compensated to some extent. To obtain high efficiency, other approaches include the application of nontraditional activation methods, such as microwave irradiation and photochemistry. Especially, heterogeneous photocatalysis thus established offers numerous perspectives for the heterocycle synthesis via C–H activation processes.

In this chapter, we aim to give a picture of recent advances of C–H activation for the heterocycle synthesis enabled by heterogeneous catalysis. Scope, limitations, and problems associated with the application of heterogeneous catalysts will be covered. Mechanistic aspects of the catalysis are briefly addressed, but an extensive survey about this field is beyond the scope of this chapter. This limitation is due to the fact that the mechanisms are not yet fully understood and findings in this field are sometimes contradictory. To an extreme, in many cases it is not even clear whether the genuine catalysis is heterogeneous or homogeneous.

Transition Metal-Catalyzed Heterocycle Synthesis via C—H Activation, First Edition. Edited by Xiao-Feng Wu.
© 2016 Wiley-VCH Verlag GmbH & Co. KGaA. Published 2016 by Wiley-VCH Verlag GmbH & Co. KGaA.

14.2
Heterogeneous Pd-Catalyzed Heterocycle Synthesis via C–H Activation

As early as 1982, the first intermolecular C–H arylation catalyzed by Pd/C was reported by Nakamura and coworkers [5]. Until the 2005, Fagnou *et al.* [6] developed a versatile heterogeneous methodology to accomplish intramolecular arylation processes for the formation of five- and six-membered rings. By using the direct transformation of iodoarene to 6*H*-benzo[*c*]chromene as model reaction, the catalytic performance of different solid Pd catalysts was investigated (Scheme 14.1). $Pd(OH)_2/C$ (Pearlman's catalyst) gave an impressive activity and selectivity. PdS/C showed the similar yield but lower selectivity. Under the same conditions, the commonly used Pd/C gave poor reactivity, whereas Pd/Al_2O_3 resulted in an improved yield (53%). Other catalysts such as $Pd/BaCO_3$ and $Pd/BaSO_4$ were practically inactive. 10 mol% $Pd(OH)_2/C$ was employed as standard conditions, as irreproducible outcomes were frequently obtained at 5 mol%.

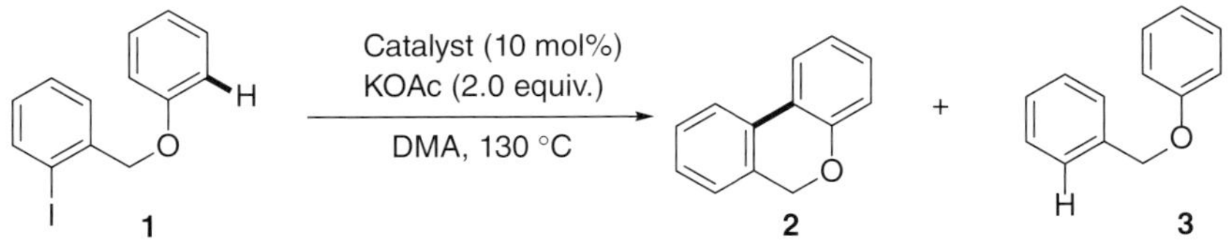

$Pd(OH)_2/C$	89%(2/3 45:1)
PdS/C	100% (2/3 8:1)
Pd/C	10%
Pd/Al_2O_3	53% (2/3 45:1)
$Pd/CaCO_3$, $Pd/BaCO_3$, $Pd/BaSO_4$	Trace

Scheme 14.1 Screening of heterogeneous catalysts.

Under the optimized conditions, the scope of intramolecular arylation was investigated (Table 14.1). Structurally diverse aryl iodide and bromide substrates underwent intramolecular arylation reactions and gave the desired products in good to excellent yield. High regioselectivity can also be observed in cases where direct arylation can occur at two chemically different arene positions as illustrated in entry 1. In this case, arylation occurs at the more sterically accessible position in a 13.5 : 1 ratio as determined by ^1H NMR analysis.

To verify whether the $Pd(OH)_2/C$ system was truly heterogeneous or not, a series of three-phase tests were carried out (Scheme 14.2). In the first case, the aryl halide was anchored to a Wang resin via an amide linker. Under standard conditions for 4.5 h, 100% conversion of supported substrate to arylated product was observed. In addition, in the presence of a silica-supported thiol-based scavenger resin (thiols are known to have a very high affinity for palladium), less than 5% of the desired product formed. These results led to the determination of

Table 14.1 Scope of $Pd(OH)_2$-C-catalyzed intramolecular direct arylation reactions.[a)]

Entry	Substrate	Product	Yield (%)[a)]
1	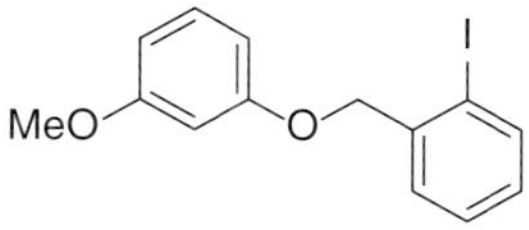	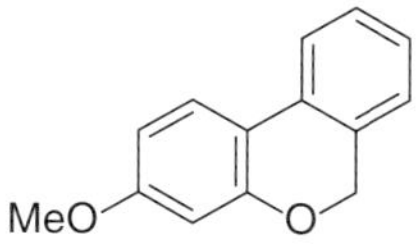	95
2	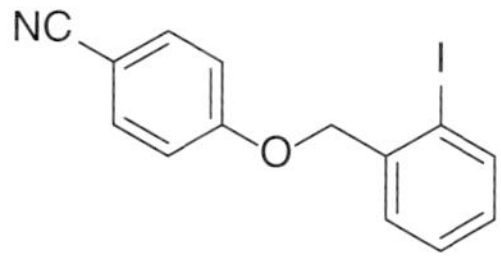	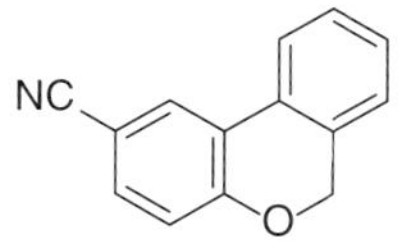	92
3	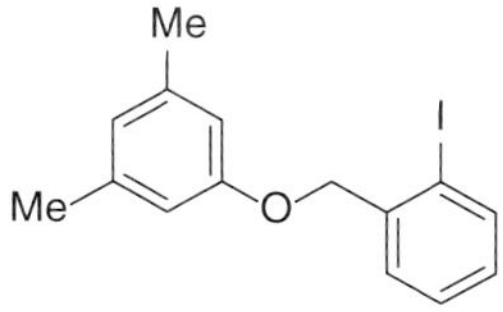	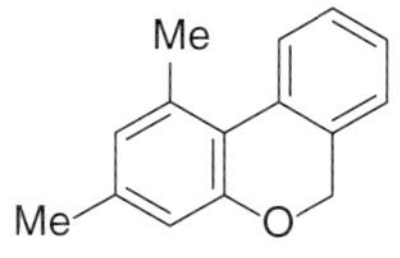	92
4	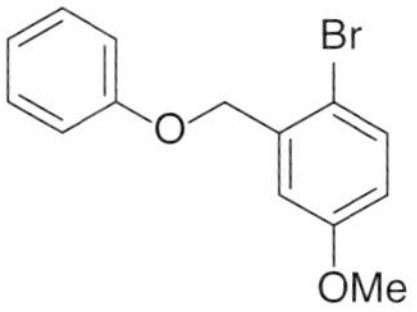	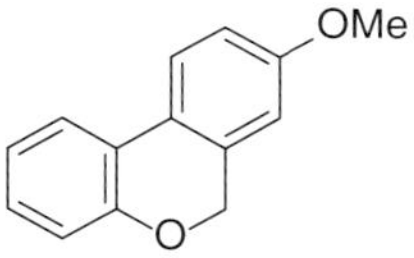	69[c)]
5	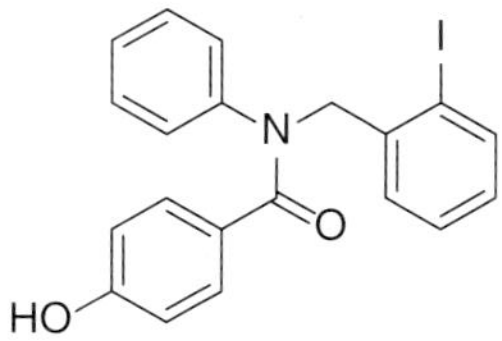	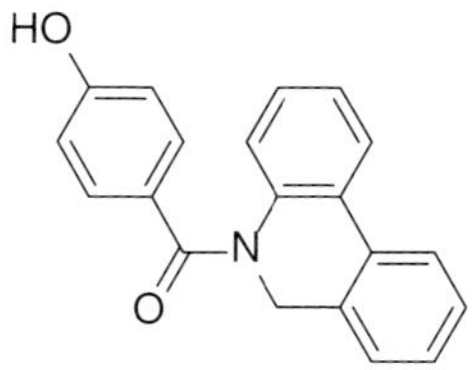	83
6	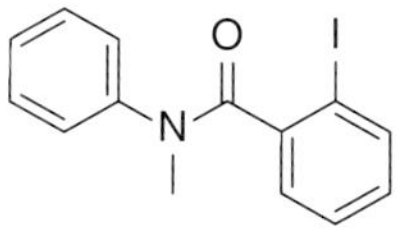	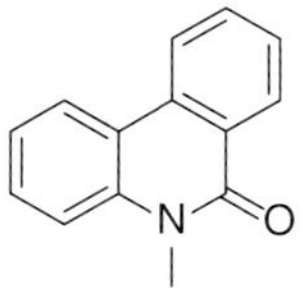	80
7	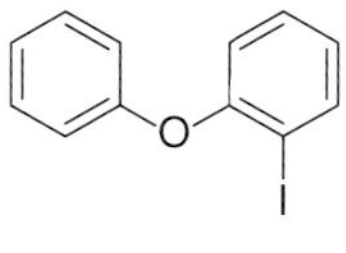	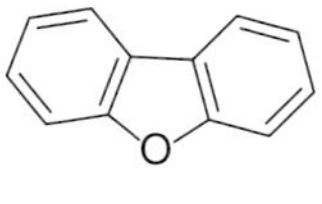	92
8	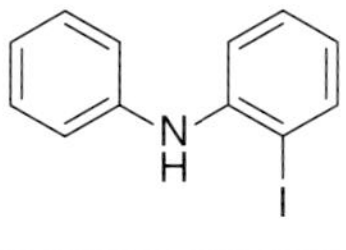	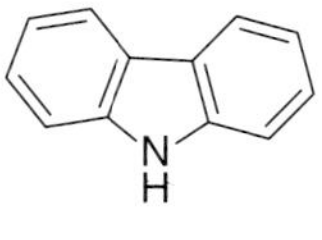	84

a) Conditions: Substrate, KOAc (2 equiv.), $Pd(OH)_2/C$ (10 mol%) added to a screw cap vial followed by DMA (0.2 M) and heating to 140 °C for 12–24 h.

b) Isolated yields. see ref. 6.

c) Twenty percent of the hydrodebrominated product was also obtained.

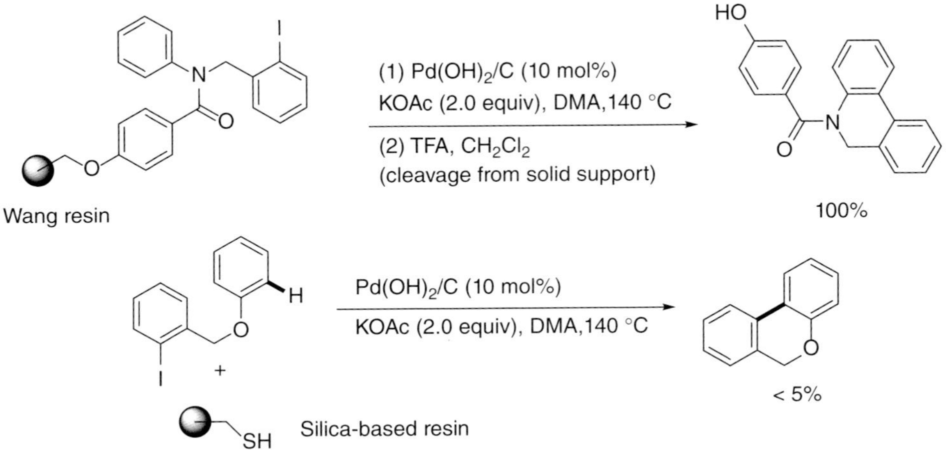

Scheme 14.2 Three-phase tests.

a mechanism in which leached homogeneous Pd species were responsible for the catalytic activity.

By contrast, Glorius and coworkers [7] have claimed that a truly heterogeneous catalysis was confirmed in the enantioselective C–H arylation of carbonyl compounds (including intramolecular version to afford indanone products in up to 85% ee, Scheme 14.3). In the arylation reaction of 2-methyl-1-tetralone and bromobenzene, Pd/Fe_3O_4/L (L = chiral *N*-heterocyclic carbine, Figure 14.1) performed as a versatile chiral nanocatalyst, furnishing the corresponding arylated product in 72% yield and with pronounced selectivity of 48% ee. Under same conditions, several homogeneous Pd–L complexes gave dramatically lower *ee* (Scheme 14.4). Standard Hg (0) poisoning test was also positive for

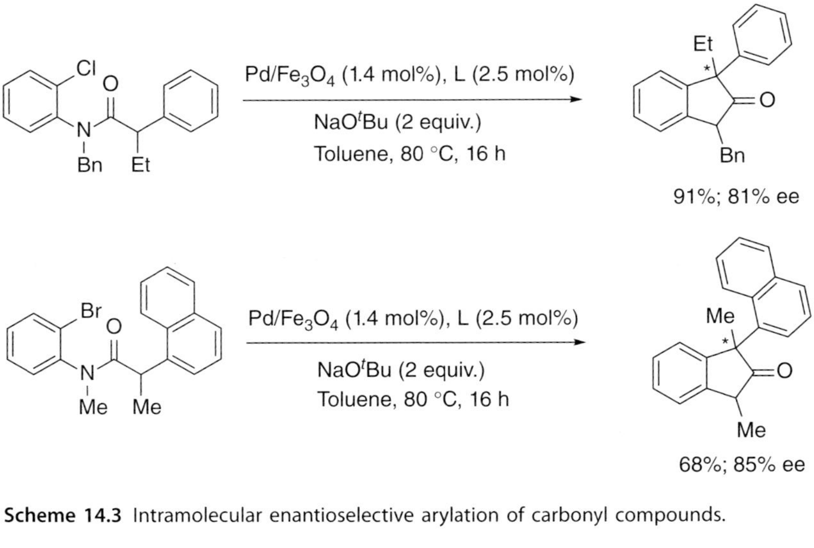

Scheme 14.3 Intramolecular enantioselective arylation of carbonyl compounds.

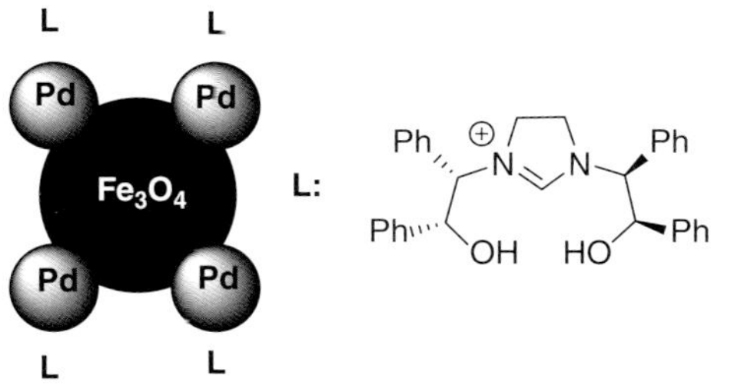

Figure 14.1 The $Pd/Fe_3O_4/L$ system.

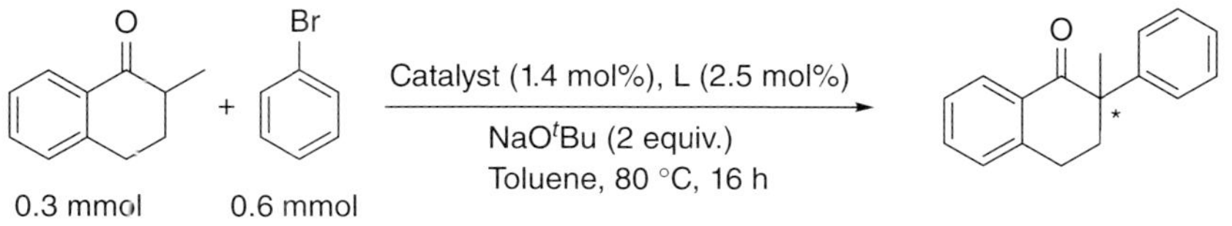

$Pd/Fe_3O_4/L$	72% yield, 48% ee
$Pd(OAc)_2/L$	52% yield, 1% ee
$Pd_2(dba)_3/L$	30% yield, 2% ee
$[Pd(allyl)Cl]_2/L$	76% yield, 1% ee

Scheme 14.4 Comparison $Pd/Fe_3O_4/L$ with homogeneous Pd–L complexes.

heterogeneous catalysis. Moreover, the leaching of the catalytically active $Pd/Fe_3O_4/L$ was negligible (0.232 ppm), whereas the leaching of Pd of the Pd/Fe_3O_4 into the solution was significant (16.6 ppm). The stabilizing effect of ligand to allow a heterogeneous mechanism was confirmed by only slightly catalytic activity loss of $Pd/Fe_3O_4/L$ in the fifth run. Although these methods evaluating the homo/heterogeneous nature of an active catalyst cannot be considered completely definitive, the results of experimental tests and literature precedent for the conclusions drawn strongly indicated that the active catalytic species was highly likely to be heterogeneous in nature.

Besides C–H arylation, intramolecular C–H amination has also emerged as an efficient synthetic method for the construction of heterocycles. The use of heterogeneous catalysts such as Pd nanomaterials for the oxidative amination has been reported by Ying group [8]. Eight supported palladium catalysts with different size and morphology were tested in the transformation of 2-acetaminobiphenyl to the corresponding carbazole. Using O_2 gas as the oxidant, the excellent isolated yield (92%) of the desired product was obtained at 120 °C (37 h) in the presence of 5 mol% Pd/C catalyst (Table 14.2, entry 4). Other mineral supports, such as SiO_2, CeO_2, TiO_2, and Al_2O_3 resulted in moderate to good yields of 78–86% (Table 14.2, entries 5–8). The Ag-Pd/C, Ag@Pd/C, and Ag_2S@Pd/C tested were not active for this reaction (Table 14.2, entries 1–3). DMSO was the solvent of choice. When the solvent was changed to other solvents DMF, DMA, 1,4-dioxane, and toluene, no product was obtained (Table 14.2, entries 9–12).

Table 14.2 C–N bond formation via C–H functionalization catalyzed by Pd nanomaterials.[a)]

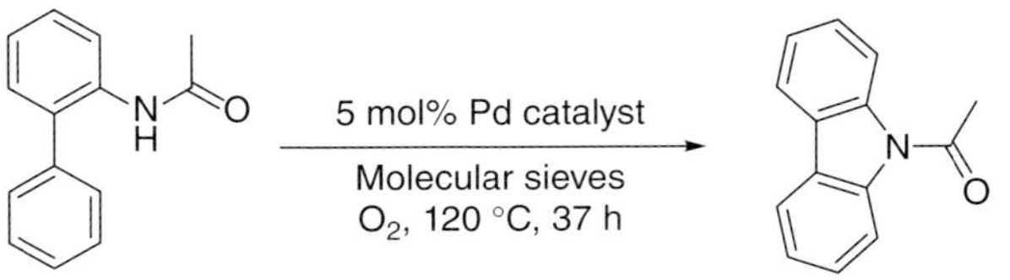

Entry	Catalyst	Pd loading (wt%)	Solvent	Yield[b)] (%)
1	Ag–Pd/C	4.0	DMSO	0
2	Ag@Pd/C	4.0	DMSO	0
3	Ag_2S@Pd/C	4.0	DMSO	0
4	Pd/C	5.6	DMSO	92
5	Pd/CeO_2	5.6	DMSO	86
6	Pd/TiO_2	5.6	DMSO	78
7	Pd/Al_2O_3	5.6	DMSO	82
8	Pd/SiO_2	5.6	DMSO	53
9	Pd/C	5.6	DMF	<1
10	Pd/C	5.6	DMA	<1
11	Pd/C	5.6	1,4-Dioxane	0
12	Pd/C	5.6	Toluene	<1

a) Conditions: 2-Acetaminobiphenyl (0.2 mmol), 4 Å molecular sieves (0.080 g), solvent (1.0 ml).
b) Isolated yields.

The recyclability of the Pd/C catalyst was also investigated. After one successive cycle, the catalytic activity decreased by 12% (80% isolated yield of product) and after two successive cycles, the catalytic activity decreased by 50%. TEM analysis showed that the fresh 2–5 nm Pd nanoparticles have increased to 20–30 nm after third run. To determine the active catalytic species in the C–H amination, a hot filtration test was performed. The reaction mixture was hot filtered at 37% yield of the product at 120 °C. Further stirring of the filtrate under the Table 14.2, in footnote (a) reaction conditions under O_2 for 14 h gave a final yield of 74%. In addition, inductively coupled plasma mass spectrometry (ICP-MS) analysis showed that 42% of the Pd was leached into the DMSO solution (probably as Pd–DMSO complexes) after heating the Pd/C catalyst at 120 °C under O_2 for 4 h. These results, together with similar performances of $Pd(OAc)_2$ catalysts, suggesting the observed catalysis is homogeneous.

A mechanism for the C–H amination catalyzed by the Pd/C nanomaterial was proposed (Scheme 14.5). Initially, soluble Pd(II) species could be leached into the solution from the surface defect sites of the Pd nanoparticles, which could be facilitated by the coordinating DMSO solvent and oxidation to Pd(II) by O_2 gas. Then, activation of the C–H bond adjacent to the amide went through directing group participation. After reductive elimination, the desired product is obtained and

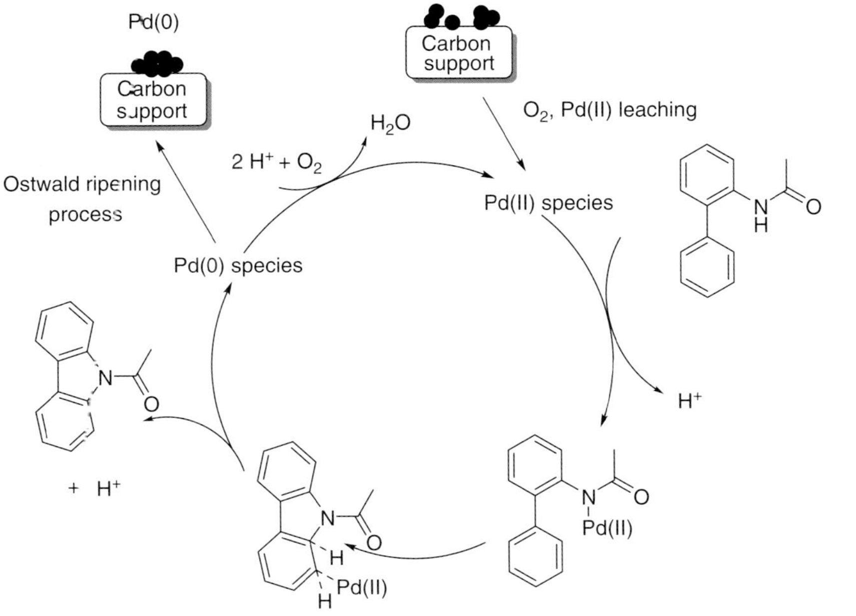

Scheme 14.5 Proposed mechanism for the C–N bond formation via C–H functionalization.

Pd(0) species can be reoxidized by O_2 to regenerate the Pd(II) species for the next catalytic cycle. Meantime, the dissolved Pd species would reattach to the existing Pd nanoparticles to form larger clusters via Ostwald ripening process.

The reaction was remarkably selective for the synthesis of a variety of carbazoles, regardless of the presence of electron-donor or electron-acceptor substituents (Scheme 14.6). Heteroaromatic groups were also investigated. However, these substrates could not be coupled to give the desired cyclized product.

Applying Pd/C as a heterogeneous catalyst, Wang *et al.* [9] developed the direct synthesis of isoquinolones from benzamides and alkynes through C–H activation (Scheme 14.7). A broad range of substituted benzamides and alkynes were used, and the products were obtained in good yields with excellent regioselectivities.

At last, recycling of the Pd/C catalyst was investigated. Desired product 2-methoxy-3,4-diphenylisoquinolin-1(2*H*)-one was obtained in yields of 92, 87, and 85% after the first, second, and third runs, respectively. To further demonstrate the efficiency and practicality of this reaction, a scale-up reaction was performed. Thus, gram-scale synthesis of 2-methoxy-3,4-diphenylisoquinolin-1(2*H*)-one was achieved in 88% yield.

Recently, Wu group synthesized a series of novel *N*-(2-pyridyl)indoles from alkynes and aniline derivatives via a heterogeneous Pd-catalyzed oxidative C–H activation reaction [10]. In the presence of Pd/CeO_2 catalyst, only catalytic amounts of Cu(II) and air as co-oxidant were required, delivering the desired indoles in moderate to excellent isolated yields (Scheme 14.8). As shown, both

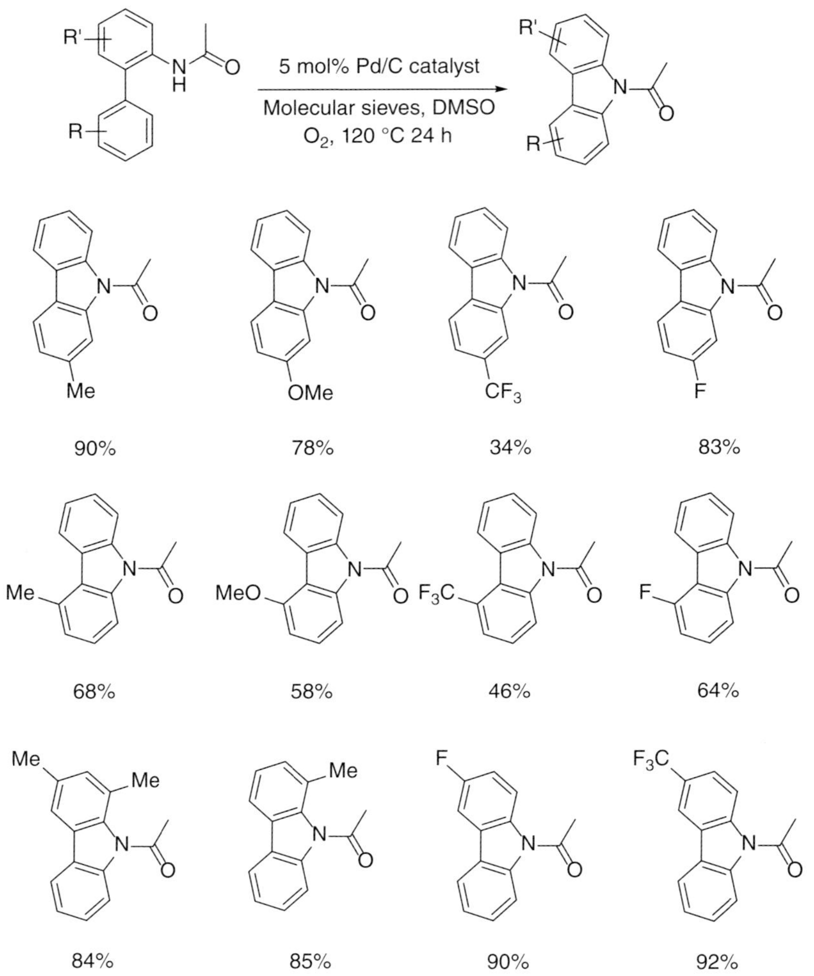

Scheme 14.6 Scope of C–N bond formation via C–H functionalization catalyzed by Pd/C.

electron-donating and electron-withdrawing groups are tolerated. Notably, bromide, a widely applied group in Pd-catalyzed coupling chemistry, can be kept intact and 89% of the corresponding indole was obtained. Furthermore, aliphatic alkynes were tested and good yields were achieved. Successively, diphenylacetylene was chosen to test different *N*-arylpyridin-2-amines. Methyl, methoxy, and methylthio as representative examples of electron-donating groups can be tolerated. Several electron-withdrawing substituted *N*-phenylpyridin-2-amines were used as well. In general, moderate to excellent yields (48–99%) of the desired products can be isolated. For meta-substituted substrates, poor yields were generally observed. When *N*-phenylpyrimidin-2-amine was tested with diphenylacetylene instead of *N*-phenylpyridin-2-amine, only a trace of the desired product was formed.

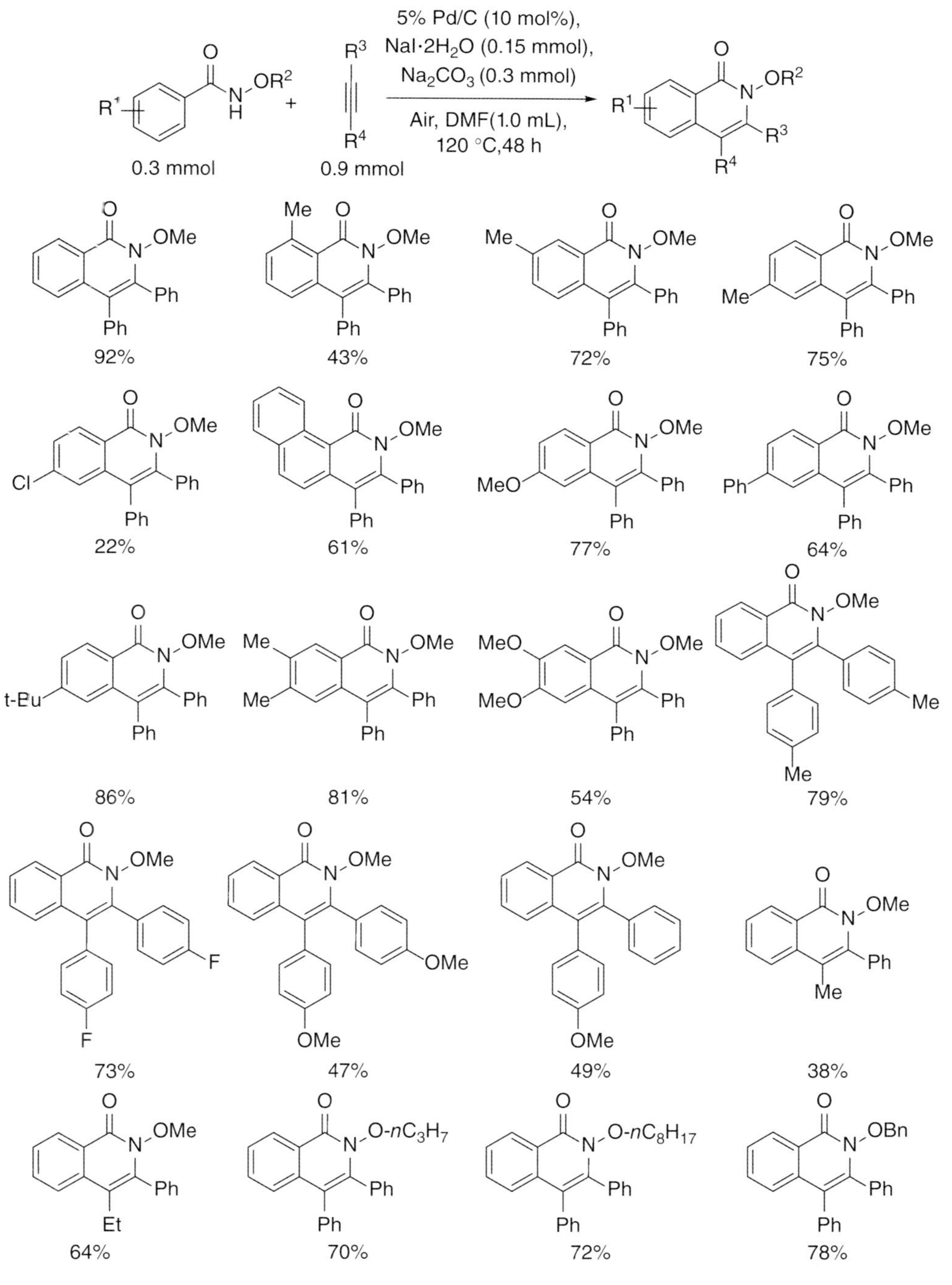

Scheme 14.7 Direct synthesis of isoquinolones from benzamides and alkynes.

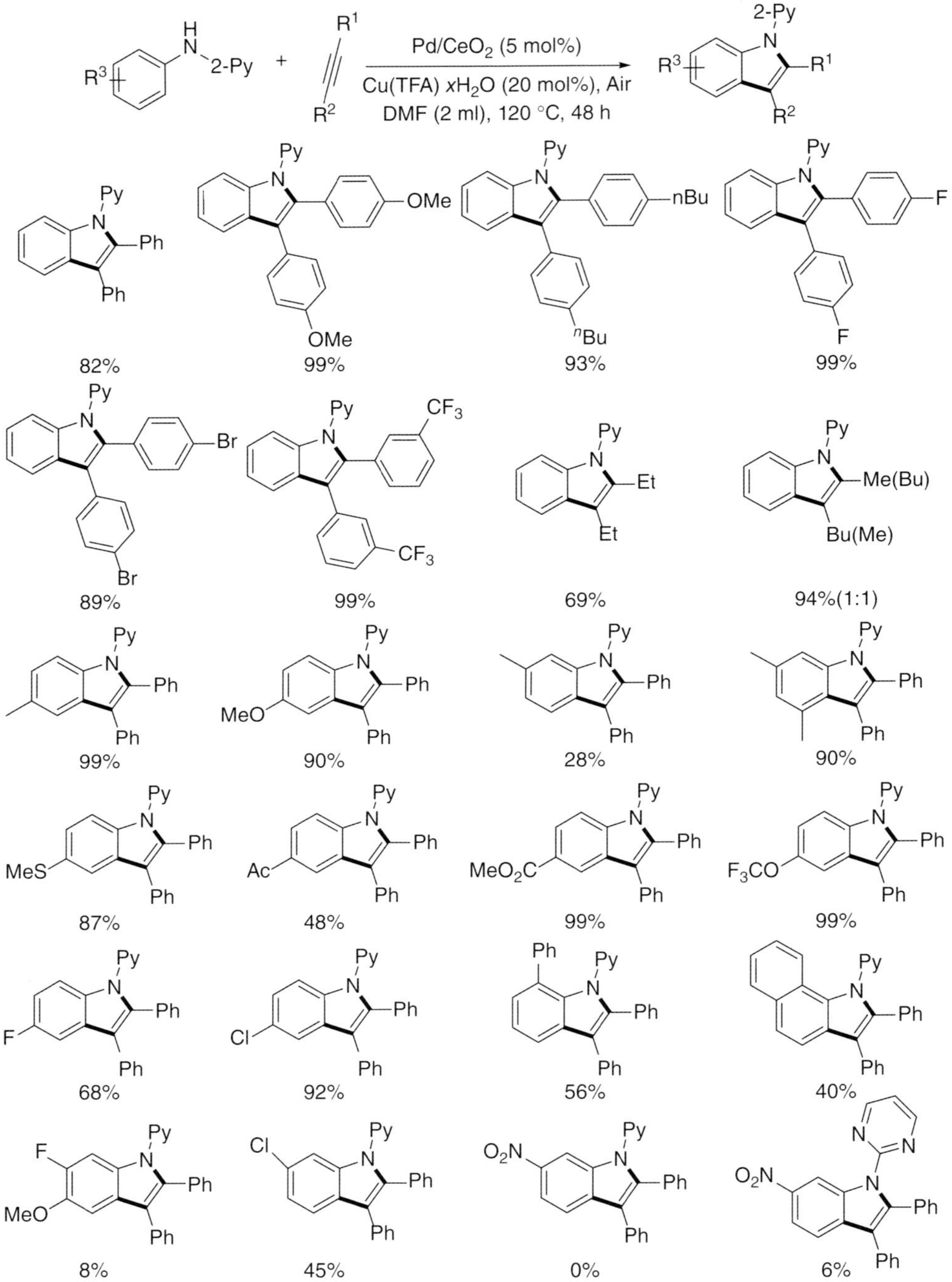

Scheme 14.8 Pd/CeO_2-catalyzed oxidative synthesis of indoles.

To get more understanding of the reaction, several control experiments were performed. In a Hg-poisoning experiment, the yield of the corresponding indole was decreased from 88% (without Hg addition) to 20% (with Hg addition). In addition, the recyclability of Pd/CeO_2 was also investigated. Unfortunately, a decrease in yield was obtained for the second (36%) run. It seems hard to say that the observed catalysis is heterogeneous or not. In kinetic isotope effect study, k_H/k_D was 1.0 (Scheme 14.9), suggesting that C–H activation might not act as the rate-determining step for the transformation.

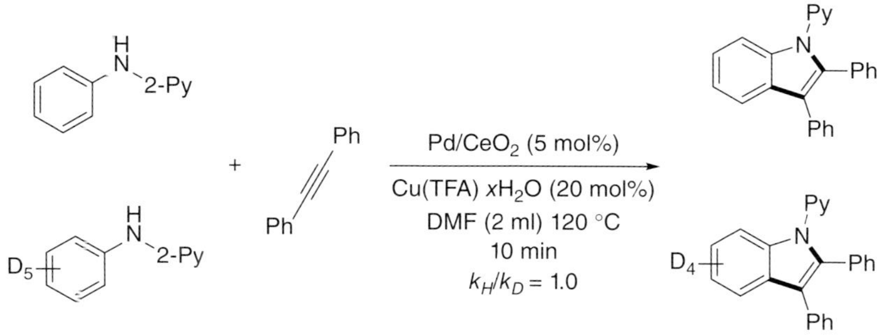

Scheme 14.9 Direct synthesis of isoquinolones from benzamides and alkynes.

With microwave-assisted technology, Porcheddu and colleagues [11] prepared benzimidazoles via one-pot C–H activation of tertiary amines/oxidative cyclization with 1,2-phenylenediamine in the presence of Pd/C (Scheme 14.10). To ascertain the heterogeneous character of the catalytically active species in this process, further leaching studies were performed. Both "Sheldon's hot filtration test" and atomic absorption spectroscopy analysis (ICP-MS) of the filtrate detected no significant quantities of leached Pd in solution, suggesting that the reaction was a heterogeneous process.

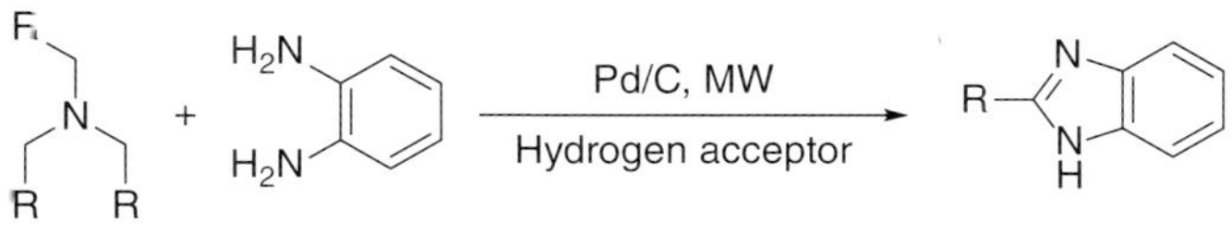

Scheme 14.10 Preparation of benzimidazole derivatives.

Proposed mechanism for the Pd/C-catalyzed oxidative synthesis of 2-substituted benzimidazoles was depicted in Scheme 14.11. The initial step seems to be the insertion of palladium into a C–H adjacent to the nitrogen, leading to a highly active intermediate complex of an iminium ion. The proposed catalytic cycle involves the removal of a hydride from tertiary amine by a metal catalyst to form (putative) oxidized iminium cation species, which undergoes a transamination reaction with o-phenylenediamine to generate intermediate imine, with generation of 1 mol equivalent of H_2. Subsequent formation of

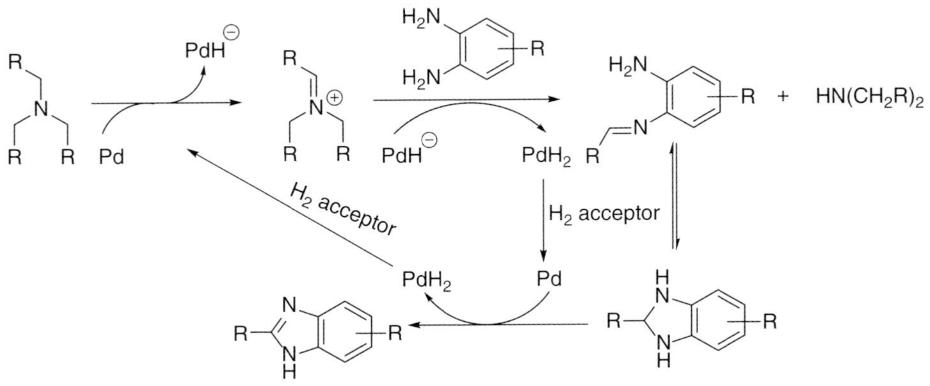

Scheme 14.11 Proposed mechanism for the oxidative synthesis of two-substituted benzimidazoles.

dihydrobenzazole and oxidation in the presence of a hydrogen acceptor generate desired heterocycles with regeneration of the Pd catalyst and formal overall transfer of two hydrogen molecules to crotonitrile.

14.3 Heterogeneous Photocatalysis for the Heterocycle Synthesis via C–H Activation

The combination of photocatalytic redox-cyclization reaction may help to develop a new strategy toward the heterocycle synthesis via C–H bond activation [12, 13]. TiO_2 as a redox active heterogeneous photocatalyst has gathered much attention because of its easy availability, high photochemical and thermal stability, and nontoxicity [14]. However, as TiO_2 has a band gap of ~3.2 eV, UV excitation was indispensable [15]. This is one of the main reasons why TiO_2 has not gained wide application in organic synthesis as transition metal complexes [16] and organic dyes [17] that have their absorption maxima in the visible region of the electromagnetic spectrum. To shift the band gap of TiO_2, modification of TiO_2 with metal oxide nanoclusters has been investigated. Tada and coworkers [18–20] found that surface modification of TiO_2 with such metal oxides as NiO, FeO_x, or CuO could greatly enhance the visible-light photocatalytic activity by narrowing the band gap and reducing the hole–electron pair recombination. Taking advantages of this, Shen and coworkers [21] used surface-modified titanium dioxide (TiO_2/NiO) as an efficient visible-light responsive photocatalyst in the direct cyclization between various tertiary anilines with maleimides to give tetrahydroquinoline products in moderate to high yields at ambient temperature (Table 14.3).

Experimentally, the catalyst is easily removed from the system after reaction by a simple centrifugal separation, and the catalyst was reused without further treatment. Remarkably, the photocatalyst could be efficiently recycled in nine consecutive catalytic cycles without significant loss of reactivity and selectivity.

Table 14.3 Visible-light-mediated cyclization of tertiary anilines and maleimides with TiO_2/NiO as photocatalyst.[a)]

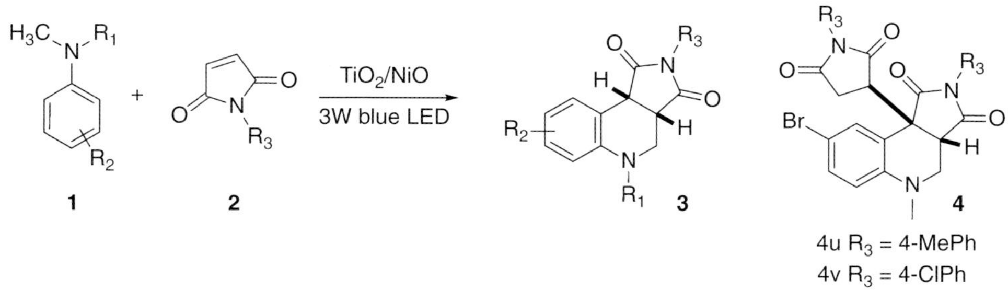

Entry	R_1	R_2	1	R_3	2	3	Yield[b)] (%)
1	Me	H	1a	Ph	2a	3a	75
2	Me	H	1a	4-MeOPh	2b	3b	64
3	Me	H	1a	4-MePh	2c	3c	89
4	Me	H	1a	4-FPh	2d	3d	81
5	Me	H	1a	4-ClPh	2e	3e	62
6	Me	H	1a	Me	2f	3f	67
7	Me	H	1a	t-Bu	2g	3g	69
8	Me	H	1a	Bn	2h	3h	68
9	Me	4-Me	1b	Ph	2a	3i	85
10	Me	4-Me	1b	4-MeOPh	2b	3j	68
11	Me	4-Me	1b	4-MePh	2c	3k	93
12	Me	4-Me	1b	4-FPh	2d	3l	86
13	Me	4-Me	1b	4-ClPh	2e	3m	73
14	Me	4-Me	1b	Me	2f	3n	73
15	Me	4-Me	1b	Bn	2h	3o	67
16	Me	4-MeO	1c	Ph	2h	3p	69
17	Me	2-Me	1d	Ph	2a	3q	64
18	Me	4-F	1e	Ph	2a	3r	86
19	Me	4-Cl	1f	Ph	2a	3s	75
20	Me	4-Br	1g	Ph	2a	3t	61
21	Me	4-Br	1g	4-MePh	2c	3u	68[c)]
22	Me	4-Br	1g	4-ClPh	2e	3v	52[d)]

a) Conditions: 1.25 mmol of 1a, 0.625 mmol of 2a, and 1 mol% TiO_2/NiO, in 25 ml of DMF, 3 W blue LED lamp, 12 h.
b) Yield after chromatographic purification.
c) A secondary product 4u was also formed in 10% yield.
d) A secondary product 4v was also formed in 8% yield.

Compared the used catalyst with the fresh one, no significant morphological changes with regard to the shape and structure of the catalyst were found.

A plausible reaction pathway for this visible-light-induced oxidative cyclization reaction catalyzed by surface-modified TiO_2 is proposed in Scheme 14.12. Single-electron transfer (SET) from the tertiary amine to the hole of the excited semiconductor catalyst leads to the amine cation radical A, while SET from the conducting band of the catalyst to ground state oxygen generates superoxide anion radical. Proton transfer from the amine cation radical A to the superoxide anion radical gives the α-aminoalkyl radical B and hydroperoxy radical. Radical addition of B to the maleimide, followed by intramolecular cyclization, leads to radical C. SET oxidation of C by oxygen or by the excited semiconductor catalyst to a carbocation, followed by deprotonation, furnishes the product. Meanwhile, the

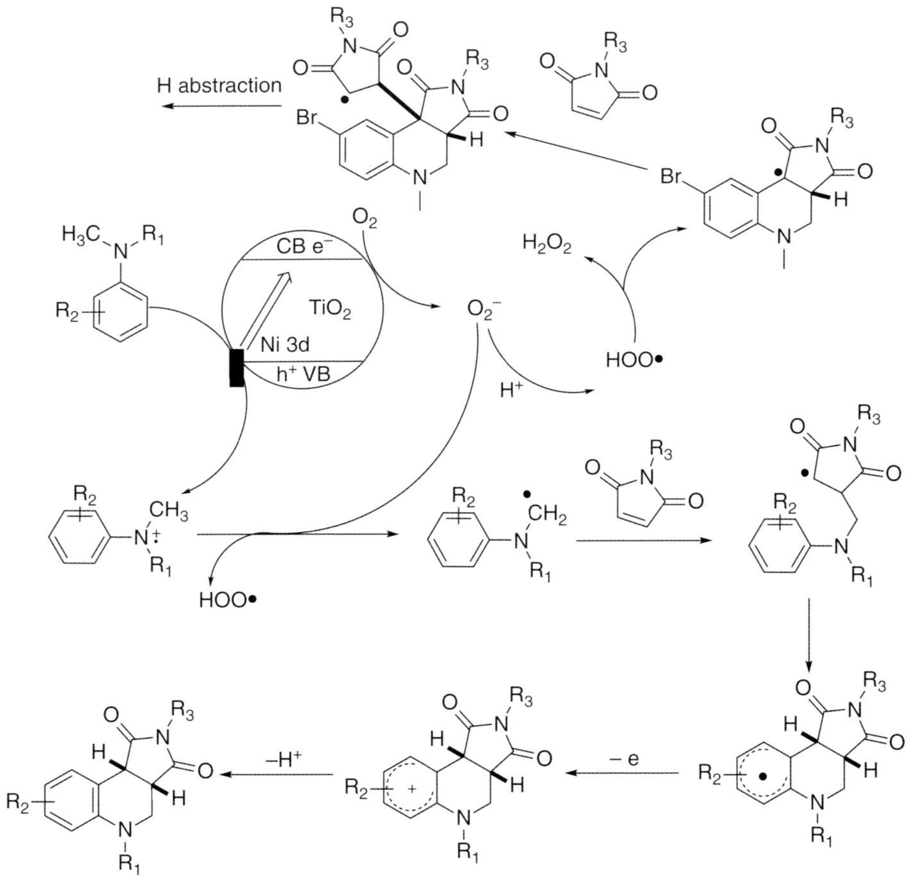

Scheme 14.12 Proposed reaction mechanism for the photoredox of tertiary anilines and maleimides with surface-modified TiO_2.

•OOH radical may also abstract a hydrogen atom from *N,N*-dimethylaniline to contribute another pathway to the formation of the α-aminoalkyl radical B.

Through a combined photoredox reaction, Swaminathan *et al.* [22] have demonstrated that Au–TiO_2 can enable efficient quinoline production from nitroarenes and ethanol. Scheme 14.13 provided a tentative overview of plausible reaction mechanisms leading to the cyclization products identified by GC-MS. The nitroaromatic compound was first reduced to aniline by photogenerated conduction band electron on the Au–TiO_2 surface. Simultaneously the alcohol is oxidized to the corresponding aldehyde consuming the photogenerated valance band holes on TiO_2. Pathways A and B suggest the formation of an imine (Schiff base) via the reaction between the aniline and acetaldehyde, whereas pathway C requires the formation of crotonaldehyde, the product of an aldol condensation of two acetaldehyde units. Since crotonaldehyde was produced in low concentration and consumed readily, it could not be detected by GC-MS under the experimental conditions. In fact, when crotonaldehyde was made to react with *N*-ethylideneaniline thermally, the reaction was completed within 4 h. More evidence for the reaction pathway B is the formation of tetrahydroquinoline, which is also detected by GC-MS. 2-Methyl-*N*-phenyl-1,2,3,4-tetrahydroquinolin-4-amine can be formed via cycloaddition of the imine to its enamine tautomer. Subsequent elimination of an aniline molecule yields the dihydroquinoline as intermediate. Further dehydrogenation gave the quinoline products.

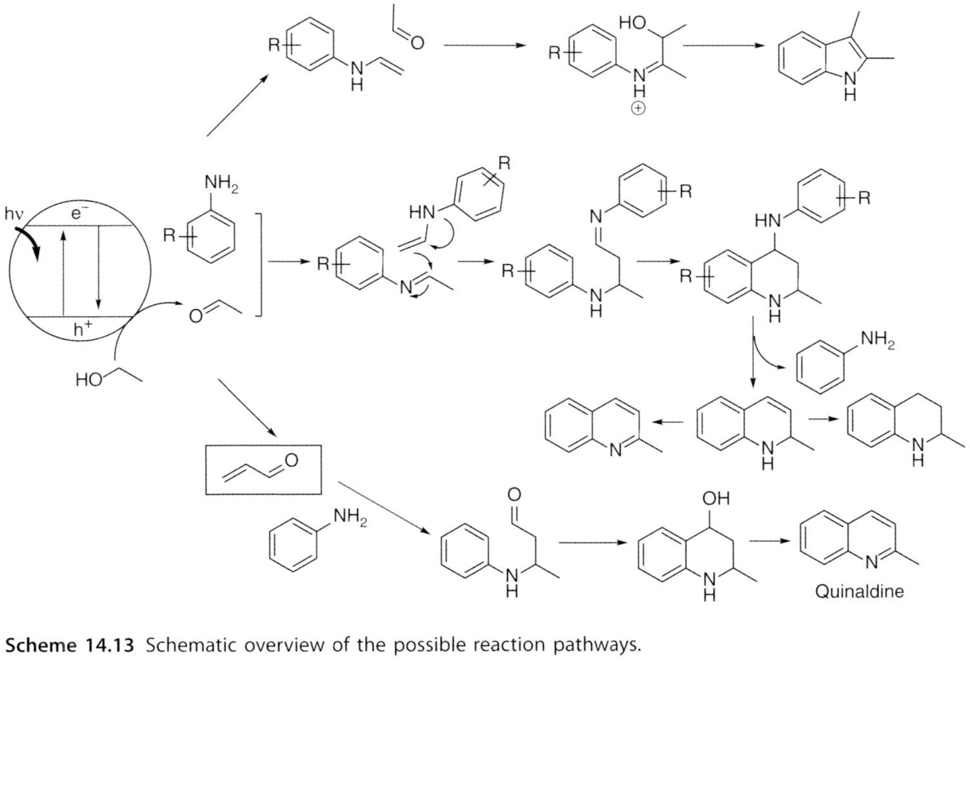

Scheme 14.13 Schematic overview of the possible reaction pathways.

14.4 Summary

This chapter featured recent developments in the discovery of heterogeneous catalytic procedures (mainly, palladium ones) for the synthesis of heterocycles via C–H activation. This research area, while mainly developed during the last 10 years, is still in its infancy and studied by only few research groups. So far, the majority of catalysts applied is commercial Pd/C catalysts. Other supported catalysts, which can allow tuning of the properties of the catalyst, have to be prepared by the user. We hope that this short chapter will stimulate further works in this direction. Future research in the detailed mechanistic studies is also desired to understand the sometimes contradictory results in this field. It has been increasingly accepted that most heterogeneous catalysts catalyze in a homogeneous way by leached metal species. However, general conclusions are difficult to attain because in most cases the mechanism of a certain reaction is affected by the conditions.

As radical chemistry is going through its renaissance, there is also a modern trend in heterogeneous photocatalyzed transformation involving C–H activation. To increase reaction rates, several studies devoted to elucidate the mechanistic details of photocatalytic reactions are essential and they are giving important information. As to the concern of the process economy, employment of sunlight is of course the ideal option.

References

1. Horvath, I.T. and Anastas, P.T. (2007) *Chem. Rev.*, **107**, 2167.
2. Alberico, D., Scott, M.E., and Lautens, M. (2007) *Chem. Rev.*, **107**, 174.
3. Yu, J.-Q. and Shi, Z.-J. (2010) *C-H Activation*, Springer, Berlin.
4. Djakovitch, L. and Felpin, F.-X. (2014) *ChemCatChem*, **6**, 2175.
5. Nakamura, N., Tajima, Y., and Sakai, K. (1982) *Heterocycles*, **17**, 235.
6. Parisien, M., Valette, D., and Fagnou, K. (2005) *J. Org. Chem.*, **70**, 7578.
7. Ranganath, K.V.S., Kloesges, J., Schäfer, A.H., and Glorius, F. (2010) *Angew. Chem. Int. Ed.*, **49**, 7786.
8. Chng, L.L., Yang, J., Wei, Y.-F., and Ying, J.Y. (2014) *Chem. Commun.*, **50**, 9049.
9. Shu, Z., Li, W., and Wang, B.Q. (2015) *ChemCatChem*, **7**, 605.
10. Chen, J., He, L., Natte, K., Neumann, H., Beller, M., and Wu, X.-F. (2014) *Adv. Synth. Catal.*, **356**, 2955.
11. De Luca, L. and Porcheddu, A. (2011) *Eur. J. Org. Chem.*, **2011**, 5791.
12. Prier, C.K., Rankic, D.A., and MacMillan, D.W.C. (2013) *Chem. Rev.*, **113**, 5322.
13. Narayanam, J.M.R. and Stephenson, C.R.J. (2011) *Chem. Soc. Rev.*, **40**, 102.
14. Lang, X., Chen, X., and Zhao, J. (2014) *Chem. Soc. Rev.*, **43**, 473.
15. Hoffman, M.R., Martin, S.T., Choi, W., and Bahnemann, D. (1995) *Chem. Rev.*, **95**, 69.
16. Xuan, J. and Xiao, W.-J. (2012) *Angew. Chem. Int. Ed.*, **51**, 6828.
17. Ravelli, D. and Fagnoni, M. (2012) *ChemCatChem*, **4**, 169.

18. Iwaszuk, A., Nolan, M., Jin, Q., Fujishima, M., and Tada, H. (2013) *J. Phys. Chem. C*, **117**, 2709.
19. Jin, Q., Ikeda, T., Fujishima, M., and Tada, H. (2011) *Chem. Commun.*, **47**, 8814.
20. Fujishima, M., Tanaka, K., Sakami, N., Wada, M., Morii, K., Hattori, T., Sumida, Y., and Tada, H. (2014) *J. Phys. Chem. C*, **118**, 8917.
21. Tang, J., Grampp, G., Liu, Y., Wang, B.-X., Tao, F.-F., Wang, L.-J., Liang, X.-Z., Xiao, H.-Q., and Shen, Y.-M. (2015) *J. Org. Chem.*, **80**, 2724.
22. Selvam, K. and Swaminathan, M. (2011) *Catal. Commun.*, **12**, 389.

15
Transition Metal-Catalyzed Carbonylative Synthesis of Heterocycles via C–H Activation

Jianbin Chen and Xiao-Feng Wu

15.1
Introduction

Carbonylation reactions have gained numerous attentions during the past decades since the pioneering work by Heck in 1974 with palladium catalyst [1]. Carbonylation reaction is the most potent methodology to produce carbonyl-containing compounds and increase the carbon chain at the meantime. For most of the cases, interests are mainly focusing on the utility of aryl halide or pseudohalides (ArX, X = I, Br, OTf, N_2BF_4, etc.) as the starting material for the formation of the carbon–metal bond via oxidative addition of metal catalyst to the C–X bond [2]. However the prefunctionalization of arene for the synthesis of the aromatic halides causes additional energy and time. At this point, the straightforward usage of C–H bond will be interesting.

Going through literature, the direct activation of the C–H bond has also experienced impressive improvements in methodology development [3]. Based on these achievements, the C–H activation under carbonylative conditions has also been exploited recently [4]. Compared to the well-defined carbonylation reaction of aryl halides, the carbonylative C–H bond functionalization still faces many challenges: (i) the reduction of Pd(II) to Pd(0) is more easy to occur under the CO atmosphere which has an excellent reducing ability, (ii) the strong binding ability of CO might inhibit the electronic attachment of Pd(II) toward the aromatic C–H bonds because of their occupancy at the active sites, and (iii) CO as an excellent π-acid behaving the unique π-backbonding might also decrease the electron density on palladium center. Efforts and solutions to solve those challenges will be discussed in the chapter.

Additionally, heterocyclic molecules are an important class of compound that is frequently encountered in the medicinal and pharmaceutical chemistry. And numerous efforts have been made to synthesize the core heterocycles in the recent years [2b, 5]. Carbonylative reactions have been applied in the synthesis of heterocycles as well [2b]. Here we wish to discuss the main progresses in the carbonylative synthesis of heterocycles via the C–H bond functionalization. To make

Transition Metal-Catalyzed Heterocycle Synthesis via C—H Activation, First Edition. Edited by Xiao-Feng Wu.
© 2016 Wiley-VCH Verlag GmbH & Co. KGaA. Published 2016 by Wiley-VCH Verlag GmbH & Co. KGaA.

it easier for the reader to understand about this chemistry, we compile this chapter in respect to the metal catalysts and then the size of heterocycle.

15.2
Cobalt-Catalyzed Heterocyclic Synthesis via Carbonylative C–H Activation

15.2.1
Five-Membered Ring Synthesis

Takahashi *et al.* reported that acetylenes could be transformed into indan-1-ones under water gas shift condition with cobalt as the catalyst in 1990 [6]. Nevertheless, the reaction proceeded at harsh conditions (100 atm of CO and 220 °C), and eight examples were provided in moderate to good yields (Table 15.1).

In order to obtain some detail about this transformation, a series of deuterium isotope-labeling experiments were conducted. By the treatment of diphenylacetylene with D_2O instead of H_2O, the deuterated indanone was

Table 15.1 Scope of the cyclocarbonylation of acetylene[a].

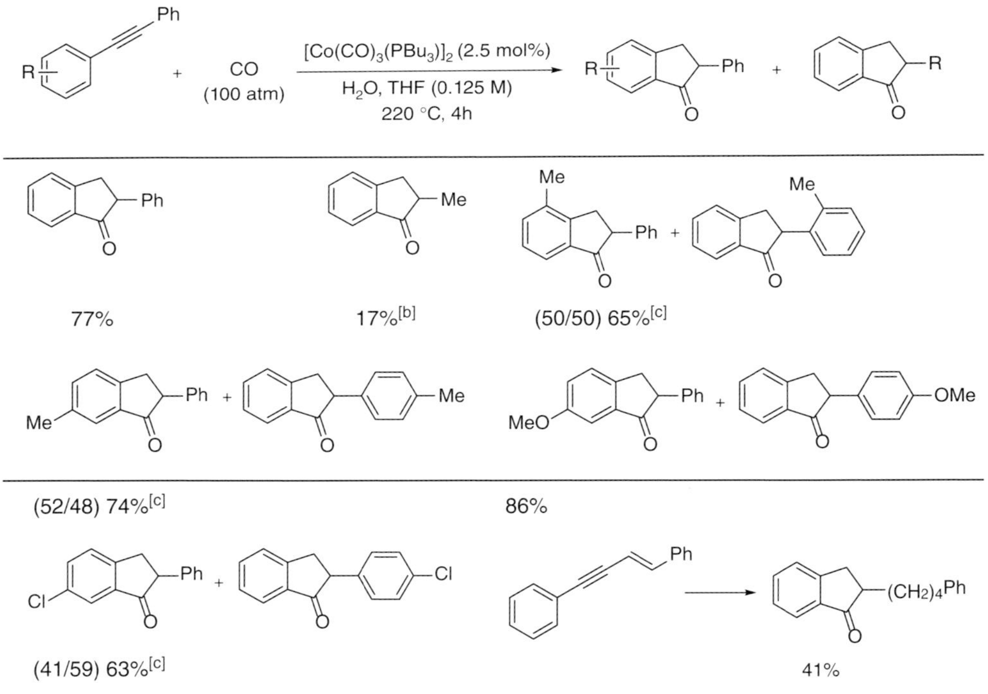

Reaction conditions: [a] **1** alkyne (5 mmol), $[Co(CO)_3(PBu_3)]_2$ (2.5 mol%), H_2O (11 equiv.), THF (40 ml), CO (100 atm), 220 °C, 4 h, isolated yields. [b] Determined by GLC. [c] Determined by ^1H NMR spectroscopy.

obtained indicating that hydrogen comes from water (Eq. (1)). The ratio of the H/D was in good agreement with the results of the reaction of deuterated phenylacetylene-*d*10 with water (Eq. (2)). Those data suggested that the hydrogen of aromatic C–H bond migrated to the acetylenic carbon of the diphenylacetylene during this cyclocarbonylation. By using molecular hydrogen as the hydrogen source, no desired indanone was obtained, whereas only dibenzyl was formed in good yield, thus indicating the water gas condition is essential for this transformation. When indanone was treated with the standard condition in the presence of D_2O, hydrogen in the methyne group of indanone was almost changed to the deuterium, while the methylene group was intact (Eq. (3)). Indenone also gave the deuterated indanone by the same treatment. The ^1H NMR spectrums show that the H/D ratio in indanone was 0.62:1.38, and the methyne group was entirely exchanged by deuterium (Eq. (4)) (Scheme 15.1).

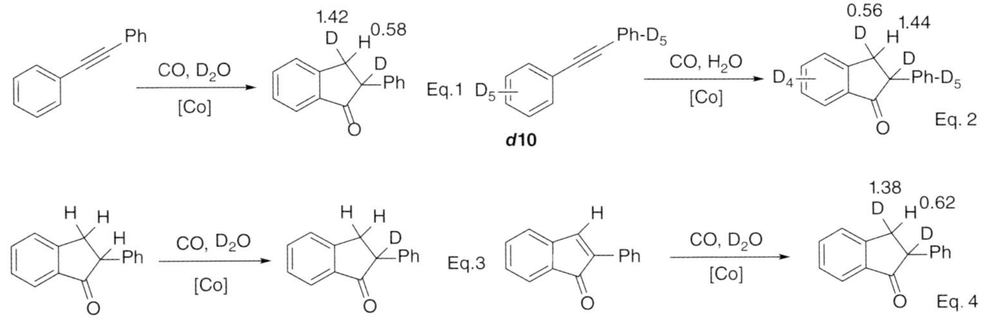

Scheme 15.1 Deuterium-labeling experiments.

A possible mechanism was outlined by the authors. Coordination of diphenylacetylene to the cobalt complex could induce an electronic attack of the central cobalt metal on the aromatic ring and migration of the aromatic hydrogen to the acetylenic carbon initialing the C–H metalation step. Insertion of CO into the cobaltocycle complex gives the indenone. The resultant indenone could be hydrogenated by the [Co]–H species formed under water gas shift reaction conditions to furnish the indanone formation as discussed previously.

The formation of *N*-alkyl isoindolin-1-one through the carbonylative C–H activation of *N*-alkyl benzaldimines with cobalt as the catalyst was achieved by Sen and coworkers in 2006 (Table 15.2) [7]. The author wants to exploit the formation of polypeptides from aldimines with CO; after the analysis of the products with GCMS, NMR, and X-ray, it turns out to be the isoindolin-1-one scaffolds. Reactions run under high temperature and pressure (100 °C and 69 bar of CO), and deuterium experiments indicate that the [Co]–H species was generated by the C–H metalation step of the aromatic ring, followed by the insertion to the C=N bond of the benzaldimine affording the final products.

In 2014, a cobalt-catalyzed carbonylative synthesis of phthalimide motifs from benzamides was demonstrated by Daugulis and coworker [8]. Here it uses the 8-aminoquinoline as the *N*,*N*-bidentate directing system. Reactions

Table 15.2 Scope of the carbonylation of aldimines[a].

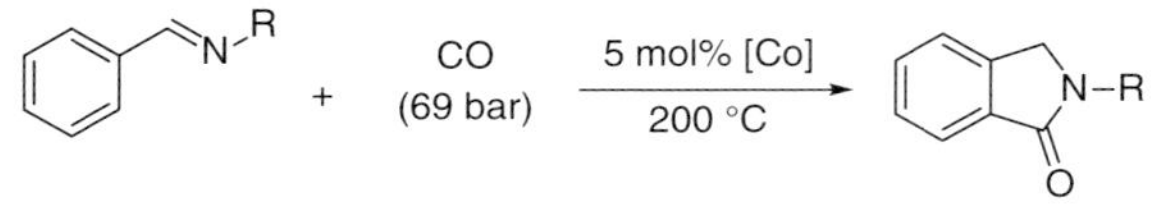

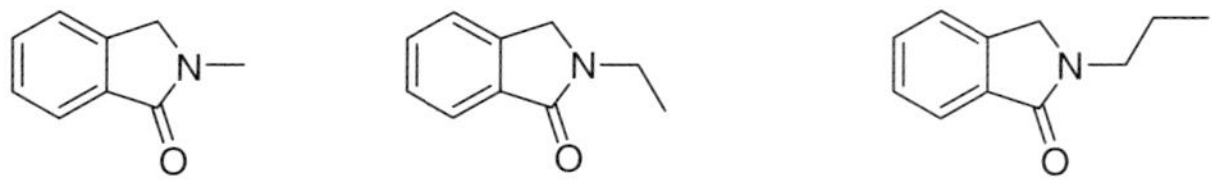

71% 48 h | 71% 48 h (92%[b]) | 62% 24 h

Reaction conditions: [a] *N*-alkylbenzaldimine (0.4 mmol), [Co(^{13}MeCO(CO)$_3$(P(*o*-tol)$_3$)] (5 mol%), (D$_8$)-1,4-dioxane (2 ml), CO (69 bar), 200 °C, noted time, yields were determined by ^{1}H NMR spectroscopy with 1,3,5-trioxane as an external standard in CDCl$_3$. [b] *N*-Alkylbenzaldimine (1.2 mmol), Co$_2$(CO)$_8$ 5 mol%, in C$_6$D$_6$ (2 ml), 6 h.

Table 15.3 Scope of the carbonylation of amides[a].

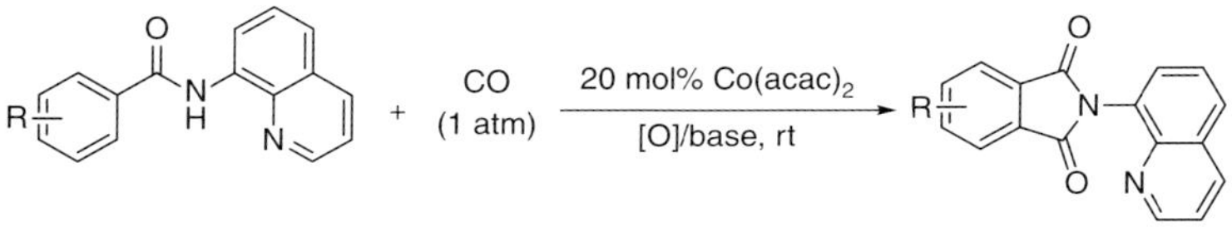

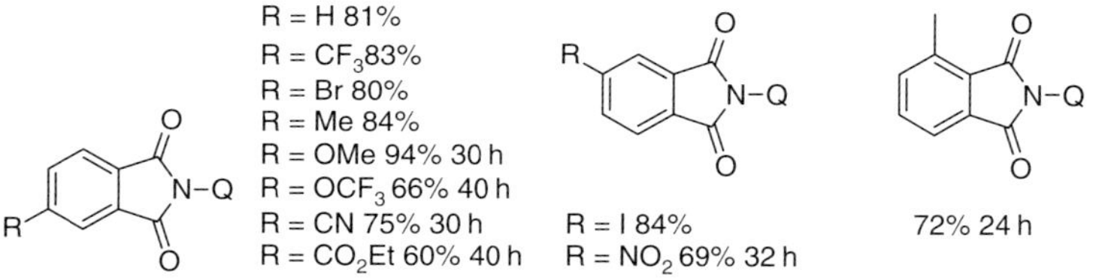

R = H 81%
R = CF$_3$ 83%
R = Br 80%
R = Me 84%
R = OMe 94% 30 h
R = OCF$_3$ 66% 40 h
R = CN 75% 30 h
R = CO$_2$Et 60% 40 h

R = I 84%
R = NO$_2$ 69% 32 h

72% 24 h

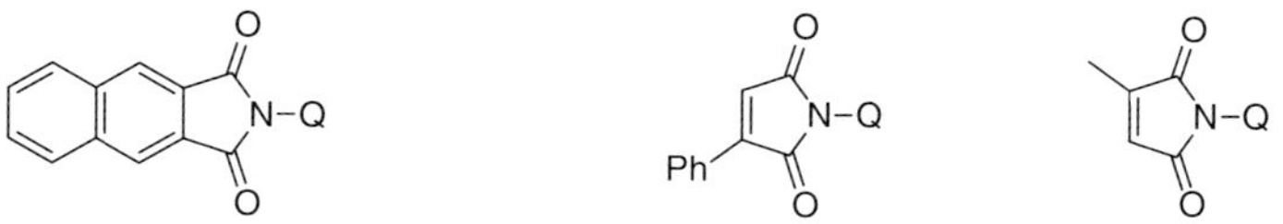

77% 30 h | 60% 24 h | 33% 40 h

Reaction conditions: [a] amide (0.5 mmol), Co(acac)$_2$ (20 mol%), Mn(OAc)$_3$ · H$_2$O (1.0 equiv.), NaOPiv (2.0 equiv.), TFEtOH (5 ml), CO (1 atm, balloon), air, rt, 16 h unless otherwise noted, Q = 8-quinolinyl.

run at room temperature under 1 atm of CO in trifluoroethanol as the solvent, use oxygen from air as the oxidant, and use Mn(OAc)$_3$·H$_2$O as the additive (Table 15.3). A wide range of functionality groups including methyl, methoxy, cyano, ester, trifluoromethyl, and trifluoromethoxyl can be compatible well and gave the desired products in good yields. Notably, the nitro group which could be easily reduced under CO conditions was incorporated into the products without any problem. Interestingly, the substrates bearing the halogen group such as bromo and iodo gave the corresponding products in 80 and 84% yield, respectively. Control experiments indicate that oxygen from air serves as the oxidant (without air, only 20% yield of the product was obtained). The author stated that $Mn(OAc)_3{\cdot}H_2O$ play the role of cocatalyst.

15.3 Rhodium-Catalyzed Heterocyclic Synthesis via Carbonylative C–H Activation

15.3.1 Five-Membered Ring Synthesis

In 2011, an efficient approach for the formation of phthalimides through the oxidative carbonylation of aromatic amides was developed by Rovis and coworkers in the presence of rhodium catalyst [9]. In this context, 16 examples were synthesized (Table 15.4). Different kinds of *N*-protecting groups such as alkyl (methyl, *n*-butyl, *n*-hexyl, and isopropyl), benzyl, phenyl ethyl, and CH_2CO_2Me can be utilized with moderate to good efficiency. Substrates bearing *para* or *meta* electron-donating groups with the example of methyl, methoxy, and phenyl

Table 15.4 Represented examples of the carbonylation of amides toward phthalimides[a].

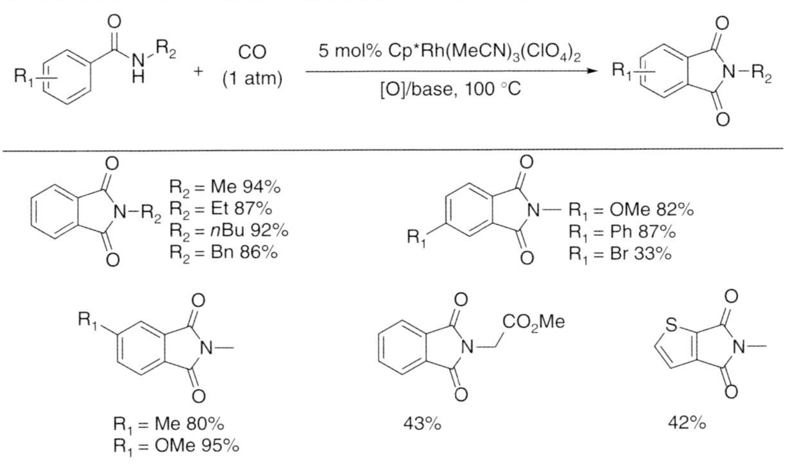

Reaction conditions: [a] aromatic amide (0.2 mmol), $Cp^{*}Rh(MeCN)_3(ClO_4)_2$ (5 mol%), Ag_2CO_3 (2.0 equiv.), KH_2PO_4 (2.0 equiv.), *t*-AmOH (0.37 ml), CO (1 atm, balloon), 100 °C, 24 h.

on the aromatic ring provided the corresponding phthalimide products in high yields. Notably, the bromo functionality was successfully untouched under this condition although in 33% yield. Interestingly, heteroaryl amide runs well in this system and gave the corresponding phthalimide in moderate yield. Reactions show a preference for the electron-rich aromatic amides in the C–H activation step, and the deuterium experiments suggest that C–H metalation step is largely irreversible.

15.4 Ruthenium-Catalyzed Heterocyclic Synthesis via Carbonylative C–H Activation

15.4.1 Five-Membered Ring Synthesis

Murai *et al.* in 1997 developed the carbonylation of aromatic imines for the preparation of indenones in the presence of $Ru_3(CO)_{12}$ [10]. This process runs under 7 bar of ethylene and 5 atm of CO at 160 °C (Table 15.5). Control experiments demonstrated that one-pot two-step manner is involved: first, the carbonylation of aromatic imines with CO and ethylene giving the keto imine and then the intramolecular aldol-type condensation of ketoimine would readily undergo to generate **1** *in situ*. Treatment of the crude reaction mixture with silica gel resulted in the elimination of *tert*-butylamine from **1** and deliverance of the final product.

The *ortho*-substituted group can be varied with electron-donating group such as methoxy or with electron-withdrawing group such as trifluoromethyl and fluoro in good yields. The reaction of *N*-benzylidene-*tert*-butylamine afforded two products among which the second acylation took place giving the side-product in 30% yield. The heteroaromatic compound only generates the corresponding acylated ketone; no further aldol-type condensation took place under this standard condition perhaps because the keto and imino group are located too far from one to another. Olefins such as the 1-hexene, styrene, and methyl acrylate failed, whereas the *tert*-butylethylene and trimethylvinylsilane underwent smoothly to produce the final products in 41 and 64% yield, respectively.

Murai and Chatani extend the method from aromatic imines to α,β-unsaturated imines in 2002. Very interestingly, no formation of the anticipated cyclopentadienones was detected, yet β,γ-unsaturated γ-butyrolactams were formed (Table 15.6) [11].

The reaction proceeds via a two-step sequence involving the initial carbonylation at the β-olefinic C–H bond of α,β-unsaturated imines forming the ketone derivatives, followed by the intramolecular nucleophilic attack of the imine nitrogen on the ketonic carbon to generate a tetrahedral intermediate, which then undergoes a 1,2-alkyl migration. Compared to the electron-donating group, the trifluoromethyl as an example of electron-withdrawing group resulted in

Table 15.5 Selected examples for the preparation of indenones from aromatic imines[a].

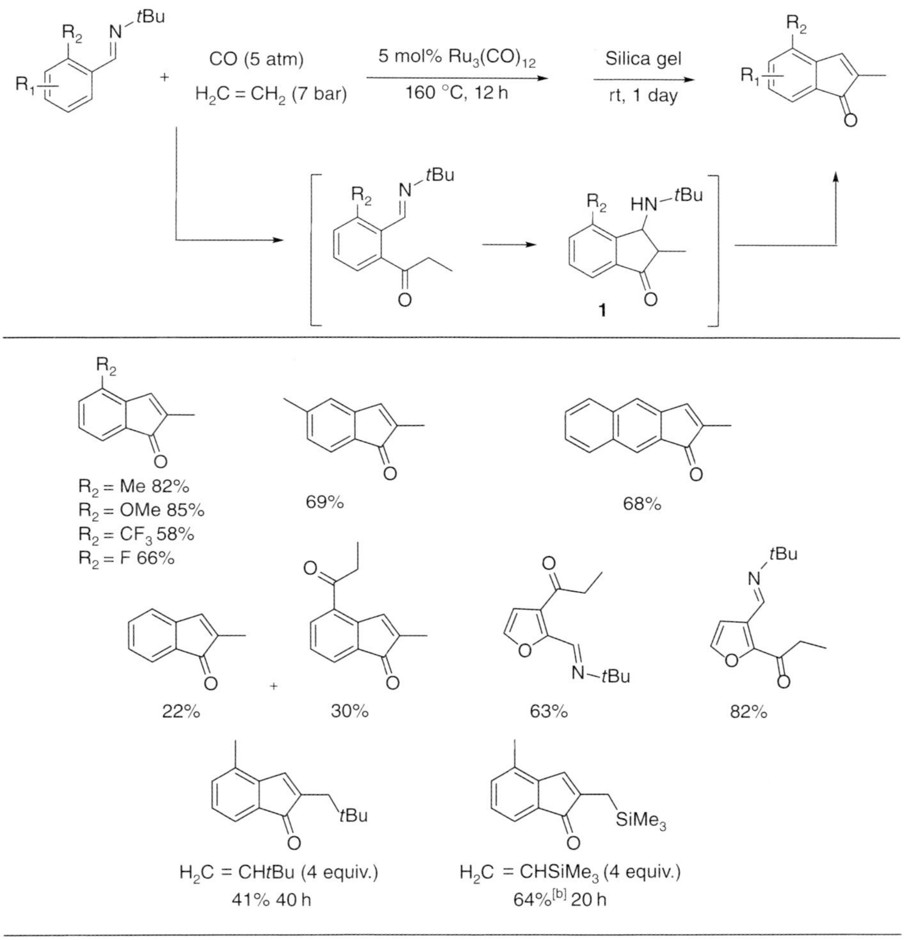

Reaction conditions: [a] aromatic imines (2 mmol), $Ru_3(CO)_{12}$ (5 mol%), toluene (6.0 ml), CO (5 atm), ethylene (7 bar), 160 °C, 12–40 h. [b] With 14% of product formed.

the lower yield. Notably, imines contained an alkyl group at the β-position also can be applied to this condition albeit in low yield. Terminal alkene such as the 1-hexene failed, while the vinylsilane and norbornene as the alkene partners run smoothly forming the lactam products in 72 and 44% yield, respectively. In this case, direct carbonylation at the olefinic C–H bonds is the key step. The author proposed a mechanism in which 1,2-hydride shifts and 1,2-alkyl migration are involved for this transformation.

Chatani *et al.* exploded a bidentate-chelating system in 2009 in the presence of ruthenium catalyst, in which the carbonylation took place at the *ortho*-C–H bonds

Table 15.6 Selected products for the carbonylation of α,β-unsaturated imines[a].

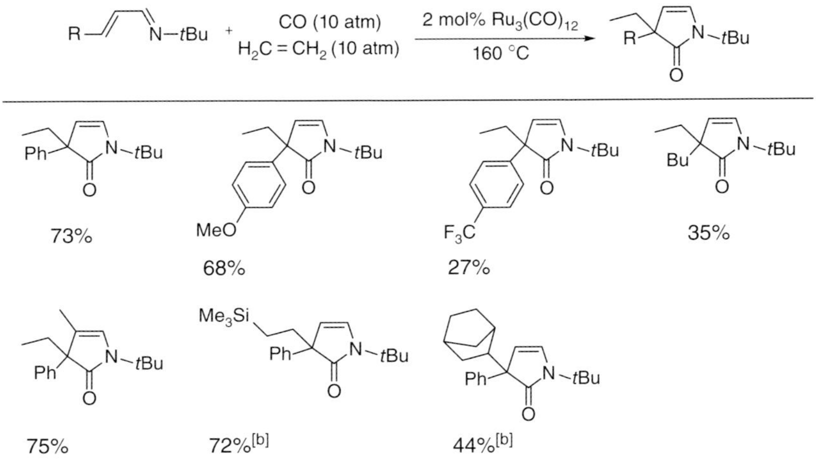

Reaction conditions: [a] imines (1 mmol), $Ru_3(CO)_{12}$ (2 mol%), toluene (2.0 ml), CO (10 atm), ethylene (10 atm), 160 °C, 20 h. [b] With 14% of the product formed.

Table 15.7 Bidentate system for the carbonylation of amides[a].

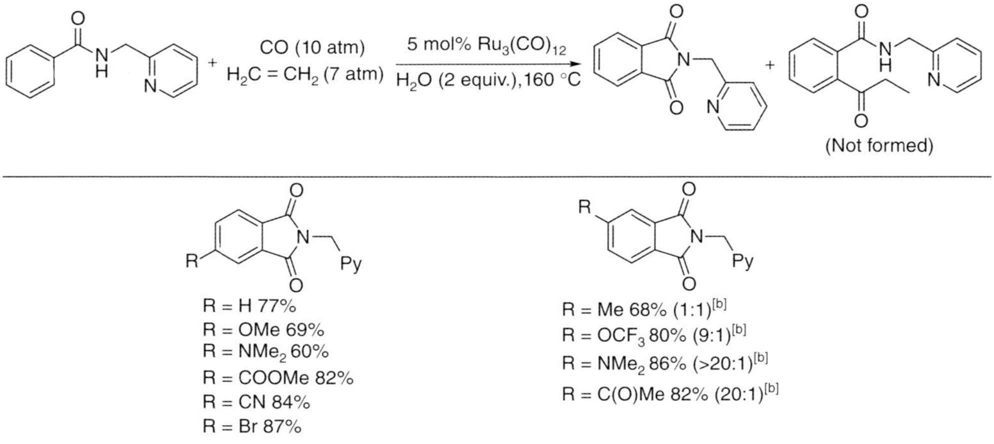

Reaction conditions: [a] amides (1 mmol), $Ru_3(CO)_{12}$ (5 mol%), H_2O (2.0 equiv.), toluene (3,0 ml), CO (10 atm), ethylene (7 atm), 160 °C, 24 h. [b] Major products were given.

in aromatic amides leading to phthalimides (Table 15.7) [12]. Intriguingly, no acylation was observed although the condition is very similar to their previous work. It should be noted that compared to the normal monodentate system, ethylene serves as a H_2 acceptor and *N,N*-bidentates are key features for the successful cyclocarbonylation as no carbonylation occurred with non-chelating substrates (Scheme 15.2).

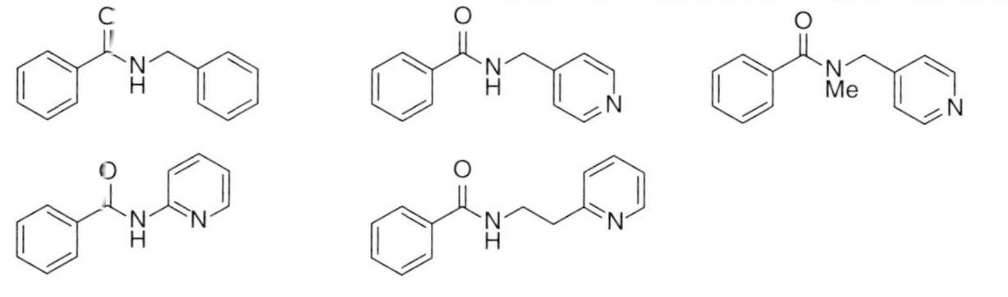

Scheme 15.2 Other amides compared to the *N,N*-bidentate system.

A wide variety of functional groups, including methoxy, amino, ester, cyano and even bromo, can be incorporated into the products. Next *meta*-substituted amides were used to test the regioselectivity. Except methyl, other functional groups such as the OCF_3, NMe_2, and ketone-substituted analogs favor the less sterically hindered position which indicates that the electron effect is not a dominant factor. Investigations demonstrate that the true active species is a dinuclear Ru(I) confirmed by X-ray crystallography. It should be kept in mind that the water is necessary in this procedure as the yield would drop sharply without water

In 2011, the group of Chatani extend the bidentate-chelation system to the inactivated C(sp^3)–H bonds in the presence of ruthenium carbonyl compounds leading to the corresponding imides (Table 15.8) [13]. Reactions proceed under similar conditions [12] and show a preference for C–H bonds of methyl groups as opposed to methylene C–H bonds. A series of aliphatic amides can be transformed into the imides. Compared to ethyl, butyl, methoxylmethyl, and cyclohexyl, the methyl group was selectively carbonylated to give the products in high yields. The steric effect was a predominant factor as exemplified by **2**, and hydrolysis of the starting material **2** is a contaminated result. To avoid hydrolysis, a more sterically bulky directing group **3** was designed. As expected, the CO was successfully incorporated into imide. Interestingly, different kinds of benzyl group were still intact under this system. The cyclopropyl group shows higher activity than the methyl group. Stoichiometric reaction of an amide with $Ru_3(CO)_{12}$ generated a dinuclear ruthenium complex which was confirmed by the X-ray crystallography. Inter- and intramolecular kinetic isotope effect (KIE) values suggest that the cleavage of C–H bond involve in the rate-determining step.

Table 15.8 Represented products for the C(sp^3)–H carbonylation[a].

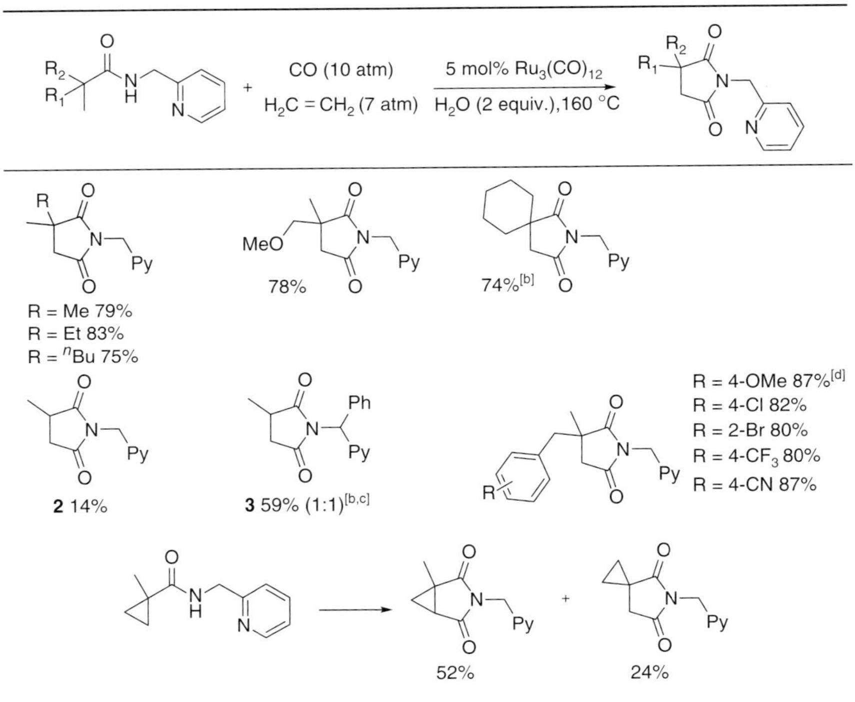

Reaction conditions: [a] amides (1 mmol), $Ru_3(CO)_{12}$ (5 mol%), H_2O (2.0 equiv.), toluene (3.0 ml), CO (10 atm), ethylene (7 atm), 160 °C, 5d. [b] $Ru_3(CO)_{12}$ (10 mol%). [c] Diastereomeric ratio. [d] Toluene (4.0 ml).

15.4.2
Six-Membered Ring Synthesis

After gaining some acknowledgement of this bidentate-chelating system, Chatani *et al.* in 2012 simply change the starting material to the arylacetamides; as expected, the isoquinoline-1,3(2*H*,4*H*)-diones could be obtained in good yield (Table 15.9) [14]. Compared to inactivated C(sp^3)–H, the C(sp^2)–H bonds of the aromatic ring more easily participated in this reaction which is understandable as the bond energy of C(sp^2)–H is far more low than the C(sp^3)–H one. In this context, except ethylene, the methyl acrylate can also serve as the hydrogen acceptor producing the desired product in 83% yield. Also the stoichiometric reaction of an arylacetamide with $Ru_3(CO)_{12}$ gave one dinuclear ruthenium complex which was also confirmed by the X-ray crystallography. Instead of $Ru_3(CO)_{12}$, this dinuclear ruthenium complex could be used as the catalyst with water as the additive.

In 2013, Kondo and Inamoto reported the carbonylative C–H cyclization of 2-arylphenol with ruthenium complex as the catalyst and *N*-heterocyclic carbene

Table 15.9 Selected examples for the carbonylative isoquinoline-1,3(2*H*,4*H*)-dione preparation[a].

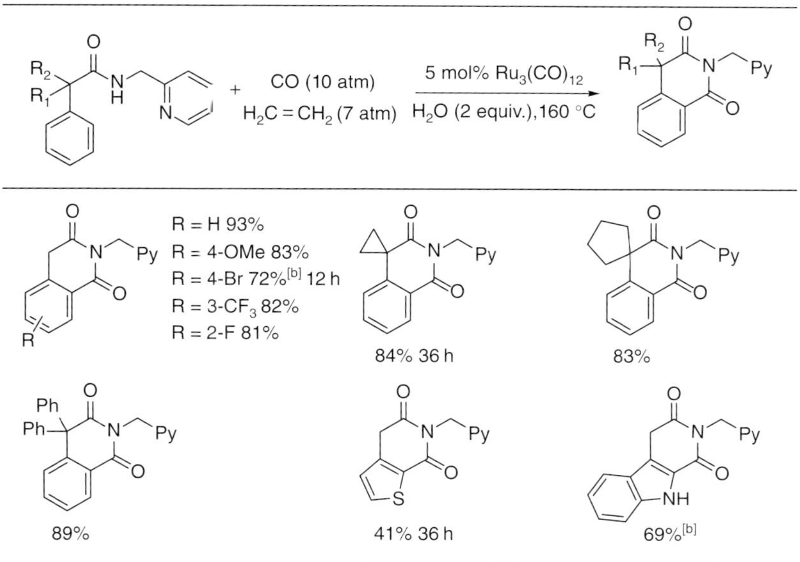

Reaction conditions: [a] amides (1 mmol), $Ru_3(CO)_{12}$ (5 mol%), H_2O (2.0 equiv.), toluene (3.0 ml), CO (10 atm), ethylene (7 atm), 160 °C, 1d. [b] Dioxane as solvent.

(*NHC*) as the ligand [15]. This approach works under balloon pressure of CO with O_2 (balloon also), producing 6*H*-dibenzo[*b*,*d*]pyran-6-ones in acceptable to good yields (Table 15.10). The electron-donating or electron-withdrawing substituents on the *ortho*-phenyl rings do not show explicit difference. For the phenol ring part, substituents on the *ortho* position impeded the reaction, while the *meta*- or *para*-substituted analogs provided the product without any problem. An intermolecular competition experiment indicates that electron-rich materials show more reactivity, and KIE results exemplify that the C–H metalation step is reversible.

15.5 Palladium-Catalyzed Heterocyclic Synthesis via Carbonylative C–H Activation

15.5.1 Four-Membered Ring Synthesis

In 2014, Lei and You demonstrated the β-lactam scaffold synthesis through the oxidative carbonylation of *N*-allylamines in the presence of palladium catalyst, thus a series of (19 examples) α-methylene-β-lactams was produced smoothly (Table 15.11) [16]. DFT calculations indicated that a four-membered-ring

Table 15.10 Represented examples for the carbonylation of *ortho*-phenyl phenol[a].

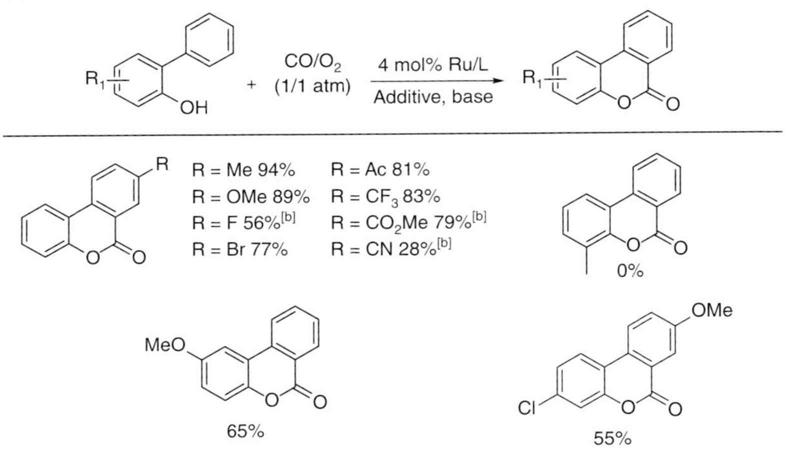

Reaction conditions: [a] *ortho*-aryl phenol, $[RuCl_2(p\text{-cymene})]_2$ (4 mol%), IPr-HCl (12 mol%), PivOH (10 mol%), Cs_2CO_3 (3.0 equiv.), mesitylene, CO (1 atm, balloon), O_2 (1 atm, balloon), 100 °C, 24 h. [b] $[RuCl_2(p\text{-Cymene})]_2$ (8 mol%), IPr-HCl (24 mol%).

Table 15.11 Selected products of α-methylene-β-lactam formation through the carbonylation of *N*-allylamines[a].

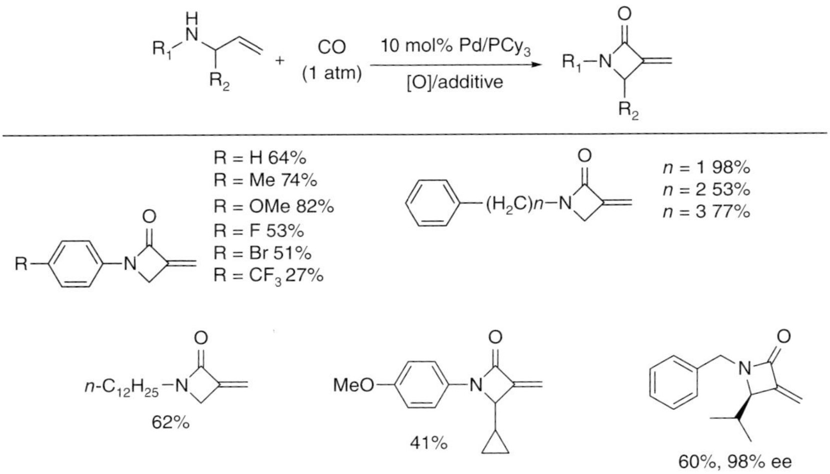

Reaction conditions: [a] *N*-allylamines (0.2 mmol), $PdCl_2$ (10 mol%), PCy_3 (20 mol%), $Cu(OPiv)_2$ (2.0 equiv.), PivOH (2.0 equiv.), DMF (2.0 ml), CO (1 atm), 100 °C, 1–24 h.

Table 15.12 Scope for the fluoren-9-one synthesis via the carbonylation of *ortho*-halobiarys[a].

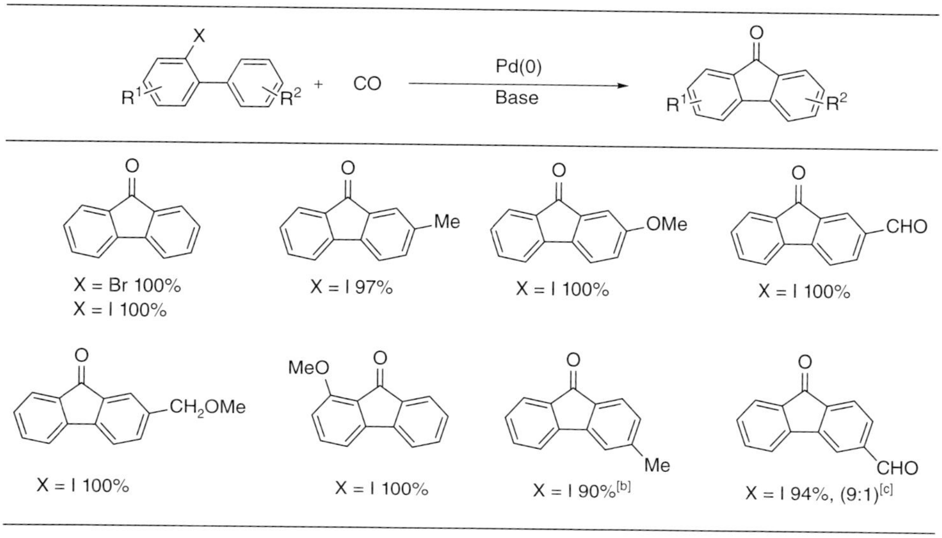

Reaction conditions: [a] ArX (0.25 mmol), $Pd(PCy_3)_2$ (5 mol%), CsPiv (2.0 equiv.), CO (1 atm), DMF (6 ml), 110 °C, 7 h. [b] With 10% isomer formed. [c] Main products; the ratio was determined by ^1H NMR spectroscopic analysis.

transition state is favorable. As we can see, *N*-allylanilines bearing electron-withdrawing or electron-donating groups on the aniline rings process well under this condition and provided the lactams in acceptable to good yields. Remarkably, the *N*-alkyl-substituted allylic amines also can be applied to this carbonylation process. Notably, no loss of enantioselectivity was observed when enantioenriched *N*-allylamine was used as the starting material. This approach presents the remarkable progress for the carbonylation and has the potential to be used in organic and medicinal chemistry.

15.5.2 Five-Membered Ring Synthesis

Larock *et al.* in 2000 developed the fluoren-9-one synthesis by the cyclocarbonylation of *ortho*-halobiarys with a commercially available $Pd(PCy_3)_2$ catalyst [17] This procedure was initiated by the palladium(0) and could be utilized to the synthesis of polycyclic and heterocyclic fluorenones containing the fused isoquinoline, indole, pyrrole, thiophene, benzothiophene, and benzofuran rings (Table 15.12).

It should be noted that this protocol could be applied to the vinylic halides, and the corresponding products were isolated in high yields (Scheme 15.3). The author proposed a Pd(0)/Pd(II) catalytic cycle.

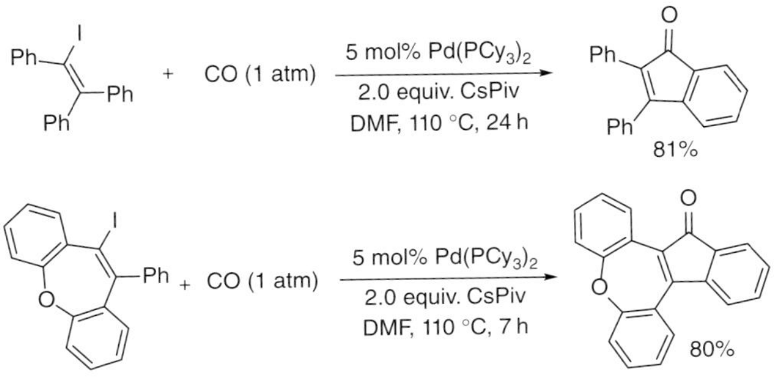

Scheme 15.3 Carbonylation of vinylic halides.

Table 15.13 Selected examples of the carbonylation of aromatic acid[a].

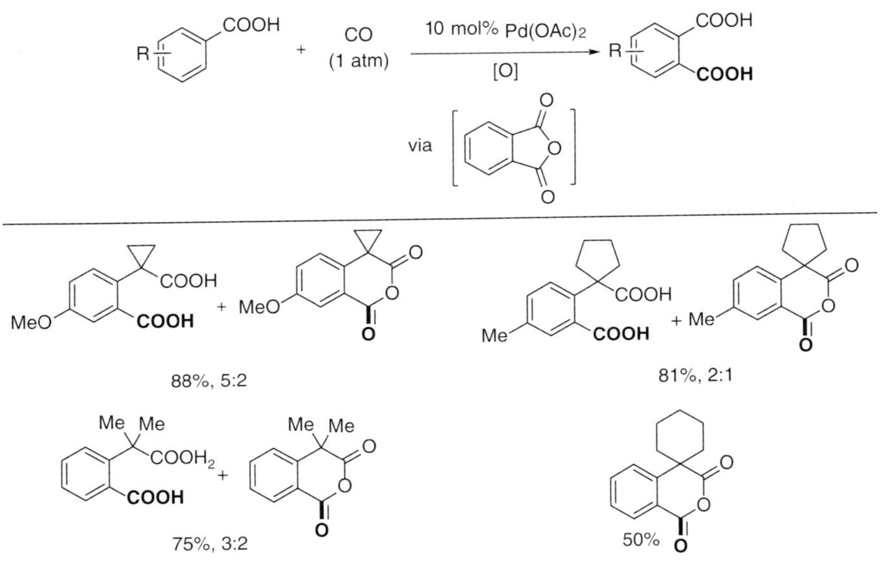

Reaction conditions: [a] using $Pd(OAc)_2$ (10 mol%), Ag_2CO_3 (2.0 equiv.), NaOAc (2.0 equiv.), CO (1 atm), 1,4-dioxane (2 ml), 130 °C, 18 h. [b] K_2HPO_4 in place of NaOAc.

Yu *et al.* in 2008 pioneered the palladium(II)-catalyzed carbonylation of aromatic or vinyl acid for the preparation of 1,2- and 1,3-dicarboxylic acid under 1 atm of CO (Table 15.13) [18]. Treatment of sodium carboxylate with $Pd(OAc)_2$ in CH_2Cl_2 or dioxane at 100 °C formed a C–H insertion intermediate complex. A suspension of this palladacycle in solvent could react stoichiometrically with CO under room temperature producing the anhydride acid quantitatively which was finally hydrolyzed to dicarboxylic acid *in situ* under the base and trace amount of water. Twenty-four examples were prepared and isolated through this protocol.

Later the same group extend this methodology to the inactivated C(sp^3)–H bonds leading to the corresponding succinimides [19]. The major obstacles for this transformation is that the excess CO inhibits the activation of the inert C(sp^3)–H by competitively occupying coordination sites on palladium(II) center, thereby preventing the requisite C–H agnostic interaction. Compared to the *N*-toluene-sulfonyl amide, the analogous acidic amides which are used as the directing group gave the corresponding succinimides in better yields. Interestingly, screening the additive reveals that a TEMPO serving as the cooxidant with AgOAc gave the full conversion. Even the precise role of TEMPO remains to be elucidated; one plausible explanation is that oxoammonium salts (the oxidization from TEMPO) reoxidize Pd(0) to Pd(II) more efficiently than solely AgOAc. Starting material (Table 15.14) bearing quaternary α-atom provides the desired products in good to excellent yields. Reactions took place favorably at the methyl C(sp^3)–H bonds instead of the benzylic C(sp^3)–H bond. The benzyl motif proved to be a better protecting group for the β-hydroxy substrates, while TBS-protected substrates failed under this condition. Notably, this protocol was effective for the methylene C(sp^3)–H bonds of cyclopropyl substrates. Intriguingly, cyclopropyl C(sp^3)–H bonds could be selectively carbonylated over a methyl C(sp^3)–H bond and an *ortho*-aryl C(sp^2)–H bond. Sixteen examples have been prepared with this methodology.

Table 15.14 Represented substrates for the carbonylation of C(sp^3)–H bonds[a].

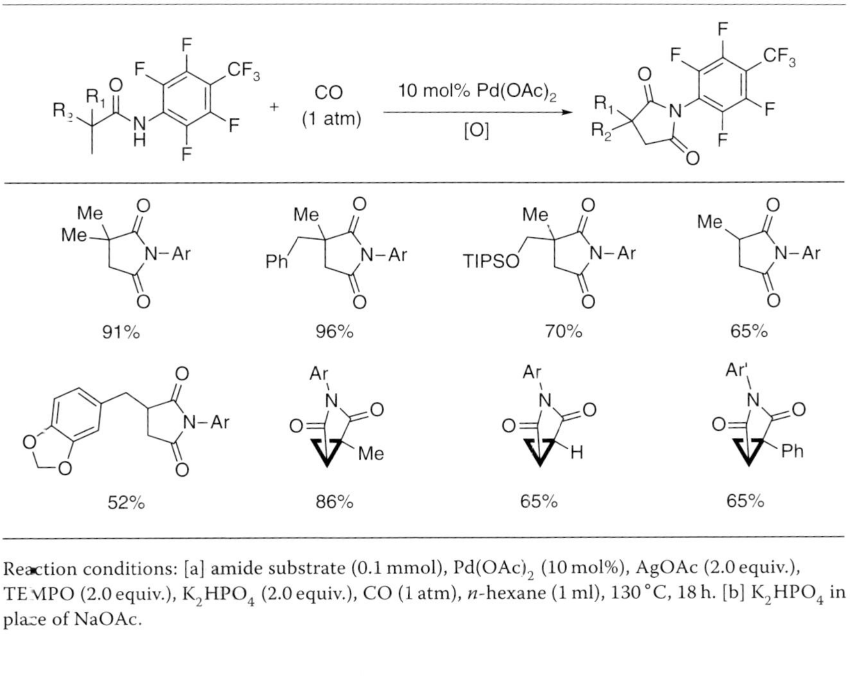

Reaction conditions: [a] amide substrate (0.1 mmol), $Pd(OAc)_2$ (10 mol%), AgOAc (2.0 equiv.), TEMPO (2.0 equiv.), K_2HPO_4 (2.0 equiv.), CO (1 atm), *n*-hexane (1 ml), 130 °C, 18 h. [b] K_2HPO_4 in place of NaOAc.

Table 15.15 Phthalimide synthesis via Pd(II)-catalyzed carbonylative C–H functionalization of *N*-alkoxybenzamides[a].

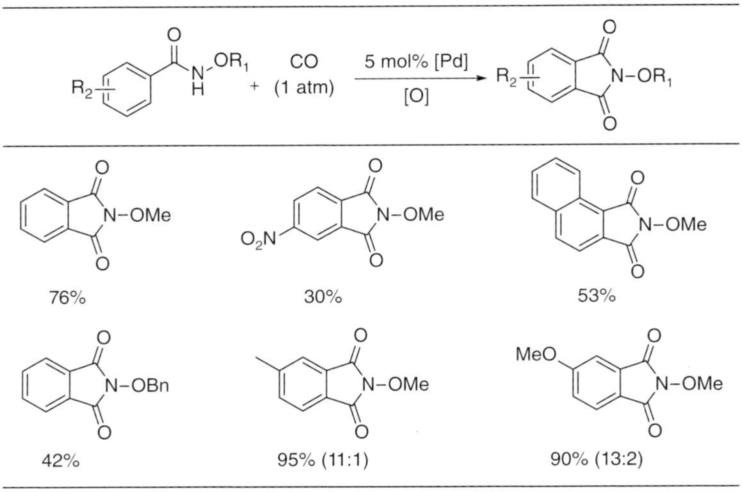

Reaction conditions: [a] *N*-methoxybenzamides (0.5 mmol), $Pd(OAc)_2$ (5 mol%), BQ (2.0 equiv.), HOAc (4.0 ml), CO (1 atm), 110 °C, 5–18 h.

Booker-Milburn and Lloyd-Jones in 2011 developed a procedure for the substituted phthalimide synthesis via the Pd(II)-catalyzed carbonylative C–H functionalization of *N*-alkoxybenzamides under 1 atm of CO [20]. A series of *N*-alkylphthalimides (10 examples) were achieved in moderate to excellent yield (Table 15.15). Notably, the nitro group was still intact albeit with low yield (30% yield). *Ortho*-substituted substrates underwent favorably the less steric site, giving the corresponding products with high selectivity.

Lei *et al.* in 2013 developed the palladium-/copper-catalyzed oxidative C–H alkenylation/*N*-dealkylative carbonylation of tertiary anilines toward the synthesis of 3-methyleneindolin-2-one derivatives (Table 15.16) [21]. For the olefin partner, electron-rich and electron-poor substrates worked well under this condition, giving the products in moderate to good yields. Anilines with electron-withdrawing or electron-donating substituents also run smoothly, providing the desired products in good yield without any problem. Notably, the bromo group was still untouched under this system, thus providing the reaction site for further manipulations. Interestingly, *N*-ethyl-*N*-methyl aniline generates two types of products in 13 and 48% yield, respectively, suggesting that the C–N bond cleavage favored the less sterically hindered alkyl group (Scheme 15.4). Control experiments show that the dealkylation step took place through copper-/oxygen-mediated C–N bond cleavage of tertiary amines. Intermolecular or intramolecular KIE experiments and *in situ* IR spectroscopy reaction rates of

Table 15.16 The palladium-/copper-catalyzed oxidative C–H alkenylation/*N*-dealkylative carbonylation of tertiary anilines[a].

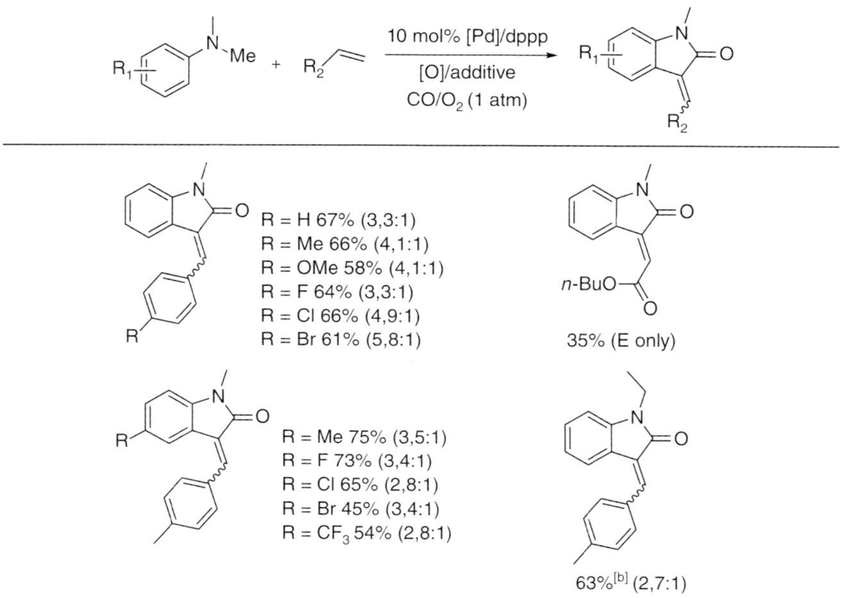

Reaction conditions: [a] *N*,*N*-dialkylanilines (2.0 equiv.), olefins (0.2 mmol), $Pd(PPh_3)_2Cl_2$ (10 mol%), dppp (10 mol%), $Cu(OAc)_2 \cdot H_2O$ (30 mol%), HOAc (50 mol%), toluene/DMF (3.0 : 0.3 ml), CO/O_2 (1 atm, 7 : 1), 100 °C, 24 h. The ratio of *E*/*Z* was determined by ^1H NMR spectroscopy. [b] Using *N*,*N*-diethylaniline as the starting material.

N,*N*-dimethylaniline with $[D]_5$-*N*,*N*-dimethylaniline indicated that the C–H cleavage might be involved in the rate-determining step.

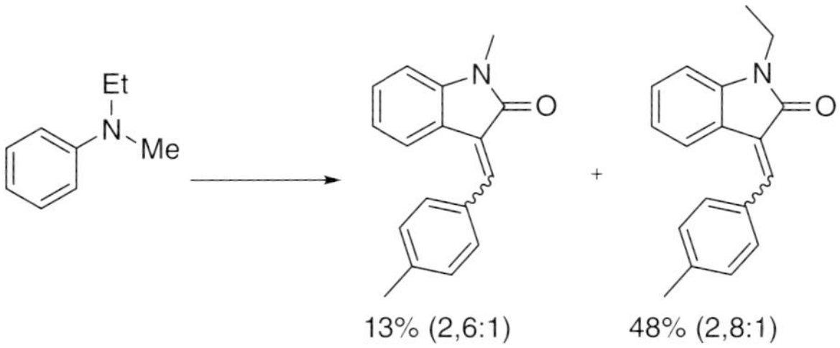

Scheme 15.4 *N*-Ethyl-*N*-methyl aniline for the C–N cleavage.

Xu *et al.* in 2013 exploited the synthesis of tetrasubstituted furan carboxylates using a tandem cycloisomerization/carbonylation sequence of cyclopropenes [22]. In the presence of palladium and copper, this tandem metal relay catalysis

(TMRC) pathway accounted for the high efficiency of this transformation. The key step is the ring-opening reaction of the cyclopropene in the presence of copper catalyst and subsequent transmetalation to palladium generating the key C–Pd bonds, which could undergo the carbonylative steps providing the desired products.

Lei *et al.* in 2014 pioneered the isatin synthesis from readily available anilines via the palladium-catalyzed double carbonylation of C–H bonds [23]. Reactions proceed under atmospheric pressure of CO without any additives, and 24 examples have been given in this context (Table 15.17). A lot of *N*-methyl anilines proceed smoothly under this oxidative condition affording the isatins in moderate to good yields, and the *N*-substituted allyl and cyclic tetrahydroquinoline also gave the corresponding isatin products in 56 and 38%, respectively. Intriguingly, no benzolactam [24] (in sharp contrast to Orito's work) was detected under this system for the *N*-benzyl anilines and selectively underwent the double carbonylation reactions, affording the corresponding isatin products in moderate yields. This is the

Table 15.17 The palladium-catalyzed double carbonylation of aniline C–H bonds to prepare isatin[a].

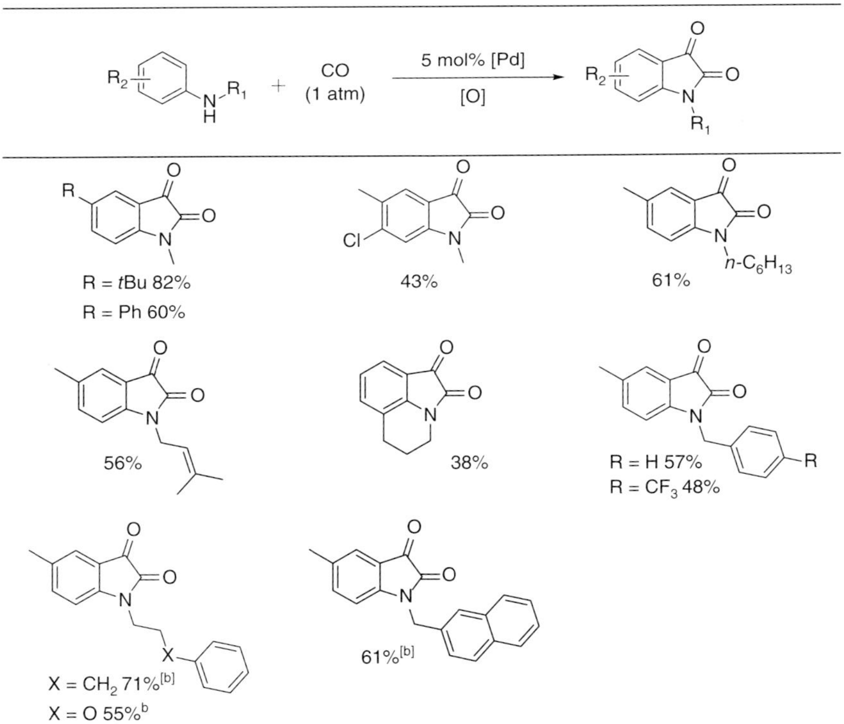

Reaction conditions: [a] *N*-alkylanilines (0.2 mmol), $Pd(PPh_3)_2Cl_2$ (5 mol%), $Cu(OPiv)_2$ (2.0 equiv.), toluene/DMSO (0.5 : 0.5 ml), CO (1 atm), 100 °C, 24 h. [b] With pivalic acid (2.0 equiv.).

first example to exploit the C–H oxidative double carbonylation for the various heterocyclic compound syntheses.

15.5.3 Six-Membered Ring Synthesis

Orito *et al.* in 2004 pioneered the carbonylation of secondary amines leading to benzolactams in the presence of $Pd(OAc)_2$ and $Cu(OAc)_2$ under 1 atm of CO containing air [24, 25]. Thirty examples have been tested giving moderate to good yield with good functional tolerance (Table 15.18). Generally speaking, the benzylic amines underwent carbonylation at a rate much faster than that of corresponding phenethylamines. Rate of the five-membered ring formation was 11 times greater than that of the six-membered ring formation.

Table 15.18 Selected benzolactam synthesis from carbonylation of secondary amines[a].

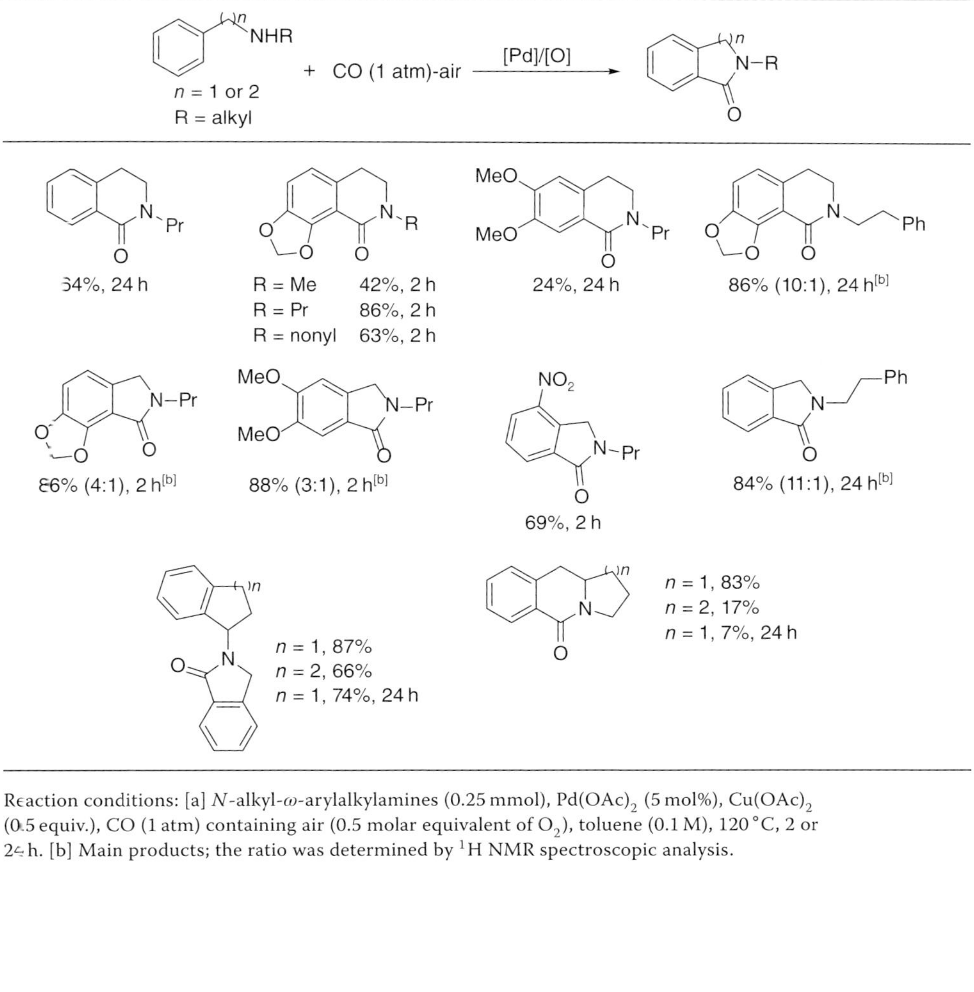

Reaction conditions: [a] *N*-alkyl-*ω*-arylalkylamines (0.25 mmol), $Pd(OAc)_2$ (5 mol%), $Cu(OAc)_2$ (0.5 equiv.), CO (1 atm) containing air (0.5 molar equivalent of O_2), toluene (0.1 M), 120 °C, 2 or 24 h. [b] Main products; the ratio was determined by ^1H NMR spectroscopic analysis.

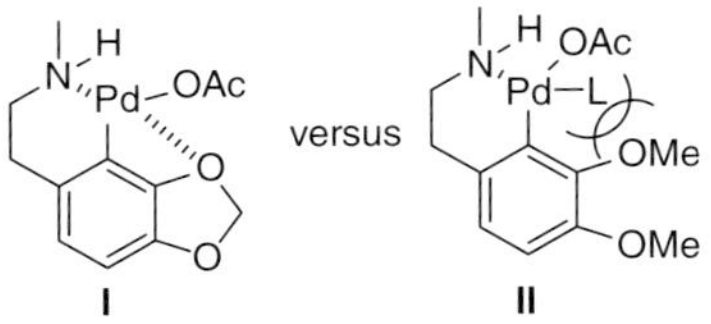

Scheme 15.5 Explanation of intermediates.

Scheme 15.5 explains the difference of the yields, in which the intermediate **II** is a more steric hindrance. Control experiments suggested that $Cu(OAc)_2$ served as not only an oxidant but also the ligand role. Only *N*,*N'*-dialkylureas were generated if the primary amine was applied as the starting material. In 2007 and 2009, Orito applied this methodology to the preparation of *N*-protected staurosporines or benzolactams such as phenanthro[9,10-*b*]indolizidin-9-ones and phenanthro[9,10-*b*]quinolizidin-9-ons, respectively [26].

Vicente and Saura-Llamas *et al.* in 2007 reported the *ortho*-palladation of L-phenylalanine methyl ester and then the carbonylation of this cyclopalladated complex under 1 atm of CO delivering the corresponding lactams [27]. It should be kept in mind that it's a stoichiometric procedure of palladium for the carbonylation of phenylalanine methyl ester. So Granell and Garcia *et al.* in 2011 developed a catalytic process for the same substrates, *N*-unprotected arylethylamines, with $Pd(OAc)_2$ as the catalyst and benzoquinone (BQ) as the oxidant under 1 atm of CO in refluxing HOAc (Table 15.19) [28]. Thus a series of quaternary α-amino acid esters could be transformed into the corresponding benzolactams. Sixteen examples have been isolated in moderate to good yield. Interestingly, secondary amine also could be used as the substrate and dimethylphenethylamine provided lactam in 73% yield. The hydroxymethyl is still intact under this oxidative condition. The substituted group on the aromatic ring such as the hydroxyl and nitro group could be tolerant in moderate to excellent yield. Unfortunately, only one example for the five-membered benzolactam was achieved with triphenylamine as the starting material. Compared to Orito's procedure [24], with $Cu(OAc)_2$ as the cooxidant, in which the five-membered ring analogs were clearly favored for the carbonylation of secondary amines, this protocol with BQ as the oxidant more favors the six-membered benzolactam.

A further experiment was especially significant (Scheme 15.6). Treatment of the substrate with the standard conditions only gave the six-membered ring lactam. In contrast, the first cyclometalation of substrate gave the favored five-membered palladacycle which was applied to the carbonylation with CO in the absence of BQ, affording the five-membered lactam in 86% yield as well as a minor amount (6%) of the isomer. The steric hindrance around the amino group is pivotal for the success of this transformation because the full substitution of the carbon in α-position of the amino esters plays an important role in their cyclopalladation step.

Gaunt *et al.* in 2011 demonstrated the carbonylation of the secondary amine with *N*-bearing aryl group in the presence of $Pd(OAc)_2$ under 1 atm of CO and

Table 15.19 Carbonylation of quaternary α-amino acid esters toward benzolactam synthesis[a].

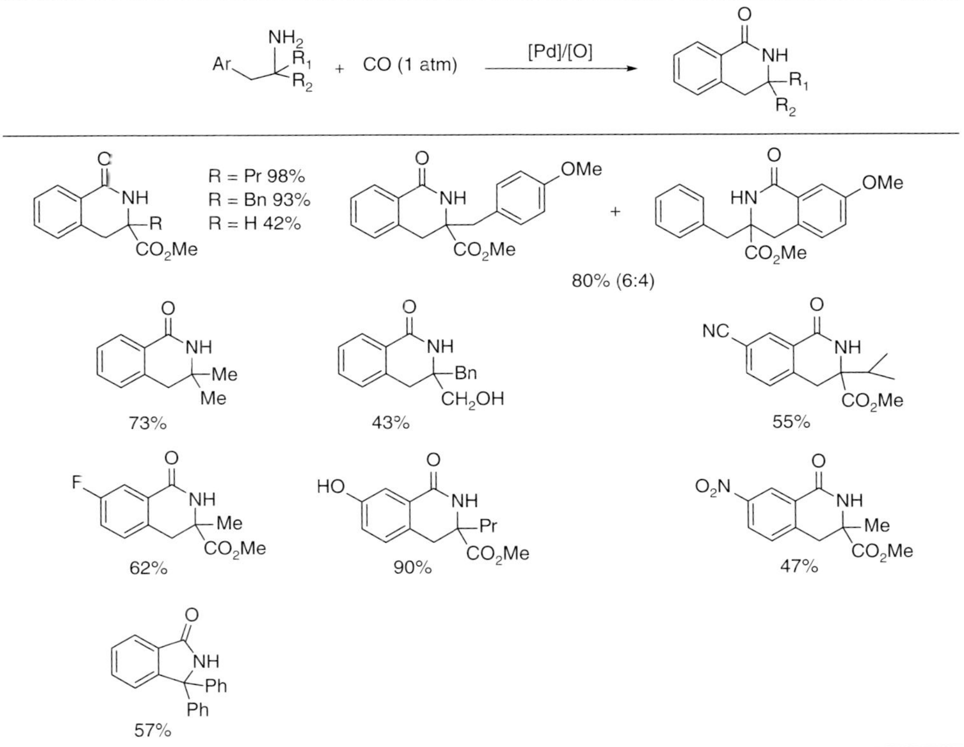

Reaction conditions: [a] arylethylamines, $Pd(OAc)_2$ (5 mol%), BQ (2.0 equiv.), CO (1 atm), HOAc as solvent, reflux, 6 h.

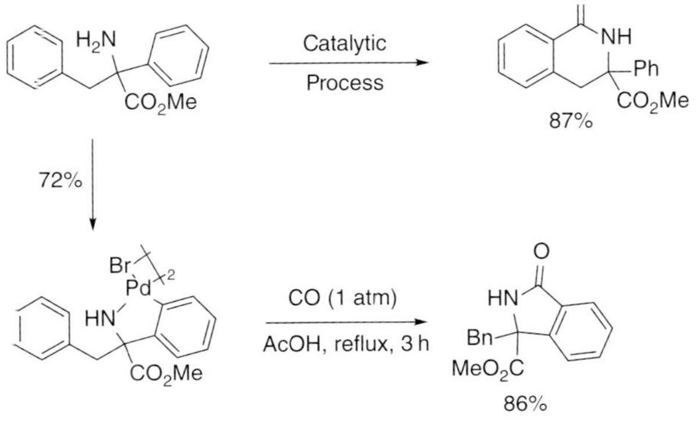

Scheme 15.6 Catalytic process versus stepwise pathway.

Table 15.20 Selected examples for the carbonylation of secondary amine[a].

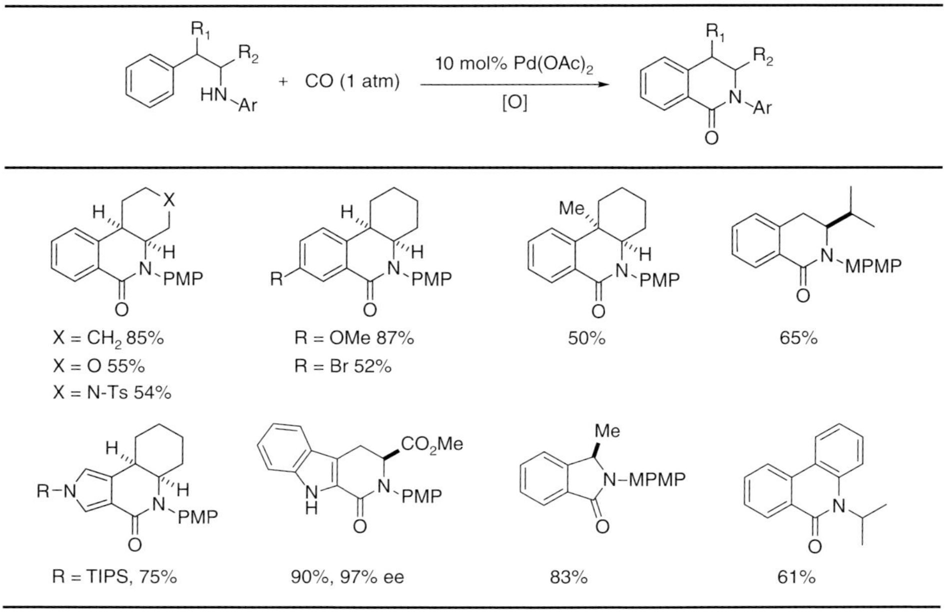

Reaction conditions: [a] *N*-arylethylamines (0.2 mmol), $Pd(OAc)_2$ (10 mol%), BQ (2.0 equiv.), t-BuO_2Ac (2.0 equiv.), NaOAc (6.0 equiv.), CO (1 atm) containing O_2, HOAc (2 ml) as solvent, room temperature. PMP = *para*-methoxyphenyl; MPMP = 4-methoxy-2-methylphenyl.

O_2 *at room temperature* providing the corresponding lactams (Table 15.20) [28a]. Reactions are tolerant of delicate stereogenic centers and functionality. The aryl group on the nitrogen atom as the protecting group is necessary, and 15 examples have been prepared in moderate to good yields. Tetrahydropyran and piperidine ring system run smoothly, and piperidine in particular represents an attractive scaffold for the medicinal chemistry purpose. The nature of arene could be varied with naphthalene-, pyrrole-, and indole-derived heteroarenes. The tryptophan derivative could be obtained without erosion of enantiomeric purity highlighting its potential in complex molecular architecture application. A more steric bulk *N*-aryl substituent was required when the chain amines were used as the substrates to suppress the addition of amine to BQ. It should be noted that carbopalladation proceeds at room temperature generating a dimeric complex 1a·Pd and O_2, BQ, and *t*-BuO_2Ac, all playing the role of oxidant. The solvent HOAc could appear to suppress the CO-mediated reduction of the Pd(II) to palladium black.

Alper and Yu in 2007 developed the palladium-catalyzed highly substituted endocyclic enol lactone synthesis via the carbonylation of terminal alkyne and 1,3-diketones in an ionic liquid system (Table 15.21) [29]. Thirteen examples have been isolated under this process. It should be noted that this catalytic system can be recycled five times with only modest loss of its catalytic activity.

Table 15.21 Carbonylation for the highly substituted endocyclic enol lactone synthesis[a].

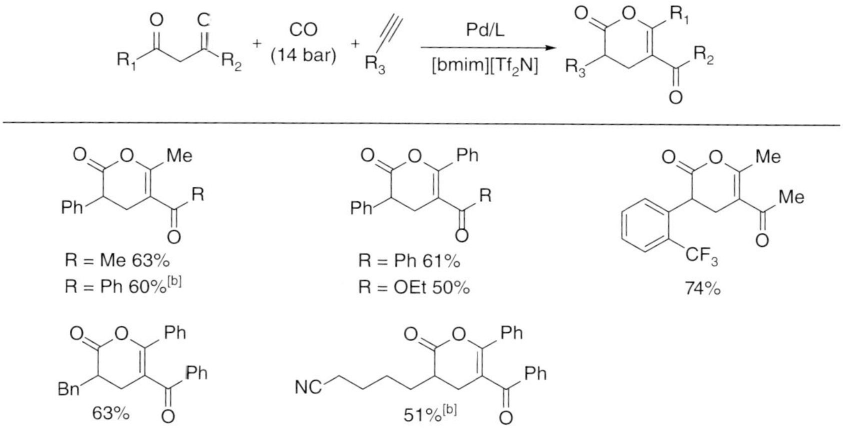

Reaction conditions: [a] alkyne (2 equiv.), 1,3-dicarbony compound (1 mmol), $Pd(PPh)_3Cl_2$ (3 mol%), dppp (6 mol%), CO (14 bar), $[bmim][Tf_2N]HOAc$ (2 g) as solvent, 110 °C, 24 h. [b] (±)-BINAP instead of dppp.

Yu *et al.* in 2011 revealed an approach to 1-isochromanones via hydroxyl-directed carbonylative C–H bond of tertiary alcohol enabled by mono-*N*-protected amino acid ligand [30]. The author tested several *N*-Boc-protected amino acids as the ligand, among them Boc-Leu-OH·H_2O increased the yield to 66% and (+)-menthyl(O_2C)-Leu-OH also gave the similar efficacy. Perhaps this kind of amino acid ligands can reduce the reduction rate of Pd(II) to palladium black under CO atmosphere. About 20 examples have been synthesized using this method (Table 15.22). In general, substrates bearing electron-rich aryl group performed better than the analogous with electron-poor aryl group. It should be noted that this approach can tolerate the halide groups such as the Br. The substituent effects on the alkyl chain adjacent to the hydroxyl group were also tested by the author. Different levels of steric bulk did not show detriment to the system. Most importantly, this protocol potential application was highlighted by the synthesis of a biologically active molecule histamine release inhibitor.

Booker-Milburn and Lloyd-Jones in 2009 revealed the cationic palladium-catalyzed carbonylation of aryl urea derivatives *at room temperature* under 1 atm of CO forming the cyclic imidates [31]. Key features of this procedure are the utilization of the $Pd(OTs)_2(MeCN)_2$ as the catalyst precursor and the powerful activating effect of the aniline–urea moiety in contrast with acetanilide (0% yield). For the *N,N*-disubstituted substrates (Table 15.23), the yields were dropping with the steric bulk increasing, whereas for the *N*-monosubstituted analogs, the trend was opposite. Prolonged reaction time resulted in low yields which can be rationalized by the reflection that the imidate products were sensitive to the added acid.

Table 15.22 Carbonylation of tertiary alcohol[a].

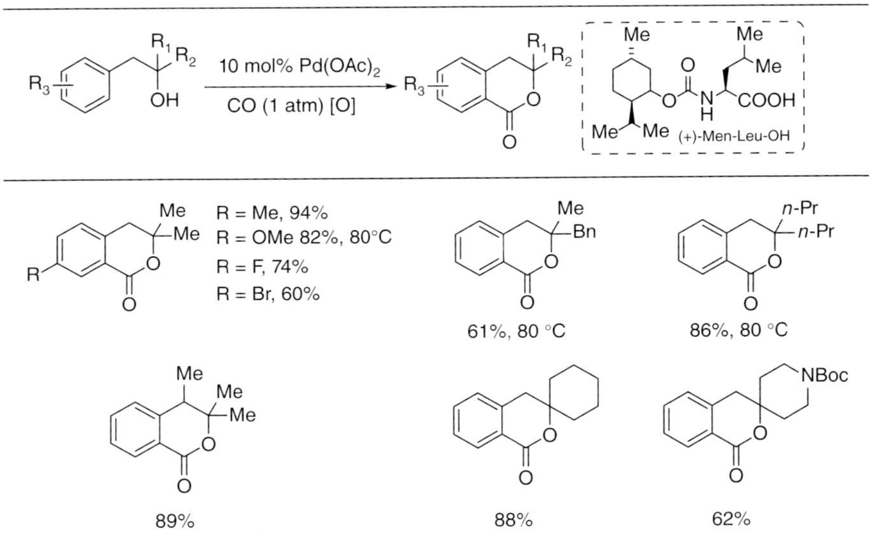

Reaction conditions: [a] phenethyl alcohol (1.0 equiv.), $Pd(OAc)_2$ (10 mol%), (+)-menthyl(O_2C)-Leu-OH (20 mol%), AgOAc (3.0 equiv.), Li_2CO_3 (1.0 equiv.), CO (1 atm), DCM, 110 °C, 48 h.

Table 15.23 Represented products for the carbonylation of aryl urea derivatives[a].

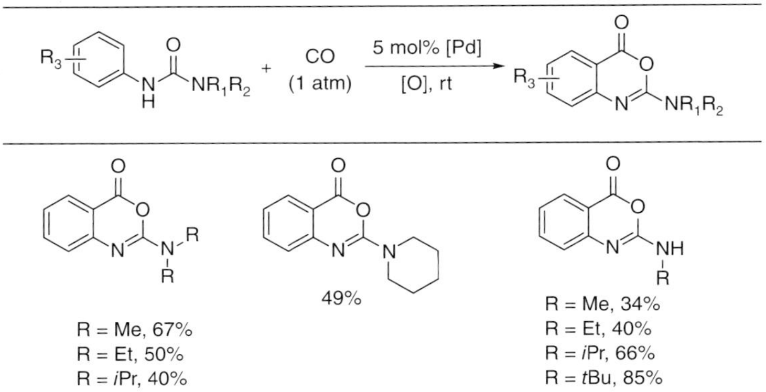

Reaction conditions: [a] aryl urea derivatives (1.0 mmol), $[(MeCN)_2Pd(OTs)_2]$ (5 mol%), BQ (2.0 equiv.), TsOH (1.0 equiv.), CO (1 atm), DCM, ambient temperature, 3–5 h.

Table 15.24 Selected examples for the carbonylative synthesis of quinazolin-4(3*H*)-ones[a].

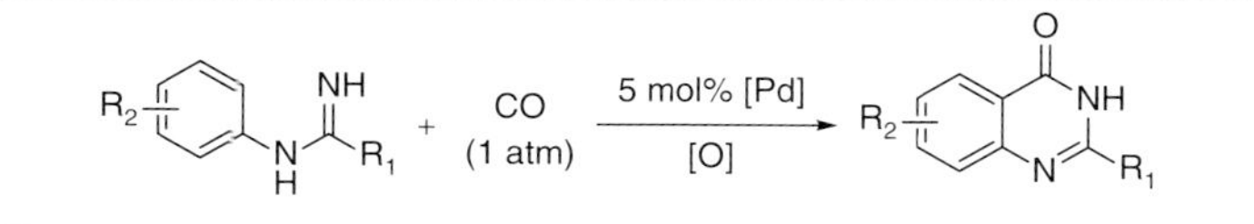

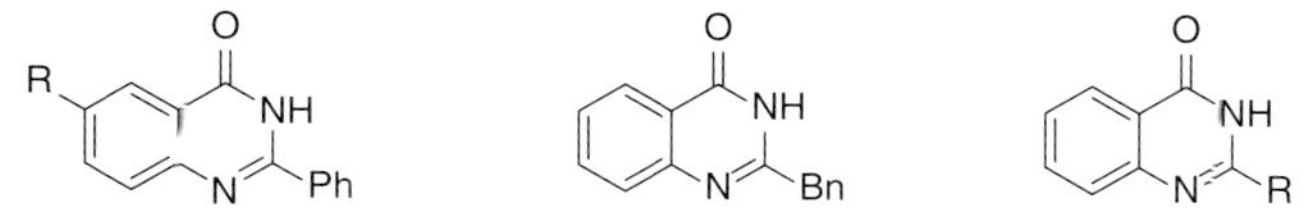

R = Me, 68%	60%	R = *i*Pr, 60%
R = OMe, 54%		R = *t*Bu, 60%
		R = Me, 81%

Reaction conditions: [a] *N*-arylamidines (0.2 mmol), $Pd(OAc)_2$ (10 mol%), CuO (1.0 equiv.), HOAc (1.5 ml), CO (1 atm, balloon), 110 °C, 23 h.

A synthetic route to the quinazolin-4(3*H*)-ones via Pd(II)-catalyzed intramolecular aminocarbonylation of *N*-arylamidines has been disclosed by Zhu *et al.* in 2011 [32]. Reactions precede in refluxing HOAc under 1 atm of CO with CuO as the oxidant. Sixteen examples have been given in moderate to good yields (Table 15.24). Deuterium experiments indicated that the C–H bond metalation step was reversible and deuterium–hydrogen scrambling occurred during the reaction.

Guan *et al.* in 2012 pioneered the directed carbonylation of the simplest *N*-alkylanilines in the presence of palladium catalyst under 1–2 atm of CO [33]. No external directing group is needed, and thus the elaboration of readily available *N*-alkylanilines resulted a number of isatoic anhydrides in good yield with good functionality tolerance under mild conditions (Table 15.25). Electron-rich anilines could run smoothly even *at ambient temperature,* while additives such as pivalic acid and oxygen were necessary for the electron-poor analogs to obtain good yields. Especially, the *meta*-nitroanilines worked well under the system. Although the indoline was almost inert to the transformation, tetrahydroquinoline was carbonylated into the tricyclic corresponding products in 75% yield. It should be noted that the substitutents on the nitrogen atom can be varied with the other alkyl group or benzyl group. Interestingly, the key intermediate for this transformation has been isolated and characterized by X-ray.

An interesting route for the synthesis of xanthone derivatives was pioneered by Lei *et al.* in 2012, thus providing an atom-economical approach to form biologically active products using the diaryl ether as the starting material [34]. Process runs under 1 atm of CO and in trifluoroacetic acid (TFA) at 50 °C (Table 15.26). Symmetrically diaryl ethers worked well forming the xanthone products in moderate to excellent yields as well as the asymmetrically analogs does. It should be noted that the bromo was untouched under these conditions which was impossible for the Pd(0)-catalyzed chemistry. Intriguingly, the biphenyl ether only generates the xanthone products in 27% yield. IR spectroscopy experiments

Table 15.25 Carbonylation of *N*-alkylanilines toward to isatoic anhydrides[a].

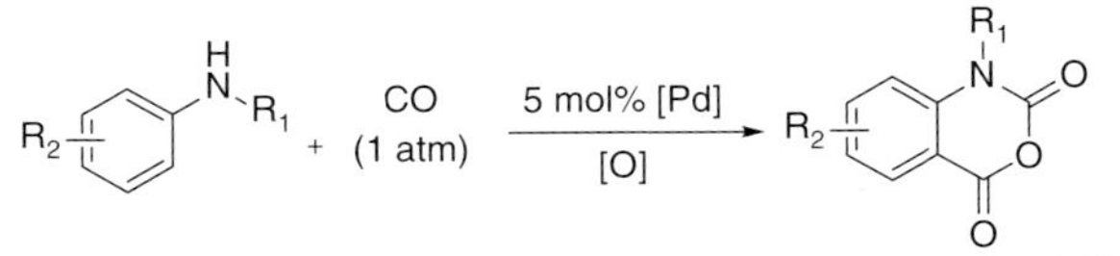

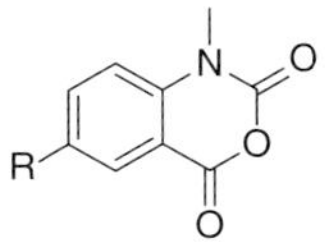

R = H 85%
R = Me 76%
R = OMe 70%[b]

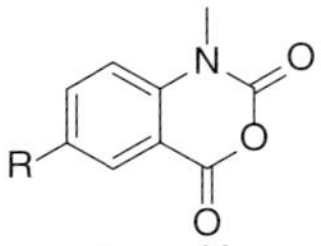

R = F 66%[c]
R = Br 86%[c]
R = CHO 77%[c]

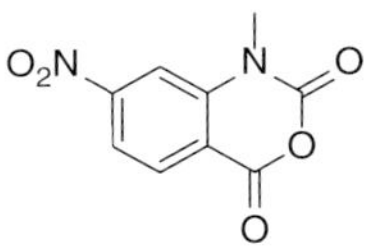

63%[c]

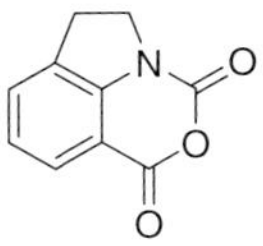

<5%

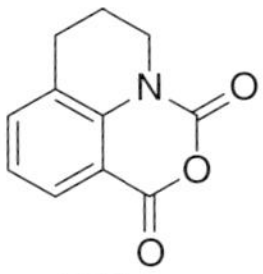

75%

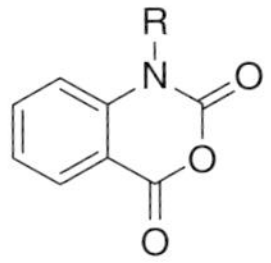

R = Et 84%
R = Pr 63%
R = Cy 51%
R = Bn 72%

Reaction conditions: [a] anilines (0.2 mmol), $Pd(OAc)_2$ (5 mol%), $Cu(OAc)_2$ (2.2 equiv.), KI (0.2 equiv.), MeCN (2.0 ml), CO (1 atm), 60 °C, 5–18 h. [b] At rt. [c] Pivalic acid (1.0 equiv.) was added, CO/O_2 (2 : 1) (3 atm).

Table 15.26 Double C–H carbonylative activation of diaryl ether for the synthesis of xanthones[a].

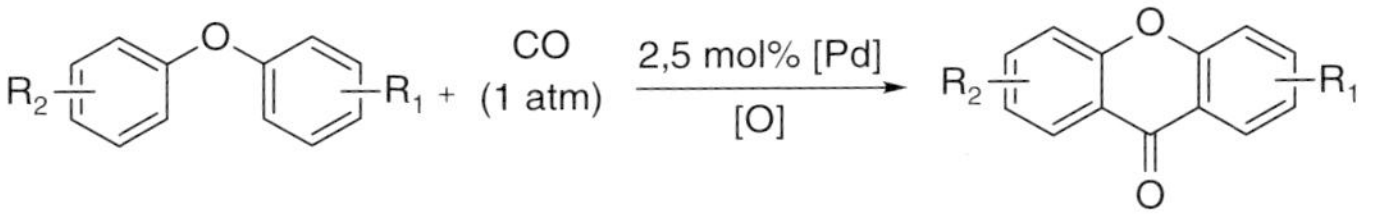

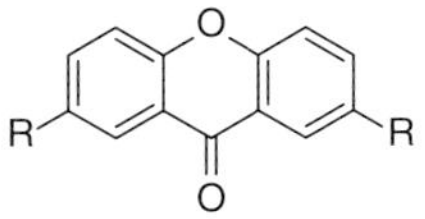

R = Me 96%
R = *t*Bu 92%
R = OMe 50%
R = Br 95%

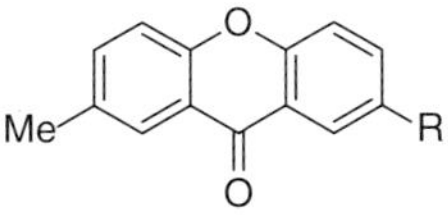

R = CF_3 56%
R = CO_2Et 51%
R = COMe 58%

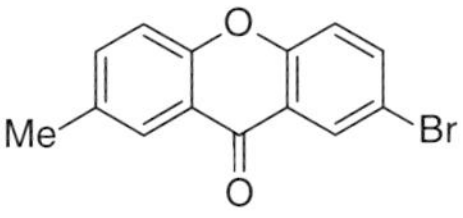

70%

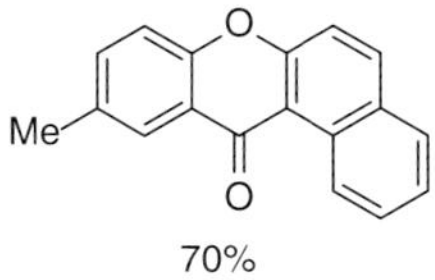

70%

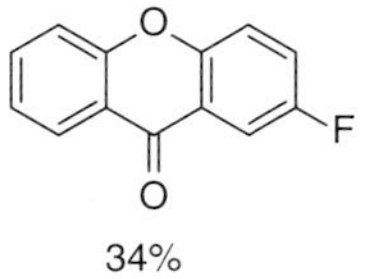

34%

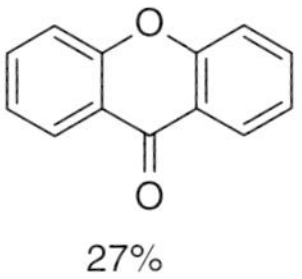

27%

Reaction conditions: [a] diaryl ethers (0.2 mmol), $Pd(OAc)_2$ (2.5 mol%), $K_2S_2O_8$ (2.0 equiv.), TFA (1.0 ml), CO (1 atm), 50 °C, 2–6 h.

show that zero-order kinetics for the carbonylation of different substrates were observed, thus indicating that the first C–H functionalization of diaryl ether step with Pd(II) was not the rate-determining step. The kinetic profiles of the different CO pressure suggest that the insertion of CO was not a rate-determining step either under the conditions.

Du and Zhang in 2014 reported another protocol for the xanthone synthesis via the palladium(0)-catalyzed carbonylation of *ortho*-diazonium salts of diaryl ether [35]. Compared to Lei's work, this transformation was initiated by Pd(0) with catalytic amount of tetrabutyl ammonium bromide as the additive. In this context, 23 examples have been prepared in accepted to good yields.

Guan *et al.* in 2013 described the synthesis of 1,3-oxazin-6-ones via the palladium-catalyzed carbonylation of alkenyl C–H bonds in the enamides [36]. Until now most of the palladium-catalyzed carbonylation of C–H bonds preferentially proceeds under the acidic conditions which may favor the generation of the cationic Pd(II) species and inhibition of the possible reduction of Pd(II) to Pd(0) by CO. But this procedure runs under basic conditions (in DMF with basic DABCO and KI as the additive, Table 15.27). Many kinds of substituents on the aromatic ring of enamides were tolerated and especially the bromo and

Table 15.27 Carbonylative synthesis of 1,3-oxazin-6-ones from palladium-catalyzed alkenyl C–H bond activation in the enamides[a].

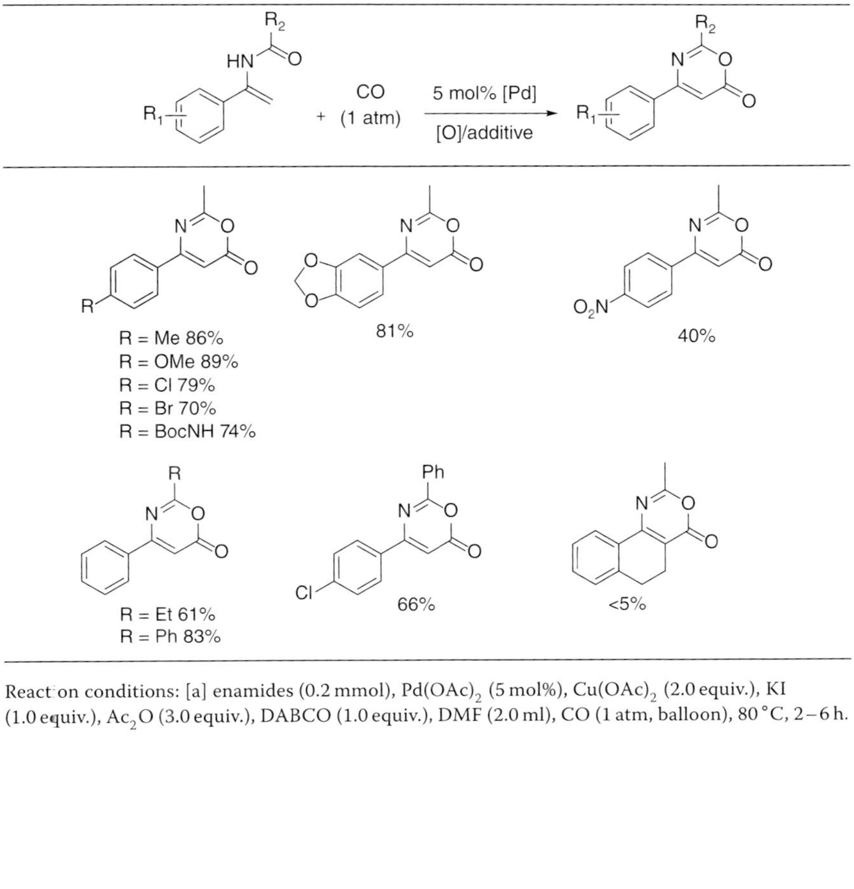

Reaction conditions: [a] enamides (0.2 mmol), $Pd(OAc)_2$ (5 mol%), $Cu(OAc)_2$ (2.0 equiv.), KI (1.0 equiv.), Ac_2O (3.0 equiv.), DABCO (1.0 equiv.), DMF (2.0 ml), CO (1 atm, balloon), 80 °C, 2–6 h.

nitro groups still intact under this system. The *tert*-butoxycarbonyl-protected amino also runs well under these conditions. Furthermore, the *N*-propionyl and *N*-benzoyl enamide analogs were also investigated to explore this carbonylation scope giving the corresponding 2-substituted 1,3-oxazin-6-ones in good yields. However, the *β*-substituted enamides were unable to generate the products. Key intermediates, six-membered vinylpalladium complex, were synthesized and treated with CO providing the final products smoothly.

Elman in 2013 reported the synthesis of naphthalic anhydride (NAn) via the $Pd(OAc)_2$-catalyzed oxidative carbonylation of aromatics with $K_2S_2O_8$ as the oxidant under 2 atm of CO *at room temperature* [37]. The subsequent alkaline hydrolysis of naphthalic anhydride and isomerization of the obtained 1,8-naphthalenedicarboxylic acid salt are known to generate the NDA.

Zhu and Zhang groups described the synthesis of phenanthridinone via the palladium-catalyzed aminocarbonylation of unprotected *ortho*-aryl anilines in 2013, independently [38]. A series of substituents can be tolerated under this procedure (Table 15.28). Notably, the free hydroxyl group is not detrimental to this procedure forming the products in good yields. It should be a reminder that the heteroaryl aniline substrates also worked well providing the products

Table 15.28 Carbonylation of *ortho*-amino biaryl[a].

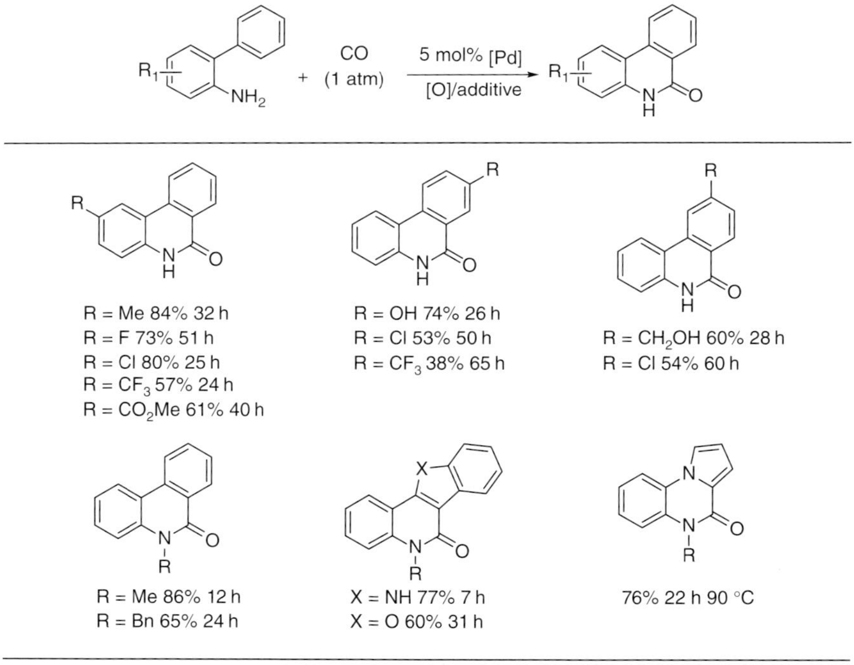

Reaction conditions: [a] *ortho*-aryl anilines (0.2 mmol), $Pd(MeCN)_2Cl_2$ (5 mol%), $Cu(TFA)_2$ (1.0 equiv.), TFA (1.0 equiv.), 1,4-dioxane (1.0 ml), CO (1 atm, balloon), 110 °C.

in good yields. Importantly, the palladacycle intermediate was isolated and further treated with CO providing the carbonylated products smoothly in 70% yield.

The acidic additive or solvent is necessary for those transformations. Zhang's process utilizes the trifluoroethanol (TFEOH) as the solvent with $Cu(TFA)_2$ as the oxidant in the presence of the $Pd(OAc)_2$ under atmospheric pressure of CO at 70 °C, and 19 examples have been prepared (Scheme 15.7).

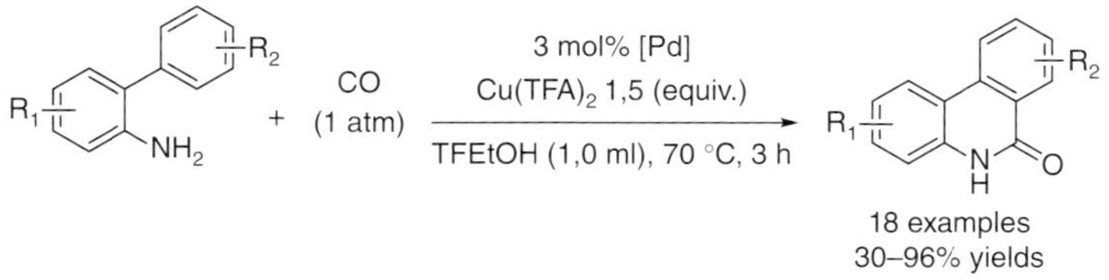

Scheme 15.7 Zhang's procedure for the synthesis of phenanthridinones.

Similarly, the group of Shi and group of Chuang and Cheng in 2013 revealed the palladium-catalyzed carbonylation of *ortho*-arylphenols leading to dibenzopyranones independently [39]. For the substrate scope (Table 15.29), a numerous functionalities can be tolerated, and about 30 examples were isolated, among which especially for the nitro group was intact albeit in low yields (31%). Remarkably, reactions proceed under basic conditions in methylene in the presence of sodium carbonate (2 equiv.) as the base and pivalic acid as the additive. Deuterium experiments indicate that the C–H metalation step might go through an S_EAr mechanism rather than the concerted metalation deprotonation (CMD) pathway.

Table 15.29 Carbonylation of *ortho*-hydroxy biaryl[a].

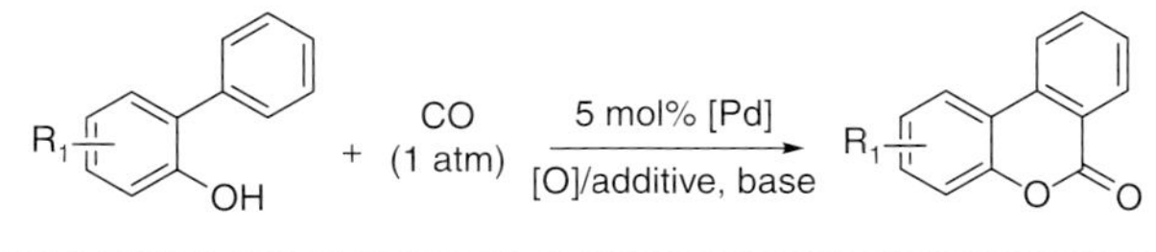

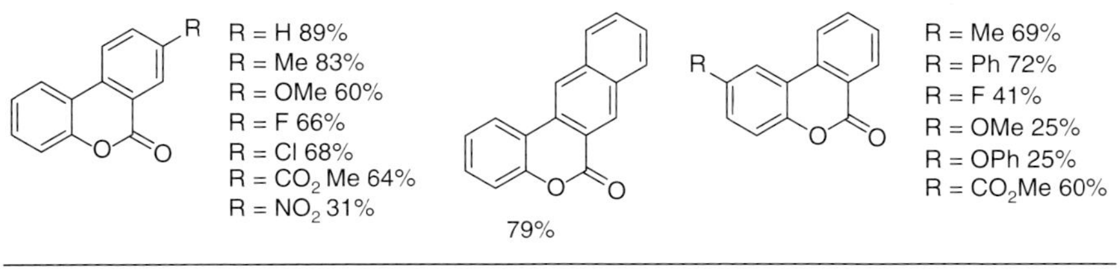

Reaction conditions: [a] *ortho*-arylphenol (0.2 mmol), $Pd(OAc)_2$ (5 mol%), $Cu(OAc)_2$ (10 mol%), Na_2CO_3 (2.0 equiv.), PivOH (0.5 equiv.), mesitylene (2.0 ml), CO (1 atm), air, 120 °C, 6 h.

Interestingly, in contrast to Shi's procedure, Chuang and Cheng's system applies the Ag(I) as the oxidant without extra acid or base, and high yield was obtained with broad substrate scope (Scheme 15.8).

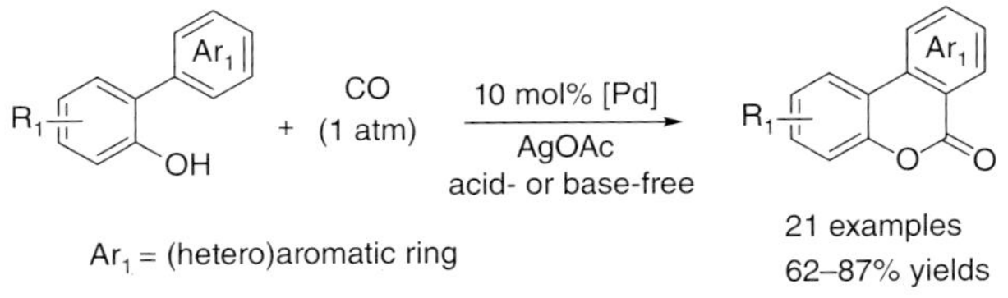

Scheme 15.8 Acid- or base-free procedure.

Our group in 2014 developed the first cabonylative [3+2+1] annulation of *N*-pyridyl anilines with internal alkynes and $Mo(CO)_6$ toward the facile synthesis of 2-quinolinones, and 19 examples were synthesized [40]. Here, the solid CO source $Mo(CO)_6$ was used instead of the CO gas which was highly toxic (Table 15.30). A series of symmetric and asymmetric internal alkynes could

Table 15.30 2-Quinolinone synthesis from carbonylative [3+2+1] annulation of *N*-pyridyl anilines with internal alkynes and $Mo(CO)_6$[a].

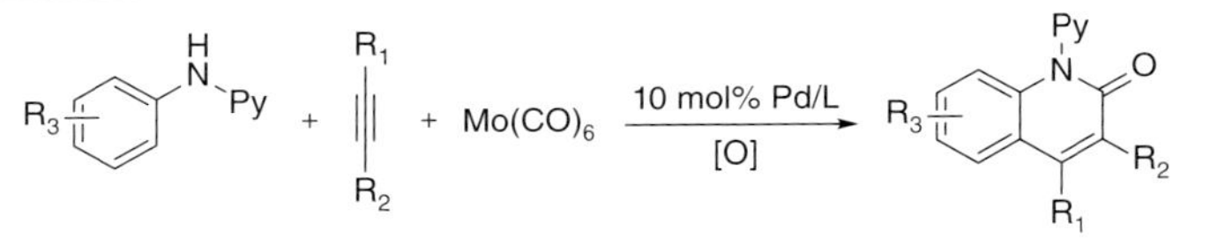

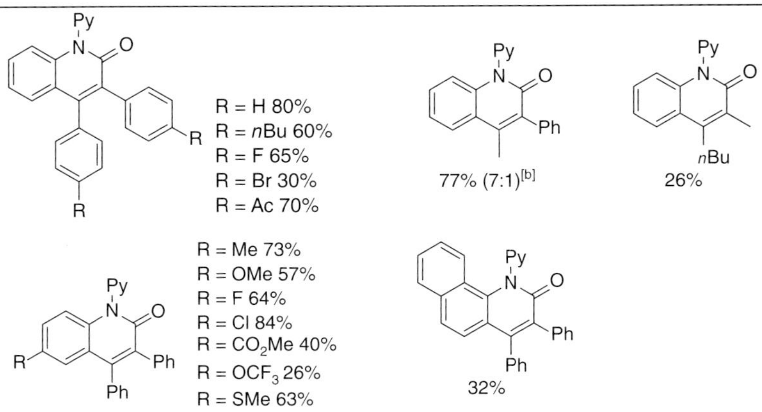

Reaction conditions: [a] *N*-pyridyl anilines (1.2 equiv.), alkyne (0.2 mmol), $Pd(OAc)_2$ (10 mol%), L-proline (20 mol%), BQ (2.0 equiv.), AgOAc (2.0 equiv.), HOAc (2 ml), $Mo(CO)_6$ (1.2 equiv.), air, 140 °C, 24 h. [b] Major products were given.

Table 15.31 Carbonylative γ-C–H functionalization of aliphatic acid[a].

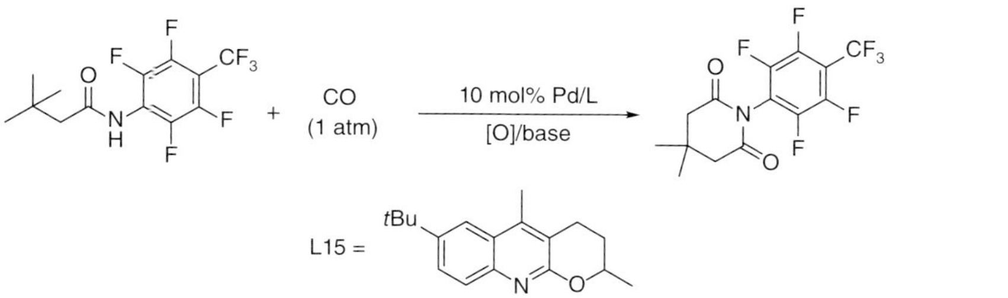

Reaction conditions: [a] aliphatic amides (0.1 mmol), $Pd(OAc)_2$ (10 mol%), ligand (20 mol%), AgOAc (2.0 equiv.), ${}^tBuOOBu^t$ (2.0 equiv.), KH_2PO_4 (2.0 equiv.), hexane (2 ml), CO (1 atm), 150 °C, 20 h.

be used as the coupling partners giving the quinolinones in moderate to good yields, and the regioselectivity was high. Substrates bearing electron-withdrawing or electron-donating groups on the aniline ring run smoothly delivering the products in moderate to good yields. KIE experiments ($k_H/k_D = 1.0$) show that the C–H bond metalation step might be not involved in the rate-determining step.

Yu *et al.* in 2014 accomplished the γ-C–H functionalization of aliphatic acid via utilizing a combination of a quinoline-based ligand and a weak coordinating directing group [41]. The corresponding lactam derivatives could be obtained in good yields which provide a new route for constructing richly functionalized all-carbon quaternary carbon centers at the β-position of aliphatic acids (Table 15.31). Only one example has been given in this context. The large-scale experiment highlighted the potential application as reaction was performed successfully in 1 g scale generating the product in 61% isolated yield.

Zhu *et al.* in 2014 reported the palladium-catalyzed pyridocarbonylation of *N*-pyridyl anilines forming 11*H*-pyrido[2,1-*b*]-quinazolin-11-one [42]. This process runs in the TFA as the solvent with $K_2S_2O_8$ as the oxidant under atmospheric pressure of CO, and 19 examples were listed (Table 15.32). Electron-donating or electron-withdrawing groups on the aromatic or the pyrido rings worked well giving the quinazolinone scaffold in good yields. For the *meta*-substituted group on the anilines, reactions favored the less sterically hindered position. Notably, the *N*-pyrimidyl aniline can provide the product in 40% yield. Here, the pyridyl group not only serves as the directing group but also as the intramolecular nucleophile for this transformation.

Lee *et al.* in 2014 exploited the phosphonic and phosphinic acid as the directing group for the carbonylative phosphaannulation in the presence of palladium catalyst, thus a series of oxaphosphorinanone oxides were obtained

Table 15.32 Selected examples for the pyridocarbonylation of *N*-pyridyl anilines to form 11*H*-pyrido[2,1-*b*]-quinazolin-11-one[a].

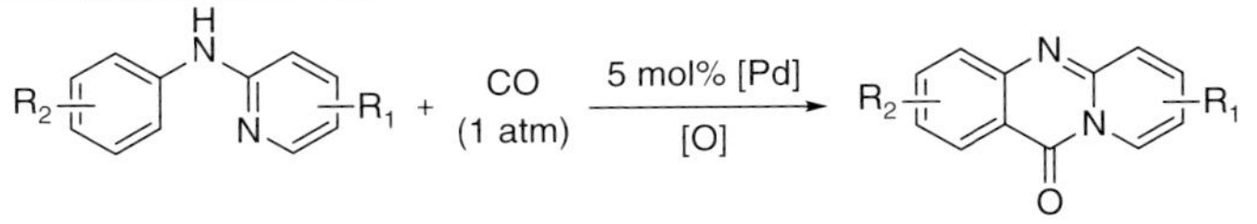

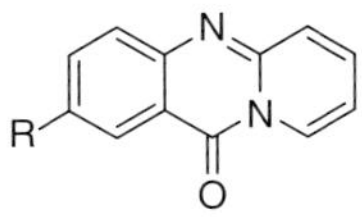

R = *t*Bu 77% 6,5 h
R = F 70% 6 h
R = Cl 77% 3 h[b]

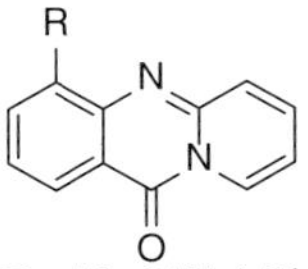

R = Me 81% 4,5 h
R = OMe 66% 3 h
R = F 77% 6 h[b]

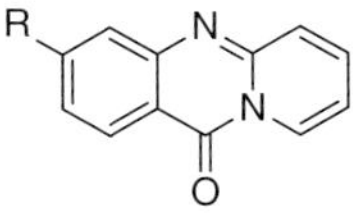

R = Me 59% 7 h
R = Ph 75% 4,5 h
R = Cl 70% 7 h[b] (10:6)[c]

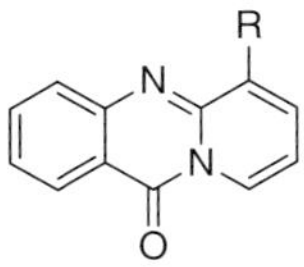

R = Me 86% 4 h
R = Cl 67% 7 h[b]

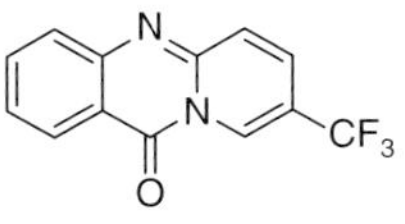

76% 1,5 h[b]

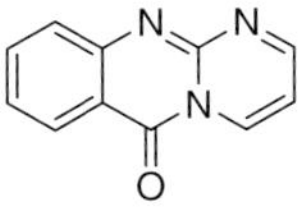

40% 4 h

Reaction conditions: [a] *N*-pyridyl anilines (0.2 mmol), $Pd(OAc)_2$ (5 mol%), $K_2S_2O_8$ (3.0 equiv.), TFA (1.0 ml), CO (1 atm; balloon), 70 °C, 24 h. [b] $K_2S_2O_8$ (5.0 equiv.). [c] Major product was given.

(Table 15.33) [43]. Effect of substituents at the α-position of the ethyl hydrogen benzylphosphonate indicates that introduction of two substituents at the α-position is essential for successful carbonylation. A series of electron-donating and electron-withdrawing groups can be tolerated and especially for the example of bromo show further manipulation opportunity. Not only the phosphonic acid but also the phosphinic acid proceeded well affording the corresponding products in 67% yield. Substrates bearing five- and six-membered spiro-type ring at the α-position are also an excellent starting material for this C–H carbonylation, delivering the desired oxaphosphorinanone oxides in good yields. KIE experiments ($k_H/k_D = 3.76$) indicated that C–H metalation at the *ortho* position of benzylphosphonic acid is most likely to be involved in the rate-determining step.

The synthesis of frutinone A and its derivatives was developed by Hong and Lee in 2014 via the palladium-catalyzed carbonylation of 2-phenolchromones [44]. Remarkably, several products were applied to test the inhibitory activities of the CYP1A2 enzyme in the nanomolar range and exerted potent efficiency.

Table 15.33 Pd(II)-catalyzed carbonylative phosphaannulation of phosphonic and phosphinic acid to prepare oxaphosphorinanone oxides[a].

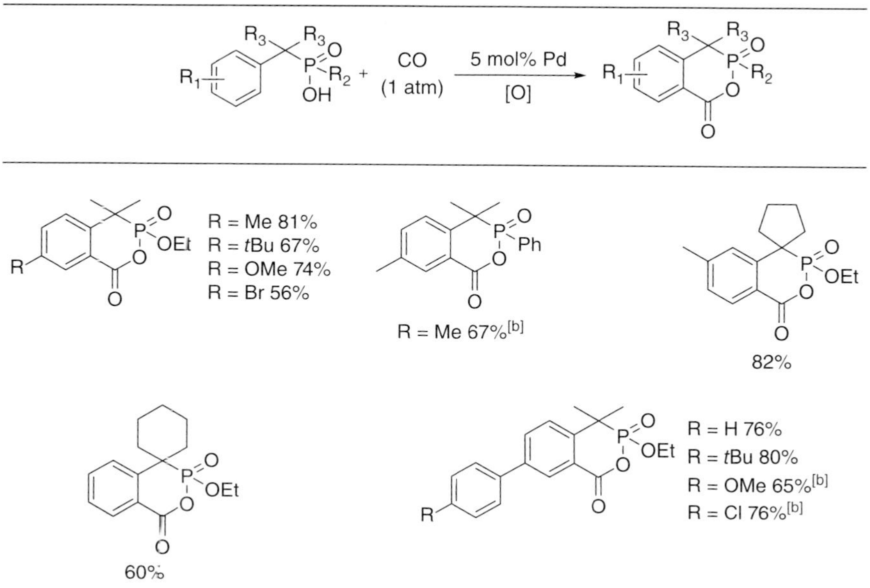

Reaction conditions: [a] ethyl hydrogen benzylphosphonates (0.2 mmol), $Pd(OAc)_2$ (5 mol%), $PhI(OAc)_2$ (1.5 equiv.), AgOAc (1.5 equiv.), DCE (2.0 ml), CO (1 atm), 60 °C, 20 h. [b] 36 h, $PhI(OAc)_2$ (2.0 equiv.) was used.

In 2015, Gevorgyan *et al.* disclosed the palladium-catalyzed silanol-directed carbonylation of phenol affording the 4*H*-benzo[*d*][1,3,2]dioxasilin-4-one which was then followed by desilylation giving the final salicylic acid [45]. Here, silanol serves as a powerful traceless directing group, thus producing a useful method for the preparation of salicylic acid from accessible phenol. Substrates containing electron-donating or electron-withdrawing groups on the *ortho*, *meta*, and *para* position of phenol are compatible (Table 15.34). Notably, α- or β-naphthalenol generates the corresponding salicylic acid in 66 or 85% yield, respectively. Interestingly, a special ether group can also work well in this system, and the heterocyclic aromatic phenol gave the product in 64% yield. The α,β-unsaturated ketonyl group at the *meta* position runs smoothly in 66% yield.

Interestingly, when the ^{18}O-label **^{18}O** was subjected to the standard conditions, results revealed the complete retention of the ^{18}O label in the silacyclic compound **^{18}O**. Thus, the intermediate should be formed in the catalytic cycles (Scheme 15.9).

Table 15.34 Selected products for silanol-directed carbonylation toward salicylic acid synthesis[a].

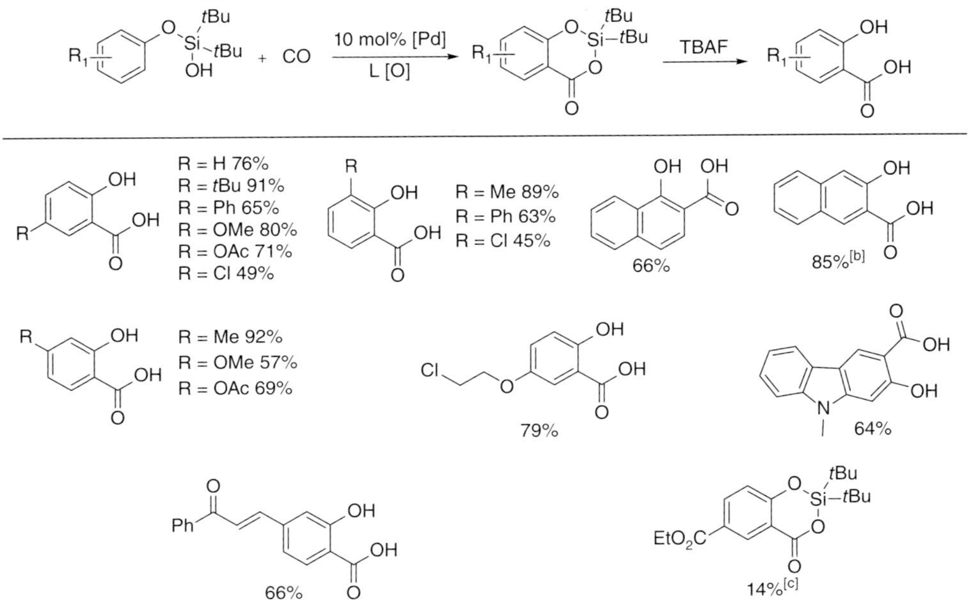

Reaction conditions: [a] substrate (0.2 mmol), $Pd(OAc)_2$ (10 mol%), Boc-Leu-OH (20 mol%), AgOAc (3.0 equiv.), CF_3CH_2OH (3.0 equiv.), DCE (0.8 ml), CO/Ar (1 : 8, balloon), 95 °C, 18 h. [b] A 2:1 ratio of regioisomers was formed; main product was shown. [c] Yield of the silacycle after the first step, as determined by NMR spectroscopy.

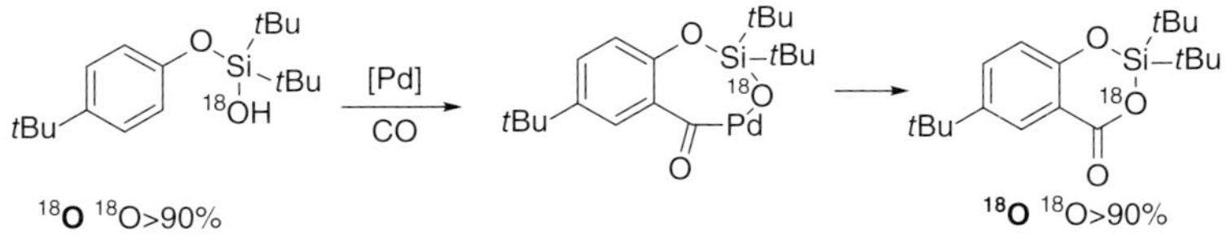

Scheme 15.9 ^{18}O-Labeling experiments.

15.6 Summary and Outlook

In this chapter, the transition metal-catalyzed carbonylative C–H functionalization toward heterocycle synthesis has been summarized. Four metal catalysts such as cobalt, ruthenium, rhodium, and palladium were explored. Among them, palladium obtained more attention than the other three metals. The main challenge for the Pd-catalyzed carbonylation is the excellent reducing ability of CO introducing the rapid reduction of Pd(II) to Pd(0). Solution to solve this problem includes the (i) utilization of unique ligand [30, 41] such as special α-amino acid, (ii) employment of cationic Pd(II) precursor [31, 46] which could enhance the nucleophilic attack of the C–H bond toward the palladium center, and (iii) usage of acid as solvent or as additive which also favors the generation of cationic palladium center [34, 40, 46, 47]. In the coming days, efforts should be focused on the following topics: (i) the development of mild reaction conditions such as reactions run at room temperature, the employment of molecular oxygen as a "green" oxidant, and acid-free conditions; (ii) the exploitation of directing group substrates; (iii) the application of double carbonylation to C–H heterocycle preparation; and (iv) the taking advantage of environmentally benign carbonyl sources for CO gas-free carbonylations.

References

1. (a) Schoenberg, A. and Heck, R.F. (1974) *J. Am. Chem. Soc.*, **96**, 7761–7764; (b) Schoenberg, A. and Heck, R.F. (1974) *J. Org. Chem.*, **39**, 3327–3331; (c) Schoenberg, A., Bartoletti, I., and Heck, R.F. (1974) *J. Org. Chem.*, **39**, 3318–3326.
2. For the recently selected reviews, please see: (a) Wu, X.F., Neumann, H., and Beller, M. (2011) *Chem. Soc. Rev.*, **40**, 4986–5009; (b) Wu, X.F., Neumann, H., and Beller, M. (2013) *Chem. Rev.*, **113**, 1–35.
3. For the recently selected reviews and examples, please see: (a) Chen, X., Engle, K.M., Wang, D.-H., and Yu, J.-Q. (2009) *Angew. Chem. Int. Ed.*, **48**, 5094–5115; (b) Tang, R.-Y., Li, G., and Yu, J.-Q. (2014) *Nature*, **507**, 215–220; (c) He, J., Li, S., Deng, Y., Fu, H., Laforteza, B.N., Spangler, J.E., Homs, A., and Yu, J.-Q. (2014) *Science*, **343**, 1216–1220; (d) Chu, L., Xiao, K.-J., and Yu, J.-Q. (2014) *Science*, **346**, 451–455; (e) Liu, Y.-J., Xu, H., Kong, W.-J., Shang, M., Dai, H.-X., and Yu, J.-Q. (2014) *Nature*, **515**, 389–393; (f) Sun, C.-L. and Shi, Z.-J. (2014) *Chem. Rev.*, **114**, 9219–9280; (g) Sun, C.-L., Li, B.-J., and Shi, Z.-J. (2010) *Chem. Rev.*, **111**, 1293–1314; (h) Ackermann, L., Vicente, R., and Kapdi, A.R. (2009) *Angew. Chem. Int. Ed.*, **48**, 9792–9826; (i) Yeung, C.S. and Dong, V.M. (2011) *Chem. Rev.*, **111**, 1215–1292; (j) Lyons, T.W. and Sanford, M.S. (2010) *Chem. Rev.*, **110**, 1147–1169.
4. For the recently selected reviews, please see: (a) Wu, X.F., Neumann, H., and Beller, M. (2013) *ChemSusChem*, **6**, 229–241; (b) Liu, Q., Zhang, H., and Lei, A. (2011) *Angew. Chem. Int. Ed.*, **50**, 10788–10799.
5. For the recently selected reviews and examples, please see: (a) Guo, X.-X., Gu, D.-W., Wu, Z., and Zhang, W. (2015) *Chem. Rev.*, **115**, 1622–1651; (b) Gulevich, A.V., Dudnik, A.S., Chernyak, N., and Gevorgyan, V. (2013) *Chem. Rev.*, **113**, 3084–3213.
6. Joh, T., Doyama, K., Fujiwara, K., Maeshima, K., and Takahashi, S. (1991) *Organometallics*, **10**, 508–513.
7. Funk, J.K., Yennawar, H., and Sen, A. (2006) *Helv. Chim. Acta*, **89**, 1687–1695.
8. Grigorjeva, L. and Daugulis, O. (2014) *Org. Lett.*, **16**, 4688–4690.
9. Du, Y., Hyster, T.K., and Rovis, T. (2011) *Chem. Commun.*, **47**, 12074–12076.
10. Fukuyama, T., Chatani, N., Kakiuchi, F., and Murai, S. (1997) *J. Org. Chem.*, **62**, 5647–5650.
11. Chatani, N., Kamitani, A., and Murai, S. (2002) *J. Org. Chem.*, **67**, 7014–7018.
12. Inoue, S., Shiota, H., Fukumoto, Y., and Chatani, N. (2009) *J. Am. Chem. Soc.*, **131**, 6898–6899.
13. (a) Hasegawa, N., Charra, V., Inoue, S., Fukumoto, Y., and Chatani, N. (2011) *J. Am. Chem. Soc.*, **133**, 8070–8073; (b)

Hasegawa, N., Shibata, K., Charra, V., Inoue, S., Fukumoto, Y., and Chatani, N. (2013) *Tetrahedron*, **69**, 4466–4472.
14. Shibata, K., Hasegawa, N., Fukumoto, Y., and Chatani, N. (2012) *ChemCatChem*, **4**, 1733–1736.
15. Inamoto, K., Kadokawa, J., and Kondo, Y. (2013) *Org. Lett.*, **15**, 3962–3965.
16. Li, W., Liu, C., Zhang, H., Ye, K., Zhang, G., Zhang, W., Duan, Z., You, S., and Lei, A. (2014) *Angew. Chem. Int. Ed.*, **53**, 2443–2446.
17. (a) Campo, M.A. and Larock, R.C. (2000) *Org. Lett.*, **2**, 3675–3677; (b) Campo, M.A. and Larock, R.C. (2002) *J. Org. Chem.*, **67**, 5616–5620.
18. Giri, R. and Yu, J.-Q. (2008) *J. Am. Chem. Soc.*, **130**, 14082.
19. Yoo, E.J., Wasa, M., and Yu, J.-Q. (2010) *J. Am. Chem. Soc.*, **132**, 17378–17380.
20. Wrigglesworth, J.W., Cox, B., Lloyd-Jones, G.C., and Booker-Milburn, K.I. (2011) *Org. Lett.*, **13**, 5326–5329.
21. Shi, R., Lu, L., Zhang, H., Chen, B., Sha, Y., Liu, C., and Lei, A. (2013) *Angew. Chem. Int. Ed.*, **52**, 10582–10585.
22. Song, C., Dong, S., Feng, L., Peng, X., Wang, M., Wang, J., and Xu, Z. (2013) *Org. Biomol. Chem.*, **11**, 6258–6262.
23. Li, W., Duan, Z., Zhang, X., Zhang, H., Wang, M., Jiang, R., Zeng, H., Liu, C., and Lei, A. (2015) *Angew. Chem. Int. Ed.*, **54**, 1893–1896.
24. Orito, K., Horibata, A., Nakamura, T., Ushito, H., Nagasaki, H., Yuguchi, M., Yamashita, S., and Tokuda, M. (2004) *J. Am. Chem. Soc.*, **126**, 14342–14343.
25. Orito, K., Miyazawa, M., Nakamura, T., Horibata, A., Ushito, H., Nagasaki, H., Yuguchi, M., Yamashita, S., Yamazaki, T., and Tokuda, M. (2006) *J. Org. Chem.*, **71**, 5951–5958.
26. (a) Wada, Y., Nagasaki, H., Tokuda, M., and Orito, K. (2007) *J. Org. Chem.*, **72**, 2008–2014; (b) Yamashita, S., Kurono, N., Senboku, H., Tokuda, M., and Orito, K. (2009) *Eur. J. Org. Chem.*, **2009**, 1173–1180.
27. (a) Vicente, J., Saura-Llamas, I., Garcıa-Lopez, J.A., and Calmuschi-Cula, B. (2007) *Organometallics*, **26**, 2768–2776; (b) Vicente, J., Saura-Llamas, I., Garcıa-Lopez, J.-A., and Bautista, D. (2009) *Organometallics*, **28**, 448–464.
28. (a) Haffemayer, B., Gulias, M., and Gaunt, M.J. (2011) *Chem. Sci.*, **2**, 312; (b) Albert, J., Ariza, X., Calvet, T., Font-Bardia, M., Garcia, J., Granell, J., Lamela, A., López, B., Martinez, M., Ortega, L., Rodriguez, A., and Santos, D. (2013) *Organometallics*, **32**, 649–659.
29. Li, Y., Yu, Z., and Alper, H. (2007) *Org. Lett.*, **9**, 1647–1649.
30. Lu, Y., Leow, D., Wang, X., Engle, K.M., and Yu, J.-Q. (2011) *Chem. Sci.*, **2**, 967–971.
31. Houlden, C.E., Hutchby, M., Bailey, C.D., Ford, J.G., Tyler, S.N., Gagne, M.R., Lloyd-Jones, G.C., and Booker-Milburn, K.I. (2009) *Angew. Chem. Int. Ed.*, **48**, 1830–1833.
32. Ma, B., Wang, Y., Peng, J., and Zhu, Q. (2011) *J. Org. Chem.*, **76**, 6362–6366.
33. Guan, Z.-H., Chen, M., and Ren, Z.-H. (2012) *J. Am. Chem. Soc.*, **134**, 17490–17493.
34. Zhang, H., Shi, R., Gan, P., Liu, C., Ding, A., Wang, Q., and Lei, A. (2012) *Angew. Chem. Int. Ed.*, **51**, 5204–5207.
35. Xu, Y., Zhou, J., Zhang, C., Chen, K., Zhang, T., and Du, Z. (2014) *Tetrahedron Lett.*, **55**, 6432–6434.
36. Chen, M., Ren, Z.-H., Wang, Y.-Y., and Guan, Z.-H. (2013) *Angew. Chem. Int. Ed.*, **52**, 14196–14199.
37. Elman, A.R. (2013) *Tetrahedron Lett.*, **54**, 5527–5531.
38. (a) Liang, Z., Zhang, J., Liu, Z., Wang, K., and Zhang, Y. (2013) *Tetrahedron*, **69**, 6519–6526; (b) Liang, D., Hu, Z., Peng, J., Huang, J., and Zhu, Q. (2013) *Chem. Commun.*, **49**, 173–175.
39. (a) Lee, T.-H., Jayakumar, J., Cheng, C.-H., and Chuang, S.-C. (2013) *Chem. Commun.*, **49**, 11797–11799; (b) Luo, S., Luo, F.X., Zhang, X.S., and Shi, Z.J. (2013) *Angew. Chem. Int. Ed.*, **52**, 10598–10601.
40. Chen, J., Natte, K., Spannenberg, A., Neumann, H., Beller, M., and Wu, X.-F. (2014) *Chem. Eur. J.*, **20**, 14189–14193.
41. Li, S., Chen, G., Feng, C.-G., Gong, W., and Yu, J.-Q. (2014) *J. Am. Chem. Soc.*, **136**, 5267–5270.
42. Liang, D., He, Y., and Zhu, Q. (2014) *Org. Lett.*, **16**, 2748–2751.
43. Shin, S., Jeong, Y., Jeon, W.H., and Lee, P.H. (2014) *Org. Lett.*, **16**, 2930–2933.

44. Shin, Y., Yoo, C., Moon, Y., Lee, Y., and Hong, S. (2015) *Chem. Asian J.*, **10**, 878–881.

45. Wang, Y. and Gevorgyan, V. (2015) *Angew. Chem. Int. Ed.*, **54**, 2255–2259.

46. Giri, R., Lam, J.K., and Yu, J.-Q. (2010) *J. Am. Chem. Soc.*, **132**, 686–693.

47. Lopez, B., Rodriguez, A., Santos, D., Albert, J., Ariza, X., Garcia, J., and Granell, J. (2011) *Chem. Commun.*, **47**, 1054–1056.

16
Synthesis of Natural Products and Pharmaceuticals via Catalytic C–H Functionalization

Junichiro Yamaguchi, Kazuma Amaike, and Kenichiro Itami

16.1
Introduction

Heterocycles are privileged structures with many interesting functions and fascinating optoelectronic or biological properties, and therefore, the construction of these scaffolds has been a topic of great importance in chemistry. Recently, the catalytic C–H functionalization of heterocycles has garnered much attention from the synthetic chemistry community as a next-generation method to construct such motifs [1]. The development of such C–H functionalization methods, particularly involving heteroaromatics with controlled regioselectivity, has been well established. However, application to the synthesis of complex and highly functionalized molecules still has considerable room for further investigation in terms of overcoming challenges in reactivity and regioselectivity, as well as rendering the reaction conditions milder. Recent developments have aimed to address these issues using the platform of complex natural product synthesis.

In this book chapter, methods in heterocycle substitution and synthesis using catalytic C–H functionalization are presented with a focus on complex-molecule synthesis, classified by heterocycle: indoles (Section 16.2.1); pyrroles (Section 16.2.2); carbazoles (Section 16.2.3); benzofurans (Section 16.2.4); imidazoles, oxazoles, and thiazoles (Section 16.2.5); quinazolines (Section 16.2.6); quinolines, isoquinolines, and phenanthridines (Section 16.2.7); pyridines (Section 16.2.8); and other heterocycles (Section 16.2.9).

16.2
Natural Products Containing Heteroaromatics

16.2.1
Indoles and Related Compounds

Indoles are commonly found in alkaloid natural products and pharmaceutical agents. Thus, tremendous efforts in the catalytic C–H functionalization reaction

Transition Metal-Catalyzed Heterocycle Synthesis via C—H Activation, First Edition. Edited by Xiao-Feng Wu.
© 2016 Wiley-VCH Verlag GmbH & Co. KGaA. Published 2016 by Wiley-VCH Verlag GmbH & Co. KGaA.

of this representative motif have been reported. These can be classified into two modes of catalytic C–H functionalization (Figure 16.1): the C–H functionalization of indoles (i.e., ring substitution) and the formation of indoles by C–H functionalization (i.e., ring synthesis). Regarding the former type, arylation (Sections 16.2.1.1 and 16.2.1.7) and alkenylation (Section 16.2.1.2) at the indole C3 position, alkenylation (Section 16.2.1.3) and alkylation (Sections 16.2.1.3 and 16.2.1.4) at the C2 position, alkylation (Section 16.2.1.5) and nitrene insertion (Section 16.2.1.7) at the C4 position, and borylation at the C7 position (Sections 16.2.1.6 and 16.2.1.7) have been reported (Figure 16.1a). Regarding the latter type, oxidative Larock indole synthesis (Section 16.2.1.8: C–H alkenylation), intramolecular oxidative α-arylation (Section 16.2.1.9: C–H alkenylation), and nitrenoid C–H insertion (Section 16.2.1.10) have been developed (Figure 16.1b). What follows are prototypical examples in the synthesis of indole alkaloids and related compounds enabled by these two modes of catalytic C–H functionalization.

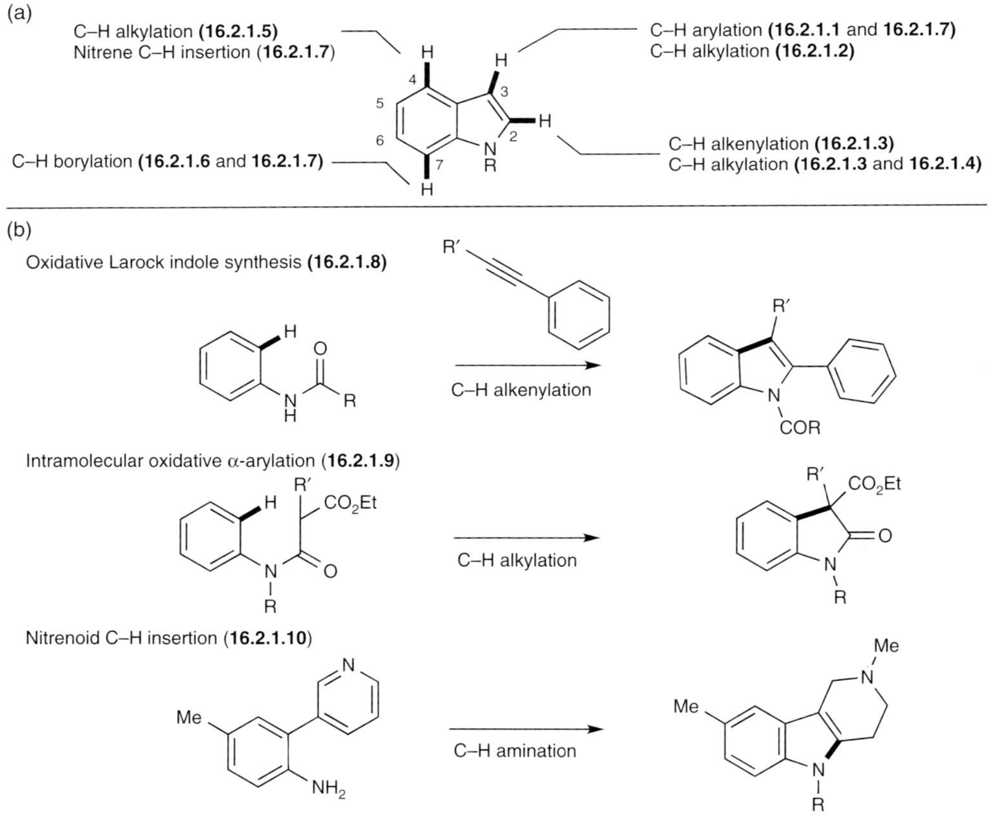

Figure 16.1 Two distinct modes of C–H functionalization in the synthesis of indole alkaloids: indole substitution and indole synthesis. (a) C–H functionalization of indoles and (b) formation of indoles by C–H functionalization.

16.2.1.1 Dragmacidin D (C–H Arylation of Indoles at the C3 Position)

In 2011, Yamaguchi, Itami, and coworkers achieved the synthesis of dragmacidin D, a marine bisindole alkaloid, using three direct C–H arylations (Scheme 16.1a) [2]. The synthesis commenced with the coupling of iodoindole **1** and 3-triisopropylsilyloxythiophene (**2**) in the presence of catalytic $Pd(OAc)_2/P[OCH(CF_3)_2]_3$ and Ag_2CO_3 to deliver thienylindole **3** in 60% yield with complete C4 regioselectivity [3]. Thienylindole **3** was then converted into ketone **4** through a three-step process. Next, C–H/C–H coupling (C–H arylation of indole at the C3 position) of **4** and pyrazine *N*-oxide (**5**) took place under palladium catalysis to afford the corresponding coupling product. Converting the resulting product to pyrazinone **6,** then an oxidative C–H/C–H coupling reaction (another example of C–H arylation of indole at the C3 position) of **6** and 6-bromoindole (**7**) with a catalytic amount of CF_3SO_3H under air afforded coupling product **8** with simultaneous removal of the two MOM (methoxymethyl) groups. The final transformation from **8** to dragmacidin D necessitated the following two steps: (i) treatment with i-Pr_2NEt/Me_3SiOTf and then *N*-bromosuccinimide and (ii) aminoimidazole formation from the resultant α-bromoketone and Boc-(tert-butylcarbonyl) guanidine, followed by removal of the Boc group. In this manner, the total synthesis of dragmacidin D was completed using C–H arylations in 15 steps total, which is 10 steps shorter than the preceding synthesis that employed Suzuki–Miyaura couplings [4]. Of note, this mode of catalytic C–H/C–H coupling was also effective in a concise total synthesis of eudistomin U (Scheme 16.1b) [5].

16.2.1.2 Clavicipitic Acid (C–H Alkenylation of Indoles at the C3 Position)

In 1995, an asymmetric total synthesis of (−)-clavicipitic acid through an intermolecular indole–alkene C–H/C–H coupling (C–H alkenylation at the C3 position of indole) was reported by Yokoyama, Murakami, and coworkers (Scheme 16.2) [6]. Their synthesis commenced with the coupling of bromoindole **11** with *N*-Boc-dehydroalanine methyl ester (**12**) by palladium catalysis. Although the previously reported conditions (1.0 equiv. of $Pd(OAc)_2$ in AcOH at 120 °C) [7] were ineffective, they found that catalytic $Pd(OAc)_2$ (25 mol%) with chloranil as an oxidant in 1,2,4-trichlorobenzene at 90 °C furnishes alkenylated product **13** in 38% yield. Asymmetric hydrogenation of **13** with a Rh–DIPAMP catalyst (94% ee), followed by vinylation under Heck conditions with **14** in the presence of Ag_2CO_3, provided coupling product **15** in 83% overall yield. After a few more steps, the synthesis of clavicipitic acid was completed. They also reported the synthesis of (−)-chanoclavine-I in a similar manner [8].

16.2.1.3 Paraherquamide B (Intramolecular C–H Alkylation of Indoles at the C2 Position)

Although a stoichiometric palladium complex was often required, palladium-catalyzed C–H alkylation of indole at the C2 position was used in syntheses of complex natural products decades ago. For example, in 1993, the total synthesis of (+)-paraherquamide B was reported by Williams and coworkers using a reductive

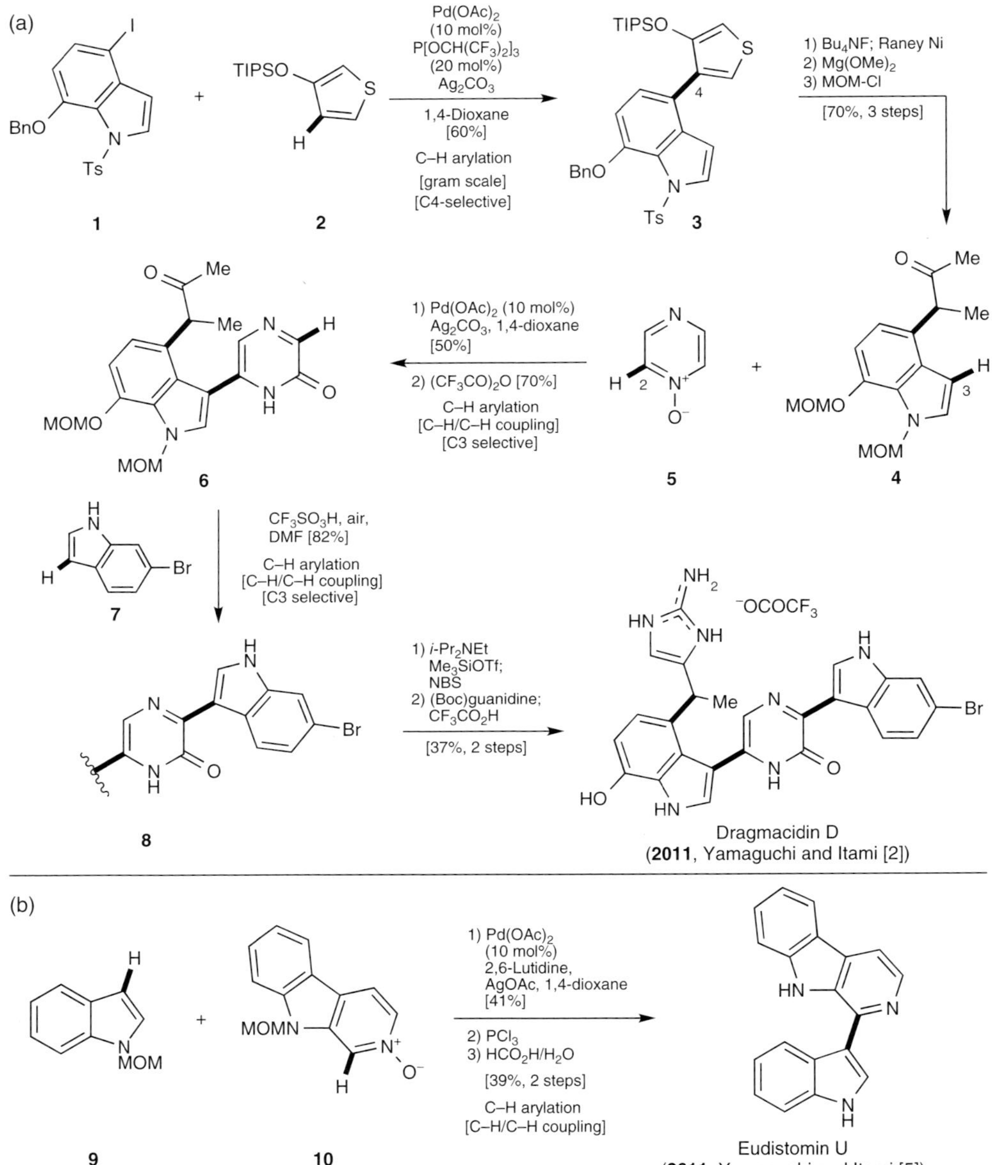

Scheme 16.1 Synthesis of dragmacidin D and eudistomin U enabled by C–H arylation of indoles at the C3 position.

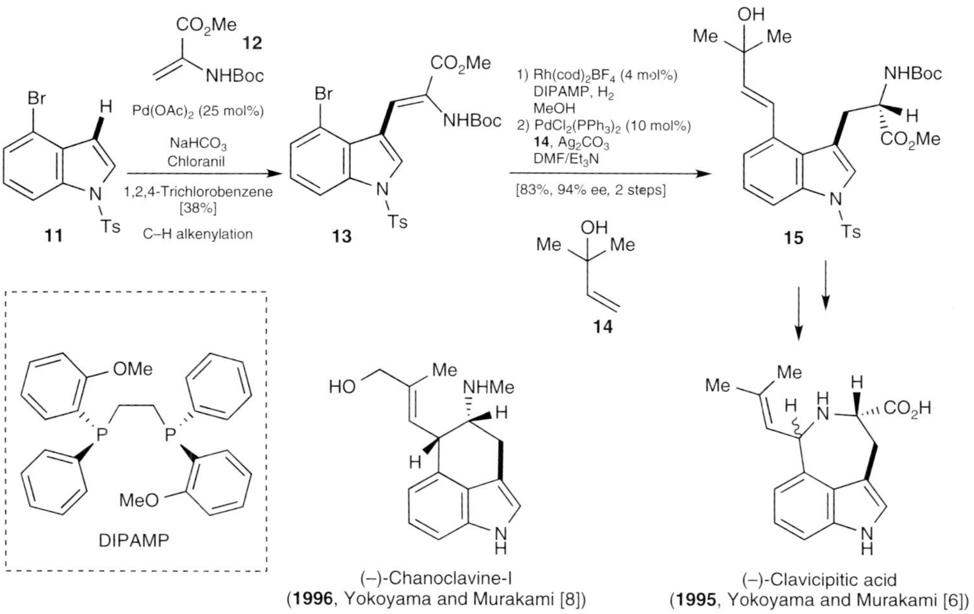

Scheme 16.2 Syntheses of clavicipitic acid and chanoclavine-I enabled by C–H alkenylation of indoles at the C3 position.

C–H alkylation strategy (Scheme 16.3a) [9]. Treatment of indole **16** with Pd(II) according to Trost's C–H alkylation procedure for the syntheses of ibogamine and catharanthine (see Scheme 16.3b) [10] afforded the corresponding heptacyclic compound. After several steps from **18,** the synthesis of (+)-paraherquamide B was accomplished. Williams and coworkers also achieved a concise, asymmetric, stereocontrolled total synthesis of stephacidins A and B and notoamide B in 2007 [11]. Additionally, in 2002 and 2003, Corey's group employed an intramolecular C–H alkylation as a key step for the total synthesis of okaramine N, austamide, hydratoaustamide, and deoxyisoaustamide (Scheme 16.3b) [12].

16.2.1.4 PKC Inhibitor (Intramolecular C–H Alkylation of Indoles at the C2 Position)

In 2006, Ellman, Bergmann, and coworkers reported the synthesis of a PKC inhibitor by using a rhodium-catalyzed intramolecular C–H alkenylation at the C2 position of the indole moiety (Scheme 16.4) [13]. Imine **19**, which was prepared from the corresponding aldehyde and a benzylamine, was treated with a rhodium complex and phosphoramidite ligand **20** in toluene, followed by hydrolysis of the imine group under acidic conditions to afford **22** in 61% yield with 90% ee (over two steps). Mechanistically, imine **19** directed the rhodium-catalyzed C–H activation via intermediate **21**, with subsequent intramolecular alkylation to give the desired product **22**. Thereafter, the synthesis of the target PKC inhibitor from **22** over several steps has been accomplished.

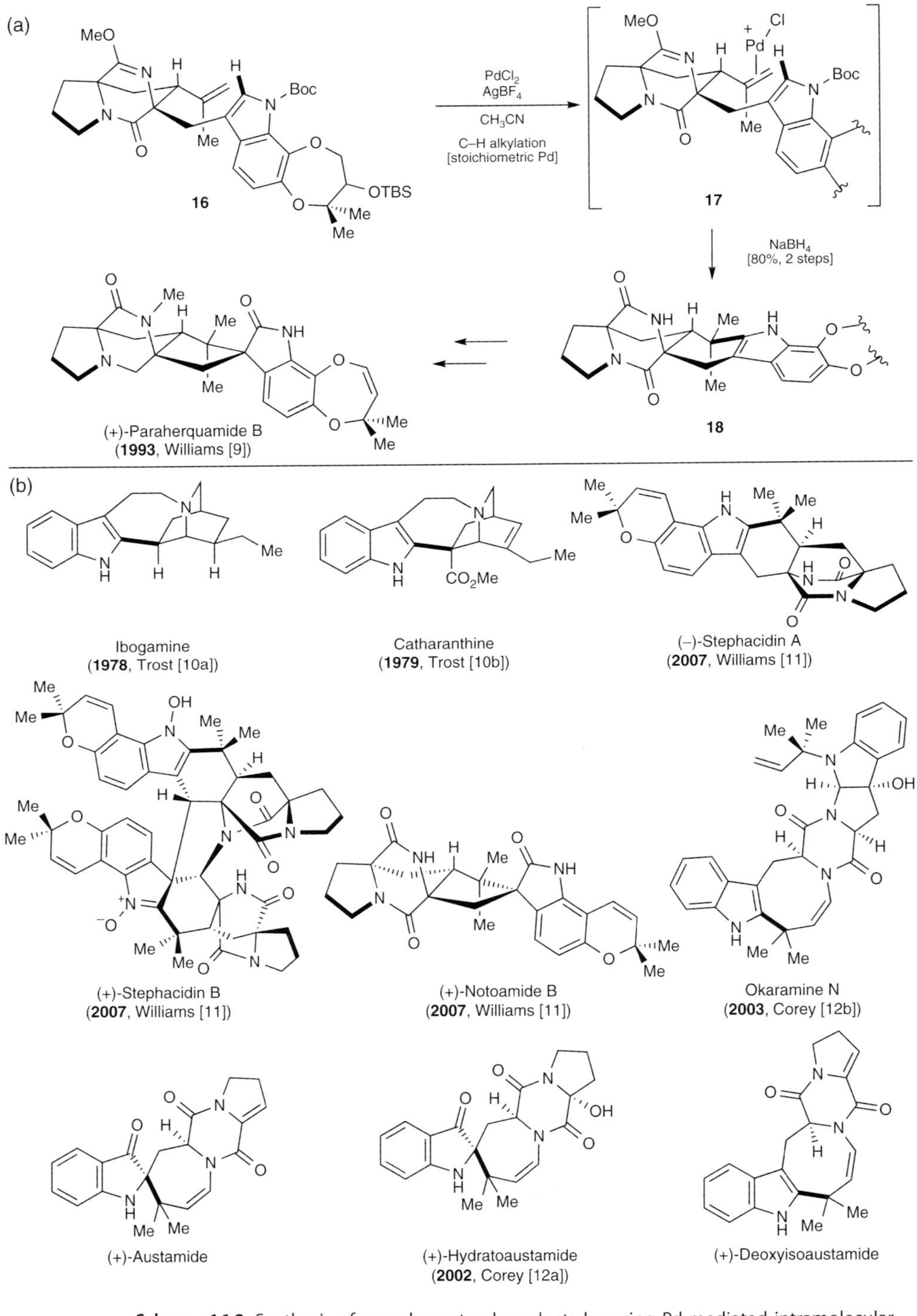

Scheme 16.3 Synthesis of complex natural products by using Pd-mediated intramolecular C–H alkylation of indoles at the C2 position.

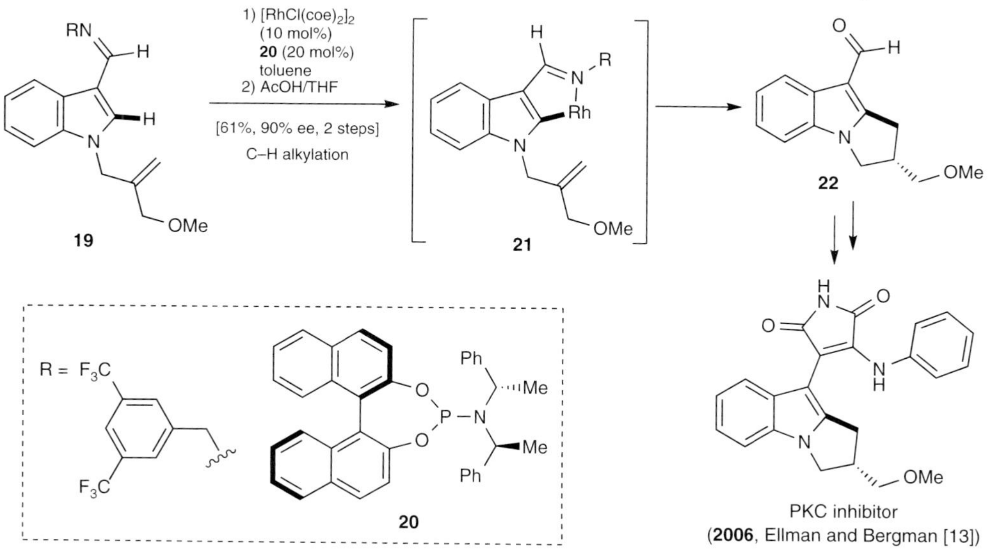

Scheme 16.4 Synthesis of a PKC inhibitor made possible by C–H alkylation of an indole at the C2 position.

16.2.1.5 Clavicipitic Acid (C–H Alkenylation of Indoles at the C4 Position)

In 2013, Jia and coworkers accomplished the total synthesis of clavicipitic acid (Scheme 16.5), which is the same natural product as shown in Section 16.2.1.3 [14]. In their synthetic strategy, regioselective C–H alkenylation at the C4 position of tryptophan derivative **23,** with a directing group on the amino unit, was designed as the key step. Indeed, C4-selective C–H alkenylation of tryptophan derivative **23** with alkene **24** in the presence of $Pd(OAc)_2$ and AgOAc proceeded to give alkenylated product **25** and cyclized product **26** in 17 and 14% yields, respectively. The reaction did not work without the triflate group on the amino unit of tryptophan **23.** Treatment with AgOAc converted **25** to desired **26** in good yield, albeit with lower diastereoselectivity. Compound **26** was then reacted with *t*-BuOK in DMSO (dimethyl sulfoxide) to give dehydrogenated product **27** in 80% yield, which was followed by several steps to give (−)-*trans*-clavicipitic acid and (+)-*cis*-clavicipitic acid.

16.2.1.6 Hippadine (C–H Borylation of Indoles at the C7 Position)

In 2010, Hartwig reported the synthesis of hippadine using an iridium-catalyzed C–H borylation (Hartwig–Miyaura–Ishiyama borylation) of indoles at the C7 position as the key step (Scheme 16.6) [15]. The reaction required a dimethylsilyl group on the *N*-substituent of indole to ensure regioselectivity for the C–H borylation step. Indole (**28**) was protected by the dimethylsilyl group, followed by C–H borylation (Ir complex and 4,4′-di-*tert*-butyl-2,2′-dipyridyl (dtbpy)) through plausible intermediate **30** to selectively afford a C7-borylated indole. The resulting 7-borylindole was coupled with aryl bromide **29** in the presence of a palladium

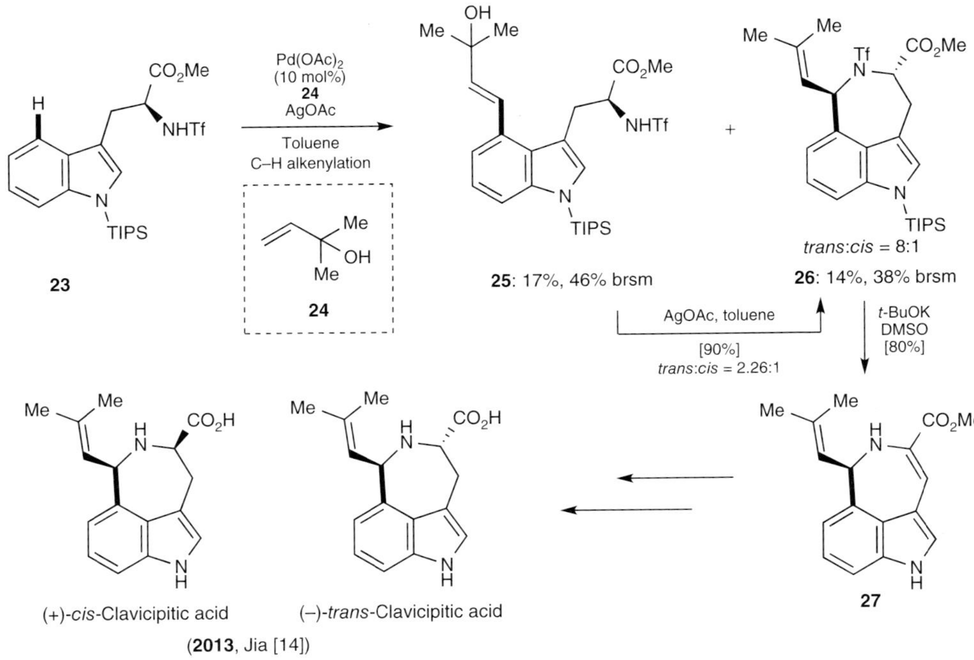

Scheme 16.5 Synthesis of clavicipitic acid using C–H alkenylation of an indole at the C4 position as the key step.

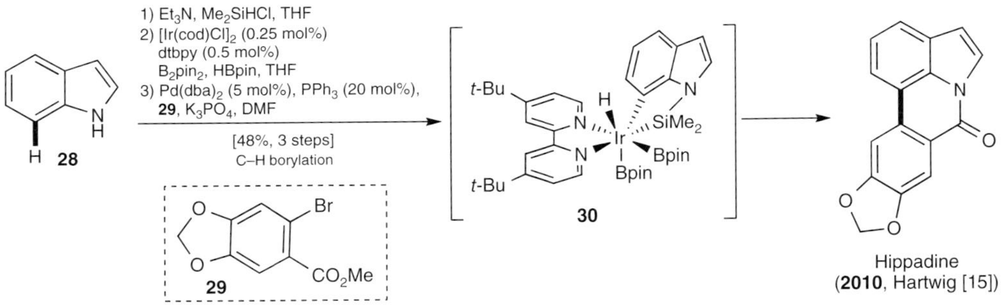

Scheme 16.6 Synthesis of hippadine enabled by C–H borylation of an indole at the C7 position.

catalyst, followed by subsequent cyclization to deliver hippadine in 48% yield over three steps.

16.2.1.7 Dictyodendrin B (C–H Arylation of Pyrroles at the C3 Position, C–H Borylation of Indoles at the C7 Position, and Nitrene C–H Insertion of Indoles at the C4 Position)

In 2015, Gaunt and coworkers reported an elegant synthesis of dictyodendrin B using three distinct types of catalytic C–H functionalization (Scheme 16.7) [16]. The first C–H functionalization involved a copper-catalyzed C–H arylation of

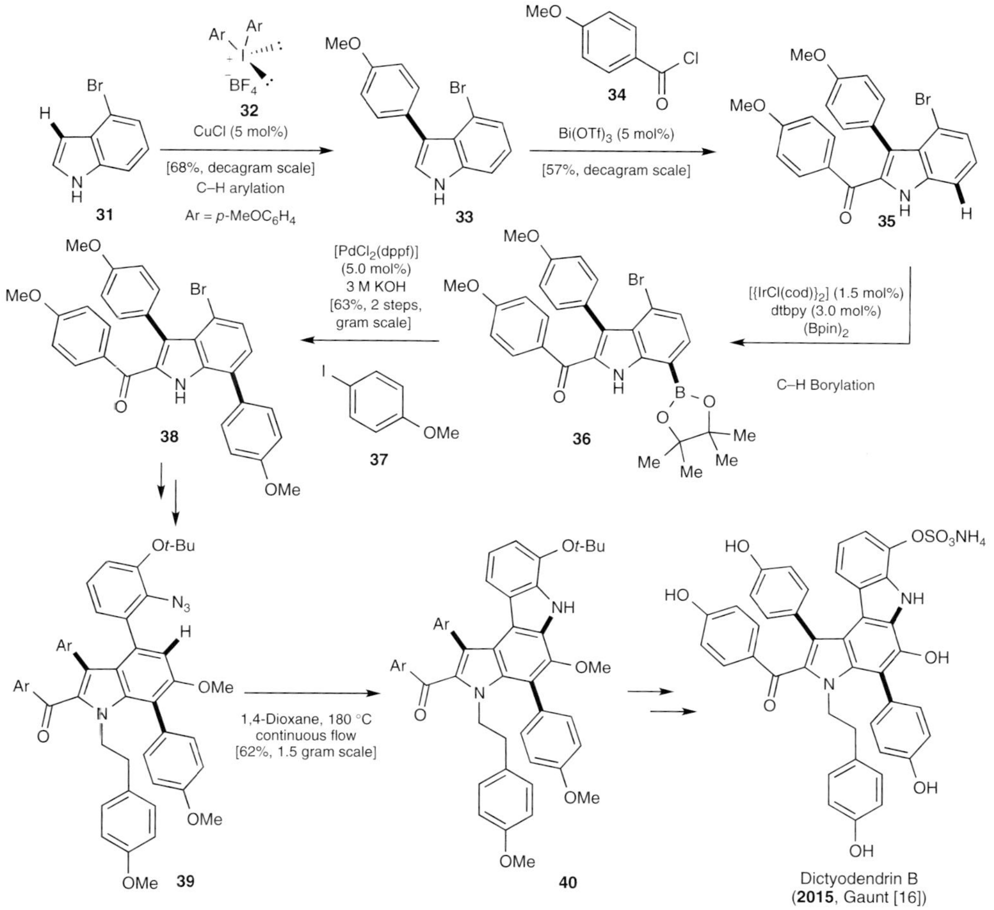

Scheme 16.7 Synthesis of dictyodendrin B showcasing three different types of heteroarene C–H functionalization.

4-bromoindole (**31**) with diaryliodonium salt **32** to afford C3-arylated indole **33** in good yield and regioselectivity (conducted on 42 g). Acylation of the resulting indole **33** with aroyl chloride **34** by using a bismuth catalyst gave C2-acylated indole **35** in moderate yield but in large quantities. The second C–H functionalization, an iridium-catalyzed borylation of **35,** proceeded at the C7 position of indole due to the effect of the indole N–H bond as a directing group, affording boronate ester **36.** Boronate **36** then underwent Suzuki–Miyaura coupling with 4-methoxyphenyl iodide (**37**) under standard palladium-catalyzed conditions to generate triarylindole **38** in 63% over two steps. After several steps from triarylindole **38** to **39**, the last C–H functionalization, a nitrene C–H insertion, was performed under light irradiation to complete the synthesis of the dictyodendrin B framework **40** (conducted on 1.5 g). After converting the *tert*-butoxy group of **40** to a sulfate salt by using a known four-step operation, the total synthesis of dictyodendrin B has been accomplished. It is of note that this synthesis achieves both indole ring substitution and indole ring synthesis using C–H functionalization strategies.

16.2.1.8 Paullone (Oxidative Larock Indole Synthesis)

In 2010, Stuart and Fagnou reported an oxidative Larock-type heterocycle synthesis using rhodium catalysis and applied it to the synthesis of paullone (Scheme 16.8) [17]. In the synthesis of paullone, acetanilide (**41**) was coupled with alkyne **42** in the presence of a rhodium catalyst and $Cu(OAc)_2{\cdot}H_2O$ as an oxidant under air to form disubstituted indole **44** via intermediate **43** in 71% yield. Removal of acetyl and Boc groups, followed by treatment with DBU (1,8-diazabicyclo[5.4.0]undec-7-ene), resulted in paullone in a total of eight steps.

16.2.1.9 Horsfiline (Indole Synthesis by Intermolecular C–H Coupling)

In 2010, Taylor and coworkers reported the formal synthesis of an anticancer and analgesic oxindole alkaloid, horsfiline, using a copper-catalyzed intermolecular C–H coupling (Scheme 16.9) [18]. They had already reported a stoichiometric approach to oxindoles using 1.0 equiv. of $Cu(OAc)_2{\cdot}H_2O$ and 1.1 equiv. of KO*t*-Bu in DMF (*N*,*N*-dimethyl formamide) at 110 °C in air [19]. After a slight change in these conditions (without base, in mesitylene, and heating at 165 °C), the reaction was rendered catalytic (5 mol% $Cu(OAc)_2{\cdot}H_2O$). With this catalytic system in hand, they targeted spirooxindole **47**. Intramolecular C–H/C–H coupling of amide **45** afforded coupling product **47** in 56% yield. The transformation from oxindole **47** to horsfiline was already known, and therefore the synthesis of horsfiline was achieved following previously reported procedures [20]. This C–H alkylation is an effective method for constructing quaternary centers without prefunctionalization, shortening the synthesis of horsfiline.

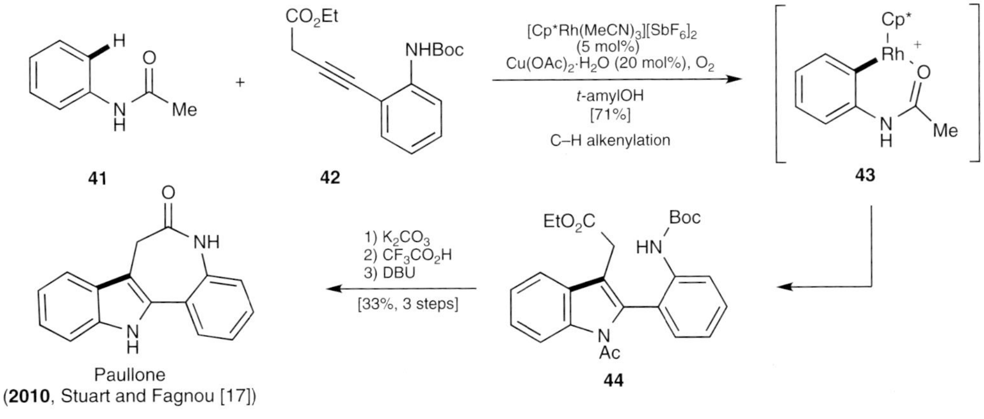

Scheme 16.8 Synthesis of paullone using oxidative Larock indole synthesis.

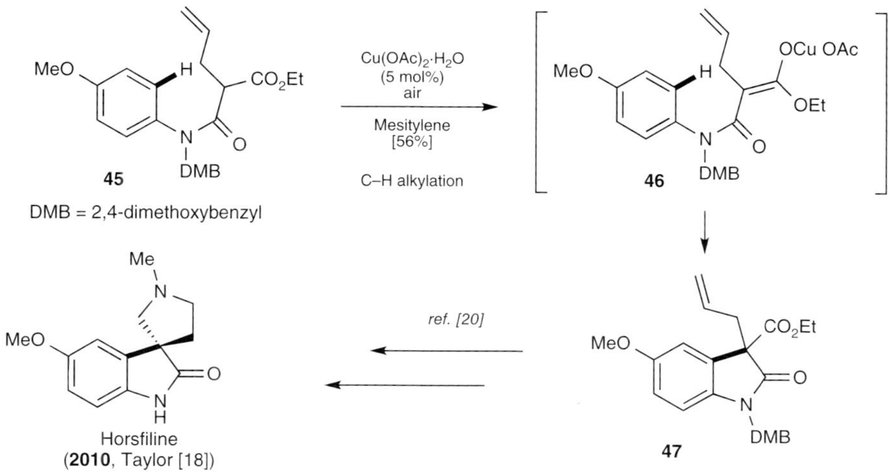

Scheme 16.9 Formal synthesis of horsfiline using intermolecular C–H coupling.

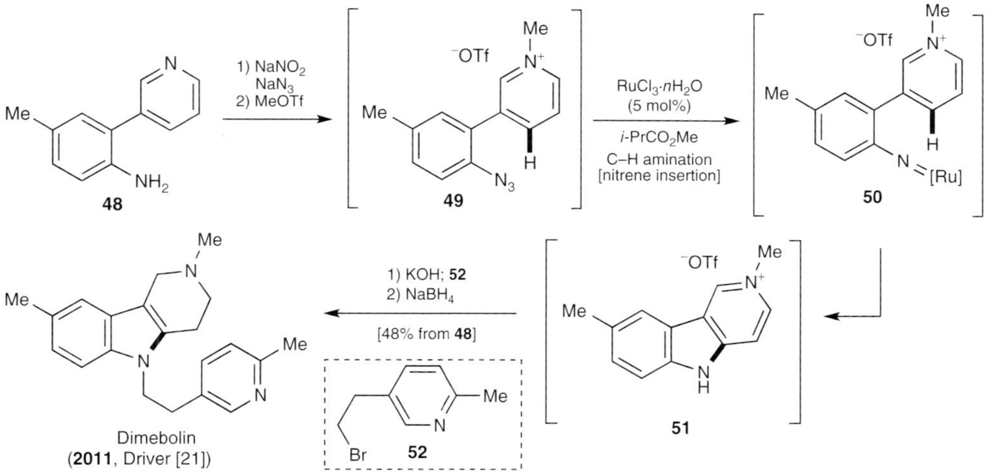

Scheme 16.10 Synthesis of dimebolin using a nitrenoid C–H insertion reaction.

16.2.1.10 **Dimebolin (Indole Synthesis by Nitrenoid C–H Insertion Reaction)**

In 2011, Driver and coworkers synthesized dimebolin by utilizing a ruthenium-catalyzed nitrenoid insertion reaction that they developed (Scheme 16.10) [21]. Azide **49** was readily prepared from **48** by a two-step sequence, consisting of treatment with sodium nitrite and sodium azide, and alkylation of pyridine. Addition of catalytic $RuCl_3$ hydrate to **49** formed putative ruthenium nitrenoid **50** [22], which then underwent regioselective C–H amination at the C4 position of the pyridinium ion to give γ-carbolinium salt **51** in 91% yield. Deprotonation of the N–H group of **51**, followed by alkylation with **52**, appended the requisite pyridine side chain. Finally, treatment of the resulting compound with sodium borohydride completed the synthesis of dimebolin in 48% yield from **48**.

16.2.2
Pyrroles and Related Compounds

In 1988, a synthesis of prodigiosin that makes use of an intramolecular C–H/C–H coupling strategy (C–H arylation of pyrroles at the C2 position) was reported by Boger and coworkers (Scheme 16.11) [23]. 1,1′-Carbonyldipyrrole derivative **53** was treated with a polymer-supported $Pd(OAc)_2$ catalyst in AcOH to afford **54** in 95% yield. This transformation most likely occurs through double C–H palladation of two pyrrole rings to Pd(II), followed by reductive elimination of **55** from intermediate **54**. After three steps, the total synthesis of prodigiosin was achieved. Using this C–H/C–H coupling strategy, they also synthesized a related family member, prodigiosene. After this early contribution, natural product synthesis using C–H functionalization of pyrroles had not been reported until 2005.

16.2.2.1 Rhazinilam (Intramolecular and Intermolecular C–H Arylation of Pyrroles at the C4 Position)

In 2005, Trauner and coworkers reported the synthesis of rhazinilam using a palladium-catalyzed intramolecular C–H arylation of a pyrrole unit with an iodoarene (Scheme 16.12a) [24]. Treatment of compound **57** with a palladium catalyst resulted in the formation of coupling product **59** through intermediate **58**. After removal of the MOM group and decarboxylation, the synthesis of

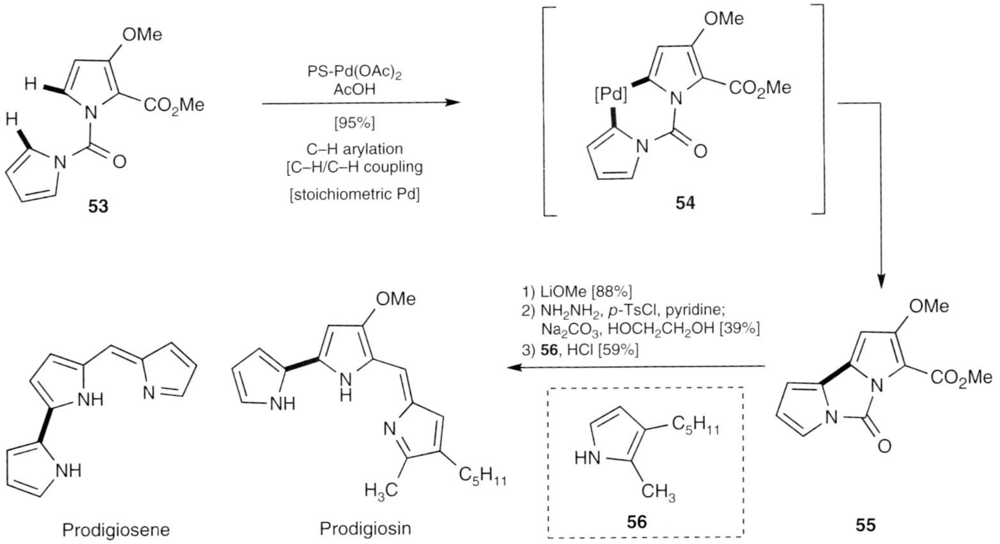

Scheme 16.11 An early contribution for the synthesis of a pyrrole-containing natural product by using an intramolecular C–H arylation strategy.

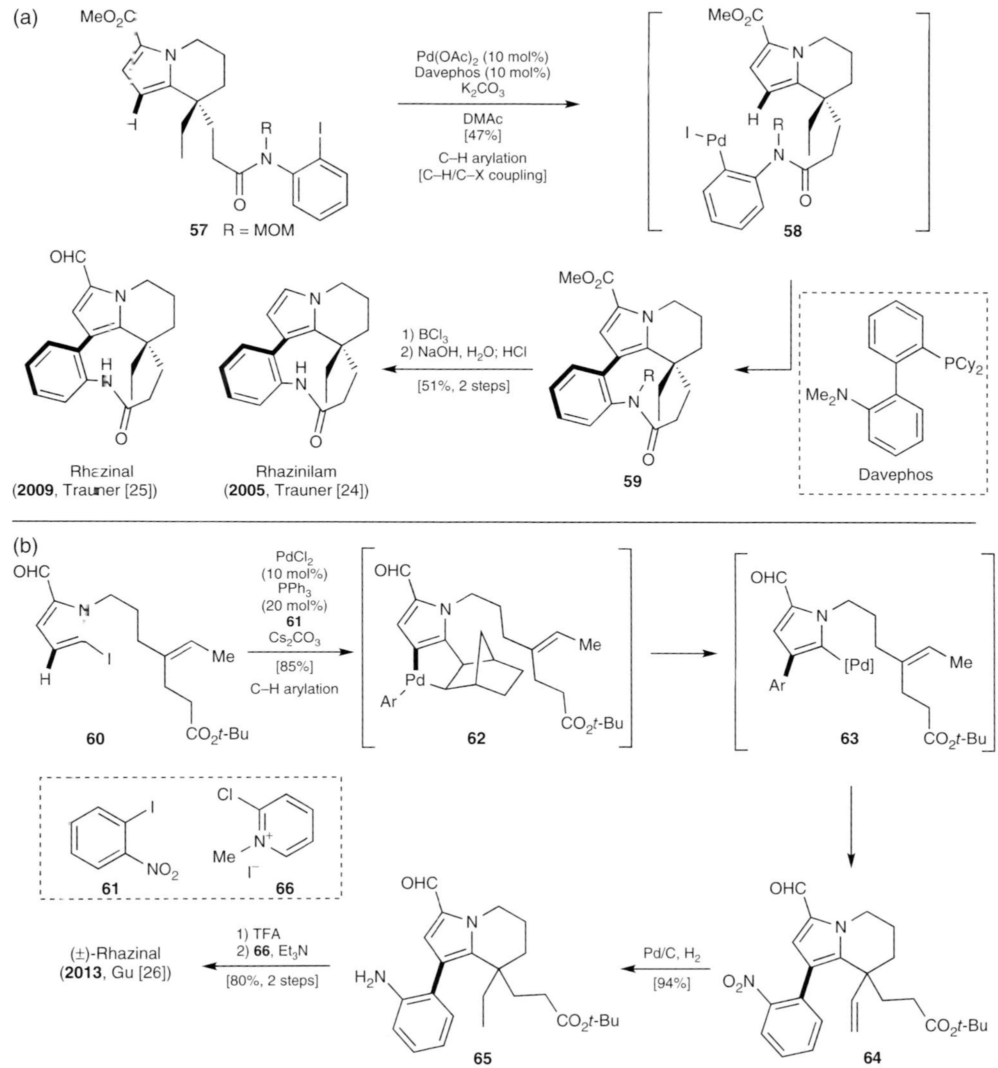

Scheme 16.12 Synthesis of rhazinal: a comparison of intramolecular and intermolecular C–H functionalization strategies.

rhazinilam was achieved. In 2009, Trauner also reported the synthesis of rhazinal, a congener of rhazinilam, by extending this intramolecular coupling strategy [25].

In contrast, Gu and coworkers achieved the synthesis of rhazinal by using an *intermolecular* C–H arylation of pyrroles at the C3 position and subsequent Heck alkylation (Scheme 16.12b) [26]. When iodopyrrole **60** was reacted with aryl iodide **61** in the presence of catalytic $PdCl_2$, PPh_3, and excess amounts of norbornene (6 equiv.), a Catellani-type reaction occurred to give intermediate **62.** This intermediate then underwent reductive elimination and norbornene release to generate intermediate **63.** Arylpalladium **63** then reacted with an internal alkene by intramolecular Heck reaction to form the desired product **64** in 85% yield. Next, hydrogenation of the double bond and the nitro group of **64** gave **65** in 94% yield. Removal of the *tert*-butyl group of **65** by treatment with trifluoroacetic acid (TFA), followed by cyclization between the generated carboxylic acid and a pendant aniline using Mukaiyama's reagent (**66**), afforded rhazinal in 80% yield over two steps. This synthesis ingeniously achieved an intermolecular C–H arylation at the C3 position of a pyrrole, as well as an alkylation at the C2 position of the pyrrole at the same time.

16.2.2.2 Rhazinilam and Aspidospermidine (C–H Borylation and C–H Alkylation of Pyrroles at the C4 and C5 Positions)

In 2012, the Gaunt group also reported the synthesis of rhazinilam by using two metal-catalyzed C–H functionalizations (C–H borylation and C–H alkylation) of pyrroles (Scheme 16.13a) [27]. Although they had already accomplished the synthesis of rhazinicine, another congener of rhazinilam [28], the same strategy was applied to the synthesis of rhazinilam and kopsiyunnanine C3 and cleverly extended to the synthesis of aspidospermidine by using a reductive transannular cyclization. Their synthesis began with Boc-pyrrole **67**, onto which Ir-catalyzed C–H borylation gave C3-borylated pyrrole **68**. This intermediate was reacted with **69** (Suzuki–Miyaura cross-coupling reaction), and the subsequent Boc removal afforded C3-arylated pyrrole **70** in 63% yield with virtually complete regioselectivity. Alkylation of the N–H group of pyrrole **70** with **71** proceeded smoothly to give the desired product **72** in 85% yield.

The resulting pyrrole **72** underwent C–H alkylation by the action of $Pd(OAc)_2$, PivOH, and NaO*t*-Bu to afford **73** in 60% yield, and this was converted to rhazinilam in a few steps. Rhazinilam was then transformed to aspidospermidine by the following four steps: (i) Boc protection of the amide group; (ii) reduction of the amide by $LiBHEt_3$, then acetylation; (iii) formation of acyliminium intermediate **74** using TFA, with one-pot reductive cyclization by treatment with $NaCNBH_3$; and finally, (iv) removal of the Boc group using TFA. This synthesis did not only produce highly functionalized pyrrole alkaloids but also gained entry into *Aspidosperma* alkaloids by reducing a pyrrole group (which is not a common retrosynthetic strategy toward this natural product family).

In 2014, Lin, Yao, and coworkers also synthesized rhazinilam by palladium-catalyzed C–H functionalization (Scheme 16.13b) [29]. In their synthesis, pyrrole **75**, which is slightly different from Gaunt's intermediate **72**, was used. The palladium-catalyzed C–H alkylation of **75** successfully occurred, and they accomplished the synthesis of rhazinilam after several more steps.

16.2.2.3 Lamellarins C and I (Inter- and Intramolecular C–H Arylation of Pyrroles at the C2, C3, and C4 Positions)

In 2014, Itami, Yamaguchi, and coworkers accomplished the synthesis of lamellarins C and I by using a C–H arylation at the C3 position and intramolecular C–H arylations at the C2 and C4 positions of pyrroles (Scheme 16.14) [30].

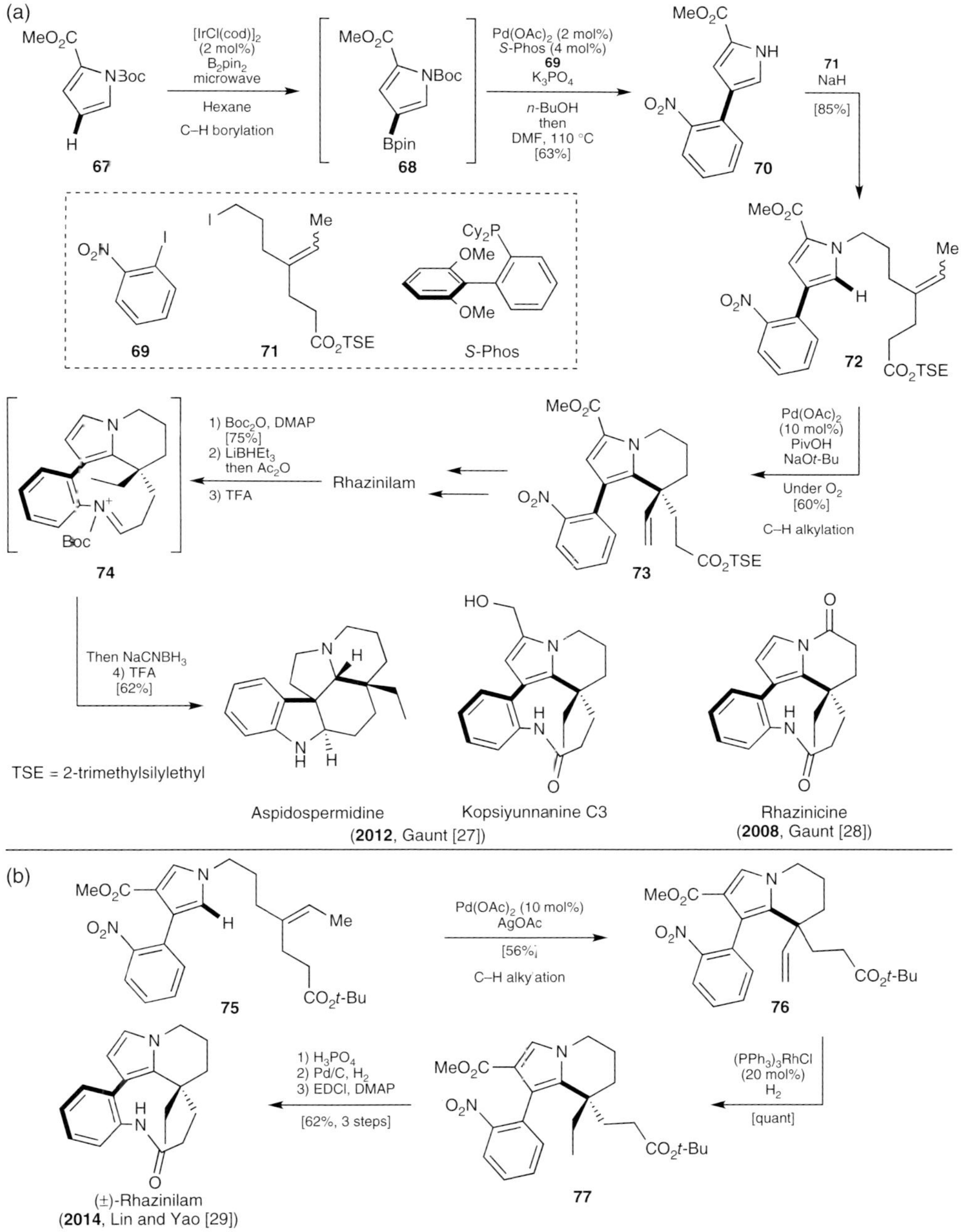

Scheme 16.13 Synthesis of rhazinilam and its congeners, kopsiyunnanine C3, and aspidospermidine.

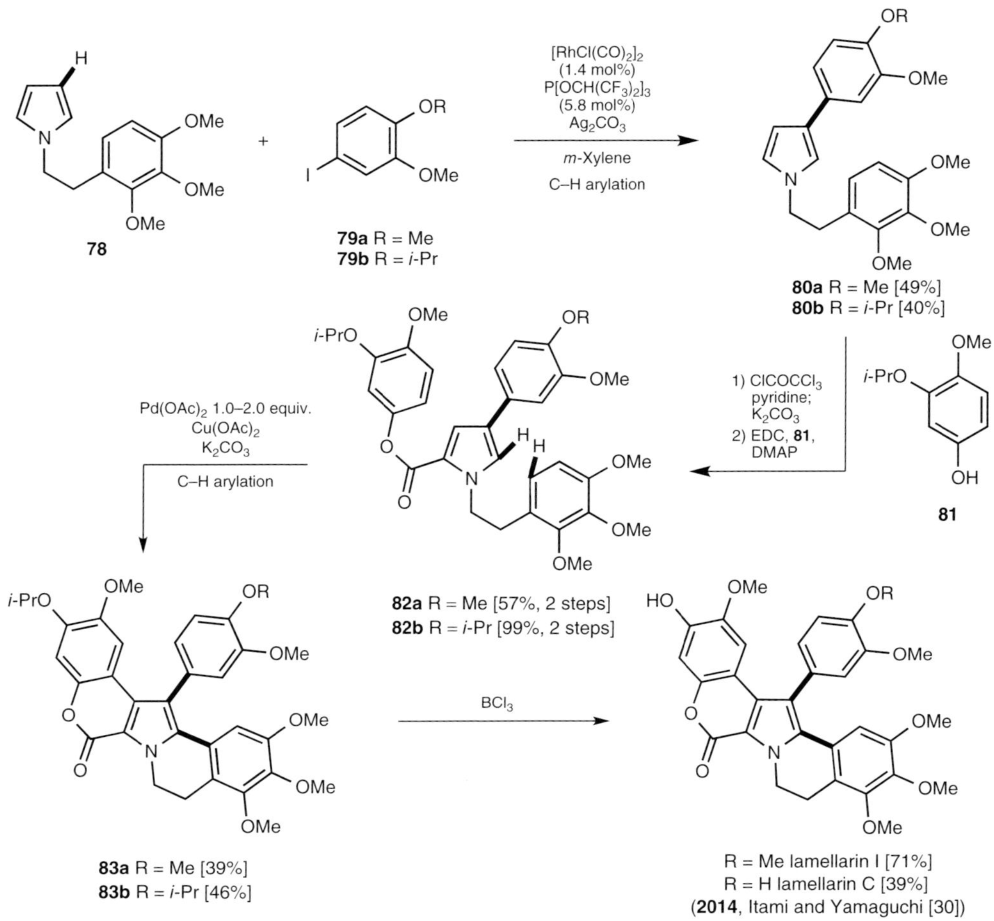

Scheme 16.14 Synthesis of lamellarins I and C accomplished by a series of C–H functionalization events.

In order to generate these natural products, they have developed a regioselective C–H arylation of pyrrole at the C3 position by using a rhodium catalyst bearing a highly electron-deficient phosphite ligand $P[OCH(CF_3)_2]_3$ [31]. The synthesis began from the coupling of pyrrole **78** and aryl iodide **79a**/**79b** in the presence of the aforementioned rhodium catalyst and Ag_2CO_3 to afford the corresponding C3-arylated pyrroles **80a** and **80b** in moderate yields with complete C3 selectivity. Trichloroacetylation of **80a** and **80b**, hydrolysis of the resultant trichloromethyl ketone, and condensation with phenol **81** gave targets **82a** and **82b** in good yield. While an intramolecular oxidative coupling between the pyrrole and the tethered arene was attempted, general oxidants such as DDQ (2,3-dichloro-5,6-dicyano-p-benzoquinone) and hypervalent iodine reagents did not result in desired product. After significant experimentation, treatment of **82** with a stoichiometric $Pd(OAc)_2$/$Cu(OAc)_2$/base system gave highly functionalized pyrroles **83a** and

83b in 39 and 46% yields, respectively. Finally, the removal of the isopropyl group from **83a** and **83b** achieved the rapid synthesis of lamellarins I and C, respectively.

16.2.2.4 Dictyodendrins A and F (C–H Arylation and C–H Insertion of Pyrroles on C2, C3, and C5 Position)

In 2015, Yamaguchi, Itami, Davies, and coworkers accomplished the synthesis of dictyodendrins A and F, pyrrole–carbazole alkaloids, by catalytic C–H functionalization (Scheme 16.15; compare to Section 16.2.1.7) [32]. This synthesis also utilized a rhodium-catalyzed β-selective C–H arylation of pyrroles with aryl iodides as the initial reaction. Pyrrole **84** was coupled with aryl iodide **85** in the presence of the same electron-deficient rhodium catalyst to afford

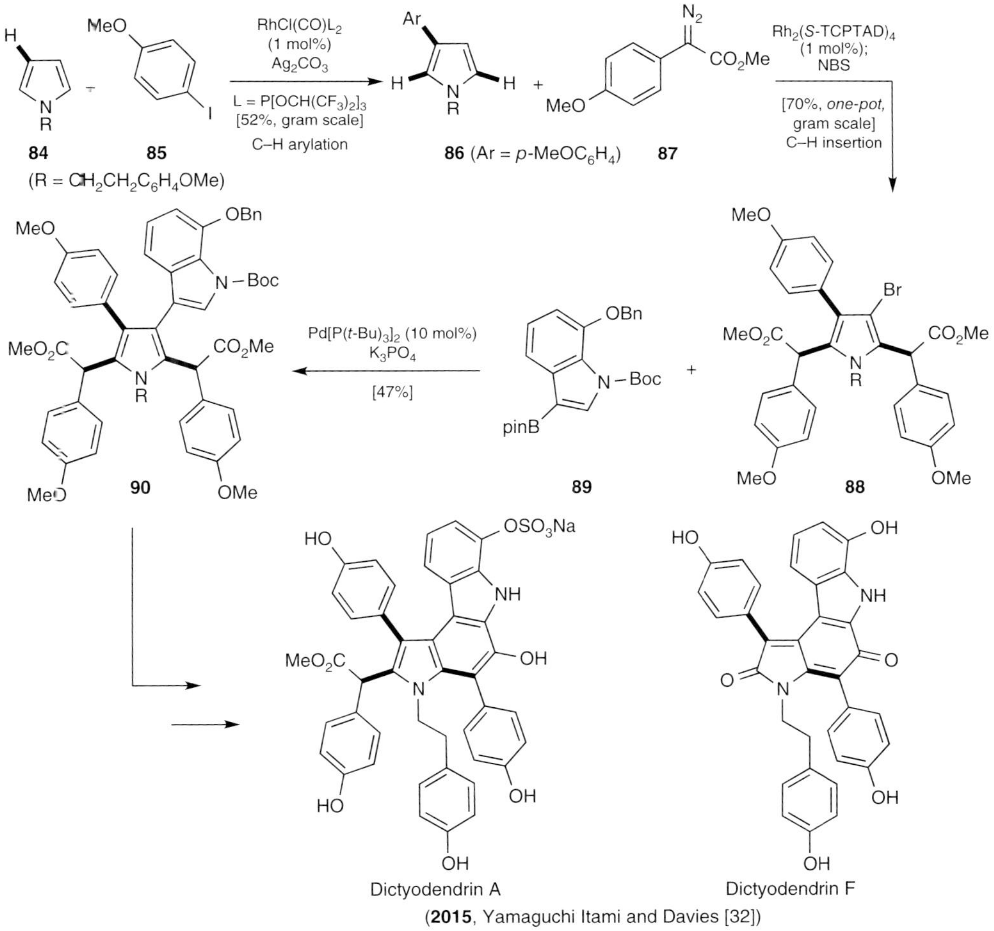

Scheme 16.15 Synthesis of dictyodendrins A and F that relies on three C–H functionalizations.

β-arylated pyrrole **86** in 52% yield on gram scale. Subsequently, C–H insertion of pyrrole **86** with diazoester **87** was conducted. Using a rhodium(II) complex, Rh_2(*S*-TCPTAD)$_4$ (S-2-adamntyl-2-(1,3-dioxoisoindolin-2-yl)acetate) [33], a double C–H insertion reaction proceeded smoothly and regioselectively at the C2 and C5 positions of pyrrole **86**, which was followed by bromination in a one-pot sequence to afford the desired product **88** in 70% yield. The obtained bromide **88** was coupled with indolyl boronic acid pinacol ester **89** by using Pd[P(*t*-Bu)$_3$]$_2$ as the catalyst and K_3PO_4 as the base to give **90** in 47% yield. Using this fully substituted pyrrole **90** as a common intermediate, the synthesis of dictyodendrins A and F has been accomplished.

16.2.3
Carbazoles and Related Compounds

16.2.3.1 Clausine P and Glycozolidine (Synthesis of Carbazoles by Intramolecular Ar–H/Ar–X Arylation)

The palladium-catalyzed intramolecular C–H arylation of aromatic C–H bonds with haloarenes (Ar–H/Ar–X coupling, where X is halogen) is effective for the synthesis of carbazole alkaloids with amine linkers, as reported by Bedford and coworkers in 2006 (Scheme 16.16a) [34]. Buchwald–Hartwig coupling of an aryl bromide and an aniline derivative generated aryl chloride **91** *in situ*, which was easily cyclized under palladium catalysis to give clausine P in 80% yield. They also

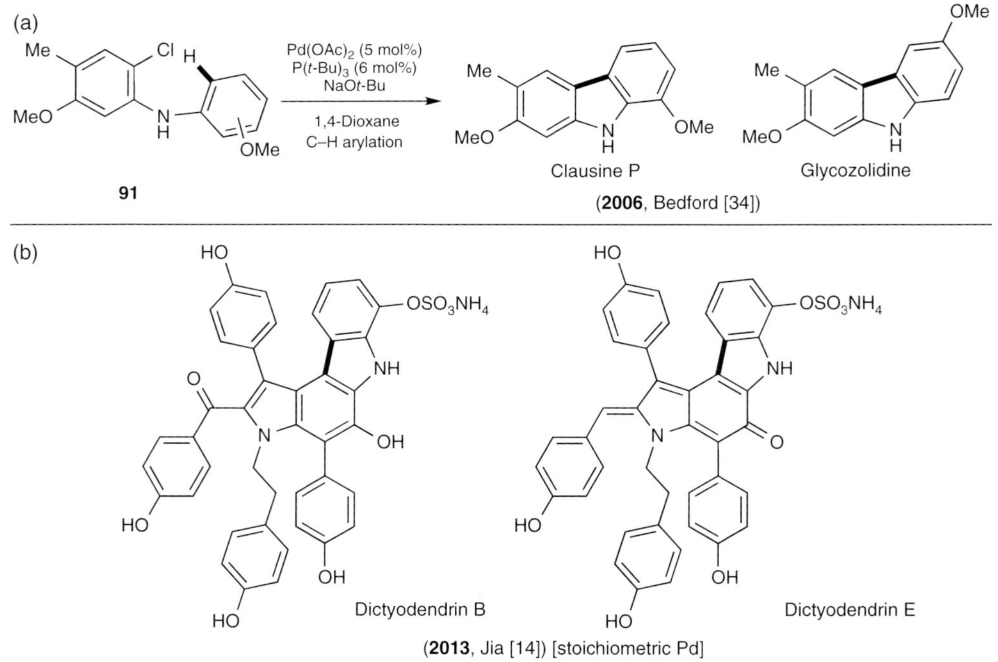

Scheme 16.16 Synthesis of carbazole-containing natural products by C–H arylation.

synthesized glycozolidine using the same reaction conditions. Furthermore, this intramolecular C–H coupling was utilized in the synthesis of the carbazole moiety of dictyodendrins B and E by Jia and coworkers, although a stoichiometric amount of palladium was required (Scheme 16.16b) [14].

16.2.3.2 Clausenine (Synthesis of Carbazoles by Intramolecular C–H/C–H Arylation)

As described earlier, carbazole synthesis can be achieved by palladium-catalyzed intramolecular Ar–H/Ar–X coupling (Scheme 16.17). However, an intramolecular oxidative coupling of two arenes (Ar–H/Ar–H coupling) is a more ideal transformation, since simple arenes can be used directly without any prefunctionalization. This strategy was successfully employed by Miller and coworkers in their synthesis of ellipticine (see Scheme 16.17b) using stoichiometric $Pd(OAc)_2$ in CF_3CO_2H and AcOH, albeit in somewhat low yield (15–25%) [35]. Then, in 2008, Fagnou and coworkers discovered that when t-$BuCO_2H$ (PivOH) was employed as an additive, the reaction yield increased dramatically in this type of reaction. For example, compound **92** was reacted under PivOH-promoted palladium catalysis to afford clausenine in 79% yield (Scheme 16.17a) [36]. They also achieved a concise synthesis of mukonine, clausine L, and murrayafoline A. The target structures of these elegant and short total syntheses are shown in

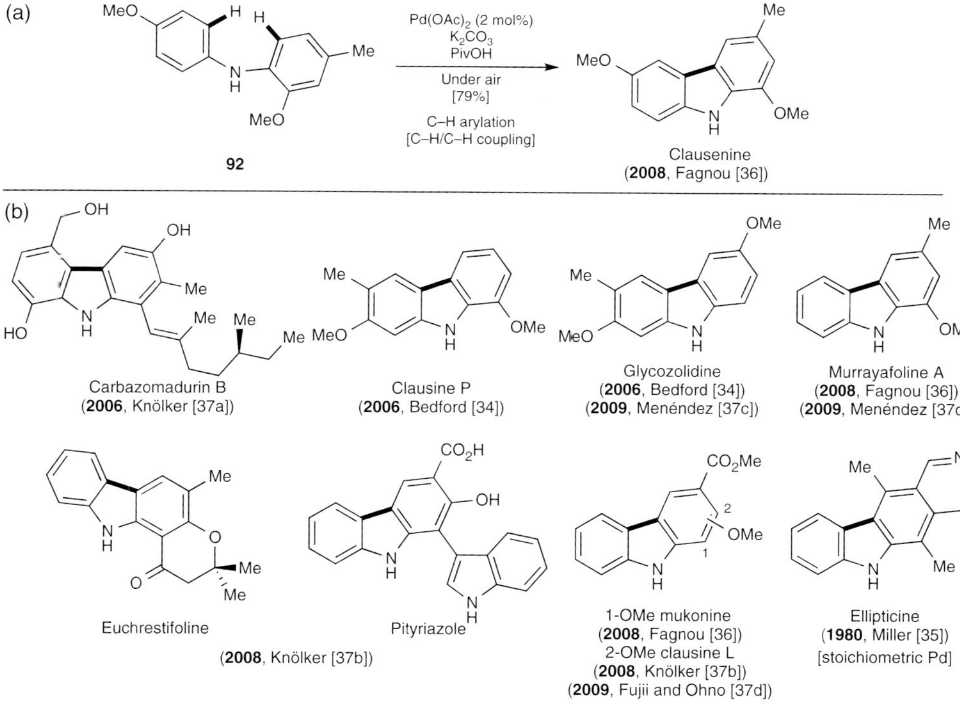

Scheme 16.17 Synthesis of carbazole alkaloids enabled by an ideal type of bond formation: C–H/C–H coupling.

Scheme 16.17b [37, 34]. Although the substrate scope of these reactions is still narrow, this intramolecular Ar–H/Ar–H coupling strategy is beginning to flourish in the synthesis of natural products and pharmaceuticals.

16.2.3.3 Clausine C and Glycozoline (Synthesis of Carbazoles by Intramolecular C–H Amination)

In an alternative approach to the synthesis of carbazole alkaloids using C–H functionalization, Youn and coworkers reported the synthesis of clausine C and glycozoline by using an intramolecular C–H amination of a biphenylamine derivative (Scheme 16.18) [38]. Treatment of tosyl-protected biphenyl 2-amine **93** with a $Pd(OAc)_2$/Oxone®/*p*-TsOH catalytic system in PivOH/DMF as a solvent afforded carbazole **94** (91% yield), followed by detosylation using tetrabutylammonium fluoride (TBAF) to give clausine C in 89% yield. Following the same strategy, glycozoline was also synthesized.

16.2.4 Benzofuran and Related Compounds

16.2.4.1 Frondosin B (C–H Alkenylation of Benzofuran)

In 2002, Trauner and coworkers reported the total synthesis of (−)-frondosin B (Scheme 16.19) [39]. In their synthesis, a palladium-catalyzed C–H alkenylation of benzofuran with an enol triflate was used for the construction of the seven-membered ring in frondosin B. Treatment of triflate **95** with catalytic $Pd(PPh_3)_4$ and *i*-Pr_2NEt in *N*,*N*-dimethylacetamide (DMAc) afforded cyclized product **97** in 70% yield. Although the mechanism of the coupling reaction is still unclear, the catalytic cycle might involve (i) C–OTf oxidative addition of **95** to Pd(0),

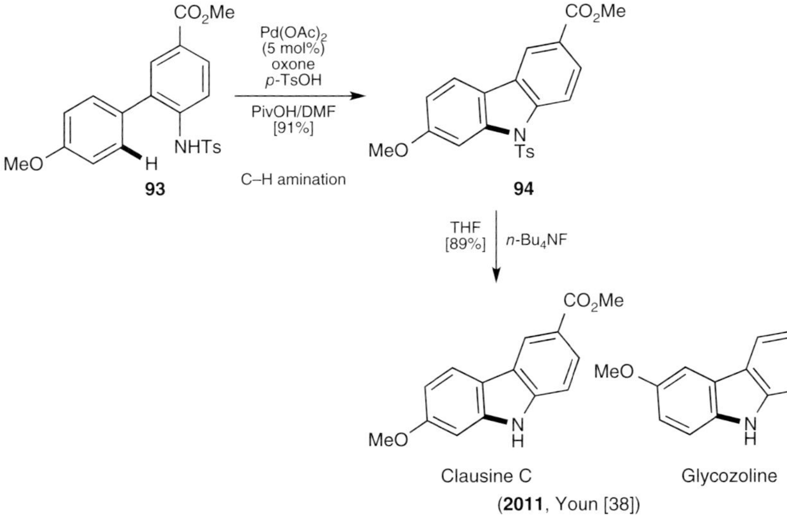

Scheme 16.18 Synthesis of clausine C and glycozoline.

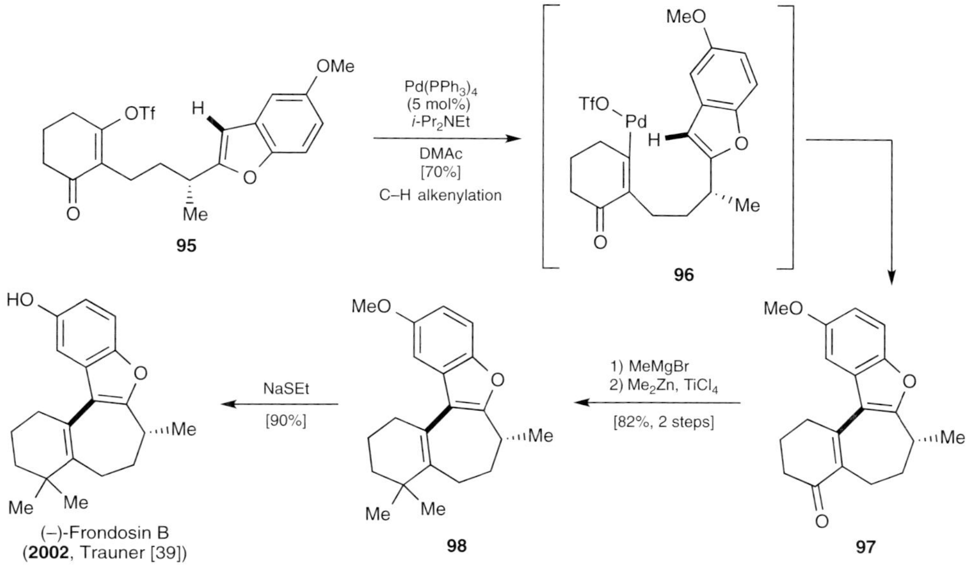

Scheme 16.19 Synthesis of frondosin B via benzofuran C–H alkenylation.

(ii) C–H palladation of the benzofuran in **96**, and finally (iii) reductive elimination and concomitant regeneration of the Pd(0) species. Transformation of the ketone to a *gem*-dimethyl group, followed by demethylation of the methoxyl moiety by NaSEt, accomplished the total synthesis of *ent*-frondosin B. This synthesis by Trauner demonstrated one of the early applications of an Ar–H/alkenyl halide-type coupling for natural products.

16.2.4.2 Diptoindonesin G (C–H Arylation of Benzofuran)

Employing a C–H arylation of benzofuran as a key step, Kim's group was able to complete the synthesis of diptoindonesin G (Scheme 16.20) under a catalytic system similar to Fagnou's work (see Section 16.2.3.2) that coupled benzofuran **77** and bromoarene **78** [40].

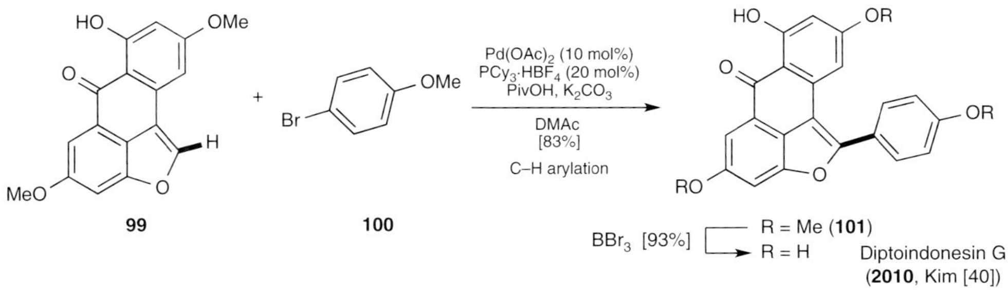

Scheme 16.20 Synthesis of diptoindonesin G using benzofuran C–H arylation as a key step.

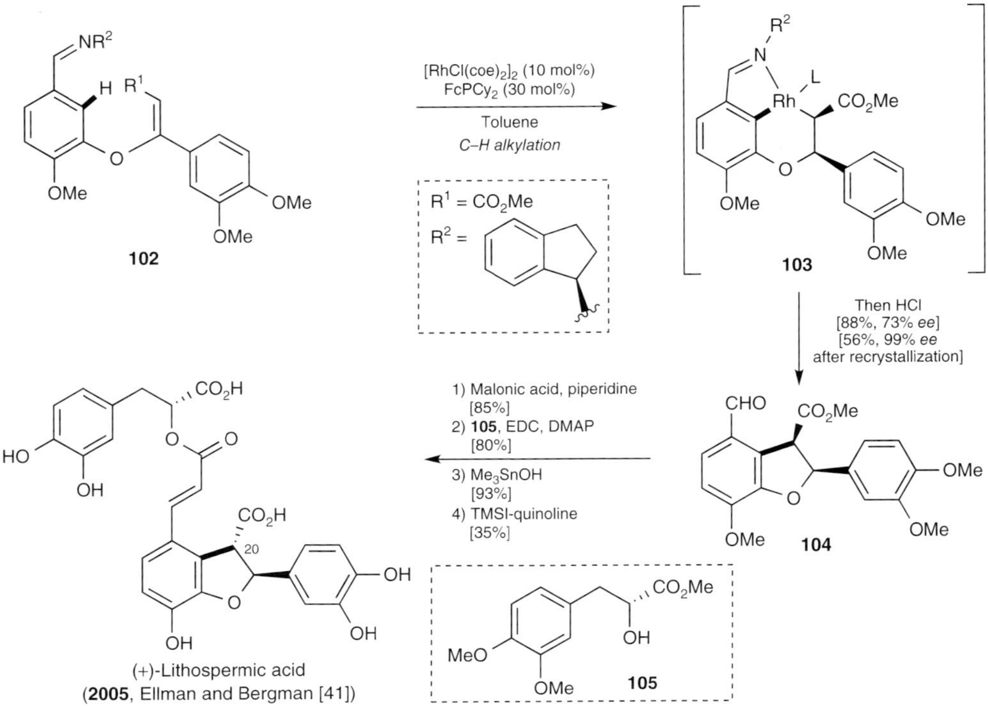

Scheme 16.21 Synthesis of (+)-lithospermic acid by the groups of Ellman and Bergman.

16.2.4.3 Lithospermic Acid (Formation of Dihydrobenzofuran Using C–H Alkylation)

In 2005, Ellman and Bergman reported the total synthesis of (+)-lithospermic acid using a diastereoselective intramolecular C–H alkylation through rhodium catalysis (Scheme 16.21) [41]. They had already developed the synthesis of chiral dihydrobenzofurans, which represent the core structure of lithospermic acid, by utilizing directing group-assisted C–H activation [42]. Imine **102**, in the presence of 10 mol% $[RhCl(coe)_2]_2$ complex and 30 mol% ferrocenyl–PCy_2 ($FcPCy_2$), yielded dihydrobenzofuran **104** in 88% yield and 73% ee, with the intermediacy of **103.** A subsequent sequence of (i) Knoevenagel condensation of **104** with malonic acid, decarboxylation, and epimerization of the C20 stereocenter, (ii) condensation with **105**, (iii) hydrolysis of the methyl ester, and (iv) demethylation completed the asymmetric total synthesis of (+)-lithospermic acid.

16.2.4.4 Lithospermic Acid (Formation of Dihydrobenzofuran Using C–H Insertion and C–H Alkenylation at the C4 Position of Dihydrobenzofuran)

In 2011, the second total synthesis of (+)-lithospermic acid was achieved by Yu and coworkers (Scheme 16.22) [43]. They employed two types of C–H functionalization; an intramolecular C–H insertion and an intermolecular C–H alkenylation to construct the dihydrobenzofuran unit and to connect the side chain. Treatment of diazo compound **106** with $Rh_2(S\text{-}DOSP)_4$ (tetrakis[1-[[4-alkylphenyl]sulfonyl]-(2S)-pyrrolidinecarboxylate) (Davies' catalyst) [44] at room temperature achieved

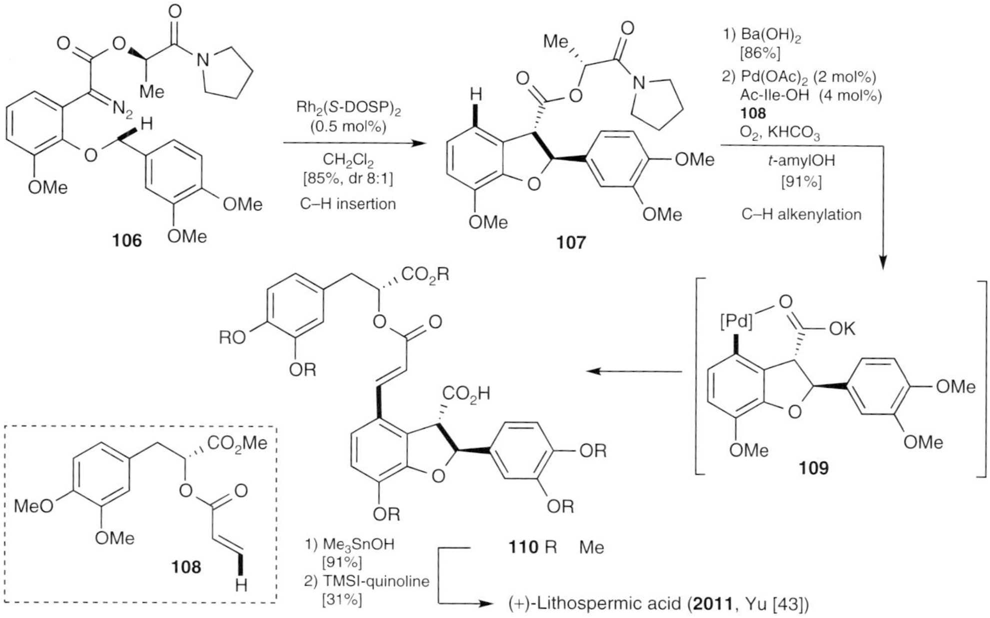

Scheme 16.22 Synthesis of lithospermic acid by Yu's group, involving two key C–H functionalization steps.

the C–H insertion to afford *trans*-dihydrobenzofuran **107** in 85% yield with high diastereoselectivity ($dr = 8:1$). After hydrolysis of the chiral auxiliary group, C–H alkenylation of the resulting carboxylic acid with acrylate ester **108** was carried out. In the presence of 2 mol% $Pd(OAc)_2$, 4 mol% Ac-Ile-OH, and 2.0 equiv. of $KHCO_3$ under an oxygen atmosphere, the corresponding coupling product **110** was synthesized in 91% yield with complete regioselectivity. As shown in intermediate **109**, the carboxylate unit in the substrate played a key role as a weakly coordinating directing group. They had already demonstrated that the key palladium-catalyzed C–H alkenylation of arenes provides the corresponding coupling product in good yield when an amino acid-based ligand and an appropriate inorganic base are used [45]. Finally, demethylation was achieved by a two-step sequence, completing the total synthesis of lithospermic acid.

Additionally, enantioselective C–H insertions using chiral rhodium catalysts have been used to forge dihydrobenzofurans. The following examples describe intramolecular C–H insertion as a strategy to create natural products containing dihydrobenzofuran units (Figure 16.2) [46].

16.2.4.5 **Morphine (Intramolecular C–H Insertion to Dihydrobenzofuran)**

In 1997, White and coworkers achieved the total synthesis of (+)-morphine (Scheme 16.23) [47]. They envisaged a C–H insertion strategy using a metal carbenoid to construct the C13 quaternary carbon in morphine. After preparing diazo ketone **111**, C–H insertion using a rhodium catalyst occurred to give

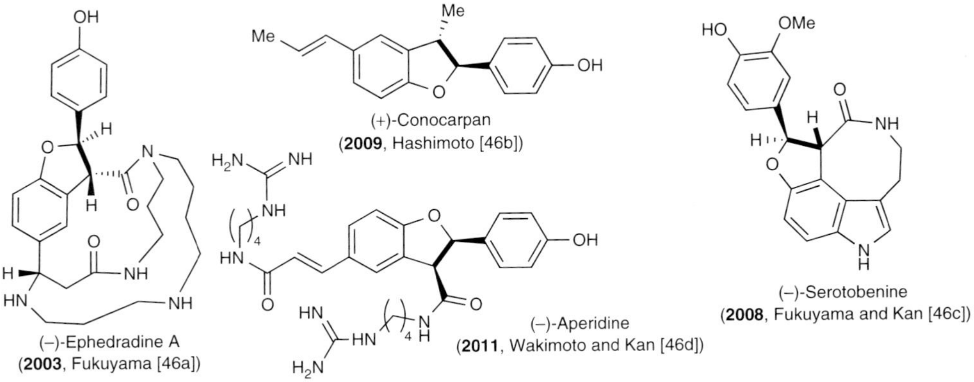

Figure 16.2 The synthesis of natural products containing dihydrobenzofuran units by intramolecular C–H insertion.

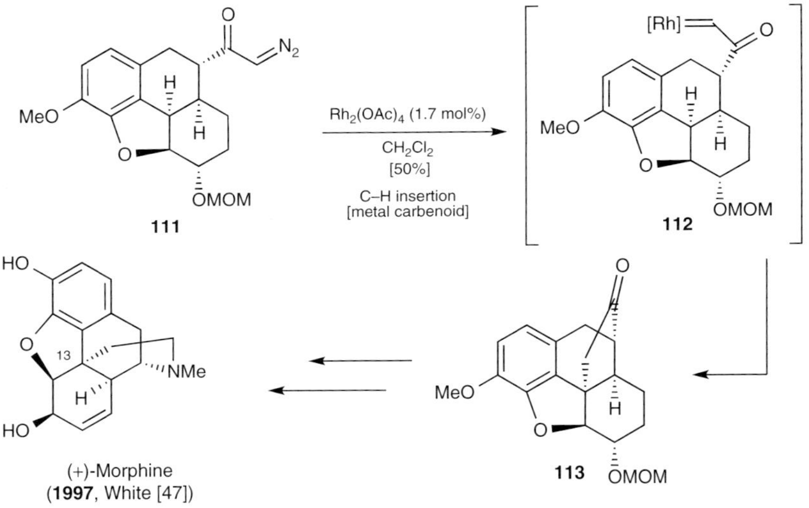

Scheme 16.23 Synthesis of morphine facilitated by a metal carbenoid C–H insertion reaction.

113 in regio- and stereoselective fashion. After a few more transformations, the synthesis of (+)-morphine, the unnatural enantiomer of morphine, was completed in a total of 28 steps.

16.2.5
Imidazoles, Oxazoles, Thiazoles, and Related Compounds

16.2.5.1 JNK3 Inhibitors (C–H Alkylation of Imidazoles)

In 2007, Ellman and Bergman have accomplished the synthesis of JNK3 inhibitors using a rhodium-catalyzed intramolecular C–H alkylation of imidazoles

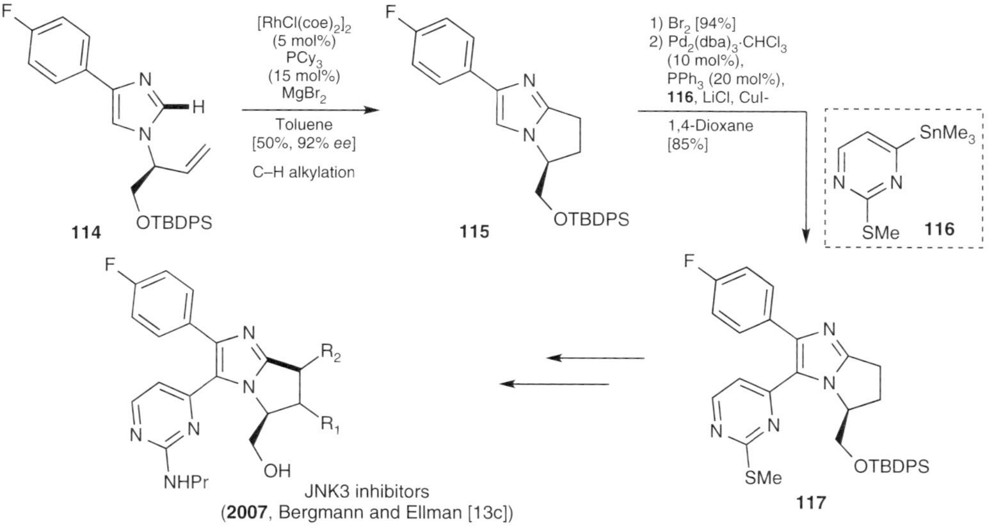

Scheme 16.24 Synthesis of JNK3 inhibitors using intramolecular C–H alkylation of imidazoles as a key step.

(Scheme 16.24) [13c]. They had already developed directing group-assisted C–H activation reactions (see Sections 16.2.1.4 and 16.2.4.3) and applied them to the synthesis of natural products. In this synthesis, they achieved C–H activation at the C2 position of imidazoles through rhodium catalysis without any directing groups. Indeed, the intramolecular C–H alkylation of imidazole **114** successfully proceeded in the presence of catalytic $RhCl(coe)_2$ and PCy_3, with $MgBr_2$ (5 mol%) as an additive, to give cyclized product **115** in 50% yield while maintaining the original enantiomeric excess. Bromination of **115**, followed by Stille coupling of the resulting bromide with **116**, afforded the coupling product in 85% yield. After several steps, the synthesis of JNK3 inhibitors has been accomplished.

16.2.5.2 Tyrosine Kinase Inhibitor (C–H Arylation of Imidazoles)

In 2009, the synthesis of Tie_2 tyrosine kinase inhibitor using multiple C–H arylations of imidazole *N*-oxide **118** was reported by the Fagnou group (Scheme 16.25a) [48]. Starting from C5-arylated imidazole *N*-oxide **118**, palladium-catalyzed C–H arylation with aryl bromide **119** occurred at 70 °C to give the coupling product **120** in 90% yield with complete regioselectivity at the C2 position of the imidazole *N*-oxide (which is also the most acidic proton in **118**). Subsequently, **120** was coupled with pyridyl bromide **121** at its C4 position in the presence of a palladium catalyst at elevated temperatures (110 °C) to deliver bis-arylated imidazole *N*-oxide **122** in 73% yield. The regioselectivity of these two sequential C–H arylation steps (C2 and C5) was controlled by the reaction temperature. Finally, reduction of *N*-oxide **122** was achieved by treatment with iron in acetic acid, completing the synthesis of tyrosine kinase inhibitor.

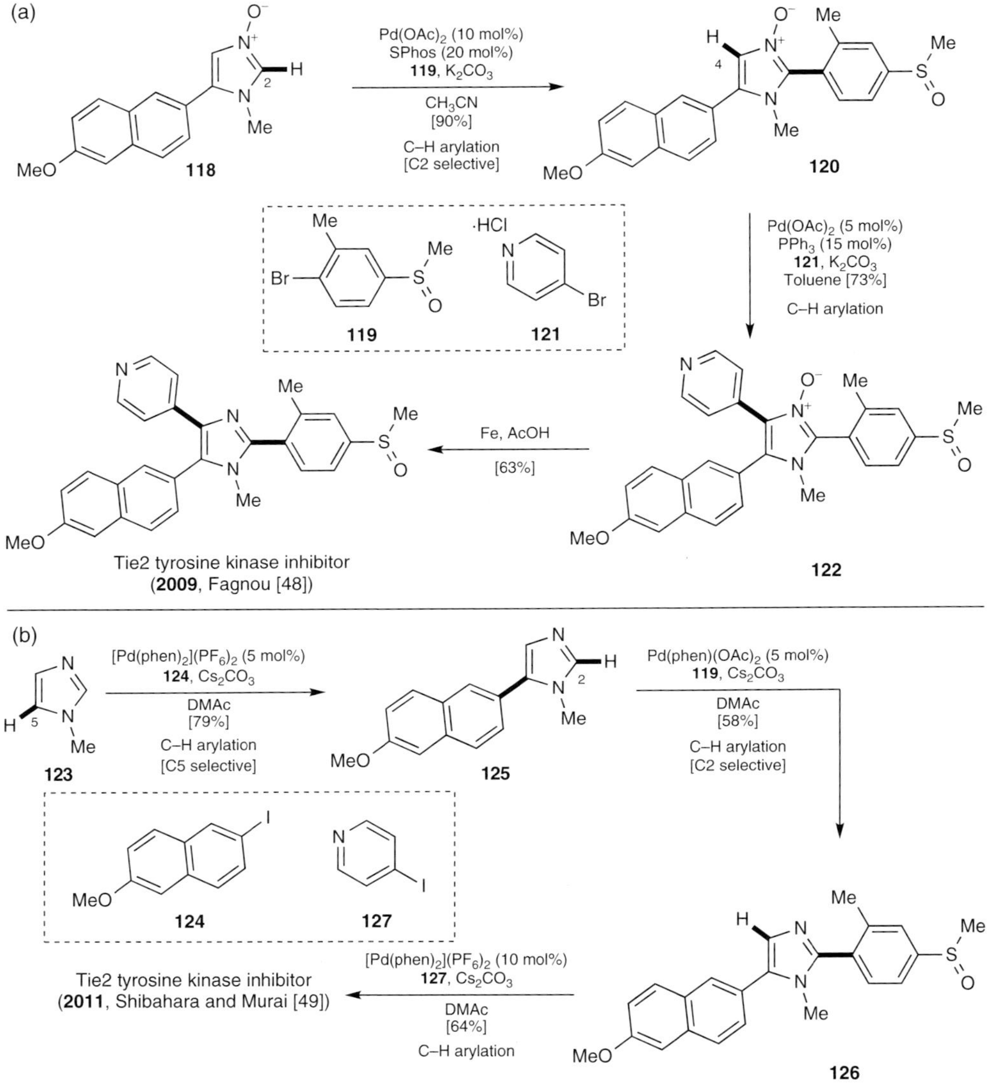

Scheme 16.25 Two syntheses of tyrosine kinase inhibitor that showcase the efficiency of C–H functionalization in polyarylated imidazole synthesis.

In 2010, Murai and Shibahara successfully synthesized the same target via sequential C–H arylation of imidazole using their own catalytic system (Scheme 16.25b) [49a]. They discovered that cationic palladium species $[Pd(phen)_2](PF_6)_2$ (phen = 1,10-phenanthroline) is more effective than neutral palladium catalysts for the C–H arylation of 1,3-azoles and, in order to demonstrate the utility of their catalyst, the synthesis of tyrosine kinase inhibitor was carried out. Their synthesis commenced with simple *N*-methylimidazole (**123**), onto which a C5-selective C–H arylation with aryl iodide **124** in the presence of $[Pd(phen)_2](PF_6)_2$ and Cs_2CO_3 gave coupling product **125** in 79% yield. Subsequent C2-selective coupling of imidazole **125** proceeded in moderate yield, and the final pyridine moiety was installed using the same cationic palladium catalyst to complete the synthesis of Tie_2 tyrosine kinase inhibitor.

16.2.5.3 Texaline, Febuxostat, and Muscoride A (C–H Arylation of Oxazoles or Thiazoles)

In 2008, Piguel, Greaney, and Hoarau independently synthesized texaline, a natural product that displays antitubercular activity, using their own palladium-catalyzed C–H coupling protocols (Scheme 16.26a). Piguel developed the coupling reaction between oxazole **128** and bromoarene **129** using a

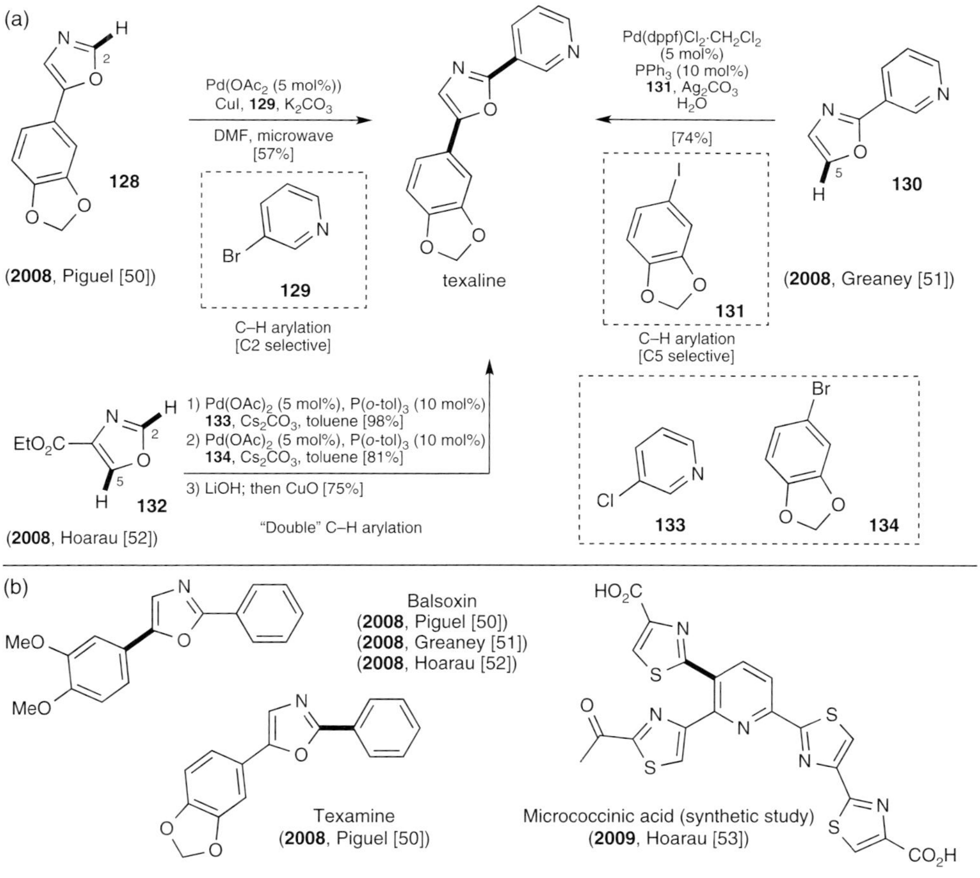

Scheme 16.26 Synthesis of oxazole-contained natural products and pharmaceuticals using C–H arylation.

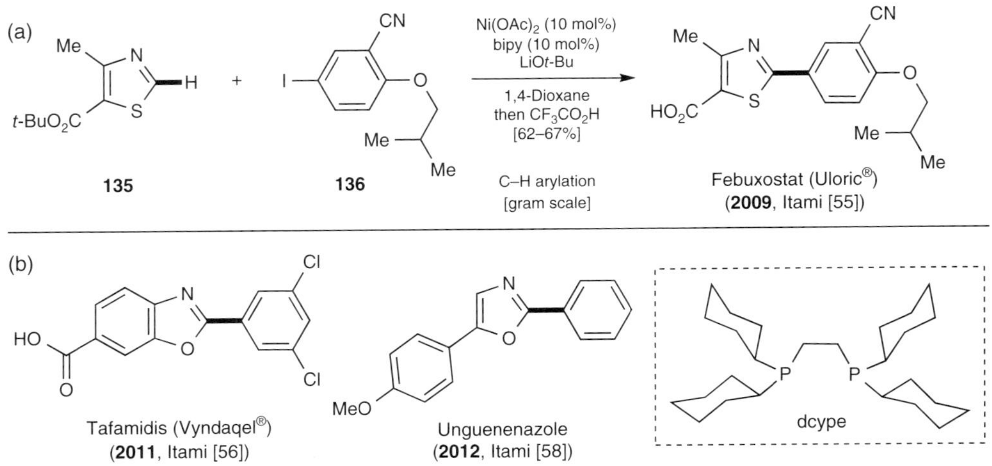

Scheme 16.27 Synthesis of oxazole-contained biologically active compounds by Ni-catalyzed C–H arylation.

$Pd(OAc)_2/CuI/K_2CO_3$ catalytic system under microwave irradiation and then applied it to the synthesis of texaline and a related compound, texamine [50]. Greaney conducted a C5-arylation of oxazole under palladium catalysis on water to access texaline. Their original method was also applied to the synthesis of balsoxin, which displays antimycobacterial activity (Scheme 16.26b) [51]. Hoarau achieved two sequential C–H arylations of oxazole **132** at its C2 and C5 positions, followed by decarboxylation to give texaline [52]. The group of Itami synthesized texaline, balsoxin, and texamine by using novel nickel-based catalytic systems (*vide infra*). Furthermore, Hoarau reported a study toward the synthesis of micrococcinic acid, in which the key step involves the construction of the thiazole–pyridine bond using a palladium-catalyzed C–H arylation method [53].

Despite the impressive utility of these catalytic reactions, palladium complexes are expensive even when used in catalytic amounts. More recently, C–H arylation methods using ubiquitous metals such as copper and iron have been reported [54]. Itami and coworkers have also developed the first, inexpensive, nickel-catalyzed C–H arylation of azoles with haloarenes in 2009 [55]. Using this nickel-based system, they achieved the rapid synthesis of febuxostat (Uloric®[1]; Scheme 16.27), which is an inhibitor of xanthine oxidase used for the treatment of gout and hyperuricemia. In the presence of catalytic $Ni(OAc)_2$/bipy and LiO*t*–Bu or $Mg(Ot\text{–}Bu)_2$, thiazole **135** and iodoarene **136** underwent C–H/C–X coupling in 1,4-dioxane, furnishing the corresponding product. Ensuing treatment with TFA afforded febuxostat in 62–67% overall yield. This methodology has also been

1) Uloric® is the trade name in the United States. It is called Adenuric® in the EU, Febutaz® in India, and Feburic® in Japan.

implemented in the synthesis of tafamidis (Vyndaqel®) and texaline (Scheme 16.27a) [56, 57]. Furthermore, the group of Itami found that a C–H arylation of azoles using phenol derivatives can be made possible with the introduction of a new catalytic system involving dcype (1,2-bis(dicyclohexylphosphino)ethane) [58]. These findings not only push the limit of biaryl coupling into a C–H/C–O manifold but should also unearth novel modes of reactivity of phenol derivatives with the $Ni(cod)_2$/dcype catalyst. They have also successfully completed the concise synthesis of oxazole alkaloids, texamine, and unguenenazole (Scheme 16.27b).

In 2012, Yamaguchi, Itami, and coworkers discovered a nickel-catalyzed decarbonylative C–H coupling between 1,3-azoles and aryl esters and applied these catalytic conditions to the synthesis of muscoride A, a natural product that displays weak antibacterial activity (Scheme 16.28) [59]. Two azole esters, **137** and **138**, were coupled under Ni/dcype catalysis to furnish the corresponding product **139** in 39% yield. As the conversion of **139** into **140** has been previously described [60], a formal synthesis of muscoride A has been completed. If the synthesis were planned and executed with typical cross-coupling substrates (aryl halides and organometallic reagents) instead, it would have become much less efficient, with many added steps.

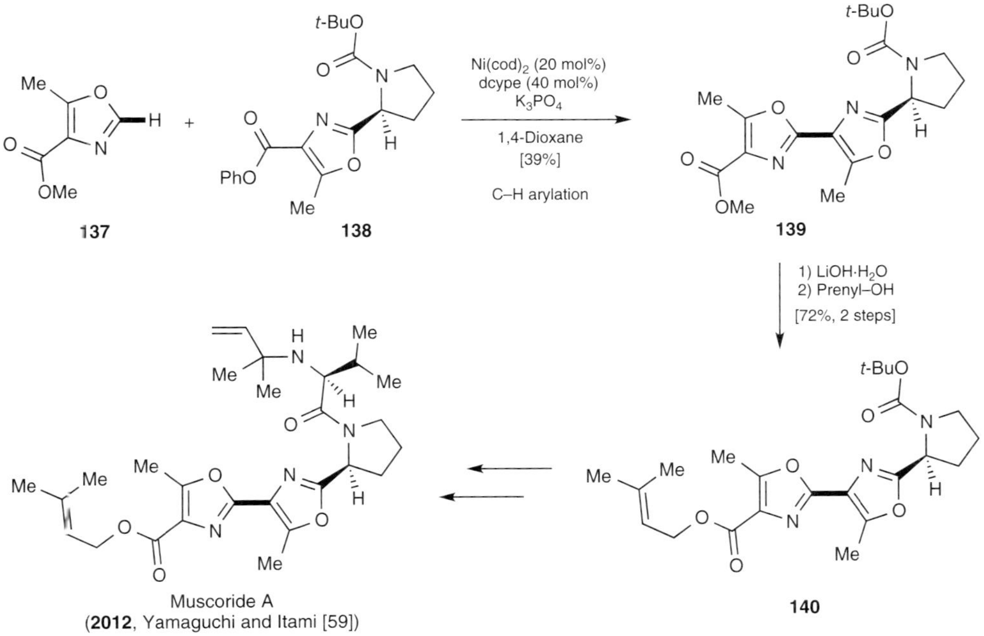

Scheme 16.28 Synthesis of muscoride A using decarboxylative C–H arylation of oxazoles as a key step.

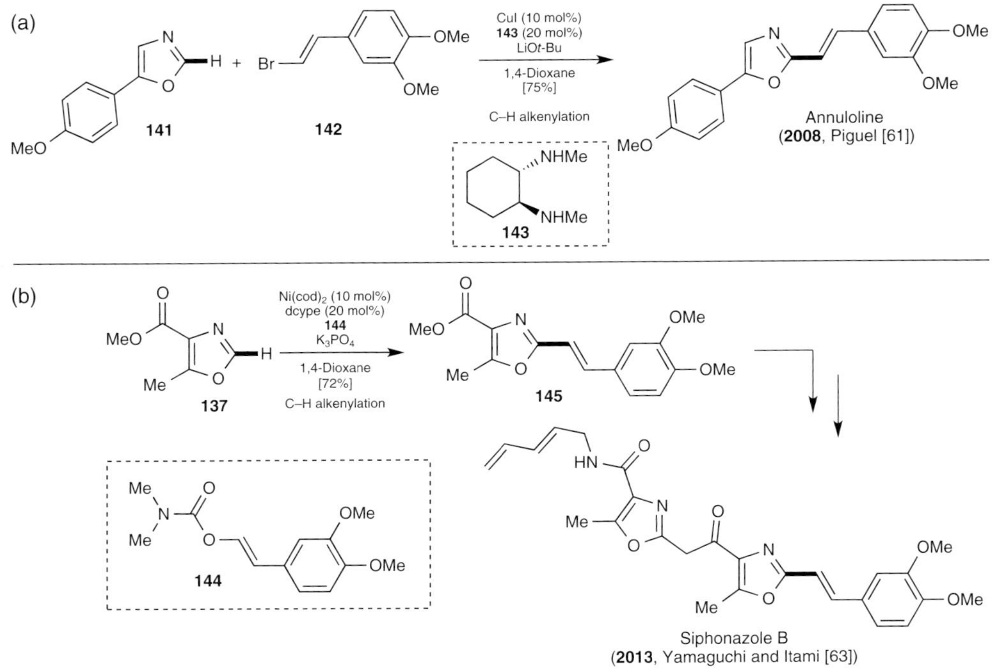

Scheme 16.29 Synthesis of annuloline and siphonazole B using C–H alkenylation of oxazoles at the C2 position.

16.2.5.4 Annuloline and Siphonazole B (C–H Alkenylation of Oxazoles at the C2 Position)

In 2008, Piguel and coworkers reported the synthesis of annuloline, the first isolated oxazole-containing natural product (Scheme 16.29a) [61]. 5-Aryloxazole **141** was coupled with bromoalkene **142** in the presence of CuI, diamine **143**, and LiO*t*–Bu to deliver annuloline in 75% yield. Although a number of similar C–H alkenylations of 1,3-azoles with alkenyl halides have been reported using transition metal catalysts [62], Piguel's synthesis is the only report of a natural product (albeit structurally simple) synthesized by intermolecular aromatic C–H alkenylation.

Meanwhile, in 2013, Yamaguchi, Itami, and coworkers reported a C–H alkenylation of 1,3-azoles with enol derivatives (C–H/C–O type alkenylation) or alkenyl esters (decarbonylative C–H alkenylation) as electrophiles and applied this strategy to the formal synthesis of siphonazole B (Scheme 16.29b) [63]. Oxazole **137** was coupled with enol derivative **144** under their nickel catalytic system using the dcype ligand to form alkenyl oxazole **145** in 72% yield by C–H alkenylation at the C2 position of oxazoles. Since oxazole **145** was a known intermediate from the previous synthesis of siphonazole B [64], a convergent formal synthesis of siphonazole B has therefore been achieved.

16.2.6
Quinazolines and Related Compounds

16.2.6.1 Luotonin B (Intramolecular C–H Arylation of Quinazoline)

In 2004, Harayama and coworkers reported an intramolecular C–H arylation between two heteroarenes (Scheme 16.30) [65]. The coupling precursor **146**, which was synthesized by *N*-alkylation of 4-quinazolinone with 2-bromo-3-(bromomethyl)quinoline, was cyclized using catalytic $Pd(OAc)_2$ and PPh_3 to give luotonin A in 66% yield. This natural product was then hydroxylated in two steps to provide luotonin B as well as rutaecarpine using the same strategy.

16.2.6.2 Vasicoline (C–H Alkylation of Quinazoline)

In 2006, the group of Ellman and Bergman applied their rhodium-catalyzed C–H activation method (see details in Section 16.2.5.1) to the synthesis of vasicoline, a quinazoline natural product (Scheme 16.31) [13a]. In this synthesis,

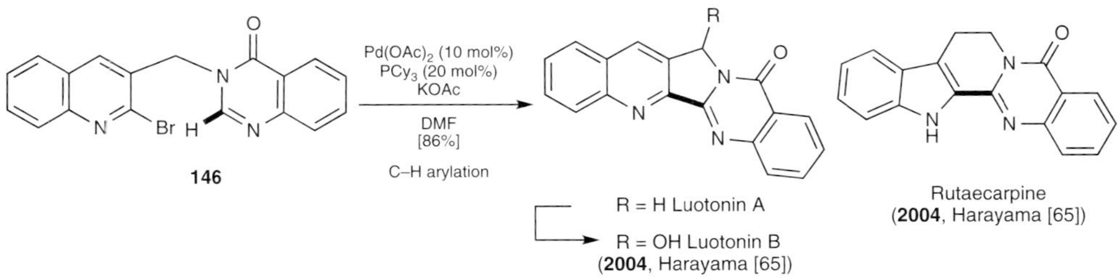

Scheme 16.30 Synthesis of luotonin B and rutaecarpine using intramolecular C–H arylation.

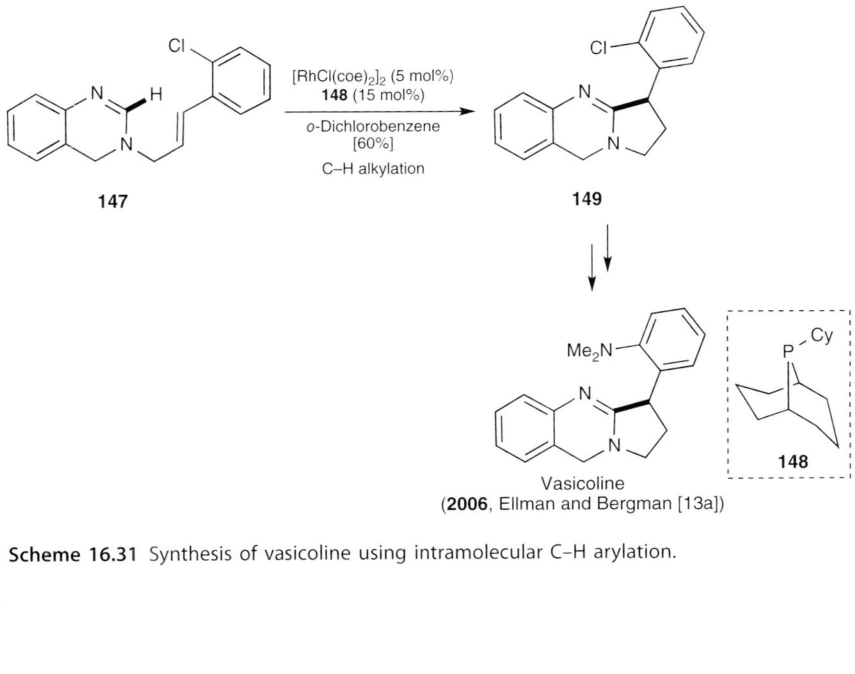

Scheme 16.31 Synthesis of vasicoline using intramolecular C–H arylation.

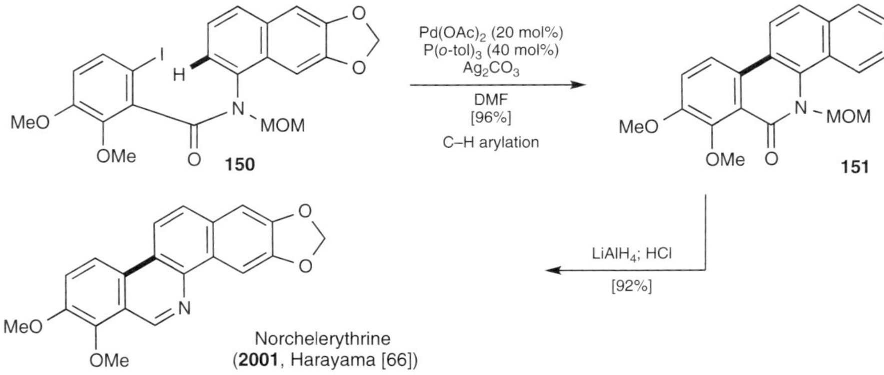

Scheme 16.32 Synthesis of norchelerythrine using intramolecular C–H arylation.

cyclohexyl-phoban ligand **148** was effective compared with PCy_3 or other ligands, and the intramolecular coupling of **147** proceeded smoothly to give coupling product **149** in 60% yield. After several steps, the synthesis of vasicoline has been achieved.

16.2.7
Quinolines, Isoquinolines, Phenanthridines, and Related Compounds

16.2.7.1 Norchelerythrine (Intramolecular C–H Arylation)

In 2001, Harayama and coworkers demonstrated an intramolecular C–H arylation (Ar–H/Ar–X coupling) using *N*-naphthylbenzamide **150** for the synthesis of norchelerythrine, a phenanthridine alkaloid (Scheme 16.32) [66]. After preparation of aryl iodide **150**, Pd-catalyzed intramolecular C–H arylation was performed to afford the coupling product **151** in 96% yield. After reduction of the amide moiety in **151** and removal of the MOM group, the total synthesis of norchelerythrine was accomplished.

16.2.7.2 Nitidine and NK 109 (Catellani-Type C–H Arylation/*N*-Arylation)

In 2011, Lautens and coworkers achieved a formal synthesis of nitidine and NK 109 through a tandem direct arylation/*N*-arylation strategy (Scheme 16.33) [67]. To this end, aryl triflate **152** and imine **153** were coupled together by the action of a palladium catalyst and norbornene to give coupling product **155**. This intriguing reaction might be undergoing a Catellani-type mechanism, involving (i) C–OTf oxidative addition of **152** to Pd(0), (ii) carbopalladation to norbornene, (iii) intramolecular arene C–H palladation, (iv) C–Br oxidative addition of **153** to the resultant palladacycle to form Pd(IV) intermediate **154**, and (v) chemoselective reductive elimination with concomitant decarbopalladation (release of norbornene).

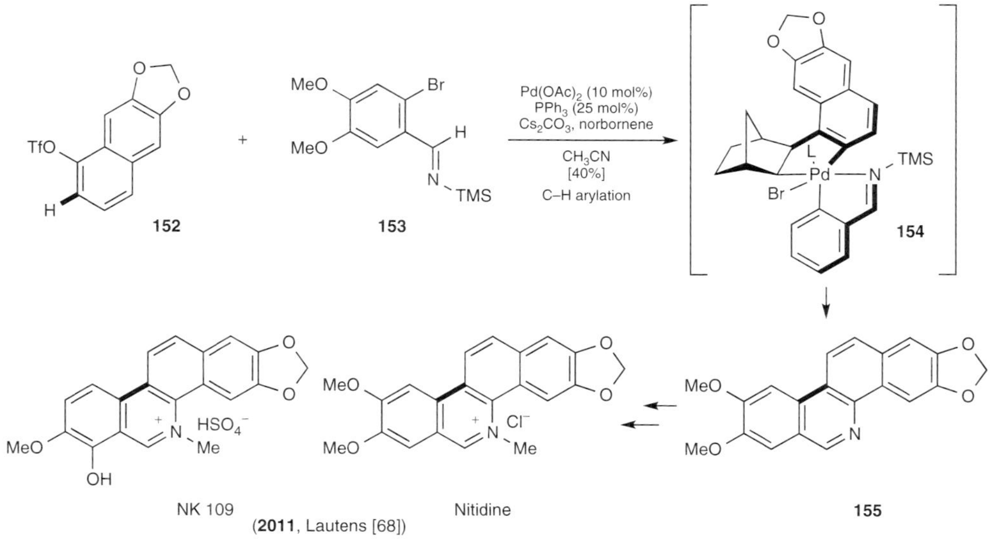

Scheme 16.33 Synthesis of nitidine and NK109 using Catellani-type tandem C–H arylation/*N*-arylation.

16.2.7.3 LTB4 Antagonist and MCH-1R Receptor Modulator (sp^3 C–H Arylation/Intramolecular C–H Amination)

In 2014, Yu and coworkers reported a triple C–H activation for the synthesis of substituted quinolines and applied it to the synthesis of two bioactive compounds (Scheme 16.34) [69]. They discovered that propionamides can react with aryl iodides in the presence of catalytic $PdCl_2$, 2,5-lutidine as an enabling ligand, and Ag_2CO_3 to afford quinolines in good to moderate yields. Thus, treatment of propionamide **156** with iodobenzene under palladium-catalyzed conditions formed intermediate **157** through sp^3 C–H arylation, dehydrogenation, and Heck reaction, followed by C–H amidation to afford quinolinone **158** in 42% yield. Removal of the 4-methoxybenzene group of **158** by treatment with BBr_3 and PIFA (bis(trifluoroacetoxy)iodo)benzene) delivered 4-phenylquinolinone **159**, and the formal synthesis of the two bioactive quinoline derivatives, LTB4 antagonist [70] and MCH-1R receptor modulator [57], have been achieved from **159.**

16.2.7.4 Tipifarnib (C–H Alkenylation and Cyclization)

In 2015, Liu and coworkers reported the formal synthesis of tipifarnib, a farnesyltransferase inhibitor, by using a palladium-catalyzed C–H alkenylation as a key step (Scheme 16.35) [71]. Aniline derivative **160** reacted with Ac_2O to form an acetanilide intermediate, followed by *ortho*-palladation under the influence of a palladium catalyst to produce compound **162.** Oxidative Heck reaction with **161** and cyclization/deacetylation then occurred to afford quinolinone **163** in 95% yield in one pot. This quinolinone was then converted to tipifarnib by the reported procedure [72].

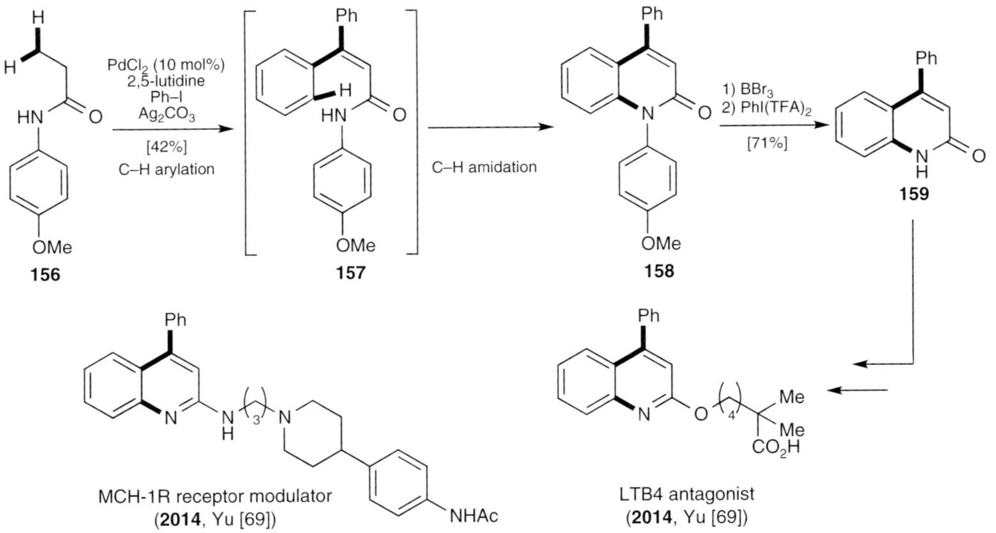

Scheme 16.34 Formal synthesis of LTB4 antagonist and MCH-1R receptor modulator using a sp^3 C–H arylation/intramolecular C–H amination cascade.

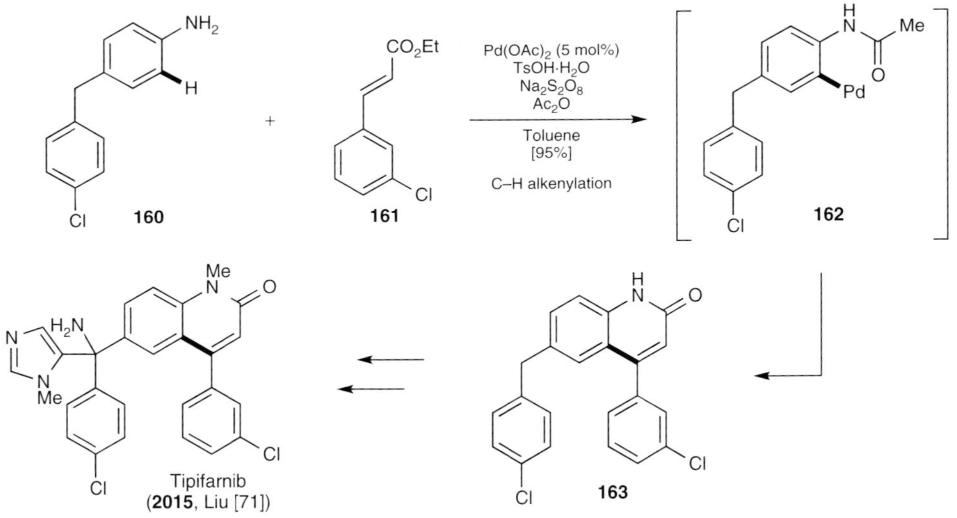

Scheme 16.35 Formal synthesis of tipifarnib: formation of quinolinones using C–H alkenylation and cyclization.

16.2.7.5 Oxychelerythrine (C–H Alkenylation and Annulation)

In 2012, Cheng and coworkers reported a one-pot synthesis of isoquinolinium salts by rhodium-catalyzed C–H alkenylation/annulation reactions [73] and applied it to the synthesis of oxychelerythrine, an isoquinoline alkaloid (Scheme 16.36) [74]. The reaction of aldehyde **164** with alkyne **165** in the presence of

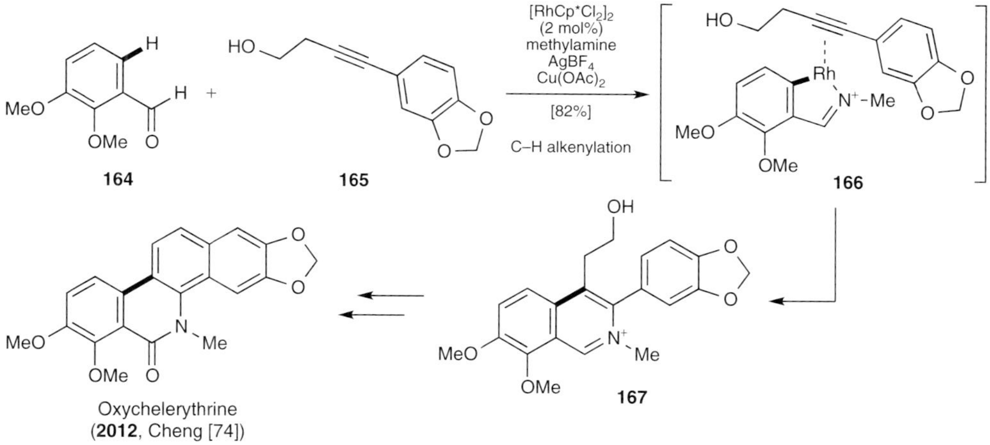

Scheme 16.36 Synthesis of oxychelerythrine using a C–H alkenylation/annulation reaction.

1.5 equiv. of methylamine, $[RhCp^*Cl_2]_2$, $AgBF_4$, and $Cu(OAc)_2$ in *t*-Amyl-OH formed five-membered rhodacycle **166**, followed by coordination of the Rh center to alkyne and annulation to afford isoquinolinium salt **167** in 82% yield in regioselective manner. This isoquinolinium salt **167** can be converted to oxychelerythrine by a three-step procedure.

16.2.8
Pyridines and Related Compounds

16.2.8.1 Sodium Channel Inhibitor and Antimalarial Agent (C–H Arylation of Pyridines at the C2 Position)

In 2009, the synthesis of a sodium channel inhibitor was accomplished by Fagnou and coworkers by employing a palladium-catalyzed C–H arylation of an azine *N*-oxide with a haloarene as the key reaction (Scheme 16.37) [48]. Pyridine *N*-oxide **169** was reacted with aryl bromide **168** using catalytic $Pd(OAc)_2$ and $P(t\text{-}Bu)_3$ to furnish coupling product **170** in 72% yield. Then, pyridine N-oxide deoxygenation and concomitant C4-chlorination, Buchwald–Hartwig amination of the generated pyridyl chloride with morpholine [75], and hydrolysis of the nitrile gave the sodium channel inhibitor in good overall yield. Notably, this pyridine C–H arylation method was also applied to the formation of an antimalarial reagent (formal synthesis) [76].

16.2.8.2 Complanadine A and B (C–H Borylation of Pyridine at C3 Position or C–H Arylation of Pyridines at C2 Position)

In 2010, Sarpong and coworkers exploited the use of C–H borylation to synthesize complanadine A, an unsymmetrical lycodine dimer (Scheme 16.38a) [77]. Disubstituted pyridine **171**, which was readily prepared from **174** by removal of the

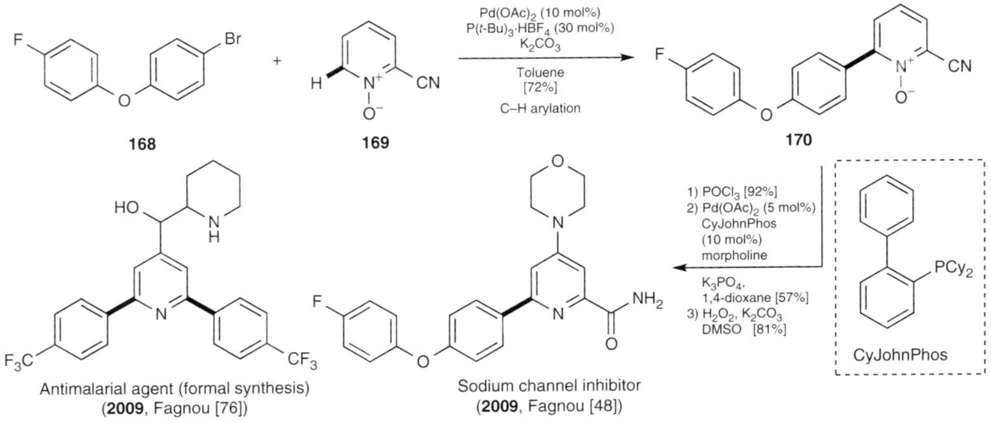

Scheme 16.37 Synthesis of a sodium channel inhibitor and an antimalarial agent using a pyridine *N*-oxide C–H arylation reaction.

triflate group, was treated with the standard Hartwig–Miyaura–Ishiyama C–H borylation conditions to afford boronate **173** as a single regioisomer in 75% yield. The Suzuki–Miyaura cross-coupling of **173** with triflate **174** then led to complanadine A in a total of eight steps. In 2013, the Sarpong group has also accomplished the synthesis of complanadine B, a congener of complanadine A, using a similar strategy [78].

In 2013, Tsukano and coworkers also reported the synthesis of complanadines A and B by using a palladium-catalyzed C–H arylation method (Scheme 16.38b) [79]. Bromide **175** and *N*-oxide **165**, which were prepared from the same intermediate in several steps, were coupled by C–H arylation under palladium catalysis ($Pd(OAc)_2$/*t*-BuDavePhos/Cs_2CO_3/PivOH) according to the method developed by Fagnou and coworkers [80] to obtain coupling product **177** in 62% yield. Reduction of *N*-oxide **177** formed complanadine A, and acetylation of *N*-oxide **177**, followed by a Claisen-type rearrangement, produced acetate **178** as a 3 : 1 epimeric mixture. Deacetylation of **178**, subsequent oxidation, and removal of the Cbz group completed the total synthesis of complanadine B.

16.2.8.3 Anabashine (C–H Arylation of Iminopyridium Ylides)

In 2008, Charette and coworkers demonstrated a direct C–H arylation of pyridine *N*-oxide (compare to Section 16.2.8.1) analogs such as iminopyridium ylides with aryl halides. Iminopyridium ylide **179** and pyridine *N*-oxide have similar abilities to furnish arylated products such as **180** by activating the C2 position using a palladium catalyst. After hydrogenation of **180**, SmI_2-mediated reductive cleavage of the nitrogen–nitrogen bond and Boc protection of the exposed secondary amine formed a Boc-protected intermediate, which was treated with TFA to give anabashine in good yield over three steps (Scheme 16.39) [81].

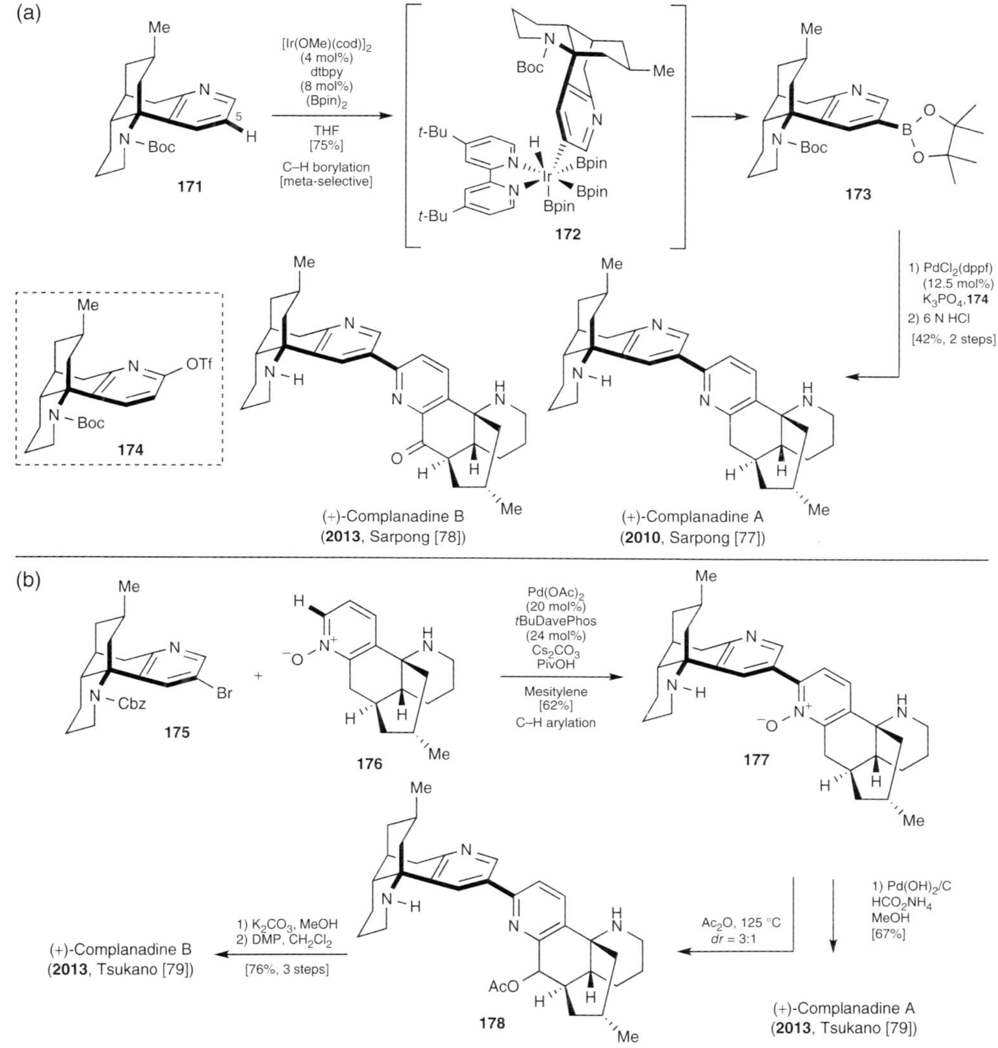

Scheme 16.38 Synthesis of complanadine A and complanadine B by C–H borylation or arylation.

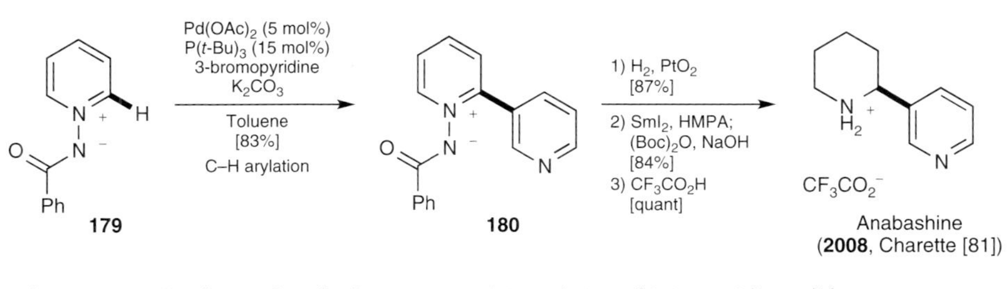

Scheme 16.39 Synthesis of anabashine using a C–H arylation of iminopyridium ylides.

16.2.8.4 Preclamol (C–H Arylation of Pyridine at the C3 Position)

For electron-deficient heteroaromatics such as pyridines, the first efficient C–H arylation was reported by Ellman and Bergman using rhodium catalyst [68], wherein C–H arylation proceeds at the C2 position of pyridine. Recently, directing group-assisted C3- or C4-arylation of pyridine was reported by Yu and Sames [82]. Later in 2011, the Yu group also demonstrated a C3-selective C–H arylation of pyridine derivatives and applied it to a gram-scale synthesis of preclamol (Scheme 16.40) [83]. The palladium-catalyzed C3-selective coupling of excess amounts of pyridine (37 equiv.) with 1-bromo-3-methoxybenzene (**181**) under a $Pd(OAc)_2$/phen/Cs_2CO_3 catalytic system afforded the desired coupling product **182** in 70% yield. A subsequent sequence of (i) *N*-alkylation of **182** with 1-bromopropane, (ii) hydrogenation of the pyridinium to piperidine using H_2/PtO_2, and (iii) demethylation afforded preclamol in good yield.

16.2.9 Other Heterocycles

16.2.9.1 Celecoxib (C–H Arylation of Pyrazoles)

In 2011, Gaulier and coworkers reported the synthesis of celecoxib (Celebrex®) via C–H arylation of pyrazoles at the C2 position (Scheme 16.41) [84]. Pyrazole **183** was coupled with aryl bromide **184** using Fagnou's catalytic system to form

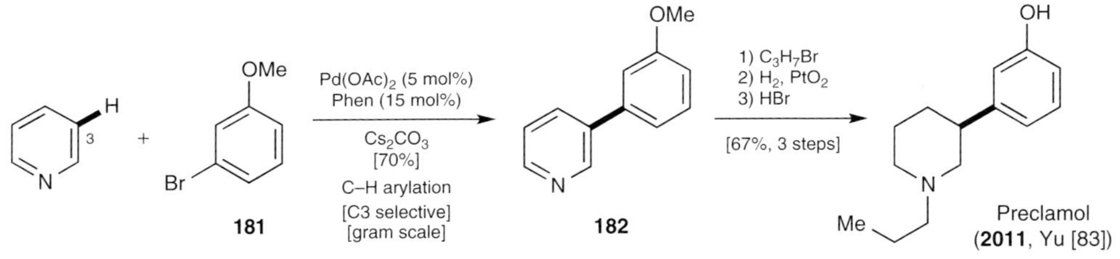

Scheme 16.40 Synthesis of preclamol employing a C3-selective C–H arylation of pyridines.

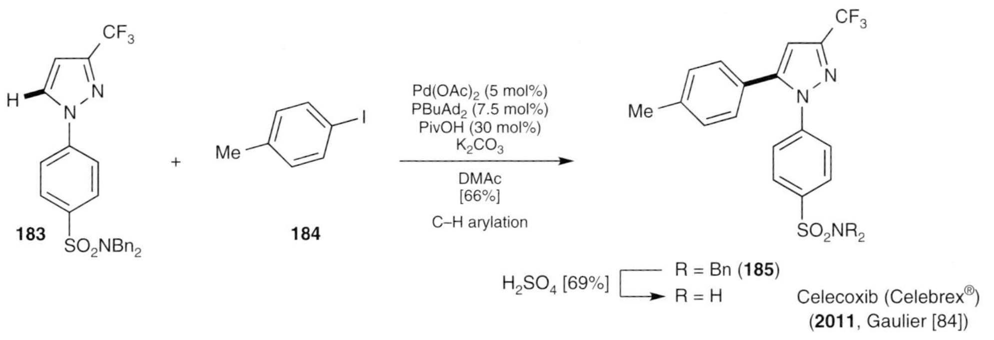

Scheme 16.41 Synthesis of celecoxib using a Pd-catalyzed C–H arylation.

the coupling product **185** in 66% yield. After removal of the benzyl group of **185**, the synthesis of celecoxib has been achieved.

16.2.9.2 GABA 2/3 Agonist (C–H Arylation of Imidazopyrimidines)

In 2005 and 2006, researchers at Merck demonstrated the application of transition metal-catalyzed intermolecular C–H arylation of arenes to the synthesis of γ-aminobutyric acid (GABA) agonists, which contains an imidazopyrimidine moiety (Scheme 16.42) [85]. In the presence of catalytic $Pd(OAc)_2/PPh_3$ and KOAc, imidazopyrimidine **186** was coupled with aryl bromide **187** to afford the GABA agonist as the sole regioisomer in excellent yield. Notably, this direct coupling reaction was accomplished on kilogram scale. The synthesis of other GABA $\alpha_{2,3}$ agonists used slightly modified protocols (with aryl chloride, XPhos as ligand, and Bu_4NHSO_4 as additive) [85c].

16.2.9.3 Nigellidine Hydrobromide, YD-3, and YC-1 (C–H Arylation of Indazoles)

In 2012, the group of Yamaguchi and Itami demonstrated the C–H arylation of indazoles, which are well-known bioisosteres of pharmaceutically important cores, and applied it to the synthesis of YD-3 (platelet antiaggregating agent) and YC-1 (an antitumor agent) (Scheme 16.43a) [86]. 1-Benzyl-1*H*-indazole (**188**) was coupled with aryl iodide **189** under a $PdCl_2$/phen catalytic system with K_3PO_4 as a base in DMAc to afford YD-3 in 63% yield. Using the same reaction conditions, the synthesis of YC-1 was also accomplished.

In 2013, Yu and coworkers also discovered the reactivity of simple 1*H*-indazoles and pyrazoles using a $Pd(OAc)_2/phen/Cs_2CO_3$ catalytic system in toluene [87].

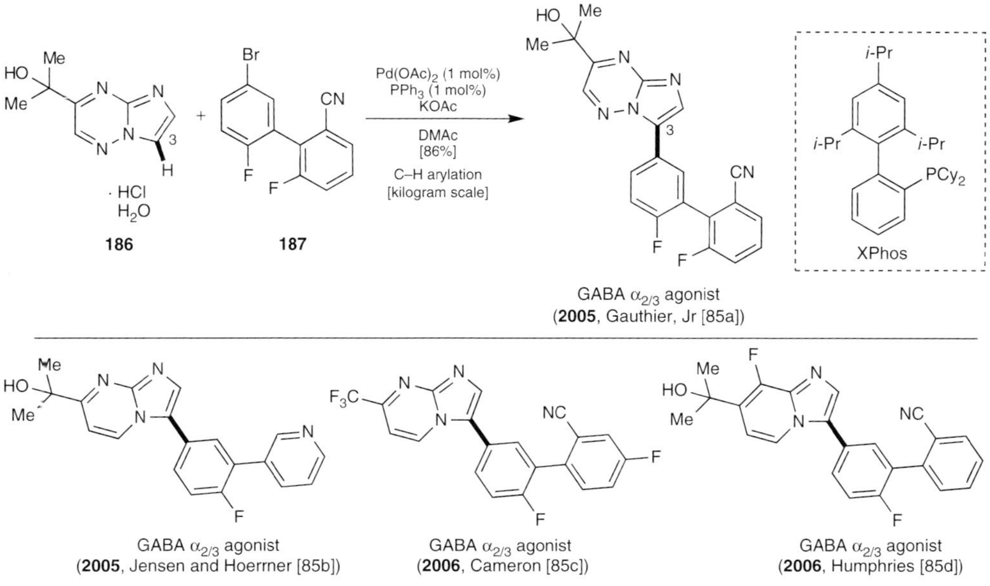

Scheme 16.42 Synthesis of GABA agonists using a C–H arylation of imidazopyrimidines.

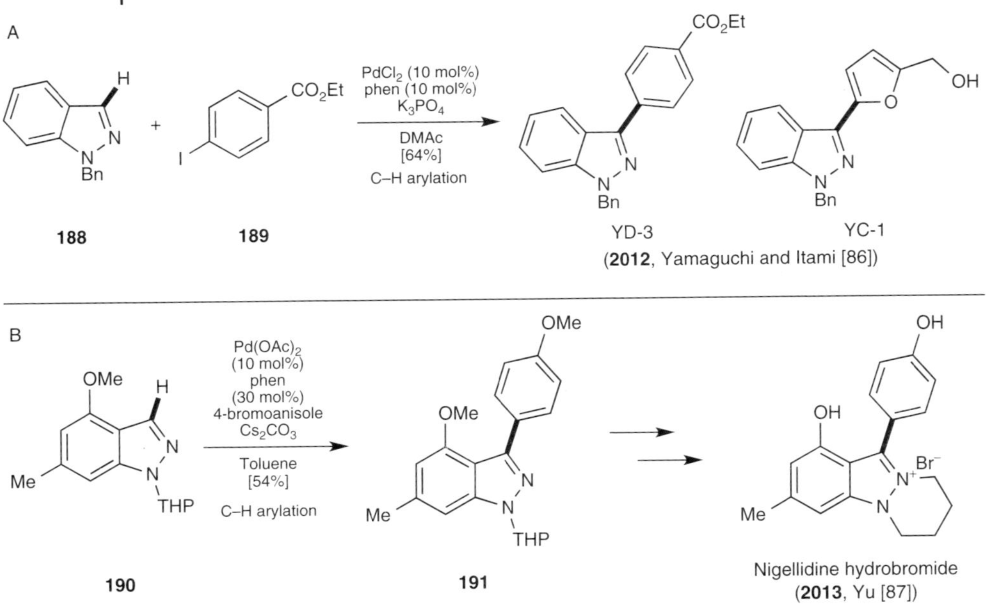

Scheme 16.43 C–H arylation of indazoles: synthesis of YD-3, YC-1, and nigellidine hydrobromide.

This catalytic reaction was successfully applied to the C–H arylation of indazole **190** with 4-bromoanisole to synthesize a biologically active compound, nigellidine hydrobromide (Scheme 16.43b).

16.2.9.4 Pseudoheliotridane (Formation of Pyrrolidines Using sp^3 C–H Insertion)

It was described earlier in this book chapter that the rhodium-catalyzed intramolecular C–H insertion reaction of metal carbenoids (mainly produced from diazocarbonyl compounds) is useful for the construction of natural products containing a dihydrofuran core (see details in Sections 16.2.4.4 and 16.2.4.5). C–H insertion using achiral or chiral rhodium(II) catalysts was also used to construct pyrrolidine frameworks, with representative applications to the synthesis of natural products and pharmaceuticals shown in Figure 16.3 [88].

The most recent example of pyrrolidine synthesis by C–H functionalization was illustrated in 2014, when Che and coworkers demonstrated the synthesis of pseudoheliotridane, a simple pyrrolidine alkaloid, by employing a ruthenium-catalyzed C–H insertion reaction (Scheme 16.44) [89]. Although diazocarbonyl compounds are generally required as the precursor of the metal carbenoid intermediate, they utilized an alkyl diazomethane as the carbene source. To this end, tosylhydrazone **192** was reacted with [Ru(TTP)(CO)] (TTP:tetra(p-tolyl)porphyrin) (1 mol%) and K_2CO_3, which generated ruthenium carbene **193** *in situ*, followed by C–H insertion to produce pseudoheliotridane in 95% yield with high diastereoselectivity.

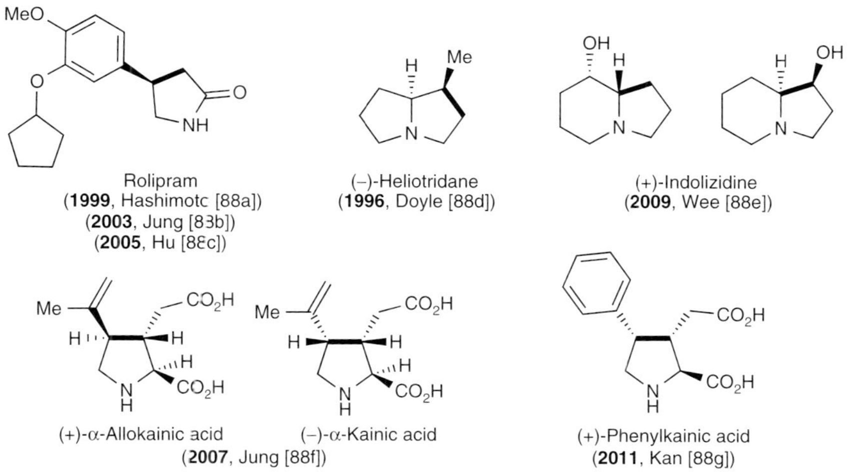

Figure 16.3 Synthesis of pyrrolidine-containing natural products and pharmaceuticals using Rh-catalyzed C–H insertion.

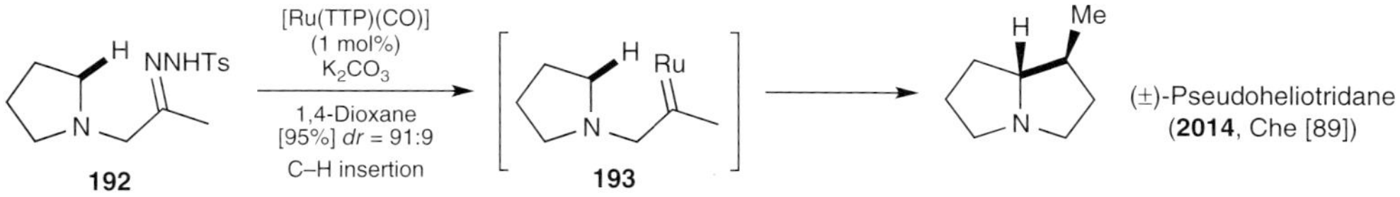

Scheme 16.44 Synthesis of pseudoheliotridane using ruthenium-catalyzed C–H insertion.

16.2.9.5 Aeruginosin (sp^3 C–H Alkenylation and Arylation)

In 2015, Baudoin and coworkers reported a scalable synthesis of aeruginosins, marine natural products, using sp^3 C–H activation reactions (Scheme 16.45) [90]. The synthesis commenced with intramolecular sp^3 C–H alkenylation of **194**, which was prepared in gram scale by a six-step operation from dihomoallylic alcohol, in the presence of $Pd(PCy_3)_2$, K_2CO_3, and t-$BuCO_2K$ as an additive in toluene to form pyrrolidine core **195** in 68–71% yield. Thereafter, **195** was derivatized to **196**, a fragment of aeruginosins. For the other component of these natural products, amino acid derivative **197**, bearing a 2-pyridinylisopropyl group as a directing group, was coupled with aryl iodide **198** in the presence of a palladium catalyst to give the coupling product via sp^3 C–H arylation in 50% yield. Removal of the directing group of **199** by $NOBF_4$ gave carboxylic acid **200**, which was then was condensed with D-allo-isoleucine methyl ester (D-allo-Ile-OMe) to form **201**. Subsequent saponification of the methyl ester and condensation with **196** afforded the aeruginosin core **202** in 60% yield over two steps. After several steps from **202**, the synthesis of aeruginosin 98B has been achieved. Additionally, the synthesis of aeruginosin 298A was accomplished using a similar operation from **200**.

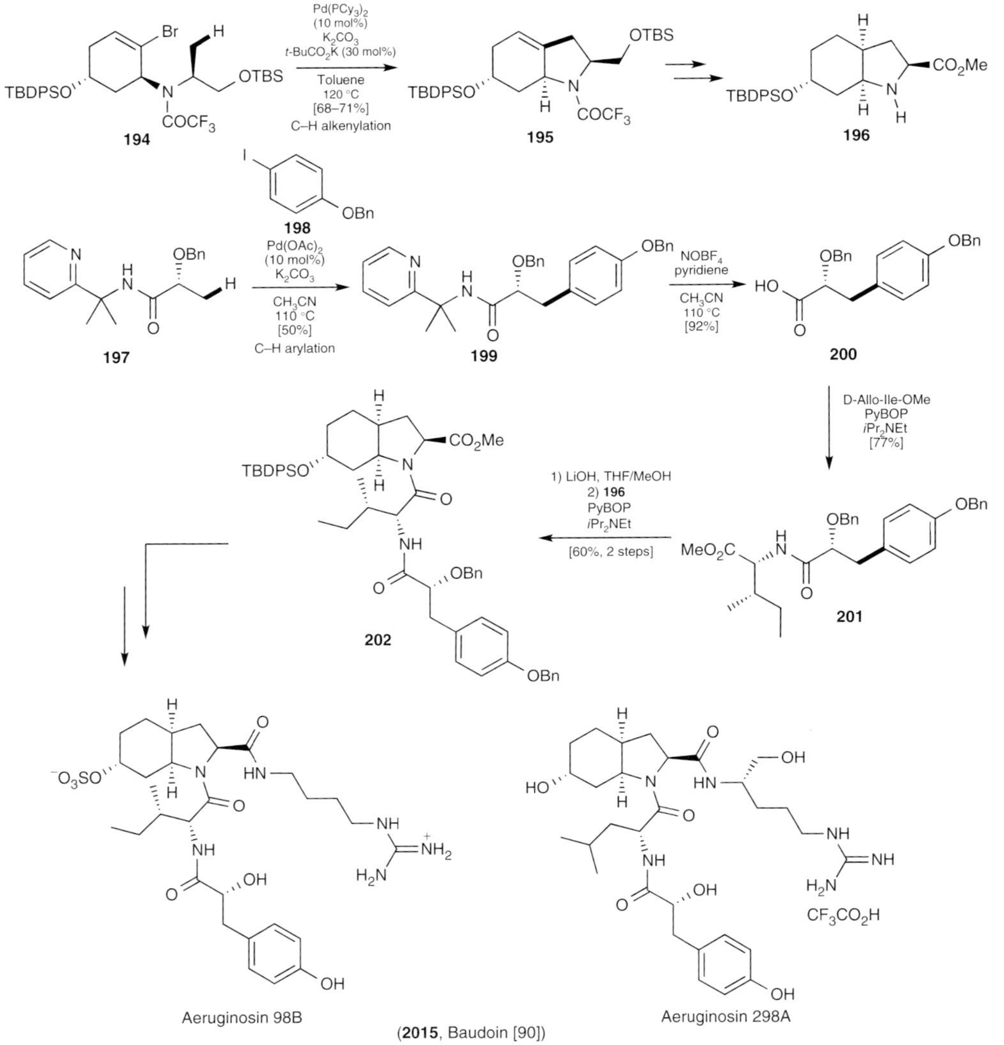

Scheme 16.45 Synthesis of aeruginosins using two sp^3 C–H functionalization reactions.

16.3 Summary

Methods in heterocycle substitution and synthesis using catalytic C–H functionalization are increasingly employed for the synthesis of complex natural products and pharmaceuticals. However, almost all of the reactions in this book chapter can be classified as sp^2 C–H functionalization, with sp^3 C–H functionalization of heterocycles, particularly when applied to highly functionalized molecules, still limited. The ultimate demonstration of versatility and applicability of a given reaction—that is, to depart from the "methodology level" of research—requires practice in complex-molecule synthesis. Pursuit of the "ideal synthesis" through C–H bond functionalization is a means for one to realize what is truly required of current synthetic methodology.

References

1. Review (a) Godula, K. and Sames, D. (2006) *Science*, **312**, 67; (b) Alberico, D., Scott, M.E., and Lautens, M. (2007) *Chem. Rev.*, **107**, 174; (c) Kakiuchi, F. and Kochi, T. (2008) *Synthesis*, **40**, 3013; (d) Davies, H.M. and Manning, J.R. (2008) *Nature*, **451**, 417; (e) Bellina, F., Cauteruccio, S., and Rossi, R. (2008) *Curr. Org. Chem.*, **12**, 774; (f) Ackermann, L., Vicente, R., and Kapdi, A.R. (2009) *Angew. Chem. Int. Ed.*, **48**, 9792; (g) Bellina, F. and Rossi, R. (2009) *Tetrahedron*, **65**, 10269; (h) Chen, X., Engle, K.M., Wang, D.-H., and Yu, J.-Q. (2009) *Angew. Chem. Int. Ed.*, **48**, 5094; (i) Thansandote, P. and Lautens, M. (2009) *Chem. Eur. J.*, **15**, 5874; (j) Ishihara, Y. and Baran, P.S. (2010) *Synlett*, 1733; (k) Satoh, T. and Miura, M. (2010) *Synthesis*, **42**, 3395; (l) Bouffard, J. and Itami, K. (2010) *Top. Curr. Chem.*, **292**, 231; (m) Gutekunst, W.R. and Baran, P.S. (2011) *Chem. Soc. Rev.*, **40**, 1976; (n) McMurray, L., O'Hara, F., and Gaunt, M.J. (2011) *Chem. Soc. Rev.*, **40**, 1885; (o) Hirano, K. and Miura, M. (2011) *Synlett*, 294; (p) Engle, K.M., Mei, T.-S., Wasa, M., and Yu, J.-Q. (2012) *Acc. Chem. Res.*, **45**, 788; (q) Brückl, T., Baxter, R.D., Ishihara, Y., and Baran, P.S. (2012) *Acc. Chem. Res.*, **45**, 826; (r) Yamaguchi, J., Yamaguchi, A.D., and Itami, K. (2012) *Angew. Chem. Int. Ed.*, **51**, 8960; (s) Wencel-Delord, J. and Glorius, F. (2013) *Nat. Chem.*, **5**, 369; (t) Segawa, Y., Maekawa, T., and Itami, K. (2015) *Angew. Chem. Int. Ed.*, **54**, 66.
2. Mandal, D., Yamaguchi, A.D., Yamaguchi, J., and Itami, K. (2011) *J. Am. Chem. Soc.*, **133**, 19660.
3. Ueda, K., Yanagisawa, S., Yamaguchi, J., and Itami, K. (2010) *Angew. Chem. Int. Ed.*, **49**, 8946.
4. Garg, N.K., Sarpong, R., and Stoltz, B.M. (2002) *J. Am. Chem. Soc.*, **124**, 13179.
5. Yamaguchi, A.D., Mandal, D., Yamaguchi, J., and Itami, K. (2011) *Chem. Lett.*, **40**, 555.
6. Yokoyama, Y., Matsumoto, T., and Murakami, Y. (1995) *J. Org. Chem.*, **60**, 1486.
7. Yokoyama, Y., Takahashi, M., Kohno, Y., Kataoka, K., Fujikawa, Y., and Murakami, Y. (1990) *Heterocycles*, **31**, 803.
8. Yokoyama, Y., Kondo, K., Mitsuhashi, M., and Murakami, Y. (1996) *Tetrahedron Lett.*, **37**, 9309.
9. (a) Cushing, T.D., Sanz-Cervera, J.F., and Williams, R.M. (1993) *J. Am. Chem. Soc.*, **115**, 9323; (b) Greshock, T.J., Grubbs, A.W., Jiao, P., Wicklow, D.T., Gloer, J.B., and Williams, R.M. (2008) *Angew. Chem. Int. Ed.*, **47**, 3573.
10. (a) Trost, B.M., Godeski, S.A., and Genèt, J.P. (1978) *J. Am. Chem. Soc.*, **100**, 3930; (b) Trost, B.M., Godeski, S.A., and Belletire, J.L. (1979) *J. Org. Chem.*, **44**, 2052.
11. Artman, G.D. III,, Grubbs, A.W., and Williams, R.M. (2007) *J. Am. Chem. Soc.*, **129**, 6336.
12. (a) Baran, P.S. and Corey, E.J. (2002) *J. Am. Chem. Soc.*, **124**, 7904; (b) Baran, P.S., Guerrero, C.A., and Corey, E.J. (2003) *J. Am. Chem. Soc.*, **125**, 5628.
13. (a) Wiedemann, S.H., Ellman, J.A., and Bergman, R.G. (2006) *J. Org. Chem.*, **71**, 1969; (b) Wilson, R.M., Thalji, R.K., Bergman, R.G., and Ellman, J.A. (2006) *Org. Lett.*, **8**, 1745; (c) Rech, J.C., Yato, M., Duckett, D., Ember, B., LoGrasso, P.V., Bergmann, R.G., and Ellman, J.A. (2007) *J. Am. Chem. Soc.*, **129**, 490.
14. Liang, J., Hu, W., Tao, P., and Jia, Y. (2013) *J. Org. Chem.*, **78**, 5810.
15. Robbins, D.W., Boebel, T.A., and Hartwig, J.F. (2010) *J. Am. Chem. Soc.*, **132**, 4068.

16. Pitts, A.K., O'Hara, F., Snell, R.H., and Gaunt, M.J. (2015) *Angew. Chem. Int. Ed.*, **54**, 5451.
17. Stuart, D.R., Alsabeh, P., Kuhn, M., and Fagnou, K. (2010) *J. Am. Chem. Soc.*, **132**, 18326.
18. Klein, J.E.M.N., Perry, A., Pugh, D.S., and Taylor, R.J.K. (2010) *Org. Lett.*, **12**, 3446.
19. (a) Perry, A. and Taylor, R.J.K. (2009) *Chem. Commun.*, **45**, 3249; (b) Pugh, D.S., Klein, J.E.M.N., Perry, A., and Taylor, R.J.K. (2010) *Synlett*, **21**, 934; Similar reaction has also been reported by Kündig; see (c) Jia, Y.-X. and Kündig, E.P. (2009) *Angew. Chem. Int. Ed.*, **48**, 1636.
20. Trost, B.M. and Brennan, M.K. (2006) *Org. Lett.*, **8**, 2027.
21. Dong, H., Latka, R.T., and Driver, T.G. (2011) *Org. Lett.*, **13**, 2726.
22. (a) Shou, W.G., Li, J., Guo, T., Lin, Z., and Jia, G. (2009) *Organometallics*, **28**, 6847; (b) Fantauzzi, S., Gallo, E., Caselli, A., Ragaini, F., Casati, N., Macchi, P., and Cenini, S. (2009) *Chem. Commun.*, **45**, 3952; (c) Milczek, E., Boudet, N., and Blakey, S. (2008) *Angew. Chem. Int. Ed.*, **47**, 6825; (d) Kawabata, H., Omura, K., Uchida, T., and Katsuki, T. (2007) *Chem. Asian J.*, **2**, 248; (e) Leung, S.K.-Y., Tsui, W.-M., Huang, J.-S., Che, C.-M., Liang, J.-L., and Zhu, N. (2005) *J. Am. Chem. Soc.*, **127**, 16629.
23. Boger, D.L. and Patel, M. (1988) *J. Org. Chem.*, **53**, 1405.
24. Bowie, A.L. Jr.,, Hughes, C.C., and Trauner, D. (2005) *Org. Lett.*, **7**, 5207.
25. Bowie, A.L. Jr., and Trauner, D. (2009) *J. Org. Chem.*, **74**, 1581.
26. Sui, X., Zhu, R., Li, G., Ma, X., and Gu, Z. (2013) *J. Am. Chem. Soc.*, **135**, 9318.
27. McMurray, L., Beck, E.M., and Gaunt, M.J. (2012) *Angew. Chem. Int. Ed.*, **51**, 9288.
28. Beck, E.M., Hatley, R., and Gaunt, M.J. (2008) *Angew. Chem. Int. Ed.*, **47**, 3004.
29. Su, Y., Zhou, H., Chen, J., Xu, J., Wu, X., Lin, A., and Yao, H. (2014) *Org. Lett.*, **16**, 4884.
30. Ueda, K., Amaike, K., Maceiczyk, R.M., Itami, K., and Yamaguchi, J. (2014) *J. Am. Chem. Soc.*, **136**, 13226.
31. Yanagisawa, S., Sudo, T., Noyori, R., and Itami, K. (2006) *J. Am. Chem. Soc.*, **128**, 11748.
32. Yamaguchi, A.D., Chepiga, K.M., Yamaguchi, J., Itami, K., and Davies, H.M.L. (2015) *J. Am. Chem. Soc.*, **137**, 644.
33. Wang, H., Guptill, D.M., Alvarez, A.V., Musaev, D.G., and Davies, H.M.L. (2013) *Chem. Sci.*, **4**, 2844.
34. Bedford, R.B. and Betham, M. (2006) *J. Org. Chem.*, **71**, 9403.
35. Miller, R.B. and Moock, T. (1980) *Tetrahedron Lett.*, **21**, 3319.
36. Liégault, B., Lee, D., Huestis, M.P., Stuart, D.R., and Fagnou, K. (2008) *J. Org. Chem.*, **73**, 5022.
37. (a) Knöll, J. and Knölker, H.-J. (2006) *Synlett*, **17**, 651; (b) Gruner, K.K. and Knölker, H.-J. (2008) *Org. Biomol. Chem.*, **6**, 3902; (c) Sridharan, V., Martín, M.A., and Menéndez, J.C. (2009) *Eur. J. Org. Chem.*, **2009**, 4614; (d) Watanabe, T., Oishi, S., Fujii, N., and Ohno, H. (2009) *J. Org. Chem.*, **74**, 4720.
38. Youn, S.W., Bihn, J.H., and Kim, B.S. (2011) *Org. Lett.*, **13**, 3738.
39. (a) Hughes, C.C. and Trauner, D. (2002) *Angew. Chem. Int. Ed.*, **41**, 1569; (b) Hughes, C.C. and Trauner, D. (2004) *Tetrahedron*, **60**, 9675.
40. Kim, K. and Kim, I. (2010) *Org. Lett.*, **12**, 5314.
41. O'Malley, S.J., Tan, K.L., Watzke, A., Bergman, R.G., and Ellman, J.A. (2005) *J. Am. Chem. Soc.*, **127**, 13496.
42. (a) Ahrendt, K.A., Bergman, R.G., and Ellman, J.A. (2003) *Org. Lett.*, **5**, 1301; (b) Thalji, R.K., Ahrendt, K.A., Bergman, R.G., and Ellman, J.A. (2001) *J. Am. Chem. Soc.*, **123**, 9692.
43. Wang, D.-H. and Yu, J.-Q. (2011) *J. Am. Chem. Soc.*, **133**, 5767.
44. (a) Davies, H.M.L. and Hansen, T. (1997) *J. Am. Chem. Soc.*, **119**, 9075; (b) Davies, H.M.L., Hansen, T., and Churchill, M.R. (2000) *J. Am. Chem. Soc.*, **122**, 3063; (c) Davies, H.M.L. and Antoulinakis, E.G. (2001) *J. Organomet. Chem.*, **617**, 47; (d) Davies, H.M.L., Cantrell, W.R. Jr.,, Romines, K.R., and Baum, J.S. (1992) *Org. Synth.*, **70**, 93.
45. (a) Wang, D.-H., Engle, K.M., Shi, B.-F., and Yu, J.-Q. (2010) *Science*, **327**, 315;

(b) Engle, K.M., Wang, D.-H., and Yu, J.-Q. (2010) *Angew. Chem. Int. Ed.*, **49**, 6169; (c) Engle, K.M., Wang, D.-H., and Yu, J.-Q. (2010) *J. Am. Chem. Soc.*, **132**, 14137.

46. (a) Kurosawa, W., Kan, T., and Fukuyama, T. (2003) *J. Am. Chem. Soc.*, **125**, 8112; (b) Natori, Y., Tsutsui, H., Sato, N., Nakamura, S., Nambu, H., Shiro, M., and Hashimoto, S. (2009) *J. Org. Chem.*, **74**, 4418; (c) Koizumi, Y., Kobayashi, H., Wakimoto, T., Furuta, T., Fukuyama, T., and Kan, T. (2008) *J. Am. Chem. Soc.*, **130**, 16854; (d) Wakimoto, T., Miyata, K., Ohuchi, H., Asakawa, T., Nukaya, H., Suwa, Y., and Kan, T. (2011) *Org. Lett.*, **13**, 2789.
47. White, J.D., Hrnciar, P., and Stappenbeck, F. (1997) *J. Org. Chem.*, **62**, 5250.
48. Campeau, L., Stuart, D., Leclerc, J., Bertrand-Laperle, M., Villemure, E., Sun, H., Lasserre, S., Guimond, N., Lecavallier, M., and Fagnou, K. (2009) *J. Am. Chem. Soc.*, **131**, 3291.
49. (a) Shibahara, F., Yamaguchi, E., and Murai, T. (2011) *J. Org. Chem.*, **76**, 2680; (b) Shibahara, F., Yamaguchi, E., and Murai, T. (2010) *Chem. Commun.*, **46**, 2471.
50. Besselièvre, F., Mahuteau-Betzer, F., Grierson, D.S., and Piguel, S. (2008) *J. Org. Chem.*, **73**, 3278.
51. Ohnmacht, S.A., Mamone, P., Culshaw, A.J., and Greaney, M.F. (2008) *Chem. Commun.*, **44**, 1241.
52. Verrier, C., Martin, T., Hoarau, C., and Marsais, F. (2008) *J. Org. Chem.*, **73**, 7383.
53. Martin, T., Laguerre, C., Hoarau, C., and Marsais, F. (2009) *Org. Lett.*, **11**, 3690.
54. For selected examples of C–H arylation using ubiquitous catalyst, see (a) Do, H.-Q. and Daugulis, O. (2007) *J. Am. Chem. Soc.*, **129**, 12404; (b) Yoshizumi, T., Tsurugi, H., Satoh, T., and Miura, M. (2008) *Tetrahedron Lett.*, **49**, 1598; (c) Do, H.-Q., Khan, R.M.K., and Daugulis, O. (2008) *J. Am. Chem. Soc.*, **130**, 15185; (d) Norinder, J., Matsumoto, A., Yoshikai, N., and Nakamura, E. (2008) *J. Am. Chem. Soc.*, **130**, 5858; (e) Yoshikai, N., Matsumoto, A., Norinder, J., and Nakamura, E. (2009) *Angew. Chem. Int. Ed.*, **48**, 2925.
55. Canivet, J., Yamaguchi, J., Ban, I., and Itami, K. (2009) *Org. Lett.*, **11**, 1733.
56. Yamamoto, T., Muto, K., Komiyama, M., Canivet, J., Yamaguchi, J., and Itami, K. (2011) *Chem. Eur. J.*, **17**, 10113.
57. Labaudiniére, R., Hendel, W., Terlain, B., Cavy, F., Marquis, O., and Dereu, N. (1992) *J. Med. Chem.*, **35**, 4306.
58. Muto, K., Yamaguchi, J., and Itami, K. (2012) *J. Am. Chem. Soc.*, **134**, 169.
59. Amaike, K., Muto, K., Yamaguchi, J., and Itami, K. (2012) *J. Am. Chem. Soc.*, **134**, 13573.
60. Wipf, P. and Venkatraman, S. (1996) *J. Org. Chem.*, **61**, 6517.
61. Besselièvre, F., Piguel, S., Mahuteau-Betzer, F., and Grierson, D.S. (2008) *Org. Lett.*, **10**, 4029.
62. For examples of alkenylation of heteroaromatics with alkenes before Piguel's 2008 report, see(a) Matsuura, Y., Tamura, M., Kochi, T., Sato, M., Chatani, N., and Kakiuchi, F. (2007) *J. Am. Chem. Soc.*, **129**, 9858; (b) Beck, E.M., Grimster, N.P., Hatley, R., and Gaunt, M.J. (2006) *J. Am. Chem. Soc.*, **128**, 2528; For the only example of alkynylation of heteroaromatics before Piguel's 2008 report, see(c) Seregin, I.V., Ryabova, V., and Gevorgyan, V. (2007) *J. Am. Chem. Soc.*, **129**, 7742.
63. Meng, L., Kamada, Y., Muto, K., Yamaguchi, J., and Itami, K. (2013) *Angew. Chem. Int. Ed.*, **52**, 10048.
64. (a) Linder, J. and Moody, C.J. (2007) *Chem. Commun.*, **43**, 1508; (b) Linder, J., Blake, A.J., and Moody, C.J. (2008) *Org. Biomol. Chem.*, **6**, 3908; (c) Zhang, J. and Ciufolini, M.A. (2009) *Org. Lett.*, **11**, 2389.
65. Harayama, T., Hori, A., Serban, G., Morikami, Y., Matsumoto, T., Abe, H., and Takeuchi, Y. (2004) *Tetrahedron Lett.*, **60**, 10645.
66. Harayama, T., Akamatsu, H., Okamura, K., Miyagoe, T., Akiyama, T., Abe, H., and Takeuchi, Y. (2001) *J. Chem. Soc., Perkin Trans. 1*, 523.
67. Blanchot, M., Candito, D.A., Larnaud, F., and Lautens, M. (2011) *Org. Lett.*, **13**, 1486.

68. Berman, A.M., Lewis, J.C., Bergman, R.G., and Ellman, J.A. (2008) *J. Am. Chem. Soc.*, **130**, 14926.
69. Deng, Y., Gong, W., He, J., and Yu, J.-Q. (2014) *Angew. Chem. Int. Ed.*, **53**, 6692.
70. Hino, K., Furukawa, K., Nagai, Y., and Uno, H. (1980) *Chem. Pharm. Bull.*, **28**, 2618.
71. Wu, J., Xiang, S., Zeng, J., Leow, M., and Liu, X.-W. (2015) *Org. Lett.*, **17**, 222.
72. Angibaud, P., Venet, M., Filliers, W., Broeckx, R., Ligny, Y., Muller, P., Poncelet, V., and End, D. (2004) *Eur. J. Org. Chem.*, **2004**, 479.
73. (a) Huang, Q., Hunter, J.A., and Larock, R.C. (2001) *Org. Lett.*, **3**, 2973; (b) Roesch, K.R., Zhang, H., and Larock, R.C. (2001) *J. Org. Chem.*, **66**, 8042; (c) Korivi, R.P. and Cheng, C.-H. (2005) *Org. Lett.*, **7**, 5179.
74. Jayakumar, J., Parthasarathy, K., and Cheng, C.-H. (2012) *Angew. Chem. Int. Ed.*, **51**, 197.
75. (a) Paul, F., Patt, J., and Hartwig, J.F. (1994) *J. Am. Chem. Soc.*, **116**, 5969; (b) Guram, A.S. and Buchwald, S.L. (1994) *J. Am. Chem. Soc.*, **116**, 7901.
76. Schipper, D.J., El-Salfiti, M., Whipp, C.J., and Fagnou, K. (2009) *Tetrahedron*, **65**, 4977.
77. Fischer, D.F. and Sarpong, R. (2010) *J. Am. Chem. Soc.*, **132**, 5926.
78. Newton, J.N., Fischer, D.F., and Sarpong, R. (2013) *Angew. Chem. Int. Ed.*, **52**, 1726.
79. Zhao, L., Tsukano, C., Kwon, E., Takemoto, Y., and Hirama, M. (2013) *Angew. Chem. Int. Ed.*, **52**, 1722.
80. (a) Lafrance, M., Gorelsky, S.I., and Fagnou, K. (2007) *J. Am. Chem. Soc.*, **129**, 14570; (b) Lafrance, M. and Fagnou, K. (2006) *J. Am. Chem. Soc.*, **128**, 16496.
81. Larivée, A., Mousseau, J.J., and Charette, A.B. (2008) *J. Am. Chem. Soc.*, **130**, 52.
82. (a) Wasa, M., Worrell, B.T., and Yu, J.-Q. (2010) *Angew. Chem. Int. Ed.*, **49**, 1275; (b) Guo, P., Joo, J.M., Rakshit, S., and Sames, D. (2011) *J. Am. Chem. Soc.*, **133**, 16338.
83. Ye, M., Gao, G.-L., Edmunds, A.J.F., Worthington, P.A., Morris, J.A., and Yu, J.-Q. (2011) *J. Am. Chem. Soc.*, **133**, 19090.
84. Gaulier, S.M., Mckay, R., and Swain, N.A. (2011) *Tetrahedron Lett.*, **52**, 6000.
85. (a) Gauthier, D.R. Jr.,, Limanto, J., Devine, P.N., Desmond, R.A., Szumigala, R.H. Jr.,, Foster, B.S., and Volante, R.P. (2005) *J. Org. Chem.*, **70**, 5938; (b) Jensen, M.S., Hoerrner, R.S., Li, W., Nelson, D.P., Javadi, G.J., Dormer, P.G., Cai, D., and Larsen, R.D. (2005) *J. Org. Chem.*, **70**, 6034; (c) Cameron, M., Foster, B.S., Lynch, J.E., Shi, Y.-J., and Dolling, U.-H. (2006) *Org. Process Res. Dev.*, **10**, 398; (d) Humphries, A.C., Gancia, E., Gilligan, M.T., Goodacre, S., Hallett, D., Merchant, K.J., and Thomas, S.R. (2006) *Bioorg. Med. Chem. Lett.*, **16**, 1518.
86. Hattori, K., Yamaguchi, K., Yamaguchi, J., and Itami, K. (2012) *Tetrahedron*, **68**, 7605.
87. Ye, M., Edmunds, A.J.F., Morris, J.A., Sale, D., Zhang, Y., and Yu, J.-Q. (2013) *Chem. Sci.*, **4**, 2374.
88. (a) Anada, M., Mita, O., Watanabe, H., Kitagaki, S., and Hashimoto, S. (1999) *Synlett*, **10**, 1775; (b) Yoon, C.H., Nagle, A., Chen, C., Gandhi, D., and Jung, K.W. (2003) *Org. Lett.*, **5**, 2259; (c) Liu, W.-J., Chen, Z.-L., Chen, Z.-Y., and Hu, W.-H. (2005) *Tetrahedron: Asymmetry*, **16**, 1693; (d) Doyle, M.P. and Kalinin, A.V. (1996) *Tetrahedron Lett.*, **37**, 1371; (e) Wee, A.G.H., Fan, G.-J., and Bayirinoba, H.M. (2009) *J. Org. Chem.*, **74**, 8261; (f) Jung, Y.C., Yoon, C.H., Turos, E., Yoo, K.S., and Jung, K.W. (2007) *J. Org. Chem.*, **72**, 10114; (g) Higashi, T., Isobe, Y., Ouchi, H., Suzuki, H., Okazaki, Y., Asakawa, T., Furuta, T., Wakimoto, T., and Kan, T. (2011) *Org. Lett.*, **13**, 1089.
89. Reddy, A.R., Zhou, C.-Y., Guo, Z., Wei, J., and Che, C.-M. (2014) *Angew. Chem. Int. Ed.*, **53**, 14175.
90. Dailler, D., Danoun, G., and Baudoin, O. (2015) *Angew. Chem. Int. Ed.*, **54**, 4919.

Index

Transition Metal-Catalyzed Heterocycle Synthesis via C—H Activation, First Edition. Edited by Xiao-Feng Wu.
© 2016 Wiley-VCH Verlag GmbH & Co. KGaA. Published 2016 by Wiley-VCH Verlag GmbH & Co. KGaA.

o

q